# Progress in Biological Control

## Volume 18

**Series Editors**

Heikki M. T. Hokkanen
Department of Environmental and Biological Sciences, University of Eastern Finland, Kuopio, Finland
Yulin Gao
Institute of Plant Protection, Chinese Academy of Agricultural Sciences, Beijing, China

Biological control of pests, weeds, and plant and animal diseases utilising their natural antagonists is a well-established but rapidly evolving field of applied ecology. Despite its documented applications and systematic development efforts for longer than a century, biological control still remains a grossly underexploited method of pest management. Its untapped potential represents the best hope to providing lasting, environmentally sound, and socially acceptable control of most problem pests in agriculture, and of invasive alien organisms threatening global biodiversity. Based on the overwhelmingly positive features of biological control, it is the prime candidate in the search for reducing dependency on chemical pesticides. Public demand for finding solutions based on biological control is the main driving force in the rapid developments in the various strategies of utilising natural enemies for controlling noxious organisms. This book series is intended to accelerate these developments through exploring the progress made within the various aspects of biological control, and via documenting these advances to the benefit of fellow scientists, students, public officials, and the public at large. Each of the books in this series is expected to provide a comprehensive, authoritative synthesis of the topic, likely to stand the test of time.

More information about this series at http://www.springer.com/series/6417

Javad Karimi • Hossein Madadi
Editors

# Biological Control of Insect and Mite Pests in Iran

A Review from Fundamental and Applied Aspects

 Springer

*Editors*
Javad Karimi
Department of Plant Protection
Ferdowsi University of Mashhad
Mashhad, Iran

Hossein Madadi
Department of Plant Protection
Faculty of Agriculture
Bu-Ali Sina University
Hamedan, Iran

ISSN 1573-5915　　　　　　　　ISSN 2543-0076　(electronic)
Progress in Biological Control
ISBN 978-3-030-63992-1　　　　ISBN 978-3-030-63990-7　(eBook)
https://doi.org/10.1007/978-3-030-63990-7

This Springer imprint is published by the registered company Springer Nature Switzerland AG
The registered company address is: Gewerbestrasse 11, 6330 Cham, Switzerland

# Foreword

Surely long before the rise of human being and oscillating steps in the development of agriculture, insects were subject to interaction with other organisms in the environment. Some of these interactions were in favor of insects (supplying nutriments, protection, or other advantageous symbiotic relationships) and some were detrimental such as competition, disease and predation. Insects as relatively vulnerable animals fell prey to all manner of predators, parasitoids, and diseases. Therefore, studies on ecology of natural populations inspired scientist to apply natural enemies to control or suppress different pest populations which is known as biological pest control.

Regarding to the history of biological control in Iran, availability of huge information on different kinds of natural enemies, and presence of a large number of young scientists, it is necessary to have special attention and support paid by relevant authorities. So that, to be capable of increasing to produce safe or organic agricultural products it is important to review and gather all information related to biocontrol to facilitate the non-chemical tactics in pest management programs.

Many thanks go to all Iranian scientists who have contributed to this valuable book. The breadth and depth of their contributions collected here in one place should go a long way to providing those scientists just starting out in working biological control disciplines and those who want to understand basic and applied biological control aspects of agricultural pests as basic information. I want to express my thanks also to Javad Karimi, the initial editor of the book, and Hossein Madadi, for inviting researchers who want to improve IPM programs based on biological control in Iran.

Professor of Entomology,                                                      Ahad Sahragard
Guilan University,
Rasht, Iran
March 13, 2020

# Preface

Biological control of plant pests has been initiated in Iran since the 1930s but recently especially in recent decades, the worldwide development of the biological control of pests has led to a considerable application of this beneficial pest management strategy in Iran. Jalal Afshar, the founder of modern Entomological science in Iran, imported Vedalia beetle, *Rodolia cardinalis* (Mulsant) (Col., Coccinellidae) to the country in 1934. He imported the beetle to Mazandaran province in Northern Iran, where an insectarium was established to mass produce this predator to suppress the cottony cushion scale, *Icerya purchasi* (Williston) (Hem., Margarodidae). In 1941, Mohammad Kosari mentioned the role and importance of the parasitic wasp *Trissolcus* spp. (Hym., Scelionidae) against the Sunn pest, *Eurygaster integriceps* Put. (Hem., Scutelleridae), a major pest of cereal crops. By 1962, the mass production and release of *Trissoclus* spp. wasps began, ultimately reaching a scale of two million wasps released per year. In the mid-1970s, efforts were made to import and release the *Trichogramma brassicae* (Hym., Trichogrammatidae) against rice stem borer, *Chilo suppressalis* (Lep., Pyralidae) in North provinces of Iran. During this decade, *Platytelenomus hylas* (Hym., Scelionida) successfully suppressed populations of the greater sugarcane borer, *Sesamia nonagrioides* (Lep., Noctuidae) in Khuzsetan Province. The establishment of insectariums as well as research and development for this agent was initiated in the mid-1980s.

In the last three decades, facilities for mass rearing of biocontrol agents were widely developed and subsequently, the development of biological control programs accelerated. Universities and research institutions also contributed to this development via fundamental and practical researches. Moreover, national meetings such as the "Iranian Plant Protection Congress", "Iranian International Congress of Entomology, "Iranian National Biocontrol Meetings" and "International Persian Congress of Acarology" were held regularly, offering opportunities for researchers in biocontrol and other fields of study to exchange the ideas and present their findings.

This book outlines basic and applied activities in Iranian biological control. The first section includes milestone events of entomology, pest control and biological control. Other parts of the book categorized into four parts. The part I deal with key insect predators including ladybird beetles (Chap. 1), mites (Chaps. 2 and 3) and lacewings (Chap. 4). The part II includes Chap. 5 Trichogrammatidae, Chap. 6 Chalcidoidea and Ichneumonoidea, Chap. 7 Platygasteroidea and Chap. 8 Aphidiidae. The following five chapters are allocated to insect pathogens as part III, including entomopathogenic bacteria (Chap. 9), entomopathogenic fungi (Chap. 10), entomopathogenic and insect parasitic nematodes (Chap. 11) and other groups of insect pathogens including viruses, microsporidians and protistans as well as endosymbiont bacterium from *Wolbachia* and other genera (Chap. 12). Part IV includes other approaches and analyses of current states of biocontrol in Iran. This part contains Chap. 13 which reviewed the major biopesticides. Chapter 14 focuses on biological control practices in greenhouses, and Chap. 15 includes biological control of medically important arthropods. The concluding chapter (Chap. 16) reviews the current opportunities, challenges and analyses the development of biocontrol in Iran.

The book is the result of 26 Iranian scientists and researchers' contribution from various Universities and institutes along with Petr Starý, with the world authority on the aphid parasitoid, and a great favor on collaborations with researcher including Iranian entomologists. Most authors are prestigious experts in their field of study and a number of early carrier young researchers have involved in contribution the chapters. This new generation promised bright future for expanding the use of biocontrol as ecofriendly tactic for pest management and in large scale, crop production and protection the environment of the Iran as well world. Both editors spent much of their time for preparing, communicating and editing these chapters. Thus, special thanks dedicated to our families who provided the best situation for work on the book.

Both of us teach insect biocontrol course from more than a decade, so we benefited from the discussions with the students in the classes. The first editor thanks from Shokoofeh Kamali for her assist during four years of book production, gathering information, and image processing and also from Reyhaneh Darsouei for her help about text format, cross check and reference style. We also would like to appreciate from Alireza Saboori and Arman Avand – Faghih for provide some images and information and Randy Gaugler for some suggestions. We appreciate from Mariska van der Stigchel for her long time patience and other staff member of book section, Springer Nature. We honoring the memory of pioneers of biocontrol in Iran, including Aziz Kharrazi- Pakdel, the emeritus professor of entomology, who was our supervisor and teacher of biocontrol and insect pathology courses in University of Tehran some years ago.

Aziz Kharazi- Pakdel, Emeritus professor of biological control, University of Tehran

The book is dedicated to farmers of the Iranian plateau, from Khuzestan to Khorasan and from Azerbaijan to Sistan and Baluchestan, from the beach of the Caspian Sea to the azure shore of the Persian Gulf which strive to provide healthy food for the nation and the country's growth and safety.

Mashhad, Iran																	Javad Karimi
Hamedan, Iran																	Hossein Madadi

# Contents

# Contributors

**Kamran Akbarzadeh** Department of Medical Entomology and Vector Control, School of Public Health, Tehran University of Medical Sciences, Tehran, Iran

**Hana Haji Allahverdipour** Biological Control Department, Iranian Research Institute of Plant Protection, Agricultural Research, Education and Extension Organization, Tehran, Iran

**Mohammad Asadi** Department of Plant Protection, University of Mohaghegh Ardabili, Ardabil, Iran

**Mina Asgari** Department of Entomology, Science and Research Branch, Islamic Azad University, Tehran, Iran

**Hassan Askary** Biological Control Department, Iranian Research Institute of Plant Protection, Agricultural Research, Education and Extension Organization, Tehran, Iran

**Mohammadreza Attaran** Biological Control Research Station, Iranian Research Institute of Plant Protection, Agricultural Research, Education and Extension Organization, Amol, Iran

**Reyhaneh Darsouei** Department of Plant Protection, Faculty of Agriculture, Ferdowsi University of Mashhad, Mashhad, Iran

**Azadeh Farazmand** Department of Agricultural Zoology, Iranian Research Institute of Plant Protection, Agricultural Research Education and Extension Organization (AREEO), Tehran, Iran

**Sepideh Ghaffari** Department of Plant Protection, Faculty of Agriculture, Ferdowsi University of Mashhad, Mashhad, Iran

**Soleiman Ghasemi** Research and Development Unit, Nature Biotechnology Company (Biorun), Karaj, Iran

**Seyed Hossein Goldansaz** Department of Plant Protection, College of Agriculture and Natural Resources, University of Tehran, Karaj, Iran

**Hamidreza Hajiqanbar** Department of Entomology, Faculty of Agriculture, Tarbiat Modares University, Tehran, Iran

**Mahnaz Hassani-Kakhki** Department of Plant Protection, School of Agriculture, Ferdowsi University of Mashhad, Mashhad, Iran

**Mahdi Hassanpour** Department of Plant Protection, University of Mohaghegh Ardabili, Ardabil, Iran

**Shahzad Iranipour** University of Tabriz, Tabriz, Iran

**Ali Jooyandeh** Department of Plant Protection, Khorasan Razavi Agriculture and Natural Resources Research and Education Center, Mashhad, Iran

**Gholamreza Salehi Jouzani** Agricultural Biotechnology Research Institute of Iran (ABRII), Agricultural Research, Education and Extension Organization (AREEO), Karaj, Iran

**Hashem Kamali** Department of Plant Protection, Khorasan Razavi Agriculture and Natural Resources Research and Education Center, Mashhad, Iran

**Shokoofeh Kamali** Department of Plant Protection, Faculty of Agriculture, Ferdowsi University of Mashhad, Mashhad, Iran

**Javad Karimi** Department of Plant Protection, Ferdowsi University of Mashhad, Mashhad, Iran

**Ayda Khorramnejad** Department of Plant Protection, College of Agriculture and Natural Resources, University of Tehran, Karaj, Iran
Instituto BioTecMed, Departamento de Genética, Universitat de València, València, Spain

**Hossein Lotfalizadeh** Plant Protection Research Department, East-Azarbaijan Agricultural and Natural Resources Research & Education Center, AREEO, Tabriz, Iran

**Hossein Madadi** Department of Plant Protection, Faculty of Agriculture, Bu-Ali Sina University, Hamedan, Iran

**Mohammad Mehrabadi** Department of Entomology, Faculty of Agriculture, Tarbiat Modares University, Tehran, Iran

**Vahe Minassian** Department of Plant Protection, Shahid Chamran University, Ahvaz, Iran

**Abbas Mohammadi-Khoramabadi** Department of Plant Production, College of Agriculture and Natural Resources of Darab, Shiraz University, Fars, Iran

**Mohammad Reza Moosavi** Plant Pathology Department, Faculty of Agriculture, Marvdasht Branch, Islamic Azad University, Marvdasht, Iran

**Ehsan Rakhshani** Department of Plant Protection, Faculty of Agriculture, University of Zabol, Zabol, Iran

**Heshmatollah Saadi** Department of Agricultural Extension Education, Bu- Ali Sina University, Hamedan, Iran

**Petr Starý** Institute of Entomology, Biology Center AVCR, České Budejovice, Czech Republic

**Ali Asghar Talebi** Department of Entomology, Faculty of Agriculture, Tarbiat Modares University, Tehran, Iran

**Zahra Tazerouni** Department of Entomology, Faculty of Agriculture, Tarbiat Modares University, Tehran, Iran

**Nahid Vaez** Azarbaijan Shahid Madani University, Tabriz, Iran

# Chapter 1
# Overview: History of Agricultural Entomology and Biological Pest Control in Iran

Javad Karimi and Shokoofeh Kamali

## 1.1 Pioneer Entomologists

Faunistic studies of insects started from the mid-eighteenth century in Iran and the beginning of the nineteenth century in Czarist Russia and some of the European countries. The specimens gathered and identified by entomologists in that period are still being kept in some of the international museums, particularly in those of Saint Petersburg, Paris, and London. At that time, entomology was not common in Iran and up to 150 years later, it remained unknown. In 1919, Jalal Afshar (Urmia 1894–Tehran 1974) returned to Iran upon completion of his higher education and started both teaching and researching on entomology and zoology at Pasteur Institute of Iran (Abivardi 2001) (Fig. 1.1).

After a while, he started to study plant pests in the Ministry of Public Welfare currently known as the Ministry of Agriculture, and at the same time teaching entomology at Falahat School (now known as College of Agriculture). In 1923, he founded a small department called "Local Pests Diagnosis and Control center". This is considered as the official beginning of research and executive work of identifying, collecting and making collections of the Insects in Iran by Iranians themselves. During this period, Jalal Afshar established an insect museum at the College of Agriculture in Karaj, which is now known as the Jalal Afshar Zoological Museum (JAZM) (https://utcan.ut.ac.ir/en/page/3135/museum) and is considered as an invaluable collection (Fig. 1.2). From 1926 onwards, he began teaching zoology,

J. Karimi (✉)
Department of Plant Protection, Ferdowsi University of Mashhad, Mashhad, Iran
e-mail: jkb@um.ac.ir

S. Kamali
Department of Plant Protection, Faculty of Agriculture, Ferdowsi University of Mashhad, Mashhad, Iran

**Fig. 1.1** Jalal Afshar.
(Courtesy of Alireza
Saboori, University of
Tehran)

**Fig. 1.2** Historial royal building of Jalal Afshar Zoological Museum (JAZM). (Courtesy of Alireza
Saboori, University of Tehran)

**Fig. 1.3** The students, Jalal Afshar (first line, fifth from left, with a hat) and other professors of Faculty of Agriculture, University of Tehran. (Courtesy of Alireza Saboori, University of Tehran)

entomology and pest control at Falahat school and he established a small entomology laboratory (Fig. 1.3) (Bagheri-Zenouz 2003).

## 1.2   The Early Persian Books on Entomology

Three of the early books compiled on entomology are mentioned here. "Anatomy of insect structure and metamorphosis" was written by Jalal Afshar in 152 pages, published by Karaj Technical Branch for Pest Control in 1937. Jalal Afshar also published the "Entomology" book (278 pages) in 1945. The third book on "Entomology" (Vol.1) was written by Mahmoud Shojaei (Fig. 1.4), published by the University of Tehran press.

Teaching Entomology, Pest Control and Zoology courses firstly began by late Jalal Afshar at the Agriculture and Rural Development College in 1927. During the same period, phytopathology course was being taught by Tajbakhsh at the same school. Later, phytopathology was being taught by Esfandiar Esfandiari from 1940 to 1945. Abbas Davachi returned to Iran after getting his PhD from France and was in charge of teaching pest control at the College of Agriculture since 1935, yet both zoology and entomology were still being taught by Jalal Afshar. A Master of Science (MSc) program on plant protection was first offered by the College of Agriculture, University of Tehran in 1967 (Fig. 1.5). This MSc program was separated into two disciplines of entomology (1973) and phytopathology (1973) and the Ph.D. program was offered in the same year (Department of Plant Protection 2019).

**Fig. 1.4** Mahmoud Shojaei. (Courtesy of Alireza Saboori, University of Tehran)

**Fig. 1.5** Main gate of the University of Tehran

University of Tabriz      **Shiraz University**

**Fig. 1.6** The first Universities offering plant protection in Iran. From the down right side are as follows: University of Tehran, Shiraz University, University of Tabriz. The top side from right: Chamran University of Ahwaz, Isfahan University of Technology, Urmia University and Ferdowsi University of Mashhad

## 1.3   Foundation of Karaj College of Agriculture

After the University of Tehran, Shiraz University (1959) and then Tabriz University (1966), Chamran University of Ahvaz (1978), Isfahan University of Technology (1979), Urmia University (1987) and Ferdowsi University of Mashhad (1990) started offering BSc degree programs in plant protection (Fig. 1.6). Nowadays, in addition to many of those national Universities, the Islamic Azad University Campuses, as well as some of the private schools, are also offering the same course of study (Bagheri-Zenouz 2003).

## 1.4   The History of Biological Pest Control in Iran

The historical records of Iranians in the biological control dates back to the controlling of locusts using starlings. They benefit from stork and hoopoe in controlling of noxious animals and termites. Also they would place a container filled with water called "Aabsaar" for the starlings to wash their beaks while hunting the locusts that would damage grasses. In 1933, importing the novius beetle, *Rodolia cardinalis,* as a predator of the cottony cushion scale, *Icerya purchasi* triggered the biological control in Iran. It proved to be one of the most successful biological control programs of the country and it is still effective in Northern gardens and the recent years in the citrus orchards of Dezful, Khuzestan province (Farahbakhsh 1961; Shirazi et al. 2011). During two decades (1966–1986), the progress of a national successful biological pest control against sunn pest, *Eurygaster integriceps* was due to the efforts of the entomologist such as Mohammad Kosari resulted in the development of mass rearing facilities for Platygasterid parasitoid, *Trissolcus grandis*. During the later years, millions of parasitic wasps were reared in several insectariums and released innundatively in the central part of the country. There is no clear reason related to putting an end to this operation. It might be due to the introduction of synthetic chemical pesticides particularly the D.D.T. into the country and its extensive use on a large scale on pest species (Heidari 2001). In 1966, to control citrus mealybug, *Planococcus citri,* a predatory ladybird of mealybugs, *Cryptolaemus montrouzieri* was imported into Iran and reared in the North Iran, Caspian Sea shores. Still, this ladybird is rearing in some insectariums and is active in the citrus gardens (mostly in Mazandaran Province as well across the country. Later, to manage the white peach scale, *Pseudaulacaspis pentagona,* an aphelinid endoparasitoid, *Prospaltella perniciosi* was imported, mass reared and applied in the infested gardens across the Northern part of the country. Afterward, in the second half of the 1970s, another species from this genus of aphelinids, *Prospaltella berlesi* was imported to control *Pseudaulacaspis pentagona*. Subsequently, a native population of *P. berlesi* was used. Due to the restrictions of the chemical pesticide utilization for controlling the berry's pests, the Silkworm Company is in charge of mass rearing and augmentation program of this agent (Abivardi 2001). When organochlorine pesticides were widely used, the progress of utilizing the natural enemies like in other parts of the world was slowed down. In the 1970s, paying attention to the biocontrol agent in the Sugarcane Agro-industrial Complex of Khuzestan agricultural farms led to mass rearing of a Scelionid wasp, *Platytelenomus hylas* as a successful parasitoid of Sugarcane borer, *Sesamia nonagrioides*. This plan regulated the pest population and there was no need for chemical treatment for several years. In 1974, two species of *Trichogramma* were imported, reared and applied against the rice stem borer, *Chilo suppressalis* in rice fields of Northern Iran. In 1988, the predatory mite, *Phytoseiulus persimilis* was imported from the Netherlands by Hooshang Daneshvar for spider mite control.

## 1.5   Hayk Mirzayans Insect Museum

In 1943, Jalal Afshar made his first efforts to create an insect collection by using two small rooms. In 1945, a few Russian entomologists (Drs. Alexandrov, Chovachin, Kiriokhin) and some young graduates of Karaj College of Agriculture who had been Afshar's students (Ghodratollah Farahbakhsh and Hayek Mirzayans) joined this laboratory for studying the locust and sun pest control and also collecting and identifying the Iranian insect fauna (Fig. 1.7). This was the cornerstone and the first step in the creation of the Hayk Mirzayans Insect Museum (HMIM).

**Fig. 1.7** Hayk Mirzayans.
(Courtesy of IRIPP)

Memories of Mirzayans, Mir Salavatian, Ghavamoddin Sharif and others indicate that those young researchers had undergone a tough time collecting excessive specimens of insects, fungi, plants and rodents in those days.

Once this unit was changed to the Insect Taxonomy Research Department (ITRD), Mirzayans was appointed as director and Mohammad Safavi acted as the Vice Dean of the department. In 1966, a national and uninterrupted plan for investigating, collecting and identifying the insect fauna of Iran was passed and implemented. As a result, department operations were unified within certain frameworks. During those years, some prestigious insect taxonomists joined this department. The outcome of such collaborations was the development of the museum with an international reputation outside the country as either the Evin Insect Museum or PPDRI Insect Museum. The headquarters and the library of the Entomological Society of Iran (ESI: http://entsoc.ir) and Iranian Phytopathological Society (IPS: http://ips.ir) are in the ITRD.

## 1.6   Iranian Research Institute of Plant Protection (IRIPP)

Laboratory of entomology and plant pest control was first founded as pests surveying department and later on, was promoted to the General Department of Pests Survey and in 1960, the name was amended to "Plant Pests and Diseases Research Institute (PPDRI)" and continued its activities under the same title until 2006 (Fig. 1.8). The Iranian Research Institute of Plant Protection (http://www.iripp.ir)

**Fig. 1.8** Iranian Research Institute of Plant Protection. (Courtesy of IRIPP)

**Fig. 1.9** Memorial stamps
for Iranian Research
Institute of Plant Protection.
(Courtesy of IRIPP)

that won this title in 2006 has currently got 10 divisions and 32 provincial divisions across the country conducting plant protection researches as a parental institute on a national scale (Fig. 1.9).

## 1.7   Department of Biological Control

The Department of Biological Control was duly established in the Iranian Research Institute of Plant Protection in 1984. The department consists of seven specialized laboratories for parasitoids, predators, beneficial microorganisms (insect pathogens and phytopathogen antagonists), radiation roles, biological materials, useful insects, molecular biology and biotechnology in the headquarters of the Institute and a laboratory in Amol city (Mazandaran Province, North Iran) which has the mandate to ensure the production of healthy food and achieving organic products.

In 1995, the approval of the National Plan for the Reduction of Chemical Pesticides usage (Development Plan for application of Biological products and Optimization of Fertilizer and Pesticide Use in Agriculture) led to a more wisely use of chemical inputs. Consequently, it primarily resulted in a reduction of chemicals pesticides usage, especially in rice fields.

During the last decades, hundreds of research projects have been carried by the Biological Control Research Division which aimed to collect, identify and implement biological control agents. The results of this division were the introduction of *Trichogramma* sp., *Bracon hebetor*, *Cryptolaemous montrouzieri*, *Chrysoperla carnea*, development of the mass-production methods, and transferring the technical knowledge to the private sections. Moreover, there are some advances in the microbial biological control agents based on the insect pathogenic bacterium, *Bacillus thuringiensis*, the entomopathogenic fungus, *Beauveria bassiana*, the entomopathogenic viruses and the phytopathogenic antagonist from *Trichoderma* genus.

The biological control division designed plans to be ready to manage obstacles related to climate change, rainfall, and newly emerged pests and simultaneously focusing on the production of healthy and sustainable crops. In line with these priorities, a mid-term plan for the development of biological control in greenhouses was presented. Providing technical knowledge for mass rearing of *Encarsia formosa* and the predatory bug, *Nesidiocoris tenuis* as important biocontrol agents for greenhouse pests are the achievements of this program. Moreover, a new biocontrol agent, the predatory muscid fly, *Coenosia attenuata* was introduced for employing in greenhouses. Upon completion of the plan, the division has issued guidelines for the implementation of native natural enemies within IPM programs, botanical products, and conservation of natural enemies for lowering the application rate of chemical pesticides. The latest research topics are attempts toward the mass production of entomopathogenic nematodes in biological control of pests.

## 1.8   Other Institutes

In addition to the Iranian Research Institute of Plant Protection (IRIPP), there are other research institutes associated with the ministry of agriculture providing departments for conducting researches on entomology and agricultural pests. Those centers are Pasteur Institute of Iran, Razi Vaccine and Serum Research Institute, Animal Science Research Institute of Iran (ASRI), Silkworm Breeding Company and the Silkworm Research Center, the Sugar Beet Seed Institute, Seed and Plant Improvement Institute, Horticultural Science Research Institute, Rice Research Institute of Iran, Cotton Research Institute, Research Institute of Forests and Rangelands, and Agricultural Biotechnology Research Institute of Iran (ABRII).

### 1.8.1   Pasteur Institute of Iran

After destructive World War I, due to famine, diseases and the need for scientific progress, the Iranian government requested assistance from the President of the Pasteur Institute of France to founding a similar facility in Iran. After an initial attempt, once World War II ended, a new round of collaborations started between the Iranian and French Pasteur Institutes. In 1968, on the occasion of the 25th anniversary of Pasteur Institute of Iran, a scientific-technical cooperation agreement was signed between the two institutes and a French microbiologist, Marcel Baltazard was appointed as President of the Pasteur Institute of Iran with the mandate to launch new public health services such as Plague control (Figs. 1.10 and 1.11) (Bagheri-Zenouz 2003).

**Fig. 1.10** WHO was meeting on plague, the research center of emerging and reemerging infectious diseases, the branch of Pasteur Institute of Iran in Akanlu, Hamedan, August 1975. Dr. Sabar Farman Farmaian, former director of Pasteur Institute of Iran in the middle. (Courtesy of Ehsan Mostafavi, Pasteur Institute of Iran)

**Fig. 1.11** Research station of Pasteur Institute of Iran in Akanloo village, Hamdean. Marcel Baltazard (second from right), Mahmoud Bahmanyar (first from right) and two unknown persons in front of the research station (1962). (Courtesy of Ehsan Mostafavi, Pasteur Institute of Iran)

The research activities conducted in Pasteur Institute of Iran included research projects on malaria, Leishmaniasis, Toxoplasmosis, Hydadits and vector-borne diseases by arthropods. Now, the Pasteur Institute of Iran (http://en.pasteur.ac.ir), after a century-old existence has become an advanced and significant center active in diverse fields such as public health and biological researches, for both research and education (Shahbaziand and Mostafavi 2018). This institute is currently accomplishing effective activities in identifying diseases such as Rabies, Tuberculosis, Malaria, Hepatitis, Aids and viral fevers and in some domains such as Rabies the institute is regarded as a reference center.

### 1.8.2 Razi Vaccine and Serum Research Institute

To prevent animal diseases and also to train some groups of veterinarians, Mostafa Qoli Bayat founded the Razi Vaccine and Serum Research Institute (http://www.rvsri. ac.ir). This institute commenced operating under the supervision of the Ministry of Agriculture and Public Interests (then Ministry of Agriculture) in 1924 by research into combating the Paramyxovirus ruminant diseases, Rinderpest and manufacturing an effective vaccine against the disease, which effectively started a new era of diseases control in the country. Soon after this success, the mission to manufacture various types of vaccines and medical serums became part of the institute's tasks. The institute's boom and progress periods dates back to 1950 when the institute was handed over to be run by Iranian specialists. A large portion of the vaccines and other biologic products have been manufactured during these years.

In the recent years, the activities of Razi Institute have been increased in both quantity and quality producing several animal and human vaccines including mumps, rubella and DTP

### 1.8.3 Animal Science Research Institute of Iran (ASRI)

By 1934, the idea of creating a center for domestic animal breeding and genetics in Iran turned into reality upon allocating the land plots of Heydarabad Village in Karaj for this purpose. Later on in 1934, Mostafa Qoli Bayat was appointed in charge of this position as well as founding the Heydar Adbad Live Stock Institute. The center is now known as the Animal Science Research Institute of Iran (http://www.asri.ir) and part of its projects aimed at honey bee and silkworm.

### 1.8.4 Silkworm Breeding Company and the Silkworm Research Center

Considering the long history of silkworm breeding in Iran and the available sources proving the Iranian root of the silkworm yellow cocoon, this industry proves to have a mysterious history in Iran. Silkworm breeding was mainly first started in Giulan province and it was also bred in Golestan, Mazandaran and Khorasan provinces. During the Safavid era, silkworm breeding was a booming business, but due to the outbreak of Pebrin disease, the government adapted regulations on the importing of healthy sericultures. These regulations were passed in March 1928 and the Sericulture Department was founded and continued its operation under various titles and names, ruled by various ministries till 1973. In 1980, a collection out of the previous units was formed under the title of "Iran Silkworm Company". Besides, a facility called" Silkworm Research Center" affiliated to the Ministry of Agriculture was organized (Iran Silk Research Center 2019).

## 1.9 Plant Protection Organization (PPO)

The Plant Protection Organization (https://ppo.ir) was formed to safeguard the country from the spread of pests and quarantine diseases as an affiliation with the Ministry of Agriculture. The PPO was duly established in 1929 when an extensive invasion of desert locust took place damaging the Iranian economy.

**Fig. 1.12** Main building of the Plant Protection Organization including its logo. (Courtesy of PPO)

The first legal plant quarantine regulation was carried out in September 1935 by the aim of eradication of pink bollworm, *Pectinophora gossypiella* outbreak in Sistan and Baluchestan province and across the shore lands of Hormozgan province. Eventually, The Department of Plant Pests Control was founded in 1941; nevertheless, it was later changed into the General Department of Pest Control. In September 1946, the legal law of plant quarantine was approved. Eventually, in 1967, upon passing the national plant protection act by the National Council, the PPO was officially and legally established (Fig. 1.12) (Zomorodi 2003). The most significant biocontrol tasks of this organization are as follows:

1. Planning and supervising the development of new and modern approaches in non-chemical and biological control.
2. Quality control of biocontrol agents and their regular monitoring.
3. Overseeing the process of agricultural produces in gardens, farms, and greenhouses.

## 1.10   Societies and Congresses

### 1.10.1   The Entomological Society of Iran (ESI)

The Entomological Society of Iran was founded on 14th. September 1968, as a result of some of the noted Iranian entomologists' efforts. The first board of directors

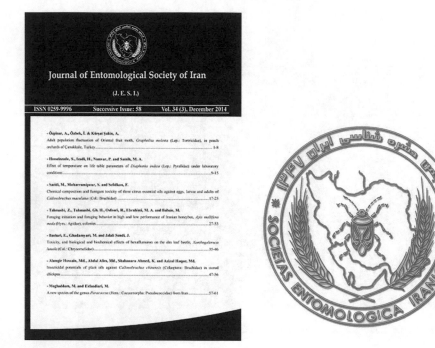

**Fig. 1.13** Formal logo of the Entomological Society of Iran (ESI) and the formal journal, JESI. (Courtesy of ESI)

consisted of Abbas Davachi, Firooz Taghizadeh, Mamoud Shojaei, Abdolghafour Mirzaei, Morteza Esmaili, Ghoratollah Farahbakhsh, Mohammad Safavi, and Hossein Sepasgozarian (Zomorodi 2003). The society has a regular journal, Journal of Entomological Society of Iran (JESI: http://jesi.areeo.ac.ir) and is among the oldest and largest national and scientific journals (Fig. 1.13). Its fiftieth anniversary was celebrated during the Plant Protection Conference held in Gorgan in 2018. The ESI has 21 boards of director members, with approximately 1500 permanent, affiliated and honorary members.

## 1.10.2  Iranian Plant Protection Congress

This congress is considered as one of the oldest scientific ones in the country and was held due to the efforts of great Iranian entomologists and phytopathologists including Abbas Davatchi, Firouz Taghizadeh, Mahmoud Shojaei, Abdolghafour Mirzaei, Ghodratollah Farahbakhsh, Mohammad Safavi, Amir Nezamoddin Ghaffari, Hossein Sepasgozarian and Morteza Esmaili, who acted as the board of directors in the first congress. The first (Fig. 1.14) and the second congresses were held at the

University of Tehran (1966 and 1968, respectively) and the 23rd congress was held at the Agriculture and Natural Resources University of Gorgan in 2018 Also, the Natinal Meeting of Biocontrol has hold biannualy since 1997 (Fig. 1.15).

**Fig. 1.14** The first Board of Directors of the Entomological Society of Iran: Top: from Right to Left: Firooz Taghizadeh, Mamoud Shojaei, Abdolghafour Mirzaei, Ghoratollah Farahbakhsh, Mohammad Safavi. Down, from right to left: Hossein Sepasgozarian, Morteza Esmaili and Abbas Davachi. (Courtesy of Alireza Saboori, University of Tehran)

**Fig. 1.15** Aziz Kharazi Pakdel, Professor of the University of Tehran, pioneer of Iranian insect pathology (2nd seated-person from left) along with a generation of his students. From left, back line: Hossein Madadi, Javad Karimi, Mohammadreza Rezapanah, Mahmoud Fazeli, Aliasghar Kousari, Shahab Manzari, Jafar Mohaghegh, Reza Talaei- Hassanlouei. Ayda Khoramnejad also visible in right corner. Front line, from left: Mojtaba Hosseini, Aziz Kharazi –pakdel, Arash Rasekh and his son, Artin. (2nd. National Meeting of Biological Control Karaj, 28 August 2013)

### 1.10.3   Iranian Acarology

Jalal Afshar (1936) published the first paper on Iranian Acari as a group of cotton pests. In the early 1940s, veterinary acarology started at the Razi Institute, Karaj, with work on ectoparasitic ticks of livestock. A list of mite as agricultural pests was prepared by Davachi (1949). Later, Farahbakhsh presented the first checklist of mites in 1961. Also, Khalil-Manesh (1969, 1973) had reports on phytophagous mites of Iran. The first paper about Iranian mites presented by Sepasgozarian in 1971 at the International Congress of Acarology in Prague. He wrote *"Until twenty years ago there was no problem of mites affecting our agricultural crops. Since then, the biological equilibrium has been disturbed and the population of useful predators decimated because of agricultural mechanization and the use of manufactured pesticides such as organochlorides. The population of mites increased gradually and this caused problems in agricultural areas* (Sepasgozarian 1973)". Sepasgozarian (1977) provided all consolidated works related to Iranian mite fauna (Cokendolpher et al. 2019). Some salient works on Iranian acarology include those by Sepasgozarian (1973), Khalil-Manesh (1969), Khalil-Manesh (1973), Daneshvar and Denmark (1982), Kamali et al. (2001) and Akrami and Saboori (2012). During new emerged generation of Iranain acarologists, Karim Kamali had a nostalgic role by supervising and teaching acarology across the universities including the University of Tehran, Chamran University of Ahwaz and Tarbiat Modares University (Fig. 1.16). The Acarological Society of Iran (ASI) (http://www.acarology.ir) established on 30 August 2008. The ASI has members from different countries and publishes a newsletter and an international journal, Persian Journal of Acarology (https://www.biotaxa.org/pja). The society arranged International Persian Congress of Acarology. The third one was held in 2017 (Hajiqanbar and Saboori 2017) (Fig. 1.17).

**Fig. 1.16**   Hossein Sepasgozarian (right) and Karim Kamali (left), pioneers of Iranian Acarology professors. (Courtesy of Alireza Saboori, University of Tehran)

**Fig. 1.17** International Persian Congress of Acarology and front cover of Persian Journal of Acarology (right corner). (Courtesy of Alireza Saboori, University of Tehran)

## 1.11   Medical Entomology

The diseases transmitted by arthropods are highly significant and diseases such as malaria and leishmania continue to remain as the first degree hygienic issues of the country; therefore, conducting applied research on the said issues is among the research objectives of the current period. Accordingly, the Department of Medical Entomology and Vector Control started its educational activities at the University of Tehran since 1953. The teaching staff of this department gravely contributed to foundation of the Medical Research Institute and thereafter to foundation of the Public Health School. Nowadays, such department is developed in more Universities and there is an association for their activities.

The Iranian Society of Medical Entomology (ISME) is related scientific society (http://issme.ir).

Iranian Journal of Arthropod-Borne Diseases is the scientific publication of the ISME (http://jad.tums.ac.ir). The society organizes a congress which 2nd International Congress of Vector-Borne Diseases and Climate Change" in conjunction with "4th National Congress of Medical Entomology" was held on Oct. 2019 (Department of Medical Entomology and Vector Control, TUMS 2019).

## 1.12   Current Biological Control Plans in Iran

Reports show that Integrated Pest Management (IPM) programs using biocontrol agents were carried out on 233,000 ha of cultivated lands and orchards. As, augmentative release of parasitoids and predators was operated over 220,000 ha, for control of pests in rice, pea, soybean, tomato, cotton, corn, cucumber, pomegranate and apple [involving species of *Trichogramma* and *Bracon hebetor* (Hym.: Braconidae), *Chrysoperla carnea* (Neu.: Chrysopidae) and phytoseiid predatory mites], sugarcane field [by *Platytelenomus hylas* (Hym.: Scelionidae)], mulberry [with *Prospaltella belesi* (Hym.: Aphelinidae)], citrus and tea [using *Cryptolaemus montrouzieri* (Col.: Cocccinellidae)] and pistachio's pests (with *C. carnea*) (Anonymous 2020). Also, about 130,000 kg of *B. thuringiensis* var. *kurstaki* (Bt) was applied on more than 13,000 ha of different crops like rice, cotton, corn and apple for control of lepidopteran pests. It is noteworthy that biological control in greenhouses which started in 2009 on 12 ha, has been extended to manage insect pests of different greenhouse crops include cucumber, tomato, strawberry, eggplant, bell pepper and

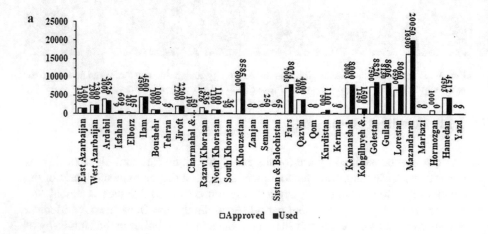

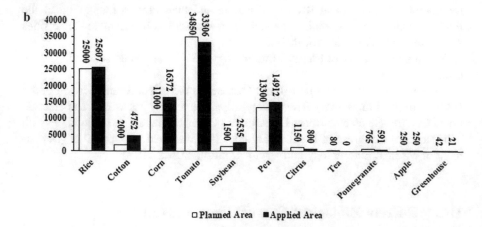

**Fig. 1.18** Information about annual mass rearing and use of biocontrol agents in various provinces of the country (**a**) and application quantity of biocontrol agents in main crops through 2018–2019 (ha) (**b**). (Data retrieved from PPO 2019)

ornamental plants (R. Marzban, Tehran, 2015, personal communication). Considering these diverse biological control plans, questions about mass production technologies, identification methods, estimation the impact of natural enemies on native/invasive pests as well as procedures for safe importation and release of natural enemies have been studied by several laboratories and facilities in the country, and private companies are working on mass rearing of parasitoids, predators and Bt. Based on the PPO report, eight macro agents used as inundative release in 27 provinces and 83 insectariums are active in 24 provinces (Personal Communication; PPO report, 2019; Department of Plant Protection 2019) (Fig. 1.18).

**Acknowledgment**  I would like to express my best appreciation to Arman Avand- Faghih (IRIPP), Ali Rezaei (PPO), Hana Haji Allahverdi Pour (IRIPP), and Ehsan Mostafavi (Pasteur Institute of Iran) for providing some images and information. We thank Alireza Saboori (University of Tehran) to check a part of the text and sharing images. Gary Dunphy (McGill University) and Ayda Khorramnejad (University of Valencia) reviewed the draft, Jafar Ershad (IRIPP) and Ahad Sahragard (Guilan University) provided some comments over the text which we would like to thanks them.

# References

Abivardi C (2001) Iranian entomology – volume 2: applied entomology. Springer Verlag, Berlin-Heidelberg-New York, pp 1033

Akrami MA, Saboori A (2012) Acari of Iran, Vol. II (Oribatid mites). University of Tehran Press, Tehran, p 281

Bagheri-Zenouz E (2003) A history of the development of agricultural sciences in Iran (from Ancient to the Present time). University of Tehran Press, Tehran, p 327

Cokendolpher JC, Zamani A, Snegovaya NY (2019) Overview of arachnids and arachnology in Iran. J Insect Biodivers Syst 5(4):301–367

Daneshvar H, Denmark HA (1982) Phytoseiids of Iran (Acarina: Phytoseiidae). Int J Acarol 8 (1):3–14

Davachi A (1949) Major economic pests of agricultural plants in Iran and their control. Chemical Institute Publication, Tehran, pp 294 [In Persian]

Department of Medical Entomology and Vector Control, School of Public Health, Tehran University of Medical Sciences. http://sph.tums.ac.ir/esph/portal/home/?47357/school-of-public-health. Accessed 30 Feb 2019

Department of Plant Protection (2019). http://www.ase.ut.ac.ir/plant-protection. Accessed 20 Feb 2020

Farahbakhsh G (1961) Family Pentatomidae (Heteroptera). In: Farahbakhsh G (ed) A checklist of economically important insects and other enemies of plants and agricultural products in Iran, vol 1. Department of Plant Protection, Ministry of Agriculture, Tehran, pp 25–28

Hajiqanbar H, Saboori A (2017) Program and Abstract book of the third International Persian Congress of Acarology. Paper presented at the 3rd International Persian Congress of Acarology, College of Science, University of Tehran, Tehran, 23–25 August 2017 [in Persian with English summary]

Heidari H (2001) Integrated production and protection management in maize and millet in north of Cameroon. Final report of kake mate NGO

Iran Silk Research Center. http://abrisham.areeo.ac.ir. Accessed 2 Feb 2019

Kamali K, Ostovan H, Atamehr A (2001) A catalog of mites and ticks (Acari) of Iran. Islamic Azad University Scientific Publication Center, Tehran, p 192

Khalil-Manesh B (1969) Tea pests in Iran. Paper presented at the 2rd Iranian Plant Protection Congress, Tehran University, Tehran [in Persian with English summary]

Khalil-Manesh B (1973) Phytophagous mite fauna of Iran. Appl Entomol Phytopathol 35:30–38 [in Persian with English summary]

Sepasgozarian H (1973) Mites and their economic importance in Iran. In: Daniel M, Rosický B (eds) Proceedings of the 3rd International Congress of Acarology held in Prague August 31– September 6, 1971. Academia, Publishing House of the Czechoslovak Academy of Sciences, pp 241–242

Sepasgozarian H (1977) The twenty years of researches in Acarology in Iran. J Iran Soc Eng 56:40–50 [in Persian]

Shahbaziand N, Mostafavi E (2018) Dr. Sabar Mirza Farman Farmaian; Benefactor and Former Director of Pasteur Institute of Iran. Iran Biomed J 22(1):1–3

Shirazi J, Attaran M, Farrokhi SH, Dadpour H, Nouri H (2011) An analytical review on the classical biological control of pests in Iran and the world. Paper presented at the 1st Biological Control Development Congress in Iran, Iranian Research Institute of Plant Protection, Tehran, 27–28 July 2011 [in Persian with English summary]

Zomorodi A (2003) History of Iranian plant protection. Agriculture Education Press, p 698

# Part I
# Predators

# Chapter 2
# Lady Beetles; Lots of Efforts but few Successes

Hossein Madadi

## 2.1 Introduction, Why Lady Beetles Are so Important?

Undoubtedly, the family Coccinellidae has been considered as important and well-known predators of sap feeder pests. This easily identified group contains 6000 described species world widely (Biranvand et al. 2016; Giorgi and Vandenberg 2009). Adults are generally convex and oval, brightly colored; their body size ranges from 0.8 to 10 mm (Triplehorn and Johnson 2005). Their tarsal formula is 3-3-3 or 4-4-4 that separate them from similar Chrysomelids (Triplehorn and Johnson 2005; Hodek et al. 2012). The life span includes seven stages, egg, four larval instars, pupa and Adult. Larvae are elongate, vigorous and their body covered by hairy tubercles or different patterns characterize them. Most ladybirds overwinter as adults in different sites. The food habits of lady beetles are various from carnivory (mostly) to herbivory (subfamily Epilachninae), and even fungivory has been identified among lady beetles (Coccinellinae: Tribe Psylloborini). Prey types are different, but most Coccinellids prefer coccids or aphids. This family has been divided into seven subfamilies; among them subfamily Coccinellinae is a very important in aphid biocontrol. One of the main species of this group is *Hippodamia variegata* (Kontodimas and Stathas, 2005). This is a Palearctic species that has been reported from different parts of the world (Franzmann 2002), including many parts of Iran (Yaghmaee and Kharazi Pakdel 1995; Lotfalizadeh 2001; Haghshenas et al. 2004; Jafari et al. 2008b; Ansari pour and Shakarami, 2012). The striking point is that this species is the dominant species of lady beetles in most parts and crops. It attacks many aphid species and also feeds on non-aphid pest species (Asghari et al. 2012). Many studies have been conducted on different biological aspects of *H. variegata* in

H. Madadi (✉)
Department of Plant Protection, Faculty of Agriculture, Bu-Ali Sina University, Hamedan, Iran
e-mail: hmadadi@basu.ac.ir

© The Author(s), under exclusive license to Springer Nature Switzerland AG 2021        23
J. Karimi, H. Madadi (eds.), *Biological Control of Insect and Mite Pests in Iran*,
Progress in Biological Control 18, https://doi.org/10.1007/978-3-030-63990-7_2

Iran. One goal of this this review is to consider those studies, their strengths and weakness.

Lady beetles can be found within many agroecosystems, orchards, greenhouses, pastures, parks, countryside and even urban environments. Because of brightly colored bodies and easy identification, most people are familiar with them and know some things about them. Most researchers consider them as a voracious predatory group that could be useful in some cases in suppressing pest populations substantially. Literately, there are some characters attributed to an efficient natural enemy. Some of the most important desirable properties enumerated for a biocontrol agent are high fecundity rate, weather concordance with prey habitat, direct numerical response to prey increase, high searching rate, ability to disperse within host habitat, patch time allocation behavior, high consumption rate, mass rearing simplicity and synchrony with host physiology, especially for parasitoids.

Lady beetles have many biological properties making them effective candidates for biological control projects. Some species (e.g., *Stethorus* species are monophagous or oligophagous and prey on a few mite species) while most others are generalist aphidophagous or coccidophagous species. It should be noted that prey types of lady beetles are not limited to mites, aphids and coccids. Spotted lady beetle, *Colemegilla maculata* (Degeer) is an effective natural enemy of eggs and larvae of Colorado potato beetle –*Leptinotarsa decemlineata* Say- in North America and Canada (Arpaia et al. 1997). Furthermore, it has been documented that other Coccinellid species frequently feed on psyllids, whiteflies, thrips, eggs and early larvae of Lepidoptera, and even alfalfa weevil, *Hypera postica* (Gyllenhal) small larvae opportunistically (Edward 2009). Many Coccinellids are voracious predators both as adults and prematures e.g., one *Hippodamia variegata* fourth instar larvae can kill up to $112.2 \pm 2.05$ and $114 \pm 2.15$ fourth instar cotton aphid and pea aphid nymphs during 24 h under laboratory conditions (Madadi et al. 2011). Consequently, the fourth instar larvae or females have been suggested as releasing stages. The life history studies of lady beetle show that most of them are active during the growing season and multivoltine. Furthermore, they could be reared under laboratory conditions on relatively simple diets. Most Coccinellids do not have special requirements for mass production. Recently, there are some attempts to rear them on artificial diets (Mirkhalilzadeh et al. 2013). Many Coccinellid species could use pollen or other non-prey foods besides natural prey allowing them to survive in the absence of their prey.

## 2.2   The Most Important Lady Beetles Species Reported from Iran

There are many studies conducted on faunistic of lady beetle species and determining the dominant species in different parts of Iran, but some of them are old conference papers or simple checklists which were not available digitally. Here we tried to address a list of described ladybird species plus their distribution and sometimes their prey as follows.

### 2.2.1   Adalia bipunctata *(Linnaeus 1758)*

This species is a common species and has been reported from different parts of Iran, including Tehran and Mazandaran provinces (Vojdani 1965), Kerman (Koohpayezadeh and Mossadegh 1991), Khorasan (Yaghmaee and Kharrazi Pakdel 1995), Golestan (Montazeri and Mossadegh 1995), Chaharmahal va Bakhtiari (Bagheri and Mossadegh 1995; Noorbakhsh 2000), Guilan (Haji-Zadeh et al. 1998), Hamedan (Ahmadi 2000), Moghan plain (Ardebill province) (Lotfalizadeh 2002; Razmjou and Hajizadeh 2000), West Azerbaijan (Akbarzadeh Shoukat and Rezvani 2000), West Khorasan (Kalantari and Sadeghi 2000), Qazvin (Mohammadbeigi 2000), Fars (Fallahzadeh et al. 2000), Isfahan (Haghshenas et al. 2004), Sistan and Baluchestan (Modarres Najaf Abadi et al. 2008a), Lorestan (Jafari and Kamali 2007; Ansari pour and Shakarami 2011; Biranvand et al. 2014), Varamin and vicinity (Samin and Shojai 2013), Kermanshah (Gholami Moghaddam et al. 2014).

### 2.2.2   Adalia decempunctata *(Linnaeus 1758)*

Golestan (Montazeri and Mossadegh 1995), Lorestan (Jafari and Kamali 2007; Ansari pour and Shakarami 2011; Biranvand et al. 2014), Arak, Markazi province (Ahmadi et al. 2012).

### 2.2.3   Adalia fasciatopunctata revelierei *Muls*

Zanjan (Mohiseni et al. 1998).

### 2.2.4   Adalia tetraspilota *(Hope 1831)*

Lorestan province (Biranvand et al. 2014).

### 2.2.5   Anisosticta bitriangularis *Say*

Khorasan (Yaghmaee and Kharazi-Pakdel 1995).

## 2.2.6   Anisosticta novemdecimpunctata *(Linnaeus 1758)*

Karaj region (Tehran province) from alfalfa fields (Vojdani 1965).

## 2.2.7   Bromus gebleri *Weise*

Fars province (Fallahzadeh et al. 2006).

## 2.2.8   Bromus octosignatus *(Gebler 1830)*

This coccidophgous lady beetle has been reported from North as well as Isfahan and Tehran provinces firstly (Vojdani 1965), then it has been reported from Kerman (Koohpayezadeh and Mossadegh 1991) Khorasan (Yaghmaee and Kharazi-Pakdel 1995), and Guilan (Haji-Zadeh et al. 1998).

## 2.2.9   Bromus undulatus

Arak, Markazi province (Ahmadi et al. 2012).

## 2.2.10   Calvia quatrodecimguttata *(Linnaeus 1758)*

Kerman (Koohpayezadeh and Mossadegh 1991).

## 2.2.11   Cheilomenes sexmaculatus *(Fabricius 1781)*

Kerman (Koohpayezadeh and Mossadegh 1991).

## 2.2.12   Chilocorus bipustulatus *(Linnaeus 1758)*

Northern part of Iran (Vojdani 1965), Kerman (Koohpayezadeh and Mossadegh 1991; Yazdani and Ebrahimi 1993), Chaharmahal va Bakhtiari Province (Bagheri and Mossadegh 1995), Golestan (Montazeri and Mossadegh 1995), Khorasan

(Yaghmaee and Kharazi Pakdel 1995), Kohkiloyeh va Boyer Ahmad (Saeedi 1998), Hamedan (Sadeghi and Khanjani 1998), Zanjan (Mohiseni et al. 1998), Mazandaran (Maafi et al. 1998), Guilan (Haji-Zadeh et al. 1998), West Khorasan (Kalantari and Sadeghi 2000), Fars (Alemansoor and 1993; Fallahzadeh et al. 2000, 2004, 2006), Fars province, Jahrom (Fallahzadeh and Hesami 2004) Alborz (Ansari pour 2012), Arak, Markazi province (Ahmadi et al. 2012), Tehran (Ghanbari et al. 2012), Varamin (South of Tehran province) (Samin and Shojai 2013), Lorestan (Biranvand et al. 2014).

### 2.2.13   Clitostethus arcuatus *Rossi*

Kerman (Koohpayezadeh and Mossadegh 1991), Chaharmahal va Bakhtiari (Bagheri and Mossadegh 1995), Golestan (Montazeri and Mossadegh 1995), Fars (Alemansoor and Ahmadi 1993; Fallahzadeh et al. 2004).

### 2.2.14   Coccinella elegantula *(Weise 1980)*

Khorasan (Yaghmaee and Kharazi Pakdel 1995), Isfahan (Bagheri and Emami 1998), Lorestan province, Khorramabad (Ansari pour and Shakarami 2011).

### 2.2.15   Coccinella magnopunctata Rybakow

Khorasan Razavi (Farahi and Sadeghi Namaghi 2011).

### 2.2.16   Coccinella novemnotata *Herbst*

Khuzestan (Ebrahimzadeh and Mossadegh 2004).

### 2.2.17   Coccinella quatrodecimpustulata *(Linnaeus 1758)*

Najafabad, Kashan (Isfahan province) (Vojdani 1965), Hamedan (Sadeghi and Khanjani 1998), Isfahan (Seifollahi et al. 2000).

## 2.2.18   Coccinella redemita *(Weise 1895)*

Chaharmahal va Bakhtiari (Bagheri and Mossadegh 1995).

## 2.2.19   Coccinella septempunctata *(Linnaeus 1758)*

This is a native and most famous species reported nearly from all faunistic studies of ladybirds in different ecosystems and crops. Vojdani (1965) firstly reported this species from Karaj vicinity. Then, it has been reported from almost all parts of Iran. Additionally, it has been reported from Kerman (Koohpayezadeh and Mossadegh 1991), Khorasan (Yaghmaee and Kharazi Pakdel 1995), Golestan(Montazeri and Mossadegh 1995), Chaharmahal va Bakhtiari (Bagheri and Mossedegh 1995; Noorbakhsh 2000), Kohkiloyeh va Boyer Ahmàd (Saeedi 1998) Hamedan (Ahmadi 2000; Sadeghi and Khanjani 1998), Zanjan (Mohiseni et al. 1998), Guilan (Haji-Zadeh et al. 1998), Moghan (Ardebill province) (Razmjou and Hajizadeh 2000); Lotfalizadeh 2002), West Azerbaijan (Akbarzadeh Shoukat and Rezvani 2000), Khuzestan (Ebrahimzadeh and Mossadegh 2004; Kalantari and Sadeghi 2000), Qazvin (Mohammadbeigi 2000), Fars (Alemansoor and Ahmadi 1993; Fallahzadeh et al. 2000), East Azerbaijan (Sadaghian et al. 2000), Isfahan (Haghshenas et al. 2004), Sistan and Baluchestan (Modarres Najaf Abadi et al. 2008a, b) on cabbage aphid and green wheat aphids respectively. This species is also dominant in Lorestan (Ansari pour and Shakarami 2011, 2012; Biranvand et al. 2014), Arak, Markazi province (Ahmadi et al. 2012), Alborz (Ansari pour 2012), Varamin and vicinity (Samin and Shojai 2013), Tehran (Ghanbari et al. 2012) and Kermanshah (Gholami Moghaddam et al. 2014).

## 2.2.20   Coccinella undecimpunctata *(Linnaeus 1758)*

Tehran, Isfahan, Golestan, Mazandaran, West Azerbaijan provinces (Montazeri and Mossadegh 1995; Vojdani 1965), Kerman (Koohpayezadeh and Mossadegh 1991), Fars (Alemansoor and Ahmadi 1993), Khorasan (Yaghmaee and Kharazi Pakdel 1995), Hamedan (Sadeghi and Khanjani 1998), Guilan (Haji-Zadeh et al. 1998), Moghan plain (Ardebill province) (Razmjou and Hajizadeh 2000), West Khorasan (Kalantari and Sadeghi 2000), Khuzestan (Ebrahimzadeh and Mossadegh 2004), Sistan and Baluchestan province (Modarres Najaf Abadi et al. 2008a, b) on cabbage aphid and green wheat aphids, Alborz (Ansari pour 2012), Varamin and vicinity (Samin and Shojai 2013).

## 2.2.21   Cryptolaemus montrouzieri *(Mulsant 1853)*

One of the most well-known and usable lady beetle species, widely mass reared and used against mealy bugs, especially in the tropical and subtropical part of Iran (North part, Guilan, Mazandaran and Golestan provinces (Montazeri and Mossadegh 1995), Lorestan province (Ansari pour and Shakarami 2011), Markazi province (Ahmadi et al. 2012). This exotic species has been introduced to Iran about 50 years before and has been used successfully.

## 2.2.22   Diloponis furschi *Ahmadi and Yazdani*

Fars province (Ahmadi et al. 1993a).

## 2.2.23   Diomus rubidus *(Motschulsky 1837)*

Koohpayezadeh and Mossadegh (1991) from Kerman province.

## 2.2.24   Exochomus flavipes *(Thunberg 1781)*

Tehran and Isfahan provinces (Vojdani 1965), Chaharmahal va Bakhtiari (Bagheri and Mossadegh 1995), Khorasan (Yaghmaee and Kharazi Pakdel 1995; Farahi and Sadeghi Namghi 2009), Lorestan (Jafari and Kamali 2007; Ansari pour and Shakarami 2012).

## 2.2.25   Exochomus illaesicollis *Roubal*

Kerman (Yazdani and Ebrahimi 1993).

## 2.2.26   Exochomus melanocephalus *(Zoubkoff 1833)*

This species has been reported from Chaharmahal va Bakhtiari (Bagheri and Mossadegh 1995), Khorasan (Yaghmaee and Kharazi-Pakdel 1995) and alfalfa fields of Lorestan province (Ansari pour and Shakarami 2012; Biranvand et al. 2014).

## 2.2.27   Exochomus nigripennis *Erichsion*

Kerman (Koohpayezadeh and Mossadegh 1991; Yazdani and Ebrahimi 1993), Golestan (Montazeri and Mossadegh 1995), Guilan (Haji-Zadeh et al. 1998), West Khorasan (Kalantari and Sadeghi 2000), Fars province on cotton whitefly (Alemansoor and Ahmadi 1993), on *Maconellicoccus hirsutus* (Fallahzadeh and Hesami 2004) and mealybugs (Fallahzadeh et al. 2006). Additionally, it was reported by Sadat Alizadeh et al. (2013) from Ahvaz, Khuzestan province and by Samin and Shojai (2013) from Varamin (Tehran province).

## 2.2.28   Exochomus nigromaculatus *(Goeze, 1777)*

Kerman (Yazdani and Ebrahimi 1993), Chaharmahal va Bakhtiari (Bagheri and Mossadegh 1995), Golestan (Montazeri and Mossadegh 1995), Hamedan (Sadeghi and Khanjani 1998), Zanjan (Mohiseni et al. 1998), Isfahan (Bagheri and Emami 1998; Haghshenas et al. 2004), Guilan (Haji-Zadeh et al. 1998), Fars (Alemansoor and Ahmadi 1993; Fallahzadeh et al. 2006; Fallahzadeh and Hesami 2004), Alborz (Ansari pour 2012), Arak, Markazi province (Ahmadi et al. 2012), Tehran (Ghanbari et al. 2012), Lorestan (Biranvand et al. 2014).

## 2.2.29   Exochomus pubescens *(Kuster 1848)*

Kerman (Yazdani and Ebrahimi 1993), Fars (Alemansoor and Ahmadi 1993), Golestan province (Montazeri and Mossadegh 1995), Isfahan (Haghshenas et al. 2004), Khuzestan (Ebrahimzadeh and Mossadegh 2004), Sistan and Baluchestan (Modarres Najaf Abadi 2008b) on green wheat aphids, Lorestan province (Jafari and Kamali 2007; Ansari pour and Shakarami 2012; Biranvand et al. 2014), Varamin and vicinity (Samin and Shojai 2013).

## 2.2.30   Exochomus quadripustulatus *(Linnaeus 1758)*

Northern part of Iran (Vojdani 1965), Kerman (Koohpayezadeh and Mossadegh 1991; Yazdani and Ebrahimi 1993), Khorasan (Yaghmaee and Kharazi Pakdel 1995) and Chaharmahal va Bakhtiari by (Bagheri and Mossadegh 1995), Kohkiloyeh va Boyer Ahmad (Saeedi 1998), Zanjan (Mohiseni et al. 1998), Isfahan (Bagheri and Emami 1998), Qazvin (Mohammadbeigi 2000), Fars (Fallahzadeh et al. 2000, 2006; Fallahzadeh and Hesami 2004), Lorestan (Ansari pour and Shakarami 2011; Biranvand et al. 2014), Alborz (Ansari pour 2012). Fallahzadeh et al. (2011)

reported this species from vine mealybug, *Planococcus ficus* (Signoret) (Hemiptera: Pseudococcidae).

## *2.2.31*   **Exochomus undulatus** *(Weise 1878)*

Kerman (Yazdani and Ebrahimi 1993), Chaharmahal va Bakhtiari (Bagheri and Mossadegh 1995), Golestan (Montazeri and Mossadegh 1995), Kohkiloyeh va Boyer Ahmad (Saeedi 1998), Isfahan (Bagheri and Emami 1998), Fars (Fallahzadeh et al. 2000), Khorasan (Farahi and Sadeghi Namghi 2009), Lorestan (Ansari pour and Shakarami 2011; Biranvand et al. 2014), Tehran (Ghanbari et al. 2012), Kermanshah (Gholami Moghaddam et al. 2014).

## *2.2.32*   **Hippodamia** *(Adonia)* **variegata** *(Goeze 1777)*

This Palearctic famous species is quietly common in different parts of Iran and mostly has been reported from Alfalfa fields. Tehran province (Vojdani 1965), Kerman (Koohpayezadeh and Mossadegh 1991), Khorasan (Yaghmaee and Kharazi Pakdel 1995), Chaharmahal va Bakhtiari (Bagheri and Mossadegh 1995), Golestan (Montazeri and Mossadegh 1995), Kohkiloyeh va Boyer Ahmad (Saeedi 1998), Hamedan (Sadeghi and Khanjani 1998), Guilan (Haji-Zadeh et al. 1998), Moghan plain (Ardebill province) (Lotfalizadeh 2002; Razmjou and Hajizadeh 2000), West Khorasan (Kalantari and Sadeghi 2000), Qazvin (Mohammadbeigi 2000), Fars (Alemansoor and Ahmadi 1993; Fallahzadeh et al. 2000), Isfahan (Haghshenas et al. 2004; Seifollahi et al. 2000), Khuzestan (Ebrahimzadeh and Mossadegh 2004), Sistan and Baluchestan (Modarres Najaf Abadi et al. 2008a, b), Lorestan (Ansari pour and Shakarami 2011, 2012; Biranvand et al. 2014) Alborz (Ansari pour 2012), Arak, Markazi province (Ahmadi et al. 2012), Tehran (Ghanbari et al. 2012), Varamin and vicinity (Samin and Shojai 2013), Kermanshah (Gholami Moghaddam et al. 2014).

## *2.2.33*   **Hippodamia tredecimpunctata** *(Linnaeus 1758)*

Chaharmahal va Bakhtiari (Bagheri and Mossadegh 1995), Kohkiloyeh va Boyer Ahmad (Saeedi 1998).

### 2.2.34    Hyperaspis duvergeri *(Fursch 1985)*

Chaharmahal va Bakhtiari (Bagheri and Mossadegh 1995).

### 2.2.35    Hyperaspis concolor *(Suffrian 1843)*

Chaharmahal va Bakhtiari (Bagheri and Mossadegh 1995).

### 2.2.36    Hyperaspis marmottani *Fairmaire*

Khorasan (Yaghmaee and Kharazi Pakdel 1995).

### 2.2.37    Hyperaspis polita *Weise*

This species reported from unknown mealybug species (Asadeh and Mossadegh 1991; Novin et al. 2000) from Khuzestan province. Moreover, it has been recorded from Kerman province (Koohpayezadeh and Mossadegh 1991) and Golestan (Montazeri and Mossadegh 1995). Furthermore, it was reported on vine mealybug, *Planococcus ficus* (Fallahzadeh et al. 2011), pink Hibiscus mealybug, *Maconellicoccus hirsutus* (Greenblatt) (Hemiptera: Pseudococcidae) (Fallahzadeh and Hesami 2004), different mealybug species (Fallahzadeh et al. 2006) from Fars province and on *M. hirsutus* from Khuzestan province (Sadat Alizadeh et al. 2013).

### 2.2.38    Hyperaspis quadrimaculata *(Redtenbacher 1844)*

Lorestan (Jafari and Kamali 2007; Ansari pour and Shakarami 2011, 2012).

### 2.2.39    Hyperaspis reppensis *(Herbst 1783)*

Alborz (Ansari pour 2012), Yazd (Zare Khormizi et al. 2014).

## 2.2.40   Hyperaspis syriaco *(Weise 1885)*

Chaharmahal va Bakhtiari (Bagheri and Mossadegh 1995).

## 2.2.41   Hyperaspis vinciqerrae *Capra*

This species was firstly reported by Sadat Alizadeh et al. (2013) as a new report on *M. hirsutus* from Ahvaz, Khuzestan Province (Southwest of Iran).

## 2.2.42   Nephus arcuatus *Kapur*

Ahvaz, Khuzestan province on pink Hibiscus mealybug *M. hirsutus* (Greenblatt) (Hemiptera: Pseudococcidae) (Sadat Alizadeh et al. 2013).

## 2.2.43   Nephus bipunctatus *(Kugelann 1794)*

Kerman (Koohpayezadeh and Mossadegh 1991), Zanjan (Mohiseni et al. 1998), Guilan (Haji-Zadeh et al. 1998) and Chahar Mahal va Bakhtiari (Esfandiari et al. 2002). On top of this, this species reported on cotton whitefly and Vine mealybug, *P. ficus* from Fars province (Alemansoor and Ahmadi 1993; Fallahzadeh et al. 2011, respectively).

## 2.2.44   Nephus biguttatus

Golestan (Montazeri and Mossadegh 1995).

## 2.2.45   Nephus bisignatus etesiacus *Fursch*

Khorasan (Yaghmaee and Kharazi Pakdel 1995).

## 2.2.46    Nephus fenestratus *Sahlberg*

Khuzestan (Novin et al. 2000), Jahrom, Fars Jahrom (Fallahzadeh et al. 2006; Fallahzadeh and Hesami 2004).

## 2.2.47    Nephus includes *Kirsch*

It has been reported as a predator of mealybugs from Khuzestan province (Asadeh and Mossadegh 1991; Novin et al. 2000), Jahrom, Fars province (Fallahzadeh et al. 2004).

## 2.2.48    Nephus nigricans *(Weise 1879)*

Lorestan (Biranvand et al. 2014).

## 2.2.49    Nephus quadrimaculatus *(Herbst 1783)*

This species formerly known as *Scymnus quadrimaculatus* has been reported on several mealybugs and Coccids from North provinces of Iran (Vojdani 1965) and Kermanshah (Gholami Moghaddam et al. 2014).

## 2.2.50    Nephus ulbrichi *(Fursch 1977)*

This species firstly reported by Jalilvand et al. (2012) on *Planococcus vovae* (Hemiptera: Pseudococcidae).

## 2.2.51    Oenopia conglobata *(Linnaeus 1758)*

Kerman (Koohpayezadeh and Mossadegh 1991), Khorasan (Yaghmaee and Kharazi Pakdel 1995), Golestan (Montazeri and Mossadegh 1995), Chaharmahal va Bakhtiari (Bagheri and Mossadegh 1995), Kohkiloyeh va Boyer Ahmad (Saeedi 1998), Hamedan (Sadeghi and Khanjani 1998), Guilan (Haji-Zadeh et al. 1998), West Khorasan (Kalantari and Sadeghi 2000), Qazvin (Mohammadbeigi 2000), Moghan region (Ardebill province) (Lotfalizadeh 2002), Fars province (Alemansoor and Ahmadi 1993; Fallahzadeh et al. 2000, 2006; Fallahzadeh and Hesami 2004), Lorestan (Jafari and Kamali 2007; Ansari pour and Shakarami 2011, 2012; Biranvand et al. 2014), Alborz (Ansari pour 2012), Tehran (Ghanbari et al. 2012), Varamin and vicinity (Samin and Shojai 2013).

### 2.2.52   Oenopia oncina *(Olivier 1808)*

This species reported from Tehran, Khuzestan, Hormozgan provinces (Vojdani 1965), Chaharmahal va Bakhtiari (Bagheri and Mossadegh 1995), Golestan (Montazeri and Mossadegh 1995), West Khorasan (Kalantari and Sadeghi 2000), Isfahan (Seifollahi et al. 2000), Fars province, Jahrom (Alemansoor and Ahmadi 1993; Fallahzadeh et al. 2006; Fallahzadeh and Hesami 2004), Lorestan Province (Ansaripour and Shakarami 2011, b, 2012; Biranvand et al. 2014), Arak, Markazi province (Ahmadi et al. 2012), Kermanshah (Gholami Moghaddam et al. 2014).

### 2.2.53   Pharoscymnus brunneosignatus *(Marder 1949)*

It has been reported from Khorasan Razavi (Ebrahimi et al. 2012).

### 2.2.54   Pharoscymnus ovoideus *Anthor,* P. arabicus *Anthor* and P. setulesus *Chevrolat*

These three species have been reported by Ahmadi and Yazdani (1991) from Fars province feeding on scale insects (Diaspididae) especially *Parlatoria blanchardi*. The first one has been reported by Haji-Zadeh et al. (1998) from Guilan.

### 2.2.55   Pharoscymnus pharoides *Marseul*

It was reported by Ahmadi et al. (1993b) from Fars province on Diaspidid scale insects firstly. Then Bagheri and Mossedegh (1995) reported it from Chaharmahal va Bakhtiari.

### 2.2.56   Pharoscymnus cf. smirnovi *(Dobzhansky 1927)*

Yazd (Zare Khormizi et al. 2014).

## 2.2.57    Propylea quatuordecimpuctata *(Linnaeus 1758)*

Chaharmahal va Bakhtiari (Bagheri and Mossadegh 1995), Khorasan (Yaghmaee and Kharazi Pakdel 1995), Golestan (Montazeri and Mossadegh 1995), Kohkiloyeh va Boyer Ahmad (Saeedi 1998), Hamedan (Sadeghi and Khanjani 1998), Mazandaran (Maafi et al. 1998), Guilan (Haji-Zadeh et al. 1998), Moghan plain (Ardebill province) (Lotfalizadeh 2002; Razmjou and Hajizadeh 2000), East Azerbaijan (Sadaghian et al. 2000), Isfahan (Haghshenas et al. (2004), Lorestan (Jafari and Kamali 2007; Ansari pour and Shakarami 2011, 2012; Biranvand et al. 2014), Alborz (Ansari pour 2012).

## 2.2.58    Psyllobora vigintiduopunctata *(Linnaeus 1758)*

Karaj, Alborz Province (Vojdani 1965), Kerman (Koohpayezadeh and Mossadegh 1991), Golestan (Montazeri and Mossadegh 1995), Golestan (Montazeri and Mossadegh 1995), Chaharmahal va Bakhtiari (Bagheri and Mossadegh 1995; Esfandiari et al. 2002), Khorasan (Yaghmaee and Kharazi Pakdel 1995), Kohkiloyeh va Boyer Ahmad (Saeedi 1998), Hamedan (Sadeghi and Khanjani 1998), Zanjan (Mohiseni et al. 1998), Guilan (Haji-Zadeh et al. 1998), West Khorasan (Kalantari and Sadeghi 2000), Fars (Fallahzadeh et al. 2000), East Azerbaijan (Sadaghian et al. 2000), Lorestan (Jafari and Kamali 2007; Ansari pour and Shakarami 2011, 2012; Biranvand et al. 2014).

## 2.2.59    Rodolia cardinalis *(Mulsant 1850)*

It is one of the most successful ladybeetle species that have been arrived in Iran in 1912. Now this species is well established in many citrus orchards (Vojdani 1965). Golestan (Montazeri and Mossadegh 1995) and Guilan (Haji-Zadeh et al. 1998) are recorded as the main sites of distribution of this predator.

## 2.2.60    Rodolia fusti *Weise*

Khorasan (Yaghmaee and Kharazi Pakdel 1995).

## 2.2.61    Scymniscus biflammatus *(Motschulsky 1837)*

Lorestan (Biranvand et al. 2014).

## 2.2.62   Scymnus argutus *Mulsant*

Chaharmahal va Bakhtiari (Bagheri and Mossadegh 1995).

## 2.2.63   Scymnus araraticus *(Iablokoff-Khnzorian 1969)*

Fars province (Yazdani and Ahmadi 1991), Guilan (Haji-Zadeh et al. 1998), Loretan (Biranvand et al. 2014).

## 2.2.64   Scymnus araxicola *(Fleischer 1900)*

Kermanshah (Gholami Moghaddam et al. 2014).

## 2.2.65   Scymnus apetzi *(Mulsant 1846)*

Fars province (Alemansoor and Ahmadi 1993; Yazdani and Ahmadi 1991), Golestan province (Montazeri and Mossadegh 1995), Chaharmahal va Bakhtiari (Bagheri and Mossadegh 1995), Mazandaran (Maafi et al. 1998), Isfahan (Bagheri and Emami 1998; Haghshenas et al. 2004), Guilan (Haji-Zadeh et al. 1998), Lorestan province (Jafari and Kamali 2007; Ansari pour and Shakarami 2011, 2012; Biranvand et al. 2014), Alborz (Ansari pour 2012).

## 2.2.66   Scymnus auritus *(Thunberg 1795)*

Fars province (Yazdani and Ahmadi 1991), Chaharmahal va Bakhtiari (Bagheri and Mossadegh 1995), Guilan (Haji-Zadeh et al. 1998).

## 2.2.67   Scymnus flavicollis *(Redtenbacher 1844)*

Fars province (Alemansoor and Ahmadi 1993; Yazdani and Ahmadi 1991), Chaharmahal va Bakhtiari (Bagheri and Mossadegh 1995), Isfahan (Bagheri and Emami 1998; Haghshenas et al. 2004), Guilan (Haji-Zadeh et al. 1998), Khuzestan (Ebrahimzadeh and Mossadegh 2004), Lorestn province (Ansari pour and

Shakarami 2011, 2012), Arak, Markazi province (Ahmadi et al. 2012), Varamin and vicinity (Samin and Shojai 2013).

### 2.2.68 Scymnus frontalis *(Fabricius 1787)*

Khorasan (Yaghmaee and Kharazi Pakdel 1995).

### 2.2.69 Scymnus impexus *(Mulsant, 1850)*

West Khorasan (Kalantari and Sadeghi 2000).

### 2.2.70 Scymnus levaillanti *(Mulsant 1850)*

This species has been reported from Kerman (Koohpayezadeh and Mossadegh 1991), Fars (Alemansoor and Ahmadi 1993; Yazdani and Ahmadi 1991), Khuzestan (Ebrahimzadeh and Mossadegh 2004).

### 2.2.71 Scymnus mongolicus *Weiser*

Isfahan (Bagheri and Emami 1998), Khuzestan (Ebrahimzadeh and Mossadegh 2004).

### 2.2.72 Scymnus nubilus *(Mulsant 1846)*

Lorestan province (Biranvand et al. 2014).

### 2.2.73 Scymnus pallipediformis *(Gunther 1958)*

Arak, Markazi province (Ahmadi et al. 2012).

## 2.2.74 Scymnus pallipes *(Mulsant 1850)*

This species was reported from Chaharmahal va Bakhtiari (Bagheri and Mossadegh 1995), Golestan (Montazeri and Mossadegh 1995), Khorasan (Yaghmaee and Kharazi Pakdel 1995), Isfahan (Bagheri and Emami 1998), and from Alfalfa fields of Lorestan province (Ansari pour and Shakarami 2012).

## 2.2.75 Scymnus quadriguttatus *(Fürsch and Kreissl 1967)*

Fars (Yazdani and Ahmadi 1991), Khorasan (Yaghmaee and Kharazi Pakdel 1995), Golestan (Montazeri and Mossadegh 1995), Guilan (Haji-Zadeh et al. 1998), Sistan and Baluchestan province (Modarres Najaf Abadi et al. 2008a).

## 2.2.76 Scymnus rubromaculatus *Goeze*

Fars (Alemansoor and Ahmadi 1993; Yazdani and Ahmadi 1991), Guilan (Haji-Zadeh et al. 1998), Isfahan (Seifollahi et al. 2000), Moghan region (Ardebill province) (Lotfalizadeh 2002), Fars province (Fallahzadeh et al. 2004), Kermanshah (Gholami Moghaddam et al. 2014), Lorestan (Biranvand et al. 2014).

## 2.2.77 Scymnus schmidt *Fürsch*

Khorasan Razavi (Ebrahimi et al. 2012).

## 2.2.78 Scymnus subvillosus *Goeze*

Kerman (Koohpayezadeh and Mossadegh 1991), Mazandaran (Agha-Janzadeh et al. 1995; Maafi et al. 1998).

## 2.2.79 Scymnus suffrianioides *(Sahlberg 1913)*

Lorestan (Biranvand et al. 2014).

## 2.2.80    Scymnus suffrianioides apetzoides *(Capra and Fursch 1967)*

Yazd (Zare Khormizi et al. 2014).

## 2.2.81    Scymnus syriacus *(Marseul 1868)*

Fars (Alemansoor and Ahmadi 1993; Yazdani and Ahmadi 1991), Chaharmahal va Bakhtiari (Bagheri and Mossadegh 1995), Golestan (Montazeri and Mossadegh 1995), Khorasan (Yaghmaee and Kharazi Pakdel 1995), Zanjan (Mohiseni et al. 1998), Isfahan (Bagheri and Emami 1998), Guilan (Haji-Zadeh et al. 1998), Moghan plain (Ardebill province) (Lotfalizadeh 2002; Razmjou and Hajizadeh 2000), Fars province (Fallahzadeh et al. 2004), Ahvaz, Khuzestan province (Ebrahimzadeh and Mossadegh 2004; Sadat Alizadeh et al. 2013), Arak, Markazi province (Ahmadi et al. 2012), Varamin and vicinity (Samin and Shojai 2013), Lorestan (Biranvand et al. 2014).

## 2.2.82    Scymnus testaceus *(Motschulsky 1837)*

Lorestan province (Biranvand et al. 2014).

## 2.2.83    Serangium montazerii *Fursch*

Golestan (Montazeri and Mossadegh 1995), Zanjan (Mohiseni et al. 1998), Guilan (Haji-Zadeh et al. 1998).

## 2.2.84    Sidis *(Nephus)* biflamulatus

Chaharmahal va Bakhtiari (Bagheri and Mossadegh 1995).

## 2.2.85    Stethorus gilvifrons *(Mulsant 1850)*

Kerman (Koohpayezadeh and Mossadegh 1991), Golestan (Montazeri and Mossadegh 1995), Khorasan (Yaghmaee and Kharazi Pakdel 1995), Chaharmahal va Bakhtiari (Bagheri and Mossadegh 1995), Isfahan (Bagheri and Emami 1998), Guilan (Haji-Zadeh et al. 1998), Moghan plain, Ardebill province (Razmjou and Hajizadeh 2000), Lorestan province (Jafari and Kamali 2007; Ansari pour and Shakarami 2011; Biranvand et al. 2014), Alborz (Ansari pour 2012), Varamin and vicinity (Samin and Shojai 2013).

## 2.2.86   Stethorus punctillum *(Weise 1891)*

Firstly, it was reported from Tehran province (Vojdani 1965) on pear trees infested to spider mites. Chaharmahal va Bakhtiari (Bagheri and Mossadegh 1995), Guilan (Haji-Zadeh et al. 1998), Lorestan (Biranvand et al. 2014).

## 2.2.87   Stethorus siphonulus *(Kapur 1948)*

It has been reported first time from Lorestan province (Ansari pour and Shakarami 2011).

## 2.2.88   Sympherobious elegans *Stephens*

Mazandaran (Maafi et al. 1998).

## 2.2.89   Synharmonia conglobata *L.*

Zanjan (Mohiseni et al. 1998).

## 2.2.90   Thea vigintiduopunctata *L.*

Hamedan (Ahmadi 2000).

## 2.2.91   Tytthaspis gebleri *(Mulsant 1850)*

Lorestan province (Biranvand et al. 2014).

## 2.2.92   Vibidia  duodecimguttata (poda, 1761)

East Azerbaijan (Mohammadi et al. 2013).

Perhaps, the outstanding publication of Iran's lady beetles written in Persian with French abstract is one of the earliest publications about the ladybird fauna of Iran (Vojdani 1965). Afterward, Many faunal studies have been conducted which there is not enough space to point to all of them. Parvizi et al. (1987) published a revision on the ladybird fauna of the West Azerbaijan province. They reported 19 species from 14 genera, which all but one (*Bulea lichatchovi* Humn.) collected species were predators of aphids, coccids, whiteflies (*Bemisia*) and herbivorous mites.

Koohpayezadeh and Mossadegh (1991), during a two-year study on lady beetle fauna of Kerman Province (South of Iran), reported 24 species. Ahmadi and Yazdani (1991) reported three new ladybird species for Iran's fauna on Diaspidid scale insects. Yazdani and Ahmadi (1991) reported eight species of genus *Scymnus* from Fars province which two species, *S. quadriguttatus* and *S. auritus* were new records for Iran's fauna of lady beetles. Yaghmaee and Kharazi Pakdel (1995) and Bagheri and Mossadegh (1995) reported 24 and 28 species in a faunistic survey on Coccinellidae from Mashhad (Khorasan) and Chaharmahal va Bakhtiari regions, respectively. From those, four species, including *Scymnus palipediformis* Gunther, *Scymnus polyps*, *Exochomus melanocephalus* Zoubk and *Rodolia fausti* Weise and five species including *Hyperaspis duvergeri* (Fursch), *H. concolor* (Suffrian), *H. syriaco* (Weiser) and *Coccinula redemita* (Weiser) were reported for the first time from Iran. In a study carried on from 1993 to 1994 in Gorgan plain (North of Iran), 27 species reported which *Serangium montazerii* Fursch was a new world record (Montazeri and Mossadegh 1995).

Some faunistic studies have been conducted on the species fauna of lady beetles in a specific crop, e.g., Sadeghi and Khanjani (1998) and Saeedi (1998), who reported 10 ladybird species of alfalfa from Hamedan and Kohkiloyeh va Boyer Ahmad (Southwest of Iran) provinces of Iran respectively. They stated the variegated lady beetle, *Hippodamia variegata* could be an important species suppressing the aphid population, especially at the second and third harvesting of alfalfa (Sadeghi and Khanjani 1998). Maafi et al. (1998) recorded five Coccinellids as natural enemies of *P. citri* from Mazandaran province. Mohiseni et al. (1998), in a four-year study, stressed on the role of some species against olive psyllid (*Euphyllura olivina* Costa) (Zanjan Province). Three species of lady beetles have been reported from Hamedan province as natural enemies of Russian wheat Aphid (Ahmadi 2000). Razmjou and Hajizadeh (2000) reported eight species in Moghan Plain (Northwest of Iran, Ardebill province) from cotton fields, of which *C. septempunctata* and *H. variegata* played an essential role in reducing cotton aphid populations. Besides aphidophagous, Coccidophagous lady beetles have been studied in different parts of Iran. Kalantari and Sadeghi (2000) listed 13 lady beetle species in dry almond orchards, while the *Adonia* (*Hippodamia*) *variegata* was the dominant species in spring and *A. bipunctata* was the most abundant in summer and autumn. Additionally, Mehrnejad (2000a) recorded the four important lady beetle species, *Oenopia conglobata, Exochomus nigripennis, Coccinella spetempunctata* and *C. undecimpunctata* as predators feeding on pistachio psylla, *Aganoscena pistaciae* in Kerman province. Among them, the *O. conglobata* was an active psyllids predator and all others preyed upon aphids and scales. Lady beetle species of other preys except aphids and coccids also have been considered e.g., Fallahzadeh et al. (2000) reported nine lady beetles preyed on olive psylla from Fars province.

In terms of habitat, most studies about lady beetle fauna have been carried on in fruit orchards and crop fields. As a rare case, Seifollahi et al. (2000) reported four lady beetle species as predators of Gaz psyllid, *Cyamophila dicora* Lognivora, from which *H. variegata* was the most dominant.

In a study on natural enemies of Ash whitefly, *Siphoninus phillyreae* conducted in Shiraz, *Nephus (Sidis) hiekei* collected and reported as a new species (Fallahzadeh et al. 2003). This species is thermophilous and has been reported previously from different countries of the Middle East (Fallahzadeh et al. 2003).

Mossadegh and Aleosfoor (2004) recorded 16 lady beetle species feeding on oleander aphid in Shiraz and Khuzestan provinces (two Southern provinces of Iran). Surprisingly, they reported aphidophagy of *Chilocorus bipustulatus, Rodolia cardinalis* and *Stethorus* sp. despite their different feeding regime. Hesami and Fallahzadeh (2004) reported eight species as natural enemies of citrus mealybug from Fars province and noted to the potential of *Hyperaspis polita* Weise and *Exochomus nigripennis* Erichson. Haghshenas et al. (2004) stated that the *H. variegata* contributed 65.8% lady beetle complex of Isfahan province wheat fields, which was a dominant species in that region.

Lady beetle faunistic of Lorestan province (West of Iran) was surveyed during 2003–2006 (Jafari and Kamali 2007). They reported 28 species from 14 different genera collectively, which *Aphidecta obliterata* (Linnacus) 1758 and *Scymnus mediterraneus* Iablokoff-Khnzorian, 1977 were new reported for the first time from Iran. (Jafari and Kamali 2007). Those species were collected on oak, spear thistle, *Cirsium vulagare* and Almond. Similar to many related studies, they reported *C. septempunctata* is the dominant among the collected species.

Modarres Najaf Abadi et al. (2008a, b) in two different studies, reported five and four lady beetle species on cabbage aphid (*Brevicoryne brassicae*) and green wheat aphid in the Sistan province while the seven spotted ladybird, *C. septempunctata* with 58.3% and 70.1% contribution was the dominant species among them.

Jafari et al. (2012) reviewed the lady beetle fauna of the Zarand Region (Kerman province, South of Iran). They recorded 13 species from nine genera from this region. They have been collected on different crops, vegetables, rosaceous fruit trees, pomegranate and pistachio. Among the collected species, *H. variegata* was a dominant species with a 52% contribution in collecting species composition.

Jafari and Kamali (2007) reported 19 ladybugs from Lorestan province, which *Hyperaspis quadrimaculata* (Redtenbacher 1843) was a new species for Iran. This species has been collected from the alfalfa field fed on cowpea aphid as prey (Ansari pour et al. 2011).

Urban lady beetles have been studied scarcely. As a rare case, 16 lady beetle species reported from Chitgar Park (Northwest of Tehran) which *O. conglobata* was the dominant species with 34.1% of species composition. All species predating on aphids, Coccids and the dominant species also feed on elm leaf beetle eggs and larvae (*Xanthogaleruca luteola* (Müller 1766)) (Coleoptera: Chrysomelidae) (Abdi et al. 2013). Mohammadi et al. (2013) reported the *Vibidia duodecimguttata* as a new record for Iranian fauna. They collected this species from East Azerbaijan province (46° 34′ 27″ N, 38° 47′ 0″ E) and 47° 21′ 58″ N, 38° 51′ 2″ E). This is a fungivorous species that feeds on powdery mildew fungi.

The *Pharoscymnus brunneosignatus* Marder and *Scymnus schmidt* Fürsch were reported for the Iran's fauna for the first time (Ebrahimi et al. 2014) in a collection program took from 2011 through 2012 on Mashhad (Northeast of Iran). This species is a coccidophagous one and was reported from France, Brazil, China, India, Pakistan, Yemen, Nepal before. In another study, *Nephus ulbrichi* Fursch was recorded on *Planococcus vovae* (Hemiptera: Pseudococcidae) on cypress trees for the first time in Iran (Jalilvand et al. 2012).

Sadat Alizadeh et al. (2013) reported the most important natural enemies of pink hibiscus mealybug (PHM), *M.hirsutus* and recorded five species which among them, the *Hyperaspis vinciquerrae* Capra was a new record for Iran. They reported that among described lady beetle species as expected, *N. arcuatus* had a high density during a relatively long period and thus, it could be the main predator of PHM in Ahvaz. Referring to the thermophilous nature of this species, this result was entirely predictable.

Most studies about the Iranian ladybird fauna are limited to recording new species or dealing with the lady beetle fauna of a specific region or province. Although recently, subfamily of Iranian Scymnine was reviewed as a checklist (Jafari et al. 2013).

In a checklist of scymninae of Iran, Jalilvand et al. (2014) surveyed natural enemies of mealybugs in Kermanshah province. Besides, Biranvand et al. (2016) published a checklist of subfamily Microweiseinae of Iran. They recorded 11 species or this subfamily and presented host plant and prey data for those species.

## 2.3   Life History and Demography

Reviewing the papers and other records published in Iran, shows that perhaps the most studied aspect of lady beetles is their life history and biology which have been studied extensively. Most of the studies have been done under laboratory conditions on one or several prey types. In most of these studies, main life table parameters like intrinsic rate of increase ($r$), net reproductive rate ($R_0$), mean generation time ($T$), finite rate of increase ($\lambda$) and gross reproductive rate (GRR) have been estimated. However, earlier studies only included simple descriptive data about reproduction or mean duration of different life stages. Recently, some researchers used life table data to show the effect of different factors on the survival and fecundity of lady beetles more efficiently. Development of age-stage, two-sex life table theory which includes the role of both males and females in population increase and consumption, helped create a progress in life table studies and many Iranian researchers rearrange their studies based on this new analysing method. However, still, there are some publications that use the traditional female-based method. In this period, many similar studies evaluated different suitability of aphids or even other prey (Pistachio psylla) (Mehrnejad et al. 2011) to the development and survival of *H. variegata*. One of the serious deficiencies of life history studies of lady beetles is that nearly all of them have been conducted under very artificial conditions, at constant temperature and humidity. Obviously, these results could not be extrapolated to real conditions and need to be validated. Most of these invesstigations have been done at $25 \pm 1$ °C, but a few researchers tested different temperatures too.

Like functional response studies, an important part of life history researches has been devoted to variegated lady beetle species as an ubiquitous dominant species in Iran. For conciseness, demographic values of different species of genus *Hippodamia* were summarized in Table 2.1 separately. Mollashahi et al. (2002) reported the life table parameters of *H. variegata* by feeding on melon aphids. Mollashahi et al. (2004a) revealed that the product index of *H. variegata* and *C. septempunctata* could be influenced by temperature and prey species; as such, this index of both species was highest at 30 °C when that were fed on *Aphis craccivora*. Studying the temperature - dependent development of *H. variegata* revealed that the minimum temperature threshold needed for growth of first instar larvae and pupae was 2.47 and 14.63, respectively. Furthermore, this species needs 346.4 degree- days (DD) to complete its development (Jafari and Vafaei Shoushatri 2010).

Stored moth egg as factitious prey has been suggested for the rearing of different lady beetle species. It has been stated that the Mediterranean flour moth was the best diet for rearing and mass production of *Pharoscymnus ovoideus* (Hajizadeh 1993). Furthermore, Tavoosi Ajvad et al. (2011) showed that *H. variagata* could complete its life cycle on Mediterranean flour moth (MFM) eggs and adult longevity on cotton aphids, pea aphids and *Ephestia kuehniella* were 53.83, 52.43 and 67.75 days. Thus the use of MFM eggs reduced the use of natural prey and labor needed.

Recently, tritrophic relationships became as a significant field of interest for Coccinellids studies and many researchers studied the basic biology of ladybirds under this context. In a tritrophic study, it was proved that host plant resistance could be effective on consumption rate of *C. septempunctata* and resulted in higher feeding rate of adults than feeding on aphids on susceptible crops (Barkhordar et al. 2012). Ghafouri Moghaddam et al. (2013) showed the premature period and survival of *H. variegata* significantly influenced by host plants. Thus, the *Sitobion avenae* Fabricius reared on wheat was more profitable for this lady beetle species. Similarly, it has been shown that Mustard plants affected on laboratory biology of this lady beetle via feeding on cabbage aphid (Nikan Jablou et al. 2013). Accordingly, the suitability of cereal aphids reared on different host plants for mass rearing of *H. variegata* has been evaluated by a few studies. Asgharian et al. (2014) showed the higher suitability of barley for this species among the four host plant tested of Russian wheat aphid. Another related research considered the effect of two resistant and susceptible wheat cultivars to *Diuraphis noxia* (Kurdjumov) on basic biology and life history of *H. variegata* (Zanganeh et al. 2015). They found a

**Table 2.1** The life history traits of *Hippodamia variegata* under different treatments

| Prey species | Developmental time (days) | Female longevity (days) | Fecundity (eggs/female) | APOP | TPOP | r (day⁻¹) | R₀ (eggs) | λ (day⁻¹) | T (days) | References |
|---|---|---|---|---|---|---|---|---|---|---|
| *Aphis gossypii* | - | - | - | - | - | 0.23 | - | - | - | Jalali et al. (2002) |
| *Aphis gossypii* | - | 53.4 ± 4.44 | 1916 ± 127 | 5.6 ± 0.27 | - | 0.254 | 387.9 | - | 23.46 | Mollashahi et al. (2004b) |
| *Aphis fabae* | - | - | 943.9 ± 53.53 | 6.2 ± 0.13 | - | 0.287 | 509 | - | 21.7 | Jafari et al. (2008a) |
| *Schizaphis graminum* | - | - | - | - | - | 0.195 | 310.5 | | 29.42 | Mollashahi and Saboori (2010) |
| *Agonoscena pistaciae* | 12.51 ± 0.4 | - | - | - | - | 0.21 | - | | - | Mehmejad et al. (2011) |
| *Aphis gossypii* | 11.69 ± 0.7 | - | - | - | - | 0.23 | - | | - | Mehmejad et al. (2011) |
| *Aphis fabae* | 16.3 ± 0.07 | 44.9 ± 3.1 | 1139.2 ± 67.8 | 3.4 ± 0.16 | 19.6 ± 0.15 | 0.203 ± 0.005 | 389.0 ± 54.0 | 1.225 ± 0.007 | 29.4 ± 0.4 | Farhadi et al. (2011) |
| *Agonoscena pistaciae* | - | - | - | 6.1 | - | 0.149 | 70.64 | 1.16 | 28.88 | Asghari et al. (2011) |
| *Aphis gossypii* | - | - | - | - | - | 0.033 ± 0.012 | 3.011 ± 1.158 | 1.033 ± 0.013 | 33.464 ± 0.894 | Mohajeri Parizi et al. (2012) |
| *Acyrthosiphon pisum* | - | - | - | - | - | 0.143 ± 0.007 | 118.511 ± 19.734 | 1.154 ± 0.009 | 33.422 ± 1.076 | Mohajeri Parizi et al. (2012) |
| *A. fabae* | - | 64.75 ± 0.87 | 799.65 ± 29.51 | 3.15 ± 0.11 | - | 0.181 ± 0.004 | 236.54 ± 15.59 | 1.198 ± 0.004 | 30.23 ± 0.39 | Golizadeh and Jafari-Behi (2012) |
| *Aphis gossypii* | - | 72.50 ± 1.61 | 587.75 ± 38.10 | 3.00 ± 0.10 | - | 0.183 ± 0.002 | 290.97 ± 12.79 | 1.201 ± 0.002 | 31.02 ± 0.26 | Golizadeh and Jafari-Behi (2012) |
| *Macrosiphum rosae* | - | 73.10 ± 1.20 | 446.95 ± 11.49 | 3.05 ± 0.14 | - | 0.156 ± 0.003 | 183.23 ± 7.52 | 1.169 ± 0.004 | 33.37 ± 0.61 | Golizadeh and Jafari-Behi (2012) |
| *Aphis fabae* | - | 60.86 ± 3.0 | 709.3 ± 59.69 | 3.53 ± 0.15 | 18.03 ± 0.18 | 0.18 ± 0.007 | 232.49 ± 39.44 | 1.20 ± 0.008 | 29.03 ± 0.49 | Rahmani and Bandani (2013a) |
| *Aphis gossypii* | 13.65 ± 0.09 | 94.31 ± 2.04 | 1332.2 ± 57.6 | 5.67 ± 1.19 | 19.89 ± 1.19 | 0.187 ± 0.009 | 621.09 ± 116.56 | 1.2056 ± 0.011 | 34.39 ± 1.26 | Davoodi Dehkordi et al. (2013b) |
| *Aphis gossypii* | - | - | - | - | - | 0.178 ± 0.005 | 262.66 ± 39.67 | 1.195 ± 0.006 | 31.33 ± 0.44 | Bigdelou (2012) |

| | | | | | | | | | | |
|---|---|---|---|---|---|---|---|---|---|---|
| *Agonoscena pistaciae* (Hexaflumuron treatment) | – | – | – | – | – | 0.086 ± 0.006 | 23.17 ± 0.346 | 1.09 ± 0.007 | 39.25 ± 2.46 | Alimohammadi Davarani et al. (2013) |
| *Agonoscena pistaciae* (Spirodiclofen treatment) | – | – | – | – | – | 0.108 ± 0.004 | 53.23 ± 0.51 | 1.11 ± 0.005 | 37.57 ± 1.525 | Alimohammadi Davarani et al. (2013) |
| *Aphis fabae* (thiamethoxam-LC$_{10}$) | | 58.15 ± 6.32 | 656.5 ± 80.38 | 2.92 ± 0.076 | 18.846 ± 0.337 | 0.15 ± 0.012 | 114.60 ± 31.57 | 1.16 ± 0.014 | 30.56 ± 1.51 | Rahmani and Bandani (2013b) |
| *Aphis fabae* (thiamethoxam-LC$_{30}$) | – | 56.58 ± 7.135 | 524.8 ± 93.81 | 3.400 ± 0.221 | 19.10 ± 0.458 | 0.13 ± 0.013 | 68.51 ± 21.58 | 1.14 ± 0.015 | 31.24 ± 1.58 | Rahmani and Bandani (2013b) |
| *Diuraphis noxia* (on Omid wheat c.v.) | – | – | 1716 | – | – | 0.24 ± 0.01 | 399.35 ± 53.01 | 1.27 ± 0.01 | 24.67 ± 0.28 | Zanganeh et al. (2015) |
| *Diuraphis noxia* (on Sardari wheat c.v.) | – | – | 1480 | – | – | 0.20 ± 0.01 | 221.56 ± 34.68 | 1.23 ± 0.01 | 26.50 ± 0.41 | Zanganeh et al. (2015) |
| *Sitobion avenae* (on Wheat) | 14.89 ± 0.13 | 48.40 ± 1.60 | 470.34 ± 19.08 | 4.32 ± 0.28 | 19.36 ± 0.35 | 0.181 ± 0.006 | 235.05 ± 36.15 | 1.199 ± 007 | 29.95 ± 0.41 | Ghafouri Moghaddam et al. (2016) |
| *Sitobion avenae* (Barley) | 16.31 ± 0.15 | 44.40 ± 1.27 | 432.79 ± 14.38 | 4.27 ± 0.24 | 20.73 ± 0.33 | 0.162 ± 0.006 | 183.97 ± 33.37 | 1.176 ± 0.007 | 32.03 ± 0.44 | Ghafouri Moghaddam et al. (2016) |
| *Sitobion avenae* (Corn) | 15.89 ± 0.14 | 53.28 ± 1.31 | 587.3 ± 15.71 | 3.83 ± 0.21 | 19.76 ± 0.23 | 0.179 ± 0.006 | 291.14 ± 46.66 | 1.197 ± 0.008 | 31.48 ± 0.40 | Ghafouri Moghaddam et al. (2016) |
| *Sitobion avenae* (Sorghum) | 16.02 ± 0.09 | 39.10 ± 1.54 | 362.47 ± 16.42 | 4.63 ± 0.29 | 20.68 ± 0.33 | 0.163 ± 0.006 | 134.30 ± 23.66 | 1.177 ± 0.008 | 29.87 ± 0.33 | Ghafouri Moghaddam et al. (2016) |

significant positive relationship between wheat resistance to Russian wheat aphid and effect on developmental time, fecundity, survival and predation rate of *H. variegata*. Suppose this synergistic effect is shown to be occurred and confirmed by more real field experiments, in that case, there is a big step towards combining biological control and host plant resistance against Russian wheat aphid. Correspondingly, the effect of four host plants of English grain aphid, *S. avenae* on life table parameters of *H .variegata* has been assessed (Ghafouri Moghaddam et al. 2016). They suggested that the wheat (c.v. Tajan) was the most suitable among four host plants tested while the barley (c.v. dasht) was the least suitable for *H. variegata* as a host plant of *S. avenae*. This difference could be chemical-based; indeed semiochemicals of different host plants or even different cultivars might affect growth, development and reproduction of lady beetles indirectly.

One of the major applications of life table is the mass rearing of predator and parasitoids. Mass rearing of lady beetles, especially *H. variegata* have been noted in many Iranian kinds of literature. Most of those studies have been carried out by using natural aphid species as prey. Talebi et al. (2014) tested the basic biology of *H. variegata* on three cereal species and in terms of female body weights, the *S. avenae* was the most profitable species to feed. As a rare study, it has been shown that the use of canola pollen with black bean aphid significantly reduced the larval period and improved the rearing of *H. variegata* (Mirkhalilzadeh Ershadi et al. 2013). Besides, they tested the effect of different artificial diets on survival and fecundity of *H. variegata* and concluded that it completed life cycle on some artificial diet albeit with a lower rate (Mirkhalilzadeh Ershadi et al. 2013). However, they did not assess the effect of feeding from artificial diets on fecundity of *H. variegata*. It seems that there is still a big gap between suitability of natural prey species and artificial diets in terms of *Hippodmia* fecundity. As a novel case study, Davoodi Dehkordi et al. (2013) proved that with increasing prey densities, the fecundity and intrinsic rate of increase of *H. variegata* raised while the developmental time and preoviposition period decreased. This study stressed the role of prey biomass quantity that a lady beetle receives to reach the development and reproduction potentials. Optimization of lady beetle sex ratio in mass rearing project is an ignored aspect which was addressed (Aldaghi et al. 2013). They showed that despite common belief, the fecundity of *H. variegata* is not reduced with decreasing male to female ratio; therefore they recommended sex ratio could be manipulated in laboratory rearings of variegated lady beetle in such a way that saves labor and maximizes reproduction output.

Demographic toxicology studies have been started recently in Iran. Some researchers studied the effects of different pesticides and acute or sublethal effects of many compounds with different modes of action assessed on life history and reproduction of different lady beetles especially *Hippodamia variegata* (Table 2.1). Possible use of *H. variegata* with pesticide application surveyed by evaluating the sublethal effect of pesticides on life history and demography of lady beetles. (Alimohammadi Davarani et al. 2013; Rahmani and Bandani 2013b). For instance, the safety of Spirodiclofen to larval stages of *H. variegata* was revealed; however, it was not safe for eggs (Alimohammadi Davarani et al. 2014). This information certainly would improve the concomitant use of insecticides and lady beetles under an IPM framework.

Another important lady beetle species that has been noted largely by Iranian researchers is *C. septempunctata*. The demography and life table parameters of this species similar to *H. variegata* have been reported under different conditions and presented briefly in Table 2.1. As a case study, some life history parameters of this species have been addressed on several combinations of mustard aphid, *Lipaphis erysimi* (Kaltenbach 1843) (Hemiptera: Aphididae) and two-spotted spider mite (Kianpour et al. 2011). Their study showed that diet including two-spotted spider mite only could not be used as an effective alternative diet for seven spotted lady beetle adults; however life table analysis would give a better insight.

Many Iranian pertinent studies considered the effect of different temperatures, and some others took into account the effects of quantity and quality of prey on ladybird biology. Hajizadeh et al. (1995) concluded that the 25–30 °C was the optimum range for activity and reproduction of an acarophagous lady beetle, *Stethorus givifrons* on two-spotted spider mite, but at 40 °C, the reproduction of *S. gilvifrons* was stopped. Moreover, the effect of seven constant temperatures on development and life table parameters of *S. gilvifrons* has been reported (Taghizadeh et al. 2008a, b). In this regard, they estimated 222.72 degree-days and 12. 47 °C as thermal constant and low temperature threshold for *S. gilvifrons* entire developmental time on *T. urtice*. This is not surprising because it is a thermophile species and has been collected from Iran's hot and arid regions. Jalali et al. (2002) reported theoretical lower threshold and thermal constant for *H. variegata*, *C. undecimpunctata*, *O. conglobata contaminata* and *E. nigripennis*. They also reported the intrinsic rate of increase of those four lady beetle species reared on psyllid or aphid prey in another study. The degree-days and lower temperature threshold for the complete life cycle of *O. conglobata* was estimated as 270.32 DD and 10 °C (Mojib Hagh Ghadam et al. 2004). These values have been reported as 285.71 DD and 9.34 °C respectively by feeding on pomegranate green aphid (Rounagh et al. 2014). Nazari et al. (2004) reported the lowest developmental threshold of *Exochomus nigromaculatus* as 10.87 for fourth instar larvae. The optimum temperature for *O. conglobata contaminata* rearing was estimated between 25–30 °C (Mokhtari and Samih 2014) and 27.5 to 32.5 °C (Rounagh et al. 2014) by feeding on *M. persicae* and *A. punicae*, respectively. Their study showed the degree-day and thermal thresholds could be impressed by prey species (Mokhtari and Samih 2014; Rounagh et al. 2014). The effects of five constant temperatures intervaled by 5 °C on demography and population growth of *Nephus arcuatus* has been presented at 25 °C (Table 2.1). Of five temperature tested, the highest life table parameters (except mean generation time) was highest at 30 °C. Thus, it has been suggested that this species shows the best efficacy at 30 °C and active in hot regions when all other lady beetles may not efficient (Zarghami et al. 2014b). In another study with cotton mealybugs, the 35 °C has been reported as an optimum temperature for growth and reproduction of *N. arcatus* (Table 2.1). The survey of the life table of *C. montrouzieri* at three temperatures showed the maximum and minimum intrinsic rate of increase (*r*) and mean generation time (T) at 27 °C and confirmed the earlier results. Therefore, 27 °C as an optimum temperature for mass production of this species was recommended (Mortazavi Malekshah et al. 2015).

Apart from *H. variegata* and *C. septempunctata*, life table and population parameters of other ladybird species have been studied. Two-spotted lady beetle, *Adalia bipunctata* is one of efficient species against different aphids, especially in orchards. The reproduction parameters of this predator was evaluated against *A. punicae* (Dehghan Dehnavi et al. 2008). Based on developmental time, immature survival and life table parameters of *A. bipunctata revelierei* Mulsant *E. kuehniella* eggs was the inferior diet compared to green peach aphid and pistachio psylla (Table 2.2) (Mehrnejad et al. 2015). Meanwhile, it has been suggested that *A. pistaciae* was an essential prey for *A. bipunctata* and this lady beetle could be used as a biocontrol candidate against pistachio psylla. Emami et al. (1998) showed that the total and daily fecundity of *Scymnus syriacus* increased linearly with temperature. Another Coccinellid which is well studied relatively is *O. conglobata*. Mojib Hagh Ghadam et al. (2002a) in surveying the effect of three different aphid species on laboratory biology of *O. conglobata* suggested the positive impact of suitable prey on predator and prey populations. Additionally, they reported the least preimaginal development rate, highest total fecundities and ovipositional duration of *O. conglobata* feeding upon poplar shoot aphid (*Chaitophorus populeti* (Panz)) at 30 °C among three distinct temperatures (20, 25 and 30 °C) tested (Mojib Hagh Ghadam et al. 2002b). The degree-day of this species was recorded as 217.3 DD. They also studied the biology of this species on *Timocallis saltans* (Nev) (Mojib Hagh Ghadam et al. 2002c). Poplar aphid, *Chaitophorus leucomelas* (Koch) was another aphid that serves as prey to study the laboratory biology of *O. conglobata*. The highest mortality was occurred at the egg stage and afterward, the mortality has been reduced. They also showed that fourth instar larvae was the most voracious stage in terms of daily feeding rate (Sadeghi et al. 2004). Most studies of ecology and predation rate of *O. conglobata* have been done in Rafsanjan region (Kerman province), the main center of pistachio and pomegranate production of Iran, where this species is one of the most dominant predators of pomegranate aphid (*Aphis punicae* Pass) and pistachio psylla. As one of the earliest study, they estimated descriptive biological parameters, prey preference and consumption rates of *O. conglobata* on pistachio psylla, confirmed the superiority of pistachio psylla than cotton aphid for development and reproduction of this species, however, it should be considered the cotton aphid as an alternative prey when pistachio psylla is unavailable (Mehrnejad and Jalali 2004). This study suggested that besides aphids, *O. conglobata* could be considered as a main predator of other target pests like psyllids. In this regard, the life table output of this predator showed the possible efficiency of *O. conglobata contaminata* to bring pomegranate green aphid under control (Table 2.2) (Rounagh and Samih 2014). The biology and consumption rate of *O. conglobata* on *Rhopalosiphum padi*, *Macrosiphum rosae* and *C. populi* showed that feeding of *R. padi* increased fecundity of *O. conglobata* significantly (Ajam Hassani 2015). The effect of three diets including pistachio psylla, cotton aphid and Meditteranean flour moth eggs on life table parameters and fitness of *O. conglobata* and *Cheilomenes sexmaculata* (Fabricius, 1781) showed the best

**Table 2.2** The life history traits of lady beetle species reared on various preys or treated with different pesticides

| Predator (prey species) | Developmental time (days) | Female longevity (days) | Fecundity (eggs/female) | APOP | TPOP | $r$ (day$^{-1}$) | $R_0$ (Eggs) | $\lambda$ (day$^{-1}$) | T (days) | References |
|---|---|---|---|---|---|---|---|---|---|---|
| *Oenopia conglobata contaminata* (Cotton Aphid) | – | – | – | – | – | 0.18 | – | – | – | Jalali et al. (2002) |
| *Coccinella undecimpunctata* (Cotton Aphid) | – | – | – | – | – | 0.23 | – | – | – | Jalali et al. (2002) |
| *Elytroleptus nigripennis* (Cotton Aphid) | – | – | – | – | – | 0.13 | – | – | – | Jalali et al. (2002) |
| *Coccinella septempunctata* (*Aphis gossypii*) | – | – | 1267 ± 149 | – | – | 0.159 | 373.9 | – | 37.25 | Mollashahi et al. (2004b) |
| *Oenopia conglobata* (*Aganoscena pistaciae*) | – | – | – | – | – | 0.16 | – | 1.8 | – | Hassani et al. (2004) |
| *Oenopia conglobata contaminata* (*Agonoscena pistaciae*) | – | – | 310 ± 16.6 | 5.7 ± 0.2 | – | 0.19 | – | – | – | Mehmejad and Jalali (2004) |
| *Oenopia conglobata contaminata* (*Aphis gossypii*) | – | – | – | 5.8 ± 0.12 | – | 0.18 | – | – | – | Mehmejad and Jalali (2004) |
| *Exochomus quadripustulatus* (*Planococcus vovae*) | – | 181.17 ± 4.16 | 468.35 ± 1.63 | – | – | 0.092 ± 0.00007 | 291.31 ± 1.01 | 1.09 ± 0.0001 | 61.906 ± 0.045 | Ameri et al. (2006) |
| *Stethorus gilvifrons* (*Olionychus afrasiaticus*) | – | 44.22 ± 0.361 | – | – | – | 0.189 ± 0.002 | 70.01 ± 4.72 | 1.2 ± 0.003 | 22.38 | Matin et al. (2008) |
| *Stethorus gilvifrons* (*Euetranychus orientalis*) | – | 45.05 ± 3.36 | 318.00 ± 32.57 | 2.88 ± 0.21 | – | 0.221 ± 0.011 | 154.08 ± 30.56 | 1.247 ± 0.014 | 22.83 ± 0.84 | Imani et al. (2009) |

(continued)

**Table 2.2** (continued)

| Predator (prey species) | Developmental time (days) | Female longevity (days) | Fecundity (eggs/female) | APOP | TPOP | $r$ (day$^{-1}$) | $R_0$ (Eggs) | $\lambda$ (day$^{-1}$) | T (days) | References |
|---|---|---|---|---|---|---|---|---|---|---|
| *Stethorus gilvifrons* (*Tetranychus turkestani*) | – | 58.00 ± 0.53 | 175.14 ± 3.19 | 2.85 ± 0.09 | – | 0.171 ± 0.007 | 97.6 ± 15.86 | 1.186 ± 0.008 | 26.76 ± 0.62 | Imani et al. (2009) |
| *Coccinella septempunctata* (*Aphis gossypii*) | | – | 1267 ± 14.0 | – | – | 0.159 | 373.91 | 1.4 | 37.25 | Mollashahi et al. (2009) |
| *Clistothetus arcuatus* (*Trialeurodes vaporariorum*) | 27.68 ± 0.31 | – | – | 6.86 ± 0.35 | – | 0.063 ± 0.002 | 224.6 | 1.065 ± 0.002 | 43.4 ± 1.1 | Yazdani et al. (2010) |
| *Cryptolaemus montrouzieri* (*Planococcus maritimus*) | 30.76 ± 0.123 | 41.24 ± 1.93 | 449.482 ± 0.99 | – | – | 0.13 | 336.67 ± 0.74 | 1.14 | 44.97 | Keshtkar et al. (2010) |
| *Cryptolaemus montrouzieri* - Abamectin treated | – | 17.1 ± 1.32 | 41.2 ± 5.1 | – | – | 0.286 ± 0.0015 | – | – | – | Ahmadi et al. (2010) |
| *Cryptolaemus montrouzieri* – Imidacloprid treated | – | 32.5 ± 5.73 | 86.4 ± 10.69 | – | – | 0.296 ± 0.0009 | – | – | – | Ahmadi et al. (2010) |
| *Coccinella undecimpunctata* (*Agonoscena pistaciae*) | 12.38 ± 0.19 | – | – | – | – | 0.22 | – | – | – | Mehmejad et al. (2011) |
| *Coccinella undecimpunctata* (*Aphis gossypii*) | 11.39 ± 0.17 | – | – | – | – | 0.23 | – | – | – | Mehmejad et al. (2011) |
| *Exochomus nigripennis* (*Agonoscena pistaciae*) | – | – | – | – | – | 0.12 | – | – | – | Mehmejad et al. (2011) |

| Species | | | | | | | | | | Reference |
|---|---|---|---|---|---|---|---|---|---|---|
| Exochomus nigripennis (Aphis gossypii) | – | – | – | – | – | 0.13 | – | – | – | Mehmejad et al. (2011) |
| Oenopia conglobata contaminata (hexaflumuron treated) | – | – | – | – | – | 0.10 ± 0.005 | 15.67 ± 2.68 | 1.10 ± 0.005 | 28.49 ± 1.51 | Baniasadi et al. (2012) |
| Oenopia conglobata contaminata (thiamethoxam treated) | – | – | – | – | – | 0.08 ± 0.003 | 8.68 ± 0.99 | 1.09 ± 0.003 | 26.03 ± 0.77 | Baniasadi et al. (2012) |
| Coccinella elegantula (Agonoscena pistaciae) | – | – | – | – | – | 0.093 | 61.54 | 1.09 | 43.9 | Parish et al. (2012b) |
| Coccinella elegantula (Aphis craccivora) | – | – | – | – | – | 0.07 | 39.36 | 1.11 | 44.73 | Parish et al. (2012b) |
| Oenopia conglobata (Myzus persicae) | – | 61 ± 3.3 | 1104 ± 34.8 | 3.2 ± 0.12 | – | – | – | – | – | Ajam Hassani (2013) |
| Nephus arcuatus (Nipaecoccus viridis) | – | 116.52 ± 6.56 | 660.47 ± 37.97 | 3.8 ± 0.8 | 21.2 ± 0.2 | 0.165 ± 0.0069 | 198.14 ± 32.3 | 1.179 ± 0.0083 | 32.12 ± 0.7 | Zarghami et al. (2013) |
| Serangium montazerii (Diaphorina citri on Page tangerine) | 20.84 ± 0.302 | 87 ± 0.79 | 457.14 ± 11.10 | – | – | 0.126 ± 0.006 | 160 ± 35.12 | 1.14 ± 0.08 | 40.57 ± 0.46 | Fotukkiaii et al. (2013) |
| Serangium montazerii (Diaphorina citri on Thompson navel orange) | 19.27 ± 0.251 | 85.57 ± 0.55 | 528.57 ± 12.37 | – | – | 0.134 ± 0.063 | 185 ± 40.59 | 1.15 ± 0.08 | 39.03 ± 0.5 | Fotukkiaii et al. (2013) |
| Cryptolaemus montrouzieri (Pulvinaria auranti on mandarin) | 27.48 ± 0.09 | 80.19 ± 0.39 | 801.0 ± 7.12 | 4.11 ± 0.06 | 31.78 ± 0.11 | 0.122 ± 0.003 | 309.02 ± 46.68 | 1.129 ± 0.004 | 47.1 ± 0.3 | Bozorg-Amirkalaee et al. (2014) |
| Cryptolaemus montrouzieri | 28.28 ± 0.12 | 80.55 ± 0.61 | 752.55 ± 5.11 | 4.25 ± 0.1 | 32.7 ± 0.21 | 0.11 ± 0.004 | 214.21 ± 40.79 | 1.116 ± 0.002 | 48.76 ± 0.66 | Bozorg-Amirkalaee et al. (2014) |

(continued)

**Table 2.2** (continued)

| Predator (prey species) | Developmental time (days) | Female longevity (days) | Fecundity (eggs/female) | APOP | TPOP | $r$ (day$^{-1}$) | $R_0$ (Eggs) | $\lambda$ (day$^{-1}$) | $T$ (days) | References |
|---|---|---|---|---|---|---|---|---|---|---|
| *(Pulvinaria auranti* on Sour orange) | | | | | | | | | | |
| *Nephus arcuatus* (*Nipaecoccus viridis*) | 29.5 ± 0.2 | 93.8 ± 10.7 | 415.2 ± 57.8 | 5.8 ± 0.3 | 35.5 ± 0.3 | 0.088 ± 0.005 | 95.4 ± 21.6 | 1.0918 ± 0.005 | 51.5 ± 1.1 | Zarghami et al. (2014b) |
| *Coccinella septempunctata* (*Macrosiphum rosae*) | – | – | – | – | – | 0.17 | 187.55 | 1.18 | 30.78 | Aminafshar et al. (2014) |
| *Oenopia conglobata* (*Aphis punicae*) | – | – | 592.11 ± 72.7 | – | – | 0.18 ± 0.009 | 251.65 ± 55.86 | 1.98 ± 0.011 | 30.5 ± 0.75 | Rounagh and Samih (2014) |
| *Cryptolaemus montrouzieri* (*Planococcus citri*) | 36.43 | 183.53 ± 5.78 | 278.62 ± 0.21 | 7.66 ± 0.10 | – | 0.081 ± 0.001 | 103.86 ± 5.73 | 1.085 ± 0.001 | 57.11 ± 0.75 | Abdollahi Ahi et al. (2015) |
| *Cryptolaemus montrouzieri* (*Pseudococcus viburni*) | – | 78.50 ± 2.1 | 385.33 ± 0.84 | 6.3 ± 0.16 | – | 0.094 ± 0.001 | 169.27 ± 5.98 | 1.099 ± 0.001 | 54.57 ± 0.14 | Abdollahi Ahi et al. (2015) |
| *Cryptolaemus montrouzieri* (*Planococcus citri-* 20 °C) | – | – | – | – | – | 0.045 ± 0.0005 | 77.52 ± 5.73 | 1.04 ± 0.0006 | 96.94 ± 1.91 | Mortazavi Malekshah et al. (2015) |
| *Cryptolaemus montrouzieri* (*Planococcus citri-* 25 °C) | – | – | – | – | – | 0.084 ± 0.0009 | 208.48 ± 9.12 | 1.14 ± 0.055 | 61.99 ± 0.74 | Mortazavi Malekshah et al. (2015) |
| *Cryptolaemus montrouzieri* (*Planococcus citri-* 27 °C) | – | – | – | – | – | 0.085 ± 0.0009 | 147.87 ± 6.97 | 1.08 ± 0.00125 | 57.37 ± 0.94 | Mortazavi Malekshah et al. (2015) |

| Species (prey) | | | | | | | | | | Reference |
|---|---|---|---|---|---|---|---|---|---|---|
| Cheilomenes sexmaculata (Ephestia eggs) | 9.76 ± 0.08 | – | 164.27 ± 35.67 | – | 4.39 ± 0.84 | 0.209 ± 0.001 | 55.332 ± 1.098 | 1.232 ± 0.002 | – | Mirhosseini et al. (2015) |
| Oenopia conglobata (Ephestia eggs) | 16.3 ± 0.10 | – | 192.8 ± 18.82 | – | 5.47 ± 0.38 | 0.159 ± 0.0003 | 77.317 ± 0.504 | 1.172 ± 0.0003 | – | Mirhosseini et al. (2015) |
| Adalia bipunctata (Agonoscena pistaciae) | 13.7 ± 0.08 | – | – | – | 6.13 ± 0.15 | 0.172 | 230.32 | – | 31.6 | Mehrnejad et al. (2015) |
| Adalia bipunctata (Myzus persicae) | 13.5 ± 0.15 | – | – | – | 6.67 ± 0.14 | 0.191 | 596.41 | – | 33.5 | Mehrnejad et al. (2015) |
| Adalia bipunctata (Ephestia kuehniella) | 16.1 ± 0.13 | – | – | – | 8.61 ± 0.16 | 0.131 | 81.40 | – | 59.5 | Mehrnejad et al. (2015) |
| Nephus arcuatus (Phenacoccus solenopsis) | – | 61.46 ± 4.26 | 125.0 ± 14.42 | – | – | 0.150 ± 0.00 | 35.809 ± 4.10 | 1.162 ± 0.00 | 23.761 ± 0.51 | Foruzan et al. (2016) |
| Oenopia conglobata contaminata (Agonoscena pistacia) | – | – | – | – | – | 0.187 ± 0.0012 | 565.31 ± 18.21 | 1.206 ± 0.0015 | 23.67 ± 0.07 | Mokhtari and Samih (2016) |
| Oenopia conglobata contaminata (Myzus persicae) | – | – | – | – | – | 0.175 ± 0.002 | 576.63 ± 21.33 | 1.191 ± 0.002 | 36.13 ± 0.13 | Mokhtari and Samih (2016) |

diets among these three is pistachio psylla for both species. However, using moth eggs could be useful specifically for rearing premature life stages therefore, reducing costs and labor needed (Mirhosseini et al. 2015). Furthermore, the age-stage, two-sex life table of this species on green peach aphid and pistachio psylla showed no difference between these two prey species in terms of suitability for reproduction and growth of *O. conglobata contaminata* (Mokhtari and Samih 2016).

Regarding aphidophagous or cocidophagous lady beetles, the life table and descriptive biological parameters of Acarophagous lady beetle have been studied scarcely. Eslami Zadeh and Pourmirza (1998) reviewed the fecundity and predation rate of *Stethorus punctillum* on red spider mite. Preadult duration and daily fecundity of *S. gilvifrons* on the old world date mite, *Oligonychus afrasiaticus* (McGregor) was reported as 18 days and five eggs at 27 °C, respectively. The survey of basic biology and predation rate of *S. gilvifrons* on two-spotted spider mite showed this predator kills 212.34 mites during its premature development averagely. The suitability of two Tetranychid mite species, *Tetranychus turkestani* Ugarov and Nycolsky and *Eutetranychus orientalis* Klein for development and reproduction of *S. gilvifrons* revealed that both species could be used as essential and complete prey. Although, the citrus brown mite seems more suitable than *T. turkestani* slightly (Imani et al. 2009)

Similar to aphidophagous lady beetles, Coccidophagous species have been studied fairly detailed. Mehdian et al. (1998) considered the rearing of *Chilocorus bipustulatus* L. on Dictyospermum scale, black scale, the Mediterranean flour moth eggs and the Angoumois grain moth eggs (*Sitotroga cerealella* (Olivier 1789)). Among them, *C. bipustulatus* did not complete its life cycle on *S. cerealella*. Although, lack of significant differences between daily fecundity and preadult development period of lady beetles reared on natural and factitious prey proved that the *E. kuehniella* eggs could be used as alternative prey for mass rearing this species. However, Mehrnejad (2000b) showed significant differences in pupal mortality of *Exochomus nigripennis* reared on *S. cerealella* eggs and citrus mealy-bugs, *Pseudococcus citri*. The natural prey made them significantly larger and heavier than those reared on angoumois grain moth eggs. Similarly, Emmai et al. (2002) showed the inadequacy of *Ephestia* and *Sitotroga* eggs compare to green citrus aphid or Spirea aphid for reproduction and development of *S. syriacus*. Ghanadamooz et al. (2010) reported the suitability of mulberry scale, *Pseudaulacapis pentagona* Targioni cultivated on pumpkin and potato for rearing of *C. bipustulatus*. Lotalizadeh et al. (2000) studied *E. quadripustulatus* (L.) biology on cypress tree mealybug and reported total developmental time on *Planococcus vovae* (Nasanov) was 34.47 days.

Perhaps the mealybug destroyer, *Cryptolaemus montrouzieri* Mulsant, 1850 is the most known and efficient Coccidophagous lady beetle in Iran. Therefore, the laboratory biology, life table parameters and consumption rate of *C. montrouzieri* Mulsant, 1850 have been studied extensively on different prey species or at various circumstances (Table 2.2). It has been revealed that this species is one of the main predators of coccids and feeds on *Pulvinaria* ovisacs, especially on citrus trees. In a choice test, the fourth instar larvae prefer cottony camellia scale, *P. floccifera*

Westwood ovisacs to *P. aurantii* Cockerell ones (Khazaeipool et al. 2008). The potential effects of Imidacloprid and Abamectin on life table and demography of *C. mountrouzieri* showed that these two pesticides had some adverse effects on fecundity and life table parameters of this lady beetle and could not be used together (Ahmadi et al. 2010). Crypt ladybird is an Australian species well accommodated with Mediterranean climate of Northern parts of Iran. Therefore, the thermal requirements and zero developmental temperature of *C. montrouzieri* was estimated as 441.09 DD and 10.46 °C for immature stages (Mortazavi Malekshah et al. 2010). It also seems that this species tolerates critical high temperatures in such a way that adult emergence and egg mortality (%) were estimated as $82 \pm 2.58$ and $10 \pm 2.0$ at 37.5 ° C. (Saberi et al. 2013). Tritrophic surveys included crypt ladybeetle, orange pulvinaria scale and citrus trees showed that the Grapefruit (c.v. Red blush) was the best host tree for this scale in terms of population parameters (Bozorg-Amirkalaee et al. 2015). Life table study also suggested that the obscure mealybug *Pseudococcus viburni* Signoret is a more suitable diet than the citrus mealybug (*Planococcus citri* Risso) as the *C. montouzieri* had higher rate of increase and net reproduction when fed by obscure mealybugs (Abdollahi Ahi et al. 2015).

## 2.4  Miscellaneous Studies

Besides aphids, cooccids, and mites, as mentioned earlier, psyllids constitute a major food item in the lady beetle food list. The thermal requirements and life table of *Coccinula elegantula* by feeding on *A. pistaciae* was evaluated (Parish et al. 2012a, b). According to results, the theoretical lower threshold and thermal constant (°D) required for complete premature development was estimated 14.9 °C and 256.4 DD (Parish et al. 2012a). The basic biology of *C. bipustulatus* on pistaschio psylla has been surveyed repeatedly. (Atrchian et al. 2014). In a tritrophic study, it has been shown that host plant species did not have any influence on life history and demography of *Serangium montazerii* feeding on citrus whitfly unexpectedly (Fotukkiaii et al. 2013). Although Bozorg-Amirkalaee et al. (2014) revealed the superiority of clementine mandarin to Sour orange as a host plant of orange *pulvinaria* scale, *Pulvinaria auranti* to survival and fecundity of *C. montrouzieri* (Table 2.2).

Collectively, numerous studies about life table have been done in Iran. In most cases, the effect of one or more factors on the population parameters of a few lady beetle species under completely artificial conditions have been investigated. Interestingly, from an extensive review of life table studies it has been cleared that from 16 studies of *H.variegata* in seven and five cases, cotton aphid and black bean aphid have been used respectively (Table 2.1). This may impose repetition of results without any further innovations. The other problem relates to data analysis. Many of the earlier studies used traditional female-based life table, while in recent years, the age-stage, two-sex life table have been used more frequently. This method estimates population parameters more accurately however, the differences between these two methods seem negligible, still, because of the unrealistic assumptions of traditional life table analysing method, the estimated outputs are unreliable. Another deficiency of conducted studies is the small cohort size, which is so small that it might produce an incorrect estimation of population parameters.

## 2.5 Predator-Prey Dynamics

### 2.5.1 Functional Response of Most Important Coccinellid Species to Different Prey Types

The functional response (F.R.) assessment is one of the most commonly used criteria evaluating predator efficiency. This is a quantitative description of a predator's ability to consume prey. Although Solomon (1949) introduced it initially, Holling (1959) developed the concept in details and described quantitative methods for measuring the functional response. It describes the relation of predation with increasing prey densities. Basically, there are three types of functional response to initial prey densities, the linear (type I), the curvilinear (type II) and the sigmoidal (type III). The output parameters of functional response experiments are search rate and handling time. Instantaneous search rate (or attack constant) is a function of encounter rate between predator and prey. It could be defined as the extent of area covered in a time unit by a predator (Holling 1961; Rogers 1972). Handling time is the amount of time spent on non-searching activities (subduing, killing, eating a prey and resting) (Juliano 2001). The type of functional response and values of related parameters of different species of lady beetles, especially variegated lady beetle, *Hippodamia variegata* towards different prey species has been addressed in Iranian studies extensively (Table 2.3).

One of the earliest functional response studies of variegated lady beetle done in Iran dates back to Jafari and Goldasteh (2009) who reported the functional response of *H. variegata* females and males to different densities of black bean aphid. Farhadi et al. (2010) considered the functional response of all life stages of *H. variegata* on the same aphid species *Aphis fabae*. Both researchers reported a type II functional response for *H. vareigata*. In spite of the shared prey type and similar equation used for analysis, their results were very different (Table 2.3). They reported type II response for all life stages, however, searching efficiency of females was more than 100 times higher than that reported by Jafari and Goldasteh (2009) although, the handling time showed lesser difference (two times). Based on their results, males and secondly, the fourth instar larvae have the highest searching efficiency among life stages and the females and fourth instar larvae showed the shortest handling time. Therefore, it seems that females are the most voracious stage. This has been confirmed by Madadi et al. (2011), who reported the fourth instar larvae has the lowest handling time compared to other stages tested in their study on the effects of prey types and experimental set-ups on the functional response type of *H. variegata*.

**Table 2.3** Functional response type and parameters of *Hippodamia variegata* to different aphid species

| Aphid species | Predator life stage | Type | Search rate (h$^{-1}$) | Handling time (h) | Arena | Reference and model used |
|---|---|---|---|---|---|---|
| *Aphis fabae* | Female | II | 0.00078 ± 0.0000111 | 0.1774 ± 0.0004 | Plexiglas cage (6*11*23 cm) | Jafari and Goldasteh (2009) (Disc equation) |
| *Aphis fabae* | Female | II | 0.00093 ± 0.000201 | 0.1999 ± 0.0094 | Plexiglas cage (6*11*23 cm) | Jafari and Goldasteh (2009) (Random predator equation) |
| *Aphis fabae* | 1st instar larvae | II | 0.0634 ± 0.0164 | 6.9332 ± 0.6657 | Petri dish | Farhadi et al. (2010) |
| *Aphis fabae* | 2nd instar larvae | II | 0.0596 ± 0.0103 | 3.3433 ± 0.24 | Petri dish | Farhadi et al. (2010) |
| *Aphis fabae* | 3rd instar larvae | II | 0.1031 ± 0.0293 | 1.9099 ± 0.1272 | Petri dish | Farhadi et al. (2010) |
| *Aphis fabae* | 4th instar larvae | II | 0.1138 ± 0.0223 | 0.4547 ± 0.023 | Petri dish | Farhadi et al. (2010) |
| *Aphis fabae* | Females | II | 0.0926 ± 0.0212 | 0.4098 ± 0.048 | Petri dish | Farhadi et al. (2010) |
| *Aphis fabae* | Males | II | 0.1589 ± 0.0435 | 1.1945 ± 0.0691 | Petri dish | Farhadi et al. (2010) |
| *Aphis nerii* | 4th instar larvae | II | 0.862 | 0.289 | Petri dish | Esmaeily et al. (2011) |
| *Aphis gossypii* | 3rd instar larvae | II | 0.033 ± 0.006 | 0.074 ± 0.041 | Three- dimensional-set-up | Madadi et al. (2011) (Disc equation) |
| *Aphis gossypii* | 4th instar larvae | II | 0.039 ± 0.002 | 0.023 ± 0.01 | Three- dimensional-set-up | Madadi et al. (2011) (Disc equation) |
| *Aphis gossypii* | Females | II | 0.043 ± 0.002 | 0.102 ± 0.013 | Three- dimensional-set-up | Madadi et al. (2011) (Disc equation) |
| *Aphis gossypii* | 3rd instar larvae | II | 0.099 ± 0.014 | 0.057 ± 0.023 | Cucumber Leaf disc | Madadi et al. (2011) (Random predator equation) |
| *Aphis gossypii* | 4th instar larvae | II | 0.122 ± 0.014 | 0.06 ± 0.016 | Cucumber Leaf disc | Madadi et al. (2011) (Random predator equation) |
| *Aphis gossypii* | Females | II | 0.094 ± 0.002 | 0.112 ± 0.018 | Cucumber Leaf disc | Madadi et al. (2011) (Random predator equation) |
| *Acyrthosiphon pisum* | 3rd. instar larvae | II | 0.066 ± 0.009 | 0.074 ± 0.031 | Petri dish | Madadi et al. (2011) |
| *Acyrthosiphon pisum* | 4th instar larvae | II | 0.06 ± 0.005 | 0.002 ± 0.024 | Petri dish | Madadi et al. (2011) |

(continued)

**Table 2.3** (continued)

| Aphid species | Predator life stage | Type | Search rate ($h^{-1}$) | Handling time (h) | Arena | Reference and model used |
|---|---|---|---|---|---|---|
| *Acyrthosiphon pisum* | Females | II | $0.099 \pm 0.006$ | $0.086 \pm 0.01$ | Petri dish | Madadi et al. (2011) |
| *Aphis gossypii* | 4th instar larvae | III | $0.0106 \pm 0.00383$ | $0.5040 \pm 0.0189$ | Cucumber leaf discs | Bigdelou (2012) |
| *Aphis gossypii* | Females | II | $0.1208 \pm 0.0162$ | $0.1843 \pm 0.017$ | Cucumber leaf discs | Bigdelou (2012) |
| *Aphis gossypii* | Males | II | $0.1340 \pm 0.0218$ | $0.6234 \pm 0.0422$ | Cucumber leaf discs | Bigdelou (2012) |
| *Aphis gossypii* | Female (One predator individual) | II | $0.122 \pm 0.021$ | $0.105 \pm 0.048$ | Black eyed bean leaf | Davvodi Dehkordi et al. (2012) |
| *Aphis gossypii* | Female (Two predator individuals) | II | $0.003 \pm 0.000424$ | $0.106 \pm 0.00441$ | Black eyed bean leaf | Davvodi Dehkordi et al. (2012) |
| *Aphis gossypii* | Female | II | $0.083 \pm 0.011$ | $0.197 \pm 0.040$ | Open Patch design. ($18 \times 23 \times 5$ cm) | Davvodi Dehkordi and Sahragard (2013a) |
| *Aphis gossypii* | 4th instar larvae | II | $0.1780 \pm 0.0592$ | $0.392 \pm 0.0337$ | Cucumber Leaf disc-20 °C | Ebrahimi Arfaa (2014) |
| *Aphis gossypii* | Females | III | $0.00809 \pm 0.00233$ | $0.4647 \pm 0.0225$ | Cucumber Leaf disc-20 °C | Ebrahimi Arfaa (2014) |
| *Aphis gossypii* | Males | II | $0.1264 \pm 0.0494$ | $0.5067 \pm 0.0864$ | Cucumber Leaf disc-20 °C | Ebrahimi Arfaa (2014) |
| *Aphis gossypii* | 4th instar larvae | II | $0.5134 \pm 0.2191$ | $0.3070 \pm 0.0154$ | Cucumber Leaf disc-25 °C | Ebrahimi Arfaa (2014) |
| *Aphis gossypii* | Females | II | $0.0856 \pm 0.0147$ | $0.0975 \pm 0.0212$ | Cucumber Leaf disc-25 °C | Ebrahimi Arfaa (2014) |
| *Aphis gossypii* | Males | II | $0.1347 \pm 0.0528$ | $1.1192 \pm 0.1063$ | Cucumber Leaf disc-25 °C | Ebrahimi Arfaa (2014) |
| *Aphis gossypii* | 4th instar larvae | II | $0.0463 \pm 0.00508$ | $0.0683 \pm 0.0183$ | Cucumber Leaf disc-30 °C | Ebrahimi Arfaa (2014) |
| *Aphis gossypii* | Females | II | $0.0484 \pm 0.0057$ | $0.0861 \pm 0.0199$ | Cucumber Leaf disc-30 °C | Ebrahimi Arfaa (2014) |

| Prey | Stage | | | | Food/condition | Reference |
|---|---|---|---|---|---|---|
| *Aphis gossypii* | Males | II | 0.0554 ± 0.00445 | 0.2034 ± 0.0247 | Cucumber Leaf disc-30 °C | Ebrahimi Arfaa (2014) |
| *Myzus persicae* | Females | II | 0.1335 ± 0.0171 | 0.1935 ± 0.0228 | Leaf disc | Hassankhani and Allahyari (2013) |
| *Myzus persicae* | Males | II | 0.1657 ± 0.0207 | 0.3339 ± 0.0237 | Leaf disc | Hassankhani and Allahyari (2013) |
| *Diuraphis noxia* | 3rd instar larvae | II | 0.0285 ± 0.00309 | 0.00791 ± 0.0309 | Potted wheat plant (Sardari c.v.) | Behnazar and Madadi (2015) |
| *Diuraphis noxia* | 4th instar larvae | II | 0.0561 ± 0.00204 | $1 \times 10^{-8} \pm 0$ | Potted wheat plant (Sardari c.v.) | Behnazar and Madadi (2015) |
| *Diuraphis noxia* | Females | II | 0.059 ± 0.0051 | 0.0168 ± 0.019 | Potted wheat plant (Sardari c.v.) | Behnazar and Madadi (2015) |
| *Diuraphis noxia* | Males | II | 0.0412 ± 0.00419 | 0.0173 ± 0.0176 | Potted wheat plant (Sardari c.v.) | Behnazar and Madadi (2015) |
| *Diuraphis noxia* | 3rd instar larvae | II | 0.0475 ± 0.00486 | 0.0782 ± 0.0185 | Potted wheat plant (Back cross c.v.) | Behnazar and Madadi (2015) |
| *Diuraphis noxia* | 4th instar larvae | II | 0.0753 ± 0.00749 | 0.0529 ± 0.0132 | Potted wheat plant (Back cross c.v.) | Behnazar and Madadi (2015) |
| *Diuraphis noxia* | Females | II | 0.0516 ± 0.00416 | 0.0077 ± 0.0123 | Potted wheat plant (Back cross c.v.) | Behnazar and Madadi (2015) |
| *Diuraphis noxia* | Males | II | 0.045 ± 0.00358 | 0.00604 ± 0.0132 | Potted wheat plant (Back cross c.v.) | Behnazar and Madadi (2015) |

They showed that prey type did not have any impact on searching rate and handling time of fourth instar larvae but did on third instar larvae and females. They postulated the fourth instar larvae is the most voracious stage of this predator and in addition to females it could be considered as releasing stages in an inundative biocontrol project against cotton aphid. This speculation was confirmed again by using Russian wheat aphid (RWA) as prey for *H. variegata* (Behnazar and Madadi 2015). Their results supported that fourth instar larvae and female adults had the highest searching efficiency on Sardari c.v. of wheat, but instead, male adults was the most voracious stage on Back cross c.v. of wheat (Table 2.3).

Virtually all of these F.R. studies on Coccinellids used the two-step analysis method developed by Juliano (2001). This method determines the type of response precisely and produces acceptable estimates of attack constant and handling time. Collectively, studies conducted on functional response of *H. variegata* in Iran showed different values for attack constant and handling time that makes comparison difficult. The first point is that most of these studies (except three cases) have been done on small artificial arenas like leaf disc or even more simple medium like Petri dish that increases predator-prey encounter rate and therefore type II functional response has been reported. However, large-scale experiments did not change the response type (Madadi et al. 2011; Davoodi Dehkordi and Sahragard 2013a). Indeed, small scale experiment only gives an insight into the predator's maximum predation potential and does not produce actual value of predation rate. Additionally, extrapolating the results of these studies to real conditions is difficult. This point is one issue that should be considered in future experiments and as possible as Petri dish experiments should be discarded. Most of these studies exhibited no difference in F.R. response type. Furthermore, searching efficiency and handling time did not deviate from corresponding values of other studies substantially. Although, as a novelty, Davoodi Dehkordi and Sahragad (2013a) used an open patch design to study the functional response of *H. variegata* to the cotton aphid. The design included the black eyed bean leaf's petiole wrapped with a wet cotton wool and was placed in an open larger plastic box (18 × 23 × 5 cm) which unlike Petri dish arena allow the lady beetles to freely move across the experimental unit. Another point is that many F.R. studies have been done with only one prey stage (mostly fourth instar nymphs), while it is obvious that predators usually encounter with a mixture of prey stages simultaneously while searching in nature. This deficiency might be attributed to difficulties in data analysis that the number of killed prey from different stages could not be incorporated in the analysis. The development of multistage functional response models could be useful for analyzing this kind of data. Similar to life table studies, among aphids, only five species have been used in F.R. evaluations and cotton aphid was used more than other preys (Table 2.3), meanwhile few fruit tree aphid species have been used as prey. This might be due to difficulties in rearing fruit tree aphid pests under laboratory conditions, but certainly, this is an apparent defect of F.R. studies of *H. variegata* conducted to date in Iran.

All life stages of *H. variegata* have predatory habits and kill their prey, but in many of F.R. studies, only a single life stage of *H. variegata* has been used individually. Although, this is a good approach to quantifying the predation potential of a single stage of a species, all life stages should be employed to reach a right conclusion. It has rarely been seen in nature that only one predatory stage kills the prey.

The other artifact of this kind of experiment is that it does not permit detection of mutual interference on number of prey killed. Actually, only in one case, two predator individuals have been used in F.R. experimental unit. The results showed it decreased searching efficiency but not handling time relative to the classic F.R. experiments used only one predator individual. Moreover, in most cases, a complex of aphidophagous predators present in a prey patch and compete with each other for shared prey. Under experimental condition, the effects of intra- and interspecific competition could not be clarified. Solving this problem could be done with prey labeling, i.e., if possible, prey labeled with traceable colors, proteins, or radioactive compounds in a way that number of killed prey by each stage could be estimated. This design also improves the reality of experiments because the effect of mutual interference among different predatory stages could be evaluated. Finally, the spatial distribution of predator and prey is a crusial factor affecting the predation rate that could not be simulated in Petri dish based experiments. Table 2.3 shows there are some variations in searching efficiency and handling time with regard to prey species and experimental arena. As would be expected, the immature stages of *H. variegata* influenced more intensively by small experimental arena than adults. This might be because adult body size is too large to be influenced by leaf area extent or host plant physical traits. Only in one case (Jafari and Goldasteh 2009) the searching rate was very small compared to other studies. This difference might be attributed to size of experimental unit. In their studies the volume of three-dimensional experimental arena searched by females was 1518 $cm^3$ while in most two dimensional experiments (leaf discs) the searching area volume was not more than about 80 $cm^3$ approximately.

Another F.R. parameter (the handling time) shows high variations among different studies. It ranges from low and biologically unrealistic values as $1 \times 10^{-8}$ h to high value as 1.9 h. This variation seems inconceivable; because it means that one third instar larva spent about 114 min in non-searching activities (pursuing, feeding, digesting, grooming and resting) for a single prey. Conversely, in another study, fourth instar larvae spent nearly zero time for handling the prey that clearly it is unrealistic and resulting from artificial data gathering and analyzing. The handling time of *H. variegata* female showed less variations and it was more congruent.

The one point that tended to be overlooked in some F.R. experiments is that some lack any control to correct treatment mortality. Of course, control mortality is negligible but considering control mortality would be increased results accuracy.

Other issue that should not be ignored in comparing the results of functional response studies is that those studies conducted at different parts of Iran and lady beetles used have been collected from different elevations and climates. It has been shown that variegated lady beetle or spotted amber ladybird shows variations in their wing size regarding their origins (Abdollahi Mesbah et al. 2015); therefore it is possible that local populations composed of different morphs with diverse biological properties that may also vary in terms of searching efficiency or handling time. However, this hypothesis needs to be validated.

The effect of different factors on searching efficiency and handling time of *H. variegata* has been evaluated. Prey types, developmental stages of predator (Bigdelou 2012; Farhadi et al. 2010; Madadi et al. 2011), experimental scale, host cultivar properties (Behnazar and Madadi 2015), temperature (Asghari et al. 2012; Ebrahimi Arfaa 2014), predator sex (Farhadi et al. 2010; Hassankhani and Allahyari 2013) and even some pesticides (Alimohammadi Davarani et al. 2012) are among the factors that have been surveyed but still it seems that the influence of many other traits (e.g. moisture, host plant qualities, feeding history and plant structure) on functional response of *H. variegata* should be addressed.

*H. variegata*, as a dominant lady beetle species of most parts of Iran, attracted most of attention to itself and it is more studied than any native lady beetle species. In addition to this species, the sporadic researches have been done on functional response of other species more or less. Similar to most *H. variegata* studies, it has been reported that *A. bipunctata* (Linnaeus 1758) exhibited type II response, while the fourth instar larvae had higher searching efficiency and lower handling time than females to pomegranate aphids, *A. punicae* (Dehghan Dehnavi et al. 2007).

There are a few studies dealing with the functional response of lady beetles to herbivorous mites. Most of these studies have been focused on functional response studies with *S. gilvifrons* (Mulsant), a major predator of different aphid species especially in South of Iran, mainly found in sugarcane and castor bean fields (Hajizadeh 1995; Modarres Awal 2001; Afshari et al. 2007; Mehrkhou et al. 2008) (Table 2.4). All such studies except Sohrabi and Shishehbor (2007), exhibited a type II response for *S. gilvifrons*. This could be doubtful, mainly because they did not specify the analysis method used for determining the type of response. The noticeable point in *S. gilvifrons* studies is low attack constant and long handling time to strawberry spider mite (Sohrabi and Shishehbor 2007). Imani et al. (2009) reported the type II functional response of *S. gilvifrons* Mulsant females to different life stages of citrus brown mite, *Eutetranychus orientalis* Klein on leaf discs. Accordingly, they reported that among stages, ladybeetle females had the lowest handling time for citrus brown mite larvae. In another study, functional response of all predatory stages of *S. gilvifrons* to *E. orientalis* (Klein) eggs assessed. Similarly, using the same prey, the fourth instar larvae of *S. gilvifrons* showed the maximum predation rate relative to females and third instar larvae. This was not unpredictable, but the striking point in their results was that the instantaneous searching rate of *S. gilvifrons* female was lower than all other tested stages, even the first instar larvae (Imani and Shishehbor 2011). Of course, they did not compare the results statistically, but 95% confidence intervals of searching efficiency were overlapped, which might imply nonsignificant difference in results. In terms of handling time, the fourth instar larvae and females have the most voracious life stages of *S. gilviforns* similar to aphidophagous lady beetles. Approximately all of these studies have been conducted under artificial laboratory conditions. This might be inevitable because the size of predator and prey mites is tiny and recording the number of killed prey is not possible easily without examining under stereomicroscope. It should be noted that apart from *S. gilvifrons*

**Table 2.4** Functional response type and parameters of acarophagous lady beetle, *Stethorus gilvifrons* to different mite species

| Predator species | Prey species | Type | Search rate (h$^{-1}$) or b (type III) | Handling time (h) | Arena | Study |
|---|---|---|---|---|---|---|
| *Stethorus gilvifrons* (Female) | *Tetranychus turkestani* | III | 0.0012 | 0.415 | Cowpea leaf disc | Sohrabi and Shishehbor (2007) |
| *Stethorus gilvifrons* (Female) | *Oligonychus afrasiaticus* | II | 0.085 | 0.138 | Petri Dish | Matin et al. (2010) |
| *Stethorus gilvifrons* (1st instar larvae) | *Eutetranychus orientalis* | II | 0.065 ± 0.10 | 0.207 ± 0.036 | Castor bean leaf arena (16 cm$^2$) | Imani and Shishehbor (2011) |
| *Stethorus gilvifrons* (2nd instar larvae) | *Eutetranychus orientalis* | II | 0.089 ± 0.014 | 0.166 ± 0.029 | Castor bean leaf arena (16 cm$^2$) | Imani and Shishehbor (2011) |
| *Stethorus gilvifrons* (3rd instar larvae) | *Eutetranychus orientalis* | II | 0.143 ± 0.033 | 0.150 ± 0.029 | Castor bean leaf arena (16 cm$^2$) | Imani and Shishehbor (2011) |
| *Stethorus gilvifrons* (4th instar larvae) | *Eutetranychus orientalis* | II | 0.125 ± 0.028 | 0.134 ± 0.034 | Castor bean leaf arena (16 cm$^2$) | Imani and Shishehbor (2011) |
| *Stethorus gilvifrons* (female) | *Eutetranychus orientalis* | II | 0.046 ± 0.004 | 0.082 ± 0.02 | Castor bean leaf arena (16 cm$^2$) | Imani and Shishehbor (2011) |
| *Stethorus gilvifrons* | *Eotetranychus hirsti* (Eggs) | II | 0.13 | 0.196 | Petri Dish | Sharifnia (2012) |
| *Stethorus gilvifrons* | *Eotetranychus hirsti* (Female) | II | 0.14 | 0.193 | Petri Dish | Sharifnia (2012) |
| *Stethorus gilvifrons* | *Tetranychus urticae* | II | 0.038 | 0.139 | Not mentioned- 15 °C | Mehrkhou et al. (2008) |
| *Stethorus gilvifrons* | *Tetranychus urticae* | II | 0.053 | 0.053 | Not mentioned- 35 °C | Mehrkhou et al. (2008) |

other acarophagous lady beetle species have not been studied so far and this should be considered in future researches in Iran.

In addition to those researches presented here, some studies have been done on the functional response of *S. gilvifrons* to different prey types or under different conditions but most of them mainly were unpublished MSc. thesis, so that access to their results or experimental details is difficult or impossible.

Except those studies mentioned above, there are some sporadic investigations about the functional response of other native or exotic lady beetles such as *Exochomus nigromaculatus* (Goeze), *Adalia bipunctata* (Linnaeus 1758), *Nephus arccuatus* Kapur, *Serangium montazerii* Fürsch, *Cryptolaemus montrouzieri* Mulsant and *Scymnus syriacus* Marseul with different prey species (Table 2.5). In most of these studies, the different life stages of ladybird showed type II that is common among related documents (Table 2.5). Different aphid species have been used as prey in most of F.R. studies. Below, we will review some of the most relevant ones.

*Exochomus nigromaculatus* Goeze is one of the important but less known lady beetles species showing type II functional response to two aphid species, oleander aphid, *A. nerii* Boyer de Fonscolombe and cowpea aphid, *A. craccivora* Koch, 1854 (Nazari et al. 2005). They reported the higher suitability of oleander aphid relative to cowpea aphid for *E. nigromaculatus* feeding according to searching efficiency.

Dehghan Dehnavi et al. (2007) again showed the superiority of fourth instar larvae of *A. bipunctata* in predation of their prey relative to females. However, their study duration was only six hours which makes the comparison of outcomes be more complex, because typical F.R. experiments last about 24 hours.

Another native lady beetle species, *Scymnus syriacus* Marseul adults, exhibited an inverse density- dependent response to black bean aphid density. As expected, the females have higher instantaneous searching efficiency and lower handling time than males (Sabaghi et al. 2011a ). Although, this species showed type III functional response to increasing prey densities of *A. craccivora*. They showed the inverse relationship of arena size and increasing prey densities with the searching efficacy of *S. syriacus*. The handling time ranged from 0.494 to 0.687, depending on arena size (Sabaghi et al. 2011b).

Fotukkiaii and Sahragard (2012, 2013) reported type II response of whitefly lady beetle, *S. montazerii* females to different densities of citrus whitefly eggs, while the kind of patch design (closed or open) did not have any influence on functional response type. They also showed the patch time residence of *S. montazerii* was inversely dependent on prey density (Fotukkiaii and Sahragard 2013). They stressed higher efficiencies of fourth instar larvae than females of *S. montazerii* against citrus whitefly that have been confirmed with another lady beetle species. Although, they implied their control potential against *D. citri*.

The functional response of *Cryptolaemus montrouzieri* Mulsant (fourth Instar larvae and females) to different stages of citrus mealybug *Planococcus citri* (Risso) (second, third and females) was also investigated (Abdollahi Ahi et al. 2012). They showed that fourth instar larvae of *C. montrouzieri* showed a type III response to the second instar nymph of citrus mealybug suggesting that the earlier release of

**Table 2.5** Functional response type and parameters of different lady beetles to various kinds of prey

| Predator species | Prey species | Type | Search rate (h$^{-1}$) or b (Type III) | Handling time ($t_h$) | Arena | Study (Model) |
|---|---|---|---|---|---|---|
| Oenopia conglobata | Aganoscena pistaciae | II | 0.0469 | 0.0152 | Petri dish (Pistachio leaf disc) | Hassani et al. (2004) |
| Exochomus nigromaculatus (4th instar larvae) | Aphis nerii | II | 1.24 | 0.006 | Petri dish (Host plant leaf) | Nazari et al. (2005) Disc equation |
| Exochomus nigromaculatus (Female) | Aphis nerii | II | 1.12 | 0.007 | Petri dish (Host plant leaf) | Nazari et al. (2005) (Disc equation) |
| Exochomus nigromaculatus (4th instar larvae) | Aphis craccivora | II | 1.17 | 0.009 | (Host plant leaf) Petri dish | Nazari et al. (2005) (Disc equation) |
| Exochomus nigromaculatus (Female) | Aphis craccivora | II | 1.01 | 0.008 | (Host plant leaf) Petri dish | Nazari et al. (2005) (Disc equation) |
| Adalia bipunctata (4th instar larvae) | Aphis punicae | II | 0.22 ± 0.037 | 0.073 ± 0.021 | Pomegranate leaf disc | Dehghan Dehnavi et al. (2007) |
| Adalia bipunctata (Female) | Aphis punicae | II | 0.144 ± 0.031 | 0.084 ± 0.037 | Pomegranate leaf disc | Dehghan Dehnavi et al. (2007) |
| Scymnus syriacus (Female) | Aphis fabae | II | 0.123 ± 0.007 | 0.4349 ± 0.012 | Plastic container (15x13 cm). | Sabaghi et al. (2011a) |
| Scymnus syriacus (Male) | Aphis fabae | II | 0.115 ± 0.008 | 0.5145 ± 0.0169 | Plastic container (15x13 cm). | Sabaghi et al. (2011a) |
| Scymnus syriacus (Female) | Aphis craccivora | III | Different values | – | Broad bean leaf at five different arenas (195, 247, 304, 385 and 650 cm$^2$) | Sabaghi et al. (2011b) |

(continued)

**Table 2.5** (continued)

| Predator species | Prey species | Type | Search rate (h$^{-1}$) or b (Type III) | Handling time (h) | Arena | Study (Model) |
|---|---|---|---|---|---|---|
| *Oenopia conglobata contaminata* | *Brevicoryne brassicae* | II | 0.041 ± 0.012 | 1.3867 ± 0.164 | Petri Dish (Cabbage leaf) | Rounagh and Samih (2012) |
| *Cryptolaemus montrouzieri* (female) | *Planococcus citri* (3rd instar nymphs) | II | 0.053 ± 0.008 | 0.4934 ± 0.054 | Petri Dish (Coleus leaves) | Ghorbanian, et al. (2010) |
| *Cryptolaemus montrouzieri* (male) | *Planococcus citri* (3rd instar nymphs) | II | 0.0403 ± 0.0056 | 0.4625 ± 0.0568 | Petri Dish (Coleus leaves) | Ghorbanian et al. (2010) |
| *Cryptolaemus montrouzieri* (4th instar larvae) | *Planococcus citri* (2nd instar nymphs) | III | 0.0126 ± 0.004 | 0.701 ± 0.033 | Petri dish | Abdollahi Ahi et al. (2012) |
| *Cryptolaemus montrouzieri* (4th instar larvae) | *Planococcus citri* (3rd. instar nymphs) | III | 0.0165 ± 0.0067 | 1.012 ± 0.0504 | Petri dish | Abdollahi Ahi et al. (2012) |
| *Cryptolaemus montrouzieri* (4th instar larvae) | *Planococcus citri* (female) | II | 0.002 ± 0.0032 | 2.05 ± 0.059 | Petri dish | Abdollahi Ahi et al. (2) (Disc equation) |
| *Cryptolaemus montrouzieri* (female) | *Planococcus citri* (2nd instar nymphs) | II | 0.048 ± 0.003 | 0.177 ± 0.033 | Petri dish | Abdollahi Ahi et al. (2) (Disc equation) |
| *Cryptolaemus montrouzieri* (female) | *Planococcus citri* (3rd instar nymphs) | III | 0.013 ± 0.0064 | 0.843 ± 0.058 | Petri dish | Abdollahi Ahi et al. (2) |
| *Cryptolaemus montrouzieri* (female) | *Planococcus citri* (female) | II | 0.098 ± 0.0186 | 2.794 ± 0.123 | Petri dish | Abdollahi Ahi et al. (2) (Disc equation) |
| *Rodolia cardinalis* (females and 4th instar larvae) | *Icerya purchase* (Maskell) | II | Not available | Not available | Petri dish (Bitter orange and Spurge laurels) | Bazyar et al. (2012) (Disc equation) |

| | | | | | | |
|---|---|---|---|---|---|---|
| Serangium montazerii (4th instar larvae) | Dialeurodes citri | II | 0.2540 ± 0.0587 | C.3715 ± 0.0174 | Petri dish (on Citrus leaf) | Fotukkiaii and Sahragard (2013) |
| Serangium montazerii (Female) | Dialeurodes citri | II | 0.1614 ± 0.0456 | 0.4641 ± 0.0377 | Petri dish (on Citrus leaf) | Fotukkiaii and Sahragard (2013) |
| Serangium montazerii | Dialeurodes citri | II | 0.0421 ± 0.00945 | 0.0896 ± 0.0362 | Open patch design (transparent plastic container 12 × 9 × 5 cm) | Fotukkiaii and Sahragard (2013) |
| Oenopia conglobata contaminata (Female) | Myzus persicae | II | 0.063 ± 0.0089 | 0 1425 ± 0.029 | Petri Dish (on Cabbage leaf disc) | Samih and Mokhtari (2014) |
| Nephus arcuatus | Nipaecoccus viridis (Eggs) | III | 0.00311 ± 0.00267 | 0.2819 ± 0.0107 | Plastic container (9 × 7 × 3) | Zarghami et al. (2014a) |
| Oenopia conglobata contaminata (3rd instar larvae) | Aphis punicae | II (at 25 °C) | 0.092 ± 0.0194 | 0.5008 ± 0.0428 | Petri dish | Rounagh and Samih (2015) |
| Oenopia conglobata contaminata (4th instar larvae) | Aphis punicae | III (at 25 °C) | 0.0124 ± 0.0596 | 0.3389 ± 0.0410 | Petri dish | Rounagh and Samih (2015) |
| Oenopia conglobata contaminata (Female) | Aphis punicae | II (at 25 °C) | 0.1276 ± 0.0379 | 0.3965 ± 0.0452 | Petri dish | Rounagh and Samih (2015) |
| Cryptolaemus montrouzieri (3rd instar larvae, female and male) | Planococcus citri | II (except for 3rd instar larvae at 40 °C) | Different values | Different values | Petri dish | Mohasesian et al. (2015) |

fourth instar larvae could be more profitable. The authors used Holling's disc equation and Roger's random predator equation for estimating searching efficiency and handling time for type II and III functional responses of *C. montrouzieri*, respectively. The selection of each these two models for the data analysis does not depend on the type of F.R. exhibited by the predator but the random predator equation should have been used whenever there is prey depletion (Juliano 2001).

In a detailed study, the functional response of *C. montrouzieri* (third instar larva, female and male) to citrus mealybug (second and third instar nymphs) at seven different temperatures ranged from 15 to 40 °C with 5 degrees intervals was investigated (Mohasesian et al. 2015). They also showed type II response for all predator stages except the third instar larvae at 40 °C exhibited type III. Moreover, there is an increase in searching efficiency (from $0.004 \pm 0.015$ to $0.008 \pm 0.031$ $h^{-1}$ for females) and decrease in handling time (from $0.279 \pm 2.314$ to $0.216 \pm 1.821$ h for females) of *C. montrouzieri* associated with increasing temperature. This issue implies the greater potential of this predator at higher temperatures, especially against mealybugs (Mohasesian et al. 2015).

Samih and Mokhtari (2014) worked on functional and numerical responses and the predation rate of *O. conglobata contaminata* recommended use of this lady beetle against green peach aphid in orchards.

As a case study, native lady beetle *Nephus arcuatus* Kapur functional response to mealybug *Nipaecoccus viridis* (Newstead) matched Holling's type III response (Zarghami et al. 2014a, 2015). This result suggests this thermophilic species, mainly found in South and Southwestern of Iran could regulate the citrus mealybug populations. High consumption rate, long oviposition period, density-dependent functional response, direct numerical response, high reproductive potential and ability to survive under hot temperatures all contribute to its potential to reduce the mealybug population.

It must be remembered that there were some other functional response studies with different lady beetle species that did not contain specific values for searching efficiency or handling time, so we could not consider them here (Bazyar et al. 2012).

Most of the functional response studies with different lady beetle species have been carried on to crop pests. As an exception, Rounagh and Samih (2015) reported the type of functional response of third and fourth instars larvae and females of *O. conglobata contaminata* (Menetries) to pomegranate green aphids, *A. punicae* at two different temperatures. They recorded type II response for all tested stages except fourth instar larvae, which showed type III at 25 °C (Rounagh and Samih 2015). They showed the effect of increasing temperatures on searching efficiency and handling time. As in many other studies, attack constant and handling time demonstrate that release of fourth instar larvae of *O. conglobata contaminata* at 27.5 °C would be the most useful approach against pomegranate aphid. However, referring to rearing storing and releasing difficulties of a large number of fourth instar larvae, the females seem to be the most appropriate releasing stage.

A common deficiency of some of the above F.R. studies is the low numbers of used replicates. In some experiments, just five or six replicates have been used while obtaining precise estimates of search rate and handling time needs at least

15 replications or even more, especially at low initial densities of prey offered. The low number of replicates can increase variations and in turn, standard error of mean among different treatments lead to incorrect estimations of searching efficiency or handling time. Additionally, again, the control treatment has not been used to correct observed mortality. Moreover, in some outdated investigations, only the Holling's disc model was used before testing the Roger's random predator equation, which is more realistic and considers prey depletion during the experimental period.

## 2.5.2   Numerical Response

The numerical response is another crucial aspect of predation. By definition, the numerical response is an increase in the number of predator densities in response to increasing prey density (Holling 1959). It has three primary forms, direct, no response and inverse (Solomon 1949).

Compare to functional response, there are very few numerical response studies and these kinds of surveys have been done scarcely in Iran. Partially, this due to complicated set-ups of numerical response experiments, which is more time consuming than a short-term experiment. Another point is that there is not a standard set-up or even standard data analysis method for these kinds of investigations. A brief review revealed puzzling variations in data-analysis and set-ups. Most carried out together with functional response studies concomitantly. Many researchers used the same set-up of functional response with slight modifications (i.e., they used a longer experimental duration for numerical response experiments instead of 24 h commonly used for functional response (Zarghami et al. 2015). Actually, the functional and numerical respons experimental set-ups are similar to each other apparently, but the different point is that in numerical response experiments, the number of eggs laid by predator in response to increasing prey densities is recorded. In contrast, the functional response considers the number of killed prey in response to initial different prey densities. Even though, there were more sophisticated studies with up to two weeks or longer as experimental duration (Mokhtari and Madadi 2013; Zarghami et al. 2015). Generally, at nearly all numerical response experiments, it has been exhibited that oviposition increases with increasing prey density (Sabaghi et al. 2011a ; Mokhtari and Madadi 2013). It has been suggested that the number of eggs laid by *H . variegata* to *A. fabae* increases curvilinearly up to an asymptote and then levelled up at density of 96 black bean aphids (Mokhtari and Madadi 2013). Moreover, there was an inverse density-dependent relationship between increasing prey density and premature developmental time. The numerical response of *Nephus arccuatus* to *Nipaecoccus viridis* (Newstead) eggs was also curvilinear reached a plateau at 115 spherical mealybugs (Zarghami et al. 2015). This value was 80 for the reproductive numerical response of *S. syriacus* to black bean aphid (Sabaghi et al. 2011a ). Sohrabi and Shishehbor (2007) showed a linear numerical response between *Tetranychus turkestani* density and *S. gilvifrons* oviposition and the lower egg production threshold was eight *T. turkestani* females.

## 2.6 Population Fluctuations of *Hippodamia variegata* and Other Lady Beetle Species

Insect populations like other organisms in response to different factors increase, decline and oscillate profoundly. Clearly, different factors such as climatic factors, competition, host quality and quantity, diseases, natural enemies and others might be influential and arise or dampen insect populations. Population dynamics attempt to find the causes of those fluctuations, their periods and strength. Although it seems that many questions have not been answered yet. Undoubtedly, knowledge of the population dynamics of aphids and their natural enemies leads us to a better understanding of their associations and integrated pest management (Rakhshani et al. 2009).

There are few studies that dealt with the population structure and fluctuation of lady beetle in Iran. Usually, the population oscillations of aphids and lady beetle have been presented across one or two consequent years. Sometimes, the relation of different abiotic or biotic factors with prey and lady beetle populations has been correlated. It has been expressed that *H. variegata* was an abundant species in central plain of Iran and its population had two peaks yearly (May–July and early of autumn) while the seven spotted lady beetle had just one dominant peak during autumn (Rakhshani et al. 2009). The interesting point was that the lady beetle population was positively correlated with aphid populations. This synchrony increases the effectiveness of the lady beetle to suppress the aphid population. Similarly, the main population peaks of these two lady beetles reached on May–July and September in Alfalfa fields of Hamedan province (Soleimani and Madadi 2015; Tavoosi Ajvad et al. 2012). The *H. variegata* population varies significantly with increasing temperatures, but unlike *C. septempunctata*, it was not correlated with pea aphid population significantly. Farsi et al. (2010) reported the peak of *C. septempunctata* in the middle of March following canola aphids rising in Ahvaz (Southern Iran). Furthermore, Afshari et al. (2000) reported the highest density of *S. gilvifrons* and *Oligonychus sacchari* McGregor, 1942 in late July and their populations were positively correlated. This species emerged on sugarcane in early summer and overwintered as an adult out of the sugarcane fields (Narrei et al. 2005).

## 2.7 Intraguild Predation Studies

Intraguild predation or predatory interference is a relatively new field in insect ecology. This term was coined by Polis et al. (1989) firstly and afterward became prevalent quickly. By definition, intraguild predation (IGP) is a combination of predation and competition, i.e. two or more competitors that compete for a shared prey involved in predation. In this interaction, one of the competitors is superior and named IG or top predator while, other is inferior, called IG prey. The shared herbivore prey often is extraguild prey (EXG). The outcome of IGP based on extraguild prey density, might be synergistic, additive, non-additive or even antagonistic Aphidophagous predators are a common guild which IGP occurs among them frequently. Most IGP studies have been conducted by different member of this guild. There are few published studies about intraguild predation in Iran, and many of them have been undertaken with predatory mites as intraguild predators and preys. In one of the earliest and most relevant records, the effect of IGP occurrence between *H. variegata* (H.v.) and *Episyrphus balteatus* De Geer (Diptera: Syrphidae) on cotton aphid population have been evaluated on a microcosm scale (Tavoosi Ajvad et al. 2014). According to their results, the IGP between third instar larvae of *E. balteatus* and *H. variegata* second instar larvae was asymmetrical, and the former acts as IG predator against the second instar larvae of *H. variegata*. They showed that the interaction of these two predators on the cotton aphid population was non-additive or even antagonistic i.e. using both predators did not suppress aphid population more efficiently than a single application of each predator stage.

Different attributes of predator and environment affect IGP outcome. It has been proposed that density of extraguild prey substantially influences of IGP occurrence (Hatami et al. 2013). However, the IGP interaction between *H. variegata* different life stages and *Aphidoletes aphidimyza* (Rondani 1847) two-days old larvae was antagonistic at highest density of extraguild prey (Hatami et al. 2013).

## 2.8 Applied Studies Employing Different Coccinellid Species in Greenhouse and Field Environment

Unfortunately, most Iranian researches on lady beetles have been restricted to laboratorical studies. Therefore, it is not easy to find practical studies with coccinellids in a small field or even greenhouse scale. They have been done under net covered cages which produce the illusive outcomes. It has been proposed that releasing 20 seven spotted lady beetle females per $m^2$ suppressed the cereal aphid population up to 45% (Haghshenas et al. 2006). The possibility of *Vedalia* beetle establishment in south citrus orchards despite hot summers considered (Eslami Zadeh and Barzkar 2006; Mossadegh et al. 2008b). Among the lady beetles, "Crypts" or "Mealybug destroyers" (*C. montrouzieri*) has been employed practically more than any other species to control mealybugs as far as more than 6,000,000 and 4,000,000 individuals were released in 2008 and 2009 (Malkeshi et al. 2010). In this regard, the possible application of this species against citrus spherical mealybug was prospective, although, hot summers suppressed activity on July (Mossadegh et al. 2008a). Of course, keep in mind that this species is sensitive to low temperatures and could not endure long-term storage (Shahriari et al. 2016).

**Acknowledgement** The special thanks dedicated to Prof. Dr. Steven Juliano for critically reading the part of this manuscript and sending his valuable comments.

# References

Abdi AR, Sadeghi SE, Talebi AA et al (2013) Coccinellid fauna of Chitgar park and determination of dominant species. Iran J Forest Range Prot Res 10(2):135–152. (In Persian with English Abstract)

Abdolahi Mesbah R, Nozari J, Dadgostar S (2015) A geometric morphometric study on geographical populations of *Hippodamia variegata* (Goeze, 1777) (Coleoptera: Coccinellidae) in some parts of Iran. J Crop Prot 4(2):207–215

Abdollahi Ahi GA, Afshari A, Baniameri V et al (2012) Functional response of *Cryptolaemus montrouzieri* Mulsant (Col.; Coccinellidae) to Citrus mealybug, *Planococcus citri* (Risso) (Hom.; Pseudococcidae) under laboratory conditions. J Plant Prot 35(1):1–14. (In Persian with English Abstract)

Abdollahi Ahi G, Afshari A, Baniameri V et al (2015) Laboratory survey on biological and demographic parameters of *Cryptolaemus montrouzieri* (Mulsant) (Coleoptera: Coccinellidae) fed on two mealybug species. J Crop Prot 4(3):267–276

Afshari A, Mossadegh MS, Kamali K (2000) Population dynamics of *Stethorus givifrons* ( Mulsant) and *Oligonychus sacchari* McGre. in sugarcane fields of Khuzestan. Paper presented at the 14th Iranian Plant Protection Congress, Isfahan University of Technology, Isfahan 5–8 September 2000

Afshari A, Mossadegh MS, Kamali K (2007) Seasonal changes and spatial distribution of sugarcane mite, *Oligonychus sacchari* (Prostigmata: Tetranychidae )and predatory ladybird, *Stethorus gilvifrons* (Mulsant) (Coleoptera: Coccinellidae) in sugarcane fields of Ahwaz. Sci J Agric 30 (1):135–147

Agha-Janzadeh S, Rasolian GR, Rezwani A, Esmaili M (1995) Identification of the aphids attacking citrus trees in West-Mazandaran and their population dynamics. Paper presented at the Proceedings of the 12th Iranian Plant Protection Congress, Karaj Junior College of Agriculture, Karaj, 2–7 September 1995

Ahmadi R (2000) Natural enemies of Russian wheat aphid (*Diuraphis noxia* Mordvilko) in Hamedan province. Paper presented at the 14th Iranian Plant Protection Congress, Isfahan University of Technology, Isfahan 5–8 September 2000

Ahmadi AA, Yazdani A (1991) First record of three species of *Pharoscymnus* Redel. From Iran. Paper presented at the 10th Iranian Plant Protection Congress, University of Shahid Bahonar, Kerman, 1–5 September 1991

Ahmadi AA, Lachinani P, Yazdani A (1993a) Biology of *Diloponis furschi* Ahmadi and Yazdani (Col.: Coccinellidae), a predator of Diaspidid scales in Fars province. Paper presented at the 11th Iranian plant protection congress, Guilan university, Rasht, 28 August-2 September 1993

Ahmadi AA, Alichi M, Yazdani A (1993b) A new records of *Pharoscymnus pharoides* Mars. from Iran. Paper presented at the 11th Iranian Plant Protection Congress, Guilan University, Rasht, 28 August-2 September 1993

Ahmadi F, Khani A, Ghadamyari M, Nouri-Ganbalani G (2010) Side-effects of abamectin and imidacloprid insecticides on life table parameters of *Cryptolaemus montrouzieri* Mulsant (Col. Coccinellidae). Paper presented at the 19th Iranian Plant Protection Congress, Iranian Research Institute of Plant Protection, Tehran, 31 July-3 August, 2010

Ahmadi A, Jafari R, Vafai R (2012) The faunistic survey of ladybird (Col., Coccinellidae) in orchards and crops Arak and shrub. Paper presented at the 20th Iranian Plant Protection Congress, Shiraz University, Shiraz, 25–28 August 2012

Ajam Hassani M (2013) Biology of *Oenopia conglobata* (Col.: Coccinellidae) on *Myzus persicae* (Hem.: Aphididae). In: Talaei-Hassanloui R (ed) Proceedings of the Conference of Biological Control in Agriculture and Natural Resources, College of Agriculture and Natural Resources, Karaj, 2013

Ajam Hassani M (2015) Study on Some biological characteristic of *Oenopia conglobata* by feeding on three aphids, *Rhopalosiphum padi*, *Macrosiphum rosae* and *Chaitophorus populi*. Biocontrol Plant Prot 3(1):101–105. (In Persian with English Abstract)

Akbarzadeh Shoukat G, Rezvani A (2000) Studies on apple aphids and their natural enemies in Urmia apple orchards. Paper presented at the 14th Iranian Plant Protection Congress, Isfahan University of Technology, Isfahan 5–8 September 2000

Aldaghi M, Allahyari H, Talaee-Hasanlouei R (2013) Effect of male to female ratio on fecundity and fertility of *Hippodamia variegata* (Col.: Coccinellidae). Biol Control Pests Plant Dis 2 (2):123–127. (In Persian with English Abstract)

Alemansoor H, Ahmadi AA (1993) Natural enemies of cotton whitefly, *Bemisia tabaci* (Gennadius) in Fars province. Paper presented at the 11th Iranian Plant Protection Congress, Guilan University, Rasht, 28 August-2 September 1993

Alimohammadi Davarani N, Samih MA, Izadi H (2012) Effects of Hexaflumuron and spirodiclofen on functional response of *Hippodamia variegata* at different densities of *Agonoscena pistaciae*. Biol Control Pests Plant Dis 1(1):1–10. (In Persian with English Abstract)

Alimohammadi Davarani N, Samih MA, Izadi H (2013) Effect of Hexaflumuron and Spirodiclofen on demography of *Hipodamia variegata* (Goez)(Col:Coccinellidae) predator of *Agonoscena pistaciae* Burckhardt and Lauterer under laboratory conditions. J Plant Prot 26(4):424–436. (In Persian with English Abstract)

Alimohammadi Davarani N, Samih MA, Izadi H et al (2014) Developmental and biochemical effects of hexaflumuron and spirodiclofen on the ladybird beetle, *Hippodamia variegata* (Goeze) (Coleoptera: Coccinellidae). J Crop Prot 3(3):335–344

Ameri A, Talebi AA, Fathipour Y, Zamani AA (2006) Life table and population growth parameters of *Exochhomus quadripustulatus* predator of cypress mealybug, *Planococcus vovae* in laboratory conditions. Paper presented at the 17th Iranian Plant Protection Congress, College of Agriculture and Natural Resources, University of Tehran, Karaj, 2–5 September 2006

Aminafshar E, Khanjani M, Zahiri B (2014) Evaluation of life table parameters of *Coccinella septempunctata* (L.) fed on *Macrosiphum rosae* (L.). Paper presented at the 21th Iranian Plant Protection Congress, Urmia University, Urmia, 23–26 August 2014

Ansari pour A (2012) Identification of some of the fauna ladybirds (Col.: Coccinellidae) in Alborz province. Paper presented at the 20th Iranian Plant Protection Congress, Shiraz University, Shiraz, 25–28 August 2012

Ansari pour A, Shakarami J (2011) Study of ladybirds (Col: Coccinellidae) in Khorramabad district and the first report of *Hyperaspis quadrimaculata* (Redtenbacher 1844) for Iranian fauna. Life Sci J 8(3):488–495. https://doi.org/10.7537/marslsj080311.76

Ansari pour A, Shakarami J (2012) Recognition of ladybird fauna (Col: Coccinellidae) in the alfalfa fields of Khorramabad. J Anim Plant Sci 22(4):939–943

Arpaia S, Gould F, Kennedy R (1997) Potential impact of *Coleomegilla maculata* predation on adaptation of *Leptinotarsa decemlineata* to Bt-transgenic potatoes. Entomol Exp Appl 82 (1):91–100. https://doi.org/10.1046/j.1570-7458.1997.00117.x

Asadeh GhA, Mossadegh MS (1991) An investigation of the mealybugs (*Pseudococcus* spp.) natural enemies fauna in the Khuzestan's province. Paper presented at the 10th Iranian Plant Protection Congress, University of Shahid Bahonar, Kerman, 1–5 September 1991

Asghari F, Samih MA, Izadi H (2011) Demography of *Hippodamia variegata* (Coleoptera: Coccinellidae) feeding on *Agonoscena pistaciae* (Hem.: Aphalaridae) under laboratory condition. J Plant Prot 34(2):75–88. (In Persian with English Abstract)

Asghari F, Samih MA, Mahdian K et al (2012) Predatory efficiency of *Hippodamia variegata* (Col.: Coccinellidae) on common pistachio psylla, *Aganoscena pistachiae* (Hem. Aphalaridae), under laboratory conditions. J Entomol Soc Iran (JESI) 32(1):37–58. (In Persian with English Abstract)

Asgharian N, Golizadeh A, Hassanpour M, Fathi SAA (2014) The effect of host plants of *Diuraphis noxia* (Hemi.: Aphididae) on biology of predator coccinellid, *Hippodamia variegata* (Col.: Coccinellidae). Paper presented at the 21th Iranian Plant Protection Congress, Urmia University, Urmia, 23–26 August 2014

Atrchian, H. Mahdian, K., Shahidi, S. Rahimy, R. (2014, August). Developmental duration and survival of *Chilocorus bipustulatus* L. (Col.:Coccinellidae) on common pistachio psylla

*Agonoscena pistaciae* (Hem.: Psyllidae). Paper presented at the 21th Iranian Plant Protection Congress, Urmia University, Urmia, 23–26 August 2014

Bagheri MR, Emami MS (1998) Introduction of ten new species of Coccinellidae for Isfahan province. Paper presented at the 13th Iranian Plant Protection Congress, Karaj Junior College of Agriculture, Karaj, 23–27 August 1998

Bagheri MR, Mossadegh MS (1995) Faunistic studies of Coccinellidae in Chaharmahal va Bakhtiari province. Paper presented at the 12th Iranian Plant Protection Congress, Karaj Junior College of Agriculture, Karaj, 2–7 September 1995

Baniasadi M, Hassani MR, Basirat M, Olyaie Torshiz A, Ghorbani H (2012) Side-effects of hexaflumuron and thiamethoxam on life table parameters of *Oenopia conglobata contaminata* (Col.: Coccinellidae). Paper presented at the 20th Iranian Plant Protection Congress, Shiraz University, Shiraz, 25–28 August 2012

Barkhordar B, Khalghani J, Salehi Jouzani Gh, Nouri Ganbalani G, Shojai M, Karimi E, Soheilivand S (2012) Impact of host plant resistance on the tritrophic interactions between wheat genotypes, *Schizaphis graminum* (Rondani)(Hom.: Aphididae) and *Coccinella septempunctata* (Col.: Coccinellidae) using molecular methods. Paper presented at the 20th Iranian Plant Protection Congress, Shiraz University, Shiraz, 25–28 August 2012

Bazyar M, Alichi M, Minaei K, Hamzezarghani H, Saharkhiz M J (2012) Functional response of *Rodolia cardinalis* (Mulsant) (Col.: Coccinellidae) to different densities of *Icerya purchasi* (Maskell) (Hem. Margarodidae) on two plant host *Citrus aurantium* and *Daphne giraldii* Paper presented at the 20th Iranian Plant Protection Congress, Shiraz University, Shiraz, 25–28 August 2012

Behnazar T, Madadi H (2015) Functional response of different stages of *Hippodamia variegata* (Col.: Coccinellidae) to *Diuraphis noxia* (Hemiptera: Aphididae) on two wheat cultivars. Biocontrol Sci Tech 25(10):1180–1191. https://doi.org/10.1080/09583157.2015.1040374

Bigdelou B (2012) Life table and predation capacity of *Hippodamia variegata* (Col.: Cocinellidae) feeding on *Aphis gossypii* (Hem.:Aphididae). MSc. Dissertation, University of Tehran

Biranvand A, Jafari R, Zare Khormizi M (2014) Diversity and distribution of Coccinellidae (Coleoptera) in Lorestan Province, Iran. Biodivers J 5(1):3–8

Biranvand A, Nedved O, Tomaszewska W et al (2016) An annotated checklist of Microweiseinae and Sticholotidini of Iran (Coleoptera, Coccinellidae). Zookeys 587(2):37–48. https://doi.org/10.3897/zookeys.587.8056

Bozorg-Amirkalaee M, Fathi SAA, Golizadeh A et al (2014) Life table of *Cryptolaemus montrouzieri* fed on ovisacs of *Pulvinaria aurantii* on clementine mandarin and sour orange. Iranian J Plant Prot Sci 45(1):161–170. (In Persian with English Abstract)

Bozorg-Amirkalaee M, Fathi SAA, Golizadeh A et al (2015) Performance of *C. montrouzieri* feeding on *Pulvinaria aurantii* ovisacs on citrus plants. Biocontrol Sci Tech 25(2):207–222. https://doi.org/10.1080/09583157.2014.968521

Davoodi Dehkordi S, Sahragard A, Hajizadeh J (2012) Comparison of functional response of two and one individual female predator, *Hippodamia variegata* Goeze (Coleoptera: Coccinellidae) to different densities of *Aphis gossypii* Glover (Hemiptera: Aphididae) under laboratory conditions. Munn Entomol Zool 7(2):998–1005

Davoodi Dehkordi S, Sahragard A, Hajizadeh J (2013) The effect of prey density on life table parameters of *Hippodamia variegata* (Coleoptera: Coccinellidae) fed on *Aphis gossypii* (Hemiptera: Aphididae) under laboratory conditions. ISRN Entomol. https://doi.org/10.1155/2013/281476

Davvodi Dehkordi S, Sahragard A (2013a) Functional response of *Hippodamia variegata* (Coleoptera: Coccinellidae) to different densities of *Aphis gossypii* (Hemiptera: Aphididae) in an open patch design. J Agric Sci Techn-Iran 15(4):651–659

Davvodi Dehkordi S, Sahragard A (2013b) The Effect of prey density on life table parameters of *Hippodamia variegata* (Coleoptera: Coccinellidae) fed on *Aphis gossypii* (Hemiptera: Aphididae) under laboratory conditions. Hindawi Pub Corp. https://doi.org/10.1155/2013/281476

Dehghan Dehnavi L, Samih MA, Talebi AA et al (2007) Functional response of *Adalia bipunctata* (Col.: Coccinellidae) reared on *Aphis punicae* (Hom., Aphididae) in laboratory conditions. Quarterly new findings Agric 1(3):215–223. (In Persian with English Abstract)

Dehghan Dehnavi L, Samih MA, Talebi AA, Goldasteh Sh (2008) Life table and reproduction parameters of *Adalia bipunctata* (Col.: Coccinellidae) on *Aphis punicae* (Hem. Aphididae) in laboratory conditions. Paper presented at the 18th Iranian Plant Protection Congress, Bu-Ali Sina University, Hamedan, 25–28 August 2008

Ebrahimi Arfaa M (2014) Effect of temperature on interference and functional response of *Hippodamia variegata* (Col. : Coccinellidae) on *Aphis gossypii* (Hem.: Aphididae). MSc. Dissertation, University of Tehran

Ebrahimi S, Modarres-Awal M, Karimi J, Fekrat L, Nedved O (2014) Two new records of ladybirds (Col.: Coccinellidae) for the Iranian beetle fauna. J Entomol Soc Iran (JESI) 34(2):11–12

Ebrahimi S, Karimi J, Modarres-Awal M (2012) The first record of the ladybird, *Pharoscymnus brunneosignatus*, Mader,1949 (Col., Coccinellidae) for Iran. Paper presented at the 20th Iranian plant protection congress, shiraz Univeristy, shiraz, 25–28 August 2012

Ebrahimzadeh P, Mossadegh MS (2004) The coccinellids and aphids of alfalfa fields in Khoozestan. Paper presented at the 16th Iranian Plant Protection Congress, Tabriz University, Tabriz, 28 August-1 September 2004

Edward WE (2009) Lady beetles as predators of insects other than Hemiptera. Biol Control 51 (2):255–267. https://doi.org/10.1016/j.biocontrol.2009.05.011

Emami MS, Sahragard A, Hajizadeh J (1998) Observation on mating and oviposition behaviours of *Scymnus syriacus* Marseul and the effect of different temperatures on developmental stages of predator under laboratory conditions. Paper presented at the 13th Iranian Plant Protection Congress, Karaj Junior College of Agriculture, Karaj, 23–27 August 1998

Emami MS Sahragard A, Hajizadeh J (2002) Studies on mass rearing possibility of *Scymnus syriacus* predator of *Aphis spiraecola*. Paper presented at the 15th Iranian Plant Protection Congress, Razi University of Kermanshah, Kermanshah, 7–11 September 2002

Esfandiari H, Rajabi G, Bagheri M, Barari H (2002) Predator Coccinellids of *Sphaerolecanium prunastri* Fonsc. (Hom. Coccidae) on almond and investigation on population changes of dominant species. Paper presented at the 15th Iranian Plant Protection Congress, Razi University of Kermanshah, Kermanshah, 7–11 September 2002

Eslami Zadeh R, Barzkar M (2006) Survey of stable of ladybird *Rodolia cardinalis* (Nab) in citrus orchard of Khuzestan province. Paper presented at the 17th Iranian Plant Protection Congress, College of Agriculture and Natural Resources, University of Tehran, Karaj, 2–5 September 2006

Eslami Zadeh R, Pourmirza AA (1998) Surveys on the biology and efficiency of the species *Orius minutes* (L.) and *Stethorus punctillum*Wies by feeding in the red spider mites *Panonychus ulmi* Koch. Under lab. conditions. Paper presented at the 13th Iranian Plant Protection Congress, Karaj Junior College of Agriculture, Karaj, 23–27 August 1998

Esmaeily S, Samih MA, Jafarbeigi F, Zarabi M (2011) Functional response of *Hippodamia variegata* to different densities of Oleander aphid, *Aphis nerii* under laboratory conditions. Paper presented at 1st Biological Control Development Congress in Iran, Iranian Research Institute of Plant Protection, Tehran, 27–28 July 2011

Fallahzadeh M, Hesami Sh (2004) Study of the natural enemies of *Maconellicoccus hirsutus* (Homoptera: Pseudococcidae) in Jahrom region. Paper presented at the 16th Iranian Plant Protection Congress, Tabriz University, Tabriz, 28 August-1 September 2004

Fallahzadeh M, Eghtedar E, Ebrahimi E (2000) Natural enemies of olive psylla, *Euphyllura olivina* Costa in Fars province. Paper presented at the 14th Iranian Plant Protection Congress, Isfahan University of Technology, Isfahan 5–8 September 2000

Fallahzadeh M, Hesami S, Fursch H (2003) Report of *Nephus* (*Sidis*) *hiekei* (Col.: Coccienllidae) in Iran. J Entomol Soc Iran (JESI) 22(2):81–82

Fallahzadeh M, Hesami Sh, Alemansoor H (2004) Citrus whiteflies and their natural enemies in
    Jahrom region of Fars province. Paper presented at the 16th Iranian Plant Protection Congress,
    Tabriz University, Tabriz, 28 August-1 September 2004
Fallahzadeh M, Shojai M, Ostovan H, Kamali K (2006) An investigation of the mealybug predators
    fauna in fars province- Iran. Paper presented at the 17th Iranian Plant Protection Congress,
    College of Agriculture and Natural Resources, University of Tehran, Karaj, 2–5 September
    2006
Fallahzadeh M, Japoshvili G, Saghaei N et al (2011) Natural enemies of *Planococcus ficus*
    (Hemiptera: Pseudococcidae) in Fars Province vineyards, Iran. Biocontrol Sci Tech 21
    (4):427–433. https://doi.org/10.1080/09583157.2011.554801
Farahi S, Sadeghi Namaghi H (2011) Report of *Coccinella magnopunctata* (Col.: Coccinellidae)
    from Iran. J Entomol Soc Iran (JESI) 30(2):79–80. (In Persian with English Abstract)
Farhadi R, Allahyari H, Juliano SA (2010) Functional response of larval and adult stages of
    *Hippodamia variegata* (Coleoptera: Coccinellidae) to different densities of *Aphis fabae*
    (Hemiptera: Aphididae). Environ Entomol 39(5):1586–1592. https://doi.org/10.1603/EN09285
Farhadi R, Allahyari H, Chi H (2011) Life table and predation capacity of *Hippodamia variegata*
    (Coleoptera: Coccinellidae) feeding on *Aphis fabae* (Hemiptera: Aphididae). Biol Control 59
    (2):83–89. https://doi.org/10.1016/j.biocontrol.2011.07.013
Farsi A, Kocheyli F, Soleymannejadian E et al (2010) Population dynamics of Canola aphids and
    their dominant natural enemies in Ahvaz. Plant Prot (Sci J Agric) 32(2):55–65
Farahi S, Sadeghi Namghi H (2009) Fauna of aphids and their coccinellid predators of wheat fields
    in Mashhad region ( Razavi Khorasan province). Plant Prot 23(2):89–95. (In Persian with
    English Abstarct)
Fotukkiaii SM, Sahragard A (2012) Functional response of fourth instar larvae and the female
    *Serangium montazerii* Fursch (Coleoptera: Coccinellidae) to different densities of *Dialeurodes
    citri* (Ashmead) (Hemiptera: Aleyrodidae) under laboratory conditions. J Entomol Res Soc 14
    (2):1–7
Fotukkiaii SM, Sahragard A (2013) Functional response of *Serangium montazerii* (Col.:
    Coccinellidae) to different densities of *Dialeurodes citri* (Hem.: Aleyrodidae): an open-patch
    approach. J Entomol Soc Iran 33(2):1–7
Fotukkiaii SM, Sahragar A, Halajisani MF (2013) Comparing demographic parameters of
    *Serangium montazerii* (Coleoptera: Coccinellidae) on citrus whitefly, *Dialeurodes citri*
    (Hemiptera: Aleyrodidae) fed on two host plants. J Crop Prot 2(1):51–61
Franzmann BA (2002) *Hippodamia variegata* (Goeze) (Coleoptera: Coccinellidae), a predacious
    ladybird new in Australia. Aust J Entomol 41:375–377
Frouzan A, Shishehbor P, Esfandiari M et al (2016) Biological characteristics and life table
    parameters of coccinelid *Nephus arcuatus* feeding on *Phenacoccus solenopsis* at different
    temperatures. Plant Prot (Sci J Agric) 39(1):75–85. (In Persian with English Abstract)
Ghafouri Moghaddam M, Golizadeh A, Hassanpour M, Rafiee H, Razmjou J (2013) The effect of
    host plants of *Sitobion avenae* (Hemi.: Aphididae) on biology of predatory coccinellid,
    *Hippodamia variegata* (Col.: Coccinellidae). In: Talaei-Hassanloui R (ed) Proceedings of the
    Conference of Biological Control in Agriculture and Natural Resources, College of Agriculture
    and Natural Resources, Karaj, 2013
Ghanadamooz S, Malkeshi SH, Sahragard A (2010) Investigation of biological factors and deter-
    mine appropriate host in rearing of coccinellid (*Chilocorous bipustulatus* L.) for controlling
    agricultural pests. Paper presented at the 19th Iranian Plant Protection Congress, Iranian
    Research Institute of Plant Protection, Tehran, 31 July-3 August, 2010
Ghanbari A, Sadeghi SE, Ladan Moghadam AR, Fakhredini M (2012) An Investigation on
    predaceous coccinellid's fauna, their distribution and determining dominant species on shading
    trees and shrubs in green spaces and parks of 16th Tehran municipal zone. Paper presented at the
    20th Iranian Plant Protection Congress, Shiraz University, Shiraz, 25–28 August 2012
Ghafouri Moghaddam M, Golizadeh A, Hassanpour M, Rafiee-Dastjerdi H, Razmjou J (2016)
    Demographic traits of *Hippodamia variegata* (Goeze) (Coleoptera: Coccinellidae) fed on
    *Sitobion avenae* Fabricius (Hemiptera: Aphididae). J Crop Prot 5(3):431–445

Gholami Moghaddam S, Vahedi HA, Heydarzade A, Ebrahimi S (2014) Faunal Study of ladybirds (Col.:Coccinellidae) in Kermanshah province and five new record for Kermanshah province. Paper presented at the 21th Iranian Plant Protection Congress, Urmia University, Urmia, 23–26 August 2014

Ghorbanian S, Ghajarieh H, Ranjbar Aghdam H, Malkeshi H (2010) Functional response of predator *Cryptolaemus montrouzieri* Mulsant to the mealy bug *Planococcus citri* (Risso) on *Solenostemon scutellarioides* (L.) Codd. Paper presented at the 19th Iranian Plant Protection Congress, Iranian Research Institute of Plant Protection, Tehran, 31 July-3 August, 2010

Giorgi A, Vandenberg N (2009) Coccinellidae. Lady beetles, ladybird beetles, ladybugs. http://tolweb.org. Accessed 20 June 2016

Golizadeh A, Jafari-Behi V (2012) Biological traits and life table parameters of variegated lady beetle, *Hippodamia variegata* (Coleoptera: Coccinellidae) on three aphid species. Appl Entomol Zool 47 (3): 199–205. doi: doi:https://doi.org/10.1007/s13355-012-0108-8

Haghshenas AR, Malkeshi SH, Bagheri MR (2004) The fauna of coccinellids in cereal aphids and investigation on population fluctuation of dominant species in Isfahan province. Paper presented at the 16th Iranian Plant Protection Congress, Tabriz University, Tabriz, 28 August-1 September 2004

Haghshenas AR, Malkeshi SH, Mahlooji M (2006) An investigation of efficiency of *Coccinella septempunctata* L.(Col.: Coccinellidae) on cereal aphids in wheat fields of Esfahan province. Paper presented at the 17th Iranian Plant Protection Congress, College of Agriculture and Natural Resources, University of Tehran, Karaj, 2–5 September 2006

Hajizadeh J (1993) Studies on the mass rearing possibility of *Pharoscymnus ovoideus* (Coccinellidae) predator of the date palm scale *Parlatoria blanchardi*. Paper presented at the 11th Iranian Plant Protection Congress, Guilan University, Rasht, 28 August-2 September 1993

Hajizadeh J (1995) Identification of Stethorus coccinellid beetles in Tehran province and study on biology, and possibility of production of *Stethorus gilvifrons* Mulsant. Dissertation, Tarbiat Modares University (TMU)

Hajizadeh J, Kamali K, Assadi HB (1995) The effect of different temperatures on developmental stages of *Stethorus gilvifrons* Mulsant (Col. Coccinellidae). Paper presented at the 12th Iranian Plant Protection Congress, Karaj Junior College of Agriculture, Karaj, 2–7 September 1995

Hajizadeh J, Jalali J, Peyrovy H (1998) A part of coccinellids (Col. Coccinellidae) fauna of Guilan province. Paper presented at the 13th Iranian Plant Protection Congress, Karaj Junior College of Agriculture, Karaj, 23–27 August 1998

Hassani MR, Mehrnejad MR, Ostovan H (2004) Determination of life table and functional response parameters of *Oenopia conglobata* L. (Col. : Coccinellidae) in laboratory conditions. Paper presented at the 16th Iranian Plant Protection Congress, Tabriz University, Tabriz, 28 August-1 September 2004

Hassankhani K, Allahyari H (2013) Functional response of adult male and female of *Hippodamia variegata* Goeze (Col.: Coccinellidae) on peach aphid. Biol Control Pests Plant Dis 2(1):65–70. (In Persian with English Abstract)

Hatami N, Allahyari H, Hosseini M (2013) Simultaneous use of *Hippodamia variegata* and *Aphidoletes aphidimyza* on Cotton aphid, *Aphis gossypii*. Biol Control Pests Plant Dis 1 (2):87–94. (In Persian with English Abstract)

Hesami S, Fallahzadeh M (2004) Study on the natural enemies of citrus mealybug, *Nipaecoccus viridis* (Homoptera: Pseudococcidae), in Jahrom, Fars province. Paper presented at the 16th Iranian Plant Protection Congress, Tabriz University, Tabriz, 28 August-1 September 2004

Hodek I, van Emden HF, Honek A (2012) Ecology and behaviour of the ladybird beetles. Wiley-Blackwell, UK

Holling CS (1959) The components of predation as revealed by a study of small-mammal predation of the European pine sawfly. Can Entomol 91(5):293–320. https://doi.org/10.4039/Ent91293-5

Holling CS (1961) Principles of insect predation. Annu Rev Entomol 6(1):163–182

Imani Z, Shishehbor P (2011) Functional response of *Stethorus gilvifrons* (Coleoptera: Coccinellidae) to different developmental stages of *Eutetranychus orientalis*. J Entomol Soc Iran (JESI) 31(1):29–40. (In Persian with English Abstract)

Imani Z, Shishehbor P, Sohrabi F (2009) The effect of *Tetranychus turkestani* and *Eutetranychus orientalis* (Acari: Tetranychidae) on the development and reproduction of *Stethorus gilvifrons* (Coleoptera: Coccinellidae). J Asia Pac Entomol 12(4):213–216. https://doi.org/10.1016/j. aspen.2009.05.004

Jafari R, Goldasteh S (2009) Functional response of *Hippodamia variegata* (Goeze) (Coleoptera: Coccinellidae) on *Aphis fabae* (Scopoli) (Homoptera: Aphididae) in laboratory conditions. Acta Entomol Serbica 14 (1): 93–100

Jafari R, Kamali K (2007) Faunistic study of Ladybird (Col.: Coccinellidae) in Lorestan province and report of new records in Iran. New Findings Agric 1(4):349–359. (In Persian with English Abstract)

Jafari R, Vafaei Shoushtari R (2010) Effect of different temperatures on life developmental stages of *Hippodamia variegata* Goeze (Col., Coccinellidae), feeding on *Aphis fabae* Scopoli (Hem., Aphididae). J Entomol Res 1 (4): 289–297. (In Persian with English Abstract)

Jafari R, Zarei Jallalabad N, Vafaei Shoushtari R (2011) The faunestic survey on Coccinellids in Zarand zone. J Entomol Res 3(4):277–284. (In Persian with English Abstarct)

Jafari R, Kamali K, Shojai M et al (2008a) Life table parameteres of *Hippodamia variegata* (Col.: Coccinellidae) on *Aphis fabae* (Hom.: Aphididae) under laboratory condition. Agroecol J 4 (10): 17–25. (In Persian with English Abstract)

Jafari R, Kamali K, Ostovan H (2008b) The faunistic survey of ladybirds (Coleoptera, Coccinellidae) in Lorestan province, Iran. Paper presented at the 18th Iranian Plant Protection Congress, Bu-Ali Sina University, Hamedan, 25–28 August 2008

Jafari R, Zarei Jallalabad N, Vafaei Shoushtari R (2012) The faunestic survey on Coccinellids in Zarand Zone. J Entomol Res 3 (4): 277–284. (In Persian with English Abstract)

Jafari R, Fursch H, Zarei M (2013) A checklist of the Scymninae (Coleoptera: Coccinellidae) of Iran. Int J Sci Basic Appl Res (IJSBAR) 4(12):4055–4061

Jalali MA, Mehrnejad MR, Asadi GH, Sadeghi SE (2002) Compiling a life table for the psyllaephagous lady birds, the natural enemies of the common pistachio psylla, *Aganoscena pistaciae* in the controlled conditions. Paper presented at the 15th Iranian Plant Protection Congress, Razi University of Kermanshah, Kermanshah, 7–11 September 2002

Jalilvand K, Fallahzadeh M, Vahedi HA, Shirazi M (2012) The first record of two predator species Coccinellidae & Nitidulidae (Coleoptera) associated with scale insects (Hemiptera, Coccoidea) from Iran. Paper presented at the 20th Iranian Plant Protection Congress, Shiraz University, Shiraz, 25–28 August 2012

Jalilvand K, Shirazi M, Fallahzadeh M et al (2014) Survey of natural enemies of mealybug Species (Hemiptera, Pseudococcidae) in Kermanshah province, Western Iran to Inform Biological Control Research. J Entomol Res Soc (JESI) 16(3):1–10

Juliano SA (2001) Nonlinear curve fiting: predation and functional response curves. In: Scheiner SM, Gurevitch J (eds) Design and analysis of ecological experiments. Chapman & Hall, New York, pp 178–196

Kalantari AA, Sadeghi A (2000) Investigation survey of ladybirds and determination of prevalent species in dry orchard almond in west Khorasan province. Paper presented at the 14th Iranian Plant Protection Congress, Isfahan University of Technology, Isfahan 5–8 September 2000

Keshtkar M, Goldasteh S, Rafiei-Karahroodi Z, Bagherzadeh Z (2010) A study on the life cycle of *Cryptolaemus montrouzieri* Mulsant (Col. Coccinellidae) in laboratory conditions. Paper presented at the 19th Iranian Plant Protection Congress, Iranian Research Institute of Plant Protection, Tehran, 31 July-3 August, 2010

Khazaeipool A, Fathi SAA, Davari M, Aghajanzadeh S (2008) Prey preference of the fourth instar larvae of *Cryptolaemus montrouzieri* Mulsant from *Pulvinaria aurantii* Cockerell and *Pulvinaria floccifera* (Westwood). Paper presented at the 18th Iranian Plant Protection Congress, Bu-Ali Sina University, Hamedan, 25–28 August

Kianpour R, Fathipour Y, Kamali K, Omkar (2011) Effects of mixed prey on the development and demographic attributes of a generalist predator, *Coccinella septempunctata* (Coleoptera:

Coccinellidae). Biocontrol Sci Tech 21(4): 435–477. doi: doi:https://doi.org/10.1080/09583157.2011.554800

Kontodimas DC, Stathas GJ (2005) Phenology, fecundity and life table parameters of the predator *Hippodamia variegata* reared on *Dysaphis crataegi*. BioControl 50(2):223–233. https://doi.org/10.1007/s10526-004-0455-7

Koohpayezadeh N, Mossadegh MS (1991) Some of the ladybirds (Coccinellidae) fauna of Kerman's province. Paper presented at the 10th Iranian Plant Protection Congress, University of Shahid Bahonar, Kerman, 1–5 September 1991

Lotfalizadeh H (2001) Sex determination in some ladybirds (Col.: Coccinellidae) fauna of Moghan region. J Entomol Soc Iran (JESI) 21(1):69–88. (In Persian with English abstract)

Lotfalizadeh H (2002) Natural enemies of cotton Aphids in Moghan region, NorthWest of Iran. Paper presented at the 15th Iranian Plant Protection Congress, Razi University of Kermanshah, Kermanshah, 7–11 September 2002

Lotfalizadeh H, Hatami B, Khalghani J (2000) Biological study *Exochomus quadripustulatus* (L.) on cypress tree mealybug, *Planococcus vovae* (Nasanov)(Hom. Pseudococcidae) in Shiraz. Paper presented at the 14th Iranian Plant Protection Congress, Isfahan University of Technology, Isfahan 5–8 September 2000

Maafi Sh, Rajabi Gh, Jafari ME (1998) Introduction of the natural enemies of *Planococcus citri* in Mazandaran. Paper presented at the 13th Iranian Plant Protection Congress, Karaj Junior College of Agriculture, Karaj, 23–27 August 1998

Madadi H, Mohajeri Parizi E, Allahyari H et al (2011) Assessment of the biological control capability of *Hippodamia variegata* (Col.: Coccinellidae) using functional response. J Pest Sci 84(4):447–455. https://doi.org/10.1007/s10340-011-0387-9

Malkeshi S H, Dadpour Moghanloo H, Askary H, Rezapanah M, Alinia F, Gholami M, Fatemi A, Hadayegh M, Hasanzadeh M, Shokri R (2010) Mass rearing and releasing of *Cryptolaemus montrouzieri* with the farmers participation for biological control of *Pseudococcus viburni* in tea orchards. Paper presented at the 19th Iranian Plant Protection Congress, Iranian Research Institute of Plant Protection, Tehran, 31 July-3 August, 2010

Matin M, Nouri-Ghanbalani G, Mossadegh MS, Shishehbor P (2008) Life table and population growth parameters of *Stethorus gilvifrons*, feeding on date dust mite, *Oligonychus afrasiaticus* in laboratory conditions Paper presented at the 18th Iranian Plant Protection Congress, Bu-Ali Sina University, Hamedan, 25–28 August 2008

Matin M, Nouri-Ghanbalani G, Mossadegh MS, Shishehbor P (2010) Functional and numerical response of *Stethorous gilvifrons* to densities of the *Oligonychus afrasiaticus* in laboratory conditions. Paper presented at the 19th Iranian Plant Protection Congress, Iranian Research Institute of Plant Protection, Tehran, 31 July-3 August, 2010

Mehdian K, Sahragard A, Haji-Zadeh J (1998) Studies on mass rearing of *Chilocorous bipustulatus* L. (Col. Coccinelliae). Paper presented at the 13th Iranian Plant Protection Congress, Karaj Junior College of Agriculture, Karaj, 23–27 August 1998

Mehrkhou F, Fathipour Y, Talebi AA (2008) Temperature-dependent functional response of *Stethorus gilvifrons* (Coleoptera: Coccinellidae) on *Tetranychus urticae* (Acari: Tetranychidae) in constant temperatures. Paper presented at the 18th Iranian Plant Protection Congress, Bu-Ali Sina University, Hamedan, 25–28 August 2008

Mehrnejad MR (2000a) Four ladybirds, as important predators of the common pistachio psylla, *Aganoscena pistaciae*. Paper presented at the 14th Iranian Plant Protection Congress, Isfahan University of Technology, Isfahan 5–8 September 2000

Mehrnejad MR (2000b) Rearing media for the ladybird, *Exochomus nigripennis*, biocontrol agent of key pistachio pests. Paper presented at the 14th Iranian Plant Protection Congress, Isfahan University of Technology, Isfahan 5–8 September 2000

Mehrnejad MR, Jalali MA (2004) Life History Parameters of the Coccinellid beetle, *Oenopia conglobata contaminata*, an important predator of the common Pistachio Psylla, *Agonoscena pistaciae* (Hemiptera: Psylloidea). Biocontrol Sci Tech 14(7):701–711. https://doi.org/10.1080/09583150410001682377

Mehrnejad MR, Jalali MA, Mirzaei R (2011) Abundance and biological parameters of psyllophagous coccinellids in pistachio orchards. J Appl Entomol 135(9):673–681. https://doi.org/10.1111/j.1439-0418.2010.01577.x

Mehrnejad MR, Vahabzadeh N, Hodgson CJ (2015) Relative suitability of the common pistachio psyllid, *Agonoscena pistaciae* (Hemiptera: Aphalaridae), as prey for the two-spotted ladybird, *Adalia bipunctata* (Coleoptera: Coccinellidae). Biol Control 80:128–132. https://doi.org/10.1016/j.biocontrol.2014.10.005

Mirhosseini MA, Hosseini MR, Jalali MA (2015) Effects of diet on development and reproductive fitness of two predatory coccinellids (Coleoptera: Coccinellidae). Eur J Entomol 112 (3):446–452. https://doi.org/10.14411/eje.2015.051

Mirkhalilzadeh Ershadi SR, Allahyari H, Nozari J et al (2013) Rearing larval stages of *Hippodamia variegata* Goeze (Col.: Coccinellidae) on artificial diet. Arch Phytopathol Plant Prot 46 (7):755–765. https://doi.org/10.1080/03235408.2012.751286

Modarres Awal M (2001) List of agricultural pests and their natural enemies in Iran. Ferdowsi University Press, Mashhad

Modarres Najaf Abadi SS, Arjmandi Nejad AR, Hanifian S (2008a) The study on population fluctuation of cabbage aphid (*Brevicoryne brassicae*) and identification of it's natural enemies in Sistan region. Paper presented at the 18th Iranian Plant Protection Congress, Bu-Ali Sina University, Hamedan, 25–28 August 2008

Modarres Najaf Abadi SS, Arjmandi Nejad AR, Hejazi R (2008b) Seasonal population changes of wheat green aphid (*Schizaphis graminum*) and introduction of its natural enemies in Sistan region. Paper presented at the 18th Iranian Plant Protection Congress, Bu-Ali Sina University, Hamedan, 25–28 August 2008

Mohajeri Parizi E, Madadi H, Allahyari H, Mehrnejad MR (2012) A Comparison of life history parameters of *Hippodamia variegata* feeding on either *Aphis gossypii* Glover or *Acyrthosiphon pisum*. Iranian J Plant Prot Sci 43(1):73–81. (In Persian with English Abstract)

Mohammadbeigi A (2000) Natural enemies of the walnut aphids in Qazvin region. Paper presented at the 14th Iranian Plant Protection Congress, Isfahan University of Technology, Isfahan 5–8 September 2000

Mohammadi M, Hajizadeh J, Lotfalizadeh HA (2013) First report of *Vibidia duodecimguttata* (Poda, 1761) (Coleoptera: Coccinellidae) from Iran. Plant Pests Res 2(4):67–70. (In Persian with English Abstract)

Mohasesian M, Ranjbar Aghdam H, Pakyari H (2015) Temperature-dependent functional response of mealybug destroyer, *Cryptolaemus montrouzieri* on citrus mealybug, *Planococcus citri*. Biocontrol Plant Prot 2(2):1–11. (In Persian with English Abstract)

Mohiseni AA, Keyhanian AA, Taghaddosi MV, Boroomand H, (1998) A study of coccinellid fauna in alfalfa fields in Hamadan. Paper presented at the 13th Iranian Plant Protection Congress, Karaj Junior College of Agriculture, Karaj, 23–27 August 1998

Mojib Hagh Ghadam Z, Jalali Sendi J, Sadeghi SE, Hajizadeh J (2002a) The effect of type and the amount of prey consumption on different developmental stages and oviposition of adult lady beetle *Oenopia conglobata* L. (Col.: Coccinellidae) in laboratory conditions. Paper presented at the 15th Iranian Plant Protection Congress, Razi University of Kermanshah, Kermanshah, 7–11 September 2002

Mojib Hagh Ghadam Z, Jalali Sendi J, Sadeghi SE, Hajizadeh J (2002b) The effect of temperature on the duration of developmental periods and the amount of oviposition of the *Oenopia conglobata* L. (Col.: Coccinellidae) in laboratory conditions. Paper presented at the 15th Iranian Plant Protection Congress, Razi University of Kermanshah, Kermanshah, 7–11 September 2002

Mojib Hagh Ghadam Z, Jalali Sendi J, Sadeghi SE, Hajizadeh J (2002c) Biology of lady-bird *Oenopia conglobata* L. (Col.: Coccinellidae) on the aphid *Timocallis saltans* in laboratory conditions. Paper presented at the 15th Iranian Plant Protection Congress, Razi University of Kermanshah, Kermanshah, 7–11 September 2002

Mokhtari F, Madadi H (2013) Numerical response and impact of *Aphis fabae* densities on some biological traits of *Hippodamia variegata*. In: Talaei-Hassanloui R (ed) Proceedings of the Conference of Biological Control in Agriculture and Natural Resources, College of Agriculture and Natural Resources, Karaj, 2013

Mokhtari B, Samih MA (2014) Effect of temperature on some biological characteristics of *Oenopia conglobata contaminata* (Menteries) in feeding on the green peach aphid, *Myzus persicae* (Sulzer) in laboratory conditions. Agric Pest Manag 1(2):1–12. (In Persian with English Abstract)

Mokhtari B, Samih MA (2016) Two-sex Life Table of *Oenopia conglobata contaminata* (Mentries) Feed on *Myzus persicae* (Sulzer) and *Agonoscena pistacia* Burkhardt and Lauterer under laboratory condition. J Plant Prot 30(1):54–62. (In Persian with English Abstract)

Mollashahi M Saboori H (2010) Demography of lady beetle *Hippodamia vriegata* with feeding wheat green aphid *Schizaphis graminum*, under laboratory conditions. Paper presented at the Proceedings of the 1st Iranian Pest Management Conference (IPMC), University of Shahid Bahonar, Kerman, 29 June-1 July 2010

Mollashahi M, Sahragard A, Hosseini R (2002) Determination of population growth indices of lady beetle, *Hippodamia variegata* (Col.: Coccinellidae) under laboratory conditions. Paper presented at the 15th Iranian Plant Protection Congress, Razi University of Kermanshah, Kermanshah, 7–11 September 2002

Mollashahi M, Sahragard A, Hosseini R, Mollashahi E (2004a) Determination of product index of lady beetles *Hippodamia variegata* (Goeze) and *Coccinella septempunctata* (L.) under laboratory condition. Paper presented at the 16th Iranian Plant Protection Congress, Tabriz University, Tabriz, 28 August- 1 September 2004

Mollashahi M, Sahragard A, Hosseini R, Mollashahi E (2004b) Determination of population growth indices of lady beetle *Coccinella septempunctata* (L.) under laboratory condition. Paper presented at the 16th Iranian Plant Protection Congress, Tabriz University, Tabriz, 28 August- 1 September 2004

Mollashahi M, Sahragard A, Hosseini R (2009) A comparative study on the population growth parameters of *Coccinella septempunctata* (Col.: Coccinellidae) and melon aphid, *Aphis gossypii* (Hem.: Aphididae) under laboratory conditions. J Entomol Soc Iran (JESI) 29(1):1–12. (In Persian with English Abstract)

Montazeri MM, Mossadegh MS (1995) The coccinelids (Coleoptera) fauna of Gorgan plain and Gonbad. Paper presented at the 12th Iranian Plant Protection Congress, Karaj Junior College of Agriculture, Karaj, 2–7 September 1995

Mortazavi Malekshah SA, Khalghani J, Ranjbar Aghdam H, Rezapanah MR (2010) Estimation of thermal requirements and zero developmental temperature of *Cryptolaemus montrouzieri* Mulsant (Coleoptera: Coccinellidae) by using degree-days linear model. Paper presented at the 19th Iranian Plant Protection Congress, Iranian Research Institute of Plant Protection, Tehran, 31 July-3 August, 2010

Mortazavi Malekshah SA, Ranjbar Aghdam H, Khalghani J, Rezapanah M (2015) Effect of temperature on life table parameters of *Cryptolaemus montrouzieri* Mulsant feeding on citrus mealybug, *Planococcus citri* (Risso). J Appl Res Plant Protect 4(2):145–160. (In Persian with English Abstract)

Mossadegh MS, Aleosfoor M (2004) *Aphis nerii* Boyer de Fonscolombe and its predatory natural enemies in Shiraz and Khuzestan. Paper presented at the 16th Iranian Plant Protection Congress, Tabriz University, Tabriz, 28 August- 1 September 2004

Mossadegh MS, Eslamizadeh R, Esfandiari M (2008a) Biological study of mealybug *Nipaecoccus viridis* (New.) and possibility of its biological control by *Cryptolaemus montrouzieri* Mul. In citrus orchards of North Khuzestan. Paper presented at the 18th Iranian Plant Protection Congress, Bu-Ali Sina University, Hamedan, 25–28 August 2008

Mossadegh MS, Eslamizadeh R, Esfandiari M (2008b) Biological control of *Icerya purchasi* Maskell by *Rodolia cardinalsi* Mulsant in citrus orchards of North Khuzestan. Paper presented at the 18th Iranian Plant Protection Congress, Bu-Ali Sina University, Hamedan, 25–28 August 2008

Narrei A, Askarianzadeh AR, Taherkhani K (2005) Study of activity trend ladybird beetle, *Stethorus gilvifrons* (Mulsant) (Col.: Coccinellidae) on sugarcane in Khuzestan province. Paper presented at the 17th Iranian Plant Protection Congress, College of Agriculture and Natural Resources, University of Tehran, Karaj, 2–5 September 2006

Nazari A, Sahragard A, Hajizadeh J (2004) Biology of *Exochomus nigromaculatus* Goeze (Col.: Coccinellidae) in relation with different temperatures. Paper presented at the Proceedings of the 16th Iranian Plant Protection Congress, Tabriz (pp 23–23), University of Tabriz

Nazari A, Sahragard A, Hajizadeh J (2005) Functional response of *Exochomus nigromaculatus* (Col.: Coccinellidae) to different densities of *Aphis nerii* and *Aphis craccivora*. Appl Entomol Phytopathol 72(2):85–94. (In Persian with English Abstract)

Nikan Jablou S, Golizadeh A, Hassanpour M, Nouri Ganbalani G, Naseri B, Ghafouri Moghaddam M, (2013) Biology of predator coccinellid, *Hippodamia variegata* (Col.: Coccinellidae) when feeding on cabbage aphid, *Brevicoryne brassicae* (Hemi.: Aphididae) on turnip and radish. In: Talaei-Hassanloui R (ed) Proceedings of the Conference of Biological Control in Agriculture and Natural Resources, College of Agriculture and Natural Resources, Karaj, 2013

Noorbakhsh SH (2000) The natural enemies of almond aphids in Chaharmahal va Bakhtiari province. Paper presented at the 14th Iranian Plant Protection Congress, Isfahan University of Technology, Isfahan 5–8 September 2000

Novin M, Mossadegh MS, Karami Nejad, M, Ghasemi Nejad M (2000) Natural enemy of *Nipaecoccus viridis* (Newstead) in the north of Khuzestan. Paper presented at the 14th Iranian Plant Protection Congress, Isfahan University of Technology, Isfahan 5–8 September 2000

Parish H, Mehrnejad MR, Basirat M, Fallahzadeh M (2012a) Thermal constant and lower threshold of *Coccinula elegantula* as a predator for the common pistachio psyllid, *Agonoscena pistaciae*. Paper presented at the 20th Iranian Plant Protection Congress, Shiraz University, Shiraz, 25–28 August 2012

Parish H, Mehrnejad MR, Basirat M, Fallahzadeh M (2012b) Life table parameters of *Coccinula elegantula* predator of *Agonoscena pistaciae* in laboratory condition. Paper presented at the 20th Iranian Plant Protection Congress, Shiraz University, Shiraz, 25–28 August 2012

Polis GA, Myers CA, Holt RD (1989) The ecology and evolution of intraguild predation: potential competitors that eat each other. Annu Rev Ecol Syst 20(1):297–330

Rahmani S, Bandani AR (2013a) Demographic traits of *Hippodamia variegata* (Goeze) (Coleoptera: Coccinellidae) reared on *Aphis fabae* Scopoli (Hemiptera: Aphididae). Arch Phytopathol Plant Prot 46(12):1393–1402. https://doi.org/10.1080/03235408.2013.768061

Rahmani S, Bandani AR (2013b) Sublethal concentrations of thiamethoxam adversely affect life table parameters of the aphid predator, *Hippodamia variegata* (Goeze) (Coleoptera: Coccinellidae). Crop Prot 54:168–175. https://doi.org/10.1016/j.cropro.2013.08.002

Rakhshani H, Ebadi R, Mohammadi AA (2009) Population dynamics of alfalfa Aphids and their natural enemies, Isfahan, Iran. J Agric Sci Techn-Iran 11(5):505–520

Razmjou J, Hajizadeh J (2000) The coccinellids fauna of cotton fields in Moghan region. Paper presented at the 14th Iranian Plant Protection Congress, Isfahan University of Technology, Isfahan 5–8 September 2000

Rogers D (1972) Random predator search and insect population models. J Anim Ecol 41 (2):369–383. https://doi.org/10.2307/3474

Rounagh H, Samih MA (2012) Functional response of *Oenopia conglobata cantaminata* (Menetries) to different densities of *Brevicoryne brassicae* L. under laboratory conditions. Paper presented at the 20th Iranian Plant Protection Congress, Shiraz University, Shiraz, 25–28 August 2012

Rounagh H, Samih MA (2014) The two-sex life table and predation rate of *Oenopia conglobata contaminata* (Col.: Coccinellidae) feeding on pomegranate green aphid, *Aphis punicae* (Hem.:

Aphididae), under laboratory conditions. J Entomol Soc Iran (JESI) 34(1):59–72. (In Persian with English Abstarct)

Rounagh H, Samih MA, Mhdian K (2014) Effect of temperature on biological parameters of *Oenopia conglobata contaminata* (Menetries) by feeding on pomegranate green aphid. *Aphis punicae* Pass under laboratory conditions Plant Pests Res 4(3):25–38. (In Persian with English Abstract)

Rounagh H, Samih MA (2015) Functional response of *Oenopia conglobata* contaminata (Col.: Coccinellidae) feeding on pomegranate green aphid, *Aphis punicae* (Hem.: Aphididae). Plant Prot (Sci J Agric) 38(1):51–65. (In Persian with English Abstract)

Sabaghi R, Sahragard A, Hosseini R (2011a) Functional and numerical responses of *Scymnus syriacus* Marseul (Coleoptera: Coccinellidae) to the black bean aphid, *Aphis fabae* Scoploi (Hemiptera: Aphididae) under laboratory conditions. J Plant Prot Res 51(4):423–428. https://doi.org/10.2478/v10045-011-0070-4

Sabaghi R, Sahragard A, Hosseini R (2011b) Area dependent searching efficiency of *Scymnus syriacus* (Col.: Coccinellidae) feeding on *Aphis craccivora* (Hem.: Aphididae). J Entomol Soc Iran (JESI) 31(1):1–16

Saberi S, Talebi AA, Fathipor Y (2013) Effects of high temperature stresses on preimaginal mortality of the Mealy bug ladybird *Cryptolaemus montrouzieri* (Col.: Coccinellidae). In: Talaei-Hassanloui R (ed) Proceedings of the Conference of Biological Control in Agriculture and Natural Resources, College of Agriculture and Natural Resources, Karaj, 2013

Sadat Alizadeh M, Mossadegh MS, Esfandiari M (2013) Natural enemies of *Maconellicoccus hirsutus* (Greenblatt) (Hemiptera: Pseudococcidae) and their population fluctuations in Ahvaz, Southwest of Iran. J Crop Prot 2(1):13–21

Sadaghian B, Nikdel M, Dordaei AA (September 2000) (2000) Faunistic study on Coleoptera order in Arasbaran. Paper presented at the 14th Iranian plant protection congress, Isfahan University of Technology. Isfahan:5–8

Sadeghi SE, Khanjani M (1998) A study of coccinellid fauna in alfalfa fields in Hamadan. Paper presented at the 13th Iranian Plant Protection Congress, Karaj Junior College of Agriculture, Karaj, 23–27 August 1998

Sadeghi SE, Mojib Hagh Ghadam Z, Jalali Sendi J, Haji Zadeh J (2004) Investigation on the biology of lady beetle *Oenopia conglobate* (L.) on poplar aphid *Chaitophorus leucomelas* (Koch) in laboratory conditions. Pajouhesh-va-Sazandegi 62:20–24. (In Persian with English Abstract)

Saeedi K (1998) The coccinellid fauna of alfalfa fields in Boyere Ahmad region. Paper presented at the 13th Iranian Plant Protection Congress, Karaj Junior College of Agriculture, Karaj, 23–27 August 1998

Samih MA, Mokhtari B (2014) Efficiency and pradatory of *Oenopia conglobata contaminata* (Menetries) feeding on green peach aphid *Myzus persicae* (Sulzer) under laboratory conditions. Biol Control Pests Plant Dis 3(1):53–65. (In Persian with English Abstract)

Samin N, Shojai M (2013) A study on Coccinellidae (Coleoptera: Cucujoidea) from Varamin and vicinity, Iran. Linzer Biol Beitr 45(2):2121–2126

Seifollahi AR, Ebadi R, Sadeghi SE (2000) Natural enemies of Gaz psyllid (*Cyamophila dicora* Lognivora) in Isfahan province. Paper presented at the 14th Iranian Plant Protection Congress, Isfahan University of Technology, Isfahan 5–8 September 2000

Shahriari N, Afshari A, Dadpour Moghanlo H, Nadimi A (2016) Cold storage possibility of adult *Cryptolaemus montrouzieri* (Col.: Coccinellidae). In: Karimi J (ed) Proceedings of the 3rd National Meeting on Biological Control in Agriculture and Natural Resources, Ferdowsi University of Mashhad, Mashhad, 2016

Sharifnia A (2012) Functional Response and life table parameters of *Stethorus gilvifrons* (Mulsant) feeding from fig mite, *Eotetranychus hirsti* Pritchard & Baker (Acari: Tetranychidae) under laboratory conditions. MSc. Dissertation, Shiraz University

Sohrabi F, Shishehbor P (2007) Functional and numerical responses of *Stethorus gilvifrons* Mulsant feeding on strawberry spider mite, *Tetranychus turkestani* Ugarov and Nikolski. Pak J Biol Sci 10(24):4563–4566. https://doi.org/10.3923/pjbs.2007.4563.4566

Soleimani S, Madadi H (2015) Seasonal dynamics of: the pea aphid, *Acyrthosiphon pisum* (Harris), its natural enemies the seven spotted lady beetle *Coccinella septempunctata* Linnaeus and variegated lady beetle *Hippodamia variegata* Goeze, and their parasitoid *Dinocampus coccinellae* (Schrank). J Plant Prot Res 55(4):421–428. https://doi.org/10.1515/jppr-2015-0058

Solomon ME (1949) The natural control of animal populations. J Anim Ecol 18(1):1–35. https://doi.org/10.2307/1578

Taghizadeh R, Fathipour Y, Kamali K (2008a) Temperature-dependent development of Acarophagous ladybird, *Stethorus gilvifrons* (Mulsant) (Coleoptera: Coccinellidae). J Asia Pac Entomol 11(3):145–148. https://doi.org/10.1016/j.aspen.2008.07.001

Taghizadeh R, Fathipour Y, Kamali K (2008b) Influence of temperature on life-table parameters of *Stethorus gilvifrons* (Coleoptera: Coccinellidae) fed on *Tetranychus urticae*. Paper presented at the 18th Iranian Plant Protection Congress, Bu-Ali Sina University, Hamedan, 25–28 August 2008

Talebi AA, Jaryani R, Allahyari H (2014) Some biological characteristics of *Hippodamia variegata* (Col.: Coccinellidae) rearing on three species of wheat Aphids. J Anim Res (Iran J Biol) 27 (2):260–269

Tavoosi Ajvad F, Madadi H, Sobhani M, Kazzazi M (2011) Biology of *Hippodamia variegata* Goeze (Col.: Coccinellidae) on three different diets under laboratory conditions. Paper presented at the Proceedings of the 1st Iranian Pest Management Conference (IPMC), University of Shahid Bahonar, Kerman, 29 June-1 July 2010

Tavoosi Ajvad F, Madadi H, Sobhani M et al (2012) Seasonal changes of *Hippodamia variegata* populations and its parasitism by *Dinocampus coccinellae* in alfalfa fields of Hamedan. Biol Control Pests Plant Dis 1(1):11–18. (In Persian with English Abstract)

Tavoosi Ajvad F, Madadi H, Gharali B (2014) Influence of intraguild predation between *Episyrphus balteatus* and *Hippodamia variegata* on their prey. Arch Phytopathol Plant Prot 47(1):106–112

Triplehorn CA, Johnson NF (2005) Borror and Delong's introduction to the study of insects. Thomson Brooks/Cole, USA

Vojdani S (1965) Les Coccinellides utiles nuisibles de l'Iran. University of Tehran, Tehran. (In Persian with French Abstract)

Yaghmaee F, Kharrazi Pakdel A (1995) A faunistic survey of coccinellis in Mashhad region. Paper presented at the Proceedings of the 12th Iranian Plant Protection Congress, Karaj Junior College of Agriculture, Karaj, 2–7 September 1995

Yazdani A, Ahmadi AA(1991) New records of eight Coccinellid species of genus *Scymnus* from Fars province. Paper presented at the 10th Iranian Plant Protection Congress, University of Shahid Bahonar, Kerman, 1–5 September 1991

Yazdani A, Ebrahimi J (1993) Predators of scale insects on pistachio trees in the Kerman Province. Paper presented at the 11th Iranian Plant Protection Congress, Guilan University, Rasht, 28 August-2 September 1993

Yazdani M, Zarabi M, Samih MA (2010) Study on life table parameters of *Clitostethus arcuatus* (Rossi), feeding on *Trialeurodes vaporariorum*. Paper presented at the 19th Iranian Plant Protection Congress, Iranian Research Institute of Plant Protection, Tehran, 31 July-3 August, 2010

Zanganeh L, Madadi H, Allahyari H (2015) Demographic parameters of *Diuraphis noxia* (Hemiptera: Aphididae) and *Hippodamia variegata* (Coleoptera: Coccinellidae) recorded in the context of *D. noxia* infesting resistant and susceptible cultivars of wheat. Eur J Entomol 112 (3):453–459. https://doi.org/10.14411/eje.2015.053

Zare Khormizi M, Ostovan H, Fallahzadeh M, Mossadegh, MS (2014) Report of Three Ladybird Beetles (Coleoptera:Coccinellidae) from Iran. Paper presented at the 21th Iranian Plant Protection Congress, Urmia University, Urmia, 23–26 August 2014

Zarghami S, Mossadegh MS, Kocheili F, Allahyari H, Rasekh A (2013) Biology and life table parameters of *Nephus arcuatus* on spherical mealy bug, *Nipaecoccus viridis*. In: Talaei-Hassanloui R (ed) Proceedings of the Conference of Biological Control in Agriculture and Natural Resources, College of Agriculture and Natural Resources, Karaj, 2013
Zarghami S, Mossadegh MS, Kocheili F et al (2014a) Prey stage preference and functional response of the Coccinellid, *Nephus arcuatus* Kapur in response to *Nipaecoccus viridis* (News.). Plant Pests Res 4(3):73–86. (In Persian with English Abstract)
Zarghami S, Mossadegh MS, Kocheili F et al (2014b) Effect of temperature on population growth and life table parameters of *Nephus arcuatus* (Coleoptera: Coccinellidae). Eur. J Entomol 111 (2):199–206. https://doi.org/10.14411/eje.2014.017
Zarghami S, Mossadegh MS, Kocheili F et al (2015) Functional and numerical responses of *Nephus arcuatus* Kapur feeding on *Nipaecoccus viridis* (Newstead). Agric Pest Manag 2(1):48–59. (In Persian with English Abstract)

# Chapter 3
# Biological Control of Pests by Mites in Iran

Hamidreza Hajiqanbar and Azadeh Farazmand

## 3.1 Introduction

Acari (ticks and mites) along with spiders are the largest and most successful groups of arachnids (Arthropoda: Arachnida) occupying a wide spectrum of habitats, similar to insects. Nowadays, Acari are considered as rivals of insects in global diversity and abundance, however, owing to their tiny size, are less known and studied. The importance of mites is not hidden for anybody. A few taxa (some tetranychoids and eriophyoids) imposed economic damage to agricultural crops, some are useful natural enemies against pest insects and mites, some others are animal (including human) parasites and many of them are fungivorous, saprophages and decomposers (Krantz and Walter 2009; Rahmani et al. 2012).

Biological control has made great advances in the last decades in the identification of natural enemies and development of commercial products, leading to the major component of integrated pest management (Dogramaci et al. 2011). In biological control point of view, predatory and parasitic mites can affect host populations. Predatory mites are important agents in integrated management of phytophagous mites and some insects for example thrips (Rahmani et al. 2009a; Madadi et al. 2009), whiteflies, mealybugs and, also weeds (Gerson et al. 2003). Predators are more effective than parasites particularly in classical biological control programs however, role of parasites in natural control of their hosts is undeniable. The use of mites for pests control was reviewed by Gerson et al. (2003), Gerson and

H. Hajiqanbar (✉)
Department of Entomology, Faculty of Agriculture, Tarbiat Modares University, Tehran, Iran
e-mail: hajiqanbar@modares.ac.ir

A. Farazmand
Department of Agricultural Zoology, Iranian Research Institute of Plant Protection, Agricultural Research Education and Extension Organization (AREEO), Tehran, Iran
e-mail: afarazmand@iripp.ir

Weintraub (2007, 2012), Gerson (2014), McMurtry et al. (2015), Hajizadeh and Faraji (2016) and Fathipour and Maleknia (2016). Generally, predatory mites of the family Phytoseiidae have received considerable attention because of their potential as biological control agents of spider mites and other microarthropods (Daneshvar and Denmark 1982; McMurtry and Croft 1997).

The following mite lineages have an association with invertebrates including arthropod pests: the mesostigmatic, prostigmatic and astigmatine mites, the latter not so professionally studied in Iran. Therefore, in this chapter, we discuss predatory and parasitic Mesostigmata and Prostigmata. Regarding predatory mites, we mostly deal with non-soil dwelling families in order to focusing on more effective and applicable mites used as biocontrol agents. Systematic concepts follow those of Krantz and Walter (2009), and a list of families are arranged based on their evolutionary trend used in Lindquist et al. (2009) and Walter et al. (2009). The Parasitengonina are treated as parasitic mites because of their parasitic life in the larval stage.

This chapter discusses mesostigmatic and prostigmatic families of predatory and parasitic mites that are known or postulated to have an adverse effect on pests including insects, mites and weeds in Iran and is focused on fundamental (faunistic) studies, and if available, laboratory works and has been compiled mostly recent works related to the application of predatory mites including behavioral aspects, life table parameters, the effect of host plants on predatory mites, the side effect of acaricides on biocontrol agents, interactions with other biocontrol agents and mass rearing. We hope that this chapter will promote further researches on exploring, developing and realizing the potential of mites in pests control. The bibliography provided in this chapter is through 2016, and some selected 2017 and 2018 references have been included.

## 3.2   Predatory Mites

### 3.2.1   Order Mesostigmata

#### 3.2.1.1   Dinychidae

Like many uropodine mites, members of the family Dinychidae usually occupy decaying organic material, some nidicolous and some others living in soil or humus. Mites of the genus *Uroobovella* are mainly fungivorous and sometimes, along with other uropodines, colonize in stored products however the *Uroobovella marginata* (C.L. Koch) (formerly placed in *Fuscouropoda*) is the most important species of the genus in biological control point of view. This mite may attack nematodes, other mites and dipteran larvae including house flies (Gerson et al. 2003; Lindquist et al. 2009). *U. marginata* can also feed on the epidermis of slugs, causing necrotic and finally killing the host (Raut and Panigrahi 1991). This useful mite is also recorded from different parts of Iran available in soil, litter, humus, cow and sheep dung, manure, and stored products (Kazemi and Rajaei 2013; Nemati et al. 2018) and although may not be considered as an effective biocontrol agent, more evaluation in distributed regions would clear its role in control of mentioned pests.

### 3.2.1.2  Parasitidae

Thirty five genera and approximately 430 species of the Parasitidae are free-living predators mostly on the ground, feeding on small arthropods and nematodes (Lindquist et al. 2009). The genus *Pergamasus* prey actively upon *Tyrophagus* mites (Acaridae) and springtails in soil habitats. Some *Parasitus* spp. (e.g. *P. bituberosus* Karg, *P. coleoptratorum* L., *P. gregarious* Ito) are listed as natural enemies of *Rhizoglyphus robini* Claparédè (Bulb mite) in soil and, early instars of cecidomyiid and sciarid larvae in cultivated mushroom (Gerson et al. 2003; Castilho et al. 2015). Several authors (e.g. Brown and Wilson 1992; Schwarz and Muller 1992; Gasperin and Kilner 2015) have repeatedly studied dynamics of symbiotic relationships between carrion-feeding burying beetles of the genus *Necrophorus* (Silphidae) and mites of the *Poecilochirus* spp. Typically, beetles carry mite deutonymphs to the food resource (carrion) in order to mite feeding on nematodes and fly larvae that are rivals of the beetles. This positive effect may be balanced by predatory of mites on host beetles progeny. All of the genera mentioned above and ten other ones have been recorded from Iran (Kazemi et al. 2013; Kazemi and Rajaei 2013; Nemati et al. 2018) however, their efficiency has yet to be done.

### 3.2.1.3  Macrochelidae

The cosmopolitan and speciose family Macrochelidae consists of about 20 genera and more than 470 species, are primarily free-living predators attacking nematodes and, eggs and early stages of small invertebrates including Diptera and other arthropods in various habitats especially dung and manure (Lindquist et al. 2009; Krant 2018). Their development is almost rapid and one generation is usually completed in one week. It is well documented that some mites of the genus *Macrocheles* such as *M. muscaedomesticae* (Scopoli) and *M. merdarius* (Berlese) have a significant role in the control of synanthropic filth flies (*Musca domestica* L., *Musca autumnalis* DeGeer and *Stomoxys calcitrans* L.) (Gerson et al. 2003). However, these two useful species have been recorded from various regions of Iran, unfortunately their effect on aforementioned nuisance flies has not yet evaluated. Many other macrochelid genera and species are distributed in the country (Kazemi and Rajaei 2013; Babaeian et al. 2015; Nemati et al. 2018).

### 3.2.1.4  Phytoseiidae

The phytoseiid mites are a large family of mesostigmatic mites that can feed on small arthropods including mite groups, thrips, whiteflies and, as well as nematodes. Also, they can feed on fungi, plant exudates, pollen grains and some have the ability to extract liquid from leaf cells (McMurtry et al. 2013). This family has received the most attention since the 1950s when it became clear that they have economic importance as natural predators of phytophagous mites and small insects, and therefore are useful in the biological control of crop pests (Swirski et al. 1967). Most phytoseiids usually develop within a week at 25 °C which make them effective natural enemies of the pests (Hoy 2011). Adult females can oviposit throughout the year in tropical and subtropical areas and also in greenhouses in the temperate area (Zhang 1963). This family are now valued and 20 species of phytoseiid mites have been commercially reared and sold in more than 50 countries all over the world (Zhang 1963; Gerson et al. 2003) that provide effective pest control in greenhouses and on agricultural crops (Bjornson 2008).

In the initial stage of biological control programs of pest mites with predatory mites, Chant (1961) explained when *Phytoseiulus persimilis* Athias-Henriot was periodically released in a greenhouse experiment, had ability to control *T. urticae*. Similar results were obtained by Hussey et al. (1965) on greenhouse cucumbers and Oatman and McMurtry (1966) as well as Oatman et al. (1977) in field-grown strawberry plants. McMurtry and Croft (1997) demonstrated periodic releases of predatory mites on cotton, hops and mint. In addition to types I and II predators, other predators i.e. types III and IV can be effective in controlling the target pests. McMurtry and Scriven (1971) demonstrated predation capacity of *Amblydromalus limonicus* (Garman and McGregor), a type III predator, on *Oligonychus punicae* (Hirst). Several studies have been shown the effect of *Typhlodromus pyri* Scheuten on *Panonychus ulmi* (Koch) (Nyrop 1988) and *Euseius hibisci* (Chant), a type IV predator, is the most abundant predaceous mite in the strawberry-growing area. It is commonly associated with *T. urticae*, both on strawberries and castor bean (*Ricinus communis*) surrounding the strawberry plants. This predaceous mite also feeds on pollen and other insect pests such as various instars of whiteflies (Badii et al. 2004).

In 1988, a strain of *P. persimilis* was introduced into Iran from the Netherlands (Department of Entomology, Wageningen Agricultural University) (Daneshvar 1989) and was a beneficial biocontrol agent in greenhouses and outdoors (Daneshvar and Abaii 1994) and during the following three decades, other exotic species were employed. In the last decade, more researches have dealt with life history, foraging behavior and interactions and population dynamisms in laboratory conditions.

3.2.1.4.1 Introduced Phytoseiid Mites in Iran

**Phytoseiulus persimilis** According to McMurtry et al. (2013), this species is classified as a specialized predator. This species has the ability to rapidly increase and overcome outbreak populations of two-spotted spider mites (McMurtry and Croft 1997). This predatory mite owing to its specificity on spider mites with commercial name Spidex® is mass produced and is used in Iran and around the world. Most early studies focused on the ability of this predaceous mite to rapidly increase and overcome outbreak populations of spider mites (McMurtry and Croft 1997).

**Neoseiulus californicus (McGregor)** *Neoseiulus californicus* is placed in type II lifestyle because it is associated with tetranychid mites which produce heavy webbing and also has the ability to feed and reproduce on eriophyids, tarsonemids, tydeids, as well as pollen (McMurtry et al. 2013) and thrips (Rahmani et al. 2009a). This species is introduced in Iran with Spical® name and can survive and be efficient at high temperatures and low humidity (Weintraub and Palevsky 2008; Ahn et al. 2010).

**Amblyseius swirskii Athias-Henriot** The mites of the genus *Amblyseius* including *A. swirskii* are classified as type III predators, which feed on the whiteflies *Bemesia tabaci* and *Trialeurodes vaporariorum*, western flower thrips, *Frankliniella occidentalis*, the broad mite, *Polyphagotarsonemus latus* and eggs of moths as well as pollen (McMurtry et al. 2013; Nguyen et al. 2013). One advantage of this predator is its ability to develop and reproduce on various kinds of pollen, which allows the population of this predator to develop on plants before the appearance of pests (Calvo et al. 2014). This species has been commercialized with Swiriskii-mite® name and released as a biological control agent in more than 50 countries (Park et al. 2010; Calvo et al. 2012).

### 3.2.1.4.2 Native Phytoseiid Mites

Demite et al. (2014) in a phytoseiid database presented 76 species of phytoseiid mites available in Iran. Some selected species are mentioned below.

**Neoseiulus barkeri Hughes** *Neoseiulus barkeri* is a worldwide species that is reported from Asia, predator which feeds on storage mites, spider mites, thrips, broad mites and whiteflies eggs, as well as plant pollen (Fan and Petit 1994). This predator is an indigenous species in Iran that is reported by Kamali et al. (2001), Hajizadeh et al. (2009) and Mahjoori et al. (2015). This species can be mass-reared in wheat bran with storage mites and has been used for augmentative biological control of *Thrips tabaci* Lind. (Hansen 1988).

**Typhlodromus (Anthoseius) bagdasarjani Wainstein and Arutunjan** According to McMurtry et al. (2013), species of the genus *Typhlodromus* are considered as type III predators. *Typhlodromus bagdasarjani* is a species native to the Middle East with high abundance in orchards of Iran (Kamali et al. 2001). It feeds on spider mites, eriophyoids, tydeids, thrips and whiteflies (Daneshvar 1993). It is an effective biocontrol agent in high temperature regions (Ganjisaffar et al. 2011a) and has a population growth rate equal or higher than its prey, *T. urticae* (Khanamani et al. 2014). Generally, species in type III require more prey to complete development than for species in type II (McMurtry and Croft 1997).

**Phytoseius plumifer (Canestrini and Fanzago)**   The mites of the genus *Phytoseius* are classified as type III predators. *Phytoseius plumifer* is an important generalist indigenous predator of tetranychid mites and is widely found on various crops in Iran (Kamali et al. 2001; Hajizadeh et al. 2002). Generalists in the Phytoseiinae have a lower mean reproductive potential than in the Amblyseiinae and are very small. Issa et al. (1974) studied on the successive release of this predator for controlling spider mites on the fig seedlings. Rasmy and El-Banhawy (1974) studied some diets including *Aceria ficus* Cotte (Acari: Eriophyidae), *Tetranychus arabicus* Attiah and the pollen grains of Caster-oil (*Ricinus communis* L.) on the development and reproduction of *P. plumifer*. Prey consumption of *P. plumifer* on *Rhyncaphytoptus ficifoliae* at six constant temperatures (15, 20, 25, 30, 35 and 37 °C) was investigated and the highest and lowest values of prey consumption by adult female were observed at 25 and 37 °C, respectively (Louni et al. 2014a).

**Amblyseius herbicolus Chant**   Species of the genus of *Amblyseius* are considered as type III predators (McMurtry et al. 2013). *A. herbicolus* is a native species to Iran with high abundance in Guilan province (Northern Iran) which feeds on spider mites, eriophyoids, teniupalpids and thrips (Argov et al. 2002). In a survey, Notghi Moghadam et al. (2010) studied the influence of three diets on development and oviposition of *A. herbicolus* under laboratory conditions and concluded that date palm pollen can be an alternative food in absence of main prey. In another study, Notghi Moghadam et al. (2014) studied biology and prey preference of this predatory mite on *T. urticae* and showed that developmental time of egg, larva, protonymph and deutonymph were $1.73 \pm 0.11$, $1.4 \pm 0.12$, $1$ and $1.13 \pm 0.09$ days per female, respectively. Mean generation time (egg to egg) was $6.86 \pm 0.16$ days and mean adult female longevity was 30.46 days. During oviposition period (20.33 days), the average of fecundity was calculated 1.8 eggs per day per female. In the prey stage preference experiment, predatory mites preferred immature stages of two spotted spider mite.

### 3.2.1.4.3   Behavioural Aspects

**Functional Response and Mutual Interference**
Before starting a biological control program, the study of behavioural aspects of natural enemies is essential. The performance of a predator depends on several specification, two of them are functional response and mutual interference (Fathipour et al. 2006). The functional response illustrates the relationship between an individual's consumption rate and prey density (Solomon 1949) and has been classified into three types by Holling (1959). In terms of biological control, predators which show the type III functional response are usually regarded as efficient biological control agents (Fernandez-Arhex and Corely 2003; Pervez and Omkar 2006; Xiao and Fadamiro 2010). However, there are some examples of natural enemies with the type II functional response which has been successfully released

as biocontrol agents (De clercq et al. 2000; Reis et al. 2003; Badii et al. 2004; Xiao and Fadamiro 2010).

The functional response of phytoseiids is influenced by a number of factors such as temperature (Gotoh et al. 2004; Kouhjani Gorji et al. 2009; Jafari et al. 2012), host plant (Cédola et al. 2001; Ahn et al. 2010), insecticides (Poletti et al. 2007), prey stage (Farazmand et al. 2012) and age of predator (Fathipour et al. 2017, 2018). Functional response can explain search efficiencies and predation rates of predators and the evaluation of this behaviour of predators is a critical first step in determining their ability to regulate the prey (Ahn et al. 2010; Xiao and Fadamiro 2010). Native and exotic phytoseiids show type II functional response. Functional response parameters of some native and introduced phytoseiid mites are given in Table 3.1.

One of the most important aspects of interactions among multiple conspecific predators is known as direct mutual interference. Hassell and Varley (1969) and Hassell and May (1974) described the importance of this phenomenon. This behaviour results in a reduction in searching efficiency because of time wasted when conspecifics encounter each other rather than handling (capturing) prey (Henne and Johnson 2010). However, mutual interference can assist mass rearing of predatory mites in laboratory conditions and also simplify the explanation of observed outcomes in the field. This phenomenon has been studied in native phytoseiid mite, *Typhlodromus bagdasarjani* and *P. plumifer* (Farazmand et al. 2012; Khodayari et al. 2016) and an introduced mite *Neoseiulus californicus* (Farazmand et al. 2012). In a survey, the effect of prey (spider mite) density on developmental time of *P. persimilis* and *N. californicus* was investigated and demonstrated that increasing the prey density did not effect on the development period of *P. persimilis* but there was significant difference for *N. californicus* (Nadeali et al. 2012a, b). Also, Zahedi et al. (2012) investigated the effect of prey density on oviposition rate of *P. persimilis* and stated that there is a direct relation between prey density and oviposition rate.

**Prey Stage Preference**
Prey stage preference may influence prey-predator population dynamics if the prey stage affects the development and reproduction of the predator (Pandey and Singh 1999). Understanding the prey-stage preferences of a biological control agent is necessary for the success of biological control programs as it assists with mass rearing efforts and can facilitate prey-predator population dynamics in the field (Pasandideh et al. 2015). Xiao and Fadamiro (2010) showed *P. persimilis*, *G. occidentalis* and *N. californicus* preferred nymphs to eggs of *Panonychus citri* (McGregor). Blackwood et al. (2001) stated adult females of *P. persimilis* preferred eggs of *T. urticae* over the larvae, while *G. occidentalis* and *N. californicus* showed no prey-stage preference. In another study, Badii et al. (2004) indicated that *Euseius hibisci* (Chant) consumed significantly more prey eggs than other prey stages. In contrast, Xiao et al. (2013) reported that *N. californicus* and *A. swirskii* equally preferred to feed on both eggs and nymphs of *T. urticae*.

**Table 3.1** The functional response parameters of some phytoseiid mites

| Species | Prey | Prey stage | $T$ (°C) | $a$ | $T_h$ | References |
|---|---|---|---|---|---|---|
| Amblyseius swirskii | Trialeurodes vaporariorum | Nymph | 25 | $0.142\ \mathrm{h}^{-1}$ | 2.310 h | Farhadi et al. (2015) |
| A. swirskii | T. urticae | Egg | 25 | $0.030\ \mathrm{h}^{-1}$ | 0.395 h | Fathipour et al. (2017) |
| Phytoseiulus persimilis | T. urticae | Egg | 25 | $0.130\ \mathrm{h}^{-1}$ | 0.494 h | Fathipour et al. (2018) |
| Euseius finlandicus | T. urticae | Larva | 24 | $0.0456\ \mathrm{h}^{-1}$ | 0.45 h | Shirdel et al. (2006) |
| Typhlodromus kettanehi | T. urticae | Larva | 24 | $0.0466\ \mathrm{h}^{-1}$ | 0.30 h | Shirdel et al. (2004) |
| Neoseiulus barkeri | T. urticae | Nymph | 20 | $0.036\ \mathrm{day}^{-1}$ | 0.921 day | Jafari et al. (2012) |
| N. barkeri | T. urticae | Nymph | 25 | $0.064\ \mathrm{day}^{-1}$ | 0.824 day | Jafari et al. (2012) |
| N. barkeri | T. urticae | Nymph | 30 | $0.073\ \mathrm{day}^{-1}$ | 0.597 day | Jafari et al. (2012) |
| N. californicus | T. urticae | Egg | 25 | $0.0936\ \mathrm{h}^{-1}$ | 1.64 h | Farazmand et al. (2012) |
| N. californicus | T. urticae | Larva | 25 | $0.0693\ \mathrm{h}^{-1}$ | 1.73 h | Farazmand et al. (2012) |
| T. bagdasarjani | T. urticae | Egg | 25 | $0.0893\ \mathrm{h}^{-1}$ | 1.80 h | Farazmand et al. (2012) |
| T. bagdasarjani | T. urticae | Larva | 25 | $0.0473\ \mathrm{h}^{-1}$ | 2.39 h | Farazmand et al. (2012) |
| A. swirskii | T. urticae | Nymph | 25 | $0.113\ \mathrm{h}^{-1}$ | 0.33 h | Khanamani et al. (2017c) |
| Phytoseius plumifer | T. urticae | Nymph | 15 | $0.027\ \mathrm{h}^{-1}$ | 0.492 h | Kouhjani Gorji et al. (2009) |
| Ph. plumifer | T. urticae | Nymph | 20 | $0.037\ \mathrm{h}^{-1}$ | 0.506 h | Kouhjani Gorji et al. (2009) |
| Ph. plumifer | T. urticae | Nymph | 25 | $0.059\ \mathrm{h}^{-1}$ | 0.651 h | Kouhjani Gorji et al. (2009) |
| Phytoseiulus persimilis | T. urticae | Larva | 25 | $0.114\ \mathrm{h}^{-1}$ | 3.15 h | Seiedy et al. (2012) |
| T. bagdasarjani | Panonychus ulmi | Adult female | 24 | $0.0652\ \mathrm{h}^{-1}$ | – | Shirdel and Arbabi (2012a, b) |
| T. bagdasarjani | P. ulmi | Larva | 24 | $0.0511\ \mathrm{h}^{-1}$ | – | Shirdel and Arbabi (2012a, b) |
| Eu. finlandicus | P. ulmi | Larva | 24 | $0.0384\ \mathrm{h}^{-1}$ | 0.84 h | Shirdel and Arbabi (2012a, b) |
| A. swirskii | T. urticae | Egg | 30 | $0.43\ \mathrm{h}^{-1}$ | 0.48 h | Rafizadeh Afshar and Latifi (2017) |

A few studies have been carried out in Iran regarding prey stage preference. For example, Moghadasi et al. (2014a, b) demonstrated all stage of *T. bagdasarjani* significantly preferred eggs of *T. urticae* over larvae and protonymphs. Comparing the preference indices of *Amblyseius swirskii*, a study indicated a significant preference of the predator on eggs and second instar nymphs of *T. urticae* than *Bemesia tabaci* (Gennadius) (Soleymani et al. 2016). Other study, (Khodayari et al. 2016), showed *P. plumifer* in no-choice tests consumed egg, larva, protonymph and male stages of *T. urticae* more than deutonymph and female but in choice tests, the predator significantly preferred immature stages of its prey. Prey preference of *A. swirskii* on *Trialeurodes vaporariorum* and *T. urticae* were studied showing that this predator had a significant preference for *T. urticae* (Heydari et al. 2016).

**Olfactory Response**

Plants infested and damaged with spider mites produce a variety of volatile chemicals and release them from flowers, fruits, and foliage. The production of mite-induced plant volatiles apparently can change based on the genetic characteristics of the spider mite species, the plant cultivar, and the genetics of the predators. Predatory mites are able to perceive these volatile chemical cues and even learn about them (Hoy 2011). There are studies about the role of these chemical cues and the responses by phytoseiids (Dicke and Sabelis 1988; Dicke et al. 1990; Janssen et al. 1998; Maeda et al. 2001; De Boer and Dicke 2004). Shimoda et al. (2005) showed that *N. californicus* responded to five volatiles produced by spider-mite-infested plants, including linalool, methyl salicylate, (Z)-3-hexen-1-ol, (E)-2-hexenal, and (Z)-3-hexenyl acetate. In another study, *N. womersleyi* responded to mixtures of three synthetic compounds produced by tea plants infested by *T. kanazwai* (Ishiwari et al. 2007). In Iran, Seiedy et al. (2013) showed *P. persimilis* was able to discriminate between untreated and *Beauveria bassiana*-treated *T. urticae* and proposed avoidance of the predator from the fungus *B. bassiana* would reduce the impact of intra-guild interactions. Other study conducted by Malcknia et al. (2014b) demonstrated that host plant experience and different hunger periods affect the olfactory response of *P. persimilis*. Khosravi Shastestani et al. (2014) stated the role of methyl salicylate as an important part of volatile blends in searching behaviour of the predatory mite, *P. persimilis*. This compound attracted the predatory mites (1 h starved) in a dose range of 0.02–20μg. Bohloolzadeh et al. (2013) showed that the rearing condition of *A. swirskii* females can affect their olfactory response. Mohammadi et al. (2012) compared the odor responses of two different populations of *P. persimilis* on cucumber and bean plants infested by *T. urticae* and showed a significant preference towards bean plants. In a study, an olfactory response of *N. californicus* to the strawberry plants infested by *T. urticae* and western flower thrips, *Frankliniella occidentalis* Pergande was compared and showed that this predatory mite had not ability to identify volatiles from strawberry infested with thrips (Rezaie et al. 2018).

**Mating and Oviposition Behavior**

Defensive behaviors of prey can affect the success of a predator (Sabelis and Dicke 1985). Parents can use different strategies for protection of their offspring such as change oviposition site, absorb the egg material or retain the eggs inside their body and depositing eggs away from risky patches (Montserrat et al. 2007). In a study, Askarieh Yazdi et al. (2015) used females of two forms of *T.urticae*, red and green forms, to study whether they retain eggs in response to exposure to the predatory mites *P. persimilis* and *A. swirskii* and showed egg development time of the green form was significantly shorter than that of the red form after receiving cues related to *P. persimilis*-prey interactions. The rate of predation, oviposition, and development of the predatory mite *N. californicus* by Karami Badrbani et al. (2015) was studied and showed that preying on red *T. urticae* and mixed (red+green) could be resulted in the slowest development of the protonymph and deutonymph stages of *T. urticae* than green form. Dehghani-Tafti et al. (2015) surveyed the effect of mating experience, on male individuals in *P. persimilis* and *N.californicus* and stated both naive and experienced males were able to mate females equally.

**Learning Behaviour**

Learning is defined as changed behavior following experience and could affect many important activities such as foraging, reproducing and social interactions and is considered as an optimizing activity which assists the organism behaviors regarding the varying environment (Walzer and Schausberger 2011; Strodl and Schausberger 2012). This phenomenon has been studied in predatory mites of the family Phytoseiidae (Drukker et al. 2000; Rahmani et al. 2009b; Schausberger et al. 2010). Experienced predators wasted less energy in finding, attacking, and handling prey and thus could retain their energy to egg production (Christiansen et al. 2016). Also, they showed larvae and early protonymphs of *A. swirskii* like other predatory mites, e.g. *Neoseiulus californicus* in foraging context and *Phytoseiulus persimilis* in social and cannibalism contexts can learn.

Effects of immature experience on handling time of females of the predatory mite *N. californicus* and *P. persimilis* were evaluated in laboratory conditions and have been showed experience with two-spotted spider mite larvae in the immature stages had no effect on the handling time of the predatory females on their prey. But experience with heterospecific larvae of *P. persimilis* significantly decreased handling time of *N. californicus* femalses (Asadi et al. 2014a, b). In another study, the effect of rearing conditions on the leaving or residence tendency of *N. barkeri* on bean leaf disc was investigated and described experienced predatory mites having tendency to reside on bean leaf (Bohloolzadeh et al. 2014a, b). In a study, Mohammadi et al. (2012) concluded that by increasing the reared generations of *P. persimilis* on spider mite infested plants, the higher fraction of predators had the ability to recognize their food patch. Rahmani and Hosseini (2012) demonstrated social learning in noncolonial predatory mite, *N. californicus* and stated that experienced predators attacked *T. urticae* more three times earlier than neutral predators do. Borji et al. (2014) investigated repeatability of aggressiveness against con-and heterospecific prey in *N. californicus* and declared this mite was more aggressive against both *Th. tabaci* and *Te. urticae* in choice tests than no-choice tests. Jalili Zennozi et al. (2016) studied the effect of familiarity on group formation and predation rate of *A. swirskii*. They declared that familiar individuals tend more to create group and predation rate than unfamiliars.

#### 3.2.1.4.4 Life Table Parameters

Life table is a useful method for summarizing the survival and reproductive potential of a population, and its parameters can be used in the studies of population ecology (Sakai et al. 2001), conservation (Wilcox and Murphy 1985), demographic ecotoxicology (Stark and Banks 2003), harvesting theory (Chi 1994) and pest control timing (Chi 1990). In the female age specific life table, only female individuals are involved. Since most species are bisexual and there are many differences between male and female individuals in longevity, survival rate, predation rate and pesticide susceptibility, and neglecting the variable developmental rate and male population, may cause errors in calculating demographic parameters (Chi and Liu 1985). Therefore, an age-stage, two-sex life table theory was developed to resolve these problems by including the stage separation and male population (Chi and Liu 1985; Chi 1988). The two-sex life table can calculate the age and stage structure of a two-sex population and helpful in timing pest management decisions (Chi 1990). Demographic parameters are the best indicators of fitness of a population and suitable criteria for comparing physiological states of different species, populations, etc. or even as bioclimatic or nutritional indices (Dent and Walton 1997). Also, in biological control programs, population growth rate is an essential criterion for preliminary screening and choice of potential biocontrol agents (van Lenteren and Woets 1988). As showed in Table 3.2, life table parameters can be affected by several factors, such as temperature (Jafari et al. 2010; Ganjisaffar et al. 2011b; Kouhjani-Gorji et al. 2012), host plant (Khanamani et al. 2015), type of prey (Rahmani et al. 2009a; Khodayari et al. 2013) and sublethal effects of acaricides (Hamedi et al. 2010, 2011; Alinejad et al. 2014, 2016; Maroufpoor et al. 2016a, b).

#### 3.2.1.4.5 The Effect of Host Plants on Predatory Mites

Plants are the natural habitat where herbivores live, feed and meet their natural enemies. The leaf surface in term of having domatia, trichomes, hairs and also presence of volatile substances and toxic compounds can directly affect a natural enemy's success in searching for and controlling the herbivores (Hunter 2003; Ode et al. 2004; Soufbaf et al. 2012) and interactions between phytophagous arthropods and their biological control agents (including predators) can be impressed by chemical and physical characteristics of their host plants (Cédola et al. 2001; Madadi et al. 2008; Soufbaf et al. 2012; Khanamani et al. 2015; Alipour et al. 2016; Bahari et al. 2018). The secondary metabolites in plant tissues can directly influence the growth of herbivores and they can indirectly affect the performance of

**Table 3.2** The life table parameters of some phytoseiid mites

| Species | Prey | $T$ (°C) | $r$ (day$^{-1}$) | $R_0$ | $T$ (day) | References |
|---------|------|----------|------------------|-------|-----------|------------|
| *Neoseiulus barkeri* | *Tetranychus urticae* | 15 | 0.036 | 5.510 | 47.18 | Jafari et al. (2010) |
| *N. barkeri* | *T. urticae* | 25 | 0.221 | 22.02 | 13.59 | Jafari et al. (2010) |
| *N. barkeri* | *T. urticae* | 35 | 0.247 | 14.26 | 10.47 | Jafari et al. (2010) |
| *Phytoseius plumifer* | *T. urticae* | 15 | 0.056 | 8.7011 | 48.75 | Kouhjani-Gorji et al. (2012) |
| *Ph. plumifer* | *T. urticae* | 25 | 0.187 | 24.25 | 17.06 | Kouhjani-Gorji et al. (2012) |
| *Ph. plumifer* | *T. urticae* | 35 | 0.244 | 19.98 | 11.02 | Kouhjani-Gorji et al. (2012) |
| *Typhlodromus bagdasarjani* | *T. urticae* | 20 | 0.065 | 7.101 | 28.80 | Ganjisaffar et al. (2011a) |
| *T. bagdasarjani* | *T. urticae* | 25 | 0.129 | 13.60 | 19.41 | Ganjisaffar et al. (2011b) |
| *T. bagdasarjani* | *T. urticae* | 30 | 0.156 | 13.04 | 16.60 | Ganjisaffar et al. (2011b) |
| *T. bagdasarjani* | *T. urticae* | 25 | 0.143 | 12.91 | 17.81 | Moghadasi et al. (2014b) |
| *T. bagdasarjani* | *T. urticae* | 25 | 0.153 | 7.550 | 12.97 | Khanamani et al. (2015) |
| *Ph. plumifer* | Corn pollen | 27 | 0.112 | 4.40 | 12.99 | Khodayari et al. (2013) |
| *Amblyseius swirskii* | *T. urticae* | 25 | 0.147 | 7.520 | 13.59 | Alipour et al. (2016) |
| *Phytoseiulus persimilis* | *T. urticae* | 25 | 0.217 | 20.45 | 13.83 | Alipour et al. (2016) |
| *P. persimilis* | *T. urticae* | 25 | 0.238 | 33.30 | 14.72 | Bahari et al. (2018) |
| *N. californicus* | *Thrips tabaci* | 25 | 0.041 | 1.950 | 18.62 | Rahmani et al. (2009a) |
| *P. persimilis* | *T. urticae* | 25 | 0.297 | 46.79 | 12.91 | Saemi et al. (2014) |
| *Ph. plumifer* | *T. urticae* | 27 | 0.11 | 7.75 | 18.62 | Moghadasi et al. (2011) |
| *Ph. plumifer* | *Rhyncaphytoptus ficifoliae* | 25 | 0.154 | 17.99 | 18.78 | Louni et al. (2014a, b) |

their natural enemies, therefore herbivores on different host plants often differ in their susceptibility to natural enemies (Lill et al. 2002; Zvereva and Rank 2003). Combining biological control and host plant resistance can be a promising pest management strategy (Fathipour and Sedaratian 2013).

There are some studies about the importance of host plants for spider mites and their predators in Iran. For example, Khanamani et al. (2015) compared the efficacy of *T. bagdasarjani* against spider mites on resistant and susceptible cultivars of egg-plant and found that the performance of this predator on resistant cultivar was more than that on the susceptible one. The effect of susceptible and resistant rose cultivars on demographic parameters of *T. urticae* and their predators *P. persimilis* and *A. swirskii* was determined and showed the antibiotic resistance of the rose cultivar negatively affect the population growth parameters of *T. urticae* and their both predators (Alipour et al. 2016). *Neoseiulus bakeri* showed significant preference to the cucumber leaves with fewer trichomes in order to lay their eggs and search for the prey (Maleknia et al. 2014a). In a survey, life table parameters of *P. persimilis* were determined on *T. urticae* reared on the resistant (HED) and susceptible (Beth-Alfa) cultivars of greenhouse cucumber during the first and tenth generations and demonstrated that long term feeding can decrease performance of *P. persimilis* during ten generations on resistant cucumber cultivar than susceptible one (Bahari et al. 2018). Hence, the study of interactions between host plant cultivars and biocontrol agents is important in integrated pest management programs (Fathipour and Sedaratian 2013).

### 3.2.1.4.6 The Side Effects of Acaricides on Biocontrol Agents

An understanding of the sublethal effects of pesticides on biological control agents is important to successful augmentation and conservation programs (Hamedi et al. 2009). Lord (1949) and Lord and MacPhee (1953) were the first ones to assay the effect of various spray programs on predatory mites. In the early 1990s, neonicotinoid insecticides were introduced into the market (Nauen et al. 2001). Among those, low toxicity of imidacloprid has been reported on adult females of several species of phytoseiid mites such as *Neoseiulus collegae* (DeLeon), *Phytoseiulus macropilis* (Banks), *Proprioseiopsis mexicanus* (Garman) (Mizell and Sconyers 1992); *Amblyseius womersleyi* Schicha (Leicht 1993); *Typhlodromus doreenae* Schicha, *Typhlodromus dossei* Schicha (James and Vogele 2001) and *Neoseiulus cucumeris* (Oudemans) (Sangsoo et al. 2005). Although some studies have indicated that imidacloprid is a low-risk pesticide to phytoseiid mites, James (2003) demonstrated that this product to be extremely toxic to *Galendromus occidentalis* Nesbitt and *Neoseiulus fallacies* Garman. Also, Poletti et al. (2007) showed imidacloprid significantly affected the functional response parameters of *N. californicus*. Many studies have demonstrated the adverse effect of abamectin on fecundity of phytoseiid predators (Zhang and Sanderson 1990; Ibrahim and Yee 2000; Bostanian and Akalach 2006; Nadimi et al. 2009).

The effects of pesticides and their sublethal have been studied on native and introduced predatory mites in Iran. For example, side effects of three commonly used pesticides (eptenophos, malathion and primiphos-methyl) were evaluated on an introduced predatory mite, *Phytoseiulus persimilis* and have been showed that heptenophos was more selective for *P. persimilis* than the two other pesticides tested and it could be used in integrated control programs (Kavousi and Talebi 2003). In a study by Noii et al. (2008), the side effects of three pesticides (abamectin, malathion and phosalone) were evaluated on *P. plumifer* in the laboratory and was showed these three pesticides cause 100% mortality within 24 h and were classified as harmful for the predator.

Nadimi et al. (2008, 2009) showed hexythiazox is a harmless miticide to *P. plumifer* and *P. persimilis* but fenpyroximate and abamectin were found harmful to both predatory mites at the highest field recommended concentrations. They suggested acaricide hexythiazox could be an appropriate substitute to Fenpyroximate and abamectin in integrated pest management (IPM) programs. Also, Sanatgar et al. (2011) expressed hexathiazox can be used against *T. urticae* without causing the adverse effect on population growth parameters of its predator, *P. persimilis*. While frequent applications of bifenazate influenced on biological parameters of this predatory mite (Sanatgar et al. 2012). In a research, lethal and sublethal doses of kingbo and azoxystrobin did not show a significant effect on fecundities of *P. persimilis* and percentage of egg hatchability increased in recommended dose of floramite in comparison with control treatment (Khajavi et al. 2011).

Two studies carried out by Hamedi et al. (2010, 2011) demonstrated that sublethal concentrations of abamectin and fenpyroximate can significantly reduce the population growth of *P. plumifer* and this should be considered in integrated pest management programs. Also, in another study, Hamedi et al. (2009) showed prey consumption of female mites was strongly affected by all concentrations of the two acaricides. One study has been stated fenazaquin should not be used with predatory mite *A. swirskii* in integrated management of *T. urticae* (Alinejad et al. 2014) while in another study, Alinejad et al. (2016) showed sublethal dosage of spirodiclofen has not adverse effect on population parameters of *A. swirskii*. Ghaderi et al. (2013) exhibited fenpyroximate had a negative effect on life table parameters of *P. persimilis* and should not be used simultaneously in integrated pest management programs. Kaveh et al. (2013) investigated the effect of three botanical insecticides on *P. persimilis* and showed these insecticides can be considered in the IPM of the two-spotted spider mite. Aflaki et al. (2016) evaluated effects of two compounds including azadirachtin and spirodiclofen on some biological parameters and predation rate of *N. californicus* and showed these pesticides did not have significant adverse effects on the predatory mite and especially spirodiclofen could be used in IPM programs. In another study, the effect of these both pesticides on biological and predation characteristics of *T. bagdasarjani* was studied and demonstrated that Azadirachtin was safer than spirodiclofen for this species (Seidpisheh et al. 2016). Therefore, integration of predatory mites with compatible acaricides could reduce environment hazards, which is an objective of IPM programs (Hamedi et al. 2011).

### 3.2.1.4.7    Interactions among Biocontrol Agents

**Intraguild Predation**
When several predator species compete for the same prey, intraguild predation (IGP) takes place (Gardiner and Landis 2007). IGP is seen as a common phenomenon

among arthropod food webs (Polis et al. 1989) and may be unidirectional or bidirectional. In unidirectional, one species feeds on the other and in bidirectional, both predators feed on each other (Madadi et al. 2008).

IGP can be seen in a wide range of biological control agents, for example, anthocorids and encyrtids (Erbilgin et al. 2004), coccinellids and coccinellids (Noia et al. 2008), coccinellids, chrysopids and cecidomyiids (Gardiner and Landis 2007), phytoseiids and phytoseiids (Walzer and Schausberger 1999a, b; Hatherly et al. 2005; Meszaros et al. 2007; Cakmak et al. 2009), phytoseiids and anthocorids (Madadi et al. 2008; Chow et al. 2010), and phytoseiids and thripids (van der Hoeven and van Rijn 1990; Faraji 2001; Janssen et al. 2002; Walzer et al. 2004; Magalhaes et al. 2005; Walzer and Schausberger 2009) and can be impressed by several factors such as environmental conditions, host plant characteristics (Madadi et al. 2008), mobility of prey (Provost et al. 2006), vulnerability of prey (Noia et al. 2008), feeding specificity (Farazmand et al. 2015a) and presence of extraguild (EG) prey (Lucas 2005; Farazmand et al. 2015a; Maleknia et al. 2016; Moghadasi and Allahyari 2017).

Several studies have been shown the effectiveness of releasing two or more predatory mite species versus single predatory mite species to control tetranychid mites (Schausberger and Walzer 2001; Walzer et al. 2001; Barber et al. 2003; Rhodes et al. 2006; Fitzgerald et al. 2007). The outcome of such interactions depends on the competing species, such as their feeding types (generalist against specialist). Recently, in Iran, some studies have been done about intraguild predation between introduced phytoseiids and native ones (Bohloolzadeh et al. 2013; Farazmand et al. 2015a; Haghani et al. 2015; Rahmani et al. 2015; Ghasemloo et al. 2016; Maleknia et al. 2016), phytoseiid and predatory thrips (Farazmand et al. 2015a, b), phytoseiid and predatory bugs (Madadi et al. 2009). Their results showed introduced phytoseiids were much more prone to IGP than native phytoseiids (Farazmand et al. 2015a; Rahmani et al. 2015; Maleknia et al. 2016). In these interactions, predation preferences of *N. californicus* and *T. bagdasarjani* on *Scolothrips longicornis* Pergande and heterospecific phytoseiids have been studied and both species tended to prey more on first instar larvae of the thrips compared with the heterospecific phytoseiid (Farazmand et al. 2013). Haghani et al. (2015) calculated the predation preference index (Manly's index) of three phytoseiid species (*N. californicus*, *A. swirskii* and *P. persimilis*) and indicated *A. swirskii* and *N. californicus* were able to recognize con-heterospecific larva and preferred to feed on heterospecific larvae but *P. persimilis* showed no preference between con- and heterospecific larvae (Farazmand et al. 2013). In another study, Rahmani et al. (2015) showed that intraguild predation is a weak force among *N. californicus*, *T. bagdasarjani* and *P. plumifer* and at least in the presence of *T. urticae*, elimination of one by another did not occur. Afshari et al. (2014) described larvae of *A. andersoni* had more nutritional value than conspesific larvae for *N. californicus* females and they laid more eggs when consumed on heterospecific larvae, while for *A. andersoni* females, the eggs of *N. californicus* had more nutritional value than conspesific eggs.

Borji et al. (2014) determined the level of *P. persimilis* aggressiveness against eggs of *T. urticae* (Extraguild prey) after feeding on eggs of the predatory mite

*N. californicus* (intraguild prey). They showed that type of prey before adulthood in *P. persimilis* did not have an effect on the level of its aggressiveness on spider mites eggs. Also, in another study, the effect of starvation on aggressiveness against spider mites and conspecific in *N. californicus* was studied and declared starved predators attacked *T. urticae* after $30 \pm 0$ min whereas satiated predators attacked only after $68 \pm 8$ min. Also, attack on conspecific larvae was shorter in starved than satiated predators (Borji et al. 2011).

**Cannibalism**

Cannibalism is defined as the consumption of conspecific individuals that was seen in the animal kingdom (Schausberger 2003). In the case of prey scarcity, this phenomenon can assist in conservation of predator population and has been shown to occur in phytoseiidae (Schausberger and Croft 2000; Walzer et al. 2001; Schausberger 2003; Zannou et al. 2005). The specialist predators act less as cannibal in comparison with generalist predator (Walzer and Schausberger 1999b; Farazmand et al. 2014; Ghasemloo et al. 2016). Schausberger and Croft (2000) demonstrated that cannibalistic individuals have the ability to discriminate between kin and non-kin and reported the specialist phytoseiids *P. persimilis* and *P. macropilis* are able to discriminate between related and unrelated larvae and preferentially consumed unrelated larvae when given a choice. Also, the effect of kinship on the mating preference of *P. persimilis* by Dehghani Tafti et al. (2013) was studied and described *P. persimilis* was not able to discriminate between kin and non-kin males. In another study, the effect of kinship on oviposition patterns of *P. persimilis* was discussed and showed that kinship decreased the distance between first and second eggs (Zeraatkar et al. 2013).

3.2.1.4.8    Mass Rearing

Mass rearing and releasing predatory mites, mostly phytoseiids, are one of the goals of biological control programs (McMurtry and Croft 1997). Methods for mass rearing of phytoseiids were proposed by McMurtry and Scriven (1971) and Gerson et al. (2003). Specialist phytoseiids require spider mites for efficient mass-rearing (McMurtry et al. 2015) and it is necessary to have a pure spider mite culture (Morales-Ramos and Rojas 2014), while pollen and honeydew produced by greenhouse whitefly can keep the generalist predators population in the absence of their pests (Mortazavi et al. 2018). Recently, Nguyen et al. (2015) studied population growth of four phytoseiids (*N. californicus, N. cucumeris, A. andersoni* and *A. limonicus*) on natural and artificial diets and showed growth rate was lower on artificial diet than on natural food, but survival was similar. These artificial diets may be useful as food supplements to protect predator populations after release in the crop (Wade et al. 2008). One of the cheap and effective sources for rearing some phytoseiids is pollen (McMurtry et al. 2015). Shirdel et al. (2002) compared rearing of *Typhlodromus kettanehi* Dosse and *Euseius finlandicus* (Oudemans) on two types of arenas (plastic or bean leaf) and three diets (*T. urticae*, almond pollen + *T. urticae* and almond pollen) and demonstrated that *T. bagdasarjani* had the ability to grow on both arenas and three diets but rearing of *E. finlandicus* inclusively on bean leaf and pollen or mixture of pollen+*T. urticae* was possible.

In a study, Hajmohammadloo and Shirdel (2010) investigated the effect of feeding of five pollen sources, apple, pear, almond, apricot and walnut on biological parameters of *T. bagdasarjani*, and showed the pollen grains can be alternative foods for development and reproduction of the predatory mite. The effect of three diets including (maize pollen; maize pollen and *T. urticae* eggs; maize pollen, active stages of *T. urticae* and all active stages of *T. urticae*) on the predation rate of *A. swirskii* and *P. persimilis* in three patches (4, 9 and 16 cm$^2$) was investigated and demonstrated that patch size and diet could significantly impact the predation rate of both mites (Sharifian et al. 2014). The nutritional value of seven different pollen grains (almond, castor-bean, date-palm, maize, bitter-orange, sunflower and mixed bee pollen) as a supplementary food source was evaluated on life table parameters of *N. californicus* and revealed almond pollen (and after that the maize pollen) is a more suitable diet than the others (Khanamani et al. 2017a). In this regard, the life table parameters of *N. californicus* were assessed after 20 generations and showed an extraordinary performance for this predator (Khanamani et al. 2017c). On the other hand, high performance on almond pollen have also been proved for *A. swirskii* and *T. bagdasarjani* (Riahi et al. 2016, 2017a, 2018a, b). Also, another related study by Khanamani et al. (2017b) showed that an artificial diet enriched with arthropods (*Ephestia kuhniella* Zeller or *Artemia franciscana* Kellogg cyst) as well as maize pollen or bull sperm could improve population growth parameters of *N. californicus*. Likewise, Riahi et al. (2017b) demonstrated adding *Artemia* cysts and maize pollen to the same basic artificial diet was suitable for mass production of *A. swirskii* while factitous preys (*E. kuhniella, A. franciscana* and *Tyrophagus putrescentiae*) and artificial diets were not appropriate for mass rearing of *T. bagdasrajani* (Riahi et al. 2018a, b). In other studies, pistachio pollen created a supplementary diet for mass production of *N. californicus* (Soltaniyan et al. 2018) and, Rezaei and Askarieh (2016) declared corn, sunflower and date pollen grains are an alternative food for *N. barkeri*.

There are limited studies on releasing method and long-term storage of the phytoseiids in Iran. In a research under microcosm conditions on cucumber, predator: prey release ratios of 1:4, 1:10 and 1:20 of *P. persimilis* and *T. bagdasarjani* were studied and observed the best ratio is 1:4 for two predators in controling of *T. urticae* (Moghadasi and Allahyari 2018).

### 3.2.1.5  Blattisociidae

With 14 genera and about 370 species, blattisociids occupy a wide range of terrestrial and semiaquatic habitats (Lindquist et al. 2009; de Moraes et al. 2016). The genus *Blattisocius* is one of potential genera being important in biocontrol point of view. Mites of this genus are usually found in stored products preying upon available arthropods including moths, beetles and mites of the family Acaridae. One generation is mostly raised in one week (Gerson et al. 2003). The *B. tarsalis* (Berlese) is a wide spread species with considerable control effects on some major stored product pests. Another useful species of the genus is *B. keegani* Fox. Experimental data on life cycle and predation potential of this mite, provisioned by eggs of *Amyelois transitella* (Walker) (Lepidoptera: Pyralidae), revealed about 9-day development time at 25 °C and 50–60% relative humidity with a maximum of three eggs consumed during 24 h (Thomas et al. 2011). About five species of this genus including *B. tarsalis* and *B. keegani* are found in Iran (Kazemi and Rajaei 2013; Shams et al. 2016; Nemati et al. 2018). Furthermore, the genus *Lasioseius* is another beneficial genus considered to prey on small arthropods and nematodes. Of 22 reported species, the *L. youcefi* Athias-Henriot is the most abundant one in the country.

It is noteworthy that the genera *Proctolaelaps* (Melicharidae) with 12 Iranian species and *Arctoseius* (Ascidae) with seven species are known to be a predator (sometimes predator-fungivores) and can feed on nematodes and some microarthropods, nevertheless their impact on host populations has not been yet evaluated.

### 3.2.1.6   Laelapidae

Many members of the ubiquitous and speciose family Laelapidae (with about 144 genera and more than 1300 species) are parasites of small mammals or associated with birds, arthropods and their nests. However, the subfamily Hypoaspidinae often encompasses free-living aggressive predators including some biological control agents of the genera *Stratiolaelaps* and *Gaeolaelaps* (Lindquist et al. 2009). *Stratiolaelaps scimitus* (Womersley) (wrongly known as *Hypoaspis* or *Stratiolaelaps miles*) and *Gaeolaelaps aculeifer* (Canestrini) (wrongly placed in *Hypoaspis s.l.*) are two species that attracted the attentions to be used as effective biocontrol agents and even mass reared commercially against many soil-living pests of greenhouses such as thrips, fungus gnats, acarid mites and nematodes (Walter and Proctor 1999; Gerson and Weintraub 2007). Both mentioned species have been found in Iran. In a very few limited studies carried out on the evaluation of *G. aculeifer* as a biocontrol agent, this mite alone and in combination with disinfected corms of gladiolus have had significant control results against *Rhizoglyphus echinopus* that is an important mite pest of this ornamental plant (Hosseininia et al. 2012, 2014). In another study, Amin et al. (2014) assessed effects of eight constant temperatures on development and fecundity of *G. aculeifer* feeding on *Rh. echinopus* showing the average of female life span as 102.40 days at 16 °C and 37.21 days at 32.5 °C. Also, the mean number of eggs laid per female per day increased from 0.49 at 16 °C to 3.76 at 30 °C and then decreased to 1.88 at 32.5 °C.

Some authors have reported that some species of *Hypoaspis s.s.* or *Coleolaelaps* are parasites of the eggs or larvae of plant-feeding Scarabaeidae, and therefore may have potential as biological control agents (Khanjani et al. 2013). According to Damghani (2001), *Hypoaspis elegans* Joharchi, Ostovan and Babaeian was reared in the laboratory on all stages of *Oryctes elegans* Prell. for one year, however, no heavy mortality observed during any stages of the host beetle. There was some mortality (mortality rate not stated) of pupae, and Damghani (2001) suggested that this mortality of pupae was caused by the mites' feeding on exudates of the pupae's body. This has not been established experimentally, and it will be necessary to do feeding experiments to establish the true ecological role of these mites (Joharchi et al. 2014).

More than 20 genera and 100 laelapid species have been recorded from Iran (for examples, see Kazemi and Rajaei 2013; Joharchi and Babaeian 2015; Nemati and Gwiazdowicz 2016; Nemati et al. 2018).

## 3.2.2  Suborder Prostigmata

### 3.2.2.1  Bdellidae

This family, containing about 16 genera and more than 278 species, has a wide distribution preying on nematodes, microarthropods or their eggs. Bdellids, also known as snout mites due to conical rostrum, can produce silk strands in order to secure prey (usually by circling the host and entangling body and the appendages with silken net) and moulting (Walter et al. 2009; Hernandes et al. 2016). Some species can be affect spider mites and springtails populations. For example, *Bdella depressa* Ewing preys on *Bryobia praetiosa* Koch and, *Bdellodes lapidaria* (Kramer) and *Neomolgus capillatus* Kramer on lucerne flea, *Sminthurus viridis* L. (Gerson et al. 2003). Likewise, *Bdella tropica* Atyeo is evaluated as an effective natural enemy of *Xenylla longauda* Folsom (Collembola; Hypogastruridae) which attacks edible fungi (Ji et al. 2007). Many genera (e.g. *Cyta*, *Neomolgus*, *Hexabdella*, *Bdella*, *Bdellodes*, *Biscirus*, *Spinibdella*) and species have been described and recorded in Iran (see Ostovan and Kamali 1995; Eghbalian et al. 2014, 2016; Paktinat-Saeej et al. 2015a, b; Ueckermann et al. 2007) but their biology, host range, and impact on host populations have yet to be studied in the country.

### 3.2.2.2  Cunaxidae

Similar to its sister family Bdellidae, cunaxids (with 27 genera and about 330 species) attack small arthropods and nematodes. They occupy various habitats including soil and aerial parts of plants (Walter et al. 2009). Among six subfamilies, members of the subfamily Cunaxoidinae occupy more habitats on plants than soil (Castro 2008; Skvarla et al. 2014). There are two strategies for prey capture, ambush (Cunaxinae) and cruise (Coleoscirinae and Cunaxoidinae) ones, both observed in soil and foliar taxa (Walter and Proctor 1999). One species of the genus *Coleoscirus* feeds on *Meloidogyne* spp. in sugarbeet fields and can complete one generation in about two weeks (Rahmani et al. 2012). Some species are recorded as predators of scale insects, eriophyids and spider mites. For example, *Cunaxa setirostris* (Hermann) is considered as a potential biocontrol agent of *Tetranychus ludeni* Zacher. This beneficial mite can complete its life cycle in about four weeks and consumes averagely 287 mobile preys during 87% of the total lifespan predation (Arbabi and Singh 2000). The most comprehensive taxonomic studies on Iranian cunaxids are available in Den Heyer et al. (2011a, b, 2013), providing many descriptions and records along with keys to Iranian species of the Cunaxoidinae, Coleoscirinae and Cunaxinae. Their efficiency in the control of other small arthropods is however unclear. It has to be noted that cannibalism is a negative factor that can affect this family to be used as a biocontrol agent and, in turn, mass production (Hernandes et al. 2015).

### 3.2.2.3 Tydeidae

The widely distributed members of the family Tydeidae with 30 genera and approximately 340 species are predominantly not specialized feeders and, predatory is mostly opportunistic in this family. The predators mainly prey on eriophyid mites. In this regard, some *Tydeus* and *Orthotydeus* could be generic examples. Although the fungivory is the main diet for this family, some may exhibit nematophagy (Santos et al. 1981). One generation usually takes 2–3 weeks. It is proven that some tydeids can play their role as alternative food for a number of predators particularly phytoseiid mites (Hoy 2011).

Tydeids are one of the major arthropods residing in leaf domatia (sort of physical and environmental structures on plants particularly leaves occupied by mostly useful microarthropods). The domatia exploited by mites are referred as Acarodomatia that mostly includes small invaginations and tufts of nonglandular trichomes at the junctions of primary and secondary veins on the undersides of leaves in woody dicots (Pemberton and Turner 1989; Walter and Proctor 1999). Predatory (such as Phytoseiidae, Stigmaeidae and Bdellidae) and mycophagous (such as Tydeidae) mites are frequently settled in Acarodomatia (Pemberton and Turner 1989; Norton et al. 2000). Some references mention to a number of fungivorous tydeids reducing the density of powdery mildews on plants like cucumber and grape for instance, the mite *Orthotydeus lambi* (Baker) reported decreasing the population of *Uncinula necator* (Schwein) in grapes by eating from its hyphae (English-Loeb et al. 1999). High populations of *Tydeus kochi* Oudemans is observed in sugarbeet leaves contaminated with powdery mildews in northwestern Iran (H. Hajiqanbar, pers. obs.). More faunistic studies on Iranian wild and agricultural plant ecosystems are highly recommended to access voracious fungivorous tydeids and evaluation of their efficiency in consuming and controlling pestiferous fungi.

Sepasgosarian (1997) provided a list of subfamilies, genera and species of the family Tydeidae. In recent years, many genera and species of Iranian tydeids are recorded from soil and aerial parts of plants (e.g. Khanjani and Ueckermann 2003; Andre et al. 2010; Darbemamieh et al. 2010; Sadeghi et al. 2012; Akbari et al. 2015). For example, it is documented that *T. caudatus* (Duges) is very abundant and important species in Iranian orchards including Kermanshah province (Darbemamieh et al. 2010; M. Darbemamieh pers. com.). This species is the most common tydeid species in Italian vineyards (Castagnoli 1984), associated with grape eriophyids during late winter and reported as a predator of *Colomerus vitis* (Pagenstecher) in the laboratory (Camporese and Duso 1995). However, more targeted studies are required to access beneficial predator and mycophagous Iranian tydeid mites and, to evaluate their efficacy.

### 3.2.2.4 Iolinidae

The mostly omnivorous members of the family Iolinidae include 36 genera and more than 125 species. There is an omnivorous species, *Pronematus ubiquitus* (McGregor), reported as a predator on some eriophyids including *Eriophyes ficus* Cotte and *Aculops lycopersici* (Massee) and tenuipalpids (Hernandes et al. 2015). This predator species, along with *Neopronematus neglectus* (Kuznetzov), are recorded as natural enemies of tetranychid mites on leaves of apple orchards in northwestern Iran (Khodayari et al. 2008). Also, in this region, Shemshadian et al. (2012) reported this mite as a potential predator of *Panonychus ulmi* (Koch). In another study conducted on fig trees in central parts of Iran, maximum abundance of *P. ubiquitus* populations recorded in October and December, simultaneous with reduction of prey mobility, temperature and photoperiod (Baradaran and Arbabi 2008). Many *Neopronematus* species recorded in Iran from orchards (e.g. Sadeghi et al. 2012; Darbemamieh et al. 2015) but there is no exact study on their diet and foraging material.

Another well-known species is *Homeopronematus anconai* (Banker) feeding on tomato russet mite (*Aculops lycopersici*) that can reduce population frequency of its prey (Gerson et al. 2003; Hernandes et al. 2015). This species is also recorded from Iran (Sadeghi et al. 2012; Darbemamieh et al. 2015). Some taxonomic studies on this family have recently been initiated in Iran (for example, see Darbemamieh et al. 2015, 2016).

### 3.2.2.5 Anystidae

Fast moving and relatively large mites of the Anystidae including 18 genera and about 110 species are general predators of many arthropods in soil and plant habitats. The most important potentially biocontrol agent is the genus *Anystis*, members of which attacking phytophagous insects and mites in many parts of the world (Gerson et al. 2003; Walter et al. 2009). Many pests including tetranychid mites are being preyed upon by familiar species *Anystis baccarum* (L.) and *A. agilis* (Banks). Unfortunately, long development time, no prey preference, the low intrinsic rate of increase and a predilection for cannibalism negatively affect the efficacy of these mites to being a suitable candidate as a biocontrol agent. Some ubiquitous mites of this family including *A. baccarum* have been recorded from various Iranian habitats that some of them are as follows. Mehrnejad and Ueckermann (2001) found *A. baccarum* in colonies of *Agonoscena pistaciae* Burckhardt and Lauterer (Hemiptera: Psyllidae) on the foliage of Iranian pistachio orchards during May to September. Khodayari et al. (2008) collected two anystids namely *A. baccarum* and *Erythracarus pyrrholeucus* (Hermann) from leaves of apple orchards in northwestern Iran. The *A. baccarum* and *A. wallacei* Otto were sampled from raspberry shrubs in northern parts of the country (Tajmiri 2013).

### 3.2.2.6 Stigmaeidae

The worldwide family Stigmaeidae, in addition to soil and litter, is associated with plants as active predators. After phytoseiid mites, the most important groups of predacious mites are stigmaeids so that 35% of species are active predators on plants (Fan and Flechtmann 2015). Also, these beneficial mites can assist phytoseiids in suppressing pest mites, especially in unsprayed environments (Hoy 2011). The family encompasses 34 genera and about 580 species (Fan et al. 2016) among which the genera *Zetzellia*, *Agistemus* and *Mediolata* embrace the species with the ability to prey on phytophagous mites in agricultural ecosystems especially orchards (Walter et al. 2009). The wide spread *Z. mali* Ewing feeds on spider mites and eriophyids in apple orchards (Gerson et al. 2003) however, in some parts of Iran (Zanjan province) it has been evaluated as a not strong predator for spider mites (Rahmani et al. 2010). This species has been recorded from various parts of Iran and its different aspects have been examined in the country. Field observations in apple orchards of Karaj, during 1999–2000 by Jamali et al. (2001) indicated that *Z. mali* produces three generations per year and adult females lay one egg per day and, survive without any other food on leaf discs for 10 days. Zahedi-Golpayegani et al. (2007) stated that this mite is an important predator of the hawthorn spider mite, *Amphitetranychus viennensis* (Zacher) (Acari: Tetranychidae) in black-cherry orchards in Baraghan, and has ability to distinguish odors in olfactometers, with a positive response to odors from *A. viennensis* and a negative one to odors containing a conspecific predator *Z. mali*, implying that this predator may avoid competition with other *Z. mali*.

Khodayari et al. (2008) showed that this species can develop and reproduce on *Tetranychus urticae* Koch eggs in laboratory conditions. Spatial distribution pattern of *Z. mali* and its preys *Eotetranychus frosti* (McGregor) and *Tydeus longisetosus* Kuznetzov and Zapletina in unsprayed apple orchards of Maragheh, northwestern Iran, were studied and revealed the peak density of *E. frosti* and *T. longisetosus* in late August and mid-September. In addition, the peak density of their predator (10.34) occurred in early July (Khodayari et al. 2010).

Some selected taxonomical studies on Iranian stigmaeids are mentioned below. Khanjani and Ueckermann (2003) described seven new species and documented some other records from Iran. Noei et al. (2007) found four species of Stigmaeidae associated with stored rice and decayed rice bran in northern Iran. Beyzavi et al. (2013) in their catalogue of Iranian Raphignathoidea and Tetranychoidea listed

11 genera and 69 species of the family Stigmaeidae. Hajizadeh et al. (2013) following an investigation in Guilan province (Northern Iran) increased the Iranian stigmaeids to 76 species. Two species described from soil in Isfahan province (Bagheri et al. 2014) and two others from the soil and litter in Kurdistan province (Khanjani et al. 2014). From northwestern Iran, Fakhari et al. (2015) described a species of the genus *Stigmaeus*, members of which are usually predators of scale insects.

### 3.2.2.7 Cheyletidae

This family with 77 genera and over 440 species includes parasitic (mostly on vertebrates) and free-living predator taxa, the latter feeding on small arthropods and dwelling in soil, plants, nest of vertebrates and stored products. They usually ambush their preys and some species exhibit subsocial mode of life. The subfamily Cheyletinae contains primarily predators of many microarthropods preferably Astigmatina. Some genera such as *Cheyletus*, *Cheletomorpha*, *Cheyletia* and *Acaropsis* are predators of astigmatan mites in grain products and storages. Grain mites e.g. *Acarus siro* L. were reported to be attacked by the cosmopolite species *Cheyletus eruditus* (Schrank) (Walter et al. 2009).

A summary of the selected taxonomic studies on Iranian Cheyletidae is shown below. Kamali et al. (2001) listed 14 genera and about 22 species from various regions and habitats. Four years later, Bochkov et al. (2005) reviewed Iranian cheyletids and provided a key to 18 genera and 28 recorded species in the country at that time.

Many cheyletid predators are recorded from stored products in Iran. For instance, Ardeshir et al. (2000) in a survey on mites of stored grain in northern Iran, claimed *Cheyletus malaccensis* Oudemans as the most common prostigmatic mites. Following a study on mites associated with stored wheat in Tehran, the *Acaropsellina sollers* (Kuzin) recorded as the most abundant predator mite (Ardeshir et al. 2008b). Hajizadeh et al. (2011) collected six species including *C. eruditus* from stored rice and decayed rice bran in Guilan province in northern Iran. The prey species of cheyletid predators were a cluster of acarid mites such as *Acarus siro*, *Tyrophagus putrescentiae* (Shrank) and *Glycyphagus destructor* (Schrank) to be included as destructive stored products pests. In the current study, the *Cheyletus malaccensis* was the most abundant species. During an investigation on mites of the factories producing food for domesticated animals in Karaj, four cheyletids were collected among which the *Acaropsellina sollers* and *Cheyletus carnifex* Zachvatkin constituted the most abundant species after *Tyrophagus longior* (Gervais) (Seiedy et al. 2012a, b). These two cheyletids were also collected in flour mills of Karaj (Seiedy et al. 2009). In a recent study, Salarzehi et al. (2018) described one new species of the genus *Cheyletus* collected from stored material in Guilan province.

Some other cheyletid mites have been documented living on plants. Mehrnejad and Ueckermann (2001) in their study of mites associated with pistachio trees in Iran declared that the *Cheletogenes ornatus* (Canestrini and Fanzago) was encountered in scale insects colonies on twigs. Jalaeian et al. (2005) collected *Cheletogenes ornatus* from orchards in central parts of Iran. Tajmiri (2013) mentioned *C. malaccensis*, *C. eruditus* and *Cheletomorpha lepidopterorum* (Shaw) from raspberry shrubs in Guilan province. Jalilirad et al. (2013) reported *Cheletogenes ornatus*, *C. berlesi* Oudemans and *Hemicheyletia wellsi* Summer and Price from citrus leaves in Guilan province.

Some others have been documented as associated species of insects that feed on eggs and early stages of the hosts or other microarthropods dwelling in host habitats. For instance, Ostovan and Kamali (1997) collected three cheyletid species (*Chelacheles michalskii* Samsinak, *Acaropsella kulagini* (Rohdendorf) and *Hemicheyletia wellsi*) from galleries of *Scolytus multistriatus* (Marsham) from parks of Tehran. Ahadiyat et al. (2004) recorded *Cheletogenes ornatus* and one species of *Caudacheles* associated with *Scolytus amygdali* Geurin-Meneville in Karaj. Haghighi et al. (2011) collected *Caudacheles khayae* Gerson from galleries of elm bark beetles in Fars province, southern Iran.

It is conceivable that conservation of free-living predatory cheyletids in their various habitats can increase their efficiency as natural biocontrol agents.

## 3.3   Parasitic Mites

### 3.3.1   Order Mesostigmata

#### 3.3.1.1   Otopheidomenidae

Contrary to its predatory sister family (Phytoseiidae), mites of the family Otopheidomenidae are permanent parasites of insects in temperate and tropical regions of the world. Ten genera and about 28 species are classified into three subfamilies based on host preference and morphological traits parasitizing moths (in wing bases and tympanic recesses), bugs (in the dorsum of the thorax and beneath the hemielytra), katydids (in thoracic tracheae), termites and grasshoppers (Lindquist et al. 2009). Under laboratory conditions, some species of the genus *Hemipteroseius* showed unfavorable effects on the feeding and life history parameters, and even death of the hemipteran hosts (Costa 1968; Banerjee and Datta 1980). One of the species of the *Hemipteroseius* has been recently found in Iran (Joharchi and Saboori 2010) that could be promising regarding more systematical and biological studies on these mites in future in the country.

### 3.3.2   Suborder Prostigmata

#### 3.3.2.1   Eriophyidae

Eriophyoid mites feed on plant surfaces and cause stunting, brooming, leaf rolling, gall formation and damage to vegetative and reproductive organs (Gerson et al. 2003). Nevertheless, some species of these mites have been considered to have a high potential for using as classical biological control agents of weeds and inflict significant damage on their host plants or serving as prey for phytoseiid predators (Smith et al. 2009; Skoracka et al. 2010). The role of some species of eriophyid mites in control of weeds has been considered by Gerson et al. (2003) and Hoy (2011).

Lotfollahi et al. (2013) enumerated two genera and 12 species of eriophyid mites associated with Compositae plants in Iran, some of which treated as weeds. Asadi et al. (2013) showed *Aceria chondrillae* (Canestrini) fed and reproduced on chondrilla plant. In a study based on material collected from Iran, *Aceria acroptiloni* Shevchenko and Kovalev could imposed significant impact on the aboveground biomass and particularly on the reproductive output of the invasive plant *Rhaponticum repens* and reduced the biomass of *R. repens* shoots by 40–75% (Asadi et al. 2014a, b). In the current study, infested seed heads from *R. repens* plants were collected in a field margin 5 km NE of Shirvan, northeastern Iran. Kamali (2016) listed 13 species of eriophyoids on various weeds collected from different parts of Iran.

### 3.3.2.2  Erythraeidae

The large and ubiquitous family Erythraeidae with seven subfamilies, more than 55 genera and 770 species is typically parasite of many arthropods in larval stage and predator in deutonymph and adult stages. Recently, Mąkol and Wohltmann (2012, 2013) provided an annotated checklist of the world terrestrial Parasitengona including Erythraeidae. This family as other parasitengones is categorized as protelean parasites meaning that they are parasitic only in one of the immature stages, i.e. larva (Walter and Proctor 1999). They are usually red and live on the ground or aerial parts of the plants. Some larvae are free-living and rarely parasite of vertebrates. Most erythraeid species are uni- or bivoultine (Walter et al. 2009). Although the larvae do not kill the host and mostly have no host specificity, parasitized insect host gets weakened and can be easier preyed by other predators like birds and other insectivores. There are several genera and species of this terrestrial parasitengone family attacking some important insect and mite pests (see Gerson et al. 2003). For instance, *Balaustium putmani* Smiley, 1968 in Nearctic realm is an active predator of many arthropod pests like *Panonychus ulmi*, *Aculus schlechtendali* (Nalepa) (Acari: Tetranychidae, Eriophyidae), *Quadraspidiotus perniciosus* (Comstock) and *Aphis pomi* De Geer (Hemiptera: Diaspididae, Aphididae) (Putman 1970; Cadogan and Laing 1977; Childers and Rock 1981). In Iran, among larvae of *Leptus* parasitizing a wide range of insects and arachnids, *L. esmailii* Saboori and Ostovan has been described as an ectoparasite of *Eurygaster integriceps* Puton, a serious pest of cereal fields (Saboori and Ostovan 2000). Also, larvae of *L. zhangi* Saboori and Atamehr are parasite of *Hyponomeuta malinella* Zeller, a pest of Iranian apicultures (Saboori and Atamehr 1999). Table 3.3 provides some selective larval species of the Iranian erythraeids parasitizing insect pests. The efficiency of these mites on their hosts is not evaluated but conservation efforts could be recommended to increase the efficacy of their natural control ability.

**Table 3.3** Some selective larval species of the Iranian erythraeids parasitizing insect pests

| Erythraeid mite | Insect host | References |
|---|---|---|
| *Abalakeus gonabadensis* Ahmadi et al. (2012) | *Aphis craccivora* Koch (Hemiptera: Aphididae); *Dociostaurus* cf. *tartarus* (Stshelkanovtzev) (Orthoptera: Acrididae) | Ahmadi et al. (2012) |
| *Charletonia baluchestanica* Tashakor et al. (2015) | *Ochrilidia* sp. (Orthoptera: Acrididae) | Tashakor et al. (2015) |
| *Charletonia behbahanensis* Haitlinger and Saboori 2008 | *Dociostaurus maroccanus* (Thunberg) (Orthoptera: Acrididae) | Haitlinger and Saboori (2008) |
| *Charletonia damavandica* Karimi Iravanlou et al. 2002a | *Acrotylus insubricus* (Scopoli); *Mioscirtus wagneri* (Eversmann) (Orthoptera: Acrididae) | Karimi Iravanlou et al. (2002a) |
| *Charletonia mehranensis* Haitlinger and Saboori (2007) | *Anacridium aegyptium* (L.) (Orthoptera: Acrididae) | Haitlinger and Saboori (2007) |
| *Charletonia nazeleae* Karimi Iravanlou (2002a) | *Heliopteryx humeralis* (Kuthy); *Ramburiella turcomana* (Fischer von Wald heim) (Orthoptera: Acrididae) | Karimi Iravanlou et al. (2002a) |
| *Charletonia saboorii* Karimi Iravanlou et al. (2002a) | *Heteracris littoralis* (Rambur) (Orthoptera: Acrididae) | Karimi Iravanlou et al. (2002a) |
| *Erythraeus* (*Erythraeus*) *populi* Khanjani et al. (2012) | *Stephanitis pyri* (F.) (Heteroptera: Tingidae) | Khanjani et al. (2012) |
| *Erythraeus* (*Erythraeus*) *shojaii* Saboori and Babolmorad (2000) | *Monosteira unicostata* (Mulsant and Rey) (Hemiptera: Tingidae) | Saboori and Babolmorad (2000) |
| *Iraniella moharramipouri* Karimi Iravanlou et al. (2002b) | *Oedipoda miniata* (Pallas) (Orthoptera: Acrididae) | Karimi Iravanlou et al. (2002b) |
| *Lasioerythraeus saboorii* Khanjani et al. (2011) | *Aphis punicae* Passerini (Hemiptera: Aphididae) | Khanjani et al. (2011) |
| *Leptus kamalii* Karimi Iravanlou and Saboori (2001) | *Dociostaurus hauensteini* (Bolivar) (Orthoptera: Acrididae) | Karimi Iravanlou and Saboori (2001) |
| *Momorangia binaloudensis* Noei et al. (2015) | *Apamea impedita* (Christoph) (Lepidoptera: Noctuidae) | Noei et al. (2015) |

### 3.3.2.3 Trombidiidae

The worldwide family Trombidiidae includes three subfamilies, 25 genera and more than 253 species (Zhang et al. 2011). As other Parasitengonina, the larvae are a parasite of many arthropods (protelean parasites) while, usually red and hypertrichous deutonymphs and adults are predators of small invertebrates in various habitats (Walter et al. 2009). Many parasitengone mites including this family are usually long-lived so that one generation is often completed in one year or more. A list of trombidiids and their hosts was prepared by Welbourn (1983) and later updated by Zhang (1998). Makol and Wohltmann (2012, 2013) provided annotated checklists of three superfamilies of terrestrial Parasitengona including

Trombidioidea. Saboori and Hakimitabar (2013) provided a checklist for Iranian Trombidioidea and mentioned to five families, 22 genera and 32 species. Some cosmopolite genera such as *Trombidium* and *Parathrombium* attack wide range of arthropods but, undoubtedly, the genus *Allothrombium* is potentially the most economic biocontrol agent in the family (Zhang 1991). One of the species, *A. pulvinum* Ewing, parasitizes aphids in the larval stage and prey spider mites in postlarval stages (Zhang 1992).

In Iran, there are some non-taxonomical studies on two representatives of the *Allothrombium* namely *A. shirazicum* Zhang and *A. pulvinum*. Biology of *Allothrombium shirazicum* as a potential biocontrol agent of *Aphis punicae* Passerini and some other arthropods has been studied in central parts of Iran (Saboori and Kamali 1999). The mite raises one generation per year. The eggs hatch in spring and larvae usually do not show superparasitism. Nymphs emerge in summer and adults appear in autumn when they prey on microarthropods and their eggs. Adults and eggs can hibernate, and at late winter overwintered females produce eggs, mostly hatched in spring. Regarding *A. pulvinum*, in Mazandaran province (northern Iran), larvae emerge in mid-spring and reach to the most abundant population in late spring (Saboori and Zhang 1996). Saboori et al. (2007) stated that larvae of *A. shirazicum* constitute about 69% of natural enemies of pomegranate aphid in mid-spring with maximum 13% parasitism in presence of coccinellids, syrphids, hymenopterans and spiders.

*A. pulvinum* larvae prefers *Hyalopterus amygdali* (Blanchard) in presence of two other aphids, *Aphis gossypii* Glover and *Macrosiphum rosae* (L.), showing that the parasitic larvae mostly prefer host aphids with less defence ability, less activity, more mildew productivity, thinner cuticle and more attractive body odor (Hosseini et al. 2002). In northern Iran, *A. pulvinum* adults can prey on two pests of citrus trees, *Planococcus citri* (Risso) and *Pulvinaria aurantii* Cockerell, with a preference on *P. citri* (Saboori et al. 2003a, b). Results of experimental studies revealed that each *A. pulvinum* adult can feed on 4.2, 2.8, 1.3, 0.8 and 0.8 of *H. amygdali*, *Aphis gossypii*, *Macrosiphum rosae*, *A. fabae* Scopoli and *T. urticae* per day, respectively (Hosseini et al. 2004). Experimental surveys conducted by Hosseini et al. (2005) showed that *A. pulvinum* deutonymph can consume 6.5 and 2.2 eggs of *T. urticae* and *Amphitetranychus vienensis* per day, respectively. In addition, 2.8 and 1.2 adult females of the mentioned tetranychids could be feed per day, respectively. The results of host preference studies proved *T. urticae* is preferred prey of *A. pulvinum* deutonymphs. In current study, authors demonstrated type III functional response of *A. pulvinum* deutonymphs on different densities of adult female individuals of *T. urticae* and density-dependent in low densities.

The following table (Table 3.4) listed some selective larval species of the Iranian trombidiids attacking insect pests. Conservation measures are also recommended as erythraeids.

**Table 3.4** Some larval species of the Iranian trombidiids attacking insect pests

| Trombidiid mite | Insect host | References |
|---|---|---|
| *Allothrombium shirazicum* Zhang and Rastegari (1996) | *Forda marginata* Koch (Hemiptera: Aphididae) | Zhang and Rastegari (1996) |
| *Cicaditrombium weni* Saboori and Lazarboni 2008 | *Cicadatra ochreata* Melichar (Homoptera: Cicadidae) | Saboori and Lazarboni (2008) |
| *Iranitrombium miandoabicum* Saboori et al. (2003a, b) | *Thrips tabaci* Lindeman (Thysanoptera: Thripidae) | Saboori et al. (2003a, b) |
| *Monotrombium simplicium* Zhang and Norbakhsh (1995) | *Schizaphis graminum* (Rondani); *Metopolophium dirhodum* (Walker); Sitobion avenae (F.); *Forda marginata* Koch (Hemiptera: Aphididae) | Zhang and Norbakhsh (1995) |
| *Oskootrombium prasadi* Saboori et al. (2006) | *Hyalopterus pruni* (Geoffroy) (Hemiptera: Aphididae) | Saboori et al. (2006) |

### 3.3.2.4 Other Terrestrial Parasitengonina

There are some representatives of other parasitengone families in Iran with well distribution and prevalence. One of them is a specified subelytral parasite *Neosilphitrombium tenebrionidum* Saboori, Hajiqanbar and Hakimitabar (Neothrombiidae) parasitizing *Opatroides punctulatus* Brullé (Coleoptera: Tenebrionidae) (Saboori et al. 2011). The host beetle is a broadly distributed insect in Europe, Africa and Asia, causing damage to some agricultural crops and the parasitic mite, at least in Iran, could be a promising potential natural enemy.

Likewise, many larval species of the *Eutrombidium* (Microtrombidiidae) are known to be included as parasites of some pestiferous grasshoppers or gryllids in Iran (Saboori et al. 2007).

Recently a monogeneric family, Achaemenothrombiidae (Trombidioidea), with three species has been described from Iran, two of which parasitizing noctuid genera *Catocala* and *Euxoa* (Saboori et al. 2010, 2013). Mites of this family could be potentially important as parasites of mentioned plant-feeding insects.

### 3.3.2.5 Aquatic Parasitengonina (Hydrachnidiae)

More than 6000 true water mite species, mostly freshwater, are classified in subcohort Hydrachnidiae. Like other mites of the cohort Parasitengonina, they have a life cycle with parasitic larvae and predatory post-larval stages (Walter et al. 2009). In the larval stage, these mites parasitize the aquatic insects, some of which economically and or medically significant such as Culicidae and Chironomidae (Gerson et al. 2003). For example, Arrenuridae, Hydryphantidae, Limnesiidae and Pionidae are among water mites considered as natural enemies of mosquitoes and midges. In Iran, in addition to mentioned families, 21 other familes (encompassing 43 genera and 186 species) have also been reported and distributed in most provinces, however, this number is far from available water mites in the country (Pešić et al. 2014). It seems water mites are well distributed in the country and could play their role in the natural control of the hosts. Although the effect of larva on the host can not be usually evaluated in natural habitat, conservation of mites can increase their impact on the host insects.

### 3.3.2.6  Trochometridiidae

Mites of the family Trochometridiidae are associated with three orders of the insects, including Coleoptera, Dermaptera and Hymenoptera, typically apoid bees of the families Halictidae and Andrenidae (Cross and Bohart 1979; Lindquist 1985; Hajiqanbar et al. 2009; Mortazavi et al. 2011). A study on the life cycle of one of the species of the family, *Trochometridium tribulatum* Cross, showed that despite fungivory of the mites, their impact on some hosts, especially the bees of families Halictidae and Andrenidae is like a parasitoid. It occurs during a triple symbiotic relationship among mite, fungus and bee host. The mites cause the death of the eggs and young larvae of the host by carrying the fungus spores (via pairs of pouch-like structures between base of the legs III and IV, named sporothecae) to the nest of the host and contribute to the growth of the fungus. In other words, the next generation of these mites feed on fungus mycelia that grow on the corpses of the host eggs and young larvae (Cross and Bohart 1979; Lindquist 1985; Kaliszewski et al. 1995).

Hitherto, the family encompasses two genera, *Neotrochometridium* with one species and *Trochometridium* with 6 species (Loghmani et al. 2014), all reported from Iran except *Trochometridium tribulatum* that is distributed in Nearctic and Afrotropical realms. The trochometridiids have not been demonstrated to be strong biocontrol agents, however, at least they have a role in the natural control of their insect hosts that their spectrum has been increased in recent years. In any case, further collections and life history investigations are needed to clarify rate of their mortality upon the host populations.

### 3.3.2.7  Pyemotidae

As in its sister family, Acarophenacidae, mites of the family Pyemotidae are the parasitoid of eggs and immatures of many insect orders including Coleoptera, Lepidoptera, Hymenoptera and Diptera. These mites have high fecundity due to relatively short life cycle, and development of the many same aged progeny inside mother's physogastric body on one host. Some species like *Pyemotes tritici* (LaGrèze-Fossat and Montagné) show one of the highest reproductive rates among animals (Walter and Proctor 1999). Since only one host individual is enough for the development of a single mite stage, and eventually host dies, Lindquist (1983) considered these mites as a parasitoid.

The family is currently consist of one genus, *Pyemotes*, with about 25 species that are classified to two species groups, *ventricosus* and *scolyti*, based on their toxicity to the host and degree of host specificity. *P. ventricosus* group has wide host spectrum and occupies many habitats. Females are extremely venomous and attack all host instars. Their injected toxin causes paralysis and death of the host, a factor that makes these mites a pest in insect stalk cultures and stored products. It can also cause grain itch, pruriginous dermatitis in people who have continuous contact with these mites (Walter et al. 2009). The cosmopolitan *P. tritici* with a high reproductive rate is a mortality factor of lepidopteran larvae and pupae, and other stored products pests. This mite attack over 150 insect species belonging to various orders and can be easily reared in the laboratory. It has a control effect on *Lasioderma serricorne* (F.) (Cigarette beetle), *Anagasta kuehniella* (Zeller) (Flour moth), *Pectinophora gossypiella* (Saunders) (Pink bollworm), *Vespa* spp. and *Solenopsis* spp. but unfortunately, unfavorable effects of the toxin of *ventricosus* species group on human is a limited factor in order to use these mites as a biocontrol agent (Gerson et al. 2003).

*P. scolyti* group are restricted to their host habitat and are usually associated with bark beetles, feeding host eggs, larvae and pupae. Females are not venomous and so, do not make dermatitis and other health problems in people and animals come in contact. Mites of this species group are one of the most successful natural enemies of the bark beetles, sometimes eradicating host populations (Gerson et al. 2003).

Both mentioned species groups have been recorded in Iran (Kamali et al. 2001). *P. ventricosus* (Newport) from xylophagous insects (Radjabi 1969), *P. tritici* from *Scolytus amygdali* Guerin-Méneville, (Ostovan 1997; Ahadiyat et al. 2004), and *P. herfsi* (Oudemans) from stored wheat (Mirfakhraii 1994; Ardeshir et al. 2008a) are collected from various parts of Iran showing that these mites have a proper dispersal in storages. It has to be noted that some mentioned species impose damage to insectaria and workers with which they are in contact. In addition, *P. scolyti* (Oudemans) is the only representative of the *scolyti* group reported from some bark beetles (such as *Scolytus amygdali*) and their galleries (Ostovan and Kamali 1996; Ostovan 1997; Ahadiyat et al. 2004). Bioecology and population fluctuation of this mite studied in some parks of Tehran by Ostovan and Kamali (2001). The authors considered the *P. scolyti* as a potential biocontrol agent for *S. amygdali* owing to being oligophage, having a short life cycle (8–10 days) with high reproductive rate, an about 90% female-based progeny and ~ 50% immature parasitation. Despite some limitations to utilize as biocontrol agents, members of this family need to be conserved as natural control agents of some important pests.

### 3.3.2.8  Acarophenacidae

Members of the small family Acarophenacidae with about 35 species are regarded as egg parasitoids of various insects. Among six extant genera, species of the genera *Aethiophenax, Paracarophenax*, and some *Acarophenax* have been documented to be associated with several beetle families such as Cerambycidae, Cucujidae, Dermestidae, Nitidulidae, Curculionidae (subfamily Scolytinae) and Tenebrionidae

(Magowski 1994; Goldarazena et al. 1999; Krantz and Walter 2009). All described species of the genus *Adactylidium* are not only egg parasitoid of tubuliferous thrips (Thysanoptera: Tubulifera) but some observations upon their penetrated mouthpart into the body of their adult hosts' bodies and feeding on haemolymph, suggest that they can be also considered as ectoparasites (Goldarazena et al. 2001; Gerson et al. 2003; Khaustov 2007). Host association of other genera is unknown (Rady 1992; Goldarazena et al. 1999).

In spite of the paucity of studies on the biology of most species of the family Acarophenacidae and their capability in controlling their host population, the inevitable potential of some acarophenacids as useful biological control agents in closed poultry houses and stored product situations, have provoked researchers to pay proportionately more attention to this family (Rack, 1959; Bruce and LeCato 1980; Steinkraus and Cross 1993; Faroni et al. 2000; Gerson et al. 2003). However, compared with the family Pyemotidae, acarophenacid mites feed only on the eggs of their insect hosts and have no toxic effects on people or domesticated animals (Steinkraus and Cross 1993). In one of the studies carried out by Steinkraus and Cross (1993), the *Acarophenax mahunkai* Steinkraus and Cross, an egg parasitoid of the tenebrionid pest *Alphitobius diaperinus* Panzer, imposed two-thirds mortality in eggs. Also, Faroni et al. (2000) stated that *Acarophenax lacunatus* (Cross and Krantz), an egg parasitoid of bostrichid major pest *Rhyzopertha dominica* F., could be a useful biocontrol agent specially in tropical regions.

Hitherto, four species of the genus *Acarophenax*, two species of the *Aethiophenax* and one species of *Adactylidium* have been found or described from Iran (Ostovan and Saboori 1999; Ardeshir et al. 2008b; Katlav et al. 2015; Rahiminejad and Hajiqanbar 2015; Lotfollahi et al. 2016; Arjomandi et al. 2017). These mites have been collected from storage or forest habitats on beetles of the families Nitidulidae, Sphididae and Tenebrionidae (including *Tribolium confusum* Jacquelin du Val and *Alphitobius diaperinus* that are two important pests of stored products), or on thrips. Further researches upon the biology and life history of these mites and their hosts are necessary to utilize them as effective and applicable biocontrol agents.

### 3.3.2.9  Podapolipidae

Mites of the highly specialized family Podapolipidae are external (and occasionally internal) parasites of many insect orders including Coleoptera, Orthoptera, Blattodea, Hymenoptera and Hamiptera. These mites are observed beneath elytra and in the genital tract of the beetles, in the trachea and air sacs of Orthoptera and on the body surface of grasshoppers and cockroaches (Kaliszewski et al. 1995; Walter et al. 2009). Many genera of this family are host-specific. For example, mites of the genus *Coccipolipus* exclusively parasitize ladybirds of the family Coccinellidae, a factor that can affect feeding and reproductive efficiencies of the hosts. Podapolipid mites have recently received considerable attention owing to the prospect of utilizing some species as biological control agents of some pests in agricultural ecosystems (Kenis et al. 2008; Rhule et al. 2010; Seeman and Nahrung 2013) like *Coccipolipus epilachnae* Smiley on Mexican bean beetle, *Epilachna varivestis* Mulsant (Riddick et al. 2009). The genus *Podapolipoides*, attacking grasshoppers, is another genus which is useful in the natural control of some acridid pests, mostly Oedipodinae and Cyrtacanthacridinae (Hajiqanbar and Joharchi 2011). Mites of this family are primarily transmitted during copulation and are one of the so-called Sexually Transmitted Diseases (STDs) of insects that can have a negative effect on fecundity and reproductive parameters of the host and or increase mortality during hibernation (Knell and Webberley 2004). Presently, this family encompasses 32 genera and 262 species that hitherto, eight genera (*Eutarsopolipus, Dorsipes, Ovacarus, Regenpolipus, Coccipolipus, Podapolipoides, Podapolipus* and *Tarsopolipus*) and 18 species have been reported from Iran on beetles of the families Carabidae, Scarabaeidae, Coccinellidae, Tenebrionidae and grasshoppers of the Acrididae (Hajiqanbar et al. 2007, 2008; Hajiqanbar and Khoshnevis 2011; Hajiqanbar and Mortazavi 2012; Mortazavi and Hajiqanbar 2012; Hajiqanbar 2013). The efficiency of mites of the family Podapolipidae on their hosts is mostly unclear so, evaluation of this aspect and biological studies on these mites and their hosts can elucidate further their ability as biocontrol agents in future.

## 3.4 Conclusion

The mesostigmatic and prostigmatic mites include the most successful acarine lineages in biological control programs, the reason these two main groups are focused in this chapter. In Mesostigmata, undoubtedly, the family Phytoseiidae encompasses the most economic and effective biocontrol agents particularly in classical programs because of their recognized potential to feed on mites as well as thrips and whiteflies. Hence, this family has been the subject of intensive studies about biology, behavior, ecology and interactions with other natural enemies and recently efforts are doing for mass-rearing methods and finding appropriate strategies for conservation of native species. There is a need to continue the further identification of native species and extend studies about releasing methods in the fields and particularly greenhouses. Some members of the family Laelapidae can be used as augmentative releases. The impact of other species of mentioned mesostigmatic mites has to be more studied regarding their biological and behavioral traits. In Prostigmata, the family Stigmaeidae is potentially useful in orchards specially unsprayed ones. Some families like Tydeidae and Iolinidae are partly predators and also can play their role as alternative prey and sanitizing agents. Usefulness of some families such as Anystidae has a limited value. Some other species or families in the country require more studies like Trombidiidae, Podapolipidae and Acarophenacidae. Generally, basic researches including faunistic, as well as biological and ecological studies are needed to find and elucidate more effective and promising natural enemies.

**Acknowledgements**   We thank the following persons for review and useful comments on some parts related to their expertise: Drs. Alireza Saboori (many Prostigmata including Parasitengonina), Maryam Darbemamieh (Tydeoidea) and Omid Joharchi (Laelapidae).

# References

Aflaki F, Rahmani H, Kavousi A (2016) Effects of Azadirachtin and Spirodiclofen on some biological traits and predation rate of *Neoseiulus californicus* (Acari: Phytoseiidae). In: Proceeding of biocontrol meeting. Ferdowsi University of Mashhad, Mashhad, 2–3 February 2016

Afshari N, Rahmani H, Movahhedi Fazel M (2014) Intra- and interspecific predation in the predatory mites *Amblyseius andersoni* and *Neoseiulus californicus* (Acari: Phytoseiidae). In: 21th Iranian plant protection Congress, Urmia University, Urmia, 23–26 August 2014

Ahadiyat A, Ostovan H, Saboori A (2004) Mites associated with *Scolytus amygdali* Guerin-Meneville, 1847 in Karaj region. In: 16th Iranian Plant Protection Congress, University of Tabriz, Tabriz, 28–1 August–September 2004

Ahmadi S, Hajiqanbar H, Saboori A (2012) A new species of the genus *Abalakeus* (Acari, Erythraeidae) from Iran. Acta Zool Acad Sci Hung 58(2):169–176

Ahn JJ, Kim KW, Lee JH (2010) Functional response of *Neoseiulus californicus* (Acari: Phytoseiidae) to *Tetranychus urticae* (Acari: Tetranychidae) on strawberry leaves. J Appl Entomol 134:98–104

Akbari A, Haddad Irani Nejad K, Khanjani M, Arzanlou M, Kazmierski A (2015) *Tydeus shabestariensis* sp. nov. and description of the male of *Neopronematus sepasgosariani* (Acari: Tydeoidea), with a key to the Iranian species of *Tydeus*. Zootaxa 4032(3):264–276

Alinejad M, Kheradmand K, Fathipour Y (2014) Sublethal effects of fenazaquin on life table parameters of the predatory mite *Amblyseius swirskii* (Acari: Phytoseiidae). Exp Appl Acarol 4:361–373

Alinejad M, Kheradmand K, Fathipour Y (2016) Assessment of sublethal effects of spirodiclofen on biological performance of the predatory mite *Amblyseius swirskii* (Acari: Phytoseiidae). Syst Appl Acarol 21:361–373

Alipour Z, Fathipour Y, Farazmand A (2016) Age-stage predation capacity of *Phytoseiulus persimilis* and *Amblyseius swirskii* (Acari: Phytoseiidae) on susceptible and resistant rosecultivars. Int J Acarol 42:224–228

Amin MR, Khanjani M, Zahiri B (2014) Preimaginal development and fecundity of *Gaeolaelaps aculeifer* (Acari: Laelapidae) feeding on *Rhizoglyphus echinopus* (Acari: Acaridae) at constant temperatures. J Crop Prot 31(3):581–587

André HM, Ueckermann E, Rahmani H (2010) Description of two new species closely related to *Tydeus spathulatus* (Acari: Tydeidae) from Zimbabwe and Iran. J Afrotrop Zool 6:111–116

Arbabi M, Singh J (2000) Studies on biological aspects of predaceous mite *Cunaxa setirostris* on *Tetranychus ludeni* at laboratory condition in Varanasi, India. J Agric Rural Dev 2(1/2):13–23

Ardeshir F, De Saint Georges-Gridelet D, Grootarert P, Tirry L, Wauthy G (2000) Preliminary observation on mites associated with stored grain in Iran. Belg J Entomol 2:287–293

Ardeshir F, Kamali H, Ranji H (2008a) Comparison of stored mite fauna in Khorasan and West Azarbaijan provinces. In: Proceeding of the 18th Iranian plant protection congress, University of Hamedan, Hamedan, 24–27 August 2008

Ardeshir F, Yousefi Porshokouh A, Saboori A (2008b) A faunistic study and population fluctuations of mites associated with stored wheat in Tehran region, Iran. J Entomol Soc Iran 27 (2):17–28

Argov Y, Amitai S, Beattie GAC, Gerson U (2002) Rearing, release and establishment of imported predatory mites to control citrus rust mite in Israel. BioControl 47:399–409

Arjomandi E, Hajiqanbar H, Joharchi O (2017) *Aethiophenax mycetophagi* sp. nov. (Acari: Trombidiformes: Acarophenacidae), an egg parasitoid of *Mycetophagus quadripustulatus* (Coleoptera: Mycetophagidae) from Iran. Syst Appl Acarol 22:541–549

Asadi G, Khorramdel S, Ghorbani R (2013) Preliminary evaluation of some eriophyid mites as biological control agents of some invasive weed in Northeastern Iran. Agroecology 5 (3):299–307

Asadi G, Ghorbani R, Cristofare M, Chetverikov P, Retanovic R, Vidovic B, Schaffner U (2014a) The impact of the flower mite *Aceria acroptiloni* on the invasive plant Russian knapweed, *Rhaponticum repens*, in its native range. BioControl 59:367–375

Asadi N, Rahmani H, Movahedi Fazel M (2014b) Effect of immature feeding experience on handling time of the predatory mite *Neoseiulus californicus*. In: 21st Iranian plant protection Congress, Urmia University, Urmia, 23–26 August 2014

Askarieh Yazdi S, Zahedi Golpayegani A, Saboori A, Karami Badrbani F (2015) Different forms of *Tetranychus urticae* Koch and their plasticity in retaining eggs in the presence of predatory mites, *Amblyseius swirskii* and *Phytoseiulus persimilis*. Persian J Acarol 4(3):319–327

Babaeian E, Halliday B, Saboori A (2015) A new species of *Geholaspis* Berlese (Acari: Mesostigmata: Macrochelidae) from Northern Iran. Zootaxa 3925(1):422–430

Badii MH, Hernandez-ortiz E, Flores AE, Landeros J (2004) Prey stage preference and functional response of *Euseius hibisci* to *Tetranychus urticae* (Acari: Phytoseiidae, Tetranychidae). Exp Appl Acarol 34:263–273

Bagheri M, Jafari S, Saboori A (2014) Two new species of the family Stigmaeidae (Acari: Trombidiformes) from Iran. Int J Acarol 40(2):152–159

Bahari F, Fathipour Y, Talebi AA, Alipour Z (2018) Long term feeding on greenhouse cucumber affects life table parameters of two-spooted spider mite and its predator *Phytoseiulus persimilis*. Syst Appl Acarol 23(12):2304–2316

Banerjee P, Datta S (1980) Biological control of red cotton bug, *Dysdercus koenigii* Fabricius by mite, *Hemipteroseius indicus* (Krantz and Khot). Indian J Entomol 42:265–267

Baradaran P, Arbabi M (2008) Population abundance of *Pronematus ubiquitus* (McGregor, 1932) (Acari: Tydeidae) on different fig varieties in Saveh region. J Entomol Res 1(2):177–183

Barber A, Campbell CAM, Crane H, Lilley R, Tregidga E (2003) Biocontrol of two-spotted spider mite *Tetranychus urticae* on dwarf hops by the phytoseiid mites *Phytoseiulus persimilis* and *Neoseiulus californicus*. Biocontrol Sci Tech 13:275–284

Beyzavi G, Ueckermann EA, Faraji F, Ostovan H (2013) A catalog of Iranian prostigmatic mites of superfamilies Raphignathoidea & Tetranychoidea (Acari). Persian J Acarol 2(3):389–474

Bjornson S (2008) Natural enemies of mass-reared predatory mites (Phytoseiidae) used for biological pest control. Exp Appl Acarol 46(1–4):299–306

Blackwood JS, Schausberger P, Croft BA (2001) Prey stage preferences in generalist and specialist phytoseiid mites (Acari: Phytoseiidae) when offered *Tetranychus urticae* (Acari: Tetranychidae) eggs and larvae. Environ Entomol 30:1103–1111

Bochkov AV, Hakimitabar M, Saboori A (2005) A review of the Iranian Cheyletidae (Acari Prostigmata). Belg J Entomol 7:99–109

Bohloolzadeh M, Babaeian E, Karami F, Askarieh Yazdi S, Mohammadi H, Zahedi-Golpayegani A (2013) The effect of rearing conditions on the olfactory responses of *Amblyseius swirskii* (Athias-Henriot) (Acari: Phytoseiidae). In: 2nd international Persian Congress of acarology, University of Tehran, Karaj, 29–31 August 2013

Bohloolzadeh M, Zahedi-Golpayegani A, Mohammadi H, Askarieh Yazdi S, Karami F (2014a). The effect of rearing conditions on the leaving or residence tendency of *Neoseiulus barkeri* Hughes (Acari: Phytoseiidae) on bean leaf. In: 21st Iranian plant protection Congress, Urmia University, Urmia, 23-26 august 2014, p 931

Bohloolzadeh M, Zahedi-Golpayegani A, Saboori A, Allahyari H, Nazeri M (2014b) Aggressiveness of adult female predatory mites *Amblyseius swirskii*, *Neoseiulus barkeri* and *Phytoseiulus persimilis* (Acari: Phytoseiidae) towards their heterospecific larva in the guild. In: 21th Iranian plant protection Congress. Urmia University, Urmia, 23–26 August 2014

Borji F, Jalili P, Jafari S, Rahmani H (2011) Effect of starvation on aggressiveness against spider mites and conspecifics in the predatory mite *Neoseiulus californicus* (Acari: Phytoseiidae). In: First Persian Congress of acarology, international center for science, high technology & environmental sciences, Kerman, 22–23 December 2011

Borji F, Jalili P, Jafari S, Rahmani H (2014) Effect of feeding on intraguild predator on aggressiveness of *Phytoseiulus persimilis* against *Tetranychus urticae*. In: 3rd Integrated Pest Management Conference (IPMC), University of Kerman, Kerman, 21–22 January 2014

Bostanian NJ, Akalach M (2006) The effect of indoxacarb and five other insecticides on *Phytoseiulus persimilis* (Acari: Phytoseiidae), *Amblyseius fallacies* (Acari: Phytoseiidae) and nymph of *Orius insidiosus* (Hemiptera: Anthocoridae). Pest Manag Sci 62:334–339

Brown JM, Wilson DS (1992) Local specialization of phoretic mites on sympatric carrion beetle hosts. Ecology 73:463–478

Bruce WA, LeCato GL (1980) *Pyemotes tritici*: a potential new agent for biological control of the red imported fire ant, *Solenopsis invicta* (Acari: Pyemotidae). Int J Acarol 6:271–274

Cadogan BL, Laing JE (1977) A technique for rearing the predaceous mite *Balaustium putmani* (Acarina: Erythraeidae), with notes on its biology and life history. Can Entomol 109:1535–1544

Cakmak I, Janssen A, Sabelis MW, Baspinar H (2009) Biological control of an acarine pest by single and multiple natural enemies. Biol Control 50:60–65

Calvo FJ, Bolckmans K, Belda JE (2012) Biological control-based IPM in sweet pepper greenhouses using *Amblyseius swirskii* (Acari: Phytoseiidae). Biocontrol Sci Tech 22(12):1398–1416

Calvo FJ, Knapp M, van Houten YM, Hoogerbrugge H, Belda JE (2014) *Amblyseius swirskii*: what made this predatory mite such a successful biocontrol agent? Exp Appl Acarol 65(4):419–33

Camporese P, Duso C (1995) Life history and life table parameters of the predatory mite *Typhlodromus talbii*. Entomol Exp Appl 77:149–157

Castagnoli M (1984) Contributo alla conoscenza dei Tideidi (Acarina: Tydeidae) delle piante coltivate in Italia [in Italian with English summary]. Redia 67:307–322

Castilho RC, Venancio R, Narita JPZ (2015) Mesostigmata as biological control agents, with emphasis on Rhodacaroidea and Parasitoidea. In: Carrillo D, de Moraes GJ, Peña JE (eds) Prospects for biological control of plant feeding mites and other harmful organisms. Springer International, Cham, pp 1–31

Castro TMMG (2008) Estudos taxonômicos e biológicos de Cunaxidae (Acari: Prostigmata) do Brasil. Thesis, University of São Paulo

Cédola CV, Sánchez NE, Liljesthröm GG (2001) Effect of tomato leaf hairiness on functional and numerical response of *Neoseiulus californicus* (Acari: Phytoseiidae). Exp Appl Acarol 25:819–831

Chant DA (1961) An experiment in biological control of *Tetranychus telarius* (L.) (Acarina: Tetranychidae) in a greenhouse using the predacious mite *Phytoseiulus persimilis* (Phytoseiidae). Can Entomol 93:437–443

Chi H (1988) Life-table analysis incorporating both sexes and variable development rate among individuals. Environ Entomol 17(1):26–34

Chi H (1990) Timing of control based on the stage structure of pest populations: a simulation approach. J Econ Entomol 83(4):1143–1150

Chi H (1994) Periodic mass rearing and harvesting based on the theories of both the age-specific life table and the age-stage, two-sex life table. Environ Entomol 23(3):535–542

Chi H, Liu H (1985) Two new methods for the study of insect population ecology. Bull Inst Zool Acad Sinica 24(2):225–240

Childers CC, Rock GC (1981) Observations on the occurrence and feeding habits of *Balaustium putmani* (Acari: Erythraeidae) in North Carolina apple orchards. Int J Acarol 7:63–68

Chow A, Chau A, Heinz KM (2010) Compatibility of *Amblyseius* (Typhlodromips) *swirskii* (Athias-Henriot) (Acari: Phytoseiidae) and *Orius insidiosus* (Hemiptera: Anthocoridae) for biological control of *Frankliniella occidentalis* (Thysanoptera: Thripidae) on roses. Biol Control 53:188–196

Christiansen IC, Sandra S, Schausberger P (2016) Benefit-cost trade-offs of early learning in foraging predatory mites *Amblyseius swirskii*. Sci Rep 6:23571

Costa M (1968) Notes on the genus *Hemipteroseius* Evans (Acari: Mesostigmata) with the description of a new species from Israel. J Nat Hist 2:1–15

Cross EA, Bohart GA (1979) Some observations of the habits and distribution of *Trochometridium* Cross, 1965 (Acarina: Pyemotidae). Acarologia 20:286–293

Damghani R (2001) Investigation on biology and some control methods of *Oryctes elegans* Prell. in Bam region. Thesis, science and research branch, Islamic Azad University

Daneshvar H (1989) The introduction of *Phytoseiulus persimilis* A.H. (Acari; Phytoseiidae) as active predator in Iran. In: 9th Iranian plant protection Congress, Ferdowsi University of Mashhad, Mashhad, 9–14 September 1989

Daneshvar H (1993) Distribution of two predatory mite *Amblydromella kettanehi* and *Euseius libanesi* (Acari: Phytoseiidae) in Iran. In: 11th Iranian plant protection Congress, University of Guilan, Rasht

Daneshvar H, Abaii MG (1994) Efficient control of *Tetranychus turkestani* on cotton, soybean and bean by *Phytoseiulus persimilis* (Acari: Tetranychidae, Phytoseiidae) in pest foci. Appl Entomol Phytopath 61(1, 2):61–75

Daneshvar H, Denmark HA (1982) Phytoseiids of Iran (Acarina: Phytoseiidae). Int J Acarol 8 (1):3–14

Darbemamieh M, Kamali K, Fathipour Y (2010) First report of *Tydeus caudatus* (Acari: Tydeidae) from Iran. J Entomol Soc Iran 30(1):63–65

Darbemamieh M, Hajiqanbar H, Khanjani M, Kazmierski A (2015) New species and records of *Neopronematus* (Acari: Iolinidae) from Iran with a key to world species. Zootaxa 3990 (2):235–246

Darbemamieh M, Hajiqanbar H, Khanjani M, Kazmierski A (2016) *Paurotyndareus*, a new genus of the family Iolinidae (Acari: Prostigmata), with the description of a new species from Iran. Syst Appl Acarol 21(4):398–404

De Boer JG, Dicke M (2004) The role of methyl salicylate in prey searching behavior of the predatory mite *phytoseiulus persimilis*. J Chem Ecol 30(2):255–271

De Clercq P, Mohaghegh J, Tirry L (2000) Effect of host plant on the functional response of the predator *Podisus nigrispinus* (Heteroptera: Pentatomidae). Biol Control 18(1):65–70

de Moraes GJ, Britto EPJ, de Mineiro JLC, Halliday B (2016) Catalogue of the mite families Ascidae Voigts & Oudemans, Blattisociidae Garman and Melicharidae Hirschmann (Acari: Mesostigmata). Zootaxa 4112:1–299

Dehghani Tafti H, Zahedi-Golpaygani A, Saboori A (2013) The effect of kinship on mating preference: Does *Phytoseiulus persimilis* avoid mating with kin? In: 2nd international Persian Congress of Acarology, University of Tehran, Karaj, 29–31 August 2013

Dehghani-Tafti H, Zahedi-Golpaygani A, Saboori A, Krey KL (2015) The effect of mating experience, age and territoriality on the male mating competition in *Phytoseiulus persimilis* and *Neoseiulus californicus* (Acari: Phytoseiidae). Persian J Acarol 4(1):111–123

Demite PR, McMurtry JA, De Moraes GJ (2014) Phytoseiidae database: a website for taxonomic and distributional information on phytoseiid mites (Acari). Zootaxa 3795:571–577

Den Heyer J, Ueckermann EA, Khanjani M (2011a) Iranian Cunaxidae (Acari: Prostigmata: Bdelloidea). Part I. Subfamily Coleoscirinae. Int J Acarol 27(2):143–160

Den Heyer J, Ueckermann EA, Khanjani M (2011b) Iranian Cunaxidae (Acari: Prostigmata: Bdelloidea): Part 2. Subfamily Cunaxinae. J Nat Hist 45(27–28):1667–1678

Den Heyer J, Ueckermann EA, Khanjani M (2013) Iranian Cunaxidae (Acari: Prostigmata: Bdelloidea). Part III. Subfamily Cunaxoidinae. J Nat Hist 47:31–32

Dent DR, Walton MP (1997) Methods in ecological and agricultural entomology. CAB International, Wallingford

Dicke M, Sabelis MW (1988) Infochemicals terminology: based on cost-benefit analysis rather than origin of compounds? Funct Ecol 2:131–139

Dicke M, van Beek T, van Posthumus MA, Ben Dom N, van Bokhoven H, de Groot AE (1990) Isolation and identification of volatile kairomone that affects acarine predator prey interactions: involvement of host plant in its production. J Chem Ecol 16:381–396

Dogramaci M, Arthurs SP, Chen J, McKenzie C, Irrizary F, Osborne L (2011) Management of chilli thrips *Scirtothrips dorsalis* (Thysanoptera: Thripidae) on peppers by *Amblyseius swirskii* (Acari: Phytoseiidae) and *Orius insidiosus* (Hemiptera: Anthocoridae). Biol Control 59:340–347

Drukker B, Bruin J, Jacobs G, Kroon A, Sabelis MW (2000) How predatory mites learn to cope with variability in volatile plant signals in the environment of their herbivorous prey. Exp Appl Acarol 24:881–895

Eghbalian AH, Khanjani M, Safaralizadeh MH, Ueckermann EA (2014) Two new species of *Cyta* (Acari: Prostigmata: Bdellidae) from Western Iran. Zootaxa 3847(4):567–575

Eghbalian AH, Khanjani M, Safaralizadeh MH, Ueckermann EA (2016) New species of *Hexabdella* and *Neomolgus* (Acari: Prostigmata: Bdellidae) from Iran. Zootaxa 4072 (2):291–300

English-Loeb GM, Norton AP, Gadoury DM, Seem RC, Wilcox WF (1999) Control of powdery mildew in wild and cultivated grapes by a tydeid mite. Biol Control 14:97–103

Erbilgin N, Dahlsten DL, Chen P (2004) Intraguild interactions between generalist predators and an introduced parasitoid of *Glycaspis brimblecombei* (Homoptera: Psylloidea). Biol Control 31:329–337

Fakhari N, Khanjani M, Rahmani H, Khanjani M (2015) *Stigmaeus jalili* sp. n. (Acari: Stigmaeidae) from Zanjan Province (Iran) and description of *S. haddadi* male. Biologia 70(6):782–787

Fan Q-H, Flechtmann CH (2015) Stigmaeidae. In: Carillo D, Moraes GJ, Peña JE (eds) Prospects for biological control of plant feeding mites and other harmful organisms. Springer, Berlin

Fan YQ, Petitt FL (1994) Biological control of broad mite, *Polyphagotarsonemus latus* (Banks), by *Neoseiulus barkeri* Hughes on pepper. Biol Control 4:390–395

Fan QH, Flechtmann CHW, de Moraes GJ (2016) Annotated catalogue of Stigmaeidae, with a pictorial key to genera. Zootaxa 4176:1–199

Faraji F (2001) How counter-attacking prey influence foraging and oviposition decisions of a predatory mite. Dissertation, University of Amsterdam

Farazmand A, Fathipour Y, Kamali K (2012) Functional response and mutual interference of *Neoseiulus californicus* and *Typhlodromus bagdasarjani* (Acari: Phytoseiidae) on *Tetranychus urticae* (Acari: Tetranychidae). Int J Acarol 38:369–376

Farazmand A, Fathipour Y, Kamali K (2013) Predation preference of *Neoseiulus californicus* and *Typhlodromus bagdasarjani* on heterospecific phytoseiid and *Scolothrips longicornis* in presence and absence of *Tetranychus urticae*. Persian J Acarol 2(1):181–188

Farazmand A, Fathipour Y, Kamali K (2014) Cannibalism in *Scolothrips longicornis* (Thysanoptera: Thripidae), *Neoseiulus californicus* and *Typhlodromus bagdasarjani* (Acari: Phytoseiidae). Syst Appl Acarol 19:471–480

Farazmand A, Fathipour Y, Kamali K (2015a) Intraguild predation among *Scolothrips longicornis* (Thysanoptera: Thripidae), *Neoseiulus californicus* and *Typhlodromus bagdasarjani* (Acari: Phytoseiidae) under laboratory conditions. Insect Sci 22:263–272

Farazmand A, Fathipour Y, Kamali K (2015b) Control of the spider mite *Tetranychus urticae* using phytoseiid and thrips predators under microcosm conditions: single-predator versus com- bined-predators release. Syst Appl Acarol 20:162–170

Farhadi R, Allahyari H, Chi H (2015) Functional response and mutual interference of *Amblyseius swirskii* (Acari: Phytoseiidae) on greenhouse whitefly *Trialeurodes vaporariorum* on cucumber. J Plant Protect 38(2):37–48

Faroni LRD'A, Guedes RNC, Matioli AL (2000) Potential of *Acarophenax lacunatus* (Prostigmata: Acarophenacidae) as a biological control agent of *Rhyzopertha dominica* (Coleoptera: Bostrichidae). J Stored Prod Res 36:55–63

Fathipour Y, Maleknia B (2016) Mite predators. In: Omkar (ed) Ecofriendly pest management for food security. Elsevier, San Diego

Fathipour Y, Sedaratian A (2013) Integrated management of *Helicoverpa armigera* in soybean cropping systems. In: El-Shemy H (ed) Soybean–pest resistance. InTech, Rijeka

Fathipour Y, Hosseini A, Talebi AA (2006) Functional response and mutual interference of *Diaeretiella rapae* (Hymenoptera: Aphidiidae) on *Brevicoryne brassicae* (Homoptera: Aphididae). Entomol Fennica 17:90–97

Fathipour Y, Karimi M, Farazmand A, Talebi AA (2017) Age-specific functional response and predation rate of *Amblyseius swirskii* (Phytoseiidae) on two-spotted spider mite. Syst Appl Acarol 22:159–116

Fathipour Y, Karimi M, Farazmand A, Talebi AA (2018) Age-specific functional response and predation capacity of *Phytoseiulus persimilis* (Phytoseiidae) on the two-spotted spider mite. Acarologia 58:31–40

Fernandez-Arhex V, Corley JC (2003) The functional response of parasitoids and its implications for biological control. Biocontrol Sci Tech 13:403–413

Fitzgerald J, Pepper N, Easterbrook M, Pope T, Solomon M (2007) Interactions among phytophagous mites, and introduced and naturally occurring predatory mites, on strawberry in the UK. Exp Appl Acarol 43:33–47

Ganjisaffar F, Fathipour Y, Kamali K (2011a) Effect of temperature on prey consumption of *Typhlodromus bagdasarjani* (Acari: Phytoseiidae) on *Tetranychus urticae* (Acari: Tetranychidae). Int J Acarol 37:556–560

Ganjisaffar F, Fathipour Y, Kamali K (2011b) Temperature-dependent development and life table parameters of *Typhlodromus bagdasarjani* (Phytoseiidae) fed on two-spotted spider mite. Exp Appl Acarol 55:256–272

Gardiner MM, Landis DA (2007) Impact of intraguild predation by adult *Harmonia axyridis* (Coleoptera: Coccinellidae) on *Aphis glycines* (Hemiptera: Aphididae) biological control in cage studies. Biol Control 40:386–395

Gasperin DG, Kilner RM (2015) Friend or foe: inter-specific interactions and conflicts of interest within the family. Ecol Entomol 40(6):787–795

Gerson U (2014) Pest control by mites (Acari): present and future. Acarologia 54:371–394

Gerson U, Weintraub P (2007) Mites for the control of pests in protected cultivation. Pest Manag Sci 63(7):658–676

Gerson U, Weintraub P (2012) Mites (Acari) as a factor in greenhouse management. Annu Rev Entomol 57:229–247

Gerson U, Smiley RL, Ochoa R (2003). Mites (Acari) for Pest Control. Blackwell Science, Oxford, UK, pp. 539

Ghaderi S, Minaei K, Kavousi A, Akrami MA, Aleosfoor M, Ghadamyari M (2013) Demographic analysis of the effect of Fenpyroximate on *Phytoseiulus persimilis* Athias-Henriot (Acari: Phytoseiidae). Entomol Generalis 34(3):225–233

Ghasemloo Z, Pakyari H, Arbab A (2016) Cannibalism and intraguild predation in the phytoseiid mites *Phytoseiulus persimilis* and *Typhlodromus bagdasarjani* (Acari: Phytoseiidae). Int J Acarol 37:556–560

Goldarazena A, Ochoa R, Jordana R (1999) Revision of the genus *Paradactylidium* Mahunka (Acari: Heterostigmata). Int J Acarol 25:91–99

Goldarazena A, Ochoa R, Jordana R, OConnor BM (2001) Revision of the genus *Adactylidium* Cross (Acari: Heterostigmata: Acarophenacidae), mites associated with thrips (Thysanoptera). Proc Entomol Soc Wash 103:473–516

Gotoh T, Nozawa M, Yamaguchi K (2004) Prey consumption and functional response of three acarophagous species to eggs of the two-spotted spider mite in the laboratory. Appl Entomol Zool 39(1):97–105

Haghani S, Zahedi-Golpayegani A, Saboori A, Allahyari H (2015) Aggressiveness and predation preference of predatory mites *Amblyseius swirskii* (Athias-Henriot), *Neoseiulus californicus* (McGregor) and *Phytoseiulus persimilis* (Athias-Henriot) (Acari: Phytoseiidae) towards to heterospecific larvae. Ecol Monogr 3:46–55

Haghighi R, Ostovan H, Hesami S (2011) The first report of *Caudacheles khayae* (Acari: Cheyletidae) associated with elm bark beetles in Iran. Plant Prot J 3(1):79–84

Haitlinger R, Saboori A (2007) Two new larval ectoparasitic *Charletonia* Oudemans (Acari: Prostigmata: Erythraeidae) found on Orthoptera (Insecta), and the first record of *Charletonia krendowskyi* (Feider) in Iran. Pol J Entomol 76:61–71

Haitlinger R, Saboori A (2008) *Charletonoia behbahanensis* sp. n. and *C. bojnordensis* sp. n. from Iran (Acari: Prostigmata: Erythraeidae). Ze Nauk Uniw Przyrod Wroc Biol Hod Zwi 56 (566):73–80

Hajiqanbar H (2013) *Podapolipus khorasanicus* n. sp. (Acari: Podapolipidae), an ectoparasite of *Opatroides punctulatus* (Coleoptera: Tenebrionidae) with notes on world distribution and host range of the beetle-assocites of *Podapolipus* spp. Ann Entomol Soc Am 106(2):181–188

Hajiqanbar H, Joharchi O (2011) World distribution and host range of *Podapolipoides* spp. (Acari: Heterostigmatina: Podapolipidae), with the description of a new species. Syst Parasitol 78:151–162

Hajiqanbar H, Khoshnevis M (2011) New records of the genus and species *Regenpolipus madrasensis* (Acari: Heterostigmata: Podapolipidae), the ectoparasite of carabid beetles from Iran. J Entomol Soc Iran 31:91–93

Hajiqanbar H, Mortazavi A (2012) First record of the myzus species group (Acari: Podapolipidae: *Eutarsopolipus* Berlese, 1911) from Asia, with the description of two new species parasitising carabid beetles. Syst Parasitol 83:189–202

Hajiqanbar H, Husband RW, Kamali K, Saboori A, Kamali H (2007) *Ovacarus longisetosus* n. sp. (Acari: Podapolipidae) from *Amara* (*Paracelia*) *saxicola* Zimm. (Coleoptera: Carabidae) and new records of *Coccipolipus*, *Dorsipes*, *Eutarsopolipus* and *Tarsopolipus* from Iran. Int J Acarol 33:241–244

Hajiqanbar H, Husband RW, Kamali K, Saboori A, Kamali H (2008) *Dorsipes saxicolae*, a new species of mite (Acari: Podapolipidae), an ectoparasite of *Amara* (*Paracelia*) *saxicola* Zimm. (Coleoptera: Carabidae) from Iran. Int J Acarol 34:85–90

Hajiqanbar H, Khaustov A, Kamali K, Saboori A, Kamali H (2009) New taxa of the family Trochometridiidae (Acari: Heterostigmata) associated with insects from Iran. J Nat Hist 43 (41–44):2701–2722

Hajizadeh J, Faraji F (2016) Identification guide and diagnosis key for predatory mites of the family Phytoseiidae of Iran. University of Guilan Press, Guilan

Hajizadeh J, Hosseini R, McMurtry JA (2002) Phytoseiid mites (Acari: Phytoseiidae) associated with eriophyid mites (Acari: Phytoseiidae) in Guilan province of Iran. Int J Acarol 28:373–377

Hajizadeh J, Faraji F, Rafati Fard M (2009) Predatory mites of the family Phytoseiidae of Iran. University of Guilan Press. 282p

Hajizadeh J, Noei J, Salehi L, Ostovan H (2011) Cheyletid mites associated with stored rice in Iran; the first record of *Chelacheles strabismus* from Iran and a key for their identification. J Entomol Soc Iran 30(2):85–88

Hajizadeh J, Khanjani M, Faraji F, Ueckermann EA (2013) Stigmaeid mites of Guilan Province of Iran with description of a new species and a checklist for Iranian stigmaeid mites (Prostigmata: Stigmaeidae). Int J Acarol 39(7):571–579

Hajmohammadloo E, Shirdel D (2010) Comparative study of feeding on some pollens on biology of *Typhlodromus bagdasarjani* Arutunjian & Wainstein (Acari: Phytoseiidae). J Agric Sci, 4:85–96

Hamedi N, Fathipour Y, Saber M, Sheikhi-Garjan A (2009) Sublethal effects of two common acaricides on the consumption of *Tetranychus urticae* (Prostigmata: Tetranychidae) by *Phytoseius plumifer* (Mesostigmata: Phytoseiidae). Syst Appl Acarol 14:197–205

Hamedi N, Fathipour Y, Saber M (2010) Sublethal effects of fenpyroximate on life table parameters of the predatory mite *Phytoseius plumifer*. BioControl 55:271–278

Hamedi N, Fathipour Y, Saber M (2011) Sublethal effects of abamectin on the biological performance of the predatory mite, *Phytoseius plumifer* (Acari: Phytoseiidae). Exp Appl Acarol 53:29–40

Hansen LS (1988) Control of *Thrips tabaci* (Thysanoptera: Thripidae) on glasshouse cucumber using large introductions of predatory mites *Amblyseius barkeri* (Acarina: Phytoseiidae). Entomophaga 33:33–42

Hassell MP, May RM (1974) Aggregation of predators and insect parasites and its effect on stability. J Anim Ecol 43(2):567–594

Hassell MP, Varley GC (1969) New inductive population model for insect parasites and its bearing on biological control. Nature 223:1113–1137

Hatherly IS, Bale JS, Walters KFA (2005) Intraguild predation and feeding preferences in three species of phytoseiid mite used for biological control. Exp Appl Acarol 37:43–55

Henne DC, Johnson SJ (2010) Laboratory evaluation of aggregation, direct mutual interference, and functional response characteristics of *Pseudacteon tricuspis* Borgmeier (Diptera: Phoridae). Biol Control 55:63–71

Hernandes FA, Castro TMMG, Venancio R (2015) Prostigmata (Acari: Trombidiformes) as biological control agents. In: Carillo D, Moraes GJ, Peña JE (eds) Prospects for biological control of plant feeding mites and other harmful organisms. Springer, Berlin

Hernandes FA, Skvarla MJ, Fisher JR, Dowling APG, Ochoa R, Ueckermann EA, Bauchan GR (2016) Catalogue of snout mites (Acariformes: Bdellidae) of the world. Zootaxa 4152:1–83

Heydari S, Allahyari H, Zahedi Golpayegani A, Farhadi R (2016) Prey preference and switching behavior of *Amblyseius swirskii* (Acari: Phytoseiidae) on greenhouse whitefly and two-spotted spider mite. In: Proceeding of biocontrol meeting, Ferdowsi University of Mashhad, Mashhad, 2–3 February 2016

Holling CS (1959) Some characteristics of simple types of predation and parasitism. Can Entomol 91:385–398

Hosseini M, Hatami B, Saboori A (2002) Host preference by *Allothrombium pulvinum* Ewing (Acari: Trombidiidae) larvae on aphids: *Macrosiphum rosae*, *Aphis gossypii* and *Hyalopterus amygdali* (Homoptera: Aphididae). Exp Appl Acarol 27:297–302

Hosseini A, Hatami B, Seyedoleslami H, Saboori A (2004) Preference of *Allothrombium pulvinum* Ewing (Acari: Trombidiidae) among various preys and its functional response to preferred prey. In: Abstracts of the 15th international plant protection congress, Beijing, China

Hosseini A, Hatami B, Saboori A, Allahyari H, Ashouri A (2005) Predation by *Allothrombium pulvinum* on the spider mites *Tetranychus urticae* and *Amphitetranychus viennensis*: predation rate, prey preference and functional response. Exp Appl Acarol 37:173–181

Hosseininia A, Amin MR, Baradaran Anaraki P, Arbabi M (2012) Integrated control of bulb mite, *Rhizoglyphus echinopus* by predator mite *Hypoaspis aculeifer* on corms of Gladiolus. In: Proceedings of the 20th Iranian plant protection congress, University of Shiraz, Shiraz, 4–7 September 2012

Hosseininia A, Baradaran P, Khanjani M, Amini MR (2014) Survey of biological control on bulb mite, *Rhizoglyphus echinopus* (Fumouze & Robin), by predatory mite, *Gaeolaelaps aculeifer* (Canestrini) in greenhouse condition. In: Proceedings of the 21th Iranian plant protection Congress, Urmia University, Urmia, 4–7 September 2014

Hoy MA (2011) Agricultural acarology: introduction to integrated mite management. CRC Press, Forida

Hunter MD (2003) Effects of plant quality on the population ecology of parasitoids. Agric For Entomol 5:1–8

Hussey NW, Parrm WJ, Gould HJ (1965) Observations on the control of *Tetranychus urticae* Koch on cucumbers by the predatory mite *Phytoseiulus riegeli* Dosse. Entomol Exp Appl 8:271–281

Ibrahim YB, Yee TS (2000) Influence of sublethal exposure to abamectin on the biological performance of *Neoseiulus longispinosus* (Acari: Phytoseiidae). J Econ Entomol 93:1085–1089

Ishiwari H, Suzuki T, Maeda T (2007) Essential compounds in herbivore-induced plant volatiles that attract the predatory mite *Neoseiulus womersleyi*. J Chem Ecol 33:1670-1681

Issa GI, El-Banhawy EM, Rasmy AH (1974) Successive release of the predatory mite *Phytoseius plumifer* for combating *Tetranychus arabicus* on fig seedlings. J Appl Entomol 76:442–444

Jafari S, Fathipour Y, Faraji F, Bagheri M (2010) Demographic response to constant temperatures in *Neoseiulus barkeri* (Phytoseiidae) fed on *Tetranychus urticae* (Tetranychidae). Syst Appl Acarol 15:83–99

Jafari S, Fathipour Y, Faraji F (2012) The influence of temperature on the functional response and prey consumption of *Neoseiulus barkeri* (Phytoseiidae) on two-spotted spider mite. J Entomol Soc Iran 31(2):39–52

Jalaeian M, Saboori A, Seyedoleslami H (2005) Prostigmatid mites (Acari: Prostigmata) associated with fruit trees in the western area of Isfahan. J Entomol Soc Iran 25(1):67–68

Jalili Zenoozi P, Zahedi Golpayeghani A, Saboori A, Hekmat Z, Bigdeli N (2016) The effect of familiarity and sociality on predation of *Amblyseius swirskii* (Acari: Phytoseiidae). In: Proceeding of biocontrol meeting, Ferdowsi University of Mashhad, Mashhad, 2–3 February 2016

Jalilirad M, Hajizadeh J, Noei J (2013) Fauna of prostigmatic mites (Acari: Prostigmata) associated with citrus orchards in Guilan Province. Plant Pests Res 2(4):1–13

Jamali MA, Kamali K, Saboori A, Nowzari J (2001) Biology of *Zetzellia mali* (Ewing) (Acari: Stigmaeidae) in Karaj, Iran. Syst Appl Acarol 6(1):55–60

James DG (2003) Toxicity of imidacloprid to *Galendromus occidentalis*, *Neoseiulus fallacis* and *Amblyseius andersoni* (Acari: Phytoseiidae) from hops in Washington State, USA. Exp Appl Acarol 31:275–281

James DG, Vogele B (2001) The effect of imidacloprid on survival of some beneficial arthropods. Plant Protect Quart 16:58–62

Janssen A, Pallini A, Venzon M, Sabelis MW (1998) Behaviour and indirect food web interactions among plant inhabiting mites. Exp Appl Acarol 22:497–521

Janssen A, Faraji F, van der Hammen T, Magalhaes S, Sabelis MW (2002) Interspecific infanticide deters predators. Ecol Lett 5:490–494

Ji J, Zhang Y, Chen X, Lin J (2007) Studies on the functional responses of *Bdellodes japonicus* (Ehara) (Acari: Bdellidae) to *Hypogastrura communis* Folsom (Collembola: Hypogastruridae). Syst Appl Acarol 12(1):13–17

Joharchi O, Babaeian E (2015) A new species of *Reticulolaelaps* Costa (Acari: Laelapidae) associated with *Tapinoma* sp. (Hymenoptera: Formicidae) from Iran, with a review of the world species. Acarologia 55(1):33–44

Joharchi O, Saboori A (2010) The first record of the family Otopheidomenidae (Acarina: Mesostigmata) from Iran. In: Abstract book of 19th Iranian plant protection congress, Iranian Research Institute of Plant Protection, Tehran, Iran, 3–7 September 2010

Joharchi O, Ostovan H, Babaian E (2014) A new species of *Hypoaspis* Canestrini from Iran (Acari: Laelapidae), with a key to the species occurring in the Western Palaearctic Region. Zootaxa 3846(4):569–576

Kaliszewski M, Athias-Binche F, Lindquist EE (1995) Parasitism and parasitoidism in Tarsonemina (Acari: Heterostigmata) and evolutionary considerations. Adv Parasitol 35:335–367

Kamali H (2016) Biological control potential of weeds by gall mites (Acari: Eriophyoidea) in Iran. In: Proceedings of the 3rd National Meeting on Biocontrol in Agriculture and Natural Resource, Ferdowsi University of Mashhad, Mashhad, 2–3 February 2016

Kamali K, Ostovan H, Atamehr A (2001) A catalog of mites & ticks (Acari) of Iran. Islamic Azad University Scientific Publication Center, Tehran

Karami Badrbani F, Zahedi Golpayegani A, Saboori A, Askarieh Yazdi S (2015) Oviposition, development and predation rates of *Neoseiulus californicus* fed on red and green forms of *Tetranychus urticae*. Syst Appl Acarol 20(6):603–611

Karimi Iravanlou JS, Saboori A (2001) *Leptus kamalii* sp. nov. (Acari: Erythraeidae) from Iran. Syst Appl Acarol 6:165–169

Karimi Iravanlou JS, Kamali K, Talebi A (2002a) Three new species of larval Callidosomatinae (Acari, Prostigmata, Erythraeidae) parasitic on short horned grasshoppers (Orthoptera: Acrididae) from Varamin and Karaj, region of Iran. Appl Entomol Phytopathol 69:1–34

Karimi Iravanlou JS, Kamali K, Talebi A (2002b) A new genus and species of larval Callidosomatinae (Acari, Prostigmata, Erythraeidae) parasitic on a short-horned grasshopper,

*Oedipoda miniata* Pall. (Orthoptera: Acrididae) from Varamin, Iran. Iran J Agric Sci 33 (1):123–127

Katlav A, Hajiqanbar H, Talebi AA (2015) First record of the genus *Aethiophenax* (Acari: Acarophenacidae) from Asia, redefinition of the genus and description of a new species. J Asia Pac Entomol 15(3):389–395

Kaveh M, Khajehali J, Poorjavad N (2013) Evalution of fumigant toxicity of essential oils from *Thymus daenensis*, *Rosmarinus officinalis* and *Satureia hortensis* against *Phytoseiulus persimilis* (Acari: Phytoseiidae). In: Conference of biological control in agriculture and natural resources, University of Tehran, Karaj, 27–28 August 2013

Kavousi A, Talebi K (2003) Side-effects of three pesticides on the predatory mite, *Phytoseiulus persimilis* (Acari: Phytoseiidae). Exp Appl Acarol 31:51–58

Kazemi S, Rajaei A (2013) An annotated checklist of Iranian Mesostigmata (Acari), excluding the family Phytoseiidae. Persian J Acarol 2(1):63–158

Kazemi S, Arjomandi E, Ahangaran Y (2013) A review of the Iranian Parasitidae (Acari: Mesostigmata). Persian J Acarol 2(1):159–180

Kenis M, Roy HE, Zindel R, Majerus ME (2008) Current and potential management strategies against *Harmonia axyridis*. BioControl 53:235–252

Khajavi N, Arbabi M, Golmohammadi Gh, Baradaran P (2011) Study lethal and sub lethal dose effects of three new pesticides on *Phytoseiulus persimilis* under laboratory condition. In: First Persian Congress of Acarology, International Center for Science, High Technology & Environmental Sciences, Kerman, 22–23 December 2011

Khanamani M, Fathipour Y, Hajiqanbar H, Sedaratian A (2014) Two-spotted spider mite reared on resistant eggplant affects consumption rate and life table parameters of its predator, *Typhlodromus bagdasarjani* (Acari: Phytoseiidae). Exp Appl Acarol 63:241–252

Khanamani M, Fathipour Y, Hajiqanbar H (2015) Assessing compatibility of the predatory mite *Typhlodromus bagdasarjani* (Acari: Phytoseiidae) and resistant eggplant cultivar in a tritrophic system. Ann Entomol Soc Am 108:501–512

Khanamani M, Fathipour Y, Talebi AA, Mehrabadi M (2017a) Linking pollen quality and performance of *Neoseiulus californicus* (Acari: Phytoseiidae) in two-spotted spider mite management programmes. Pest Manag Sci 73:452–461

Khanamani M, Fathipour Y, Talebi AA, Mehrabadi M (2017b) Evaluation of different artificial diets for rearing the predatory mite *Neoseiulus californicus* (Acari: Phytoseiidae): diet-dependent life table studies. Acarologia 57(2):407–419

Khanamani M, Fathipour Y, Talebi AA, Mehrabadi M (2017c) Quantitative analysis of long-term mass rearing of *Neoseiulus californicus* (Acari: Phytoseiidae) on almond pollen. J Econ Entomol 110(4):1442–1450

Khanjani M, Ueckermann EA (2003) Four new tydeid species from Iran (Acari: Prostigmata). Zootaxa 182:1–11

Khanjani M, Raisi H, Izadi H (2011) A new record of the genus *Lasioerythraeus* Welborn & Young (Acari: Erythaeidae) from Iran and description of a new species. Int J Acarol 37:544–549

Khanjani M, Mirmoayedi A, Asali-Fayaz B, Sharifian T (2012) Two new larval species of the genus *Erythraeus* (*Erythraeus*) (Acari, Erythraeidae) from Iran. Zootaxa 3479:52–68

Khanjani M, Ghaedi B, Ueckermann EA (2013) New species of *Hypoaspis* Canestrini and *Coleolaelaps* Berlese (Mesostigmata: Laelapidae) associated with *Polyphylla olivieri* Castelnau (Coleoptera: Scarabaeidae) in Iran. Zootaxa 3745:469–478

Khanjani M, Nasrollahi S, Zamani AS, Fayaz BA (2014) *Cheylostigmaeus tarae* sp. nov. and *Stigmaeus delaramae* sp. nov. (Acari: Stigmaeidae) from Kurdistan, Iran. Zootaxa 3841 (3):364–378

Khaustov AA (2007) Two new species of mites of the family Acarophenacidae (Acari, Heterostigmata) from Crimea (Ukraine). Vest Zool 41:549–553

Khodayari S, Kamali K, Fathipour Y (2008) Biology, life table and predation of *Zetzellia mali* (Acari: Stigmaeidae) on *Tetranychus urticae* (Acari: Tetranychidae). Acarina 16(2):191–196

Khodayari S, Fathipour Y, Kamali K, Naseri B (2010) Seasonal activity of *Zetzellia mali* (Stigmaeidae) and its preys *Eotetranychus frosti* (Tetranychidae) and *Tydeus longisetosus* (Tydeidae) in unsprayed apple orchards of Maragheh, Northwestern of Iran. J Agric Sci Technol 12:549–558

Khodayari S, Fathipour Y, Kamali K (2013) Life history parameters of *Phytoseius plumifer* (Acari: Phytoseiidae) fed on corn pollen. Acarologia 53(2):185–189

Khodayari S, Fathipour Y, Sedaratian A (2016) Prey stage preference, switching and mutual interference of *Phytoseius plumifer* (Acari: Phytoseiidae) on *Tetranychus urticae* (Acari: Tetranychidae). Syst Appl Acarol 21:347–355

Khosravi Shastania S, Ziaaddinia M, Latifia M, Zahedi Golpayegani A (2014) Olfactory response of predatory mite *Phytoseiulus persimilis* to synthetic methyl salicylate. Arch Phytopathol Plant Protect 47(20):2442–2446

Knell RJ, Webberley KM (2004) Sexually transmitted diseases of insects: distribution, evolution, ecology and host behaviour. Biol Rev 79:557–581

Kouhjani Gorji M, Fathipour Y, Kamali K (2009) The effect of temperature on the functional response and prey consumption of *Phytoseius plumifer* (Acari: Phytoseiidae) on the two-spotted spider mite. Acarina 17(2):231–237

Kouhjani Gorji M, Fathipour Y, Kamali K (2012) Life table parameters of *Phytoseius plumifer* (Phytoseiidae) fed on two-spotted spider mite at different constant temperatures. Int J Acarol 38 (5):377–385

Krant GW (2018) Allogynaspis flechtmanni, a new genus and species of the subfamily Macrochelinae (Acari: Mesostigmata: Macrochelidae) from southeastern Brazil, with comments on cheliceral dentition, reproductive strategies, and postepigynal platelets. Zootaxa 4455:150–160

Krantz GW, Walter DE (eds) (2009) A manual of acarology, 3rd edn. Texas, Texas Tech University Press

Leicht W (1993) Imidacloprid: a chloronicotinyl insecticide. Pestic Outlook 4:17–24

Lill JT, Marquis RJ, Riclefs RE (2002) Host plant influence parasitism of forest caterpillars. Nature 417:170–173

Lindquist EE (1983) Some thoughts on the potential for use of mites in biological control, including a modified concept of "parasitoids". In: Hoy MA, Cunningham GL, Knutson L (eds) Biological control of pests by mites. University of California Division of Agriculture and Natural Resources, Special Publ, 3304:12–20

Lindquist EE (1985) Discovery of sporothecae in adult female *Trochometridium* Cross, with notes on analogous structures in *Siteroptes* Amerling (Acari: Heterostigmata). Exp Appl Acarol 1:73–85

Lindquist EE, Krantz GW, Walter DE (2009) Order Mesostigmata. In: Krantz GW, Walter DE (eds) A manual and acarology, 3rd edn. Texas Tech University Press, Texas

Loghmani A, Hajiqanbar H, Talebi AA (2014) An illustrated key to world species of the mite family Trochometridiidae (Acari: Prostigmata), with description of a new species and new insect host records. Can Entomol 146:471–448

Lord FT (1949) The influence of spray programs on the fauna of apple orchards in Nova Scotia. III. Mites and their predators. Can Entomol 81:202–230

Lord FT, MacPhee AW (1953) The influence of spray programs on the fauna of apple orchards in Nova Scotia. VI. Low temperatures and the natural control of the oystershell scale, *Lepidosaphes ulmi* (L.) (Homoptera: Coccidae). Can Entomol 85:282–291

Lotfollahi P, Irani-Nejad KH, Khanjani M, Moghadam M, De Lillo E (2013) Eriophyoid mites (Acari: Prostigmata: Eriophyidae) associated with Compositae in Iran. Zootaxa 3664 (3):349–360

Lotfollahi P, Tajaddod S, Houshyari F (2016) First record of the genus *Adactylidium* Cross, 1965 (Acari: Trombidiformes: Acarophenacidae) from Asia. In: 22th Iranian Plant Protection Congress, University of Tehran, Karaj, 28–31 August 2016

Louni M, Jafari S, Shakarami J (2014a) Life table parameters of *Phytoseius plumifer* (Phytoseiidae) fed on *Rhyncaphytoptus ficifoliae* (Diptilomiopidae) under laboratory conditions. Syst Appl Acarol 19(3):275–282

Louni M, Jafari S, Shakarami J (2014b) Prey consumption of *Phytoseius plumifer* (Phytoseiidae) fed on *Rhyncaphytoptus ficifoliae* (Diptilomiopidae) in laboratory condition. In: 21st Iranian plant protection Congress, Urmia University, Urmia, 23–26 August 2014

Lucas E (2005) Intraguild predation among aphidophagous predators. Eur J Entomol 102:351–364

Madadi H, Enkegaard A, Brødsgaard HF, Kharrazi-Pakdel A, Ashouri A, Mohaghegh-Neishabouri J (2008) *Orius albidipennis* (Heteroptera: Anthocoridae): Intraguild predation of and prey preference for *Neoseiulus cucumeris* (Acari: Phytoseiidae) on different host plants. Entomol Fen 19:32–40

Madadi H, Enkegaard A, Brodsgaard HF, Kharazi-Pakdel A, Ashouri A, Mohaghegh-Neishabouri J (2009) Inetraction between *Orius albidipennis* (Heteroptera: Anthocoridae) and *Neoseiulus cucumeris* (Acari: Phytoseiidae): effects of host plants under microcosm condition. Biol Control 50:137–142

Maeda T, Takabayashi J, Yano S, Takafuji A (2001) Variation in the olfactory response of 13 populations of the predatory mite *Amblyseius womersleyi* to *Tetranychus urticae*-infested plant volatiles (Acari: Phytoseiidae, Tetranychidae). Exp Appl Acarol 25(1):55–64

Magalhaes S, Janssen A, Montserrat M, Sabelis MW (2005) Prey attack and predators defend: counterattacking prey trigger parental care in predators. Proc R Soc B Biol Sci 272:1929–1933

Magowski WL (1994) Discovery of the first representative of the mite subcohort Heterostigmata (Arachnida: Acari) in the Mesozoic Siberian amber. Acarologia 35:229–241

Mahjoori M, Hajizadeh J, Abbasii Mozhdehi MR (2015) A checklist and a key for the phytoseiid and blattisociid mites (Acari: Phytoseioidea) associated with olive orchards in Guilan Province Iran. Entomofauna 36(8):97–108

Mąkol J, Wohltmann A (2012) An annotated checklist of terrestrial Parasitengona (Actinotrichida: Prostigmata) of the world, excluding Trombiculidae and Walchiidae. Ann Zool 62(3):359–562

Mąkol J, Wohltmann A (2013) Corrections and additions to the checklist of terrestrial Parasitengona (Actinotrichida: Prostigmata) of the world, excluding Trombiculidae and Walchiidae. Ann Zool 63(1):15–27

Maleknia B, Fathipour Y, Soufbaf M (2014a) Effects of prey density and plant leaf structure on predation behavior of *Neoseiulus barkeri* (Acari: Phytoseiidae). In: 21th Iranian plant protection Congress, Urmia University Urmia, 23–26 August 2014

Maleknia B, Zahedi-Golpayegani A, Saboori A, Mohammadi H (2014b) Olfactory responses of the predatory mite *Phytoseiulus persimilis* (Acari: Phytoseiidae) to rose leaves: starvation and previous host plant experience. Persian J Acarol 3(1):77–90

Maleknia B, Fathipour Y, Soufbaf M (2016) Intraguild predation among three phytoseiid species *Neoseiulus barkeri*, *Phytoseiulus persimilis* and *Amblyseius swirskii*. Syst Appl Acarol 21 (4):417–426

Maroufpoor M, Ghoosta Y, Pourmirza AA, Lotfalizadeh H (2016a) The effects of selected acaricides on life table parameters of the predatory mite, *Neoseiulus californicus*, fed on European red mite. NW J Zool 12(1):1–7

Maroufpoor M, Ghoosta Y, Pourmirza AA (2016b) Toxicity of Diazinon and Acetamipridon on life table parameters of the predatory mite, *Neoseiulus californicus* (Acari: Phytoseiidae) when fed on the European red mite *Panonychus ulmi* (Koch). Egypt J Biol Pest Control 26(1):15–19

McMurtry JA, Croft BA (1997) Life-styles of phytoseiid mites and their roles in biological control. Annu Rev Entomol 42:291–321

McMurtry JA, Scriven GT (1971) Predation by *Amblyseius limonicus* on *Oligonychus punicae*: effects of initial predator-prey ratios and prey distribution. Ann Entomol Soc Am 64:219–224

McMurtry JA, De Moraes GJ, Sourassou NF (2013) Revision of the lifestyles of phytoseiid mites (Acari: Phytoseiidae). Syst Appl Acarol 18:297–320

McMurtry JA, Sourassou NF, Demite PR (2015) Mesostigmata as biological control agents, with emphasis on Rhodacaroidea and Parasitoidea. In: Carrillo D, de Moraes GJ, Peña JE (eds)

Prospects for biological control of plant feeding mites and other harmful organisms. Springer International, Berlin

Mehrnejad MR, Ueckermann EA (2001) Mites (Arthropoda: Acari) associated with pistachio trees (Anacardiacae) in Iran (I). Syst Appl Acarol Spec Publ 6:1–12

Meszaros A, Tixier MS, Cheval B, Barbar Z, Kreiter S (2007) Cannibalism and intraguild predation in *Typhlodromus exhilaratus* and *T. phialatus* (Acari: Phytoseiidae) under laboratory conditions. Exp Appl Acarol 41:37–43

Mirfakhraii Sh (1994) Faunistic study on house mites and biology of important species in Urmieh. Thesis, Tarbiat Modares University

Mizell RF, Sconyers MC (1992) Toxicity of imidaclopride to selected arthropod predators in laboratory. Fla Entomol 75:277–280

Moghadasi M, Allahyari H (2017) Effect of Prey and pollen on interactions between *Typhlodromus bagdasarjani* and *Phytoseiulus persimilis* (Acari: Phytoseiidae) on cucumber (Cucurbitaceae). Can Entomol 149:581–591

Moghadasi M, Allahyari H (2018) Effect of different predator: prey release ratios of *Phytoseiulus persimilis* and *Typhlodromus bagdasarjani* (Acari: Phytoseiidae) on reduction of *Tetranychus urticae* (Acari: Tetranychidae) on cucumber under microcosm conditions. J Agric Sci Technol 20:509–519

Moghadasi M, Hajizadeh H, Saboori A, Nowzari J (2011) Life table parameters of *Phytoseius plumifer* (Canestrini & Fanzago) (Acari: Phytoseiidae) feeding on *Tetranychus urticae* Koch (Acari: Tetranychidae). First Persian Congress of Acarology, Internatoional Center for science, High Technology & Enviromental Sciences, Kerman, 22–23 December 2011

Moghadasi M, Saboori A, Allahyari H, Zahedi Golpayegani A (2014a) Prey stages preference of different stages of *Typhlodromus bagdasarjani* (Acari: Phytoseiidae) to *Tetranychus urticae* (Acari: Tetranychidae) on rose. Persian J Acarol 2(3):531–538

Moghadasi M, Saboori A, Allahyari H, Zahedi Golpayegani A (2014b) Life table and predation capacity of *Typhlodromus bagdasarjani* (Acari: Phytoseiidae) feeding on *Tetranychus urticae* on rose. Int J Acarol 40:501–508

Mohammadi H, Saboori A, Zahedi Golpayegani A (2012) *Phytoseiulus persimilis* (Acari: Phytoseiidae) attraction towards infested plant volatiles intensifies by experience. In: 20th Iranian Plant protection Congress, Shiraz University, Shiraz, 26–29 August 2012

Montserrat M, Bas C, Magalhães S (2007) Predators induct egg retention in prey. Behav Ecol 150:699–705

Morales-Ramos JA, Rojas MG (2014) A modular cage system design for continuous medium to large scale in vivo rearing of predatory mites (Acari:Phytoseiidae). Psyche 2014:1–8

Mortazavi A, Hajiqanbar H (2012) A new podapolipid species (Acari) on *Scarabaeus* (*Scarabaeus*) *acuticollis* (Insecta: Coleotera: Scarabaeidae) from Iran. J Parasitol 98:746–753

Mortazavi A, Hajiqanbar H, Saboori A (2011) A new species of the family Trochometridiidae (Acari: Heterostigmatina) associated with *Paulusiella* sp. (Coleoptera: Elateridae) from Iran. Zootaxa 2746:57–68

Mortazavi N, Fathipour Y, Talebi AA (2018) The efficiency of *Amblyseius swirskii* in control of *Tetranychus urticae* and *Trialeurodes vaporariorum* is affected by various factors. Bull Entomol Res:1–11. https://doi.org/10.1017/S0007485318000640

Nadeali T, Zahedi Golpayegani A, Saboori A (2012a) Prey *Tetranychus urticae* Koch (Acari: Tetranychidae) density and its effect on the predator *Neoseiulus californicus* (McGregor) (Acari: Phytoseiidae) developmental time. In: 20th Iranian Plant Protection Congress, Shiraz University, Shiraz, 26–29 August 2012

Nadeali T, Zahedi Golpayegani A, Saboori A (2012b) Prey *Tetranychus urticae* Koch (Acari: Tetranychidae) density and its effect on the predator *Phytoseiulus persimilis* Athias-Henriot (Acari: Phytoseiidae) developmental time. 20th Iranian Plant Protection Congress, Shiraz University, Shiraz, 26–29 August 2012

Nadimi A, Kamali K, Arbabi M, Abdoli F (2008) Selectivity of three miticides to spider mite predator, *Phytoseilus persimilis* (Acari: Phytoseiidae) under laboratory conditions. Munis Entomol Zool 3(2):556–567

Nadimi A, Kamali K, Arbabi M, Abdoli F (2009) Selectivity of three miticides to spider mite predator, *Phytoseius plumifer* (Acari: Phytoseiidae) under laboratory conditions. Agric Sci China 8:326–331

Nauen R, Ebbinghaus-Kintscher U, Schmuck R (2001) Toxicity and nicotinic acetylcholine receptor interaction of imidacloprid and its metabolites in *Apis mellifera* (Hymenoptera: Apidae). Pest Manag Sci 57:577–586

Nemati A, Gwiazdowicz DJ (2016) A new genus and species of Laelapidae from Iran with notes on *Gymnolaelaps* Berlese and *Laelaspisella* Marais & Loots (Acari, Mesostigmata). Zookeys 549:23–49

Nemati A, Riahi E, Khalili-Moghadam A, Gwiazdowicz DJ (2018) A catalogue of the Iranian Mesostigmata (Acari): additions and updates of the previous catalogue. Persian J Acarol 7:115–191

Nguyen DT, Vangansbeke D, Lu X, De Clercq P (2013) Development and reproduction of the predatory mite *Amblyseius swirskii* on artificial diets. BioControl 58:369–377

Nguyen DT, Vangansbeke D, De Clercq P (2015) Performance of four species of phytoseiid mites on artificial and natural diets. Biol Control 80:56–62

Noei J, Hajizadeh J, Salehi L, Ostovan H, Faraji F (2007) Stigmaeid mites associated with stored rice in northern Iran (Acari: Stigmaeidae). Int J Acarol 32(2):153–156

Noei J, Rabieh MM, Saboori A (2015) First record of *Momorangia* (Acari: Erythraeidae) from Asia with description of a new species. Syst Appl Acarol 20(7):789–798

Noia M, Borges I, Soares AO (2008) Intraguild predation between the aphidophagous ladybird beetles *Harmonia axyridis* and *Coccinella undecimpunctata* (Coleoptera: Coccinellidae): the role of intra and extraguild prey densities. Biol Control 46:140–146

Noii S, Talebi K, Saboori A, Allahyari H, Sabahi Q, Ashouri A (2008) Study on the side-effects of three pesticides on the predatory mite, *Phytoseius plumifer* (Canestrini & Fanzago) (Acari: Phytoseiidae) under laboratory conditions. Pestic Benefits Org 35:141–151

Norton AP, English-Loeb G, Gadoury D, Seem RC (2000) Mycophagous mites and foliar pathogens: leaf domatia mediate tritrophic interactions in grapes. Ecology 81(2):490–499

Notghi Moghadam BA, Hajizadeh J, Jalali Sendi J, Rafati Fard M (2010) Influence of three diets on development and oviposition of *Amblyseius herbicolus* under laboratory conditions. J Entomol Soc Iran 30(1):51–68

Notghi Moghadam BA, Hajizadeh J, Jalali Sendi J, Rafati-fard M (2014) Biology and prey preference of predatory mite *Amblyseius herbicolus* Chant (Acari: Phytoseiidae) feeding on *Tetranychus urticae*. In: 3rd Integrated Pest Management Conference (IPMC), Shahid Bahonar University of Kerman, Kerman, 21–22 January 2014

Nyrop JP (1988) Spatial dynamics of an acarine predatory-prey system: *Typhlodromus pyri* (Acari: Phytoseiidae) preying on *Panonychus ulmi* (Acari: Tetranychidae). Environ Entomol 17:1019–1031

Oatman ER, McMurtry JA (1966) Biological control of the two-spotted spider mite on strawberry in southern California. J Econ Entomol 59:433–439

Oatman ER, McMurtry JA, Gilstrap FE, Oatman ER, McMurtry JA, Gilstrap FE, Voth V (1977) Effect of *Amblyseius californicus*, *Phytoseiulus persimilis* and *Typhlodromus occidentalis* on the two spotted spider mite on strawberry in southern California. J Econ Entomol 70:45–47

Ode PJ, Berenbaum MR, Zanger AR, Hardy ICW (2004) Host plant, chemistry and the polyembryonic parasitoid *Copidosoma sosares*: indirect effects in a tritrophic interaction. Oikos 104:388–400

Ostovan H (1997) Mites associated with elm bark beetle (Col.: Scolytidae) and biocenotic aspects of *Pyemotes scolyti* (Oud.) (Acari: Pyemotidae) as biocontrol agent for *Scolytus multistriatus* (Marsham). Dissertation, Islamic Azad University

Ostovan H, Kamali K (1995) Some snout mites (Acari: Bdellidae) from Iran and a key for their identification. J Agric Sci 1:29–43

Ostovan H, Kamali K (1996) First record and biology of *Pyemotes scolyti* (Oud.) (Acari: Pyemotidae) an important parasite of elm bark beetle *Scolytus multistriatus* in Tehran parks. J Agric Sci 2(7/8):5–14

Ostovan H, Kamali K (1997) Biodiversity of mites (Acari) associated with elm bark beetle *Scolytus multistriatus* (Marsh.) (Coleoptera: Scolytidae) in parks of Tehran. J Agric Sci 3(11&12):23–67

Ostovan H, Kamali K (2001) Bioecology of *Pyemotes scolyti* (Acari: Pyemotidae) as biocontrol agent for *Scolytus multistriatus* (Marsham) (Col.: Scolytidae). In: Proceeding of the 4th Asia Pacific conference of entomology, Kuala Lumpur

Ostovan H, Saboori A (1999) Some mites of the families Podapolipidae, Acarophenacidae and Podocinidae in Iran. J Agric Sci Azad Univ 5:81–90

Paktinat-Saeej S, Bagheri M, Saboori A, Ahaniazad M (2015a) Two new Bdellidae (Trombidiformes: Bdelloidea) from Iran and the status of *Neobiscirus* Gomelauri, 1963. Zootaxa 4013(4):519–530

Paktinat-Saeej S, Bagheri M, Saboori A, Seilsepour N, Ueckermann EA (2015b) A new snout mite, *Spinibdella tabarii* sp. nov. (Trombidiformes: Bdellidae) from Iran, with a summary of *Spinibdella* distributions worldwide. Syst Appl Acarol 20(6):693–706

Pandey S, Singh R (1999) Host size induced variation in progeny sex ratio of an aphid parasitoid *Lysiphlebia mirzai*. Entomol Exp Appl 90:61–67

Park HH, Shipp L, Buitenhuis R (2010) Predation, development, and oviposition by the predatory mite *Amblyseius swirkii* (Acari: Phytoseiidae) on tomato russet mite (Acari: Eriophyidae). J Econ Entomol 103(3):563–569

Pasandideh A, Talebi AA, Hajiqanbar H, Tazerouni Z (2015) Host stage preference and age-specific functional response of *Praon volucre* (Hymenoptera: Braconidae, Aphidiinae) a parasitoid of *Acyrthosiphon pisum* (Hemiptera: Aphididae). J Crop Protect 4(4):563–575

Pemberton RW, Turner CE (1989) Occurrence of predatory and fungivorous mites in leaf domatia. Am J Bot 76:105–112

Pervez A, Omkar (2006) Ecology and biological control application of multicoloured Asian ladybird, *Harmonia axyridis*: a review. Biocontrol Sci Tech 16(2):111–128

Pešić V, Smit H, Saboori A (2014) Checklist of the water mites (Acari, Hydrachnidia) of Iran: second supplement and description of one new species. Ecol Mont 1(1):30–48

Poletti M, Maia AHN, Omoto C (2007) Toxicity of neonicotinoid insecticides to *Neoseiulus californicus* and *Phytoseiulus macropilis* (Acari: Phytoseiidae) and their impact on functional response to *Tetranychus urticae* (Acari: Tetranychidae). Biol Control 40:30–36

Polis GA, Myers CA, Holt RD (1989) The ecology and evolution of intraguild predation: potential competitors that eat each other. Annu Rev Ecol Syst 20:297–330

Provost C, Lucas E, Coderre D (2006) Prey preference of *Hyaliodes vitripennis* as an intraguild predator: active predator choice or passive selection? Biol Control 37:148–154

Putman WL (1970) Life history and behavior of *Balaustium putmani* (Acarina: Erythraeidae). Ann Entomol Soc Am 63(1):76–81

Rack G (1959) *Acarophenax dermestidarum* sp. n. (Acarina, Pyemotidae), ein Eiparasit von Dermestes-Arten. Z. f. Parasite 19:411–431

Radjabi Gh (1969) Some bioecological studies on a newly known mite, parasitic on xylophagous insects in Iran. In: Proceeding of the 2nd Iranian plant protection congress, Tehran University, Tehran

Rady GH (1992) New genus and species of Acarophenacidae (Acari: Tarsonemina) from Egypt. Ann Agric Sci 30:1129–1135

Rafizadeh Afshar F, Latifi M (2017) Functional response and predation rate of *Amblyseius swirskii* (Acari: Phytoseiidae) at three constant temperatures. Persian J Acarol 6(4):299–314

Rahiminejad V, Hajiqanbar H (2015) A new species of the genus *Acarophenax* (Acari: Heterostigmatina: Acarophenacidae) associated with *Sphindus* sp. (Coleoptera: Sphindidae) from Iran. Persian J Acarol 4(3):277–286

Rahmani H, Hosseini MS (2012) Social learning in the nonclonal predatory mite, *Neoseiulus californicus* (Acari: Phytoseiidae), decreases cost of forgoing activities. In: 20th Iranian plant protection Congress. Shiraz University, Shiraz, 26–29 August 2012

Rahmani H, Fathipour Y, Kamali K (2009a) Life history and population growth parameters of *Neoseiulus californicus* (Acari: Phytoseiidae) fed on *Thrips tabaci* (Thysanoptera: Thripidae) in laboratory condition. Syst Appl Acarol 14:91–100

Rahmani H, Hoffmann D, Walzer A, Schausberger P (2009b) Adaptive learning in the foraging behavior of the predatory mite *Phytoseiulus persimilis*. Behav Ecol 20:946–950

Rahmani H, Fathipour Y, Kamali K (2010) Spatial distribution and seasonal activity of *Panonychus ulmi* (Acari: Tetranychidae) and its predator *Zetzellia mali* (Acari: Stigmaeidae) in Apple Orchards of Zanjan, Iran. J Agric Sci Technol 12(2):155–165

Rahmani H, Saboori A, Hajiqanbar H (2012) Acarology. University of Zanjan Press, Zanjan. (in Farsi)

Rahmani H, Daneshmandi A, Walzer A (2015) Intraguild interactions among three spider mite predators: predation preference and effects on juvenile development and oviposition. Exp Appl Acarol 67:493–505

Rasmy AH, El-Banhawy EM (1974) Behaviour and bionomics of the predatory mite, *Phytoseius plumifer* (Acarina: Phytoseiidae) as affected by physical surface features of host plant. Entomophaga 19:255–257

Raut SK, Panigrahi A (1991) The mite *Fuscuropoda marginata* (C.L. Koch) for the control of pest slugs *Laevicaulis alte* (Ferussac). Mod Acarol 12:683–687

Reis PR, Sousa EO, Teodoro AV, Neto MP (2003) Effect of prey densities on the functional and numerical response of two species of predaceous mites (Acari: Phytoseiidae). Neotrop Entomol 32:461–467

Rezaei M, Askarieh S (2016) Effect of different pollen grains on life table parameters of *Neoseiulus barkeri* (Acari: Phytoseiidae). Persian J Acarol 5(3):239–253

Rezaie M, Saboori A, Baniameri V (2018) The effect of strawberry cultivars, infested with *Tetranychus urticae* (Acari: Tetranychidae), on the olfactory response of the predatory mite *Neoseiulus californicus* (Acari: Phytoseiidae). J Berry Res 8:71–80

Rhodes EM, Liburd OE, Kelts C, Rondon SI, Francis RR (2006) Comparison of single and combination treatments of *Phytoseiulus persimilis*, *Neoseiulus californicus*, and Acramite (bifenazate) for control of two-spotted spider mites in strawberries. Exp Appl Acarol 39:213–225

Rhule EL, Majerus ME, Jiggins FM, Ware RL (2010) Potential role of the sexually transmitted mite *Coccipolipus hippodamiae* in controlling populations of the invasive ladybird *Harmonia axyridis*. Biol Control 53:243–247

Riahi E, Fathipour Y, Talebi AA, Mehrabadi M (2016) Pollen quality and predator viability: life table of *Typhlodromus bagdasarjani* on seven different plant pollens and two-spotted spider mite. Syst Appl Acarol 21:1399–1412

Riahi E, Fathipour Y, Talebi AA, Mehrabadi M (2017a) Linking life table and consumption rate of *Amblyseius swirskii* (Acari:Phytoseiidae) in presence and absence of different plant pollens. Ann Entomol Soc Am 110:244–253

Riahi E, Fathipour Y, Talebi AA, Mehrabadi M (2017b) Attempt to develop cost-effective rearing of *Amblyseius swirskii* (Acari: Phytoseiidae): assessment of different artificial diets. J Econ Entomol 110:1525–1532

Riahi E, Fathipour Y, Talebi AA, Mehrabadi M (2018a) Interactions among food diets and rearing substrates sffect development and population growth rate of *Typhlodromus bagdasarjani*. Syst Appl Acarol 23(9):1845–1856

Riahi E, Fathipour Y, Talebi AA, Mehrabadi M (2018b) Factitious prey and artificial diets: do they all have the potential to facilitate rearing of *Typhlodromus bagdasarjani* (Acari: Phytoseiidae)? Int J Acarol 44(2–3):121–128

Riddick EW, Cottrell TE, Kidd KA (2009) Natural enemies of the Coccinellidae: parasites, pathogens, and parasitoids. Biol Control 51:306–312

Sabelis MW, Dicke M (1985) Long range dispersal and searching behaviour. In: Helle W, Sbelis MW (eds) Spider mites, their biology, natural enemies and control. Elsevier, Amsterdam

Saboori A, Atamehr A (1999) A new larval *Leptus* (Acari: Erythraeidae) from Iran. Syst Appl Acarol 4:159–163

Saboori A, Babolmorad M (2000) A new larval mite (Acari: Erythraeidae) ectoparasitic on *Monosteira unicostata* (Hemiptera: Tingidae) from Iran. Syst Appl Acarol 5:119–123

Saboori A, Hakimitabar M (2013) A checklist of the Trombidioidea (Acari: Prostigmata) of Iran. J Crop Protect 2(1):33–42

Saboori A, Kamali K (1999) Biology of *Allothrombium shirazicum* Zhang (Acari: Trombidiidae) in Garmsar, Semnan Province, Iran. Syst Appl Acarol 4:199–200

Saboori A, Lazarboni H (2008) A new genus and species of Trombidiidae (Acari: Trombidioidea) described from larvae ectoparasitic on *Cicadatra ochreata* Melichar (Homoptera: Cicadidae) from Iran. Zootaxa 1852:50–58

Saboori A, Ostovan H (2000) A new species of the genus *Leptus* Latreille, 1796 (Acari: Erythraeidae) ectoparasitic on sun pest, *Eurygaster integriceps* Puton (Hemiptera: Scutelleridae) from Iran. Syst Appl Acarol 5:143–147

Saboori A, Zhang Z-Q (1996) Biology of *Allothrombium pulvinum* Ewing (Acari: Trombidiidae) in West Mazandaran, Iran. Exp Appl Acarol 20:137–142

Saboori A, Hajiqanbar H, Haddad Irani-nejad K (2003a) A new genus and species of mite (Acari: Trombidiidae) ectoparasitic on thrips in Iran. Int J Acarol 29(2):127–132

Saboori A, Hosseini M, Hatami B (2003b) Preference of adults of *Allothrombium pulvinum* Ewing (Acari; Trombidiidae) for eggs of *Planococcus citri* (Risso) and *Pulvinaria aurantii* (Cockerell) on citrus leaves in the laboratory. Syst Appl Acarol 8:49–54

Saboori A, Bagheri M, Haddad Irani-nejad K, Kamali K, Khanjani M (2006) A new genus and species of Trombidiidae (Acari: Trombidioidea) described from larvae ectoparasitic on *Hyalopterus pruni* (Aphidoidea, Aphididae) from Iran. Syst Appl Acarol 11:247–252

Saboori A, Hosseini A, Asadi M (2007) Acari of Iran, Vol. 1, Parasitengone mites. University of Tehran Press, Tehran

Saboori A, Wohltmann A, Hakimitabar M (2010) A new family of trombidioid mites (Acari: Prostigmata) from Iran. Zootaxa 2611:16–30

Saboori A, Hajiqanbar H, Hakimitabar M (2011) First Iranian species of *Neosilphitrombium* (Acari: Prostigmata: Neothrombiidae) with a key to world species. Zootaxa 2738:60–68

Saboori A, Wohltmann A, Hakimitabar M, Shirvani A (2013) A new species of the genus *Achaemenothrombium* (Acari: Achaemenothrombiidae) from Iran. Zootaxa 3694:143–152

Sadeghi H, Łaniecka I, Kaźmierski A (2012) Tydeoid mites (Acari: Triophtydeidae, Iolinidae, Tydeidae) of Razavi Khorasan Province, Iran, with description of three new species. Ann Zool 62(1):99–114

Saemi S, Rahmani H, Kavousi A (2014) Age-stage, two-sex life table and predation rate of the predatory mite *Phytoseiulus persimilis* (Acari: Phytoseiidae) feeding on *Tetranychus urticae* (Acari: Tetranychidae) eggs. In: 21th Iranian plant protection Congress, Urmia University, Urmia, 23–26 August 2014, p 936

Sakai AK, Allendorf FW, Holt JS et al (2001) The population biology of invasive species. Annu Rev Ecol Syst 32:305–332

Salarzehi S, Hajizadeh J, Ueckermann EA (2018) A new species of *Cheyletus* Latreille (Prostigmata: Cheyletidae) from Iran and a key to the Iranian species. Acarologia 58:640–646

Sanatgar E, Vafaei Shoushtari R, Zamani AA, Arbabi M, Soleyman Nejadian E (2011) Effect of frequent application of Hexythiazox on predatory mite *Phytoseiulus persimilis* Athias – Henriot (Acari: Phytoseiidae). Acad J Entomol 4(3):94–101

Sanatgar E, Vafaei Shoushtari R, Zamani AA, Arbabi M, Soleyman Nejadian E (2012) Effect of frequent application of bifenazate on biology and biological parameters of predatory mite *Phytoseiulus persimilis* Athias-Henriot (Acari: Phytoseiidae). In: 20th Iranian plant protection Congress, University of Shiraz, Shiraz, 26–29 August 2012

Sangsoo K, Sangi S, Jongdae P, Seongon K, Doik K (2005) Effects of selected pesticides on the predatory mite, *Amblyseius cucumeris* (Acari: Phytoseiidae). J Entomol Sci 40:107–114

Santos PE, Phillips J, Whitford WG (1981) The role of mites and nematodes in early stages of buried litter decomposition in a desert. Ecology 62:664–669

Schausberger P (2003) Cannibalism among phytoseiid mites: a review. Exp Appl Acarol 29:173–191

Schausberger P, Croft BA (2000) Cannibalism and intraguild predation among phytoseiid mites: are aggressiveness and prey preference related to diet specialization? Exp Appl Acarol 24:709–725

Schausberger P, Walzer A (2001) Combined versus single species release of predaceous mites: predator-predator interactions and pest suppression. Biol Control 20:269–278

Schausberger P, Walzer A, Hoffmann D, Rahmani H (2010) Food imprinting revisited: early learning in foraging predatory mites. Behaviour 147:883–897

Schwarz HH, Müller JK (1992) The dispersal behaviour of the phoretic mite *Poecilochirus carabi* (Mesostigmata, Parasitidae): adaptation to the breeding biology of its carrier *Necrophorus vespilloides* (Coleoptera, Silphidae). Oecologia 89:487–493

Seeman OD, Nahrung HF (2013) Two new species of *Chrysomelobia* Regenfuss, 1968 (Acariformes: Podapolipidae) from *Paropsis charybdis* Stal (Coleoptera: Chrysomelidae). Syst Parasitol 86:257–270

Seidpisheh E, Rahmani H, Kavousi A (2016) Biological and feeding attributes of the native predatory mite, *Typhlodromus bagdasarjani* (Acari: Phytoseiidae) under Azadirachtin and Spirodiclofen treatments. In: Proceeding of biocontrol meeting. Ferdowsi University of Mashhad, Mashhad, 2–3 February 2016

Seiedy M, Saboori A, Kamali K, Kharazi Pakdel A (2009) Mites (Acari) found in flour mills in the Karaj region of Iran. Syst Appl Acarol 14:191–196

Seiedy M, Saboori A, Allahyari H (2012a) Preliminary observations on mites found in domesticated animal food factories in Karaj, Iran. Persian J Acarol 1(2):119–125

Seiedy M, Saboori A, Allahyari H, Talaei-Hassanloui R, Tork M (2012b) Functional response of *Phytoseiulus persimilis* (Acari: Phytoseiidae) on untreated and Beauveria bassiana- treated adults of *Tetranychus urticae* (Acari: Tetranychidae). J Insect Behav 25:543–553

Seiedy M, Saboori A, Zahedi-Golpayegani A (2013) Olfactory response of *Phytoseiulus persimilis* (Acari: Phytoseiidae) to untreated and *Beauveria bassiana*- treated *Tetranychus urticae* (Acari: Tetranychidae). Exp Appl Acarol 60:219–227

Sepasgosarian H (1997) The world genera and species of the family Tydeidae (Actinedida: Acaridia). J Entomol Soc Iran 7:1–54

Shams MH, Kazemi S, Saboori A (2016) A new species of the genus *Blattisocius* Keegan (Acari: Mesostigmata: Blattisociidae) from Iran. Syst Appl Acarol 21(1):139–145

Sharifian A, Zahedi Golpayegani A, Ahadiyat A, Saboori A (2014) The effect of patch size, previous diet and their interaction on the predation rate of *Amblyseius swirskii* Athias-Henriot and *Phytoseiulus persimilis* Athias-Henriot on *Tetranychus urticae* Koch eggs and larvae. In: 21th Iranian plant protection Congress, Urmia University Urmia, 23–26 August 2014

Shemshadian A, Talaei Hassanloei R, Hakimitabar M (2012) Report of *Pronematus ubiquitus* (Acari: Iolinidae) as a potential predator for European red mite *Panonychus ulmi* (Acari: Tetranychidae) in western Azarbayjan. In: 20th Iranian plant protection Congress, University of Shiraz, Shiraz, 25–28 August 2012

Shimoda T, Ozawa R, Sano K, Yano E, Takabayashi J (2005) The involvement of volatile infochemicals from spider mites and from food-plants in prey location of the generalist predatory mite *Neoseiulus californicus*. J Chem Ecol 31:2019–2032

Shirdel D, Arbabi M (2012a) Functional response of the predatory mite, *Euseius finlandicus* (Oudemans) (Acari: Phytoseiidae) to European red mite. In: 20th Iranian plant protection Congress, University of Shiraz, Shiraz, 25–28 August 2012

Shirdel D, Arbabi M (2012b) Functional response of the predatory mite, *Typhlodromus bagdasarjani* (Wainstein and Artunjan) (Acari: Phytoseiidae) to European red mite. In: 20th Iranian plant protection Congress, University of Shiraz, Shiraz, 25–28 August 2012

Shirdel D, Kamali K, Ostovan H, Arbabi M (2002) Comparison of rearing methods of two predatory mites *Typhlodromus kettanehi* Dosse and *Euseius finlandicus* (Oudemans) (Acari: Phytoseiidae). Appl Entomol Phytopathol:101–120

Shirdel D, Kamali K, Ostovan H, Arbabi M (2004) Functional response of the predatory mite, *Typhlodromus kettanehi* Dosse (Acari:Phytoseiidae) on two-spotted spider mite. In: 16th Iranian Plant protection congress, University of Tabriz, Tabriz, 28–1 August–September 2004

Shirdel D, Kamali K, Ostovan H, Arbabi M (2006) Functional response of the predatory mite, *Euseius finlandicus* (Oudemans) (Acari:Phytoseiidae) on two-spotted spider mite. In: 17th Iranian plant protection Congress, University of Tehran, Karaj, 2–5 September 2006

Skoracka A, Smith L, Oldfield G, Cristofaro M, Amrine JW (2010) Host-plant specificity and specialization in eriophyoid mites and their importance for the use of eriophyoid mites as biocontrol agents of weeds. Exp Appl Acarol 51:92–113

Skvarla MJ, Fisher JR, Dowling APG (2014) A review of Cunaxidae (Acariformes, Trombidiformes): histories and diagnoses of subfamilies and genera, keys to world species, and some new locality records. ZooKeys 418:1–103

Smith L, de Lillo E, Amrine JW Jr (2009) Effectiveness of eriophyid mites for biological control of weedy plants and challenges for future research. Exp Appl Acarol 51:115–149

Soleymani S, Hakimitabar M, Sciedy M (2016) Prey preference of predatory mite *Amblyseius swirskii* (Acari: Phytoseiidae) on *Tetranychus urticae* (Acari: Tetranychidae) and *Bemisia tabaci* (Hemiptera: Aleyrodidae). Biocontrol Sci Tech. https://doi.org/10.1080/09583157.2015.1133808

Solomon M (1949) The natural control of animal populations. J Anim Ecol 18:1–35

Soltaniyan A, Kheradmand K, Fathipour Y, Shirdel D (2018) Suitability of pollen from different plant species as alternative food sources for *Neoseiulus californicus* (Acari: Phytoseiidae) in comparison with a natural prey. J Econ Entomol 111(5):2046–2052

Soufbaf M, Fathipour Y, Zalucki MP, Hui C (2012) Importance of primary metabolites in canola in mediating interactions between a specialist leaf-feeding insect and its specialist solitary endoparasitoid. Arthropod Plant Interact 6:241–250

Stark JD, Banks JE (2003) Population-level effects of pesticides and other toxicants on arthropods. Annu Rev Entomol 48:505–519

Steinkraus DC, Cross EA (1993) Description and life history of *Acarophenax mahundai*, n. sp. (Acari: Tarsonemina: Acarophenacidae), an egg parasite of the lesser mealworm (Coleoptera: Tenebrionidae). Ann Entomol Soc Am 86:239–244

Strodl MA, Schausberger P (2012) Social familiarity modulates group living and foraging behaviour of juvenile predatory mites. Naturwissenschaften 99:303–311

Swirski E, Amitai S, Dorzia N (1967) Laboratory studies on the feeding, development and reproduction of the predaceous mites *Amblyseius rubini* Swirski & Amitai and *Amblyseius swirskii* Athias (Acarina: Phytoseiidae) on various kinds of food substances. Isr J Agric Res 17:101–119

Tajmiri P (2013) An introduction of the collected Prostigmatic (Acari: Trombidiformes) mites on raspberry shrubs in central area of Guilan Province. Plant Pests Res 3(1):1–9

Tashakor S, Hajiqanbar H, Saboori A, Hakimitabar M (2015) A new species of *Charletonia* from Iran (Acari, Erythraeidae). Spixiana 38(2):197–202

Thomas HQ, Zalom FG, Nicola NL (2011) Laboratory studies of *Blattisocius keegani* (Fox) (Acari: Ascidae) reared on eggs of navel orangeworm: potential for biological control. Bull Entomol Res 101:499–504

Ueckermann EA, Rastegar J, Saboori A, Ostovan H (2007) Some mites of the superfamily Bdelloidea (Acari: Prostigmata) of Karaj (Iran) with descriptions of two new species and redescription of *Bdellodes kazeruni*. Acarologia 47(3–4):127–138

van der Hoeven WAD, van Rijn PCJ (1990) Factors affecting the attack success of predatory mites on thrips larvae. Proc Exp Appl Entomol 1:25–30

Van Lenteren JC, Woets J (1988) Biological and integrated pest control in greenhouses. Annu Rev Entomol 33:239–269

Wade MR, Zalucki MP, Wratten SD, Robinson KA (2008) Conservation biological control of arthropods using artificial food sprays: current status and future challenges. Biol Control 45:185–199

Walter DE, Proctor H (1999) Mites: ecology evolution and behaviour. CAB International, Wallingford

Walter DE, Lindquist EE, Smith IM, Cook DR, Krantz GW (2009) Order Trombidiformes. In: Krantz GW, Walter DE (eds) A manual of acarology. Texas Tech University Press, Lubbock

Walzer A, Schausberger P (1999a) Predation preferences and discrimination between con- and heterospecific prey by the phytoseiid mites *Phytoseiulus persimilis* and *Neoseiulus californicus*. BioControl 43:469–478

Walzer A, Schausberger P (1999b) Cannibalism and inter- specific predation in the phytoseiid mites *Phytoseiulus persimilis* and *Neoseiulus californicus*: predation rates and effects on reproduction and juvenile development. BioControl 43:457–468

Walzer A, Schausberger P (2009) Non-consumptive effects of predatory mites on thrips and its host plant. Oikos 118:934–940

Walzer A, Schausberger P (2011) Threat-sensitive anti-intraguild predation behaviour: maternal strategies to reduce offspring predation risk in mites. Anim Behav 81:177–184

Walzer A, Blumel S, Schausberger P (2001) Population dynamics of interacting predatory mites, *Phytoseiulus persimilis* and *Neoseiulus californicus*, held on detached bean leaves. Exp Appl Acarol 25:731–743

Walzer A, Paulus HF, Schausberger P (2004) Ontogenetic shifts in intraguild predation on thrips by phytoseiid mites: the relevance of body size and diet specialization. Bull Entomol Res 94:577–584

Weintraub P, Palevsky E (2008) Evaluation of the predatory mite, *Neoseiulus californicus*, for spider mite control on greenhouse sweet pepper under hot arid field conditions. Exp Appl Acarol 45:29–37

Welbourn WC (1983) Potential use of trombidioid and erythraeoid mites as biological control agents of insect pests. In: Hoy MA, Cunningham GL, Knutson L (eds) Biological control of pests by mites. Agricultural Experiment Station, Division of Agriculture and Natural Resources, University of California, Berkeley

Wilcox BA, Murphy DD (1985) Conservation strategy: the effects of fragmentation on extinction. Am Nat 125(6):879–887

Xiao Y, Fadamiro HY (2010) Functional response and prey-stage preferences of three species of predacious mites (Acari:Phytoseiidae) on citrus red mite, *Panonychus citri* (Acari: Tetranychidae). Biol Control 53:345–352

Xiao Y, Osborne LS, Chen J, McKenzie CL (2013) Functional responses and prey-stage preferences of a predatory gall midge and two predacious mites with two spotted spider mites, *Tetranychus urticae*, as host. J Insect Sci 13(8):1–12

Zahedi Golpayegani A, Nadeali T, Saboori A (2012) The effect of prey, *Tetranychus urticae* Koch (Acari: Tetranychidae) density on oviposition rate of the predatory mite, *Phytoseiulus persimilis* Athias-Henriot (Acari: Phytoseiidae). In: 20th Iranian plant protection Congress, University of Shiraz, Shiraz, 25–28 August 2012

Zahedi-Golpayegani A, Saboori A, Sabelis MW (2007) Olfactory response of the predator *Zetzellia mali* to a prey patch occupied by a conspecic predator. Exp Appl Acarol 43:199–204

Zannou ID, Hanna R, De Moraes GJ, Kreiter S (2005) Cannibalism and interspecific predation in a phytoseiid predator guild from cassava fields in Africa: evidence from the laboratory. Exp Appl Acarol 37:27–42

Zeraatkar A, Zahedi Golpayegani ˙A, Saboori A (2013) Kin recognition in three samples of *Phytoseiulus persimilis* (Acari: Phytoseiidae). Persian J Acarol 2(2):311–319

Zhang ZQ (1963) Mites of greenhouses identification, biology and control. CABI Publishing, Wallingford

Zhang Z-Q (1991) Biology of mites of Allothrombiinae (Acari: Trombidiidae) and their potential role in pest control. In: Dusbabek F, Bukva V (eds) Modern acarology. SPB Academic Publishing, Prague

Zhang Z-Q (1992) Functional response of *Allothrombium pulvinum* deutonymphs (Acari: Trombidiidae) on twospotted spider mites (Acari: Tetranychidae). Exp Appl Acarol 15:249–257

Zhang Z-Q (1998) Biology and ecology of trombidiid mites (Acari: Trombidioidea). Exp Appl Acarol 22:139–155

Zhang Z-Q, Norbakhsh H (1995) A new genus and three new species of mites (Acari: Trombidiidae) described from larvae ectoparasitic on aphids from Iran. Eur J Entomol 92:705–718

Zhang Z-Q, Rastegari N (1996) Larval mites (Acari: Trombidiidae) parasitic on aphids in Iran: key, a new species and new record. Tij Entomol 139:91–96

Zhang ZQ, Sanderson JP (1990) Relative toxicity of Abamectin to the predatory mite *Phytoseiulus persimilis* (Acari: Phytoseiidae) and two-spotted spider mite (Acari: Tetranychidae). J Econ Entomol 83:1783–1790

Zhang Z-Q, Fan Q-H, Pesic V, et al (2011) Order Trombidiformes Reuter, 1909. In: Zhang Z-Q (ed) Animal biodiversity: an outline of higher-level classification and survey of taxonomic richness. Zootaxa, 3148:129–138

Zvereva EL, Rank NE (2003) Host plant effects on parasitoid attack on the leaf beetle *Chrysomela japponica*. Oecology 135:268–267

# Chapter 4
# Applied Ecology of Some Predacious Mites in Iran

Hossein Madadi and Hashem Kamali

## 4.1  Introduction

Acarology – the study of mites and ticks- is a young field of science especially in Iran. History review of predatory mite researches in Iran shows that most studies have been done from the perspective of faunistic. Therefore, this chapter was prepared from scientific resources available regarding the use of predatory mites as biological control agents chiefly in greenhouses. Herbivorous mites mostly belong to the superorder Acariformes, order Trombidiformes and suborder Prostigmata (Lindquist et al. 2009b). The most important plant-feeding pest mites are spider mites (family Tetranychidae), false spider mites (family Tenuipalpidae), Acarid mites (family Acaridae), Tarsonemids (family Tarsonemidae) and eriophyid mites (superfamily Eriophyoidea) that injure vegetables and ornamental plants (Zhang 2003).

Two-spotted spider mite, (*Tetranychus urticae* Koch) (TSSM) is undoubtedly the most invasive species in Iran and worldwide (Zhang 2003). This species has a wide host range and more than 1200 species have been recorded to serve as host plants (Zhang 2003), posing a major threat in Iran and cause yield reduction in all agricultural and greenhouse products (Arbabi 2010). For many years, the primary tactic for controlling herbivorous mites relies on synthetic chemical pesticides (Hoy 2011). Although especially after the Second World War, it has been confirmed that this tactic did not work well and mites should be controlled using a rationale framework named integrated pest management (IPM), employed multiple

H. Madadi (✉)
Department of Plant Protection, Faculty of Agriculture, Bu-Ali Sina University, Hamedan, Iran
e-mail: hmadadi@basu.ac.ir

H. Kamali
Department of Plant Protection, Khorasan Razavi Agriculture and Natural Resources Research and Education Center, Mashhad, Iran

© The Author(s), under exclusive license to Springer Nature Switzerland AG 2021
J. Karimi, H. Madadi (eds.), *Biological Control of Insect and Mite Pests in Iran*,
Progress in Biological Control 18, https://doi.org/10.1007/978-3-030-63990-7_4

compatible tactics (Hoy 2011). IPM could be defined as a systemic approach in which interacting components act together to reduce pest population to tolerable levels and minimizing the disadvantages of pest control programes (Fathipour and Sedaratian 2013; Pedigo and Rice 2015). Among the different controlling methods, biological control using natural enemies has a unique position. Although biological control will not eradicate all pest individuals and actually, it is not the goal of biological control, it could be served as a foundation for a more comprehensive pest control strategy, i.e. IPM, which combines a variety of pest control methods. Biological control can be effective, economical, and safe and it should be more widely used than it is today (Mahr et al. 2008).

During the past few years, considerable researches on many aspects of predatory mites have been carried on by Iranian scientists however, a few of them were developed and put into practice. To improve biological control measures using predatory mites, attempts should be made to eliminate constraints and strengthen local research programs. The results of researches with the aim of using predatory mites in greenhouse products primarily were reviewed in this chapter.

## 4.2 Important Mite Pests

During the recent half century, mites damages have become gradually a problem in the qualitative and quantitative production of agricultural products in Iran, but still, there is no clear estimation of yield loss by mite pests. In the next parts, we tried to cite some of the studies carried on by Iranian researchers on these important pests.

### 4.2.1 Two- Spotted Spider Mite

Because of the short life cycle under optimal temperature, this species has a quick developmental resistance capability against pesticides than any other greenhouse pest. This necessitates the higher application of acaricide dosages or frequency, which in turn causes other associated hazards like pest resistance, pest resurgence, toxic residual contaminations on fresh greenhouse products, human poisonings and mortality of natural enemies and pollinators. Therefore, these disadvantages make other environmentally safe control tactics are necessary for effective management of *T. urticae*, especially within greenhouses (Zhang 2003).

Many different natural enemies are associated with *T. urticae* under field and greenhouse conditions. These enemies are either predators or pathogens, but there are no known parasitoids of spider mites so far (McMurtry et al. 1970). There are two categories of predacious species in greenhouses that feed on two-spotted spider mites; those which occur naturally and those which are artificially introduced. The predacious phytoseiid mite, especially *Phytoseiulus persimilis* Athias-Henriot is the major specialist species used to control two- spotted spider mites in greenhouses.

However, some other predatory mites and insects have been used with different outcomes. Several attempts have been undertaken to manage *T. urticae* populations in agricultural crops, such as the application of new safe acaricides with the lower concentrations plus release of predacious mites such as *P. persimlis* in glasshouses on cucumber (Arbabi 2007), bean, cotton and soybean (Daneshvar and Abaii 1994).

## 4.3   Important Predatory Mites

### 4.3.1   Order Mesostigmata

#### 4.3.1.1   Family Phytoseiidae

In the nineteenth century, phytoseiid mites were considered and released locally as potential predators. Up to 1951, only 45 species were introduced for the world fauna (McMurtry et al. 2015), but especially in the past six decades; they were studied intensely which their number by April 2014 is estimated as 2709 species (Kostianinen and Hoy 1996; Arbabi et al. 2011; McMurtry et al. 2015). However, predation capacity for many species is unknown and probably only 3–5% of them can be used effectively against pest species (Sabelis 1985; McMurty 1982).

Phytoseiidae, the most well-known predatory mites has been generally considered as the most promising predators of pest mites and tiny insects on different crops and ornamentals. They feed on fungi, pollen and nectar as supplements (Helle and Sabelis 1985; Zhang 2003; Lundgren 2009). Interestingly, several species within this family may develop and reproduce by feeding exclusively on pollen (Ferragut et al. 1987; Zhimo and McMurtry 1990; Grafton-Cardwell et al. 1999; Nomikou et al. 2002; Ragusa et al. 2009; Khanamani et al. 2017). They are voracious predators and some of them have a high reproductive rate. These features collectively make them as appropriate candidates against greenhouse pests including mites and thrips (Koehler 1999). About 20 species are now mass reared commercially for biological control purposes (Gerson et al. 2003).

4.3.1.1.1   *Phytoseiulus persimilis* Athias-Henriot, 1957

One of the earliest biological control experiences on greenhouse productions was carried out by the Iranian Research Institute of Plant Protection (IRIPP) with Gyah Bazr Alvand Co's cooperation. In this project, 16 commercial natural enemies were evaluated on 12.93 ha cucumber, tomato, strawberry and bell pepper greenhouses for two cropping systems in Teheran, Isfahan, Markazi and Ghazvin provinces. Among those biological agents, the predatory mite species *Amblyseius swirskii* Athias-Henriot was used against thrips and greenhouse whitefly and two species,

*A. clifornicus* and *P. persimilis* were used against two-spotted spider mite (Baniameri and Farokhi 2011).

Compared to the other predatory mite families reported from Iran, the fauna of the Phytoseiidae is well known. According to the literature, about 85 recorded and described species are known from Iran (McMurtry 1977; Sepasgosarian 1977; Daneshvar 1978, 1980, 1987; Daneshvar and Denmark 1982; Hajizadeh et al. 2002; Kolodochka et al. 2003; Faraji et al. 2007; Rahmani et al. 2010; Zare et al. 2012; Asali Fayaz et al. 2013; Hajizadeh and Mortazavi 2015). The unique monophagous species of this family, *P. persimilis* firstly was introduced into Iran from the Netherlands (Wageningen University) in 1988 (Daneshvar 1989) after its widespread dispersal in European countries and the USA (Shirazi et al. 2011). This species was proved to be effective in controlling spider mites under greenhouses and outdoor conditions (Daneshvar and Abaii 1994). Since then, huge lots of studies considered its biology, efficiency, functional and numerical responses, pesticide tolerance, dispersal pattern, patch residence behaviour, chemical ecology, and even mass rearing requirements.

Studying the chemical ecology of *P. persimilis* and olfactory response of this predator is among the interesting research fields, which directly influence the efficiency of this biocontrol agent. Regarding this point, the attractive response of *P. persimilis* to common bean volatiles (from four cultivars) induced by *T. urticae* feedings have been evaluated and confirmed that the quality and quantity of volatiles released by different host plant species significantly influenced on attractiveness of *P. persimilis* (Tahmasebi et al. 2012; Maleknia et al. 2012). Furthermore, experienced predators showed a higher response to *T. urticae* infested red bean leaves than naïve individuals (Mohammadi et al. 2012). However, further studies still should be carried on to elucidate the role of HIPV (Herbivore-induced plant volatiles) as an attracting source of *P. persimilis* toward infested host plants (Zahedi Golpayegani et al. 2010).

This predator showed type III functional response to two- spotted spider mites; however, it has been reported that according to handling time and age-stage consumption rate, *Typhlodromus bagdasarjani* Wainstein & Arutunjan, 1967 is significantly more efficient in predation on *T. urticae* eggs than *P. persimilis* (Moghadasi et al. 2013a, b). Interestingly, despite more predation efficiency, the intrinsic rate of increase of *T. bagdasarjani* was significantly lower than the corresponding value of *P. persimilis* on spider mite eggs (Moghadasi et al. 2013c). The possible outcome of combined application of *P. persimilis* with other biocontrol agents (e.g. *Beauveria bassiana* (Bals.-Criv.) Vuill. (1912)) on two- spotted spider mite population have been studied (Seiedy and Tork 2013; Zamanpour et al. 2019) and it has been reported that in some cases they were not compatible (Seiedy and Tork 2013). This claim was proved when it has been shown that *P. persimilis* refused to attack to *Beauveria bassiana*-infected spider mites (Seiedy et al. 2012). This finding could be important in the simultaneous releasing of these two agents.

Recently, some researches studied the effect of different acaricide residues on life table parameters, fecundity and demography of *P. persimilis* e.g., it has been shown that Primphos methyl (Actellic®) did not have any hazardous effect for *P. persimilis*.

However, this insecticide is no longer used to suppress whiteflies (Kavousi et al. 2000). Moreover, tondexir® and fenpyroximate were moderately and highly toxic for *P. persimilis* according to field bioassay tests (Nadimi et al. 2006; Hamedi et al. 2010; Kabiri Raeis Abad and Zaree 2017), whereas, the Hexythiazox was found safe (Nadimi et al. 2006; Ersin and Madanlar 2006). Sanatgar et al. (2011) assessed the effect of hexythiazox spraying on the development, mortality and reproduction of *P. persimilis*. The obtained results showed that various frequent sprayings of hexythiazox (at 1, 7 and 15th generations) had no significant effect on the total pre-imaginal developmental period of *P. persimilis* among multiple generations. However, comparison of different stages revealed that frequent use of hexythiazox caused a shorter development time for larvae and protonymphs. Their results showed that frequent application of hexythiazox increased the pre-oviposition period of *P. persimilis*, but the adult longevity decreased after 7 and 15 generations (the second and third sprayings), means that the feeding period of *P. persimilis* declined. Ultimately, they concluded that frequent sprayings with hexythiazox has no negative impact on the biological characteristics of *P. persimilis* (Sanatgar et al. 2011).

In a comparative study, the effect of mineral spray oil (MSO) and imidacloprid sprayings on some phytoseiid life stages were compared and less toxic effect of MSO spraying on eggs and preimaginal stages of Phytoseiids was proved (Kord Firozjaee and Damavandian 2018). A laboratory bioassay experiment also showed the adverse impact of bifenazate on life table parameters of *P. persimilis* despite the non-significant lethal effect on adults (Sanatgar et al. 2012). Despite many laboratory bioassays, a more detailed understanding of toxicity under field conditions is required before any recommendations for their suitability or unsuitability of those acaricides in IPM programs (Nadimi et al. 2008b). Prebloom application of insecticidal oils has little or no effect on overwintered phytoseiid mites because of their habit of seeking protected sites, still it can significantly reduce overwintering European red mite (*Panonychus ulmi* (Koch, 1836)) eggs, allowing predator populations to grow concerning the pest's numbers. A single application of highly refined petroleum oil sprayed after petal fall ("summer oil"), has little impact on predators at rates of 1% or less. The combined use of this species plus Volk® oil showed their compatibility in reducing *T. urticae* population (Hosseininia and Arbabi 2008).

Most of Persian surveys on *P. persimilis* restricted to highly sophisticated studies carried on under laboratory conditions, which it is so much challenging to extrapolate their outcomes to real situations. However, few studies have carried on microcosm or more complex setups (Farid and Daneshvar 1995; Shirdel et al. 2000). As a rare case, the effect of three releasing densities of *P. persimilis* on two- spotted spider mite populations has been evaluated on rose (Tavosi et al. 2017). They concluded that 1:10 predator: prey ratio at the initial stages of spider mite infestation gave the best results and could be used to control *T. urticae* population. In this case, three predator-prey ratio of *P. persimilis* and *T. bagdasarjani* were tested on cucumber seedlings and reported that 1:4 rate (predator: prey) reduced two-spotted spider mite populations significantly (Moghadasi and Allahyari 2016). Shirdel et al. (2000) and Farid & Daneshvar (1995) reported that four and five female of *P. persimilis* per cucumber seedling (at five leaves stage) and eggplants respectively stopped the increasing trend of two- spotted spider mite (TSSM) population but they did not record the number of *T. urticae* different developmental stages. Contrary to previous documents, studying the different releasing method and predator-prey ratio of *P. persimilis* against two- spotted spider mites showed that this species even at high predator-prey ratio could not suppress *T. urticae* population on strawberry, which might be assigned to leaf trichomes of host plant (Arbabi 2015). Dispersal pattern studies on greenhouse roses showed the aggregated pattern for *T. urticae* and *P. persimilis* (Ahmadi et al. 2017). Nearly all of these researches have been done within greenhouse while as a special case, once release of *P. persimilis* plus half recommended dose of Propal acaricide produced the good results in terms of TSSM population suppression (Mohiseni et al. 2006).

*P. persimilis* is the most outstanding predatory mite species used in greenhouses, but many other Phytoseiids feed on spider mites and other pests; possibly contribute to control them in greenhouses which some species would be reviewed as follows.

### 4.3.1.1.2  *Neoseiulus* spp. Hughes, 1948

*Neoseiulus* genus sensu stricto is a large and well-known Phytoseiid group (By April 2014, 397 species had been identified, of which 355 species were valid) (Demite et al. 2014). This genus contains many species some are among the major and effective natural enemies of mites, whiteflies, and thrips. Here, we tried to address some researches carried out on different species of this genus.

There are some Persian studies on life history, mass production, prey preference, spatial distribution, feeding history, intraguild predation, predation rate and predation efficacy of this predator species commonly carried on leaf discs at unreal laboratory conditions. For instance, it has been proved that feeding premature stages of *Neoseiulus californicus* (McGregor) with different pollens and two- spotted spider mite positively influenced this predator's predation rate (Rezaie and Montazerie 2015). In a comparative survey, the superiority of *N. californicus* over *T. bagdasarjani* according to daily fecundity and population size has been recorded (Farazmand et al. 2013). Moreover, based on the attack rate and the handling time, they reported *N. californicus* was more efficient biocontrol agent of *T. urticae* on cucumber leaves than *T. bagdasarjani* (Farazmand et al. 2012a). Patch- leaving behaviour of *N. californicus* has also been studied and the positive effect of increasing prey density on patch residence and predator oviposition rate has been shown (Nadeali et al. 2012).

In an applied study, the mold mite, *Tyrophagus putrescentiae* (Schrank) as a prey and Agar oviposition substrate increased the predator's fecundity and survival. Furthermore, among pollens tested, the date palm and walnut ones fortified with spider mite eggs produced the highest longevity (Rezaie and JavanNezhad 2018).

As stated above, there are few studies on *N. cucumeris* carried on real experimental conditions. Microcosm setup, which is a semiartificial rather complex arena used to study the original behaviour of organisms while does not have many deficiencies of artificial setups showed that *N. californicus* could be used with *Scolothrips longicornis* Priesner simultaneously to suppress *T. urticae* population (Farazmand et al. 2012b).

It has been shown that strawberry cultivars did not affect type of functional response of *N. californicus* to Western flower thrips (*Frankliniella occidentalis* (Pergande)) and on all tested cultivars, the predator showed type II response (Rezaie et al. 2017). This predator was more efficient on Chandler' and 'Yalova' cultivars could be assigned to glabrous surface of leaves. This point gets more interesting when it has been reported that in olfactometer bioassay, *N. californicus* preferred Chandler than other commercial cultivars (Rezaie 2016). Host plant cultivars not only affect predation efficiency or prey preference but they are influenced on spatial distribution of *N. californicus* (Rezaie et al. 2012).

Investigating the side effects of different acaricides and insecticides through toxicological bioassays is one of common research fields in Iran as noted earlier. In a toxicological bioassay, it has been demonstrated that two common acaricides, spirodiclofen and spiromesifen residues was not harmful for *N. californicus* after 14 days past application and could be used within an IPM context against two-spotted spider mites (Sarbaz et al. 2017). Besides, spirodiclofen could be considered as a safe compound for *N. californicus* according to survival and predation rate data (Aflaki et al. 2016). However, the undesirable effects of Thiamethoxam (Actara®) sublethal doses on some demographic parameters of this predator makes it as an unsuitable candidate (Havasi et al. 2017).

*Neoseiulus barkeri* Hughes is another known spider mite predator, which has been considered e.g., it was cleared that the leaf trichomes of strawberry negatively influences some biological parameters of *N. barkeri*, like intrinsic rate of increase, generation time and net reproductive rate (Rezaie 2013). *N. barkeri* as a native predator of Tetranychids and thrips pests and as a type II lifestyle species (McMurtry et al. 2013) can survive and reproduce on alternative diets e.g. different pollens. This species certainly drew less attention than *N. californicus* and Persian investigations restricted to some laboratory life table, functional response, or predation rate studies. Unlike most phytoseiids, higher temperature does not significantly reduce the preda-tion rate of this predator and the highest daily prey consumption was recorded at 35 °C (Jafari et al. 2010a). Similarly, the upper thermal threshold and optimum temperature for the development of *N. barkeri* were estimated as 38.8 and 33.29 °C, respectively (Jafari et al. 2010b). The suitability of several different nutritional diets and pollens to *N. barkeri* has also been studied (Rezaie 2010c, 2018; Rezaie and Askari 2015). Those studies confirm that *N. barkeri* was able to complete its life cycle and reproduce on onion thrips plus corn pollen, corn, walnut and date pollens and these could be used as additional food regimes.

*Neoseiulus* (= *Amblyseius*) *cucumeris* as an important type III lifestyle predator of thrips and two- spotted spider mite (McMurtry et al. 2013) have been considered scarcely in Persian papers. To study its predation efficiency, it has been proved that this predator showed the highest performance on glabrous leaves of sweet pepper, while on hairy surfaces of cucumber it was not able to suppress pest population.

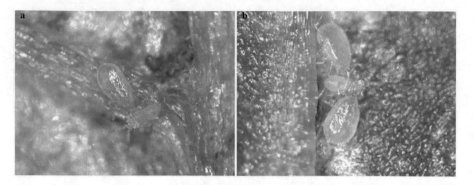

**Fig. 4.1** *Neoseiulus cucumeris* feeding from first instar larvae of *Thrips tabaci*

Accordingly, leaf trichomes reduced the searching efficiency of *N. cucumeris* on cucumber and eggplant than on sweet pepper (Madadi et al. 2007) (Fig. 4.1).

### 4.3.1.1.3   *Typhlodromus bagdasarjani* Wainstein and Arutunjan

*Typhlodromus bagdasarjani* Wainstein and Arutunjan is an widespread indigenous species of the Middle East with high abundance in orchards of Iran (Kamali
et al. 2001). It has been frequently reported on plants infested by spider mites,
eriophyids, tydeids, or insect pests such as thrips and whiteflies (Daneshvar
1993; Riahi et al. 2015). *T. bagdasarjani* is an effective biocontrol agent in greenhouses and outdoors when temperatures are above 20 °C (Ganjisaffar et al. 2011a, b)
and has a population growth rate equal to or greater than that of its prey, TSSM on
both susceptible and resistant eggplant cultivars (Khanamani et al. 2013, 2014) and
strawberry (Mortazavi et al. 2015b). According to Ganjisaffar et al. (2011a), temperature affects the feeding capacity of all life stages of *T. bagdasarjani*, except the
larvae. Food consumption of immature stages increased with increasing temperature
from 15 °C to 25 °C. Therefore, it could conclude that the optimal temperature for
predation of this predator was about 25 °C. It has been reported that this species
showed type II functional response to *Cenopalpus irani* Dosse, 1971 (Acari:
Tenuipalpidae) *P. ulmi* and TSSM eggs and it may be a potential candidate against
*C. irani* especially at low densities (Shirdel and Arbabi 2012b; Mortazavi et al.
2015a; Bazgir et al. 2019a). However, *T. bagdasarjani* exhibited type III functional
response to TSSM eggs on rose leaf discs (Moghadasi et al. 2012b), might be
influenced by the physical properties of leaves, whereas all stages of this predator
preferred *T. urticae* eggs than other stages significantly (Moghadasi et al. 2012a).
The biology or life table of *T. bagdasarjani* have not been studied extensively in
Iran; however, in an investigation about the effect of seven pollen types on biological
parameters, almond pollen has been determined as the most suitable for growth and
reproduction of this predator (Riahi et al. 2015).

According to Farazmand and Fathipour (2015), three predator species *N. californicus*, *T. bagdasarjani* and *S. longicornis*, in single-predator or combined-predator releases, affected to control TSSM on cucumber plants. They observed that TSSM control was at equivalent levels under different predator combinations. Their study showed that the predator combinations did not increase pest suppression inevitably compared with single application treatments. They observed that the densities of two phytoseiids, *N. californicus* and *T. bagdasarjani* were similar in the absence or presence of each other, suggesting that intraguild predation may not occur between the two phytoseiid species and when extraguild prey present, they did not engage in intraguild predation (Farazmand et al. 2012c). Moreover, a combined release of the two species did not significantly impact TSSM density than when each predator was released alone. It is noteworthy that *N. californicus* is a specialist predator, but *T. bagdasarjani* is a generalist (McMurtry et al. 2013). However, both species belong to Type III lifestyle and are able to survive in the absence of main prey (*T. urticae*) (Farazmand et al. 2012c).

Their results showed that all three species are effective predators of *T. urticae* under microcosm conditions and suggest that they can be used in combination without decreased efficacy through intraguild predation. This study has implications for the use of multiple species of predators in biological control programs. However, because of the short experimental period, and a practical application of their findings for biological control of TSSM, further comprehensive long-term studies in experimental greenhouses are necessary for a full evaluation of the biocontrol efficiency of the three predators (Farazmand and Fathhipour 2015).

Toxicological bioassay confirmed the negative effect of spirodilofen on survival, daily oviposition and daily predation rate while Azadirachtin seems safe to this predator (Seidpisheh et al. 2016). Besides laboratory assays, the aggregated spatial distribution of this predator in relation to *T. urticae* on apple was shown. However, a density-independence predation pattern on *T. urticae* might create suspicions about its efficacy (Darbemamieh 2008).

### 4.3.1.1.4  *Amblyseius swirskii* Athias-Henriot

Recently, this predator has been noticed as a candidate against mites, whiteflies and thrips pests in Iranian greenhouses, which among them prefers spider mite than greenhouse whitefly nymphs (Heydari et al. 2016a). However, quite unexpectedly, this species had higher predation on *T. vaporariorum* eggs than *T. urticae* eggs on leaves heavily covered by fine silk webs of *T. urticae* (Mortazavi et al. 2017b). Different studies on life table, effect of pesticides, intraguild interactions, cannibalism, predation rate, switching and prey preference, patch and oviposition behavior, predator-prey dynamics of *A. swirskii* carried out during last decade in Iran (Rezaie 2010a; Haghani et al. 2013; Askarieh Yazdi et al. 2015; Soleymani et al. 2015, Jalili Zenoozi et al. 2016; Heydari et al. 2016a, b; Mortazavi et al. 2017a; Fadaei et al. 2018). This predator showed a type II response to young instar nymphs (first and second instars) of greenhouse whitefly. Meanwhile, with increasing prey and decreasing predator densities, the predation rate increased (Farhadi et al. 2015). Functional response of *A. swirskii* to untreated and *Beauveria bassiana*-infected *F. occidentalis* larvae were surveyed under laboratory conditions after three intervals past spraying. They showed time after spraying did not influence type of functional response and at all intervals this predator exhibited type II response. However, spraying with *B. bassiana* reduced attack rate and increased handling time in 24 and 36 hours treated than control (Heidarian Dehkordi et al. 2017).

Numerical response- an increase in predator numbers in response to increasing prey densities – has been considered as an important factor affecting the number of prey killed (Holling 1961). However, because of difficulties in doing such experiments, predators' numerical response have rarely been studied compared to functional response. This is true for predatory mites. As a rare case, the numerical response of *A. swirskii* to different developmental stages of *Eotetranychus frosti* (McGregor) (Acari: Tetranychidae) showed the suitability of first instar nymphs than other prey stages for predator reproduction (Bazgir et al. 2019b).

The suitability of different species as prey has been investigated through life table studies. Accordingly, the life table of *A. swirskii* on different prey species has been studied and different values were reported for demographic parameters under various situations (Table 4.1). The non-hatchability process of mold mite eggs showed this process significantly decreased the life history parameters of *A. swirskii* (Pirayeshfar et al. 2017a) (Table 4.1). The life table of *A. swirskii* on *Phyllocoptes adalius* Keifer, an important pest of roses plants, showed that *A. swirskii* developed successfully on this pest species and could be used as a biological control agent (Maroufpoor 2016). Additionally, the temperature-dependent two-sex life table of *A. swirskii* was studied using dried fruit mite *Carpoglyphus lactis* L. (Astigmata: Carpoglyphidae). Hence, the intrinsic rate of increase (r), net reproductive rate ($R_0$) and finite rate of population increase ($\lambda$) were significantly higher at 25 °C than other temperatures and the highest population growth of *A. swirskii* occurred at this point (Jafari et al. 2016). Life table studies confirmed that cattail and olive pollens as supplemental food sources increased the fecundity and intrinsic rate of increase of *A. swirskii* (Pirayeshfar et al. 2017b).

The effect of different temperatures on oviposition and development showed the highest fecundity was occurred at 25 °C. Moreover, using linear and non-linear models, temperature threshold ($T_0$) and thermal constant (K) were estimated (Farazmand and Amir-Maafi 2019; Farazmand 2019).

Besides life table studies, the effects of different rearing beds on fecundity and developmental time showed that mulberry leaves were the most suitable for mass rearing of *A. swirskii* (Rezaie 2010b).

### 4.3.1.1.5 Other Phytoseiid Species

In addition to the above-mentioned species, there are scatter publications on other less important phytoseiid species commonly published in conference proceedings. One of the major species from this list is *Phytoseius plumifer* Canestrini and

**Table 4.1** The demographic parameters of *Amblyseius swirskii* fed by different regimes

| Prey species | Temperature (°C) | r (day$^{-1}$) | R$_0$ (offsprings) | Λ (day$^{-1}$) | References |
|---|---|---|---|---|---|
| *T. urticae* (Preimaginal stages) | 25 | 0.224 | 12.14 | 1.252 | Soleymani et al. (2015) |
| *Bemisia tabaci* (Preimaginal stages) | 25 | 0.12 | 3.90 | 1.128 | Soleymani et al. (2015) |
| *T. urticae* (without silk webs) | 25 | 0.141 | 12.66 | 1.152 | Mortazavi et al. (2017a) |
| *T. urticae* (with silk webs) | 25 | −0.0076 | 0.87 | 0.992 | Mortazavi et al. (2017a) |
| *Trialeurodes vaporariorum* (Westwood) on honey-dew free leaves | 25 | 0.132 | 8.46 | 1.141 | Mortazavi et al. (2017a) |
| *Trialeurodes vaporariorum* (Westwood) on honey-dew infested leaves | 25 | 0.171 | 18.64 | 1.186 | Mortazavi et al. (2017a) |
| *Phyllocoptes adalius* | 25 | 0.172 | 12.59 | 1.18 | Maroufpoor (2016) |
| *Carpoglyphus lactis* L. | 25 | 0.304 | 33.82 | 1.35 | Jafari et al. (2016) |
| *Tyrophagus putrescentiae* live eggs | Not indicated | 0.343 | 60.03 | 1.409 | Pirayeshfar et al. (2017a) |
| *Tyrophagus putrescentiae* non-hatchable eggs | Not indicated | 0.3 | 23.94 | 1.39 | Pirayeshfar et al. (2017a) |
| Soybean pollen | 25 | 0.09 | 2.27 | 1.09 | Fadaei et al. (2018) |

Fanzago. The temperature dependent functional response of *P. plumifer* to *Rhyncaphytoptus ficifoliae* Keife adults confirmed that temperature did not affect the type of response but did on searching efficiency and handling time (Kouhjani Gorji et al. 2009). According to functional response parameters, the optimum temperature for *P. plumifer* was 25 °C associated with the highest attack rate and lowest handling time (Louni et al. 2016), while the maximum gross fecundity and fertility values were recorded at 25 and 30 °C (Kouhjani Gorji et al. 2010). Conclusively, they recommended 25–30 °C as the best temperature for using *P. plumifer* (Kouhjani Gorji et al. 2008). Moreover, in a laboratory bioassay, according to residue tests of IOBC recommendation, proved that two acaricides fenpyroximate and abamectin was harmful for this species (Nadimi et al. 2008a). In a field research, Baradaran Anaraki and Arbabi (2008) claimed that *P. plumifer* could not control fig spider mite successfully, which could be allocated to late emergence and inability to feed on adults.

Another phytoseiid species reported from Iran is *Amblydromella kettanehi* Denmark and Daneshvar (Acari: Phytoseiidae) (Mossadegh 1995). The population fluctuations of *A. kettanehi* in response to temperature, humidity, frost duration and rainfall were studied in Tehran Province on *Pinus eldarica* Medw. The results showed the significant positive effect of increasing temperature while rain, snow and the number of freezing days influenced negatively (Arbabi and Baradaran 2001). Laboratory study showed females need 10.7 days to complete their developmental time averagely (Shirdel et al. 2006b), while Arbabi and Baradaran (2000) reported 5.5 ± 1.5 days for this period. This predator showed the type II functional response to all stages of *T. urticae* (Shirdel et al. 2004a) and by feeding on stored product mites, it might be responsible for hive hygiene (Mossadegh 1995).

*Euseius finlandicus* as a type IV lifestyle species has scarcely been considered and there is little information about its biology or efficacy, especially under field conditions. Searching into Persian documents revealed that females and males' developmental time is 7.71 and 7.21 days respectively (Shirdel et al. 2004b). In this regard, it has been shown that this predator showed type II functional response to different densities of two- spotted spider mite and European red mite larvae and females (Shirdel et al. 2006a; Shirdel and Arbabi 2012a). However, this study carried on apple leaf discs under artificial climate chamber conditions and might be difficult to extend to real conditions.

*Amblyseius herbicolus* Chant as generalist predator of *T. urticae* Koch and mulberry thrips *Pseudodendrothrips mori* Niwa has wide distribution in North parts of Iran (Guilan province) (Notghi Moghadam et al. 2008). This predator preferred *T. urticae* to mulberry thrips under laboratory conditions when encountered with these two pests simultaneously (Notghi Moghadam et al. 2010). Abbasipour et al. (2012) indicated that the predatory mites fauna associated with tea plants includes many species in Northern parts of Iran, especially Guilan and Mazandaran provinces. Among these, *A. herbicolus* was highly distributed with very diverse hosts; feeds on different phytophagous mites and small insects (Hajizadeh 2007). It might be exploited as a beneficial species in the tea gardens, but, still its rearing procedure is obscure. It is one of the beneficial species responsible for keeping scarlet tea mite, *Brevipalpus obovatus* Donnadieu populations under control in tea plantations.

In addition to the cited species, there are sporadic and preliminary documents about other less well- known phytoseiid species, which are mostly restricted to laboratory biology or bionomics. The preliminary study on *Phytoseius tropicalis* Daneshvar (Acari: Mesostigmata) showed this species completed its life cycle on TSSM eggs and larvae (Nemati et al. 2008).

According to Rafatifard et al. (2004), *Typhlodromips caspiansis* Denmark and Daneshvar is a common predator of plant mites in Guilan province and has a wide range in the whole region. Regarding daily oviposition rate, developmental time and adult longevity of *T. caspiansis* by feeding from three related prey species (*Panonychus citri* McGregor, *T. urticae* and *Oligonychus bicolor* Banks) this species can be exploited as an efficient natural enemy against spider mites, especially *O. bicolor* in Guilan province (Rafatifard et al. 2004).

A survey by Javadi Khederi and Khanjani (2014) in western Iran (mainly Hamedan Province) showed that large number of natural enemies are associated with the leaves infested by grape erineum mite, *Colomerus vitis* (Pagenstecher) (Acari: Eriophyidae). The main predators were dipteran Cecidomyiids, Tydeid and Phytoseiid mites, while *Typhlodromus* (*Anthoseius*) *khosrovensis* Arutunjan and *T. bagdasarjani* abundance was as low as 11% of the total natural enemies complex.

### 4.3.1.2  Family Tydeidae

Tydeidae family have been reported by different authors to prey on eriophyid mites, including *Tydeus caryae* Khanjani and Ueckermann on the walnut leaf gall mite, *Aceria tristriata* (Nalepa) (Khanjani and Ueckermann 2003). Reports indicating tydeids feeding on pests are rare. As a case, the spatial distribution pattern of *Tydeus longisetosus* El-Bagoury & Momen have been determined as random and aggregated on Golden delicious and Go-lab apple varieties, respectively (Khodayari et al. 2009).

### 4.3.1.3  Family Ascidae

Mites of this diverse family constitute an important group of predatory arthropods common in soil, on plants and stored products, fungi, nests of rodents, birds and even under the elytra of ground beetles. Some species are in close association with humans and animals (Hajizadeh et al. 2010a, b). They are also among the gamasine mites, which are useful as biological indicators of soil conditions and changes in soil ecosystems (Minero et al. 2009; Kazemi and Moraza 2013) and feed on a variety of nematodes, small arthropods like astigmatid mites and springtails. In the catalogue of Ascidae (Acari: Mesostigmata) from Iran (Nemati et al. 2012) based on the information gathered from the published studies, nine genera and 26 ascid species in total are known to inhabit the various parts of country. No information about their efficiency and effectivenes as biological control agents against pests is not available yet in Iran.

### 4.3.1.4  Family Laelapidae

This family belongs to superfamily Dermanyssoidea, which includes free-living and parasitic mites (Moreirea and de Moraes 2015; Lindquist et al. 2009a). Due to the predatory importance of Laelapid mites and their crucial role in the natural control of injurious edaphic pests such as arthropods and nematodes, a faunistic study was carried out for the identification of laelapid mites associated with olive orchards in Guilan province, Iran. During this study, 18 species belonging to nine genera were collected and identified. All species are new for olive orchards mite's fauna of Iran (Mahjoori et al. 2014).

There are not so many documents about the role of Laelapids in pest population suppression in Iran despite the recognized potential of those mites for the control
of soil pests. There are only some preliminary studies on basic biology and life table
of the most prominent species of this group, *Gaeolaelaps aculeifer* (Canestrini). This
species extracted and reported from many parts of Iran including saffron fields of
Khorasan Razavi. The role of commercial species *G. aculeifer* and its preference
aginst Sciarid fungus gnats (*Lycoreilla auripila* Winn., a key pest of button mush-
rooms) and its interaction with entomopathogenic fungi have been studied recently
(Tavoosi ajvad et al. 2020). Additionally, some investigations about the possiblity of
mass rearing this predator have been done (Baradaran et al. 2014). To rear this
predatory mite, four methods were used, Tashiro and Hufaker cages, cultured cells
and culture PDA. It has been confirmed that this species is nematophagous, thus as a
result, it could be breed on thin potato slices with a diameter of less than 5 mm
infected with Cephalobidae nematodes were used as a food source and oviposition
substrate (Baradaran Anaraki et al. 2013).

The single study about the biology of this species pertains to Amin et al. (2014)
who considered the life table of *Hypoaspis* (*=Gaeolaelaps*) *aculeifer* fed by bulb
mite, *Rhizoglyphus echinopus* (Fumouze & Robin, 1868). Moreover, there are some
unpublished informal documents about the introducing *G. aculeifer* to gladiola or
other bulbous host plants in Mahalat (Iran, Markazi Province).

### 4.3.1.5   Family Parasitidae

The cosmopolitan mites of the family Parasitidae are free-living predators. The
family generally feeds on a wide variety of microarthropods and nematodes, with
individual species usually having a narrower range of prey. The Parasitidae group
has poorly been studied in Iran. Up to now, 24 species have been reported. (Rahmani
2011; Kazemi et al. 2013; Nazari and Hajizadeh 2013).

## 4.3.2   Order Prostigmata

### 4.3.2.1   Family Trombidiidae

The mites of the family Trombidiidae are the large red velvet mites commonly found
in soil, litter and other terrestrial habitats.

The biology of two species of *Allothrobium* genus, *Allothrombium pulvinum*
Ewing and *A. shirazicum* Zhang, 1996 has been studied (Saboori and Kamali
1995, 1998). Research carried out in Iran showed that the *A. pulvinum* is a common
natural enemy of spider mites, aphids, and other arthropods (Saboori and Kamali
1995; Hosseini et al. 2002a). It is univoltine and overwinters as adult under the
tree barks or in the soil. After mating in spring, females lay their eggs on weeds or the
soil surface. The eggs hatch in May, larvae population reaches its peak value in

June, nymphs emerge in early summer, and adults appear in autumn under the Northern climatic conditions of Iran (Saboori and Kamali 1995). Larvae are ecto-parasites of aphids, whereas deutonymphs and adults are free-living predators of aphids and spider mites. In Northern Iran, *A. pulvinum* was found on many plants such as apple, walnut, forest trees and primarily citrus trees (Zhang and Faraji 1994). They are fast-runners and have high searching abilities (Saboori and Zhang 1996); adults are most abundant in citrus orchards from September to June (Saboori and Zhang 1996). Feeding on the eggs of *Planococcus citri* (Risso) (Hemiptera: Pseudococcidae) and *Pulvinaria aurantii* Cockerell (Hemiptera: Coccidae) by *A. pulvinum* adults observed for the first time in citrus orchards at the end of spring. The predation was mostly observed between early May and mid-June when the main prey population (nymphs and adults of aphid species, such as *Toxoptera aurantii* Boyer de Fonscolombe and *Macrosiphum rosae* L.) was low or absent in citrus orchard. Noteworthy, *P. citri* and *P. aurantii* are the most important pests of citrus in this region.

High humidity is the most critical factor for *A. pulvinum* oviposition and devel-opment (Saboori and Zhang 1996). Host species selection, host size selection, and superparasitism with mite larvae was studied with two-choice test. Three aphid species used were *Macrosiphum rosae*, *Aphis gossypii* Glover, 1877 and *Hyalopterus amygdali* (Blanchard, 1840). In multiple-choice tests, larvae of *A. pulvinum* did not show any significant preference for any aphid-plant association when confronted with *M. rosae* on rose, *A. gossypii* on cucumber and *H. amygdali* on apricot simultaneously. Two-choice tests showed that larvae preferred *H. amygdali* to *A. gossypii*, but had no preference when offered a choice between *A. gossypii* and *M. rosae* or between *H. amygdali* and *M. rosae*. In host size selection and superparasitism tests, significantly more mites selected the larger host (*M. rosae*). Furthermore, parasitized *H. amygdali* preferred to unparasitized ones (Hosseini et al. 2002b). Therefore, it seems that data on the host preferences of this predator could be helpful for predicting its potential in the biological control of those pests. Due to the wide occurrence and abundance of *A. pulvinum* in citrus orchards in the North of Iran, and because they have high searching abilities, they could be effective biocontrol agents of some citrus pests (Saboori and Zhang 1996). *P. citri* and *P. aurantii* have a synchronized presence in citrus orchards and ornamental plants in this region, and at the same time (mid to late spring, from early May to mid-June) *A. pulvinum* adults are at the end of their annual cycle, after mating and ovipositing. Thereupon, this mite can act as a biocontrol agent that could reduce the populations of these two pests by consuming their eggs. These eggs could serve as an alternate diet and may increase the adult's longevity from early May to mid-June. Because *P. citri* and *P. aurantii* are important pests of citrus trees and ornamental plants in this region, the use of this mite should be taken into account and efforts should be made to conserve its populations in natural habitats. Furthermore, some integrated pest management (IPM) programs, incorporating the use of this mite with other biocontrol agents in citrus orchards such as *Cryptolaemus montrouzieri* Mulsant and *Chilocorus bipustulatus* Gordon (Col.: Coccinellidae) to reduce the populations of these important pests recommended (Saboori et al. 2003).

**4.3.2.2 Family Cunaxidae**

The members of the family Cunaxidae are cosmopolitan and free-living predatory mites that ambush their prey; Overall, few data are available about the effectiveness of Cunaxidae to control key plant pests in Iran (Den Heyer et al. 2011a, b).

**4.3.2.3 Family Stigmaeidae**

The Stigmaeids belongs to the superorder Acariformes, order Trombidiformes, suborder Prostigmata, (Lindquist et al. 2009b). Members of this family have been collected from diverse habitats including foliage, buds and flowers, soil, litter mosses and lichens, fallen leaves, nests of birds and rats (Zhang 2003).

The genus *Stigmaeus* Koch, 1836 recorded from most parts of the country. To date, 27 species of the genus have been recorded and described from Iran (Fakhari et al. 2015). More than 35% of the species are free-living predators on plant leaves and branches. These species are mainly from two genera, *Agistemus* and *Zetzellia* (Gerson 1972). *Agistemus* and *Zetzellia* usually prefer to feed on immatures than adults of spider mites. *Zetzellia mali* (Ewing) prefered to feed on active apple rust mite, *Aculus schlechtendali* than eggs of European red mite, *Panonychus ulmi*; and eggs more than quiescent nymphs of *P. ulmi* (Clements and Harmsen 1993). It also has been recorded as a predator of *C. irani* (Rashki et al. 2004). This species showed aggregative spatial distribution pattern and density-dependent predation on *C. irani* and *Bryobia rubrioculus* Scheuten (Acari: Tetranychidae) in Kermanshah province (Darbemamieh 2008; Darbemamieh et al. 2009) but, despite of aggregated dispersal pattern, it showed density independent reaction to its prey, *P. ulmi* on apple trees (Rahmani et al. 2008).

Few number of the Iranian Stigmaeidae has been identified and for a long time, researches were restricted to *Z. mali* (Ewing) and *A. macrommatus* Gonzales. These mites can live in the soil and organic matter (Khanjani and Uckerman 2003; Khodayari et al. 2008). However, they could also survive on pollen, fungal spores and plant sap (Khodayari et al. 2008). Three species of *Zetzellia* including *Z. mali* (Ewing), *Z. kamali* Kheradmand, Fathipour and Ueckerman, and *Z. pourmirzai* Khanjani and Ueckerman are recorded from Iran. *Z. pourmirzai*, a near species to *Z. mali*, has been found in Kermanshah province (Khanjani and Ueckermann 2008). This species has been considered as a predator with the same foraging regimes (Darbemamieh 2008), and has the ability of surviving for a long time on low prey densities (Villanueva and Harmsen 1998). Quantitative knowledge of spatial distribution patterns of plant feeding mites and their natural enemies is essential for understanding their interactions and developing reliable sampling plans for monitoring pest and predator abundance (Onzo et al. 2005; Athanassiou et al. 2005). The results of a study by Darbemamieh et al. (2011) showed that the spatial distribution indices of the *Z. pourmirzai* and its two preys (*C. irani* and *B. rubrioculus*) can be used in designing sampling programs as well as in the proper estimation of their population densities in apple orchards. Although frustrating results showed the density independent predatin by *Z. mali* on two main preys *E. frosti* and *Tydeus longisetosus* (ElBagouy and Momen) in apple orchards (Khodayari et al. 2009).

The intraguild interaction response between *Z. mali* as intraguild prey and predatory thrips, *S. longicornis* as intraguild predator was studied. It has been confirmed that the simultaneous presence of thrips and hawthorn spider mite, *Amphitetranychus viennensis* (Zacher) on upwind plants led to a withdrawal of *Z. mali* toward higher parts of host trees (Zahedi Golpayegani et al. 2019).

### 4.3.2.4  Family Anystidae

A survey of mites associated with pistachio tree, *Pistacia vera* (Linnaeaus) and *Pistacia mutica* Fischer and Meyer (Sapindales, Anacardiacae) was carried on in one of the main pistachio-production region of Iran (Kerman province) (Mehrnejad and Ueckermann 2001). Fourteen mite species belonging to eleven families reported from cultivated and wild pistachio trees. One of the most abundant predator species found was *Anystis baccarum* (Linnaeus). This mite is predacious on other mites and small insects, although because of highly cannibalistic habit, it is difficult to rear. In the pistachio orchards, this species was found in psyllid colonies, *Agonoscena pistaciae* Burckhardt and Lauterer, (Hemiptera: Psyllidae). Besides, *A. baccarum* is one of the active natural enemies on Coccidae, Diaspididae and Eriococcidae (Hemiptera: Sternorrhyncha) in Kermanshah province (Jalilvand et al. 2014). This species is also active predator of grape erineum mite, *C. vitis* in western Iran (Javadi Khederi and Khanjani 2014). It has not been studied intensively and there are few documents about its predation efficiency or life history. Khanjani et al. (1995) reported a type II functional response of *A. baccarum* females to different life stages of *T. turkestani* however; they did not report any values for searching efficiency or handling time.

The other reported species of this family are *Erythracarus pyrrholeucus* (Hermann) collected on spider mites by Jalaeian et al. (2005) from Isfahan province and *A. wallacei* Otto collected on raspberry shrubs in the central area of Guilan province (Tajmiri 2013; Jalaeian et al. 2005).

### 4.3.2.5  Family Cheyletidae

This family mostly common and abundant in stored products, nests of birds and rodents, granaries, warehouses and barns. Most cheyletids are free-living predators of various micro-arthropods associated with stored-products, including insect eggs and in association with Acariformes mites (Zdarkova 1979, 1998; Hajizadeh et al. 2011). According to Ardeshir et al. (2007), they are among the dominant species of mites collected from wheat stored infected with *Acarus siro* Linnaeus, 1758, *T. putrescentiae* and *Caloglyphus* sp. *Acaropsellina sollers* and *Cheyletus malaccensis* are the beneficial dominant cheyletids in most of the silages in Iran.

## 4.4   Contribution to Research Development

The Regional IPM Programme in the Near East in collaboration with the Research Institute of Plant Protection (IRIPP) is working on IPM for reducing pesticide hazards and increasing biodiversity, established a technical training site on IPM, Biological Control and Good Agriculture Practice (GAP) at the Biological Control Research Department, IRIPP. In 2009, the Iranian Research Institute of Plant Protection (IRIPP) requested the FAO Regional IPM Programme National Project Director in Iran to set up a unit for National and Regional Training on Biological Control at Farm Level. Accordingly, the unit was designed and constructed taking into account all the basic needs for such a unit, including greenhouses, labs, preys and predators propagation sites, classrooms, and other requirements. This site has become a major reference facility conducting comprehensive research on IPM science and technology (Fathi et al. 2011).

## 4.5   How Can Improve the Biological Control by Mites in Iran?

The faunistic and floristic richness of Iran could be an excellent opportunity for the biological control. Most of the biological control agents in agricultural products are available in the Iran ecosystems. Faunistic studies in the agro-ecosystems are in progress at the fields, gardens, forests, grasslands, and pastureland ecosystems. Considering cultural and economic conditions, the employing beneficial agents in greenhouses is more feasible than in other ecosystems because in an enclosed space release of several natural enemies is more controllable. In spite of some preliminary attempts for Phytoseiids and Laelapids, predatory mite mass production has not been conducted in Iran extensively. Only some studies were conducted with personal goals or trainings in the field. The country needs to large and organized insectariums as well technical knowledge with skilled experts and equipments. They should be designed based on the biological and behavioral characteristics of the predatory mites. The last step is the evaluation and quality control methods according to international standards. Imperfect identification of predatorry mite fauna in different ecosystems and insufficient attention to the suppressing role of predatory mites in fields and greenhouses are major hindering factors that have reduced the rate of progression of biological control by predatory mites in Iran. Development of biological control requires long-term financial supported applied researches on different aspects of predatory mites, their compatibility and performance, especially relying on the results of applied researches in other countries (Ashouri 2011).

## 4.6 Conclusions

In the recent two decades, knowledge about the predatory mites' morphology and taxonomy has been increased remarkably in Iran. However, the few experiments conducted under field conditions, despite many researches on the systematic and faunistic, lead to a limited use of these mites to control plant pests. The potential use of predatory mites still needs to be investigated, although several promising native species have been identified. Most of the experimental studies were restricted to laboratory evaluations of their life cycle and little has been addressed the practical efficacy of predatory mites. To apply these mites effectively by growers, efforts should be focused on fieldworks to validate promising laboratory results. Knowledge about the diversity of predatory mites is still limited, which should be supported to increase the chances of finding species potentially useful as native biological control agents.

Currently, no native species are commercialized as biological control agent yet. The majority of imported commercial products are used in greenhouses while most of these species can not acclimatize to climatic and environmental conditions in the long term. Some biological control programs using imported *P. persimilis* to control *T. urticae* on greenhouse rose, cucumber and strawberry were carried out during past years. Generally, before any predatory species is being proposed for release, the data about its biology, ecology, response to different environmental conditions and the range of potential prey organisms in the habitat should be collected. The results of experimental studies and greenhouses should be implemented in fields and then recommended to farmers.

**Acknowledgement** The preparation of this chapter would not have been possible without the kind support and help of many Iranian individuals and organizations. We are highly indebted to Dr. Alireza Saboori, (University of Tehran); Dr. Mohammad Khanjani (Bu-Ali Sina University); Dr. Jalil Hajizadeh (University of Guilan); Dr. Farid Faraji (MITOX Consultants/Eurofins, Amsterdam); Dr. Yaghoub Fathipour (Tarbiat Modares University); Dr. Masoud Arbabi (Iranian Research Institute of Plant Protection); Dr. Shahrooz Kazemi (Institute of Science and High Technology and Environmental Sciences); Dr. Alireza Nemati (University of Shahrekord) for their guidance as well as for providing necessary information and thanks to all members of Acarological Society of Iran.

## References

Abbasipour H, Taghavi A, Rastegar F et al (2012) Phytoseiid mites (Acari: Mesostigmata) associated with tea gardens in north of Iran. Arch Phytopathol Plant Prot 45(12):1–12. https://doi.org/10.1080/03235408.2012.676238

Aflaki F, Rahmani H, Kavuosi A (2016) Effects of Azadirachtin and Spirodiclofen on some biological traits and predation rate of *Neoseiulus californicus* (Acari: Phytoseiidae). In: Karimi J (ed) Proceedings of the 3rd national meeting on biological control in agriculture and natural resources. Ferdowsi University of Mashhad, Mashhad

Ahmadi K, Fathipour Y, Bashiri M (2017) Spatial distribution and population fluctuation of *Tetranychus urticae* and its predator *Phytoseiulus persimilis* in a rose greenhouse. In: Talaei-

Hasssanloui R (ed) Proceedings of the 2nd Iranian International Congress of Entomology. Karaj, College of Agriculture and Natural Resources, University of Tehran

Amin MR, Khanjani M, Zahiri B (2014) Life table parameters of *Hypoaspis aculeifer* (Acari: Laelapidae) in feeding on *Rhizoglyphus echinopus* (Acari: Acaridae). Agric Pest Manag 1 (1):10–22. (in Persain with English abstract)

Arbabi M (2007) Study on effectiveness of *Phytoseiulus persimilis* in control of cucumber two spotted spider mite (*Tetranychus urticae* complex) in woody and iron greenhouse structures in Varamine region. Pajouhesh-va-Sazandegi 19(4):96–104

Arbabi M (2010) Review on six decades pesticides application to control mite pests in different agricultural crops of Iran. Paper presented at the 1st congress on half a century of the pesticide usage in Iran, Iranian Research Institute of Plant Protection and Iran Plant Protection Organization, Tehran, 2–3 March 2010

Arbabi M (2015) Evaluation *Phytoseiulus persimilis* A. H. in different releasing methods in control of greenhouse strawberry infested by *Tetranychus urticae*. In: Manzari S (ed) Proceedings of the 1st Iranian International Congress of Entomology, Tehran

Arbabi M, Baradaran P (2000) Study on biology of *Amblydromella kettanehi* Daneshvar and Denmark on *Pinus nigra* Arnold and *Tetranychus urticae* Koch at laboratory conditions. Paper presented at the 14th Iranian Plant Protection Congress, Isfahan University of Technology, Isfahan, 5–8 September 2000

Arbabi M, Baradaran P (2001) Study on population fluctuation of *Amblydromella kettanehi* Denmark and Daneshvar on *Pinus eldarica* Medw in Tehran and its biology on *Tetranychus urticae* Koch under laboratory condition. J Entomol Soc Iran (JESI) 20(2):1–21. (in Persain with English abstract)

Arbabi M, Daneshvar H, Shirdel D, Baradaran P (2011) Results of half century investigation of phytoseiid mite fauna in agricultural crops of Iran. Paper presented at the 1st Biological Control Development Congress in Iran, Iranian Research Institute of Plant Protection, Tehran, 2011

Ardeshir F, Yousefi Porshekoh A, Saboor A (2007) A faunistic study and population fluctuations of mites associated with stored wheat in Tehran region, Iran. J Entomol Soc Iran (JESI) 27 (2):17–28. (in Persain with English abstract)

Asali Fayaz B, Khanjani M, Tixier MS (2013) Redescription of six species of the genus *Typhlodromus* Scheuten (Acari: Phytoseiidae: Typhlodrominae) recorded from some regions of Western and North-Western Iran. Pers J Acarol 2(3):369–387

Ashouri A (2011) Production and commercial use of biological control agents in worldwide and Iran. Paper presented at 1st Biological Control Development Congress in Iran, Iranian Research Institute of Plant Protection, Tehran, 27–28 July 2011

Askarieh Yazdi S, Zahedi Golpayegani A, Saboori A, Karami Badrbani F (2015) Different forms of *Tetranychus urticae* Koch (Acari: Tetranychidae) and their plasticity in retaining eggs in the presence of predatory mites *Amblyseius swirskii* (Acari: Phytoseiidae). In: Manzari S (ed) Proceedings of the 1st Iranian International Congress of Entomology, Tehran

Athanassiou CG, Kavallieratos NG, Palyvos NE et al (2005) Spatio-temporal distribution of insects and mites in horizontally-stored wheat. J Econ Entomol 98(3):1058–1069. https://doi.org/10.1603/0022-0493-98.3.1058

Baniameri V, Farokhi S (2011) Implementation of biological control program in greenhouse crops in Iran. Paper presented at 1st Biological Control Development Congress in Iran, Iranian Research Institute of Plant Protection, Tehran, 27–28 July 2011

Baradaran Anaraki P, Arbabi M (2008) Study phenology of population abundance of *Phytoseius plumifer* (C. & F.) on different fig varieties in Saveh region. Paper presented at the 18th Iranian Plant Protection Congress, Bu-Ali Sina University, Hamedan, 25–28 August

Baradaran Anaraki P, Arbabi M, Joharchi O, Rahimi H, Hoseininia A, Ghanbari Z (2012) In vitro study biology of *Gaeolaelaps aculifer* as the most abundant of predatory mite of *Rhizoglyphus* spp. on the corm of gladiole and saffron. Paper presented at the 20th Iranian Plant Protection Congress, Shiraz University, Shiraz, 25–28 August

Baradaran Anaraki P, Arbabi M, Joharchi O (2013) Scientific knowledge of *Gaeolaelaps aculeifer* mass production on potato. Plant Protection Research Institute. Registration Number: 79342, 3, 25–26. Retrieved from http://bulletin.iripp.ir/Newsletter/1392/IRIPPNewsletter24.pdf

Baradaran P, Arbabi M, Rahimi H, Hosseininia A (2014) *Gaeolaelaps aculeifer* study the predatory mite breeding methods and the introduction of a new method to control mites gladiolus corm. Paper presented at the first National Congress Flowers and Ornamental Plants, Ornamental Plant Research Center, Karaj, 21–22 October 2014

Bazgir F, Shakarami J, Jafari S (2019a) The influence of prey density on the predacious behavior of *Typhlodromus bagdasarjani* (Acari: Phytoseiidae). In: Abstracts of the 9th National Conference on Biological Control in Agriculture and Natural Resources, Bu-Ali Sina University, Hamedan, 2019

Bazgir F, Shakarami J, Jafari S (2019b) Numerical response of *Amblyseius swirskii* (Acari: Phytoseiidae) to different densities of *Eotetranychus frosti* (Tetranychidae). In: Abstracts of the 9th National Conference on Biological Control in Agriculture and Natural Resources, Bu-Ali Sina University, Hamedan, 2019

Clements DR, Harmsen R (1993) Prey preferences of adult and immature *Zetzellia mali* Ewing (Acari: Stigmaeidae) and *Typhlodromus caudiglans* Schuster (Acari: Phytoseiidae). Can Entomol 125(5):967–969. https://doi.org/10.4039/Ent125967-5

Daneshvar H (1978) A study on the fauna of plant mites in Azarbayjan. Iran Entomol Phytopathol Appl 46(1–2):18–20

Daneshvar H (1980) Some predator mites from northern and western Iran. Iran Entomol Phytopathol Appl 48:15–17

Daneshvar H (1987) Some predatory mites from Iran, with descriptions of one new genus and six new species (Acari: Phytoseiidae, Ascidae). Appl Entomol Phytopathol 54(1–2):13–37

Daneshvar H (1989) The introduction of *Phytoseiulus persimilis* AH (Acari: Phytoseiidae) as active predator in Iran. Paper presented at the 9th Plant Protection Congress of Iran, Ferdowsi University of Mashhad, Mashhad, 9–14 September 1989

Daneshvar H (1993) Distribution of two predatory mite *Amblydromella kettanehi* and *Euseius libanesi* (Acari: Phytoseiidae) in Iran. Paper presented at the 11th Iranian Plant Protection Congress, Guilan University, Rasht, 28 August–2 September

Daneshvar H, Abaii MG (1994) Efficient control of *Tetranychus turkestani* on cotton, soybean and bean by *Phytoseiulus persimilis* (Acari: Tetranychidae, Phytoseiidae) in pest foci. Appl Entomol Phytopathol 61:61–75

Daneshvar H, Denmark HA (1982) Phytoseiids of Iran (Acarina: Phytoseiidae). Int J Acarol 8 (1):3–14. https://doi.org/10.1080/01647958208683272

Darbemamieh M, (2008) Tetranychoid mites and their acarid predators in Kermanshah orchards, spatial distribution and population dynamics of some species on apples. M.Sc. thesis, Tarbiat Modares (TMU) University

Darbemamieh M, Fathipour Y, Kamali K (2009) Bionomics of *Cenopalpus irani*, *Bryobia rubrioculus* and their egg predator *Zetzellia mali* (Acari:Tenuipalpidae,Tetranychidae, Stigmaeidae) in natural conditions. Munis Entomol Zool 4(2):341–354

Darbemamieh M, Fathipour Y, Kamali K (2011) Population abundance and seasonal activity of *Zetzellia pourmirzai* (Acari: Stigmaeidae) and its preys *Cenopalpus irani* and *Bryobia rubrioculus* (Acari: Tetranychidae) in sprayed apple orchards of Kermanshah, Iran. J Agric Sci Tech 13(2):143–154

Demite PR, McMurtry J, De Moraes GJ (2014) Phytoseiidae database: a website for taxonomic and distributional information on phytoseiid mites (Acari). Zootaxa 3795(5):571–577

Den Heyer J, Ueckermann EA, Khanjani M (2011a) Iranian Cunaxidae (Acari: Prostigmata: Bdelloidea), Part 1. Subfamily Coleoscrinae. Int J Acarol 37(2):143–160. https://doi.org/10.1080/01647954.2010.495953

Den Heyer J, Ueckermann EA, Khanjani M (2011b) Iranian Cunaxidae (Acari: Prostigmata: Bdelloidea) Part II. Subfamily Cunaxoidinae. J Nat Hist 45(27–28):1667–1678. https://doi.org/10.1080/00222933.2011.559602

Ersin F, Madanlar N (2006) Investigations on the effects of some pesticides used in greenhouse vegetables on predatory mite *Phytoseiulus persimilis* (Acarina: Phytoseiidae) in laboratory conditions. Turk Entomol Derg 30(1):67–80

Fadaei E, Hakimitabar M, Seiedy M et al (2018) Effects of different diets on biological parameters of the predatory mite *Amblyseius swirskii* (Acari: Phytoseiidae). Int J Acarol 44(7):341–346. https://doi.org/10.1080/01647954.2018.1525428

Fakhari N, Khanjani M, Rahmani H et al (2015) *Stigmaeus jalili* sp. n. (Acari: Stigmaeidae) from Zanjan Province (Iran) and description of *S. haddadi* male. Biologia 70(6):782–787. https://doi.org/10.1515/biolog-2015-0084

Faraji F, Hajizadeh J, Ueckermann EA et al (2007) Two new records for Iranian Phytoseiid mites with synonymy and keys to the species of *Typhloseiulus* Chant and McMurtry and Phytoseiidae in Iran (Acari: Mesostigmata). Int J Acarol 33(3):231–239. https://doi.org/10.1080/01647950708684527

Farazmand A (2019) Temperature-dependent development of *Amblyseius swirskii* Athias-Henriot (Acari: Phytoseiidae) feeding on immature stages of two-spotted spider mite. In: Abstracts of the 9th national conference on Biological Control in Agriculture and Natural Resources, Hamedan, 2019

Farazmand A, Amir-Maafi M (2019) Oviposition model of *Amblyseius swirski* Athias-Henriot in prey system (*Tetranychus urticae* Koch). In: Abstracts of the 9th national conference on Biological Control in Agriculture and Natural Resources, Hamedan, 2019

Farazmand A, Fathhipour Y (2015) Control of the spider mite *Tetranychus urticae* using phytoseiid and thrips predators under microcosm conditions: single-predator versus combined–predators release. Syst Appl Acarol 20(2):162–170. https://doi.org/10.11158/saa.20.2.2

Farazmand A, Fathipour Y, Kamali K (2012a) Functional response and mutual interference of *Neoseiulusn californicus* and *Typhlodromus bagdasarjani* (Acari: Phytoseiidae) on *Tetranychus urticae* (Acari: Tetranychidae). Int J Acarol 38(5):369–376. https://doi.org/10.1080/01647954.2012.655310

Farazmand A, Fathipour Y, Kamali K (2012b) Effect of *Neoseiulus californicus* (Phytoseiidae) and *Scolothrips longicornis* (Thripidae) on two-spotted spider mite population density under microcosm condition. Paper presented at the 20th Iranian Plant Protection Congress, Shiraz University, Shiraz, 25–28 August 2012

Farazmand A, Fathipour Y, Kamali K (2012c) Predation preference of *Neoseiulus californicus* and *Typhlodromus bagdasarjani* on heterospecific phytoseiid and *Scolothrips longicornis*. Paper presented at the 20th Iranian Plant Protection Congress, Shiraz University, Shiraz, 25–28 August 2012

Farazmand A, Fathipour Y, Kamali K (2013) Population growth of predatory mites, *Neoseiulus californicus* and *Typhlodromus bagdasarjani* on cucumber leaf discs. In: Talaei-Hassanloui R (ed) Proceedings of the conference of biological control in agriculture and natural resources. College of Agriculture and Natural Resources, Karaj

Farhadi R, Allahyari H, Chi H (2015) Functional response and mutual interference of *Amblyseius swirskii* (Acari: Phytoseiidae) on greenhouse whitefly *Trialeurodes vaporariorum* on cucumber. Plant Prot 38(2):37–48. (in Persian with English abstract)

Farid A, Daneshvar H (1995) *Phytoseiulus persimilis* as a predatory mite applied against *Tetranychus turkestani* in lentil and eggplant. Paper presented at the 12th Iranian Plant Protection Congress, Karaj Junior College of Agriculture, Karaj, 2–7 September 1995

Fathi H, Heidari H, Impiglia A, Fredrix M (2011) History of IPM/FFS in Iran. FAO Publication. Available via www.ipm–neareast.com

Fathipour Y, Sedaratian A (2013) Integrated management of *Helicoverpa armigera* in soybean cropping systems. In: El-Shemy H (ed) Soybean–pest resistance. In Tech, Rijeka, pp 231–280

Ferragut F, García-Marí F, Costa-Comelles J et al (1987) Influence of food and temperature on development and oviposition of *Euseius stipulatus* and *Typhlodromus phialatus* (Acari: Phytoseiidae). Exp Appl Acarol 3:317–329. https://doi.org/10.1007/BF01193168

Ganjisaffar F, Fathipour Y, Kamali K (2011a) Effect of temperature on prey consumption of *Typhlodromus bagdasarjani* (Acari: Phytoseiidae) on *Tetranychus urticae* (Acari: Tetranychidae). Int J Acarol 37(6):556–560. https://doi.org/10.1080/01647954.2010.528800

Ganjisaffar F, Fathipour Y, Kamali K (2011b) Temperature-dependent development, and life table parameters of *Typhlodromus bagdasarjani* (Phytoseiidae) fed on two-spotted spider mite. Exp Appl Acarol 55:256–272. https://doi.org/10.1007/s10493-011-9467-z

Gerson U (1972) Mites of the genus *Ledermuelleria* (Prostigmata: Stigmaeidae) associated with mosses in Canada. Acarologia 13(2):319–343

Gerson U, Smiley RL, Ochoa R (2003) Mites (Acari) for pest control. Blackwell, London

Grafton-Cardwell EE, Ouyang Y, Bugg RL (1999) Leguminous cover crops to enhance population development of *Euseius tularensis* (Acari: Phytoseiidae) in Citrus. Biol Control 16(1):73–80. https://doi.org/10.1006/bcon.1999.0732

Haghani S, Zahedi Golpayegani A, Saboori AR, Allahyari H (2013) *Amblyseius swiriskii* and *Neoseiulus californicus* aggressiveness towards *Phytoseiulus persimilis* larvae (Acari.: Phytoseiidae). In: Talaei-Hassanloui R (ed) Proceedings of the conference of biological control in agriculture and natural resources. College of Agriculture and Natural Resources, Karaj

Hajizadeh J (2007) Fauna of Phytoseiidae mites (Acari: Phytoseiidae) in Guilan Province. Final report of research project. Guilan University, Rasht, Agri Sci Coll Uni Guilan, 64 pp

Hajizadeh J, Mortazavi S (2015) The genus *Euseius* Wainstein (Acari: Phytoseiidae) in Iran, with a revised key to Iranian phytoseiid mites. Int J Acarol 41(1):53–66. https://doi.org/10.1080/01647954.2014.985712

Hajizadeh J, Hosseini R, McMurtry JA (2002) Phytoseiid mites (Acari: Phytoseiidae) associated with eriophyid mites (Acari: Eriophyidae) in Guilan province of Iran. Int J Acarol 28(4):373–378. https://doi.org/10.1080/01647950208684313

Hajizadeh J, Faraji F, Rafati Fard M (2010a) Four species of family Laelapidae (Acari: Mesostigmata), new records for the Iran. Paper presented at the 19th Iranian Plant Protection Congress, Iranian Research Institute of Plant Protection, Tehran, 31 July–3 August, 2010

Hajizadeh J, Faraji F, Rafatifard M (2010b) Ascidae (Acar: Mesostigmata) of Guilan province, A new genus and four species recorded for the Iranian mites fauna and a key to the north of Iran Ascid species. Iran J Plant Prot Sci 40(2):34–50. (in Persian with English abstract)

Hajizadeh J, Noei J, Salehi L, Ostovan H (2011) Cheyletid mites associated with stored rice in Iran, the first record of *Chelacheles strabismus* from Iran and a key for their identification. J Entomol Soc Iran (JESI) 30(2):85–88

Hamedi N, Fathipour Y, Saber M (2010) Sublethal effects of the acaricide fenpyroximate on life-table and reproduction parameter of the predatory mite, *Phytoseius plumifer* (Phytoseiidae) in laboratory conditions. Paper presented at the 19th Iranian Plant Protection Congress, Iranian Research Institute of Plant Protection, Tehran, 31 July–3 August, 2010

Havasi, M R, Kheradmand K, Mosallanejad H, Fathipour Y (2017) Demographic outcomes of *Neoseiulus californicus* McGregor treating by the thiamethoxam. In: Naeimi S (ed) Proceedings of the 8th national conference on biological control in agriculture and natural resources, Rasht

Heidarian Dehkordi M, Allahyari H, Talaei-Hasanlouie R, Parker B (2017) Functional response of *Amblyseius swirskii* (Acari: Phytoseiidae) on untreated and *Beauveria bassiana*-treated *Frankliniella occidentalis* (Thysanoptera: Thripidae). Biol Control Pests Plant Dis 6(2):245–255. (in Persain with English abstract)

Helle W, Sabelis MW (1985) World crop pests. Spider mites: their biology, natural enemies and control, 1B. Elsevier, Amsterdam

Heydari S, Allahyari H, Zahedi Golpaygani A, Farhadi R (2016a) Prey preference and switching behavior of *Amblyseius swirskii* (Acari: Phytoseiidae) on greenhouse whitefly and two-spotted spider mite. In: Karimi J (ed) Proceedings of the 3rd national meeting on biological control in agriculture and natural resources. Ferdowsi University of Mashhad, Mashhad

Heydari S, Allahyari H, Zahedi Golpaygani A, Farhadi R, Enkegaard A (2016b) Patch residence time and patch preference of the predatory mite, *Amblyseius swirskii* in relation to prey diversity.

In: Karimi J (ed) Proceedings of the 3rd national meeting on biological control in agriculture and natural resources. Ferdowsi University of Mashhad, Mashhad

Holling CS (1961) Principles of insect predation. Annu Rev Entomol 6(1):163–182

Hosseini M, Hatami B, Saboori A (2002a) Functional response of *Allothrombium pulvinum* Erwig (Acari: Trombidiidae) to *Aphis gossypii* Glov. (Homoptera: Aphididae). Paper presented at the 15th Iranian Plant Protection Congress, Razi University of Kermanshah, Kermanshah, 7–11 September 2002

Hosseini M, Hatami B, Saboori A (2002b) Host preference by *Allothrombium pulvinum* (Acari: Trombidiidae) larvae on aphids: *Macrosiphum rosae*, *Aphis gossypii* and *Hyalopterus amygdali* (Homoptera: Aphididae). Exp Appl Acarol 27:297–302. https://doi.org/10.1023/A:1023359130396

Hosseininia A, Arbabi M (2008) Integrated control of spider mite, *Tetranychus* spp. (Acari: Tetranychidae) on roses by application of predator mite, *Phytoseiulus persimilis* Athias-Henriot and biological compound (abamectin) mixed to Volk® oil. Paper presented at the 18th Iranian Plant Protection Congress, Bu-Ali Sina University, Hamedan, 25–28 August 2008

Hoy MA (2011) Agricultural acarology, introduction to integrated mite management. CRC Press, Boca Raton

Jafari S, Fathipour Y, Faraji F, Bagheri M (2010a) The influence of temperature on the functional response and prey consumption of *Neoseiulus barkeri* (Acari: Phytoseiidae) on two-spotted spider mite. Paper presented at the 19th Iranian Plant Protection Congress, Iranian Research Institute of Plant Protection, Tehran, 31 July–3 August 2010

Jafari S, Fathipour Y, Faraji F, Bagheri M (2010b) Thermal requirements for development of *Neoseiulus barkeri* (Acari: Phytoseiidae) fed on *Tetranychus urticae*. Paper presented at the 19th Iranian Plant Protection Congress, Iranian Research Institute of Plant Protection, Tehran, 31 July–3 August 2010

Jafari S, Sarraf Moayeri H, Kavousi A (2016) Temperature – dependent life table of the predatory mite, *Amblyseius swirskii* (Mesostigmata: Phytoseidae) fed on stored product mite *Carpoglyphus lactis* (Astigmata: Carpoglyphidae). J Entomol Soc Iran (JESI) 36(3):163–179. (in Persian with English abstract)

Jalaeian M, Saboori A, Seyedoleslami H (2005) Prostigmatid mites (Acari: Prostigmata) associated with fruit trees in the western area of Isfahan. J Entomol Soc Iran (JESI) 25(1):67–68

Jalili Zenoozi P, Zahedi Golpayghani A, Saboori A, Hekmat Z, Bigdeli N (2016) The effect of familiarity and sociality on predation of *Amblyseius swirskii* (Acari: Phytoseiidae). In: Karimi J (ed) Proceedings of the 3rd national meeting on biological control in agriculture and natural resources. Ferdowsi University of Mashhad, Mashhad

Jalilvand K, Shirazi M, Vahedi H et al (2014) A preliminary study on natural enemies of Coccoidea (Hemiptera, Sternorrhyncha) in Kermanshah Province, Western Iran. Acta Phyt Entomol Hun 48(2):299–308. https://doi.org/10.1556/APhyt.48.2013.2.11

Javadi Khederi S, Khanjani M (2014) Natural predatory survey on vineyards infested by grape erineum mite, *Colomerus vitis* (Pagenstecher) (Acari: Eriophyidae) in Western Iran. J Crop Prot 3(20):625–630

Kabiri Raeis Abad M, Zaree E (2017) Comparing toxicity of plant pesticides, Tondexir® and chemichal acaricides, Ortus® on two spotted spider mite *Tetranychus urticae* Koch (Acari: Tetranychidae) and its natural enemies *Phytoseiulus persimilis* Athias-Henriot (Acari: Phytoseiidae). Plant Prot 40(3):27–38. (in Persian with English abstract)

Kamali K, Ostovan H, Atamehr A (2001) A catalog of mites and ticks (Acari) of Iran. Islamic Azad University Scientific Publication Center, Tehran

Kavousi O, Talebi Kh, Arbabi M, Kharazi Pakdel A (2000) Laboratory evaluation of side effects of primphos methyl on the predatory mite, *Phytoseiulus persimilis* A.H. (Acari: Phytoseiidae). Paper presented at the 14th Iranian Plant Protection Congress, Isfahan University of Technology, Isfahan, 5–8 September 2000

Kazemi S, Moraza ML (2013) Mites of the genus *Antennoseius* Berlese (Acari: Mesostigmata: Ascidae) from Iran. Persian J Acarol 2(2):217–234. https://doi.org/10.22073/pja.v2i2.9956

Kazemi S, Arjomandi E, Ahangaran Y (2013) A review of the Iranian Parasitidae (Acari: Mesostigmata). Persian J Acarol 2(1):159–180. https://doi.org/10.22073/pja.v2i1.9951

Khanamani M, Fathipour Y, Hajiqanbar H (2013) Population growth response of *Tetranychus urticae* to eggplant quality: application of female age-specific and age-stage, two-sex life tables. Int J Acarol 39:638–648. https://doi.org/10.1080/01647954.2013.861867

Khanamani M, Fathipour Y, Hajiqanbar H et al (2014) Two-spotted spider mite reared on resistant eggplant affects consumption rate and life table parameters of its predator, *Typhlodromus bagdasarjani* (Acari: Phytoseiidae). Exp Appl Acarol 63:241–252. https://doi.org/10.1007/s10493-014-9785-z

Khanamani M, Fathipour Y, Talebi AA et al (2017) Linking pollen quality and performance of *Neoseiulus californicus* (Acari: Phytoseiidae) in two-spotted spider mite management programmes. Pest Manag Sci 73(2):452–461. https://doi.org/10.1002/ps.4305

Khanjani M, Uckerman EA (2003) The stigmaeid mites of Iran (Acari: Stigmaeidae). Int J Acarol 23(4):317–339. https://doi.org/10.1080/01647950208684309

Khanjani M, Ueckermann EA (2008) A new species of *Zetzellia oudemans* (Acari: Stigmaeidae) from Iran. Int J Acarol 34(3):237–241. https://doi.org/10.1080/01647950808684536

Khanjani M, Kamali K, Sahragard A (1995) Functional response of *Anystis baccarum* (L.) (Acari: Anystidae) on *Tetranychus turkestani* (U. and N.) (Acari: Tetranychidae). Paper presented at the 12th Iranian Plant Protection Congress, Karaj Junior College of Agriculture, Karaj, 2–7 September

Khodayari S, Kamali K, Fathipour Y (2008) Biology and predation rate of predatory mite, *Zetzellia mali* (Acari: Stigmaeidae) on two spotted spider mite in laboratory conditions. Paper presented at the 18th Iranian Plant Protection Congress, Bu-Ali Sina University, Hamedan, 25–28 August 2008

Khodayari S, Kamali K, Fathipour Y (2009) Comparison of population density and spatial distribution of *Tydeus longisetosus* El-Bagoury & Momen, *Tetranychus turkestani* (Ugarov & Nikolskii) and their predator *Zetzellia mali* (Ewing) on two apple varieties. J Entomol Res 1 (3):197–208. (in Persian with English abstract)

Koehler HH (1999) Predatory mites (Gamasina, Mesostigmata). Agric Ecosyst Environ 74:395–410. https://doi.org/10.1016/S0167-8809(99)00045-6

Kolodochka LA, Hajiqanbar H, McMurtry JA (2003) A description of unknown male and redescription of female of the rare phytoseiid mite *Neoseiulus sugonjaevi* (Wainstein & Abbasova, 1974) (Parasitiformes, Phytoseiidae) from Iran. Acarina 11(2):231–233

Kord Firozjaee Z, Damavandian MR (2018) Control of citrus leafminer using mineral oil in Mazandaran and its effects on predatory Phytoseiid mites. J Appl Res plant prot 6 (4):107–118. (In Persain with English Abstract)

Kostianinen TS, Hoy MA (1996) The Phytoseiidae, as biological control agents of pest mites and insects, a bibliography, Monograph, 17. Entomology and Nematology Department, University of Florida

Kouhjani Gorji M, Kamali K, Fathipour Y (2008) The effect of temperature on voracity of *Phytoseius plumifer* (Acari: Phytoseiidae) on two-spotted spider mite. Paper presented at the 18th Iranian Plant Protection Congress, Bu-Ali Sina University, Hamedan, 25–28 August 2008

Kouhjani Gorji M, Fathipour Y, Kamali K (2009) The effect of temperature on the functional response and prey consumption of *Phytoseius plumifer* (Acari: Phytoseiidae) on the two-spotted spider mite. Acarina 17(2):231–237

Kouhjani Gorji M, Fathipour Y, Kamali K (2010) The effect of temperature on reproduction parameters of *Phytoseius plumifer* (Acari: Phytoseiidae) on tow-spotted spider mite. Paper presented at the 19th Iranian Plant Protection Congress, Iranian Research Institute of Plant Protection, Tehran, 31 July–3 August 2010

Lindquist EE, Krantz GW, Walter DE (2009a) Order Mesostigmata. In: Krantz GW, Walter DE (eds) A manual of acarology, 3rd edn. Texas Tech University Press, Lubbock, pp 124–232

Lindquist EE, Krantz GW, Walter DE (2009b) Classification. In: Krantz GW, Walter DE (eds) A manual of acarology, 3rd edn. Texas Tech Univ Press, Lubbock, pp 97–103

Louni M, Jafari S, Shakarami J (2016) The effect of temperature on the functional response and prey consumption of *Phytoseius plumifer* (Phytoseiidae) fed on *Rhyncaphytoptus ficifoliae* (Diptilomiopidae). Plant Pest Res 6(1):71–83. (in Persain with English abstract)

Lundgren JG (2009) Relationships of natural enemies and non-prey foods, progress in biological control. Springer, Dordrecht

Madadi H, Enkegaard A, Brødsgaard HF et al (2007) Host plant effects on the functional response of Neoseiulus cucumeris to onion thrips larvae. J Appl Entomol 131(9–10). https://doi.org/10.1111/j.1439-0418.2007.01206.x

Mahjoori M, Hajizadeh J, Abbssii Mozhdehi MR (2014) Mites of the family Laelapidae (Acari: Mesostigmata) associated with olive orchards in Guilan Province Iran. Linzer Biol Beitr 46 (2):1599–1606

Mahr DL, Whitaker P, Ridgway N (2008) Biological control of insects and mites. An introduction to beneficial natural enemies and their use in pest management. Department of Entomology, University of Wisconsin, USA

Maleknia BA, Zahedi Golpayegani A, Saboori A (2012) The effect of host plant changing on the olfactory response of *Phytoseiulus persimilis* (Acari: Phytoseiidae). Paper presented at the 20th Iranian Plant Protection Congress, Shiraz University, Shiraz, 25–28 August 2012

Maroufpoor M (2016) The two-sex life table of the predatory mite *Amblyseius swirskii* fed on *Phyllocoptes adalius* in laboratory conditions. Appl Entomol Phytopathol 83(2):189–200. (in Persian with English abstract)

McMurtry JA (1977) Description and biology of *Typhlodromus persianus*, n. sp., from Iran, with notes on *T. kettanehi* (Acari: Mesostigmata: Phytoseiidae). Ann Ento Soc America:563–568. https://doi.org/10.1093/aesa/70.4.563

McMurtry JA, Huffaker CB, van de Vrie M (1970) Ecology of Tetranychid mites and their natural enemies: a review. I. Tetranychid enemies: their biological characters and the impact of spray practices. Hilgardia 40:331–390

McMurtry JA, De Moraes GJ, Sourassou NF (2013) Revision of the life style of Phytoseiid mites (Acari: Phytoseiidae) and implications for biological control strategies. Syst Appl Acarol 18:297–320. https://doi.org/10.11158/saa.18.4.1

McMurtry JA, Sourassou NF, Demite PR (2015) The Phytoseiidae (Acar: Mesostigmata) as biological control agents. In: Carrilo D, de Moraes G, Peña J (eds) Prospects for biological control of plant feeding mites and other harmful organisms, progress in biological control. Springer, Cham, pp 133–149

McMurty JA (1982) The use of Phytoseiids for biological control, prospects. In: Hoy MA (ed) Recent advance in knowledge of the phytoseiidae. Berkeley, Division of Agricultural Science, University of California, Berkeley Publishin, pp 23–48

Mehrnejad MR, Ueckermann EA (2001) Mites (Arthropoda, Acari) associated with pistachio trees (Anacardiacea) in Iran (I). Syst Appl Acarol 6:1–12. https://doi.org/10.11158/saasp.6.1.1

Minero JLDEC, Lindquist EE, De Moraes GJ (2009) Edaphic Ascid mites (Acari: Mesostigmata: Ascidae) from the state of Sao Paulo, Brazil, with description of five new species. Zootaxa 2024 (1):1–32. https://doi.org/10.11646/zootaxa.2024.1.1

Moghadasi M, Allahyari H (2016) Effect of different pretor: prey ratios of *Phytoseiulus persimilis* and *Typhlodromus bagdasarjani* (Acari: Phytoseiidae) on reduction of *Tetranychus urticae* (Acari: Tetranychidae) on cucumber under microcosm conditions. In: Karimi J (ed) Proceedings of the 3rd national meeting on biological control in agriculture and natural resources. Ferdowsi University of Mashhad, Mashhad

Moghadasi M, Saboori A, Allahyari H, Zahedi Golpayegani A (2012a) Prey-stage preference in *Typhlodromus bagdasarjani* and *Phytoseiulus persimilis* (Acari: Phytoseiidae) on *Tetranychus urticae* (Acari: Tetranychidae) on rose. Paper presented at the 20th Iranian Plant Protection Congress, Shiraz University, Shiraz, 25–28 August 2012

Moghadasi M, Saboori A, Allahyari H, Zahedi Golpayegani A (2012b) Functional response of *Typhlodromus bagdasarjani* and *Phytoseiulus persimilis* (Acari: Phytoseiidae) on eggs of

*Tetranychus urticae* (Acari: Tetranychidae) on rose. Paper presented at the 20th Iranian Plant Protection Congress, Shiraz University, Shiraz, 25–28 August 2012

Moghadasi M, Saboori A, Allahyari H, Zahedi Golpayegani A (2013a) Functional response of *Typhlodromus bagdasarjani* and *Phytoseiulus persimilis* (Acari: Phytoseiidae) feeding on *Tetranychus urticae* (Acari: Tetranychidae) on rose. Plant Pest Res 2(4):55–65. (in Persian with English abstract)

Moghadasi M, Saboori A, Allahyari H, Zahedi Golpayegani A (2013b) Predation rate of *Typhlodromus bagdasarjani* and *Phytoseiulus persimilis* (Acari: Phytoseiidae) fed on *Tetranychus urticae* (Acari: Tetranychidae) on rose. In: Talaei-Hassanloui R (ed) Proceedings of the conference of biological control in agriculture and natural resources. College of Agriculture and Natural Resources, Karaj

Moghadasi M, Saboori A, Allahyari H, Zahedi Golpayegani A (2013c) Life table parameters of *Typhlodromus bagdasarjani*and *Phytoseiulus persimilis* (Acari.: Phytoseiidae) fed on *Tetranychus urticae* (Acari.: Tetranychidae) on rose. In: Talaei-Hassanloui R (ed) Proceedings of the conference of biological control in agriculture and natural resources. College of Agriculture and Natural Resources, Karaj

Mohammadi H, Saboori A, Zahedi Golpayegani A (2012) *Phytoseiulus persimilis* (Acari: Phytoseiidae) attraction towards infested plant volatiles intensifies by experience. Paper presented at the 20th Iranian Plant Protection Congress, Shiraz University, Shiraz, 25–28 August 2012

Mohiseni AA, Soleymannejadian I, Arbabi M (2006) Integrated control of *Tetranychus urticae* Koch. by using the predatory mite *Phytoseiulus persimilis* Athias-Heenriot and chemicals in Brujerd. Paper presented at the 17th Iranian Plant Protection Congress, College of Agriculture and Natural Resources, University of Tehran, Karaj, 2–5 September 2006

Moreira GF, de Moraes G (2015) The potential of free-living Laelapid mites (Mesostigmata: Laelapidae) as biological control agents. In: Carrilo D et al (eds) Prospects for biological control of plant feeding mites and other harmful organisms, progress in biological control. Springer International Publishing, pp 77–97

Mortazavi N, Fathipour Y, Talebi AA (2015a) Functional response of *Typhlodromus bagdasarjani* (Acari: Phytoseiidae) on eggs of the two-spotted spider mite on Strawberry. In: Manzari S (ed) Proceedings of the 1st Iranian International Congress of Entomology, Tehran

Mortazavi N, Fathipour Y, Talebi AA (2015b) Biology of *Typhlodromus bagdasarjani* (Acari: Phytoseiidae) on the two-spotted spider mite. In: Manzari S (ed) Proceedings of the 1st Iranian International Congress of Entomology, Tehran

Mortazavi N, Fathipour Y, Talebi AA (2017a) Effect of spider mite-produced web and whitefly-produced honeydew on *Amblyseius swirskii* (Acari-Phytoseiidae). In: Naeimi S (ed) Proceedings of the 8th national conference on biological control in agriculture and natural resources, Rasht

Mortazavi N, Fathipour Y, Talebi AA (2017b) Prey preference of *Amblyseius swirskii* between *Tetranychus urticae* and *Trialeurodes vaporariorum* on strawberry. In: Talaei-Hasssanloui R (ed) Proceedings of the 2nd Iranian International Congress of Entomology. College of Agriculture and Natural Resources, University of Tehran, Karaj

Mossadegh MS (1995) Some predatory mite (Phytoseiidae) from honeybee *Apis mellifera* L. hives in Iran. Paper presented at the 12th Iranian Plant Protection Congress, Karaj Junior College of Agriculture, Karaj, 2–7 September 1995

Nadeali T, Zahedi Golpayegani A, Saboori A (2012) The effect of prey, *Tetranychus urticae* Koch (Acari: Tetranychidae) density on oviposition rate of the predatory mite, *Neoseiulus californicus* (McGregor) (Acari: Phytoseiidae). Paper presented at the 20th Iranian Plant Protection Congress, Shiraz Univeristy, Shiraz, 25–28 August 2012

Nadimi A, Kamali K, Arbabi M (2006) Study on the side-effects of three acarides on *Phytoseilus persimilis* under laboratory condition. Paper presented at the 17th Iranian Plant Protection Congress, College of Agriculture and Natural Resources, University of Tehran, Karaj, 2–5 September 2006

Nadimi A, Kamali K, Arbabi M, Abdoli F (2008a) Side-effects of three acarides on the predatory mite, *Phytoseius plumifer* (Acari: Phytoseiidae). Paper presented at the 18th Iranian Plant Protection Congress, Bu-Ali Sina University, Hamedan, 25–28 August 2008

Nadimi A, Kamali K, Arbabi M et al (2008b) Side-effects of three acarides on the predatory mite, *Phytoseiulus persimilis* Athias-henriot (Acari: Phytoseiidae) under laboratory conditions. Mun Entomol Zool 3(2):556–567

Nazari M, Hajizadeh J (2013) A checklist to the parasitid mites (Mesostigmata, Parasitidae) of Iran with nine new records and a key for Guilan Province Prasitidae species. Entomofauna 34 (30):397–408

Nemati A, Riahi E, Abasian S, Kamali K (2008) Preliminary study of biology of *Phytoseiulus tropicalis* (Acari: Mesostigmata). Paper presented at the 18th Iranian Plant Protection Congress, Bu-Ali Sina University, Hamedan, 25–28 August

Nemati A, Gwiazdowicz DJ, Riahi E et al (2012) Catalogue of the Iranian Mesostigmatid mites. Part 1: Family Ascidae. Intl J Agric Crop Sci 4(14):1414–1420

Nomikou M, Janssen A, Schraag R et al (2002) Phytoseiid predators suppress populations of *Bemisia tabaci* on cucumber plants with alternative food. Exp Appl Acarol 27:57–68. https:// doi.org/10.1023/A:1021559421344

Notghi Moghadam B A, Hajizadeh J, Jalali Sendi J, Rafati Fard M (2008) Investigation on biology of predatory mite *Amblyseius herbicolus* Chant (Acari: Phytoseiidae) feeding on three different foods under laboratory conditions. Paper presented at the 18th Iranian Plant Protection Congress, Bu-Ali Sina University, Hamedan, 24–27 August 2008

Notghi Moghadam BA, Hajizadeh J, Jalali Sendi J, Rafati Fard M (2010) Investigation on feeding preference from different developmental stages of two spotted spider mite and host preference between two spotted spider mite and mulberry thrips by predatory mite *Amblyseius herbicolus* Chant under laboratory conditions. Paper presented at the 19th Iranian Plant Protection Congress, Iranian Research Institute of Plant Protection, Tehran, 31 July–3 August 2010

Onzo A, Hanna R, Sabelis MW et al (2005) Temporal and spatial dynamics of an exotic predatory mite candits herbivorous mite prey on Cassava in Benin, West Africa. Environ Entomol 34:866–874. https://doi.org/10.1603/0046-225X-34.4.866

Pedigo LP, Rice ME (2015) Entomology and pest management.Waveland Press, Inc, Illinois

Pirayeshfar F, Asgari F, Sarraf Moayeri HR (2017a) Comparative assessment of some biological characteristics of the predatory mite *Amblyseius swirskii* by feeding of alive and non-hatchable eggs of *Tyrophagous putrescentiae*. In: Naeimi S (ed) Proceedings of the 8th national conference on biological control in agriculture and natural resources, Rasht

Pirayeshfar F, Sarraf Moayeri HR, Safavi SA (2017b) Application of supplementary food sources for increased performance of phytoseiid predatory mites: with introduction of Banker Food for the predatory mite, *Amblyseius swirskii*. In: Naeimi S (ed) Proceedings of the 8th national conference on biological control in agriculture and natural resources, Rasht

Rafatifard M, Hajizadeh J, Arbabi M (2004) Biology of predatory mite, *Typhlodromips caspiansis* Denmark and Daneshvar (Acari: Phytoseiidae) feeding on two spotted spider mite *Tetranychus urticae* Koch (Acari: Tetranychidae) in laboratory conditions. J Agric Sci 1(1):61–67

Ragusa E, Tsolakis H, Jordà Palomero R (2009) Effect of pollens and preys on various biological parameters of the generalist mite *Cydnodromus californicus*. Bull Insectol 62(2):153–158

Rahmani H (2011) Report of some edaphic Mesostigmatic mites (Acari) from Iran and Zanjan Province. In: Kazemi S, Saboori (eds) Proceedings of the 1st Persian Congress of Acarology, Kerman

Rahmani H, Kamali K, Fathipour Y, Faraji F (2008) Spatial distribution pattern of *Panonychus ulmi* (Tetranychidae) and its predator *Zetzellia mali* (Stigmaeidae) in Zanjan region. Paper presented at the 18th Iranian Plant Protection Congress, Bu-Ali Sina University, Hamedan, 25–28 August 2008

Rahmani H, Kamali K, Faraji F (2010) Predatory mite fauna of Phytoseiidae of Northwest Iran (Acari: Mesostigmata). Turk J Zool 34:497–508. https://doi.org/10.3906/zoo-0902-23

Rashki M, Saboori A, Nozari J (2004) Biology of *Cenopalpus irani* (Acari: Tenuipalpidae) and its predators in Mahdasht region, Karaj. Paper presented at the 16th Iranian Plant Protection Congress, Tabriz University, Tabriz, 28 August–1 September 2004

Rezaie M (2010a) The effect of cannibalism and interspecific predation on biologiy of predatory mite *Amblyseius swirskii* (Athias-Henriot) (Acari: Phytoseiidae). Paper presented at the 19th Iranian Plant Protection Congress, Iranian Research Institute of Plant Protection, Tehran, Tehran, 31 July–3 August 2010

Rezaie M (2010b) Effect of host plants on biology of predatoy mite *Amblyseius swirskii* (Athias-Henriot) (Acari: Phytoseiidae). Paper presented at the 19th Iranian Plant Protection Congress, Iranian Research Institute of Plant Protection, Tehran, Tehran, 31 July–3 August 2010

Rezaie M (2010c) Investigation on biology of predatory mite *Neoseiulus barkeri* (Hughes) (Acari: Phytoseiidae) feeding on three different food under laboratory conditions. Paper presented at the 19th Iranian Plant Protection Congress, Iranian Research Institute of Plant Protection, Tehran, Tehran, 31 July–3 August 2010

Rezaie M (2013) The effect of host plants on biology of predatory mite *Neoseiulus barkeri* (Acari.: Phytoseiidae). In: Talaei-Hassanloui R (ed) Proceedings of the conference of biological control in agriculture and natural resources. College of Agriculture and Natural Resources, Karaj

Rezaie M (2016) Comparison olfactory response of predatory mite, *Neoseiulus californicus* to two-spotted spider mite-infested and western flower thrips-infested strawberry plants. In: Karimi J (ed) Proceedings of the 3rd national meeting on biological control in agriculture and natural resources. Ferdowsi University of Mashhad, Mashhad

Rezaie M (2018) Suitability of different plant pollens as supplementary food source and natural prey for predatory mite, *Neoseiulus barkeri* Hughes (Acari: Phytoseiidae). Plant Prot 41 (4):77–90. (in Persian with English abstract). https://doi.org/10.22055/ppr.2019.14155

Rezaie M, Askari S (2015) Effect of feeding with different plant pollens on biological parameters of *Neoseiulus barkeri* (Hughes) (Acari: Phytoseiidae). In: Manzari S (ed) Proceedings of the 1st Iranian International Congress of Entomology, Tehran

Rezaie M, JavanNezhad R (2018) The effect of different factors on mass rearing of predatory mite *Neoseiulus californicus* (McGregor) (Acari: Phytoseiidae). J Appl Res Plant Protect 7(1):85–97. (in Persian with English abstract)

Rezaie M, Montazerie F (2015) Effect of feeding with different plant pollens on predation rate of *Neoseiulus californicus* (McGregor). In: Manzari S (ed) Proceedings of the 1st Iranian International Congress of Entomology, Tehran

Rezaie M, Saboori A, Baniameri V, Allahyari H (2012) Host preference of *Neoseiulus californicus* (McGregor) on different developmental stages of *Tetranychus urticae* Koch in seven strawberry cultivars. Paper presented at the 20th Iranian Plant Protection Congress, Shiraz University, Shiraz, 25–28 August 2012

Rezaie M, Baniamerie V, Saboori A (2017) Functional response and predation interference of (*Neoseiulus californicus*) (Acari: Phytoseiidae) feeding on the western flower thrips larvae on several commercial strawberry cultivars. Plant Pest Res, 6 (4): 1–15. (in Persian with English abstract)

Riahi E, Fathipour Y, Talebi AA, Mehrabadi M (2015) Comparison of biological parameters of *Typhlodromus bagdasarjani* (Acari: Phytoseiidae) feeding on different pollens. In: Manzari S (ed) Proceedings of the 1st Iranian International Congress of Entomology, Tehran

Sabelis MW (1985) Capacity of population increase, development. In: Helle W, Sabelis MW (eds) Spider mite, their biology. Natural Enemies and Control. Elsevier Publication, Amsterdam, pp 35–55

Saboori A, Kamali K (1995) Observations on biology of *Allothrombium pulvinum* Ewing (Acari: Trombidiidae) an important natural enemy of aphids and spider mites in west Mazandaran. Paper presented at the 12th Iranian Plant Protection Congress, Karaj Junior College of Agriculture, Karaj, 2–7 September 1995

Saboori A, Kamali K (1998) Some observations on biology of *Allothrombium shirazicum* Zhang 1996 (Acari: Trombidiidae) a natural enemy of *Aphis punicae* in Garmsar. Paper presented at the

13th Iranian Plant Protection Congress, Karaj Junior College of Agriculture, Karaj, 23–27 August 1998

Saboori A, Zhang ZQ (1996) Biology of *Allothrombium pulvinum* Ewing (Acari: Trombidiidae) in West Mazandran, Iran. Exp Appl Acarol 20(3):137–142. https://doi.org/10.1007/BF00051479

Saboori A, Hosseini M, Hatami B (2003) Preference of adults of *Allothrombium pulvinum* Ewing (Acari: Trombidiidae) for eggs of *Planococcus citri* (Risso) and *Pulvinaria aurantii* Cockerell on citrus leaves in the laboratory. Syst Appl Acarol 8(1):49–54. https://doi.org/10.11158/saa.8.1.5

Sanatgar E, Vafaei Shoushtari R, Zamani AA et al (2011) Effect of frequent application of hexythiazox on predatory mite *Phytoseiulus persimilis* Athias-Henriot (Acari: Phytoseiidae). Acad J Entomol 4(3):94–101

Sanatgar E, Vafaei Shoushtari R, Zamani AA, Arbabi M, Soleyman Nejadian E (2012) Effect of frequent application of bifenazate on biology and biological parameters of predatory mite *Phytoseiulus persimilis* Athias-Henriot (Acari: Phytoseiidae). Paper presented at the 20th Iranian Plant Protection Congress, Shiraz University, Shiraz, 25–28 August 2012

Sarbaz S, Goldasteh S, Zamani AA et al (2017) Side effects of spiromesifen and spirodiclofen on life table parameters of the predatory mite, *Neoseiulus californicus* McGregor (Acari: Phytoseiidae). Int J Acarol 43(5):380–386. https://doi.org/10.1080/01647954.2017.1325396

Seidpisheh E, Rahmani H, Kavousi A (2016) Biological and feeding attributes of the native predatory mite, *Typhlodromus bagdasarjani* (Acari: Phytoseiidae) under Azadirachtin and Spirodiclofen treatments. In: Karimi J (ed) Proceedings of the 3rd national meeting on biological control in agriculture and natural resources. Ferdowsi University of Mashhad, Mashhad

Seiedy M, Tork M (2013) Investigation of some behavioral parameters in *Phytoseiulus persimilis* (Acari: Phytoseiidae) feeding on untreated and *Beauveria bassiana*–treated adults of *Tetranychus urticae* (Acari: Tetranychidae). In: Talaei-Hassanloui R (ed) Proceedings of the conference of biological control in agriculture and natural resources. College of Agriculture and Natural Resources, Karaj

Seiedy M, Saboori A, Allahyari H, Talaei-Hassanloui R, Zahedi Golpayegani A, Tork M (2012) Orientation of *Phytoseiulus persimilis* (Acari: Phytoseiidae) to healthy and *Beauveria bassiana*–infected adults of *Tetranychus urticae* (Acari: Tetranychidae) in an olfactometer system. Paper presented at the 20th Iranian Plant Protection Congress, Shiraz University, Shiraz, 25–28 August 2012

Sepasgosarian H (1977) The 20 years research of acarology in Iran. J Iran Soc Eng 56:40–50

Shirazi J, Attaran M, Farrokhi S, Dadpour H, Nouri H (2011) An analytical review on the classical biological control of pests in Iran and the world. Paper presented at 1st Biological Control Development Congress in Iran, Iranian Research Institute of Plant Protection, Tehran, 27–28 July 2011

Shirdel D, Arbabi M (2012a) Functional response of the predatory mite, *Euseius finlandicus* (Oudemans) (Acari: Phytoseiidae) to European red mite. Paper presented at the 20th Iranian Plant Protection Congress, Shiraz University, Shiraz, 25–28 August 2012

Shirdel D, Arbabi M (2012b) Functional response of the predatory mite, *Typhlodromus bagdasarjani* (Wainstein and Arutunjan) (Acari: Phytoseiidae) to European red mite. Paper presented at the 20th Iranian Plant Protection Congress, Shiraz University, Shiraz, 25–28 August 2012

Shirdel D, Talebi-Chaichi P, Maleki-Milani H (2000) Study of predatory mite (*Phytoseiulus persimilis* A.-H.) different densities effect on two-spotted spider mite (*Tetranychus urticae* Koch) population control on glasshouse cucumber. Paper presented at the 14th Iranian Plant Protection Congress, Isfahan University of Technology, Isfahan, 5–8 September 2000

Shirdel D, Kamali K, Ostovan H, Arbabi M (2004a) Functional response of the predatory mite, *Typhlodromus kettanehi* Dosse (Acari: Phytoseiidae), on two-spotted spider mite. Paper presented at the 16th Iranian Plant Protection Congress, Tabriz University, Tabriz, 28 August–1 September 2004

Shirdel D, Kamali K, Ostovan H, Arbabi M (2004b) Biology of the predatory mite, *Euseius finlandicus* (Oudemans)(Acari: Phytoseiidae) on two spotted spider mite under laboratory conditions. Paper presented at the 16th Iranian Plant Protection Congress, Tabriz University, Tabriz, 28 August–1 September 2004

Shirdel D, Kamali K, Ostovan H, Arbabi M (2006a) Functional response of the predatory mite, *Euseius finlandicus* (Oudemans)(Acari: Phytoseiidae) on two spotted spider mite. Paper presented at the 17th Iranian Plant Protection Congress, Tehran University, Karaj, 2–5 September 2006, p 51

Shirdel D, Kamali K, Ostovan H, Arbabi M (2006b) Biology of the predatory mite, *Typhlodromus kettanehi* Dosse (Acari: Phytoseiidae), on two-spotted spider mite. College of Agriculture and Natural Resources, University of Tehran, Karaj, 2–5 September 2006

Soleymani S, Hakimitabar M, Seiedy M (2015) The predation rate of *Amblyseius swirskii* Athias-Henriot (Acari: Phytoseiidae) on *Tetranychus urticae* Koch (Acari: Tetranychidae) and *Bemisia tabaci* Gennadius (Hem: Aleyrodidae). In: Manzari S (ed) Proceedings of the 1st Iranian International Congress of Entomology, Tehran

Tahmasebi Z, Hosseinzadeh A, Zahedi A (2012) Effects of the herbivore induced plant volatiles of four common bean cultivars on the performance of the predatory mite, *Phytoseiulus Persimilis* (Acari: Phytoseiidae). Iran J Plant Prot Sci 43(2):275–282. (in Persian with English abstract)

Tajmiri P (2013) An introduction of the collected Prostigmatic (Acari: Trombidiformes) mites on raspberry shrubs in central area of Guilan Province. Plant Pest Res 3(1):1–9. (in Persian with English abstract)

Tavoosi Ajvad F, Madadi H, Zafari D (2020) Combined applications of an entomopathogenic fungus and a predatory mite to control fungus gnats (Diptera: Sciaridae) in mushroom production. Biol Control. https://doi.org/10.1016/j.biocontrol.2019.104101

Tavosi S, Arbabi M, Sanatgar E (2017) Different releasing ratio of *Phytoseiulus persimilis* A. H. in biological control of *Tetranychus urticae* Koch in greenhouse roses. J Entomol Res 9 (3):259–269. (in Persian with English abstract)

Villanueva RT, Harmsen L (1998) Studies on the role of stigmaeid predator *Zetzellia mali* in the acarine system of apple foliage. Proc Entomol Soc Ontario, Guelph, Ontario 129:149–155

Zahedi Golpayegani A, Saboori A, Kafil M, Yaghoubi S, Mohammadi H, Maleknia B (2010) The priority of volatiles role in interactions of biological control systems when *Amphitetranychus viennensis* (Zacher) (Acari: Tetranychidae) and *Phytoseiulus persimilis* Athias-Henriot (Acari: Phytoseiidae) are present. Paper presented at the 19th Iranian Plant Protection Congress, Iranian Research Institute of Plant Protection, Tehran, 31 July–3 August, 2010

Zahedi Golpayegani A, Saboori A, Kafil M et al (2019) Within plant migration as an alternative counter-measuring avoidance behavior in *Zetzellia mali* (Ewing) (Acari: Stigmaeidae). Biol Control Pests Plant Dis 8(1):27–35. (in Persian with English abstract)

Zamanpour M, Sedaratian-Jahromi A, Mohammadi H et al (2019) The effect of *Beauveria bassiana* on preference and switching behavior in *Phytoseiulus persimilis* (Acari: Phytoseiidae). Plant Pest Res 9(1):75–93. (in Persian with English abstract)

Zare M, Rahmani H, Faraji F, Akrami MA (2012) First description of the male of *Transeius avetianae* (Arutunjan & Ohandjanian) (Mesostigmata: Phytoseiidae) and redescription of the female. Syst Appl Acarol 17:254–260. https://doi.org/10.11158/saa.17.3.5

Zdarkova E (1979) Cheyletid fauna associated with stored products in Czechoslovakia. J Stored Prod Res 15(1):11–16. https://doi.org/10.1016/0022-474X(79)90019-5

Zdarkova E (1998) Biological control of storage mites by *Cheyletus eruditus*. Integr Pest Manag Rev 3:111–116. https://doi.org/10.1023/A:1009647814477

Zhang ZQ (2003) Mites of greenhouses: identification, biology and control. CABI Publishing, Wallingford

Zhang ZQ, Faraji F (1994) Notes on *Allothrombium pulvinum* Ewing (Acari: Trombidiidae) new to the Fauna of Iran. Acarologia 35(4):357–360

Zhimo Z, McMurtry JA (1990) Development and reproduction of 3 *Euseius* (Acari: Phytoseiidae) species in the presence and absence of supplementary foods. Exp Appl Acarol 8:233–242. https://doi.org/10.1007/BF01202134

# Chapter 5
# Lacewings: Research and Applied Aspects

Mahdi Hassanpour, Mohammad Asadi, Ali Jooyandeh, and Hossein Madadi

## 5.1 Introduction: Why Lacewings Are Important?

The Order Neuroptera encompasses 17 families, three of which including Chrysopidae (green lacewings), Hemerobiidae (brown lacewings), and
Coniopterygidae (dustywing lacewings) are considered as the most important families. These families are highly effective agents in augmentative biocontrol programs
to control a wide variety of target pests such as aphids, scales, thrips, leafhoppers,
and mites (New 1999). Based on a review of literature, 217 species from this order
are found in Iran, among which 46, 4, and 25 species belong to Chrysopidae,
Hemerobiidae, and Coniopterygidae families, respectively, covering 25 out of
31 provinces of Iran (Mirmoayedi 2002a, b; Mirmoayedi 2008; Farahi et al. 2009).
Green lacewings are an important group of insect predators. The adults of most
chrysopids are not predaceous; rather, they feed on pollen, plant nectar, and insect
honeydew (e.g., aphid or coccid honeydew; Fig. 5.1a); however, the larvae
(Fig. 5.1c) feed on a wide range of small soft-bodied insects, mites as well as eggs
and young caterpillars (Canard and Principi 1984). They have hollow, sickle-shaped
mandibles used for sucking fluids from their victims. Pupation takes place inside a
silken cocoon (Fig. 5.1d). Green lacewing eggs (Fig. 5.1b) are available from
commercial insectaries and may be purchased for augmentative releases (Uddin
et al. 2005; Woolfolk et al. 2007). The importance of green lacewings as natural

M. Hassanpour (✉) · M. Asadi
Department of Plant Protection, University of Mohaghegh Ardabili, Ardabil, Iran
e-mail: hassanpour@uma.ac.ir

A. Jooyandeh
Department of Plant Protection, Khorasan Razavi Agriculture and Natural Resources Research
and Education Center, Mashhad, Iran

H. Madadi
Department of Plant Protection, Faculty of Agriculture, Bu-Ali Sina University, Hamedan, Iran

© The Author(s), under exclusive license to Springer Nature Switzerland AG 2021          175
J. Karimi, H. Madadi (eds.), *Biological Control of Insect and Mite Pests in Iran*,
Progress in Biological Control 18, https://doi.org/10.1007/978-3-030-63990-7_5

**Fig. 5.1** Life stages of *Chrysoperla carnea*. (**a**) adult, (**b**) eggs, (**c**) larva while feeding on prey, (**d**) pupa (Permission from Jooyandeh A)

enemies of several important pests along with ease of production have led to considerable researches on this useful group within the last two decades in Iran. Lacewing populations often lag behind their prey and may not be able to satisfactorily reduce heavy infestations of pests.

## 5.2 Review on Taxonomy of Neuroptera

The Order Neuroptera represents one of the oldest lineages of endopterygota undergoing complete metamorphosis. Neuroptera comprises three suborders divided into 17 extant families containing more than 6000 species (Mansell 2002). The families Myrmeleontidae with more than 2000 species and Chrysopidae with over 1200 species are the most species-rich families, followed by Hemerobiidae (ca. 550 species) and Ascalaphidae (ca. 400 species). The suborder Nevrorthiformia with the single family of Nevrorthidae represents the most basal group, which is sporadically found in

Australia, Japan, Taiwan and Europe (Mirmoayedi 2002a). The suborder Myrmeleontiformia contains five families, namely Myrmeleontidae, Ascalaphidae, Nemopteridae, Psychopsidae, and Nymphidae. It is a well-defined group of lacewings with soil-dwelling or arboreal larvae. The suborder Hemerobiiformia contains 11 families including Ithonidae, Rapismatidae, Polystoechotidae, Osmylidae, Chrysopidae, Hemerobiidae, Coniopterygidae, Mantispidae, Rachiberothidae, Dilaridae, and Berothidae, many of which have highly specialized life cycles (New 1985; Monserrat 2002; Cannings and Cannings 2006).

## 5.3   Reports on Chrysopidae Species from Iran

Studies have been conducted to identify the Iranian green lacewings from early 1970s. The first checklist of Iranian fauna of Chrysopidae was published by Farahbakhsh in 1961 (Mirmoayedi 2008), in which the major agricultural pests of Iran have been listed.

The Chrysopidae fauna of Iran was studied by Hölzel (1966, 1967, 1982) who was the first European neuropterologist to study Iranian Chrysopidae fauna in detail. He never visited Iran and the specimens he studied were collected mainly by A. Vartian and E. Vartian, an Austrian-Armenian couple and amateur insect collectors visiting Iran as tourists. The total number of Chrysopidae from Iran reported by them and others were 45 species (Mirmoayedi 2002a). Mirmoayedi (1993) investigated Chrysopidae fauna from different regions of Kermanshah province and identified six species, including *C. carnea*, *Chrysopa septempunctata* Linnaeus, *Chrysopa viridana* Schneider, *Chrysopa dubitans* McLachlan, *Suarius nanus* (McLachlan), and *S. fedtschenkoi* (McLachlan). Later, he reported three previously unrecorded species from this province, namely *Mallada derbendica* Hölzel, *Suarius paghmana* Hölzel and *Suarius vartianae* (Hölzel) (Mirmoayedi 1995). Mirmoayedi and Yassayie (1998) studied the fauna of Neuroptera in Golestan National Park located in northeastern Iran and reported seven species of Chrysopidae, namely *Anisochrysa amseli* Brauer, *Anisochrysa flavifrons* (Brauer), *Anisochrysa prasina* Burmeister, *C. dubitans*, *C. viridana*, *S. nanus*, and *C. carnea*. Mirmoayedi (1998) published a checklist of Neuroptera in Iran comprising 39 species collected and identified between 1991 to 1996 from different regions such as Tehran, Guilan, Hormozgan, Markazi, Kermanshah, Khuzestan, and Ilam provinces. Among them, 11 species were belong to Chrysopidae (*C. dubitans*, *C. septempunctata*, *C. viridana*, *Chrysopa iranica* (Hölzel), *C. carnea*, *M. derbendica*, *Mallada prasina* (Hölzel), *S. fedtschenkoi*, *S. nanus*, *S. paghmana* and *S. vartianae* (Mirmoayedi 1998)). One year later, he published a list of seven species of Chrysopidae from Kermanshah and Kurdistan provinces (Mirmoayedi 1999a). In the same year, Mirmoayedi studied the Neuroptera fauna in Shiraz (Fars province) and identified seven species (Mirmoayedi 1999b).

Hölzel (1966) published records of 11 chrysopid species for the fauna of Iran. One year later, he published a list of Chrysopidae species comprising 23 species and described their morphological features, presented identification keys and recorded data on collecting dates as well as localities of each species (Hölzel 1967). A few years later, Mirmoayedi described two new Chrysopidae species, namely *Anisochrysa mira* (Hölzel) and *Suarius ressli* (Hölzel). In 1982, Hölzel described a new chrysopid, *Suarius laristanus* Hölzel, which was collected previously by Ressl in southern Iran. Aspöck et al. (1980) also worked on Chrysopidae fauna of Iran. In their book "Die Neuropteren Europas", Aspöck et al. mentioned four species of *Anisochrysa*, six species of *Chrysopa*, one species of *Chrysoperla,* and one species of *Suarius* as forming part of Chrysopidae fauna of Iran. Asadi (2010) reported 11 species of Chrysopidae including *C. dubitans, Chrysopa pallens* (Rambur), *C. viridana, Dichochrysa prasina* (Burmeister), *C. carnea, Chrysoperla kolthoffi* Navas, *Chrysoperla lucasina* (Lacroix), *Chrysoperla sillemi* Esben-Petersen, *I. vartianorum, M. derbendica* and *S. nanus* for Kermanshah province. Farahi et al. (2009) studied different species of Chrysopidae and Hemerobiidae from northeastern and eastern provinces of Iran and reported 12 species, seven of which were new for the investigated regions, including one new species for Iran (*Hemerobius stigma* Stephens). A brief literature review regarding Chrysopidae fauna of Iran (from 1961 to 2000) was carried out by Mirmoayedi (2002a). Mirmoayedi et al. (2014) studied new species of Neuroptera from West Azerbaijan province. They concluded that between the collected materials, there were 24 species belonging to six families of the Order Neuroptera and that all the species (except *C. carnea*) were recorded for the first time for the fauna of this province. Six species were new recordings for the fauna of Iran.

Surveys and collections of Iranian Neuroptera fauna have shown that *C. carnea* is a dominant species (Heydari 1995; Sharififard and Mossadegh 2006). However, several reports have indicated that other species such as *Chrysopa formosa* Brauer (Daniali et al. 1995; Ghadiri-Rad et al. 2004) and *C. lucasina* (Farahi et al. 2009; Kazemi and Mehrnejad 2011) were more abundant than *C. carnea* in some regions. It is clear that *C. carnea* contains several cryptic species comprising the so-called "*carnea* group" (Henry 1985; Haruyama et al. 2008). This group includes a complex of about 20 cryptic species distributed throughout the northern hemisphere and Afrotropics (Henry et al. 2002). The morphological similarities between species make it difficult to differentiate among them. The results of studies conducted on courtship behaviors of the cryptic species in "*carnea* group" have demonstrated that the unique courtship song of each cryptic species ensures reproductive isolation (Wells and Henry 1992; Henry et al. 2002). Therefore, further studies are required to identify the real species of "*carnea* complex" in Iran.

The list of Chrysopidae species of Iran is presented as follows:
*Anisochrysa amseli* (Brauer)
*Anisochrysa flavifrons* (Brauer)

*Anisochrysa genei* (Rambur)
*Anisochrysa mira* (Hölzel)
*Anisochrysa prasina* (Burmeister)
*Chrysopa caviceps* (McLachlan)
*Chrysopa derbendica* (Hölzel)
*Chrysopa dubitans* (McLachlan)
*Chrysopa formosa* (Brauer)
*Chrysopa nigricostata* Brauer
*Chrysopa pallens* (Rambur)
*Chrysopa perla* (Linnaeus)
*Chrysopa septempunctata* (Linnaeus)
*Chrysopa sogdianica* (McLachlan)
*Chrysopa viridana* (Schneider)
*Chrysopa vulgaris* (Hölzel)
*Chrysopa walkeri* (McLachlan)
*Chrysoperla carnea* (Stephens)
*Chrysoperla iranica* (Hölzel)
*Chrysoperla kolthoffi* (Navas)
*Chrysoperla lucasina* (Lacroix)
*Chrysoperla mutata* (McLachlan)
*Chrysoperla rotundata* (Navas)
*Chrysoperla sillemi* (Esben-Petersen)
*Chrysopidia ciliata* (Wesmael)
*Chrysotropia ciliata* (Wesmael)
*Cunctochrysa albolineata* (Killington)
*Dichochrysa derbendica* (Hölzel)
*Dichochrysa prasina* (Burmeister)
*Dichochrysa zelleri* (Schneider)
*Mallada derbendica* (Hölzel)
*Mallada prasina* (Hölzel)
*Italochrysa italic* (Rossi)
*Italochrysa stgmatica* (Rambur)
*Italochrysa vartianorum* (Hölzel)
*Suarius caviceps* (McLachlan)
*Suarius fedtschenkoi* (McLachlan)
*Suarius laristanus* (Hölzel)
*Suarius mongolicus* (Tjeder)
*Suarius nanus* (McLachlan)
*Suarius paghmana* (Hölzel)
*Suarius ressli* (Hölzel)
*Suarius vartianae* (Hölzel)

## 5.4    Reports on Coniopterygidae Species from Iran

Mirmoayedi (1995) recorded two Coniopterygidae species from Hormozgan province, among which *Coniopteryx deserta* Meinander was reported for the first time in Iran. Mirmoayedi (1998) also described *Coniopteryx (Xeroconiopteryx) loipetsederi* Aspöck and *Hemisemidalis pallida* (Withycombe) from Golestan National Park, of which *Co. (X.) loipetsederi* was newly recorded for Iranian fauna. To extend the knowledge of neuropteran fauna of Iran, collections were performed in the provinces of Tehran, Guilan, Hormozgan, Markazi, Kermanshah, Khuzestan, and Ilam. In this survey, the species belonging to Coniopterygidae family were *Aleuropteryx ressli* Rausch, Aspöck and Ohm (from Kermanshah and Ridjab cities), *Coniopteryx (Metaconiopteryx) lentiae* Aspöck and Aspöck, *Co. (X.) deserta*, *Hemisemidalis kasyi* (Aspöck and Aspöck), and *Nimboa vartianorum* Aspöck and Aspöck (all from Ridjab in Kermanshah province), *Co. (X.) unicef* Monserrat and *Co. (X.) venustula* Rausch and Aspöck (from Gazir in Hormozgan province), *Helicoconis pseudolutea* Ohm (from Bandar-e-Khamir in Hormozgan province and Ridjab in Kermanshah province), and *Semidalis aleyrodiformis* (Stephens) (from Minab in Hormozgan province) (Mirmoayedi 1998).

Mirmoayedi (1999b) described two new species of this family from Fars province, namely *Coniopteryx manka* Aspöck and Aspöck and *Coniopteryx farsi* Aspöck and Aspöck. These two species were reported as new records for Iranian fauna; the last species was a newly described one (Mirmoayedi 1999a). In the same year, Mirmoayedi reported *H. kasyi* and *H. pallida* from Kurdistan province. The species *Nimboa asadeva* Rausch and Aspöck, *Co. (H.) drammonti* Rousset and *Co. (X.) furcata* Meinander were reported from the provinces of Kurdistan, Hormozgan, Khorasan, and Hamedan, respectively (Mirmoayedi 1999b), and later, these species were identified as new records for Iran (Mirmoayedi 2002a). Overall, nearly 25 species of the family Coniopterygidae have been described from Iran (Mirmoayedi 2002b). *Co. (X.) furcata* and *Co. (X.) israelensis* Meinander (from Israel) and *Co. (X.) hastata* Meinander (from Iran) were described as new species, and *N. vartianorum*, *Co. drammonti*, *Co. lentiae*, *Co. (X.) unicef* and *H. kasyi* were new for the fauna of Iran (Meinander 1998). Asadi (2010) reported one species of Coniopterygidae, *A. ressli*, as a new record for fauna of Kermanshah province. Sziráki and Mirmoayedi (2012) prepared an annotated checklist of Iranian Coniopterygidae and stated that it encompassed all 24 species from Coniopterygidae family that were reported up to that time in Iran. Among them, five species were reported from the country for the first time. In addition, a number of species including *Aleuropteryx vartianorum* Aspöck and Aspöck, *Co. (C.) borealis* Tjeder, *Co. (X.) arenicola* Sziráki, *Co. (X.) kerzhneri* Meinander, *Co. (X.) latigonarcuata* Meinander and *Hemisemidalis hreblayi* Sziráki were new records for the Iranian fauna. Additionally, geographical distribution of a species (i.e. *N. asadeva*) was corrected in this report.

The list of Coniopterygidae species of Iran is given as follows:
**Subfamily: Aleuropteryginae Enderlein**
*Aleuropteryx ressli* Rausch, Aspöck and Ohm
*Aleuropteryx vartianorum* (Aspöck and Aspöck)
*Helicoconis (Ohmopteryx) pseudolutea* (Ohm)

**Subfamily: Coniopteryginae Burmeister**
*Coniopteryx (Coniopteryx) borealis* (Tjeder)
*Coniopteryx (Holoconiopteryx) drammonti* (Rousset)
*Coniopteryx farsi* Aspöck and Aspöck
*Coniopteryx (Metaconiopteryx) lentiae* (Aspöck and Aspöck)
*Coniopteryx (Xeroconiopteryx) arenicola* (Sziráki)
*Coniopteryx (Xeroconiopteryx) atlasensis* (Meinander)
*Coniopteryx (Xeroconiopteryx) deserta* (Meinander)
*Coniopteryx (Xeroconiopteryx) furcata* Meinander
*Coniopteryx (Xeroconiopteryx) hastata* (Meinander)
*Coniopteryx (Xeroconiopteryx) kerzhneri* (Meinander)
*Coniopteryx (Coniopteryx) latigonarcuata* (Meinander)
*Coniopteryx (Xeroconiopteryx) loipetsederi* (Aspöck)
*Coniopteryx (Xeroconiopteryx) manka* (Aspöck and Aspöck)
*Coniopteryx (Xeroconiopteryx) martinmeinanderi* (Sziráki)
*Coniopteryx (Xeroconiopteryx) orba* (Rausch and Aspöck)
*Coniopteryx (Xeroconiopteryx) unicef* (Monserrat)
*Coniopteryx (Xeroconiopteryx) venustula* (Rausch and Aspöck)
*Hemisemidalis hreblayi* (Sziráki)
*Hemisemidalis kasyi* (Aspöck and Aspöck)
*Hemisemidalis pallida* (Withycombc)
*Nimboa asadeva* (Rausch and Aspöck)
*Nimboa macroptera* (Aspöck and Aspöck)
*Semidalis aleyrodiformis* (Stephens )

## 5.5  Reports on Hemcrobiidae Species from Iran

Mirmoayedi recorded six species of Hemerobiidae (*Hemerobius zernyi* (Esben-Petersen), *Sympherobius pygmaeus* (Rambur), *Wesmaelius navasi* (Andreu), *Wesmaelius saudiarabicus* (Hölzel), *Hemerobius humulinus* L. and *Micromus variegatus* (Fabricius)) from different regions of Iran, among which *W. saudiarabicus* was recorded for the first time from Iran (Mirmoayedi 1998). Asadi (2010) reported three species of Hemerobiidae, i.e. *W. navasi*, *S pygmaeus*, and *Psectra maculata* (Burmeister), from Kermanshah province; *P. maculata* was recorded as a new species for fauna of Iran. Hemerobiidae species (Fig. 5.2) differ from Chrysopidae species not only in the usual color, but also in the wing venation. Hemerobiidae species have numerous long veins that are absent in chrysopids.

**Fig. 5.2** Brown lacewing
from family Hemerobiidae
(© Magnolia Press.
Reproduced with
permission from copyright
holder)

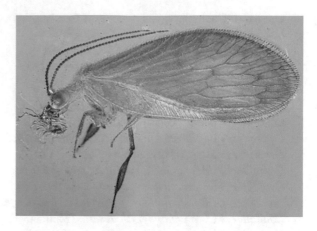

The list of Hemerobiidae species of Iran is given below:
*Hemerobius humulinus* (Linnaeus)
*Hemerobius stigma* (Stephens)
*Hemerobius zernyi* (Esben-Petersen)
*Micromus angulatus* (Walker)
*Micromus variegatus* (Fabricius)
*Psectra maculata* (Bermeister)
*Sympherobius pellucidus* (Walker)
*Sympherobius pygmaeus* (Rambur)
*Wesmaelius fassnidgei* (Andreu)
*Wesmaelius mongolicus* (Bermeister)
*Wesmaelius navasi* (Andreu)
*Wesmaelius saudiarabicus* (Hölzel)

## 5.6    Rearing of Green Lacewings

From among the several chrysopid species reported from Iran, *C. carnea* has
received the most attention as a biological control agent for research and release.
Mass rearing of *C. carnea* adults for commercial purpose to obtain eggs or larvae for
augmentative release in field or greenhouse conditions requires certain processes
such as the provision of food, harvesting of eggs, and cleaning of cages containing
*C. carnea* adults in an insectary. As a food source, the eggs of *Ephestia kuehniella*
Zeller and *Sitotroga cerealella* (Olivier) (Lepidoptera: Pyralidae) are generally used
to feed and rear larvae. However, *Ephestia* eggs have more positive effects than
*Sitotroga* eggs on larval and pupal weight of *C. carnea* (Jooyandeh 2009).

Lacewing larvae are highly predaceous and cannibalistic, making it difficult to
maintain and feed the larvae in crowd cultures. Jooyandeh (1999 and 2000) developed group rearing techniques of *C. carnea* larvae. The larval stages of *C. carnea*
were reared in rectangular feeding units (25 × 25 × 6 cm; Fig. 5.3a) where
development occurred until adult emergence. Two larval feeding methods were

**Fig. 5.3** (**a**) larvae of *Chrsysoperla carnea* collected from container, (**b**) rearing unit for *Chrysoperla carnea* adults, (**c**) ChrysoCard, (**d**) the ChrysoCard containing *Chrysoperla carnea* eggs on pear tree (Permission from Jooyandeh, A)

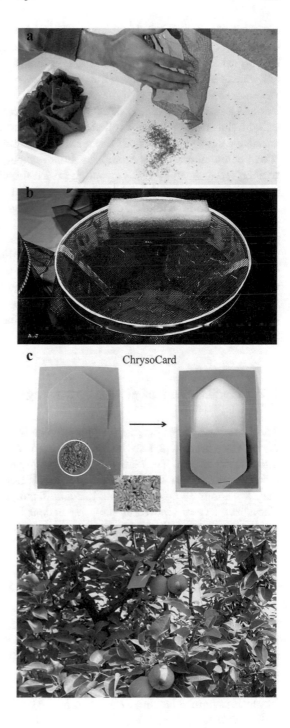

assessed. The best initial density was 600 and 1000 *C. carnea* eggs per rearing unit using the eggs and freezed adults of *S. cerealella* as food sources, respectively. A cylindrical PVC container (25 cm height and 20 cm diameter; Fig. 5.3b) was a convenient oviposition-feeding unit for *C. carnea* adults where they had access to water and a diet consisting of brewer's yeast, honey, and sugar. Water was supplied through a saturated sponge placed on top of the container. The eggs were removed using nylon netting loosely rolled into a ball and rubbed across the substrate on which the eggs were attached. The adult stage is much less expensive to produce, as it can be maintained on a mixture of honey, sugar, and brewer's yeast (1: 1: 1 by weight) (Jooyandeh 1999). Jooyandeh (1995) found that *C. carnea* females fed on yeast and sugar paste 1:1 laid on average a total of 554.6 eggs in 30 days while oviposition was high until 20 days, after which egg-laying gradually declined. In this reserach, the green lacewings were reared at 25 ± 2 °C, 60 ± 10% RH, and a photoperiod of 16: 8 (L: D) h. With some changes in the method of Vogt et al. (2000), the adult stage of *C. carnea* was reared on an artificial diet consisting of brewer's yeast, honey, and distilled water in a 4: 7: 5 ratio smeared as a paste on transparent plastic tape inserted in the cylindrical plastic containers (15.5 cm diameter; 25 cm height) (Hassanpour, 2011). He also reared *C. carnea* larvae in containers (17.5 cm diameter; 7.5 cm height with screen-covered ventilation holes in the lid), which held several layers of paper towel to reduce cannibalism among the larvae.

## 5.7 Life History of Green Lacewings

The lacewing species hibernate as pupal stage in protected areas of the soil cracks, under tree barks, and also in trash. In warm climates, the adults can be found throughout the year. They emerge in early spring and the females lay eggs on plants infected by their prey, which are commonly different species of aphids. A female may lay about 30 eggs a day and several hundred eggs during her total lifetime. They have five or six overlapping generations in each season (Uddin et al. 2005; Nasreen et al. 2007; Asadi 2010).

## 5.8 Research on Green Lacewings in Iran

Several studies on lacewings have been conducted, most of which have examined the ecological aspects of green lacewings such as demography and prey suitability, functional response, tritrophic interactions, intraguild predation, and the effects of different insecticides on life history parameters.

Fekrat et al. (2006) studied the effects of different densities of the first instar larvae of *C. carnea* feeding on the eggs of *S. cerealella* or *E. kuehniella*. Their results indicated that the searching efficiency of the first instar larvae of the predator decreased with increasing larval densities.

*Nipaecoccus viridis* (Newstead) (Hemiptera: Pseudococcidae) is a major invasive pest of citrus as well as other plants in tropical and sub-tropical regions of the world. This economically important pest was reported from Fars, Hormozgan, and Khuzestan provinces (South and Southwest of Iran). Damage of this pest severely affects citrus trees in Jahrom region from Fars province. High potential of reproduction, widespread distribution, and difficulty of chemical control demanded urgent application of natural enemies against this pest within an IPM program. Zakerin et al. (2009) studied the efficiency of *C. carnea* in control of *N. viridis* in citrus orchards of Jahrom during 2004 to 2005. First instar larvae of *C. carnea* were released based on a specific design. Moreover, movement of ants on treated citrus was banned using sticky band, and infection rate of fruits was considered as an index. The results showed that this pest can be controlled substantially by releasing the green lacewing larvae (500 larvae per citrus tree) in three releasing times at 30 April, 10 May and 20 May.

Hassani-sadi et al. (2010) studied the effects of diet and temperature regimes on the development and reproduction of green lacewing as a natural enemy of the common pistachio psyllid, *Agonoscena pistaciae* Burckhardt and Lauterer (Hemiptera: Psyllidae) and concluded that the larvae of *C. lucasina* fed 1016 fourth psyllid nymphs and 315 third instar cowpea nymphs through the whole larval period at 30 °C. The diets had a significant influence on fecundity and longevity of adult green lacewings. Furthermore, the reproduction of the green lacewing was reduced by increasing the temperature from 22.5 °C to 32.5 °C. The intrinsic rate of natural increases ($r_m$) of this insect was 0.11 and 0.09 day$^{-1}$ when it was reared using fourth instar pistachio psyllid nymphs and flour moth eggs at 25 °C through the larval stage, respectively.

Hassanpour et al. (2011) investigated the functional response of three larval instars of *C. carnea* to the eggs and first instar larvae of *H. armigera* under laboratory conditions. First and second instar larvae of predator showed type II functional responses on both prey stages. However, third instar lacewing larvae exhibited type II and III functional responses to the first instar larvae and eggs of *H. armigera*, respectively. Maximum observed predation (±SE) rate of first, second and third larval instars of *C. carnea* was 8 (±0.26), 27.5 (±1.49), and 81.6 (±1.96) eggs, and 7.5 (±0.37), 24.4 (±0.64), and 51.1 (±1.48) larvae, respectively. According to the prey consumption rate and functional response parameters, the researchers confirmed that the larvae of *C. carnea*, especially the third instar, have a high potential for control of eggs and first instar larvae of *H. armigera*. However, field-based studies are needed for a comprehensive estimation of the capacity of *C. carnea* in biological control of *H. armigera*.

Hesami et al. (2011) studied the effects of different host plants of *Sitobion avenae* (Fabricius) (Hemiptera: Aphididae) on feeding rate and longevity of *C. carnea* under laboratory conditions. There was significant difference between two host plants (wheat and oleander) in feeding rate and longevity of the green lacewing. Mean feeding rate of second instar larvae of *C. carnea* on the aphids fed on wheat and oleander plants was 40.3, and 30.6, respectively. Additionally, the developmental time of second instar larvae of the green lacewing was 3.7 and 6 days when fed on the aphids on these plants, respectively. The results indicated that the biological parameters of *C. carnea* larvae can be affected by the quality of prey.

Various species in *carnea* complex of the common green lacewing are predators of common pistachio psylla, *A. pistaciae* in cultivated as well as wild pistachio plants in Iran. Kazemi and Mehrnejad (2011) studied the seasonal occurrence and trends in the abundance of green lacewings on pistachio orchards during 2007 and 2008. Moreover, the effect of different temperatures on preimaginal development, survival rate, and prey consumption of *C. lucasina* in feeding on *A. pistaciae* nymphs was investigated. The adults appeared on pistachio trees in mid-April and were abundant in early July; subsequently, they decreased in summer and increased again in October. The relative density of green lacewing in pistachio orchards was higher when the ground was covered by weed plants. *C. lucasina* females laid about 1085 eggs during 60 days at 22.5 °C. The *C. lucasina* larvae consumed up to 1812 fourth instar nymphs of *A. pistaciae* during their total larval period at 35 °C. The net reproductive rate ($R_0$) and intrinsic rate of increase ($r_m$) of predator were estimated to be 286.90 female/female/generation and 0.11 female/female/day, respectively. Degree-day requirement for the development of *C. lucasina* (egg to adult) was estimated 385 °D.

Naderian et al. (2012) investigated the fauna of natural enemies in corn fields and surrounding grasslands of Semnan province. Among fifty-four predator and parasitoid species from 12 families, there were one species of Chrysopidae and Trichogrammatidae; two species of Chalcididae; three species of Asilidae and Eulophidae; four species of Tachinidae; five species of Coccinellidae; six species of Syrphidae; seven species of Braconidae, Carabidae and Staphylinidae and eight species of Ichneumonidae.

*Bemisia tabaci* (Gennadius) (Hemiptera: Aleyrodidae) is one of the most important pests of different crops, and lacewings are widely used as predatory species in biological control programs against this pest. Baghdadi et al. (2012) investigated the biological parameters of *C. lucasina* on this pest in laboratory. In this study, the gravid females of *C. lucasina* were collected from Sarepol-e Zahab in western Iran. The larvae were offered different diets consisting of a silver leaf whitefly *B. tabaci*, green peach aphid, *Myzus persicae* (Sulzer) (Hemiptera: Aphididae), and also lyophilized powder of drone honeybee (*Apis mellifera* L.). The results showed that the larvae had maximum duration of 27 ± 0.33 days when fed on honeybee lyophilized powder, followed by *M. persicae* (25 ± 0.27 days) and *B. tabaci* (17.9 ± 0.3 days). Furthermore, they showed a prey preference for *M. persicae* compared with *B. tabaci* as the former has a larger body size and is thus more easily found and captured by the predator larvae. There was a significant difference in the number of preys consumed by the predator larvae. The results showed that when the predator larvae had been fed on *B. tabaci*, their developmental times were shorter and their adult fertility was increased. The results indicated that the suitable prey not only accelerated developmental stages of the predator but also promoted the fecundity of the resulting adults through increasing the pupal weight.

Ahmadzadeh and Hatami (2014) investigated the predatory potential of *C. carnea* on *Trialeurodes vaporariorum* West (Hemiptera: Aleyrodidae) under laboratory conditions and concluded that a single *C. carnea* larva consumed an average of 215 different nymphal instars of *T. vaporariorum* during its total larval period. However, all of the third instar larvae died before reaching the adult stage.

Sarailoo and Lakzaei (2014) investigated the effects of artificial diets on biological parameters of *C. carnea*. The predator was reared for seven generations to provide a homogenous population under laboratory conditions. The impact of six different diets [a mixture of 30% concentrations of glucose, fructose, and sucrose (in a ratio of 1:1:1); glucose, fructose, sucrose plus extract of *S. cerealella* eggs (in a ratio of 1:1); glucose, fructose, sucrose plus extract of *Anagasta kuehniella* eggs (in a ratio of 1:1); a mixture of honey, yeast, and distilled water (in a ratio of 1:1:1); honey, yeast plus extract of *S. cerealella* eggs (in a ratio of 1:1:1), and honey, yeast plus extract of *A. kuehniella* eggs (in a ratio of 1:1:1)] was assessed on biological parameters of the predator. The results showed that the artificial diet including mixture of honey, yeast and extract of *A. kuehniella* eggs had more positive effects on biological parameters of the predator than the other diets.

The effects of different starvation levels of *C. carnea* larvae on functional response of the predator to nymphs of *Aphis fabae* (Scopoli) (Hemiptera: Aphididae) were investigated (Hassanpour et al. 2015). At three hunger levels (4, 12, and 24 h), the functional response types of first and second instar larvae of the predator were II, II, III and III, III, II, respectively. At 2, 4 and 8 h hunger levels, the third instar larvae of *C. carnea* showed types II, II and III functional response, respectively. For the first instar larvae of predator, the handling times were not significantly different among various hunger levels. However, for the second and third instar larvae, there was significant difference in handling times estimated at different hunger levels. This study indicated that predator hunger level can change its functional response type as well as the parameters. The last instar larvae of *C. carnea* killed approximately 150 third and fourth nymphal instars of *A. fabae* during 8 h, showing its high potential in biological control of the aphid.

Prey type is an important element for suitable mass rearing of the predators in IPM programs. Takalloozadeh (2015) investigated the effect of some prey species on biological attributes of *C. carnea*. Their findings revealed that pre-imaginal and total developmental times of *C. carnea* were significantly influenced by prey species. Total developmental time was estimated to be $19.63 \pm 0.125$, $20.63 \pm 0.180$, $22.06 \pm 0.183$, $22.35 \pm 0.120$, and $23.81 \pm 0.356$ days on *Aphis gossypii* (Glover), *M. persicae*, *Aphis punicae* (Passerini), *A. fabae* and *Aphis craccivora* (Koch) (Hemiptera: Aphididae), respectively. The fecundity of *C. carnea* was highest on *M. persicae* ($478.50 \pm 8.38$ eggs) and lowest on *A. craccivora* ($242.78 \pm 7.37$ eggs). The highest female longevity of the predator was on *M. persicae*. According to the results, nymphs of *M. persicae* and *Ap. gossypii* were more appropriate preys than others because the preimaginal developmental period of *C. carnea* was shorter with higher adult longevity, fecundity and survival percentage.

Farrokhi et al. (2017a) studied the effects of different artificial diets on some biological parameters of *C. carnea* adults under laboratory conditions. Their results showed that a mixture of 1:1:1:1 egg yolk, soluble vitamins, yeast, and honey positively affected the biological traits of the predator. In this research, mean male and female longevities, oviposition period, post oviposition period, and fecundity were calculated 34.86, 43.53, 26.26, 10.46 days, and 618.07 eggs, respectively. In another study, Farrokhi et al. (2017b) investigated host plant-herbivore-predator interactions in *C. carnea* and *M. persicae* systems on four plant species in laboratory conditions. Accordingly, peach plant was found to be a more appropriate host than the other plants for development and predation fitness of *C. carnea*. These results revealed that awareness of tritrophic interactions and subsequent life table evaluations of natural enemies improves IPM programs.

Moradi et al. (2019) studied the prey preference of second and third instar larvae of *C. carnea* to *Aphis spiraecola* Patch and *Ap. gossypii* on orange leaves under laboratory conditions. Their findings showed that when *A. spiraecola* or *Ap. gossypii* were separately offered to the predator larvae, the predation rate was higher on *Ap. gossypii* compared to *A. spiraecola*. The values of Manly's preference index in equal ratios of *Ap. gossypii*: *A. spiraecola* (30:30 and 40:40 for second and third instar larvae, resp.) were calculated 0.632:0.368 and 0.647:0.353, respectively, indicating the preference of *C. carnea* larvae for *Ap. gossypii*.

The influence of experimental complexity and initial prey density (*A. fabae*) on consumption rate of green lacewings was investigated by Zarei et al. (2019) who reported that a more complex microcosm arena significantly weakened the predation rate of *C. carnea*. Moreover, the predator life stages (second versus third instar larvae) differed significantly in terms of prey consumption rate.

## 5.9 Applied Studies of Lacewings for Pest Management

Despite extensive faunistic research on lacewings, the applied surveys are limited, which can be partially attributed to executive and financial problems. Some of the release programs for controlling *A. pistaciae* in Kerman province showed promising results in pistachio orchards. The use of green lacewings along with imidacloprid spray showed good ability to control *T. vaporariorum* on potato farms; nevertheless, few studies have examined techniques of applying *C. carnea* for commercial purposes. Supplementary experiments have been conducted to use the eggs and larvae of *C. carnea* for biological control of various pests in field and greenhouse (Jafari-Nadoushan 1998; Jooyandeh 1999; Rafiee-Karahroudi and Hatami 2003; Ahmadzadeh and Hatami 2006; Jooyandeh 2009).

The development of efficient methods for commercial release of natural enemies is crucial to the success of augmentative biological control. Jooyandeh (1999) reported that a mechanical applicator (atomizer sprayer model ECHO DM9) can be used for distributing the eggs and larvae of *C. carnea*. Using this method, *C. carnea* eggs in an agar solution and second larval instars mixed with sawdust were released for controlling the most important cotton pests.

Four releases of second instar larvae of *C. carnea* at release rates of 10,000–12,000/ha resulted in 63.1%, 72.0%, 82.3%, and 60.0% reduction in *Acyrthosiphon gossypii* Mordvilko (Hemiptera: Aphididae), *B. tabaci*, *Ap. gossypii*, and *Tetranychus urticae* Koch (Acari: Tetranychidae) populations, respectively. Egg releases of *C. carnea* did not contribute to satisfactory reduction in cotton pests (Jooyandeh 1999). Although lacewings are commonly released as eggs, larval releases may sometimes be more effective. However, larval release is more expensive than egg release.

*C. carnea* eggs were released to control the pear psylla, *Cacopsylla pyricola* (Foerster) (Hemiptera: Psyllidae), which is one of the most important pests of pear trees in Iran. Mature lacewing eggs were glued together with *S. cerealella* eggs on cards (Fig. 5.3c). Satisfactory reduction of pear psylla was achieved using 200 eggs/ tree (two ChrysoCards for each tree; Fig. 5.3d) at proper time points. The cards are easy and safe to install and protect the beneficial organisms against rain and radiation (Jooyandeh 2009).

In a greenhouse experiment on cucumber plants in ventilated cages, the effectiveness of two different lacewing release methods against *Ap. gossypii* were compared. A small net bag and sawdust were used as carriers of *C. carnea* eggs. The results showed that small net bags were more effective than sawdust (Rafiee-Karahroudi and Hatami 2003).

Several species of parasitoids attack *Chrysoperla* eggs and larvae, and parasitism rate can be high, especially at the end of the season. *Telenomus acrobats* Giard (Hymenoptera: Scelionidae) is a factor contributing to decreasing efficiency as well as population changes of *C. carnea* under natural conditions in Varamin region (Shahpouri-Arani et al. 2005). In some cases, release of mature eggs can greatly reduce parasitism (Tauber et al. 2000).

Pesticides play an essential role in alleviating Iranian farmer pest problems, albeit temporarily. By combining the advantages of both chemical and biological methods, greater pest suppression may be obtained. In situations where natural enemies are not present in sufficient numbers to provide adequate control, it may be helpful to release mass reared predators/parasitoids or combine them with other tactics suppressing pest arthropods more efficiently. By far, the greatest concern about the combined release of *C. carnea* and pesticide spraying is related to lethal effects (mortality assessed as $LD_{50}$, $LC_{50}$, etc.); however, few cases have considered sublethal effects such as changes in developmental rates, fecundity, and fertility.

Saber et al. (2018) stated that emamectin benzoate and spinosad at recommended field concentrations had negligible toxicity on second instar larvae of *C. carnea*. In two other stidies, the lethal effects of four insecticides (imidacloprid, lufenuron, thiamethoxam, and thiodicarb) were examined on male and female adults (Ayubi et al. 2014) as well as eggs and first instar larvae of *C. carnea* (Ayubi et al. 2013). The findings showed that males were more susceptible than females and that thiamethoxam was highly toxic on adults and immature stages whereas lufenuron proved to be the least toxic.

Sublethal effects of endosulfan, imidacloprid, and indoxacarb were investigated on *C. carnea* adults using a demographic toxicology method at concentrations of 317, 46, and 9 mg AI/lit (as $LC_{25}$), respectively. The net reproductive rate ($R_0$) and intrinsic rate of natural increase ($r_m$) of the predator were estimated 125, 93 and 61 female/female/generation, and 0.161, 0.157 and 0.136 female/female/day in imidacloprid, endosulfan and indoxacarb-treated insects, respectively. The results revealed that indoxacarb was more toxic to adult stage of this predaor than the other two compounds (Golmohammadi et al. 2013).

Ahmadzadeh and Hatami (2006) investigated the effects of single and combined use of 0.3 ml/l confidor (35% SC) and *C. carnea* against different nymphal instars of *T. vaporariorum* on tomato plants. Egg releases of the predator at a 1:1 predator-prey ratio caused inadequate reduction of the pest population. Nonetheless, pest elimination was observed after three releases of the predator beginning 20 days after spraying.

## 5.10 Conclusion

Among several species of chrysopids that have been reported from Iran, *C. carnea* has received the most attention as a biological control agent for research and practical purposes. Despite the unique potential of *C. carnea*, it has not yet been used as a commercial biological control agent in Iran. Rearing systems for *C. carnea* are kept at small-scale educational/research level; consequently, few studies have been conducted to increase *C. carnea* production commercially. Low-cost and mechanized methods for mass rearing larvae and appropriate inundative release techniques are still needed. There are not enough producers for lacewing production in Iran. Continuous production with desirable quality and quantity of lacewings and other natural enemies is necessary for successful application of these biocontrol agents in the fields and greenhouses. By employing natural enemies including *C. carnea* in IPM programs, the use of pesticides will be reduced and the utilization of higher doses of insecticides will be avoided.

## References

Ahmadzadeh Z, Hatami B (2006) Evaluation of integrated control of greenhouse whitefly, *Trialeurodes vaporariorum* West using *Chrysoperla carnea* (Stephens) and insecticide confidor in greenhouse conditions. J Sci Tech Agri Natur Res 9(4):239–251

Ahmadzadeh Z, Hatami B (2014) Feeding potential and survival of predator, *Chrysoperla carnea* on greenhouse whitefly, *Trialeurodes vaporariorum* in laboratory conditions. Plant Prot J 6 (2):113–121

Asadi M (2010) Identification of Neuroptera fauna from Eslamabad-e Gharb and Sarpol-e Zahab cities in Kermanshah province. M. Sc. Thesis, Razi University, Kermanshah, Iran

Aspöck H, Aspöck U, Hölzel H (1980) Die Neuropteran Europas. Eine Zusammenfassende Darstellung der Systematik, Okologie und Chorologie der Neuropteroidea (Megalopteraö, Raphidioptera, Planipennia) Europas. Vol. 2. Goecke & Evers, Germany

Ayubi A, Moravvej G, Karimi J et al (2014) Susceptibility of adult green lacewing, *Chrysoperla carnea* (Stephens) (Neuroptera: Chrysopidae) to some prevalent insecticides in laboratory conditions. J Entomol Res 6(4):321–332

Ayubi A, Moravvej G, Karimi J et al (2013) Lethal effects of four insecticides on immature stages of *Chrysoperla carnea* (Stephens) (Neuroptera: Chrysopidae) in laboratory conditions. Turk J Entomol 37(4):399–407

Baghdadi A, Sharifi F, Mirmoayedi A (2012) Investigation on some biological aspects of *Chrysoperla lucasina* (Chrysopidae: Neuroptera) on *Bemisia tabaci* in laboratory conditions. Commun Agric Appl Biol Sci 77(4):635–638

Canard M, Principi MM (1984) Development of Chrysopidae. In: Canard M, Semeria Y, New T (eds) Biology of Chrysopidae. Dr. Junk Publishers, The Hague, pp 57–75

Cannings RA, Cannings SG (2006) The Mantispidae (Insecta: Neuroptera) of Canada with notes on morphology, ecology, and distribution. Can Entomol 138:531–544. https://doi.org/10.4039/n06-806

Daniali M, Heydari H, Khodaman A (1995) Dominate species of lacewing in agroecosystems of Gorgan and Gonbad regions and their mass rearing. Proceedings of the 12th Iranian Plant Protection Congress, University of Tehran, Karaj, Iran. 2–7 September 1995

Farahbakhsh Gh (1961) A checklist of economically important insects and other enemies of plants and agricultural products in Iran. Dept Plant Prot, Min Agri, Tehran, Iran 1:1–153

Farahi S, Sadeghi H, Whittington AE (2009) Lacewing (Neuroptera: Chrysopidae and Hemerobiidae) from north eastern and east provinces of Iran. Mun Ent Zool 4(2):501–509

Farrokhi M, Gharekhani G, Iranipour Sh et al (2017a) Effect of different artificial diets on some biological traits of adult green lacewing, *Chrysoperla carnea* (Neuroptera: Chrysopidae) under laboratory conditions. J Entomol Zool Stud 5(2):1479–1484

Farrokhi M, Gharekhani G, Iranipour Sh et al (2017b) Host plant-herbivore-predator interactions in *Chrysoperla carnea* (Neuroptera: Chrysopidae) and *Myzus persicae* (Homoptera: Aphididae) on four plant species under laboratory conditions. J Econ Entomol 110(6):2342–2350. https://doi.org/10.1093/jee/tox268

Fekrat L, Solimannejadyan E, Jooyandeh A (2006) Mutual interference in the first instar larvae of *Chrysoperla carnea* (Stephens) (Neuroptera: Chrysopidae) in laboratory. Proceedings of the 17th Iranian Plant Protection Congress, University of Tehran, Karaj, Iran. 2–5 September 2006

Ghadiri-Rad S, Heydari H, Saghaee-Nia AM (2004) *Chrysopa formosa* Brauer (Neu.: Chrysopidae), green lacewing of winter wheat fields in Golestan province. Proceedings of the 16th Iranian Plant Protection Congress, University of Tabriz, Tabriz, Iran. 29 August– 2 September 2004

Golmohammadi G, Hejazi M, Iranipour Sh et al (2013) Sublethal effects of three insecticides on adults green lacewing *Chrysoperla carnea* (Stephens) with demographic toxicology method. J Appl Entomol Phytopath 81(1):73–82

Haruyama N, Mochizuki A, Duelli P et al (2008) Green lacewing phylogeny, based on three nuclear genes (Chrysopidae, Neuroptera). Syst Entomol 33(2):275–288. https://doi.org/10.1111/j.1365-3113.2008.00418.x

Hassani-sadi M, Mehrnejad MR, Shojaei M (2010) The effect of diet and temperature regimes on development and reproduction of green lacewing, a natural enemy of the common pistachio psyllid. J Entomol Res 2(3):179–192

Hassanpour M, Mohaghegh J, Iranipour Sh et al (2011) Functional response of *Chrysoperla carnea* (Neuroptera: Chrysopidae) to *Helicoverpa armigera* (Lepidoptera: Noctuidae): effect of prey and predator stages. Insect Sci 18(2):217–224. https://doi.org/10.1111/j.1744-7917.2010.01360.x

Hassanpour M, Maghami R, Rafiee-Dastjerdi H et al (2015) Predation activity of *Chrysoperla carnea* (Neuroptera: Chrysopidae) upon *Aphis fabae* (Hemiptera: Aphididae): effect of different hunger levels. J Asia Pac Entomol 18(2):297–302. https://doi.org/10.1016/j.aspen.2015.03.005

Henry CS (1985) Sibling species, call differences, and speciation in green lacewings (Neuroptera: Chrysopidae: *Chrysoperla*). Evolution 39(5):965–984

Henry CS, Brooks SJ, Duelli P et al (2002) Discovering the true *Chrysoperla carnea* (Stephens) (Insecta: Neuroptera: Chrysopidae) using song analysis, morphology and ecology. Ann Entomol Soc Am 95(2):172–191. https://doi.org/10.1603/0013-8746(2002)095[0172: DTTCCI]2.0.CO;2

Hesami S, Farahi S, Gheibi M (2011) Effect of different host plants of normal wheat aphid *Sitobion avenae* on the feeding and longevity of green lacewing *Chrysoperla carnea*. International Conference on Asia Agriculture and Animal, Hong Kong. 2–3 July 2011

Heydari H (1995) The list of Iranian Chrysopidae (Neuroptera). Proceedings of the 12th Iranian Plant Protection Congress, University of Tehran, Karaj, Iran. 2–7 September 1995

Hölzel H (1966) Beitrag zur kenntnis der Chrysopiden des Iran (Planipennia, Chrysopidae). Stutt Beitr Naturk 148:1–7

Hölzel H (1967) Die Neuroptera Vorderasiens, Chrysopidae. Beitr naturk Forsch SüdwDtl 26 (1):19–45

Hölzel H (1982) Redeskription von *Chysopa andresi* Navas und Beschreibung zweier neuer Arten aus Vorderasien (Planipennia, Chrysopidae). Zeitsch Arbeitsgemeinsch Osterreichisch Entomol 33:113–121

Jafari-Nadoushan A (1998) Performance evaluation of lacewing *Chrysoperla carnea* as pistachio psylla predator. MSc. Thesis, University of Tarbiat Modares, Tehran, Iran

Jooyandeh A (1995) Evaluation of suitable rearing methods of common green lacewing, *Chrysoperla carnea* (Neuroptera: Chrysopidae) and its biological aspects under laboratory conditions. MSc. Thesis, University of Tehran, Tehran, Iran

Jooyandeh A (1999) The investigation on using common green lacewing, *Chrysoperla carnea* (Neuroptera: Chrysopidae) for control of cotton pests. Final report of research project, Khorasan Razavi Agriculture and Natural Resource Research and Education Center

Jooyandeh A (2000) Mass production of common green lacewing, *Chrysoperla carnea* (Stephens) (Neu., Chrysopidae): New methods in group rearing of the larvae. Proceedings of the 14th Iranian Plant Protection Congress, Isfahan University of Technology, Isfahan, Iran. 5–8 September 2000

Jooyandeh A (2009) Study of effectiveness of common green lacewing, *Chrysoperla carnea* (Stephens) (Neu.: Chrysopidae) in control of pear psylla, *Cacopsylla pyricola* (F.) (Hem., Psyllidae). Final report of Research project, Khorasan Razavi Agriculture and Natural Resource Research and Education Center

Kazemi F, Mehrnejad MR (2011) Seasonal occurrence and biological parameters of the common green lacewing predators of the common pistachio psylla, *Agonoscena pistaciae* (Hemiptera: Psyllidae). Eur J Entomol 108(1):63–70. https://doi.org/10.14411/eje.2011.008

Mansell MW (2002) Monitoring lacewing (Insecta: Neuroptera) in Southern Africa. Acta Zool Acad Sci Hungar 48 (Suppl. 2):165–173

Meinander M (1998) Coniopterygidae (Neuroptera) from the Mediterranean region and Iran. J Neuropterol 1:23–31

Mirmoayedi A (1993) Collection and identification of species of Chrysopidae and Hemerobiidae (Neuroptera) in Kermanshah province. Proceedings of the 11th Iranian Plant Protection Congress, University of Guilan, Rasht, Iran. 28 August–2 September 1993

Mirmoayedi A (1995) A checklist of certain Neuropteroida of Hormozgan province. Proceedings of the 12th Iranian Plant Protection Congress, Karaj, Iran. 2–7 September 1995

Mirmoayedi A (1998) Neuroptera from Iran. Acta Zool Fenn 209:163–165

Mirmoayedi A (1999a) New survey on Neuropteran fauna of Kermanshah and Kurdistan provinces. Proceedings of the 8th Iranian Biology Conference, Razi University, Kermanshah, Iran. 30 August–1 September 1999

Mirmoayedi A (1999b) Lacewings of Shiraz. Proceedings of the 8th Iranian Biology Conference, Razi University, Kermanshah, Iran. 30 August–1 September 1999

Mirmoayedi A (2002a) Forty years of studies by Iranian entomologists on the Chrysopidae fauna (1961-2000) of Iran (Insecta, Neuroptera). Zool Mid East 26(1):157–162. https://doi.org/10.1080/09397140.2002.10637931

Mirmoayedi A (2002b) New records of Neuroptera from Iran. Aca Zool Acad Sci Hungar 48 (2):197–201

Mirmoayedi A (2008) An updated checklist of the neuropterans of Iran. Proceeding of the 10th International Symposium on Neuropterology, Piran, Slovenia. 22–26 July 2008

Mirmoayedi A, Yassayie A (1998) Collection and identification of Neuropteran fauna in Golestan national park. Proceedings of the 7th Iranian Biology Conference, Isfahan University, Isfahan, Iran. 22–24 Aug 1998

Mirmoayedi A, Eradati N, Adldoost H (2014) New records of Neuroptera (Insecta, Neuropterida, Neuroptera) from West Azerbaijan, Iran. Adv Biotech Patent 1:25–31

Monserrat VJ (2002) New data on the dusty wings from Africa and Europe (Insecta, Neuroptera, Coniopterygidae). Graellsia 58(1):3–19

Moradi M, Hassanpour M, Golizadeh A et al (2019) Prey preference and switching of the green lacewing Chrysoperla carnea on the citrus aphid Aphis spiraecola and the melon aphid Aphis gossypii. Plant Pest Res 8(4):43–54

Naderian H, Ghahari H, Asgari S (2012) Species diversity of natural enemies in corn fields and surrounding grasslands of Semnan province, Iran. Calodema 217:1–8

Nasreen A, Ashfaq M, Mustafa Gh et al (2007) Mortality rate of five commercial insecticides on Chrysoperla carnea (Stephens) (Chrysopidae: Neuroptera). Pak J Agri Sci 44(2):266–271

New TR (1985) A revision of the Australian Myrmeleontidae (Insecta: Myrmeleontidae) I. Introduction, Myrmeleontini, Protoplectrini. Austral J Zool (Suppl. Ser.). 104:1–90

New TR (1999) Neuroptera and biological control (Neuropterida). Stapfia 60:147–166

Rafiee-Karahroudi Z, Hatami B (2003) Comparison of two method of releasing Chrysoperla carnea (Stephens) eggs against Aphis gossypii Glover under greenhouse condition. J Sci Tech Agri Natur Res 7(2):215–224

Saber M, Vojudi S, Farrokhi M et al (2018) Effect of field recommended rates of emamectin benzoate and spinosad on biological parameters of green lacewing, Chrysoperla carnea (Stephens) (Neu: Chrysopidae) under laboratory conditions. J Appl Res Plant Prot 7(1):57–67

Sarailoo MH, Lakzaei M (2014) Effect of different diets on some biological parameters of Chrysoperla carnea (Neuroptera: Chrysopidae). J Crop Prot 3(4):479–486

Shahpouri Arani S, Talebi AA, Fathipour Y et al (2005) Comparison between the population parameters in green lacewing, Chrysoperla carnea (Steph.) (Neu: Chrysopidae) and its egg parasitoid wasp, Telenomus acrobats Giard (Hym. Scelionidae). J Agri Sci 11(1):107–116

Sharififard M, Mossadegh SMS (2006) The lacewings (Chrysopidae) fauna of Khuzestan and the effect of temperature and prey on larval voracity of dominant species. Sci J Agri 29(1):113–127

Sziráki G, Mirmoayedi A (2012) Annotated checklist of the Iranian Coniopterygidae (Neuroptera). Folia Histor Natur Mus Matr 36:45–50

Takalloozadeh HM (2015) Effect of different prey species on the biological parameters of Chrysoperla carnea (Neuroptera: Chrysopidae) in laboratory conditions. J Crop Prot 4 (1):11–18

Tauber MJ, Tauber CA, Daane KM et al (2000) Commercialization of predators: recent lessons from green lacewings (Neuroptera: Chrysopidae: Chrysoperla). Am Entomol 46(1):26–38. https://doi.org/10.1093/ae/46.1.26

Uddin J, Holliday NJ, MacKay PA (2005) Rearing lacewings, Chrysoperla carnea and Chrysopa oculata (Neuroptera: Chrysopidae), on prepupae of alfalfa leafcutting bee, Megachile rotundata (Hymenoptera: Megachilidae). Proceedings of the Entomological Society of Manitoba 61:11–19

Vogt H, Bigler F, Brown K et al (2000) Laboratory method to test effects of plant protection products on larvae of Chrysoperla carnea (Stephen) (Neuroptera: Chrysopidae). In: Condolfi MP, Blomel S, Forster R (eds) Guidelines to evaluate side effects of plant protection products to non-target arthropods. IOBC, BART, and EPPO Joint Initiative. IOBC, Gent, pp 27–44

Wells MM, Henry CS (1992) The role of courtship songs in reproductive isolation among populations of green lacewings of the genus *Chrysoperla* (Neuroptera: Chrysopidae). Evolution 46(1):31–42

Woolfolk SW, Smith DB, Martin RA et al (2007) Multiple orifice distribution system for placing green lacewing eggs into vertical larval rearing units. J Econ Entomol 100(2):283–290. https://doi.org/10.1093/jee/100.2.283

Zakerin A, Zibaiee K, Fallahzadeh M (2009) To study the biological control of *Nipaecoccus viridis* (Hem., Pseudococcidae) by *Chrysoperla carnea* (Neu., Chrysopidae) in citrus orchards of Jahrom, Fars province-Iran. J Nov Res Plant Prot 1(2):126–139

Zarei M, Madadi H, Zamani AA et al (2019) Predation rate of competing *Chrysoperla carnea* and *Hippodamia variegata* on *Aphis fabae* at various prey densities and arena complexities. Bull Insectol 72(2):273–280

# Part II
# Parasitoids

# Chapter 6
# Egg Parasitoids: Chalcidoidea with Particular Emphasis on Trichogrammatidae

**Shahzad Iranipour and Nahid Vaez**

## 6.1  Introduction

Superfamily Chalcidoidea is one of the most important and richest groups of Hymenoptera. Chalcids are often small and shiny insects. They immediately can be recognized by a reduced wing venation. Maximum four longitudinal veins are present in front wings (Fig. 6.1). Venation of platygastrids may resembles this general pattern, but chalcids may easily be recognized by quadrate pronotum in lateral view and a distinct distance between tegula and pronotum which sometimes filled by a sclerit namely prepectus or post-spiracular sclerit (Fig. 6.1). Some minute specimens like members of Trichogrammatidae and Aphelinidae are without metallic coloration and metasoma fused to mesosoma by a thorough contact. Chalcids are sometimes very strange in appearance.

Due to such vast diversity, very miscellaneous life styles have been evolved among them. Some species are phytophagous like agaonids that are pollinators of *Ficus* spp. and some eurytomids. *Eurytoma amygdali* Enderlein, *Eurytoma plotnikovi* Nikolskaya and *Megastigmus pistaciae* Walker (Hym., Torymidae) are important pests of almond and pistachio respectively (Behdad 1991). But the major fame of the superfamily is due to presence of a vast kind of parasitoid of numerous insect species. Almost all stages of insects are attacked by chalcids. Both primary parasitoids and hyperparasitoids are present in this family. They are the most important group of parasitoids in applied biological control. Many insect orders may be attacked by chalcids. In this chapter we focused only on egg parasitoids. In this regards the most important family is Trichogrammatidae. Except a few species, all remaining trichogrammatids are egg parasitoids of moths and seldom other

S. Iranipour (✉)
University of Tabriz, Tabriz, Iran

N. Vaez
Azarbaijan Shahid Madani University, Tabriz, Iran

© The Author(s), under exclusive license to Springer Nature Switzerland AG 2021
J. Karimi, H. Madadi (eds.), *Biological Control of Insect and Mite Pests in Iran*,
Progress in Biological Control 18, https://doi.org/10.1007/978-3-030-63990-7_6

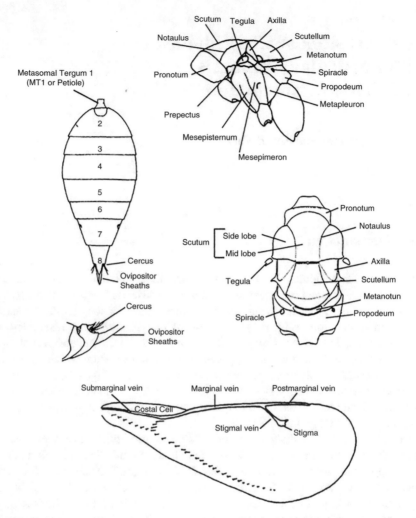

**Fig. 6.1** Structure of mesosoma, metasoma and fore wing in Chalcidoidea (http://mx.speciesfile.org/projects/77/public/site/chalcidkey/home/glossary, access date: 24 Feb. 2014) (Courtesy of USDA)

insects. There are also some egg parasitoids distributed in other families, while other members of those families are parasitoids of the other stages. In Iran, records of egg parasitoids are present from Encyrtidae, Eupelmidae, Eulophidae, Torymidae, and Mymaridae. Mymarids all are egg parasitoid of the other insects but their importance is very lower than trichogrammatids.

## 6.2   Family Trichogrammatidae Foerster, 1856

The family Trichogrammatidae involves the smallest members of the order Hymenoptera, superfamily Chalicidoidea. Adults are mate without metallic shine, often less than 1 mm in length (mostly 0.18–1.5 mm), without narrow stem between mesosoma and metasoma, and are distinguished by their 3-segmented tarsi from the other Chalcidoidea (Pinto and Stouthamer 1994). The native species of *Trichogramma* in Iranian plateau are commonly small (0.2–0.4 mm), with yellow, brown or black body. *Trichogramma* spp. have been spread in humid southern coasts of the Caspian Sea (Guilan, Mazandaran and Golestan provinces), the southern foothills of the Alborz Mountains, the Zagros mountains, central provinces and even borderlines of the central desert of Iran (Shojai 1987).

This family consists of about 800 species in ca. 83 genera which are dispersed worldwide. Most genera of the Trichogrammatidae have few species (55 genera have only five species or less) and only 14 genera have more than 10 species. Almost all species of Trichogrammatidae are primary, solitary or gregary endoparasitoids of other insects' eggs belonging to various orders. There is only one report of parasitism on a stage other than egg; that is related to Cecidomyiidae (Diptera) pupa (Viggiani 1981). The trichogrammatids have a broad host range. At least 400 species in 11 orders of insects are targets for trichogrammatids, including Homoptera (29 genera), Coleoptera (11 genera), Lepidoptera (8 genera), Hemiptera (6 genera), Orthoptera (4 genera), Diptera (3 genera), Odonata (2 genera), Hymenoptera (1 genus), Thysanoptera (1 genus) and Neuroptera (1 genus) (Hassan et al. 1988; Ciochia 1991; Nikonov et al. 1991; Pintureau 1991a; Pinto and Stouthamer 1994; Romeis and Shanower 1996; Ebrahimi 2004). Among these hosts, both agricultural pests and beneficial insects are observed (Pintureau 1991b). The most common hosts in Iran are moth eggs particularly belonging to Noctuidae, Pyralidae, Tortricidae, Pieridae, Crambidae etc. (Shojai 1987).

The species of Trichogrammatidae are used in miscellaneous agroecosystems throughout the world. They have been used against 28 insect pests in more than 30 million hectares of products such as corn, rice, sugarcane, cotton, vegetables, sugar beet, fruit trees and softwoods at 30 countries (Li 1994; van Lenteren and Bueno 2003). More recent estimates show that currently 15 million hectares are treated with *Trichogramma* spp. in globe (Bueno and van Lenteren 2010). The features that make these wasps good biocontrol agents are numerous generations, polyphagy, adaptation to various weather conditions and easy way of rearing on factitious hosts. Russia, China, Mexico, Europe, the United States and some central Asian countries, are the major consumers of trichogrammatids in the world (Li 1994; Bueno and van Lenteren 2010).

*Trichogramma* Westwood is the largest and a well-known genus of Trichogrammatidae with more than 200 species around the world (Pinto 2006). *Trichogramma* and *Oligostia* (with 110 species), include 40% members of this family. *Trichogramma* spp. are the most widely studied and successfully used natural enemies regarding their short generation time, easy mass rearing in insectaries and voracious parasitizing habit on eggs of target hosts (Cônsoli et al. 2010). More than 70 species of *Trichogramma* are used in the world, but only 20 species are mass reared to use in agroecosystems. In Europe, just five species out of 126 are artificially produced (Pintureau 1991b) while, in Iran only *T. brassicae* is augmented and limitedly used in field scale. *Trichogramma* species are used in biological control programs of some known insect pests including cotton bollworm, *Helicoverpa armigera* Hübner, European corn borer, *Ostrinia nubilalis* (Hübner), rice stem borer, *Chilo suppressalis* Walker, fall armyworm, *Spodoptera frugiperda* Smith, cactus moth, *Cactoblastis cactorum* (Berg), cabbage looper, *Trichoplusia ni* (Hübner), codling moth, *Cydia pomonella* Linnaeus, carob moth, *Spectrobates ceratoniae* Zeller and some related species (Doyon and Boivin 2005; Cônsoli et al. 2010; Altoé et al. 2012; Paraiso et al. 2012).

In Iran, extensive application of *Trichogramma* spp. was conducted at past two decades. *Trichogramma* spp. mainly used against rice stem borer, *C. suppressalis* eggs in North of Iran. An estimate of the released area by *Trichogramma* spp. in Iran is not straightforward. Based on the available data, nowadays up to 187,000 hectares of seven field crops and fruit orchards are treated by them against rice stem borer, cotton bollworm, European corn borer, carob moth and codling moth (Attaran and Dadpour-Moganlou 2011).

## 6.3 Importance of Egg Mortality

Trichogrammatidae and other egg parasitoids are idiobionts (Wajnberg and Hassan 1994). They kill their hosts at commence of development prior to any damage to vegetation. An attacked host egg lives only in early stage, and dies soon while host egg reduces to a developmental substrate for *Trichogramma* larvae. This is a big advantage for a natural enemy to reduce pest population before injury. This sequence of impact may also allow proper timing of sequential application of larvicidal insecticides and/or larvivorous predators or parasitoids following to *Trichogramma* release programs.

## 6.4 Iranian Species and Their Target Moths

Faunistic studies on *Trichogramma* in Iran began by Shojai and his colleagues in 1984 (Shojai et al. 1990). They collected and identified a rich fauna of native *Trichogramma* species in vast agricultural ecosystems such as rice, corn, cotton, pomegranate, cabbage etc. in north, center and west of Iran by an essential cooperation of Dr. Voegele. Supplementary studies on morphological and enzymatic characteristics of *Trichogramma* species was then conducted by Kahani (1998) and Ebrahimi (1991). Just four genera of Trichogrammatidae were identified in Iran. *Trichogramma* is the richest one in terms of number of species. Furthermore,

*Oligostia* Walker was reported on *Arboridia kermanshah* Dlabola eggs from Urmia, *Paracentrobia* Howard on *Sesamia nonagrioides* Lefebvre and *Uscana* Girault on *Caryedon acaciae* Gyllenhal eggs both in Khuzestan province respectively (Mostaan and Akbarzadeh-Shoukat 1995). There are evidences that show some other genera exist in Iran but because of small size and difficulty in collecting and identifying have been neglected (Ebrahimi 2004).

Unfortunately, a serious challenge in faunistic studies on Trichogrammatidae in Iran and evaluation of contribution of each species in natural control of hosts is turbulence caused by augmentative releases, which may perturb both richness and evenness of species in a region. Release programs often neglected natural role of the parasitoid reservoirs and probable impacts of the released individuals on the already established fauna. Before and after release, evaluations are needed in this regards (van Driesche and Bellows 1996).

First time, Afshar in 1938 noticed the importance of *Trichogramma* in pest control and the possibility of mass production of them in Iran (Behdad 1996). Then in the 70's, following to invasions of the *C. suppresalis* to Northern provinces, two species of *Trichogramma*, were introduced from Germany by Dr. Nikjou who neglected rich native fauna of *Trichogramma* spp. in broad Iranian plateau. Rearing efforts was begun immediately and restricted areas in Northern Iran were treated by inundative release during 1976–1978 (Ebrahimi 2004).

So far, 11 species of *Trichogramma* have been identified in Iran by using morphological characteristics as well as enzymatic analysis (Shojai et al. 1988, 1990; Ebrahimi 1996; Kahani 1998; Ebrahimi et al. 1998; Alizadeh et al. 2002; Poorjavad 2011, 2012; Mahrughan et al. 2015). The species are as below:

*Trichogramma brassicae* Bezdenko, *T. embryophagum* Hartig, *T. pintoi* Voegelé, *T. evanescens* Westwood, *T. sembilidis* (Aurivillius), *T. tsumokave* Sorokina, *T. principium* Sugonjaev & Sorokina, *T. euproctidis* Giraulf, *T. ingricum* Sorokina, *T. cacoeciae* Marchal, *T. dendrolimi* Matsumura.

Identification of Iranian trichogrammatids by foreign experts has been questionable in some cases. For example, Agamy and Hassan (see Ebrahimi 2004) have been diagnosed *T. poliae* Nagaraja and *T. minutum* Riley from Tehran, while these species are rather known as Oriental and Australian fauna and Nearctic and Oriental fauna, respectively and these diagnoses may need further revision.

The most abundant species collected in different regions is *Trichogramma brassicae* Bezdenko. In fact, anywhere one can find some *Trichogramma* specimens, this species will be present. *T. brassicae* has higher density in flat and dry areas. This species so far reported from northwestern (Zanjan, East and West Azarbaijan, Ardabil), northern (Guilan, Mazandaran, Golestan), central (Markazi, Alborz, Tehran, Isfahan, Yazd), northeastern (Khorasan) and Western (Shahr-e Kord) provinces (Shojai et al. 1990; Ebrahimi et al. 1998; Ebrahimi 1999, 2004; Fig. 6.2). Therefore, *T. brassicae* is the most important and best biocontrol agent for management of some serious insect pests in Iran (Shojai et al. 1998; Ebrahimi 2004). Ease of access to the *T. brassicae* has caused this species to be used in most augmentation and release programs. Ebrahimi et al. (1998) collected *T. brassicae* from *C. pomonella*, *Lobesia botrana* Denis and Schiffermüller (Tortricidae),

**Fig. 6.2** Provincial distribution of *Trichogramma* in Iran, br = *brassicae*, cc = *cacociae*, dr = *dendrolimi*, em = *embryophagum*, ev = *evanescense*, ig = *ingricum*, pr = *principum*, pt. = *pintoi*, sm = *semblidis*, ts = *tshumakovae*

*H. armigera*, *Mamestra brassicae* L. (Noctuidae), *Scrobipalpa ocellatella* Boyd (Gelechiidae), *Papilio demoleus* L. (Papilionidae), *O. nubilalis* and *C. suppressalis* (Crambidae) eggs.

*Trichogramma evanescens* Westwood is the second most common species that has been collected on *O. nubilalis*, *C. suppressalis*, *H. armigera*, *C. pomonella*, *Ectomyelois ceratoniae* Zeller and *Naranga aenescens* Moore eggs from Isfahan, Alborz, Markazi, East Azarbaijan, Khorasan-e-Razavi and Northern provinces (Ebrahimi et al. 1998; Ebrahimi 1999, 2004; Fig. 6.2). In different parts of Iran, *T. evanescens* is observed alongside with *T. brassicae*, with higher densities in northern slopes of central Alborz Mountains (Shirazi et al. 2010; Mahrughan et al. 2015).

*Trichogramma embryophagum* (Harting) has reported from Guilan, West and East Azerbaijan, Mazandaran, Tehran, Markazi, Isfahan and Yazd provinces on eggs of *Archips rosanus* L., *C. pomonella*, *Ocnerogyia amanda* Staudinger and *E. ceratoniae* (Ahmadi and Basiri 1991; Ebrahimi 2004; Shiri et al. 2012; Fig. 6.2).

*Trichogramma dendrolimi* Matsumura was observed first time on eggs of *C. pomonella* from northeastern, Razavi Khorasan province (Fig. 6.2). However, it may also be found in other areas (Ebrahimi et al. 1998). A thelytokous population of this species has been collected recently (Ebrahimi 2004).

*Trichogramma semblidis* (Aurivillius) was collected from apple orchards of West Azerbaijan, and on *C. suppressalis* and *O. nubilalis* eggs from rice fields of Guilan and Zanjan (Fig. 6.2). This species has the widest host range among other species and it can attack different orders of insects.

*Trichogramma principium* Sugonjaev and Sorokina was collected from West and East Azerbaijan and Mazandaran provinces on *C. suppressalis* eggs (Fig. 6.2). A broader distribution range is possible for this species in Iran.

Shojai et al. (1990) reported the *Trichogramma pintoi* Voegele from Golestan province on *H. armigera* eggs. Then Ebrahimi (1999) collected this species from Khorasan on *C. pomonella* eggs (Fig. 6.2). Subsequent investigations revealed wider host range including *Pieris brassicae* L., *Euproctis chrysorrhoea* L., *Agrotis segetum* Denis and Schiffermüller, *M. brassicae*, *Plusia* sp. Ochsenheimer, *O. nubilalis*, *L. botrana*, *Prays oleae* Bernard, *Cassida nebulosa* L. and *Acantholyda posticalis* Matsumura eggs (Ebrahimi et al. 1998; Pinto 1999; Poorjavad et al. 2012; Ranjbar-Aghdam and Ataran 2014). Surely, this species can also be found in other regions of Iran.

*Trichogramma tshumakovae* Sorokina was collected in Northern forests of Iran. This species has been reported from Iran (Ebrahimi et al. 1998) and Kyrgizstan (Sorokina 1984) on *C. suppressalis* and *M. brassicae*, respectively.

*Trichogramma ingricum* Sorokina first time recorded by Ebrahimi and Akbarzadeh-Shoukat (2008) from *C. pomonella* and *L. botrana* in West and East Azarbaijan and *C. suppressalis* in Mazandaran (Fig. 6.2). This species resembles *T. embryophagum* morphologically and some recent reports of *T. ingricum* may be related to *T. embryophagum*.

Ebrahimi (2004) collected *Trichogramma cacoeciae* Marchal only from Yazd. However, Shojai et al. (1998) reported it from Markazi, Isfahan and Yazd provinces on *E. ceratoniae* eggs in pomegranate orchards (Fig. 6.2). Furthermore, it was reported from Varamin (Tehran province). It is difficult to distinguish *T. cacoeciae* from *T. embryophagum* and some researchers considered them as synonyms (Pintureau 1997).

Poorjavad et al. (2012) and Karimi et al. (2012) collected *Trichogramma* specimens from parasitized eggs of several lepidopteran pests in Northern and Central Iran. They recognized seven *Trichogramma* species based on the morphological characteristics and some molecular facts including nucleotide sequence, size and restriction profile of the internal transcribed spacer 2 (ITS2) region of the rDNA (Table 6.1). They ascertained first record of *T. euproctidis* Girault on *Pieris brassicae* L. and *Plutella xylostella* L. eggs in Iran.

Overall, Iranian fauna of *Trichogramma* can be considered intermediate between East and West Palearctic regions, as *T. dendrolimi*, the most common species in China and widespread one in Japan and Taiwan, was collected from Khorasan as well. On the other hand, *T. brassicae*, *T. evanescens* and *T. embryophagum* was

**Table 6.1** Geographic range, host and number of *Trichogramma* specimens examined by Poorjavad et al. (2012)

| Name of species | Geographic origin (latitude, longitude) | n[a] | Host |
|---|---|---|---|
| *T. brassicae* (203)[b] | Sangtop (36.55212 N, 52.330627E) | 50 | *Ostrinia nubilalis* on *Xanthium* sp. |
| | Kasgarmahal (36.533916 N, 51.933746E) | 13 | *Chilo suppressalis* on *Oryza sativa* |
| | Lekode (36.521777 N, 52.277069E) | 6 | *O. nubilalis* on *Xanthium* sp. |
| | Rasht (37.259572 N, 49.536324E) | 3 | *C. suppressalis* on *O. sativa* |
| | Sote (36.64611 N, 52.540741E) | 9 | *C. suppressalis* on weed |
| | Tonekabon (36.815881 N, 50.873566E) | 6 | *C. suppressalis* on *O. sativa* |
| | Varaz deh (36.452218 N, 52.2000165E) | 4 | *C. suppressalis* on *O. sativa* |
| | Hosein abad (36.51957 N, 52.26059E) | 39 | *C. suppressalis* on *O. sativa* |
| | Sharam kala (36.536123 N, 52.441177E) | 51 | *C. suppressalis* on *O. sativa* |
| | Taleb amoli (36.619386 N, 52.265396E) | 7 | *C. suppressalis* on *O. sativa* |
| | Gand yab (36.539433 N, 52.036228E) | 4 | *O. nubilalis* on *Xanthium* sp. |
| | Posht nesha (37.374523 N, 49.888916E) | 2 | *C. suppressalis* on weed |
| | Velisde (36.458983 N, 52.271404E) | 6 | *C. suppressalis* on *O. sativa* |
| | Chaboksar (36.95291 N, 50.541573E) | 3 | *C. suppressalis* on *O. sativa* |
| *T. cacoeciae* (6)[b] | Qum (34.657569 N, 50.911589E) | 6 | *Ectomyelois ceratoniae* on *Punica granatum* |
| *T. evanescens* (46)[b] | Aktij mahale (36.559015 N, 52.667942E) | 12 | *O. nubilalis* on *X.* sp. |
| | dasht-naz sari (36.662636 N, 53.262749E) | 10 | *O. nubilalis* on *Zea mays* |
| | Gorgan (36.82234 N, 54.425583E) | 3 | *C. suppressalis* on *O. sativa* |
| | Salma, Gorgan (36.913666 N, 54.574585E) | 3 | *C. suppressalis* on *O. sativa* |
| | Semnan (35.57943 N, 53.387547E) | 3 | *E. ceratoniae* on *P. Granatum* |
| | Keteshest (37.213925 N, 49.850464E) | 2 | *C. suppressalis* on *O. sativa* |
| | Bishe kala (36.660432 N, 52.376289E) | 9 | *O. nubilalis* on *X.* sp. |
| | Qum (34.670488 N, 50.887642E) | 1 | *E. ceratoniae* on *P. granatum* |
| | Nokade (36.249672 N, 53.369865E) | 2 | *O. nubilalis* on *X.* sp. |
| | Shiraz (29.773914 N, 52.715149E) | 1 | *E. ceratoniae* on *P. granatum* |

(continued)

**Table 6.1** (continued)

| Name of species | Geographic origin (latitude, longitude) | n[a] | Host |
|---|---|---|---|
| T. embryophagum (25)[b] | Saryazd (31.35636 N, 54.29777E) | 6 | E. ceratoniae on P. granatum |
| | Ashkezar (31.56531 N, 54.10496E) | 6 | E. ceratoniae on P. granatum |
| | Varamin (35.357696 N, 51.992798E) | 8 | E. ceratoniae on P. granatum |
| | Neyriz (29.11293 N, 54.16297E) | 5 | E. ceratoniae on P. granatum |
| T. euproctidis (13)[b] | Nahalestan (35.483005 N, 50.58534E) | 7 | P. brassicae & P. xylostella |
| | | | on Brassica oleracea |
| | Golestanak (35.774372 N, 50.904465E) | 6 | P. brassicae & P. xylostella |
| | | | on B. oleracea |
| T. pintoi (3)[b] | Ghochhesar (35.546195 N, 51.441422E) | 2 | P. brassicae on B. oleracea |
| | Charbagh (36.032442 N, 50.565948E) | 2 | P. brassicae on B. oleracea |
| T. tshumakovae (5)[b] | Kheyrod (36.584658 N, 51.556091E) | 5 | Egg trap in forest |

Source: Data retrieved from Pourjavad et al. (2012)
[a]Number of collected wasps
[b]Total number of collected eggs parasitized by Trichogramma spp.

spread from Europe to Iran. Moreover, existence of *T. ingricum, T. tshumakovae* and *T. pintoi*, represents the relationship between Iranian fauna to Russia and Eastern Europe.

It should be noted that a high differential sensitivity of *Trichogramma* spp. either to physical conditions or hosts has led ecotypes to be evolved. For example, high ecological difference among *T. brassicae* populations in different parts of Iran has led to evolution of numerous host-related regional ecotypes (Ebrahimi 2004).

## 6.5 Laboratory Rearing on Factitious Hosts

Chambers (1977) defined mass rearing as "Production of insects competent to achieve program goals with an acceptable cost/benefit ratio and in numbers exceeding ten-thousand to one-million times of the mean productivity of the native population female". A very large number of *Trichogramma* is needed to be released in a limited area. Flanders (1927, 1929, 1930) described the first mass-production system for *Trichogramma* on *S. cerealella* eggs that was a significant step in mass production of these wasps. Initial efforts on mass rearing of *Trichogramma* in Iran, were conducted at 1981 by Iranian Research Institute of Plant Protection and Iranian Research Organization for Science and Technology. Two factitious hosts were used in this regards. Limited efforts also have been done by artificial diets (Attaran and Dadpour-Moganlou 2011). Several hosts such as Mediterranean flour moth, *Anagasta kuehniella* Zeller, angoumois grain moth, *Sitotroga cerealella* Olivier, rice moth, *Corcyra cephalonica* Stainton, Indian meal moth, *Plodia interpunctella* Hübner and greater wax moth, *Galleria mellonella* Linneaus are examined for rearing of *Trichogramma* spp. (Shojai 1987). Angoumois grain moth, *S. cerealella* eggs were used for mass-rearing of various *Trichogramma* species in Iran (Vaez 2007). For host rearing purpose, barley seeds are fumigated by Aluminum phosphide, washed and dried in room air. Then they are infested to the moth eggs (4 g *S. cereallella* eggs for 3 kg barley). The larvae penetrate into the seeds and develop for 7–10 days depending on temperature and relative humidity. Adults are transferred to funnels as soon as they begin to fly within containers. Both sides of the funnel are covered with cloth net (50 mesh) to prevent adults to escape. The funnel is inversely put on a piece of white paper. Adults lay their eggs on paper sheets where are harvested every day. Then the harvested eggs are exposed to female parasitoids.

This moth usually is grown on barley, but it may grow on other cereals as well. The barley variety can affect the quality of the produced moths, but it does not take into account in *Trichogramma* mass rearing systems because suitable varieties are not easily accessible for producers and they have to use the existing varieties that sometimes are very inappropriate (Vaez 2007). There are major defects in host rearing in insectaries. The average efficiency of *S. cerealella* production in Iran is about 4–5 g *Sitotroga* eggs per 1 g introduced to production system (Attaran and Dadpour-Moganlou 2011). This is so, while a female can lay 100–150 eggs during her life (Behdad 1996). A minor increase in efficiency from 5 to 6 will cause to a considerable saving in costs in a specified period. Indeed, price of produced agents is major limiting factor in biological control programs.

Additional to the barley varieties, seed size and problems arose by infestation of barley seeds to *Sitotroga* larvae, there are serious problems in preparation of the moth eggs. Brushing method, lace texture and the mesh size, the surface on which funnels are settled, the duration of the egg harvesting, moth density and physical conditions are factors affecting the egg production rate. Storage condition and duration of egg storage can affect the production's efficiency. Eggs of different ages often are pooled and kept in refrigerator for subsequent use. Such pooling strategy and bulking the harvested eggs (that reduces egg quality), poor storage conditions, inappropriate ventilation and fault of insectaries are the factors that affect *Trichogramma* production (Attaran and Dadpour-Moganlou 2011). Many manufacturing units in Iran are without the required standards. In a mass-rearing system all requirements cannot be met together. On the other hand, economic and cultural aspects and governmental support policy have caused the producer fails to achieve the required standards. These are true for mass rearing of *Sitotroga*, but it can be extended for *A. kuehniella* as well. The mass production system of *A. kuehniella* is more efficient than *S. cereallella* and eggs can be produced by a ratio of 1–20 per one (Attaran and Dadpour-Moganlou 2011). Because *A. kuehniella* can be easily used for mass rearing of *Habrobracon hebetor* Say as well, requires minimal rearing facilities and is suitable enough for wasps, can be considered in *Trichogramma* mass rearing systems. Nevertheless due to higher labor required, should be improved. More than 517 kg parasitized eggs of *S. cereallella* were purchased by Plant Protection Organization of Iran from manufacturers of *Trichogramma* in 2011 (Attaran and Dadpour-Moganlou 2011). In the current mass rearing system in Iran, production efficiency is three per one. In order to increase efficiency and have a cost-effective production, one have to choose the higher quality hosts and thelytokous *Trichogramma* populations with higher tolerance to temperature and pesticide-resistant ones.

## 6.6  Quality Control and Parasitoid Efficacy Studies

One challenging object in mass-rearing of *Trichogramma* spp. in Iran is quality control of the produced wasps. Quality control is a process that begins with choosing agents for initiating mass culture and continues to final steps of efficacy evaluation. The most common definition of mass production is easy way of rearing of large number of insects over the existing number to use in control unit (Shorey and Hale 1965). Several criteria are used in monitoring the quality of the produced agents to ensure continuity of production. Criteria such as viability, survival rate, adult size, density dependence, performance, sex ratio, life span and adaptation to environmental conditions. Dutton et al. (1996) divided the qualitative factors of trichogrammatid production in two categories. First-one is parameters affecting the number of released females such as percentage of adult emergence, percentage of malformed females and sex ratio. The second one is suitability and preference parameters of trichogrammatids, such as longevity and fertility both on factitious and target hosts, acceptability and suitability of target host for parasitoids that developed on factitious host and flight ability. According to Etzel and Legner (1999) the most important factor in mass rearing is fertility. The main factor limiting widespread use of *Trichogramma* on a global scale is lack of necessary information on the quality of produced Trichogrammatidae (Losey and Calvin 1995).

The International Organization of Biological Control (IOBC) has offered standards for mass rearing of *Trichogramma* spp. The sex ratio above 50%, fertility rate above 40 progeny per week, female mortality below 20% after 7d, parasitized at least 10 host eggs within 4 h, adult emergence above 80% and malformed females below 5% are standard indices of trichogrammatids in this instruction (Bigler et al. 1991; Singh et al. 2001; van Lenteren 2003). A comprehensive quality control guideline has not been published in Iran yet. However, problem is not lack of information but is rather its implementation. A qualitative evaluation of Iranian insectaries in 2006 showed that 30–40% of them are out of a standard range. An evaluation revealed that it is mainly due to low percentage of adult emergence as well as high percentage of malformed females (Shirazi et al. 2006). Low rate of adult emergence can be caused by storage system and/or mortality during transportation. Storage of *Trichogramma* spp. has received less attention than what is required. In production units, the wasps are directly transferred to refrigerator in the fifth or sixth day of their life, which causes to mortality increase and quality reduction. In traditional method, the wasps are repeatedly exposed to thermal shock and not stored at the right time. The temperature varies by insectary and therefore determining the right time of storage is difficult (Attaran and Dadpour-Moghanlou 2011). The similar problem exists in host egg storage. Abroon et al. (2014) stated that fresh eggs ($\leq$24 h) should be kept in refrigerator for 24 h. This action has no undesirable effect on parasitism while prevent host egg hatch. Lengthening the cold storage duration, will result in quality loss. Several reviews and research papers both in Iran and other countries emphasize on the importance of the quality of mass-produced *Trichogramma* in which various parameters such as longevity, survival rate, sex ratio, fecundity, life table parameters, functional response parameters, host preference etc. have been considered as quality parameters (see the next subsections).

### 6.6.1   Longevity

Female longevity is a quality indicator of *Trichogramma* spp. in field scale (Bigler 1994). It is also an important criterion in preliminary screening of ecotypes or species of *Trichogramma* in order to consider in pest management programs (Vasquez et al. 1997; Gomez et al. 1995, 2005). Longer longevity of a parasitoid provides more contact fortune to host (Steidle et al. 2001). This may increase the success of biological control (Nadeem 2010). Different scientists reported different values for longevity of *Trichogramma* females ranging from 6 to 9d. The longevity was 9.19 and 7.27d for *T. chilonis* on *C. cephalonica* and *H. armigera* respectively (Shirazi 2004), 6.79d for *T. pintoi* on *S. cerealella* eggs (Alizadeh and Ebrahimi 2004), 7, 8.12 and 9.16 for *T. brassicae* on *A. kuehniella*, *S. cerealella* and *H. armigera* respectively (Vaez et al. 2009).

### 6.6.2   Fecundity

Fecundity is an important indicator of efficiency of a parasitoid. It directly determines number of parasitized hosts (Roitberg et al. 2001). The average fecundity of *Trichogramma* spp. is recorded in laboratory and usually is used to determine quality of them prior to release. Of course, such criterion may not be adequate in field condition, because a large number of wasps may die before oviposition (Dutton et al. 1996). Shirazi (2004) stated that *T. chilonis* has a higher daily fecundity on *C. cephalonica* compared to its natural host, *H. armigera*. Fathipour and Dadpour-Moghanlou (2003) reported that the fecundity of *T. pintoi* on *A. kuehniella* and *S. cerealella* eggs were 97.3 and 71.6, respectively. Fecundity of *T. brassicae* on

*A. kuehniella, S. cerealella* and *H. armigera* were 79.17, 92.90 and 57.72 respectively (Farazmand et al. 2007; Vaez et al. 2009). Some external factors affect fecundity of these wasps. The size of host egg determines the available resources and ultimately the size, fecundity and fitness of the emerging parasitoid. On the other hand, fecundity of *Trichogramma* spp. depends on size and quality of host egg (Bulut 1990). Larger *T. brassicae* females with higher fecundity were obtained when reared on the angoumois grain moth compared to the Indian meal moth (Hosseini Bai et al. 2006). Temperature and photoperiod are the other factors that affect fecundity. Shirazi (2006) reported daily fecundity of *T. chilonis* on *C. cephalonica* at 25 and 30 °C to be 6.69 and 12.73 respectively. He also mentioned that more eggs were laid at a photoperiod of 14: 10 h (L: D). Fecundity of *T. embryophagum* was 16.61, 26.33 and 26.46 and that of the *T. pinoti* was 12.33, 30.93 and 28.50 eggs at 22, 24 and 26 °C respectively (Ranjbar-Aghdam and Attaran 2014).

Feeding prior to release, markedly increased fecundity and longevity of *Trichogramma* (Southard and Houseweart 1982; Bigler 1994; Reznik et al. 1997; Soares et al. 2012). Attaran et al. (2004) reported that fecundity of starved *T. brassicae* were 20.57 and 19.60 on *A. kuehniella* and *S. cerealella*, while that of the fed wasps by honey reached to 71.35 and 56.7 respectively. Karimi-Malati and Hatami (2005, 2010) also showed that nutrition plays an important role in increasing of longevity of *T. brassicae*, as honey supplement led to 8d-increased longevity. Yeast had no effect on longevity.

### 6.6.3  Life Table Parameters

According to Andrewartha and Birch (1954) demographic parameters are the best indicators of fitness of a population and are suitable criteria for comparing physiological states of different species, populations, etc. The most common demographic parameters are intrinsic rate of increase ($r_m$), finite rate of increase ($\lambda$), gross reproductive rate (GRR), net reproductive rate ($R_0$), mean generation time (T) and doubling time (DT) (Carey 1993; Ebert 1999). Local populations of trichogrammatids were compared in regards of life table parameters in a few studies. However, in many studies estimation of parameters such as intrinsic rate of increase or net reproductive rate was neglected due to difficulty of calculations. These parameters are used for monitoring fitness of the parasitoid at successive generations (Table 6.2).

Direct comparison among different works may be misleading because the parasitoids are very sensitive to physical conditions. On the other hand, punctuality among different researchers in treating the wasps is not similar. Data analysis methods also can cause some deviations. Disregarding such sources of variation, *T. brassicae* seems to be partially superior to the *T. pintoi* and *T. embryophagum*. This is deducible from higher intrinsic rate of increase and reproductive rates and shorter generation time. However, role of host seems to be crucial. For example, *P. interpunctella* is a poor host while *S. cereallella* is a suitable one for *T. brassicae*.

**Table 6.2** Life table parameters estimated for different species of *Trichogramma*

| *Trichogramma* species | Host | GRR | $R_0$ | $r_m$ | T | DT | $\lambda$ | References |
|---|---|---|---|---|---|---|---|---|
| *T. brassicae* | *P. interpunctella* | 39.36 | 19.26 | 0.214 | 13.75 | 3.23 | 1.242 | Iranipour et al. (2009) |
| | *A. kuehniella* | 56.23 | 45.51 | 0.291 | 13.07 | 2.38 | 1.351 | |
| | | 59.25 | 42.47 | 0.355 | 10.54 | 1.95 | 1.426 | Iranipour et al. (2010) |
| | *S. cerealella* | 77.24 | 62.31 | 0.389 | 10.61 | 1.78 | 1.476 | |
| | *H. armigera* | 55.40 | 44.68 | 0.348 | 10.87 | 1.99 | 1.416 | |
| | *E. ceratoniae* | 38.02 | 36.95 | 0.286 | 12.66 | 2.42 | 1.33 | Mohseni et al. (2016) |
| | *Plutella xylostella* | a | a | 0.296 | 12.76 | 2.34 | 1.34 | Akbari (2010) |
| | *Chilo suppressalis* | a | 37.64 | 0.442 | 8.18 | 1.57 | 1.556 | Ranjbar-Aghdam and Mahmoudian (2014) |
| *T. pintoi* | *A. kuehniella* | 46.30 | 45.30 | 0.257 | 14.81 | 2.70 | 1.29 | Dadpour Moghanlou (2002) |
| | *S. cerealella* | 49.19 | 45.68 | 0.281 | 13.59 | 2.47 | 1.32 | |
| | *E. ceratoniae* | 28.62 | 27.15 | 0.262 | 12.56 | 2.64 | 1.30 | Mohseni et al. (2016) |
| | *Plutella xylostella* | a | a | 0.343 | 11.56 | 2.02 | 1.41 | Akbari (2010) |
| *T. embryopgagum* | *A. kuehniella* | 55.24 | 44.88 | 0.238 | 16.37 | 2.91 | 1.27 | Haghani and Fathipour (2004) |
| | *S. cerealella* | 41.74 | 37.63 | 0.218 | 16.49 | 3.18 | 1.24 | |
| | *E. ceratoniae* | 39.24 | 37.20 | 0.311 | 11.64 | 2.23 | 1.36 | Mohseni et al. (2016) |
| | *Plutella xylostella* | a | a | 0.331 | 12.83 | 2.09 | 1.39 | Akbari (2010) |

[a]Data was not available

On the other hand, superiority of *S.cereallella* does not so impressive in two other species of *Trichogramma*. Similar order in host suitability is not present among various *Trichogramma* species. All these species are of applied value for integrated pest management of different pests.

## 6.6.4  Functional Response

Functional response is a key concept to understand how parasitoids/ predators influence population dynamics of their hosts/preys and how they affect the structure and hence stability of the communities in which they are living (Jervis and Kidd 1996; Reay-Jones et al. 2006; Wajnberg et al. 2008). Functional response experiments may be conducted to evaluate parasitoid/ predator potential in suppressing the different prey/host densities and can be helpful in preliminary screening of natural enemies as well as predicting the results of their release in field (Moezipour et al. 2008; Carneiro et al. 2010). Functional response of a natural enemy is indicated by two parameters: the handling time $(T_h)$, the time taken for a parasitoid to chase, overcome and handle a single host; and the attack rate or searching efficiency ($a$ or $b$), the rate of encounter to new hosts (Holling 1966; Hassell 1978). The parameter "$a$" is rather a constant value while the "$b$" shows density dependence. Estimation of these parameters has been object of a few studies in Iran (Table 6.3).

The same factors that influence life table parameters may affect functional response parameters as well. Among them age (Sahragard 1989; Asadi et al. 2012), density (Tahriri et al. 2007), strain (Farrokhi et al. 2010), host age and density (Chen et al. 2006), host developmental stage (Gonza'lez-Herna'ndez et al. 2005),

**Table 6.3** Functional response parameters of Iranian populations of *Trichogramma* spp.

| Species | Host | Shape of functional response | a or b $(h^{-1})$ | $T_h$ (h) | References |
|---|---|---|---|---|---|
| *T. pintoi* | *A. kuehniella* | III | 0.48 | 0.965 | Fathipour et al. (2002) |
| | *S. cerealella* | III | 0.16 | 1.169 | |
| *T. brassicae* | *S. cerealella* | III | 0.168 | 1.468 | Arbab Tafti et al. (2004) |
| | *A. kuehniella* | III | [a] | [a] | Lashkari et al. (2004) |
| | *S. cerealella* | III | [a] | [a] | |
| | *H. armigera* | II | 0.051 | 1.20 | |
| | *A. kuehniella* | III | 0.0042 | 0.601 | Farazmand and Iranipour (2009) |
| | *P. interpunctella* | III | 0.0022 | 1.344 | |
| | *S. cerealella* | II | 0.049 | 0.42 | Farrokhi et al. (2010) |
| | *H. armigera* | III | 0.031 | 1.134 | Vaez et al. (2013) |
| | *S. cerealella* | III | 0.0037 | 0.831 | Nikbin et al. (2014) |

[a]Data is not available

temperature and relative humidity (Parajulee et al. 2006; Shojaei et al. 2006; Moezipour et al. 2008) were considered.

Both type II and type III functional responses are seen in *Trichogramma* spp. Fathipour et al. (2002, 2003), Arbab Tafti et al. (2004), Farazmand and Iranipour (2009), Vaez et al. (2013) and Nikbin et al. (2014) reported that functional response of *Trichogramma* species to their hosts has been type-III. In contrast, in some studies (Karimian 1998; Farrokhi et al. 2010), type-II functional response also was observed in these parasitoids. In all experiments, total time of exposure was 24 h. Host and parasitoid species or populations, rearing conditions (temperature and relative humidity) etc. are the sources of variation.

Shape of the functional response curve can be affected by some factors such as temperature, species, size and physiological state of host/prey (Juliano and Williams 1985; Coll and Ridgway 1995; Runjie et al. 1996; Messina and Hanks 1998; Wang and Ferro 1998; De Clercq et al. 2000; Mohaghegh et al. 2001; Sagarra et al. 2001). Moezipour et al. (2008) reported type-II functional response of *T. brassicae* to *S. cerealella* eggs at 25 °C and type III at 20 and 30 °C. Functional response of *T. brassicae* was age-specifically changeable, as female parasitoids were more sever and more efficient during their early adult life (Nikbin et al. 2014).

## 6.7   Targets and Inundative Programs

So far, *Trichogramma* spp. were used against *C. suppressalis*, *O. nubilalis*, *S. ceratoniae*, *H. armigera* and *C. pomonella* in seven crops in Iran. Overall 187896 hectare of the mentioned crops was treated by *Trichogramma* spp. by Plant Protection Organization of Iran in 2010. The released area in mentioned crops during 1995 to 2010 is shown in Figs. 6.3, 6.4, 6.5, 6.6, 6.7, 6.8 and 6.9 (Attaran and Dadpour-Moghanlou 2011). Plant Protection Organization of Iran published an executive instruction for control of the above pests. Unfortunately, theoretical base of the following instructions is not known. Region for which this instruction was provided is also unknown. For example, date of release or number of generation differs by region, but this instruction does not provide regional information. At last, no data is available about evaluation of the results of releases. Therefore, it is likely an empirical rather than a scientific note.

### 6.7.1   Codling moth, Cydia pomonella L.

Here, biological control of codling moth by *Trichogramma* spp. has been considered.

**First generation** Release of *Trichogramma* against the first generation was recommended to be carried out in three stages by releasuing 15g parasitized eggs

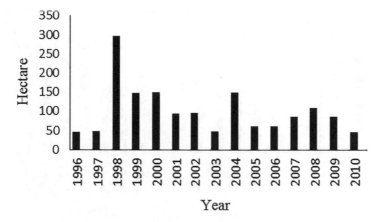

**Fig. 6.3** Release area of *Trichogramma* spp. against codling moth, *C. pomonella* in apple orchards of Iran (Attaran and Dadpour-Moghanlou 2011)

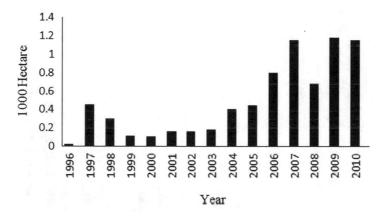

**Fig. 6.4** Release area of *Trichogramma* spp. against carob moth, *S. ceratoniae* in pomegranate orchards of Iran (Attaran and Dadpour-Moghanlou 2011)

per hectare. This is done by hanging paper sheets (Tricocards) bearing 0.01 g parasitized eggs from tree branches in three stages. The first stage will be done 48 h after oil spraying at the peak of the moth emergence with the highest release rate in this stage. The second stage will continue by releasing additional 5g parasitized eggs per hectare one week later. The final stage will continue a week later by reaching total release rate of all stages to 15 g.

**Second generation** The first phase of release is a week before the peak of adult emergence with the rate of 5g (500 Tricocard) per hectare. The exact time is determined by summing degree days accumulated for the pest. The second release will be done at the peak of the codling moth flights, with the highest release rate. Finally, one week later a supplementary release will be done to achieve 15g release rate as before.

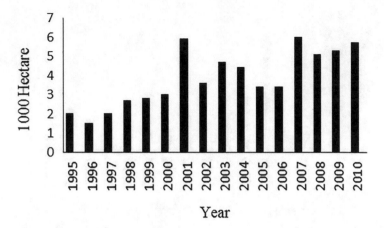

**Fig. 6.5** Release area of *Trichogramma* spp. against cotton bollworm, *H. armigera* in cotton fields of Iran (Attaran and Dadpour-Moghanlou 2011)

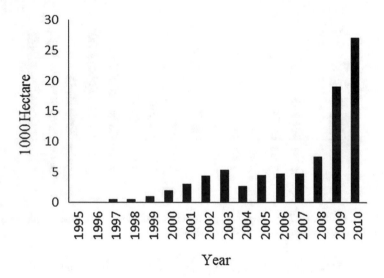

**Fig. 6.6** Release area of *Trichogramma* spp. against tomato fruit worm, *Helicoverpa zea* Boddie in tomato fields of Iran (Attaran and Dadpour-Moghanlou 2011)

**Intermediate of the second and the third generations** Depending on regional weather conditions, two additional releases between two generations may be necessary. If adult catches by traps show an increase after the second generation, which represents overlap with the next generation, one or two releases will be conducted. The first one 7d after the last release of the previous generation followed by the second one 7d later, both at the rate of 5g per hectare.

**Third generation** Three additional releases will be carried out during the third generation at a rate of 15g per hectare in three stages whether generations overlap or

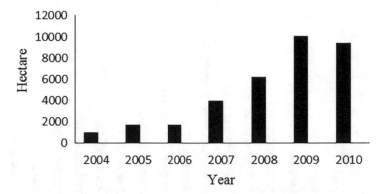

**Fig. 6.7** Release area of *Trichogramma* spp. against *H. armigera* in soybean fields of Iran (Attaran and Dadpour-Moghanlou 2011)

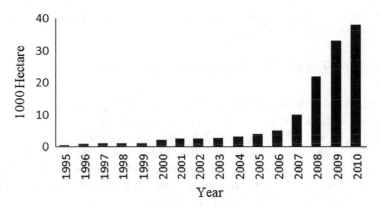

**Fig. 6.8** Release area of *Trichogramma* spp. against European corn borer, *O. nubilalis* in corn fields of Iran (Attaran and Dadpour-Moghanlou 2011)

not. In those regions where the second and third generations are overlapped, the release against the 3rd generation will continue within 7d after the second phase of the previous stage, whereas in non-overlapping ones, it will began 25-28d after the peak of the second generation relying on traps' catches. This coincides to while one week remains to peak of the third generation. The two additional releases will continue every other week. Release areas of apple orchards covered by Tricocards have been shown in Fig. 6.3.

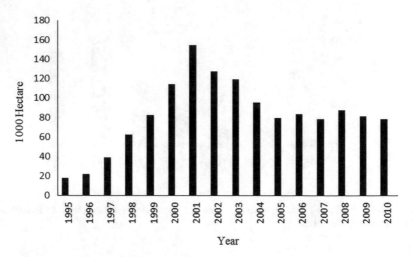

**Fig. 6.9** Release area of *Trichogramma* spp. against rice stemborer, *C. suppressalis* in rice fields of Iran (Attaran and Dadpour-Moghanlou 2011)

### 6.7.2   *Carob moth,* **Spectrobates ceratoniae** *Zeller*

An overall 8–10 times release of 600 Tricocards (0.01g parasitized eggs each) per hectare with regular intervals of 10–15 days between any two successive releases was recommended (Alizadeh et al. 2007). The first round will begin at early June and the subsequent ones will continue as follow: The 2nd time: late June, The 3rd time: early July, The 4th time: late July, The 5th time: the mid-August, The 6th time: early September, The 7th time: mid-September, The 8th time: late September, The 9th time: early October, The 10th time: mid- October.

Note: The adult emergence of overwintering generation of carob moth is gradual and occurs in a long horizon of time, with an apparent peak at mid-June when is considered the best time for the first release. Since the developmental time of carob moth at the first generation may take 50-55d, sometimes the second release is delayed and then one or two releases may be removed. Furthermore, in precocious cultivars and early harvest regions, one additional release also may be excluded. Release areas of pomegranate orchards have been shown in Fig. 6.4.

### 6.7.3   *Cotton Bollworm,* **Helicoverpa armigera** *Hübner*

#### 6.7.3.1   Release in Cotton Fields

The first release of *Trichogramma* spp. is carried out by a rate of 200 Tricocards per hectare (0.01 g parasitized egg each) when the first moths are trapped by pheromone traps in cotton fields. An additional release of 3 g (300 Tricocards) is done 5–7 days later followed by the third release of 1 g *Trichogramma* together 1000 *Habrobracon hebetor* females per hectare 5 days after that. Releases against the second generation will continue 10–15 days later at a rate of 2 g *Trichogramma* per hectare followed by another release of the same rate together 1000 *H. hebetor* females after a week. Release areas of cotton fields have been shown in Fig. 6.5.

### 6.7.3.2    Release in Tomato Fields

The first release of 1 g *Trichogramma* spp. against *H. armigera* occurs in tomato fields providing that the following requirements meet together in a field:

(a) One moth is observed in traps in tomato fields
(b) The first bollworm eggs is observed in tomato fields
(c) The tomato field is at early blooming stage

The subsequent releases continue by the rate of 2 and 1 g per hectare one week and 5 days later respectively. Release areas of tomato fields have been shown in Fig. 6.6.

### 6.7.3.3    Release in Soybean Fields

The first release of 200 Tricocards per hectare is done when the first moth is trapped within each of two traps settled in each region and the first eggs are visited on soybean leaves. Two subsequent releases will be done at the same rate 5–7 days later (Amini and Vafayi-Oskouyi 2007). Release areas of soybean fields have been shown in Fig. 6.7.

## 6.7.4    *European Corn Borer,* Ostrinia nubilalis *(Hubner)*

In corn fields of Khuzestan Province, the *Trichogramma* spp. are released at the rate of 100 Tricocards per hectare against the second generation of *Ostrinia nubilalis* when the first female was observed. The subsequent releases will be done at 50% pupation and peak of the *O. nubilalis* flights by the rates of 100 and 200 Tricocards respectively (Amir Nazari 2007; Amir Nazari et al. 2015). Release areas of corn fields have been shown in Fig. 6.8.

## 6.7.5    *Rice Stemborer,* Chilo suppressalis *Walker*

The norm of *Trichogramma* release in rice farms is 4.5 g per hectare in Mazandaran province and 4 g in Golestan province. Realese of *Trichogramma* against the first generation of *C. suppressalis* is done at the rate of 100 Tricocards per hectare at the second half of April in field or nursery based on the peak of flight. For the second generation, two releases are performed; One at the peak of pupation or as soon as the first eggs or female oviposition was recorded and the second when *C. suppressalis* catches maximize (7–10d after the first release and before dough stage). When the second and third generations overlap, release is done at the peak flights of the second generation or when the curve begins to rise at the third generation (Anonymous 2006). Release areas of corn fields have been shown in Fig. 6.9.

Above Figs. (6.3, 6.4, 6.5, 6.6, 6.7, 6.8 and 6.9) show that the wide spread use of *Trichogramma* spp. in Iran was begun 6–17 years ago depending on crop. The release area in some products such as tomato and corn have been exponentially increasing while in other crops such as rice remained constant following a sudden decline or even in some others like apple have had decreasing trend. Total release area does not exceed 0.8% of the grown lands, which distributed mainly in three Northern provinces of Iran (Attaran and Dadpour-Moghanlou 2011). This suggests a potential market for the future.

## 6.8 Comparison Between Iran and the World Status of *Trichogramma* Research and Application

Study on flight speed of *Trichogramma* spp. showed that the Iranian populations are weaker flyers than other populations. This may be an intrinsic difference or otherwise arises by poor rearing methods (Attaran 2009; Moosavian et al. 2010). Despite low price of *Trichogramma* production in Iran (260,000 IRR $\approx$ 2 US \$ per gram) the release of these wasps is still very limited. Some of the most important restrictions are lack of investments, lack of workforce, lack of organization for on-time access to required numbers of wasps, lack of governmental support, insurance, low acceptability among farmers due to inadequate control and so on.

*Trichogramma* application initially received higher attention by farmers but gradually a large number of farmers refused it and believed that biological control is not efficient enough (Attaran and Dadpour-Moghanlou 2011). Numerous reasons caused it to happen. One main challenge in *Trichogramma* use has been extra reliance on one species and neglecting local populations. Ease of access caused *T. brassicae* released in many situations and replaced to natural fauna. This is while *T. brassicae* is not suitable to control carob moth (Table 6.2). One possible solution to overcome this problem is construction of a *Trichogramma* bank of natural source. This bank may assist to conserve natural diversity and provide suitable founders for augmentative programs adapted to local conditions. The other fault in *Trichogramma* production is lack of data related to efficacy evaluation of the released wasps. This negates having a true insight about the results of our measurements. This may cause to expend unnecessary effort and cost.

More ecological studies are needed for optimum use of all biological control agents including *Trichogramma* spp. Quality control measurements also are necessary in all steps of production. Unfortunately, no serious attention was paid to this important task in Iran. Packing, storage and release methods also need to further attention. Apparently, we are at initial steps of biological control by *Trichogramma* in Iran despite two decades of application. We need to revise our past efforts once again to set up a sufficient system.

## 6.9   Other Chalcids

Among Chalcids (Superfamily Chalcidoidea) there are some other families that have members which parasitizes egg of insects. For example some members of families Encyrtidae, Eupelmidae, Eulophidae, Torymidae, and Mymaridae attack insect eggs. The other members of these families attack other stages of insects. The following species were reported as egg parasitoids in Iran (Shishehbor and Rasekh 2015):

### 6.9.1   Family Encyrtidae

*Blastothrix sericea* (Dalman) and *Metaphycus punctipes* (Dalman) are egg parasitoids of *Eulecanium coryli* L. (Hem. Coccidae) (Ebrahimi 1993). The former species has a wider range of distribution. Northern and central provinces are the regions in which this species was reported (Radjabi 1989). This species was the most abundant parasitoid on *E. coryli* with 1/3 of the collected specimens (Davoodi et al. 2004).

*Ooencyrtus* spp. egg parasitoids of bugs and moths: One species was reported as *Ooencyrtus* sp. near to *masii* Mercet upon *Ocneria terebinthina* Stgr (Lep. Lymantridae) in Fars province in southern Iran (Sabahi 1996). In addition, three species of *O. telenomicida* (Vassiljev), *O. nigerrimus* Ferriere & Voegele and *O. fecundus* Ferriere & Voegele was reported on common sunn pest *Eurygaster integriceps* Puton (Hem. Scutelleridae). All these species are gregarious parasitoids and up to four wasps can emerge from a single host egg (Iranipour 1996). In the case of *Ooencyrtus* sp. egg parasitoid of *Ocneria*, it is three wasps per a single host (Sabahi 1996). The recent species is active on all generations of the target host but parasitism rate of the first generation was very low (less than 6%). Therefor Sabahi (1996) suggested that supplementary release should be carried out at peak of the first generation of the host. Negative geotaxis, positive phototaxis, tendency to RH's around 30–40% and temperatures around 24–26 °C and finally host aggregation causes the wasps tend to crowd on upper canopy.

In a laboratory study, Safavi (1970) described oviposition and mating behaviors of *O. telenomicida* and *O. pityocampae*. He showed that *O. telenomicida* is a gregarious arrhenotokous parasitoid. Up to nine parasitoid eggs, four nymphs and three adult wasps were counted in a single host egg. Iranipour (1996) and

Ahmadpour et al. (2013) obtained four adults of both species of *O. telenomicida* and *O. fecundus* emerged from an egg. In contrast *O. pityocampae* is rather a solitary and telytokous species. Male progeny is very rare and just one male was recorded among 152 adult wasps. Also seldom more than one wasp can emerge from a single egg of *Eurygaster integriceps*. The adults of the recent wasp are larger and it may be the reason of solitary development of immature stages. Development in larger hosts has not been examined. Majority of *Ooencyrtus* species are not host-specific and can attack large number of moths' and true bugs' eggs. However, their host range is not studied enough. Hyperparasitism on *Trissolcus* species also was reported (Safavi 1970). This is considered an undesirable property that may restrict application of *Ooencyrtus* spp. against the cereal bugs. Furthermore *Ooencyrtus* species are poorer parasitoids in comparison to platygastrid egg parasitoids of cereal bugs (see Chap. 8 for comparative biostatistics of both groups). This is due to lower fecundity, slower development and polyphagy, which reduce overall impact of these wasps on target hosts particularly *E. integriceps*. Often natural parasitism of these species does not exceed 5% (Iranipour et al. 2011), although egg traps may be attacked by a higher rate late season (Shafaei et al. 2011; Nozad-Bonab and Iranipour 2011). An explanation may be role of the other hosts during massive oviposition of the target host. In other words, *Ooencyrtus* species may have other hosts that prefer them to target host and just switch to target host late season when the other hosts are going to become rare. In fact, except in West Azarbaijan province (Shafaei et al. 2011), no place else *Ooencyrtus* species was observed early season and their early season activity in West Azarbaijan province was very limited and ceased for a month. On the other hand, cereal bugs also are going to terminate their oviposition late season. Hence, only few eggs are available for parasitism.

Rafat et al. (2013a, b), Iranipour et al. (2013a, b) and Ahmadpour et al. (2013) studied effects of superparasitism and gregarious development on life history, functional response and aggregation of *O. telenomicida* and *O. fecundus* respectively. Life history studies on *O. telenomicida* revealed that two progenies within a single host develop by the same rate that one progeny develops in similar unit. They also have similar fecundity and no difference is present between twins and singletons (Rafat et al. 2013a). Ahmadpour et al. (2013) extended this arguement to higher superparasitism rates by *O. fecundus* i. e. triplets and quadruplets. Except development time and generation time, other statistics were affected by number of progenies that shared a host egg. However, no difference was observed between twins and singletons again.

Functional response studies revealed that parasitism rate was inversely density dependent in *O. telenomicida* (Rafat et al. 2013b). On the other hand, superparasitism rate decreases by increasing host density. Taking into account that four progenies can develop in a single host egg, one can expect to achieve this extreme in lower host/parasitoid rates. However, such a capacity never realizes. Safavi (1970) stated that the reason is host feeding behavior of this parasitoids that pierce egg shells for feeding and those eggs with several holes on egg shell became useless. In *O. fecundus*, Iranipour et al. (2013a) showed that parasitism rate was 100% as long as host density remained eight or less and only slightly decreased at density of 16. However, it deeply declined at higher densities. A vast number of parasitoids were permitted to oviposit on a few host egg clutches (14 eggs per clutch). Then individual eggs were separated from the clutches and developed singly in glass vials to emerge. Number of progeny per a single host egg was divided to four groups; singletons, twins, triplets, and quadruplets. Females of each group were singly confined in vials containing different number of host eggs for five hours. Parasitized eggs were maintained in laboratory and number of progeny was recorded after emergence. As expected, per capita progeny number was decreased by parent size (singletons to quadruplets) in all host densities. Nevertheless, per capita number of progeny unexpectedly did not decrease continuously. It means that number of progeny per a single egg reduced in densities below four for singleton and twin parents and remained unchanged for triplets and quadruplets. This may occur due to larval competition, host injury during handling, or interference during oviposition. Direct observations do not support contention of competing females. Difference among the parents in terms of total number of progeny was also lower than expectation. A maximum 1.74 times more progeny (10.6 vs. 6.0) was recorded at density of four host eggs in singleton mothers compared to quadruplets. This difference declined to 1.12 times at density 32. It seems that 12% difference in fertility between singleton and quadruplet wasps is insignificant. In a similar study on *O. telenomicida* both host numbers and parasitoid numbers varied (Iranipour et al. 2013b). Maximum daily fecundity was 21. So no resource limitation was expected until host/parasitoid ratio remains 20:1. It was true with one exception. In all treatments in which resource limitation was present parasitism rate was 100%. Superparasitism rate was higher than *O. fecundus*; it reached 14.7–29.4% at H/P ratios above 20:1. A sudden increase in per capita number of progeny occurred at ratios below 10:1. Maximum progeny was 3.06 per host at the lowest H/P ratio, which exceeded that of the *O. fecundus*. The results suggested that female parasitoid responds to host deficiency by reduction in fecundity rather than increasing superparasitism.

## 6.9.2   Family Eupelmidae

*Anastatus bifasciatus* Forcroy was reported on sunn pest eggs (Iranipour et al. 1998). Just males were reared in host eggs. The females also collected later from larger eggs of an unknown moth.

## 6.9.3   Family Eulophidae

*Tetrastichus xanthonelanae* Rond was collected from eggs of *Galerucella luteola* Mull. (Col. Chrysomelidae) (Azmayesh-Fard and Esmaili 1981). This is a serious pest of elm trees in landscapes of some regions. Larvae defoliate elm trees and damage is so intensive that leaves begin to drop mid-summer some years. New vegetation causes trees to weaken. A female wasp can parasitize 200 eggs. In addition, it can also feed on a few eggs prior to parasitism. In 1977 and 1978 the first adults emerged early May. This wasp as well as another pupal parasitoid *T. brevistigma* Gahan has considerable effect on the *G. luteola* of the first generation. A non-persistent insecticide application between the first and second generation satisfy pest control up to the end of season without significant effect on the parasitoid.

### 6.9.4  Family Torymidae

*Oopristus safavii* Stephan was collected from *Apodiphus amygdalii* Germar (Hem. Pentatomidae) in Tehran in 1966 (Safavi 1973). After half a century, no other report is present about this species.

### 6.9.5  Family Mymaridae

A few species of Mymaridae was reported in Iran. All of them are egg parasitoids of other insects. *Anaphes* (*Patasson*) sp. attacks eggs of *Hypera postica* (Col. Curculionidae) in Southwestern (Yasuj) alfalfa fields. Parasitism rate is as low as 3% (Saeedi 2007). *Anagrus atomus* L. is the most known species of this family. It attacks a few species of plant hoppers including *Arboridia Kermanshah* (Hem. Cicadellidae) in Isfahan (Latifian 1998; Latifian and Soleymannezhadian 2008; Hesami et al. 2001) and Khorasan vineyards (Triapitsyn 1998), *Circulifer tenellus* in sugarbeet fields of Karaj (Walker et al. 1997) and *Empoasca decipiens* in bean fields of Tehran (Naseri et al. 2007). It is also a parasitoid of a few planthoppers in Europe and Turkey. For example *Empoasca vitis* (Arno et al. 1987; Chiappini et al. 1996) *E. decipiens* (Schmidt 2000) and *Arboridia adanae* (Yigit and Erkilic 1987) are attacked by it. This parasitoid switches to alternate hosts (other planthoppers' eggs) at winter in vineyards because target host *E. vitis* hibernate as stage other than egg. This parasitoid is not able to control *E. vitis* although it has an important role in its population reduction (Latifian and Soleymannejadian, 2008). Naseri et al. (2007) observed density independent mortalirty of *E. decipiens* in four bean varieties. Those varieties with dense trichoms did not colonized by the parasitoid at all. In contrast, Agboka et al. (2004) reported an inverse density dependent relationship between mortality and host population density.

*Erythmelus panis* Enocl attacks *Stephanitis pyri* F. (Hem. Tingidae) eggs (Akbarzadeh-Shoukat 1998). This is a specific pest of pear trees with low importance. Parasitism rate achieves 6–32%.

# References

Abroon P, Mousavi SG, Ashouri A, Kishani H (2014) The impact of different qualities of flour moth eggs, *Ephestia kuehniella* on *Trichogramma brassicae* parasitism. The 1st National Conference on Stable Agriculture and Natural Resources, 30 January 2014. Mehrarvand international Institute of Technology, Abadan, pp. 1–8

Agboka K, Tounou AK, Al-Moaalem R, Poehling HM, Raupach K, Borgemeister C (2004) Life-table study of *Anagrus atomus*, an egg parasitoid of the green leafhopper *Empoasca decipiens*, at four different temperatures. BioControl 49(3):261–275

Ahmadi AA, Basiri G (1991) Biological investigations on *Trichogramma* sp., and egg parasitoid of fig defoliator, *Ocnerogyia cacoeciae* Staud. (Lep.: Lymantridae) in Fars. Proceedings of the 10th Plant Protection Congress of Iran, 1–5 September 1991, Kerman, Page 5

Ahmadpour S, Iranipour S, Asgari S (2013) Effects of superparasitism on reproductive fitness of *Ooencyrtus fecundus* Ferriere & Voegele (Hym. Encyrtidae), egg parasitoid of sunn pest, *Eurygaster integriceps* Puton (Hem. Scutelleridae). Biol Cont Pests & Plant Disease 2 (2):97–105

Akbari F (2010) Study of application of *Trichogramma* in *Plutella xylostella* (L.) control. M.Sc. Thesis of Entomology. College of Agriculture, Shahed University

Akbarzadeh-Shoukat G (1998) The first report on the occurrence of the egg parasitoid of pear lace bug in Iran. App Entomol Phytopath 66(1/2):44

Alizadeh S, Ebrahimi E (2004) Investigation on biology of *Trichogramma pintoi* (the egg parasitoid of winter moth) on laboratory host. Proceedings of the 16th Iranian Plant Protection Congress, Vol. 1: Pests. University of Tabriz; Tabriz, 28 Aug–1 Sept 2004. Page 79

Alizadeh S, Javanmoghadam H, Fotohi K, Hosseini O, Bagheri AR (2002) Comparative investigation on population density of different cutworm species in sugar beet and introduction of *Trichogramma euproctidis* Girault. The 15th Plant Protection Congress of Iran, 7–11 September 2002, Razi University, Kermanshah

Alizadeh P, Yasini A, Dashtbani K (2007) Management of trees and ornamental plants pests. Plant Protection Organization Publication. 23 pp. http://ppo.ir/LinkClick.aspx?fileticket=XSAxKp3%2bbi4%3d&tabid=874

Altoé TS, Pratissoli D, Carvalho JR, Santos Junior HJG, Paes JPP, Bueno RCOF, Bueno AF (2012) *Trichogramma pretiosum* (Hymenoptera: Trichogrammatidae) parasitism of *Trichoplusia ni* (Lepidoptera: Noctuidae) eggs under different temperatures. Ann Entomol Soc Am 105: 82–89

Amini MA, Vafayi-Oskouyi F (2007) Instructions of soybean pests, diseases and weeds management. Plant Protection Organization publication. 50 pp. http://ppo.ir/LinkClick.aspx?fileticket=FdBk2otQxDw%3d&tabid=873

Amir-Nazari M (2007) Executive instruction, control of pests, diseases and weeds of corn in Iran. Plant Protection Organization of Iran Publication. 79 pp. http://ppo.ir/LinkClick.aspx?fileticket=mSmBvHYmANc%3d&tabid=873

Amir-Nazari M, Momeni H, Arabi M, Maroof A, Nezamabadi N, Ebtali Y, Benanach K (2015) Executive instruction, control of pests, diseases and weeds of corn in Iran. Plant Protection Organization of Iran Publication. 82 pp. http://ppo.ir/LinkClick.aspx?fileticket=y5uunTv7Lbc%3d&tabid=873

Andrewartha HG, Birch LC (1954) The distribution and abundance of animals. The University of Chicago Press, Chicago

Anonymous (2006) Technical instructions of rice pests, diseases and weeds management. Plant Protection Organization publication. 51 pp. http://ppo.ir/LinkClick.aspx?fileticket=J%2fzqeNflKJE%3d&tabid=873

Arbab-Tafti R, Sahragard A, Salehi L, Asgari S (2004) Study on functional response of *Trichogramma brassicae* Bezdenko (Hym.: Trichogrammatidae) to different densities of *Sitotroga cerealella* Olivier (Lep.: Gelechiidae) eggs. J Agric Sci 1:1–8

Arno C, Alma A, Arzone A (1987) *Anagrus atomus* as egg parasite of *Typhlocybinae* (Rhynchota Auchenorrhyncha). Proceedings of the 6th Auchen. Meeting, Turin, Italy, 7–11 Sept. pp. 611–615

Asadi R, Talebi AA, Khalaghani J, Fathipour Y, Moharramipour S, Askari-Siahooei M (2012) Age-specific functional response of *Psyllaephagus zdeneki* (Hymenoptera: Encyrtidae), parasitoid of *Euphyllura pakistanica* (Hemiptera: Psyllidae). J Crop Prot 1(1):1–15

Attaran MR (2009) Investigation on flight activity of *Trichogramma brassicae* Bezd. (Hym., Trichogrammatidae) at different generations and temperatures of rearing. J Entomol Res 1 (3):229–237

Attaran MR, Dadpour-Moghanlou H (2011) An analytical review of present status and future prospective in utilization of *Trichogramma* wasps for biological control of agricultural pests in Iran. National Conference on the Development of Biological Control in Iran, Iranian Research Institute of Plant Protection,Tehran, 27–28 July 2011. Pp. 94–112

Attaran MR, Shojai M, Ebrahimi E (2004) Comparision of some quality parameters of *Trichogramma brassicae* (Hym., Trichogrammatidae). JESI 24(1):29–47

Azmayesh-Fard P, Esmaili M (1981) An investigation on the species of elm leaf beetle parasitoids and their application for population density reduction of their host in Karadj. JESI 6(1/2):21–31

Behdad E (1991) Pests of fruit crops in Iran, 2nd edn. Neshat publications, Isfahan

Behdad E (1996) Iran phytomedicine encyclopedia. Plant pests and diseases, weeds. Yadbood Publishing, Isfahan, p 3153

Bigler F (1994) Quality control in *Trichogramma* production. In: Wajnberg E, Hassan SA (eds) Biological control with egg parasitoids. CAB International, Wallingford, pp 1–36

Bigler F, Cerutti F, Laing J (1991) First draft of criteria for quality control (product control) of *Trichogramma*In: Bigler, F. (ed). Proceeding of fifth workshop of the IOBC global working group "quality control of mass reared arthropods" IOBC. Pp. 200–201

Bueno VHP, van Lenteren JC (2010) Biological control of pests in protected cultivation: implementation in Latin America and successes in Europe. Memorias, XXXVII Congreso Sociedad Colombiana de Entomologia, Bogota, D.C., 30 June – 2 July 2010: Pp. 261–269

Bulut H (1990) Investigation on the determination of host age for the egg parasitoids, *Trichogramma* spp. and some behaviour of the adults. Can Entomol 25:151–159

Carey JR (1993) Applied demography for biologists with special emphasis on insects. Oxford University Press, New York

Carneiro TR, Fernandes OA, Cruz I, Bueno RCOF (2010) Functional response of *Telenomus remus* Nixon (Hymenoptera: Scelionidae) to *Spodoptera frugiperda* (Smith) (Lepidoptera: Noctuidae) eggs: effect of female age. RBE 54:692–696

Chambers DL (1977) Quality control in mass rearing. Annu Rev Entomol 22:289–308

Chen WL, Leopold RA, Harris MO (2006) Parasitism of the glassy-winged sharpshooter, *Homalodisca coagulata* (Homoptera: Cicadellidae): functional response and superparasitism by *Gonatocerus ashmeadi* (Hymenoptera: Mymaridae). Biol Control 37:119–129

Chiappini E, Triapitsyn SV, Donev A (1996) Key to the Holarctic species of *Anagrus* (Hym. Mymaridae) with a review of the Nearctic and Palearctic (other than European) species and description of new taxa. J Nat Hist 30:551–595

Ciochia V (1991) Some aspects of the utilization of *Trichogramma* sp. in Romania. In: Wajnberg E, Vinson SB, editors. *Trichogramma* and Other Egg Parasitoids. Antibes INRA; 1991. pp. 181–182

Coll M, Ridgway RL (1995) Functional and numerical response of *Orius insidiosus* (Heteroptera: Anthocoridae) to its prey in different vegetable crops. Ann Entomol Soc Am 88:732–738

Cônsoli FL, Parra JRP, Zucchi RA (2010) Egg parasitoids in agroecosystems with emphasis on *Trichogramma*, vol 9. Springer, Verlag, Dortrecht

Dadpour-Moghanlou H (2002) An investigation on the host-parasitoid system between *Trichogramma pintoi* (Voegele) and the Mediteranean flour and angoumois grain moth, in laboratory conditions. M.Sc. Thesis of Entomology. College of Agriculture, Tarbiat Modares University

Davoodi A, Talebi AA, Rajabi G, Fathipour Y, Rezaei V, Rakhshani E (2004) An identification of parasitoids and hyperparasitoids of the most common soft scales (Hom.: Coccidae) in Tehran and Guilan provinces. Iran J Agric Sci 35(4):887–899

De Clercq P, Mohaghegh J, Tirry L (2000) Effect of host plant on the functional response of the predator *Podisus maculiventris* (Heteroptera: Pentatomidae). Biol Control 18:65–70

Doyon J, Boivin G (2005) The effect of development time on the fitness of female *Trichogramma evanescens*. J Insect Sci 5:1–5

Dutton A, Cerruti F, Bigler F (1996) Quality and environmental factors affecting *Trichogramma brassicae* efficiency under field conditions. Entomol Exp Appl 81:71–79

Ebert TA (1999) Plant and animal populations, methods in demography. Academic Press, San Diego

Ebrahimi E (1991) Morphological and enzymatic study of the genus *Trichogramma* in Iran. Ph. D. Thesis. Tarbiat Modarres University, Iran, 149 pp.

Ebrahimi E (1993) An introduction to the new six parasitoid wasps for the fauna of Iran. JESI 12-13:113

Ebrahimi E (1996) The species of *Trichogramma* Westwood in Iran. (Abstract 20–102). In: Proceeding XX International Congress of Entomology, Firenze, Italy, Page 639

Ebrahimi E (1999) The biodiversity of *Trichogramma* in Iran. Proceeding of international symposium on biological control of insect pests of agricultural crops, University of Aleppo, Syria, pp 11–19

Ebrahimi E (2004) Study and identification the species of *Trichogramma* in Iran. In: Azema M, Mirabzadeh A (eds) Issues on different aspects of applying natural enemies for biological conrtol. Sepehr Publication, Tehran, pp 5, 213 pp–71

Ebrahimi E, Akbarzadeh-Shoukat G (2008) Report of *Trichogramma ingricum* (Hym.: Trichogrammatidae) from Iran. JESI 27(2):43–45

Ebrahimi E, Pintureau B, Shojai M (1998) Morphological and enzymatic study of the genus *Trichogramma* (Hym. Trichogrammatidae) in Iran. JAEP 66:122–141

Etzel LK, Legner EF (1999) Culture and colonization. In: Bellows TS, Fisher TW (eds) Handbook of biological control. Academic Press, San Diego, pp 125–197

Farazmand A, Iranipour S (2009) Functional response of *Trichogramma brassicae* Bezdenko (Hym.; Trichogrammatidae) to different egg densities of *Anagasta kuehniella* Zell. and *Plodia interpunctella* Hubner. Iranian J Plant Protect Sci 39(1):67–72

Farazmand A, Iranipour S, Saber M, Mashhadi-Jafarloo M (2007) Comparison of some biological parameters of *Trichogramma brassicae* Bez. on *Anagasta kuehniella* Zell. and *Plodia interpunctella* Hub. in laboratory condition. J Agric Sci 17:175–185

Farrokhi S, Ashouri A, Shirazi J, Allahyari H, Huigens ME (2010) A comparative study on the functional response of *Wolbachia*-infected and uninfected forms of the parasitoid wasp *Trichogramma brassicae*. J Insect Sci 10:1–11. Available from: insectscience.org/10.167

Fathipour Y, Dadpour-Moghanlou H (2003) Comparative biology of *Trichogramma pintoi* Voegele wasps reared on two of laboratory hosts. Iran J Agric Sci 34(4):881–888

Fathipour Y, Dadpour-Moghanlou H, Attaran M (2002) The effect of the laboratory hosts on functional response of *Trichogramma pintoi* Voegele (Hym.: Trichogrammatidae). J Agric Nat Resour Sci 9:109–118

Fathipour Y, Haghani M, Attaran M, Talebi A, Moharramipour S (2003) Functional response of *Trichogramma embryophagum* (Hym.: Trichogrammatidae) on two laboratory hosts. JESI 23:41–54

Flanders SE (1927) Biological control of the codling moth (*Carpocapsa pomonella*). J Econ Entomol 20:644

Flanders SE (1929) The mass production of *Trichogramma minutum* Riley and observations on the natural and artificial parasitism of the codling moth egg. Proc 4th Int Congr Entomol Trans 2:110–130

Flanders SE (1930) Mass production of egg parasites of the genus *Trichogramma*. Hilgardia 4:465–501

Gomez LLA, Diaz AE, Lastra LA (1995) Selection of strains of *Trichogramma exigum* for controlling sugarcane borers (*Diatraea* spp.) in the Cauca valley, Colombia. Less Colloques de l'INRA 73:75–78

Gonza'lez-Hernandez H, Pandey RR, Johnson MW (2005) Biological characteristics of adult *Anagyrus ananatis* Gahan (Hymenoptera: Encyrtidae): a parasitoid of *Dysmicoccus brevipes* Cockerell (Hemiptera: Pseudococcidae). Biol Control, 35: 93–103

Haghani M, Fathipour Y (2004) Effective kind of laboratory host on population growth parameters of *Trichogramma embryophagum*. J Agric Nat Resour Sci 10:117–124

Hassan SA, Kohler E, Rost WM (1988) Erprobung verschiedener *Trichogramma* Arten zur Bekampfung des Apfelwicklers *Cydia pomonella* L. und des Apfelschalenwicklers *Adoxophyes orana* F.R. (Lep., Tortricidae). Nachrichtenbl Deut Pflanzenschutzd (Braunschweig) 40(5):71–75

Hassell MP (1978) The dynamics of arthropod predator-prey systems. Princeton University Press, Princeton

Hesami S, Seyedoleslami H, Ebadi R (2001) Morphological notes on *Anagrus atomus* (Hymenoptera: Mymaridae), an egg parasitoid of grape leafhopper, *Arboridia kermanshah* (Hom.; Cicadellidae) in Isfahan. JESI 21(1):51–67

Holling CS (1966) The functional response of invertebrate predators to prey density. Mem Entomol Soc Can, Ottawa, p 84

Hosseini-Bai S, Yazdani-Khorasgani A, Mashhadi-Jafarloo M (2006) The determination of body size of *Trichogramma brassicae* on *Sitotroga cerealella* and *Plodia interpunctella* and its relation with wasp's fecundity and longevity. Proceedings of the 17th Iranian Plant Protection Congress, Vol. 1: Pests, Campus of Agriculture and Natural Resources, University of Tehran, Karaj, 2–5 Sept. 2006, Page 17

Iranipour S (1996) A study on population fluctuation of the egg parasitoids of *Eurygaster integriceps* Put. (Heteroptera: Scutelleridae) in Karaj, Kamalabad, and Fashand. M.Sc. thesis on Agricultural Entomology, University of Tehran, Karaj, Iran. 187 pp.

Iranipour S, Kharrazi-Pakdel A, Radjabi G (1998) Introduction of a Chalcid species from Eupelmidae for Iran. 13th Iranian Plant Protection Congress, Vol.1-Pests, 23–27 August 1998, Karaj, Iran, Page 5

Iranipour S, Farazmand A, Saber M, Mashhadi-Jafarlou M (2009) Demography and life history of *Trichogramma brassicae* on two laboratory hosts, *Anagasta kuehniella* and *Plodia interpunctella*. J Insect Sci 9(51):1–8

Iranipour S, Vaez N, Nouri-Ghanbalani N, Asghari-Zakaria R, Mashhadi-Jafarloo M (2010) Effect of host change on demographic fitness of the parasitoid, *Trichogramma brassicae*. J Insect Sci 10(78):1–12

Iranipour S, Kharrazi-Pakdel A, Radjabi G, Michaud JP (2011) Life tables for sunn pest, *Eurygaster integriceps* (Heteroptera: Scutelleridae) in Northern Iran. B Entomol Res 101:33–44

Iranipour S, Ahmadpour S, Asgari S (2013a) Effects of superparasitism on searching ability of *Ooencyrtus fecundus* (Hym.: Encyrtide), egg parasitoid of sunn pest, *Eurygaster integriceps* (Hem., Scutelleridae). 2nd Global Conference on Entomology, 8–12 November 2013, Kuching Malaysia, Abstract No. 0101, Page 125

Iranipour S, Rafat A, Safavi SA (2013b) Functional and numerical response of *Ooencyrtus telenomicida* (Hym.: Encyrtidae) against sunn pest *Eurygaster integriceps* (Hem.: Scutelleridae) eggs. 2nd Global Conference on Entomology, 8–12 November 2013, Kuching Malaysia, Abstract No. 0103, Page 127

Jervis MA, Kidd NAC (1996) Insect natural enemies. Practical approaches to their study and evaluation. Chapman and Hall, London, 491 pp

Juliano SA, Williams FM (1985) On the evolution of handling time. Evolution 39:212–215

Kahani F (1998) Identification and biosystematics of *Trichogramma* in Mashhad, Nishabur and Kashmar cities. M.Sc thesis in Animal Science, Ferdowsi University of Mashhad, Iran, 244 pp.

Karimi J, Darsouei R, Hosseini M, Stouthamer R (2012) Molecular characterization of Iranian Trichogrammatids (Hymenoptera: Trichogrammatidae) and their *Wolbachia* endosymbiont. J Asia Pac Entomol 15(1):73–77

Karimian Z (1998) Biology and ecology of *Trichogramma brassicae* in rice fields of Guilan Province. M.Sc. Thesis of Entomology, University of Guilan, Guilan, Iran

Karimi-Malati A, Hatami B (2005) Effect of honey, sugar and proein diets on longevity of *Trichogramma brassicae* (Hym.: Trichogrammatidae) with and without host eggs. JESI 25 (1):1–12

Karimi-Malati A, Hatami B (2010) Effect of feeding and male presence on some biological characteristics of female *Trichogramma brassicae* (Hymenoptera: Trichogrammatidae). JESI 29(2):1–11

Lashkari AA, Talebi AA, Fathipour Y, Moharramipour S (2004) Functional response of. *Trichogramma brassicae* Bezdenko (Hym.; Trichogrammatidae) to different densities of three host eggs in laboratory condition. Proceedings of the 16th Iranian Plant Protection Congress, Vol. 1-Pests, University of Tabriz, Tabriz, 28 Aug.-1 Sept. 2004, P. 35

Latifian M (1998) Bioecology and geographical distribution of a dominant grape leafhopper species in Isfahan province. M.Sc. Thesis, Department of Entomology, Isfahan University of Technology, Isfahan, Iran

Latifian M, Soleymannezhadian E (2008) Study on the phenology of egg parasitoid *Anagrus atomus* L. (Hym.: Mymaridae) and its host (Grapevine leafhopper), *Arboridia kermashah* D. (Hom: Cicadellidae) to evaluate parasitism by recruitment method. J Agric Sci 31 (1):111–123

Li L (1994) Worldwide use of *Trichogramma* for biological control on different crops. In. Wajnberg E, Hassan SA (eds) Biological control with egg parastitoids. CAB International, Wallingford, pp 37–54

Losey JE, Calvin DD (1995) Quality assessment of four commercially available species of *Trichogramma*. J Econ Entomol 88:1243–1250

Mahrughan A, Shirazi J, Amir-Maafi M, Dadpour-Moghanlou H (2015) Dispersal of *Trichogramma brassicae* in tomato field. J Crop Prot 4(2):173–180

Messina FJ, Hanks JB (1998) Host plant alters the shape of the functional response of an aphid predator (Coleoptera: Coccinellidae). Environ Entomol 27:1196–1202

Moezipour M, Kafil M, Allahyari H (2008) Functional response of *Trichogramma brassicae* at different temperatures and relative humidities. Bull Insectology 62:245–250

Mohaghegh J, De Clercq P, Tirry L (2001) Functional response of the predators *Podisus maculiventris* (say) and *Podisus nigrispinus* (Dallas) (Hetcroptera: Pentatomidae) to the beet armyworms, *Spodoptera exigua* (Hübner) (Lepidoptera: Noctuidae): effect of temperature. J Appl Entomol 125:131–134

Mohseni E, Abbasipour H, Attaran MR, Askarianzadeh A (2016) Evaluation of the life table characteristics of three species of the genus *Trichogramma* on the carob moth, *Ectomyelois ceratoniae* under laboratory conditions. Biocont Plant Prot 3(2):47–58

Moosavian SS, Attaran MR, Shojai M, Shahrokhi S (2010) Comparison of flight ability of *Trichogramma brassicae* (Hym.: Trichogramatidae) at different generations. JAEP 78 (1):107–112

Mostaan M, Akbarzadeh-Shoukat G (1995) Studies on the egg parasitoid of the grape vine leafhopper (*Arboridia kermanshah*). *Trichogramma* and other egg parasitoids. Proceedings of the 4th International Symposium, Cairo, Egypt, 4–7 October, 1994, pp. 201–202

Nadeem S (2010) Improvement in production and storage of *Trichogramma chilonis* (Ishii), *Chrysoperla carnea* (Stephens) and their hosts for effective field releases against major insect pests of cotton. Ph.D Thesis, University of Agriculture, Faisalabad, Pakistan

Naseri B, Fathipour Y, Talebi AA (2007) Seasonal parasitism of *Empoasca decipiens* (Homoptera: Cicadellidae) by *Anagrus atomus* on four bean species in Tehran area. JAEP 75:1–11

Nikbin RA, Sahragard A, Hosseini M (2014) Age-specific functional response of *Trichogramma brassicae* (Hymenoptera: Trichogrammatidae) parasitizing different egg densities of *Ephestia kuehniella* (Lepidoptera: Pyralidae). J Agr Sci Tech 16:1205–1216

Nikonov PV, Lebedev GI, Startchevsky IP (1991) *Trichogramma* production in the USSR. In: Wajnberg E, Vinson SB (eds) *Trichogramma* and other egg parasitoids. Antibes INRA; 1991. pp. 151–152

Nozad-Bonab Z, Iranipour S (2011) Seasonal fluctuation in egg parasitoid fauna of sunn-pest *Eurygaster integriceps* Puton in wheat fields on New Bonab County, East Azarbaijan province, Iran. SAPS 20(3):71–83

Paraiso O, Hight SD, Kairo MTK, Bloem S, Carpenter JE, Reitz S (2012) Laboratory biological parameters of *Trichogramma fuentesi* (Hymenoptera: Trichogrammatidae), an egg parasitoid of *Cactoblastis cactorum* (Lepidoptera: Pyralidae). Fla Entomol 95(1):1–7

Parajulee MN, Shrestha RB, Lester JF, Wester DB, Blanco CA (2006) Evaluation of the functional response of selected arthropod predators on bollworm eggs in the laboratory and effect of temperature on their predation efficiency. Environ Entomol 35:379–386

Pinto JD (1999) Systematics of the North American species of *Trichogramma* Westwood (Hymenoptera: Trichogrammatidae). Mem Entomol Soc Wash 22:1–187

Pinto JD (2006) A review of the New World genera of Trichogrammatidae (Hymenoptera). J Hym Res 15:38–163

Pinto JD, Stouthammer R (1994) Systematics of the Trichogrammatidae with emphasis on *Trichogramma*. In: Wajenberg E, Hassan SA (eds) Biological control with egg parasitoids, CAB International. pp: 1–36

Pintureau B (1991a) Indices d'isolement reproductif entre espèces proches de Trichogrammes (Hym.: Trichogrammatidae). Ann Soc Entomol Fr 27:379–392

Pintureau B (1991b) We know *Trichogramma*, but what else in Trichogrammatidae? Insect parasitoids, 4th European Workshop. REDIA, Vol. XXIV, n.3. Appendices, pp. 375–377

Pintureau B (1997) Systematic and genetical problems revised in two closely related species of *Trichogramma, T. embryophagum* and *T. cacoeciae* (Hym., Trichogrammatidae). Misc Zool 20:11–18

Poorjavad N (2011) Morphological, molecular and reproductive compatibility studied on the systematic of the genus *Trichogramma* Westwood (Hymenoptera: Trichogrammatidae) in Tehran and Mazandaran Provinces (Iran). Ph.D. Thesis of Entomology, University of Tehran, Tehran, Iran

Poorjavad N, Goldansaz SH, Machtelinckx T, Tirry L, Stouthamer R, van Leeuwen T (2012) Iranian *Trichogramma*: ITS2 DNA characterization and natural *Wolbachia* infection. BioControl 15(2):452–459

Radjabi G (1989) Insects attacking Rosaceous fruit trees in Iran, vol 3. Plant Pests and Diseases Research Institute Publications, Tehran, Iran, Homoptera

Rafat A, Iranipour S, Safavi SA (2013a) Fecundity-life tables of *Ooencyrtus telenomicida* (Hym., Encyrtidae), egg parasitoid of *Eurygaster integriceps* (Hem., Scutelleridae). 2nd Global Conference on Entomology, 8–12 November 2013, Kuching Malaysia, Abstract No. 0095, P. 119

Rafat A, Safavi SA, Iranipour S (2013b) Density dependence of parasitic wasp *Ooencyrtus telenomicida* (Hym.: Encyrtidae) to different densities of sunn pest eggs. Proceedings of the Conference of Biological Control in Agriculture and Natural Resources. 26–27 Aug. 2013, College of Agriculture and Natural Resources, University of Tehran, Karaj, Page 14

Ranjbar-Aghdam H, Attaran MR (2014) Biological control of the codling moth by *Trichogramma embryophagum* based on Degree-Hours forecasting model. Biol Cont Pests Plant Dis 3 (2):87–96

Ranjbar-Aghdam H, Mahmoudian R (2014) Effect of different rice varieties on age specific life table and population growth parameters of *Trichogramma brassicae*, the egg parasitoid of the striped stem borer, *Chilo suppressalis*. Iranian J Plant Protect Sci 45(1):1–11

Reay-Jones FPF, Rochaf J, Goebel R, Tabone E (2006) Functional response of *Trichogramma chilonis* to *Galleria mellonella* and *Chilo sacchariphagus* eggs. Entomol Exp Appl 118:229–236

Reznik SY, Umarova TY, Voinovich ND (1997) The influence of previous host age on current host acceptance in *Trichogramma*. Entomol Exp Appl 82:153–157

Roitberg BD, Boivin G, Vet LM (2001) Fitness, parasitoids and biological control: an opinion. Can Entomol 133:1–10

Romeis J, Shanower TG (1996) Arthropod natural enemies of *Helicoverpa armigera* (Hübner) (Lepidoptera: Noctuidae) in India. Biocontrol Sci Technol 6: 481–508

Runjie Z, Heong KL, Domingo IT (1996) Relationship between temperature and functional response in *Cardiochiles philippinensis* (Hymenoptera: Braconidae), a larval parasitoid of *Cnaphalocrocis medinalis* (Lepidoptera: Pyralidae). Environ Entomol 28:1321–1324

Sabahi Q (1996) Field and laboratory studies on the biology and parasitic ability of *Ooencyrtus* prob. *masii* Mercet (Hym., Encyrtidae) egg parasitoid of *Ocneria terebinthina* (Lep. Lymantridae) and its efficiency as controlling agent in Fars province. M.Sc. Thesis, Department of Plant Protection, College of Agriculture, University of Tehran, Karaj, Iran

Saeedi K (2007) Preliminary studies on natural enemies of the alfalfa weevil *Hypera postica* (Gyllenhall), in Yasouj. J Res Agric Sci 3(1):1–13

Safavi M (1970) Study on biology of *Ooencyrtus* wasps, parasite of sunn pest egg. Third Plant Medicine Congress of Iran. 11–15 September1970, Pahlavi University, Shiraz. Iran. Pp. 249–259

Safavi M (1973) Etude bio-ecologiue des hymenopteres parasites des oeufs des punaises des cereals en Iran. Plant Pests and Diseases Research Institute, Tehran, 159 pp

Sagarra LA, Vinvent C, Stewart RK (2001) Body size as an indicator of parasitoid quality in male and female *Anagyrus kamali* (Hymenoptera: Encyrtidae). Bul Entomol Res 91:363–367

Sahragard A (1989) Biological studies on *Dicondylus indianus* (Olmi) (Hymenoptera: Drynidae) with particular reference to foraging behavior. Ph.D. Thesis, College of Cardiff, University of Wales, Wales, UK, 297 pp.

Schmidt U (2000) News on leafhoppers and their control on the island of Reichenau. Gemüse, Munchen 36(9):47–49

Shafaei F, Iranipour S, Kazemi MH, Alizadeh E (2011) Diversity and seasonal fluctuations of sunn pest's egg parasitoids (Hymenoptera: Scelionidae) in central regions of West-Azarbaijan province, Iran. J Field Crop Entomol 1(1):39–54

Shirazi J (2004) Effects of factitious host *Corcyra cephalonica* (St.) and natural host *Helicoverpa armigera* (Hub.) eggs on some important biological characters of *Trichogramma chilonis* Ishii (Hymenoptera: Trichogrammatidae). Proceeding of the 16th Iranian Plant Protection Congress, Vol. 1: Pests, 28 Aug. – 1 Sept. 2004. University of Tabriz, Tabriz, Page 14

Shirazi J (2006) Effect of temperature and photoperiod on the biological characters of *Trichogramma chilonis* Ishii (Hymenoptera: Trichogrammatidae). PJBS 9(5):820–824

Shirazi J, Attaran MR, Rezapanah MR, Farrokhi S (2006) Quality control evaluation of mass reared *Trichogramma* wasps in private insectaria of Guilan and Mazandaran provinces, Iran. Proceedings of the 17th Iranian Plant Protection Congress (Vol.1: Pests), 2–5 September 2006, Campus of Agriculture and Natural Resources, University of Tehran, Karaj, Page 21

Shirazi J, Taghizadeh M, Dadpour-Moghanlou H, Attaran MR, Zand S (2010) Investigation on the parasitism level of *Ostrinia nubilalis* (Hub.) eggs related to different densities of released *Trichogramma brassicae* Bezdenko in corn. Proceedings of the 19th Iranian Plant Protection Congress, 31 July – 3August 2010, Tehran, Page 68

Shiri H, Askari O, Arabian M, Najafi S, Ebrahimi E (2012) The report of *Trichogramma embryophagum* Hartig, the egg parasitoid of carob moth from Tarom in Zanjan province. Proceedings of 20th Plant Protection Congress of Iran, Shiraz University, Shiraz, 26–29 August 2012, Page 150

Shishehbor P, Rasekh A (2015) Taxonomic and biologic characteristic of parasitoids with an emphasis on parasitoids of Iran. Shahid Chamran University Press, Ahvaz

Shojaei S, Safaralizadeh MH, Shayesteh N (2006) Effect of temperature on the functional response of *Habrobracon hebetor* Say (Hymenoptera: Braconidae) to various densities of the host, *Plodia interpunctella* Hubner (Lepidoptera: Pyralidae). PJE 28:51–56

Shojai M (1987) Entomology (3rd ed.): social life and natural enemies. No. 1681, University of Tehran Publication, 406 pp.

Shojai M, Tirgari S, Nasrollahi A (1988) Primary report on the occurrence of *Trichogramma*. In: *Trichogramma* and other egg parasites. Les Colloques de l'INRA 43:121

Shojai M, Tirgari S, Azma M, Nasrollahi AA (1990) Faunistic study of beneficial parasitoid wasps *Trichogramma* and prospect for their application in agricultural field in Iran. IROST 9 (18):33–47

Shojai M, Ostovan H, Khodaman AR, Hosseini M, Daniali M, Seddighfar M, Nasrollahi AA, Labbafi Y, Ghavam F, Honarbakhsh S (1998) An investigation on beneficial species of *Trichogramma* spp. (Hym., Trichogrammatidae), active in apple orchards, and providing optimum conditions for mass production in laboratory cultures. J Agric Sci Islamic Azad Univ 16:5–39

Shorey HH, Hale RL (1965) Mass rearing of the larvae of nine noctuid species on a simple artificial medium. J Econ Entomol 58:522–524

Singh SP, Murphy ST, Ballal CR (2001) Augmentative biocontrol. Proceedings of the ICAR-CABI workshop, June 29th to July 1st, 2001, Project Directorate of Biological Control, Bangalore. CABI, Wallingford, 250 pp.

Soares MA, Leite GLD, Zanuncio JC, Sá VGM, Ferreira CS, Rocha SL, Pires EM, Serrão JE (2012) Quality control of *Trichogramma atopovirilia* and *Trichogramma pretiosum* (Hym.: Trichogrammatidae) adults reared under laboratory conditions. Braz Arch Biol Technol 55 (2):305–311

Sorokina AP (1984) New species of the genus *Trichogramma* Westw. (Hymenoptera, Trichogrammatidae) from the USSR. Entomologicheskoe Obozrenie 63(1):154–156

Southard SG, Houseweart MW (1982) Size differences of laboratory reared and wild populations of *Trichogramma minutum* (Hym.: Trichogrammatidae). Can Entomol 114:693–698

Steidle JLM, Rees D, Wright EJ (2001) Assessment of Australian *Trichogramma* species as control agents of stored product moths. J Stored Prod Res 37:263–275

Tahriri S, Talebi AA, Fathipour Y, Zamani AA (2007) Host stage preference, functional response and mutual interference of *Aphidius matricariae* (Hymenoptera: Braconidae: Aphidiinae) on *Aphis fabae* (Homoptera: Aphididae). J Entomol Sci 10:323–331

Triapitsyn SV (1998) *Anagrus* (Hym., Mymaridae) egg parasitoids of *Erythroneura* spp. and other leafhoppers (Hom. Cicadellidae) in North American vineyards and orchards: a taxonomic review. Trans Am Entomol Soc 124:77–122

Vaez N (2007) The effect of adaptation in *Trichogramma brassicae* Bez. reared on factitious hosts, to *Helicoverpa armigera* Hub. before release. M.Sc. Thesis of Entomology, College of Agriculture, University of Mohaghegh Ardabili, 111 pp.

Vaez N, Nouri-Ganbalani G, Iranipour S, Mashhadi-Jafarloo M, Asghari-Zakaria R (2009) Necessity of encountering *Trichogramma brassicae* Bezdenko wasps reared on alternative hosts, cereal moth and meal moth to target pest bollworm *Helicoverpa armigera* Hubner prior to release. J Agric Sci 19(1):317–332

Vaez N, Iranipour S, Hejazi MJ (2013) Effect of treating eggs of cotton bollworm with *Bacillus thuringiensis* Berliner on functional response of *Trichogramma brassicae* Bezdenko. Arch Phytopathol Plant Protect. https://doi.org/10.1080/03235408.2013.799820

van Driesche R, Bellows TS Jr (1996) Biological control. Chapman and Hall, New York

van Lenteren JC (2003) Quality control and production of biological control agents: theory and testing procedures. CAB International, Wallingford

van Lenteren JC, Buneo VHP (2003) Augmentative biological control of arthropods in Latin America. BioControl 48:123–139

Vasquez LA, Shelton AM, Hoffmann MP, Roush RT (1997) Laboratory evaluation of commercial Trichogrammatid products for potential use against *Plutella xylostella* (L.) (Lep.: Plutellidae). Biol Cont 9:143–148

Viggiani G (1981) Note su alcune speciedi *Oligosita* Walher e descrizione di auattro nouove specie. Bollettino del Laboratorio diEntomologia Agraria "Filippo Silvestri" di Portrici 38:125–132

Wajnberg E, Hassan SA (1994) Biological control with egg parastitoids. Wallingford, CAB International

Wajnberg E, Carlos Bernstein C, van Alphen J (2008) Behavioral ecology of insect parasitoids: from theoretical approaches to field applications. Wiley-Blackwell, New York, 464 pp

Walker GP, Zareh N, Bayoun IM, Triapitsyn SV (1997) Introduction of Western Asian egg parasitoids into California for biological control of beet leafhopper, *Circulifer tenellus*. Pan-Pac Entomol 73(4):236–242

Wang B, Ferro DN (1998) Functional response of *Trichogramma ostriniae* (Hymenoptera: Trichogrammatidae) to *Ostrinia nubilalis* (Lepidoptera: Pyralidae) under laboratory and field conditions. Environ Entomol 27:752–758

Yigit A, Erkilic L (1987) Studies on egg parasitoids of grape leafhopper, *Arboridia adanae* (Hom., Cicadellidae) and their effects in the region of South Anatolia. Turkiye I. Entomoloji Kongresi Bildirileri, Ege Universitesi, Bornova, Izmir, pp. 35–42

# Chapter 7
# Parasitic Wasps: Chalcidoidea and Ichneumonoidea

Hossein Lotfalizadeh and Abbas Mohammadi-Khoramabadi

## 7.1 Introduction

Chalcidoidea and Ichneumonoidea are the most diverse superfamilies of Hymenoptera. They contain the main groups of biological control agents of insect pests in both agricultural and forest ecosystems throughout the world (Quicke 2015). They affect immature stages of their hosts by using different biological traits (solitary and gregarious, endoparasitoid and ectoparasitoid, koinobiont and idiobiont).

Iran constitutes a large part of the Iranian plateau. It covers an area of 1,623,779 km$^2$ which is located between the eastern Mediterranean area and the Oriental region, and contains elements of rich fauna so the diverse topography and climate of Iran, from cool and humid mountains (Alborz in the north, Zagros in the west, North Khorasan in the northeast and Jebal Barez and Baluchistan mountains in the south and southeast of Iran) to hot and dry deserts (Zehzad et al., 2002). Iran also located at the crossroad of the Palaearctic, Afrotropical and Oriental biogeographic realms; therefore, it is one of the most diverse regions of the world that this is reflected in its insect fauna especially in parasitoid Hymenoptera or parasitoid wasps.

A review of available literature clearly shows that the majority of studies on parasitoid wasps of Iran prior to the 1960s were undertaken by foreign experts and that those by Iranian taxonomists mostly started more recently.

International projects and cooperating well-known specialists and professional laboratories accelerated the process of identifying hymenopterous parasitoids and

H. Lotfalizadeh (✉)
Plant Protection Research Department, East-Azarbaijan Agricultural and Natural Resources Research & Education Center, AREEO, Tabriz, Iran
e-mail: hlotfalizadeh@areeo.ac.ir

A. Mohammadi-Khoramabadi
Department of Plant Production, College of Agriculture and Natural Resources of Darab, Shiraz University, Shiraz, Fars, Iran

© The Author(s), under exclusive license to Springer Nature Switzerland AG 2021
J. Karimi, H. Madadi (eds.), *Biological Control of Insect and Mite Pests in Iran*,
Progress in Biological Control 18, https://doi.org/10.1007/978-3-030-63990-7_7

assessment of their potential for biological control of pests by these parasitoids. *Prospaltella berlesei* Howard (Aphelinidae) was imported and established for biological control of *Pseudaulacaspis pentagona* (Targioni) (Hem.: Diaspididae) in the north of Iran, Guilan province (Habibian 1981). An unknown Iranian species of *Gonatocerus* (Mymaridae) was introduced to the US to use it as a biocontrol agent (Triapitsyn 2013). Exploring parasitoids of *Hypera postica* (Gyllenhal) (Col.: Curculionidae) was conducted from 1973 to 1975 in Iran and nearby countries and resulted in finding two larval parasitoids of the pest, *Bathyplectes anurus* (Thomson) and *Bathyplectes curculionis* (Thomson) (Hym.: Ichneumonidae, Campopleginae) (Gonzalez et al. 1980). Abaei studied the natural enemies of *Lymantria dispar* (Lep.: Lymantriidae) a destructive pest of forest in Iran by cooperating with the parasitology laboratory of France under the supervision of Dr. Herard during 1975–1976 (Abai 2011; Herard et al. 1979).

Augmentation biological control programs by mass-rearing and releasing of some species of Braconidae and Trichogrammatidae have been started from the early recent century in Iran. *Habrobracon hebetor* (Say) (Hym.: Braconidae, Braconinae) is a polyphagous biological control agent of several crop pests (Amir-Maafi and Chi 2006; Forouzan et al. 2008, 2009; Taghizadeh and Basiri 2013). It is widely distributed in the Holarctic and Oriental regions and reported as a parasitoid of several species of Coleoptera and Lepidoptera (Yu et al. 2012).

## 7.2    The Superfamily Chalcidoidea

### 7.2.1    *Some Chalcidoids Parasitoids of Major Pests in Iran*

Agricultural stored products threaten by several storage insect pests with a loss of about 6–10% reduction in Iran (Maroof 2002). These storage pests mostly belong to Coleoptera and Lepidoptera attacked by several species of natural enemies especially hymenopterous parasitoids.

Five pteromalid species have been reared as parasitoids of stored product pests in Iran: *Anisopteromalus calandrae* (Howard), *Dinarmus vagabundus* (Timberlake), *Lariophagus distinguendus* (Förster), *Theocolax elegans* (Westwood), *Theocolax formiciformis* (Westwood). Within these parasitoids, *A. calandrae* is a well known cosmopolitan parasitoid of Anobiidae and Bruchinae (Coleoptera) associated with stored products and widely distributed in Iran (Lotfalizadeh and Hosseini 2013). It has been shown that *A. calandrae* prefers fourth instar larvae of *Callosobruchus maculatus* than other stages and exhibited type II functional response to this stage (Kazemi et al. 2004).

**Table 7.1** Chalcidoid parasitoids of Safflower seed pests in Iran (after Lotfalizadeh and Gharali 2014)

| Family | Parasitoids |
|---|---|
| Eulophidae | *Aprostocetus* sp. |
| | *Pronotalia carlinarum* (Szelényi & Erdös) |
| Eurytomidae | *Eurytoma acroptilae* Zerova |
| | *Sycophila submutica* (Thomson) |
| Ormyridae | *Ormyrus gratiosus* (Förster) |
| | *Ormyrus orientalis* Walker |
| Pteromalidae | *Colotrechnus viridis* (Masi) |
| | *Pachyneuron muscarum* (Linnaeus) |
| | *Pteromalus albipennis* Walker |
| Torymidae | *Adontomerus crassipes* (Bouček) |
| | *Microdontomerus annulatus* (Spinola) |

## 7.2.2 Parasitoids of Stored-Product Pests

Four species of fruit flies (Dip.: Tephritidae), including *Acanthiophilus helianthi* (Rossi), *Chaetorellia carthami* Stackelberg, *Terellia luteola* (Wiedemann), *Urophora mauritanica* Macquart, and the safflower gall wasp, *Isocolus tinctorius* Melika & Gharali (Hym.: Cynipidae) are serious pests of safflower fields in Iran. Twelve species of hymenopterous parasitoids of these pests were reared and some of their morphological and biological data were presented. These parasitoids belong to the families Eulophidae, Eurytomidae, Ormyridae, Pteromalidae, and Torymidae (Lotfalizadeh and Gharali 2014). *Colotrechnus viridis* (Masi), *Microdontomerus annulatus* (Bouček), *Pronotalia carlinarum* (Szelényi & Erdös), and *Pteromalus albipennis* Walker are associated with safflower fruit flies, and *Ormyrus gratiusus* (Förster) are associated with cynipid gall wasp in Iran (Table 7.1) (Lotfalizadeh and Gharali 2014).

## 7.2.3 Safflower Seed Pests Parasitoids

Study of parasitic wasps received from different localities in Iran 2008–2012, showed that there are 10 hymenopterous parasitoids on diamondback moth (DBM), *Plutella xylostella* (Linneaus) (Lep.: Plutellidae) in Iran (Lotfalizadeh et al. 2013). The larval parasitoid, *Oomyzus sokolowskii* (Kurdjumov) (Hym.: Eulophidae) is the most important parasitoid of DBM reported from Iran (Golizadeh et al. 2008a, b, c) but *Brachymeria excarinata* (Gahan) (Hym.: Chalcididae) attacks pupae (Afunizadeh et al. 2010). Two petromalid species, *Mokrzeckia obscura* (Graham) and *Pteromalus* sp. were reported as hyperparasitoids of DBM. Regarding adaptation with relatively high temperatures, it has been suggested that *O. sokolowskii* is the only chalcidoid species, which has the capacity to control DBM in tropical and semi-tropical conditions (Golizadeh et al. 2008a, b, c). Alternatively, the parasitism rate of this species in the south of Tehran was estimated too low to control the DBM efficiently (it was 7.93% and 1.8% on Boris and S-J cultivars, respectively) (Hasanshahi et al. 2012b).

### 7.2.4 Diamondback Moth Parasitoids

Study of parasitic wasps received from different localities in Iran 2008–2012, showed that there are 10 hymenopterous parasitoids on diamondback moth (DBM), *Plutella xylostella* (Linneaus) (Lep.: Plutellidae) in Iran (Lotfalizadeh et al. 2013). The larval parasitoid, *Oomyzus sokolowskii* (Kurdjumov) (Hym.: Eulophidae) is the most important parasitoid of DBM reported from Iran (Golizadeh et al. 2008a, b, c) but *Brachymeria excarinata* (Gahan) (Hym.: Chalcididae) attacks pupae (Afunizadeh et al. 2010). Two petromalid species, *Mokrzeckia obscura* (Graham) and *Pteromalus* sp. were reported as hyperparasitoids of DBM. Regarding adaptation with relatively high temperatures, it has been suggested that *O. sokolowskii* is the only chalcidoid species, which has the capacity to control DBM in tropical and semi-tropical conditions (Golizadeh et al. 2008a, b, c). Alternatively, the parasitism rate of this species in the south of Tehran was estimated too low to control the DBM efficiently (it was 7.93% and 1.8% on Boris and S-J cultivars, respectively) (Hasanshahi et al. 2012b).

### 7.2.5 Xylophagous Beetles Parasitoids

Nineteen species of Chalcidoidea have yet been recorded as parasitoids of Coleopteran xylophagous pests in Iran. They are belonging to six families including Chalcididae, Encyrtidae, Eulophidae, Eupelmidae, Eurytomidae and Pteromalidae each with 5, 1, 1, 2, 4 and 7 known species, respectively (Table 7.2) (Lotfalizadeh and Khalghani 2008; Lotfalizadeh and Jafari-Nadushan 2015). Lotfalizadeh et al. (2012) tabulated the hosts of Iranian species of the family Chalcididae and reported three species as parasitoids on Coleopterans of the family Buprestidae.

### 7.2.6 Alfalfa Leaf Miners Parasitoids

The most important alfalfa leaf miners in Iran include two dipterous and one lepidopterous species as *Liriomyza trifolii* (Burgess), *Chromatomyia horticola* (Goureau) (Dip.: Agromyzidae) and *Phyllonorycter medicaginella* (Gerasimor) (Lep.: Gracillariidae). These species were parasitized with 23 hymenopterous parasitoids including 19 eulophids and four pteromalids species (Lotfalizadeh et al. 2015) (Table 7.3).

**Table 7.2** Chalcidoid parasitoids of xylophagous beetles parasitoids (Hym.: Chalcidoidea) in Iran

| Family | Parasitoid species | Host | References |
|---|---|---|---|
| Chalcididae | *Cratocentrus tomentosus* (Nikol'skaya) | *Chrysobothris parvipunctata* (Coleoptera: Buprestidae) on *Punica granatum* | Lotfalizadeh and Jafari-Nadushan (2015) |
| | *Trigonura ninae* (Nikol'skaya) | *Chrysobothris parvipunctata* (Coleoptera: Buprestidae) on *Punica granatum* | Lotfalizadeh and Jafari-Nadushan (2015) |
| | *Trigonura ruficaudis* (Cameron) | *Chrysobothris* sp. (Col.: Buprestidae) | Lotfalizadeh et al. (2012) |
| | *Varzobia tibialis* (Nikol'skaya) | Buprestidae (Coleoptera) on pine | Lotfalizadeh et al. (2012) |
| Encyrtidae | *Heterococcidoxenus? schlechtendali* (Mayer) | – | Lotfalizadeh and Khalghani (2008) |
| Eulophidae | *Entedon ergias* (Ratzeburg) | – | Lotfalizadeh and Khalghani (2008) |
| Eurytomidae | *Eurytoma morio* (Boheman) | *Rogulascolytus mediterraneus* eggers and *Scolytochelus multistraiatus* Marsham (Col.: Scolytidae) on *Biota orientalis* | Lotfalizadeh and Khalghani (2008) |
| | *Eurytoma arctica* (Thomson) | *Smicronyx robustus* (Fst.) (Col.: Curculionidae), *Xylopertha reflexicauda* (Lesne) (Col.: Bostrychidae) and *Ruguloscolytus mediterraneus* (Col.: Scolytidae) | Lotfalizadeh and Khalghani (2008) |
| | *Eurytoma iranicola* (Zerova) | *Osphranteria coerulescens* (Col.: Cerambycidae) | Zerova et al. (2004), Lotfalizadeh and Khalghani (2008) |
| Eupelmidae | *Eupelmus muellneri* (Ruschka) | *Rogulascolytus mediteraneus* (Col.: Scolytidae) and buprestid, *Sphenoptera davatchii* Descarpentries | Lotfalizadeh and Khalghani (2008) |
| | *Eusandalum inerme* (Ratzeburg) | Buprestidae on *Ficus carica* | Lotfalizadeh and Khalghani (2008) |
| Pteromalidae | *Callocleonymus pulcher* (Masi) | *Rogulascolytus mediteraneus* (Col.: Scolytidae) and *Xylopertha reflexicauda* (Lesne) (Col.: Bostrychidae) | Lotfalizadeh and Khalghani (2008) |
| | *Chalcedectus balachowskyi* (Steffan) | *Osphranteria coerulescens* (Redtenbacher) (Col.: Cerambycidae) on *Rosa* | Lotfalizadeh and Khalghani (2008) |
| | *Cheiropachus quadrum* (Fabricius) | *Rogulascacheolytus mediteraneus* (Col.: Scolytidae) and *Xylopertha reflexicauda* (Col.: Bostrychidae) | Lotfalizadeh and Khalghani (2008) |
| | *Dinotiscus colon* (Linnaeus) | *Ruguloscolytus mediterraneus* and *Phloeosinus bicolour* (Brulle) | Lotfalizadeh and Khalghani (2008) |

(continued)

**Table 7.2** (continued)

| Family | Parasitoid species | Host | References |
|---|---|---|---|
| | *Heydenia pretiosa* (Forster) | *Ruguloscolytus mediterraneus* and *Phloeosinus bicolour* on fruit trees and *Biota orientalis* | Lotfalizadeh and Khalghani (2008) |
| | *Rhaphitelus maculatus* (Walker) | *R. mediteraneus* (Col.: Scolytidae) and *X. reflexicauda* (Col.: Bostrychidae) | Lotfalizadeh and Khalghani (2008) |
| | *Oxysychus* sp. | *Sphenoptera davatchii* and *S. kambyses* (Col.: Buprestidae) | Lotfalizadeh and Khalghani (2008) |

The diversity of hymenopterous parasitoids was studied in the northwest of Iran. This study showed the most of these parasitoids belong to the family Eulophidae (93%) including two subfamilies (Eulophinae and Entedoninae). Amongst reared parasitoids, *Diglyphus isaea* (Walker) was the most common species and had the highest frequency.

## 7.2.7   Grape Berry Moth Parasitoids

The hymenopterous parasitoids of grape berry moth *Lobesia botrana* (Denis and Schiffermüller) (Lep.: Tortricidae) in Iran were studied sparsely. This aggregation includes 12 species belong to five families Bethylidae, Braconidae, Ichneumonidae, Pteromalidae and Trichogramatidae are listed as egg (two species), larval (nine species) and pupal (four species) parasitoids (Table 7.4). During this study, two species *Dibrachys affinis* Masi (Hym.: Pteromalidae, Pteromalinae) and *Itoplectis alternans* (Gravenhorst) (Hym.: Ichneumonidae, Pimplinae) are new records for Iranian insect fauna and larval parasitoid of *L. botrana*. Finally, all of the hymenopterous parasitoids of grape berry moth in Iran were reviewed.

Noyes (2018) listed 28 chalcidoid species as reported parasitoids of grape berry moth worldwide. These parasitoids attack to different life stages of *L. botrana*. About half of these attacks to larvae and are larval parasitoid and the rest attack pupa (Akbarzadeh-Shoukat et al. 2008b). Based on Akbarzadeh-Shoukat et al. (2008a), the rate of larval parasitism of *L. botrana* varied from 1% to 16.8% (average 6.33%). This parasitism was significantly different in other localities and the pest generations, but the maximum rate of parasitism was in the first generation of the pest.

Within this list, *T. ingricum* has not been reported as a parasitoid of *L. botrana*. It is widely distributed in the north and north-west of Iran as parasitoid of *L. botrana*, *Cydia pomonella* (L.) and *Chilo supressalis* Walker (Ebrahimi and Akbarzadeh-Shoukat 2008). Suppressing potential of *T. ingricum* on *L. botrana* was assessed by about 60% in Urmia vineyards (Akbarzadeh-Shoukat and Ebrahimi 2008). They reported that it is widely distributed in this region with 40–50% parasitism rate in the first generation of *L. botrana*.

**Table 7.3** Chalcidoid parasitoids of alfalfa leaf miners in Iran

| Family | Genus | Species | Host | References |
|---|---|---|---|---|
| Eulophidae | *Baryscapus* | *Baryscapus impeditus* (Nees) | *Liriomyza trifolii* | Talebi et al. (2011) |
| | *Chrysocharis* | *Ch. albicoxis* (Erdös) | *Liriomyza trifolii* and *Chromatomyia horticola* | Lotfalizadeh et al. (2015) |
| | | *Ch. crassiscapus* (Thomson) | *Liriomyza trifolii* and *Chromatomyia horticola* | Lotfalizadeh et al. (2015) |
| | | *Ch. submutica* (Graham) | *Phyllonorycter medicaginella* | Lotfalizadeh et al. (2015) |
| | *Cirosspilus* | *C. talitzkii* (Bouček) | *Liriomyza trifolii* and *Chromatomyia horticola* | Lotfalizadeh et al. (2015) |
| | | *C. vittatus* (Walker) | *Liriomyza sativae, Liriomyza trifolii* and *Chromatomyia horticola* | Lotfalizadeh et al. (2015), Asadi et al. (2006) |
| | *Closterocerus* | *Cl. formosus* (Westwood) | *Liriomyza sativae, L. trifolii* | Asadi et al. (2006) |
| | | *Closterocerus* sp. | *Liriomyza trifolii* and *Chromatomyia horticola* | Lotfalizadeh et al. (2015) |
| | *Diglyphus* | *D. crassinervis* (Erdös) | *Liriomyza sativae, L. trifolii* | Asadi et al. (2006) |
| | | *D. isaea* (Walker) | *Liriomyza sativae, L. trifolii, Chromatomyia horticola* | Lotfalizadeh et al. (2015), Asadi et al. (2006) |
| | | *D. pachyneuron* (Graham) | *Liriomyza trifolii* and *Chromatomyia horticola* | Lotfalizadeh et al. (2015) |
| | | *Diglyphus sensilis* (Yefremova) | *Chromatomyia horticola* | Ranji et al. (2015) |
| | *Hemiptarsenus* | *H. zilahisebessi* (Erdös) | *Liriomyza sativae, L. trifolii, Chromatomyia horticola* | Zahiri et al. (2002), Asadi et al. (2006) |
| | | *H. wailesellae* (Nowicki) | *Liriomyza sativae* | Zahiri et al. (2002) |
| | | *Hemiptarsenus* sp. | *Liriomyza trifolii* and *Chromatomyia horticola* | Lotfalizadeh et al. (2015) |
| | *Pnigalio* | *P.* sp. nr. *pectinicornis* | *Liriomyza sativae, L. trifolii* | Asadi et al. (2006) |
| | *Ratzeburgiola* | *R. cristatus* (Ratzeburg) | *L. trifolii* | Dousti (2008) |

(continued)

**Table 7.3** (continued)

| Family | Genus | Species | Host | References |
|---|---|---|---|---|
| | Sympiesis | S. xanthostoma (Nees) | Liriomyza trifolii and Chromatomyia horticola | Lotfalizadeh et al. (2015) |
| | | Sympiesis sp. | Liriomyza trifolii and Chromatomyia horticola | Lotfalizadeh et al. (2015) |
| Pteromalidae | Cyrtogaster | Cyrtogaster vulgaris (Walker) | Chromatomyia horticola | Lotfalizadeh and Gharali (2008), Lotfalizadeh et al. (2015) |
| | Halticoptera | Halticoptera andriescui (Mitroiu) | Chromatomyia horticola | Ranji et al. (2015) |
| | Pteromalus | Pteromalus sp. | Chromatomyia horticola | Lotfalizadeh et al. (2015) |
| | Sphegigaster | Sphegigaster ineus (Mitroiu) | Phyllonorycter medicaginella | Lotfalizadeh et al. (2015) |

**Table 7.4** Chalcidoid parasitoids of grape berry moth in Iran

| Parasitoids | Host stage | Generation | | |
|---|---|---|---|---|
| | | 1st | 2nd | 3rd |
| **Chalcididae** | | | | |
| Brachymeria minuta (L.) | Pupae | + | + | − |
| Hockeria unicolor (Walker) | Pupae | + | + | − |
| **Pteromalidae** | | | | |
| Coelopisthia pachycera (Masi) | Pupae | + | + | − |
| Coelopisthia sp. | − | + | + | − |
| Dibrachys affinis (Masi) | Pupae | ? | ? | ? |
| Pteromalus puparum (L.) | Pupae | − | − | + |
| Homoporus sp. | Larvae | ? | ? | ? |
| **Trichogrammatidae** | | | | |
| Trichogramma embryophagum (Hartig) | Egg | + | − | − |
| Trichogramma ingricum (Sorokin) | Egg | + | − | − |

## 7.3 Recent Improvements

### 7.3.1 Chalcididae

The family Chalcididae (Hymenoptera: Chalcidoidea) is a medium-sized family within the Chalcidoidea superfamily, including more than 1500 species and 90 genera worldwide (Noyes 2018). Some species are internal and external primary parasitoids while several species are hyperparasitoids on a wide range of other insects (Habu 1960; Narendran 1986; Fry 1989; Delvare and Bouček 1992; Noyes 2018).

Chalcidids are mostly parasitoids, although a few are hyperparasitoids on a wide range of holometabolous insect hosts (Bouček 1952; Narendran 1986; Delvare and Bouček 1992; Noyes 2018). In Iran, this group has not been studied systematically, and only limited research on the taxon has been conducted (Delvare et al. 2011; Rajabi et al. 2011; Lotfalizadeh et al. 2012; Tavakoli Roodi et al. 2016; Falahatpisheh et al. 2018). The first information on Chalcididae from Iran was published by Masi (1924), who described two new species. In the following years, only a few species of this family were reported from the country (Bouček 1952; Nikol'skaya 1952, 1978). Recently, Delvare et al. (2011) described three new species from Iran. Later, Rajabi et al. (2011) and Lotfalizadeh et al. (2012) provided novel information on the Iranian chalcidid fauna together with new data on the hosts of the Chalcididae. Within five subfamilies, 68 species and 18 reported genera from Iran (Tavakoli Roodi et al. 2016; Falahatpisheh et al. 2018), the biology of 29% of species is known. Practically, this group has not been used extensively in biological control but known species from Iran were reported on Lepidoptera, Coleoptera and Diptera, respectively (Fig. 7.1). Twelve species (*Brachymeria albicrus* (Klug), *B. ceratoniae* (Delvare), *B. excarinata* Gahan, *B. femorata* (Panzer), *B. minuta* (Linnaeus), *B. rugulosa* (Förster), *B. tibialis* (Walker), *Belaspidia nigra* (Siebold), *Proconura caryobori* (Hanna), *P. nigripes* (Fonscolombe), *P. persica* (Delvare) and *P. ceratoniae* (Delvare) have been reared on lepidopterous pests. While they are mostly unknown biologically (71%) (see Fig. 7.1).

## 7.3.2   Encyrtidae

The Encyrtidae (Hym.: Chlacidoidea) comprises two subfamilies Encyrtinae and Tetracneminae with approximately 4000 nominal species in 483 genera (Noyes

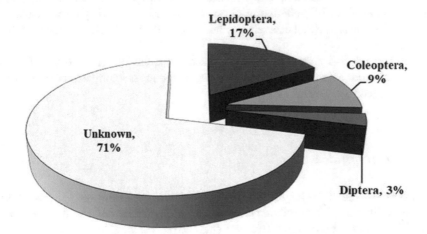

**Fig. 7.1** Biological association of the Chalcididae family in the Iranian fauna

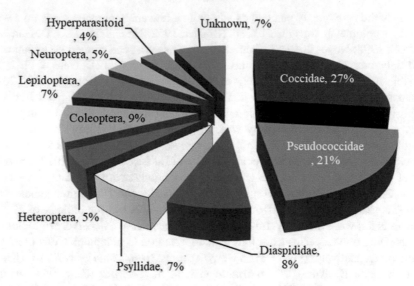

**Fig. 7.2**  Biological association of the Encyrtidae family in the Iranian fauna

2018). There are more than 1270 described species of encyrtids in the Palaearctic region (Yasnosh and Japoshvili 1999; Japoshvili 2007a, 2007b; Japoshvili and Abrantes 2006; Japoshvili and Noyes 2005, 2006). Many encyrtids have been used successfully for control of mealybugs (Pseudococcidae) and many soft scales (Coccidae).

Iranian Encyrtidae includes 149 species representing 48 genera (Fallahzadeh and Japoshvili 2010; Lotfalizadeh 2010a, 2010b; Fallahzadeh and Japoshvili 2013, 2017). Host information from Iran and distributional data are also provided.

Four new species, *Gyranusoidea iranica* Japoshvili and Fallahzadeh and *Microterys iranicus* Japoshvili and Fallahzadeh, *Metaphycus davoodii* Lotfalizadeh, *Ooencyrtus ferdowsii* Ebrahimi and Noyes are described and diagnostic characters are provided for them (Fallahzadeh and Japoshvili 2010, 2013; Lotfalizadeh 2010a; Ebrahimi et al. 2014).

Members of this family attack insect pests of the order Hemiptera (68% of hosts) and the suborder Sternorrhyncha (Hemiptera) is the most important recorded hosts with 56% records (see Fig. 7.2).

### 7.3.3  Eulophidae

Eulophidae is one of the largest families of Chalcidoidea, with over 297 genera and 4472 described species worldwide (Noyes 2018). The family has a total of 39 genera and 122 species known in Iran (Hesami et al. 2010; Talebi et al. 2011).

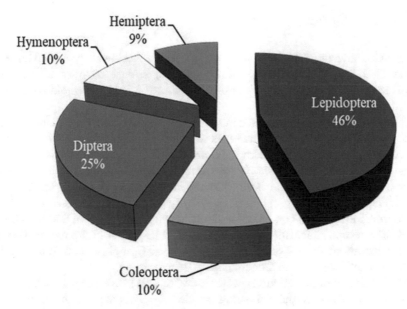

**Fig. 7.3** Biological association of the Eulophidae family in the Iranian fauna

It includes species which are parasitoids of insects from many orders and also mites. Many eulophid wasps attack pests of several forests, agricultural and fruit plants. They can regulate numbers of their hosts in natural conditions.

A review of the available literature of Iranian eulophids shows the majority of them attack lepidopterous pests (46%). Within the Lepidoptera parasitoids, 50% (25 spp.) were reared on the family Gracillariidae. While Diptera with 25% places in the second position after Lepidoptera and the dipterous leafminers, Agromyzidae is the most important host of eulophids (76% of 25%) in Iran (Lotfalizadeh et al. 2015) (Fig. 7.3). A few species attack the other arthropods.

There are not so many Persian documents about the ecology and parasitism rate of those parasitoids. Intensive searching leads us to a few documents about the life history of the most famous eulophids, *Diglyphus isaea* (Walker). It has been reported that just 3.3 days need to the population of *D. isaea* doubled (Asadi et al. 2006). The possibility of successful suppressing of cucumber leafminer, *Liriomyza* sp. by *D. isaea* has been investigated under greenhouse conditions (Dashtbani et al. 2013). It has been suggested that the activity and parasitism rate of this parasitoid increase in response to increasing temperature (Dashtbani et al. 2012, 2013). Accordingly, the maximum parasitism rate of this parasitoid reached up to 60% (Bagheri et al. 2013). Ghasemzadeh et al. (2012) indicated that there is not any synchrony between *D. isaea* and *L. trifolii* population peak, thus it is unlikely to control American serpentine leafminer relying on natural parasitism.

## 7.3.4  Eupelmidae

Eupelmidae have more than 900 species in 51 genera worldwide classified into three subfamilies: Calosotinae, Neanastatinae and Eupelminae, (Noyes 2018). These wasps are primary or secondary parasitoids and ecto or endoparasitoids of different orders of insects (Gibson 1995).

Eupelmidae of Iran were revised based on the materials collected mainly from the north-west of Iran. This family includes two subfamilies and six genera (*Anastatus* Motschulsky, *Brasema* (Cameron), *Calosota* (Curtis), *Calymmochilus* (Masi), *Eupelmus* (Dalman) and *Eusandalum* (Ratzeburg) and 29 species in Iran. These species were listed within two subfamilies Calosotinae (one species) and Eupelminae (28 species).

Eupelmidae has been considered as a relatively less well-known family in Iranian fauna. The recently faunistic study demonstrated that most of the reported species are the parasitoid of gall making wasps such as Cynipidae and Tenthridinidae (Fig. 7.4) (Lotfalizadeh and Ghadirzadeh 2016).

More than half of the reported species (56%) related to cynipid gall making wasps (Hym.: Cynipidae) as the following species were reared on them: *Eupelmus annulatus* (Nees), *E. azureus* (Ratzeburg), *E. bicolor* (Gibson & Fusu), *E. cerris* (Förster), *E. confusus* (Al khatib), *E. fulvipes* (Förster), *E. impennis* (Nikol'skaya), *E. martellii* (Masi), *E. mehrnejadi* (Gibson & Fusu), *E. urozonus* (Dalman) and *E. vesicularis* (Retzius) (Lotfalizadeh and Ghadirzadeh 2016). *Eupelmus bicolor* has been reported as ectoparasitoid of maize caterpillar (*Mythimna loreyi* Duponchel) in Iran. Its biological attributes have been studied under laboratory conditions. This parasitoid under favourable conditions has one generation per month (Naseri et al. 1998).

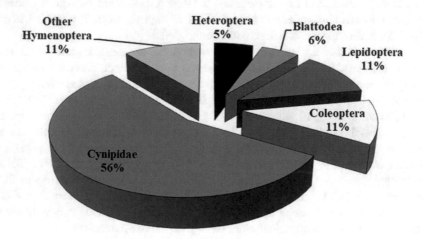

**Fig. 7.4** Biological association of the Eupelmidae family in the Iranian fauna

## 7.3.5  Eurytomidae

The family is present and common in all zoogeographical regions. Eurytomids exhibit a wide range of life histories, but the majority of the larvae are endophytic: as seed feeders, gall formers, or parasitoids of phytophagous insects. Most eurytomids are primary or secondary parasitoids, attacking eggs, larvae, or pupae of various arthropods (Diptera, Coleoptera, Hymenoptera, Lepidoptera, Orthoptera, and Araneae) (Lotfalizadeh et al. 2007a, 2007b).

Several studies on Iranian fauna, e.g. Hedicke (1921), Bouček (1952, 1970, 1977), Nikol'skaya (1952) and Burks (1971) contain some faunistic reports on Iranian Eurytomidae. Modarres-Awal (1997) cited eight valid species of eurytomids in his list of agricultural pests and their natural enemies in Iran. In recent years some faunistic studies have been done on Iranian Eurytomidae and some new species have been described (Narendran and Lotfalizadeh 1999; Zerova et al. 2004; Lotfalizadeh et al. 2007a, 2007b; Zerova et al. 2008; Saghaei et al. 2018).

This family composited 89 species belonging to 8 genera *Aximopsis*, *Bruchophagus*, *Eurytoma*, *Exeurytoma*, *Macrorileya*, *Sycophila*, *Systole* and *Tetramesa*. Of the 8 genera, *Eurytoma* (41 species, 46.06%) and *Bruchophagus* (19 species, 21.34%) are the most specious.

The majority of Iranian eurytomids are phytophagous (see Fig. 7.5) that include the genera *Bruchophagus*, *Eurytoma*, *Exeurytoma*, *Tetramesa* and *Systole*. The main phytophagous eurytomid species are seed feeder genera including *Bruchophagus*, *Eurytoma*, *Exeurytoma* and *Systole*, which are mainly pests of various Fabaceae and the genus *Tetramesa* is stem-miner in the family Poaceae.

Within phytophagous species, two species *Eurytoma amygdali* Enderlein and *E. plotnikovi* Nikol'skaya are seed feeders associated with plant species of Rosaceae and Anacardiaceae, respectively (Lotfalizadeh et al. 2007b). *Bruchophagus roddi* Gussakovsky and *B. gibbus* (Boheman) are two serious pests of alfalfa fields and

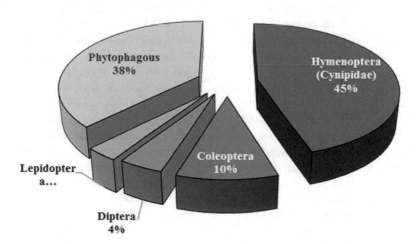

**Fig. 7.5**  Biological association of the Eurytomidae family in the Iranian fauna

these are widely distributed throughout all the alfalfa-producing areas of Iran (Arbab 2006, Eslamizadeh et al. 2008). Their damage is about 20–30% but it can reach up to 80% in a high population of the pest.

Most of the entomophagous eurytomids were reared on Cynipidae (gall maker wasps) but three orders Coleoptera, Diptera and Lepidoptera known as hosts of Eurytomidae in Iran. The main parasitic eurytomid wasps of Iran are associated with Hymenoptera, especially the family Cyipidae (about 45%), followed by Coleoptera (about 10%), Diptera (about 4%), Lepidoptera (about 3%). Most of the species (about 62%) are parasitoids associated with galls induced by cynipid wasps and tephritid flies (Saghaei et al. 2018). Nine species (about 10%) of Iranian fauna, including *Aximopsis ghazvini* (Zerova), *Bruchophagus iranicus* Ozdikmen, *B. shohadae* (Zerova), *Eurytoma iranica* Narendran and Lotfalizadeh, *E. iranicola* Zerova, *E. melikai* Zerova, *E. zerovai* Ozdikmen, *Tetramesa leucospae* Zerova and Madjdzadeh and *T. persica* (Hedicke) are originally described from Iran and are possibly endemic to the country (Saghaei et al. 2018).

## 7.3.6   Mymaridae

The family Mymaridae includes some of the smallest known insects: the combined lengths of three or even four adult individuals may not even equal 1 mm (Annecke and Doutt 1961). More than 1400 species and 100 genera are known worldwide (Noyes 2018). This family is one of the most distinctive of the superfamily Chalcidoidea, to have the antennal toruli quite far apart (3–5 times their own diameter), usually a reduced wing venation and distinctive fringed wings (Nikol'skaya 1978). Its members are abundant and easily collected using a variety of trapping methods. The hosts of Mymaridae include eggs of Hemiptera, Psocoptera, Coleoptera, Orthoptera and some other insect orders (Huber 1986), however, only about one-quarter of the genera have hosts reported for them. As in the Trichogrammatidae, several mymarids attack eggs of the aquatic insects. Huber (1986) published a comprehensive review of the known hosts of the mymarids known to that date. A few species of mymarids have been reported as responsible for successful biological control projects (Lin et al. 2007). In Iran, some biological and ecological studies were made by Hesami et al. (2004, 2009), Latifian and Soleyman-Nejadian (2009), and Akbarzadeh-Shoukat (1998). Hesami et al. (2001) studied the morphology of *Anagrus atomus* (L.) that was reared from the grape leafhopper, *Arboridia kermanshah* (Dlabola) (Hem.: Cicadellidae), in Isfahan.

Only biological information of four species of Iranian mymarids was known, that *Anagrus atomus* (L.) on *A. kermanshah* (Hem.: Cicadellidae), *Erythmelus panis* (Enock) on *Stephanitis pyri* (Fabricius) (Hem.: Tingitidae), *Gonatocerus litoralis* (Haliday) on *Zyginidia sohrab* (Zachvatkin) (Hem.: Cicadellidae) and *Gonatocerus* sp. on *Circulifer tenellus* (Hem.: Cicadellidae). The occurrence of overwintering refuges for *A. atomus* had a positive influence on the parasitism rate of grape leafhopper and earlier emergence of parasitoids in vineyards (Hesami et al. 2002, 2009).

## 7.3.7 Pteromalidae

The Pteromalid wasps are the largest family in the Chalcidoidea including over 3506 species belonging to 588 genera (Noyes 2018). The pteromalids are primary or secondary parasitoids attacking other insect groups such as Coleoptera, Diptera, Lepidoptera, Hymenoptera, Hemiptera and some Arachnida at their various stages of development (Bouček and Rasplus 1991). A few species of this family are phytophagous (Farooqi and Menon 1972). Some of them develop in the seeds of plants, or gall making and others develop as inquilines in galls created by other insects. They play an important role in the control of pest insect and several species have been employed successfully in biological control programs all over the world (Bouček and Rasplus 1991).

The first list of Pteromalidae parasitoids provided by Davatchi and Chodjai (1968) included only seven species. Goldansaz et al. (1996), Habibpour et al. (2002), Jalilvand and Gholipour (2002), Mehrnejad (2002, 2003), Rezaei et al. (2003), Sadeghi and Askary (2001), Sadeghi and Ebrahimi (2001), Sharifi and Javadi (1971a, b), Steffan (1968), Lotfalizadeh (2002a, 2002b, 2004, 2015) and Lotfalizadeh and Ahmadi (1998, 2000) added some new records to the list without providing any additional information.

Lotfalizadeh and Gharali (2008) presented the first informative checklist which reported 78 species of Pteromalidae from different parts of Iran including biological and geographical distribution data. They compared the composition of species with those of the world and the Palaearctic region. Most recently 129 species within 62 genera belonging to 11 subfamilies of Pteromalidae have been listed for Iranian fauna (Abolhassanzadeh et al. 2017). Of which biological association of 40.8% is known (Fig. 7.6).

Five species related to aphids including *Asaphes suspensus* (Nees), *Euneura lachni* (Ashmead), *Pachyneuron aphidis* (Bouché), *Pachyneuron leucopiscida* (Mani) and *Pachyneuron solitarium* (Hartig).

Fifteen species related to the beetles that 50% of them attack xylophagous beetles in Iran. Ten species were reared on flies (Diptera) of Tephritidae, Muscidae and Agromyziidae. Some biological aspects of *Spalangia endius* Walker, as a parasitoid of some dipterous families, have been studied by Behbahani et al. (1995). They revealed that the effect of temperature up to 30 °C has a positive influence on the reproduction of *S. endius*. Five species were reported on Hymenoptera, *Caenacis inflexa* (Ratzeburg), *Cecidostiba fungosa* (Geoffroy), *Hobbya stenonota* (Ratzeburg), *Mesopolobus amaenus* (Walker) and *Pteromalus bedeguaris* (Thomson) attack gall wasps (Cynipidae), and eight species attack Lepidoptera. Three pteromalids, *Eunotus nigriclavis* (Förster), *Moranila californica* (Howard) and *Pachyneuron muscarum* (L.) are related to Coccoidea (Hemiptera).

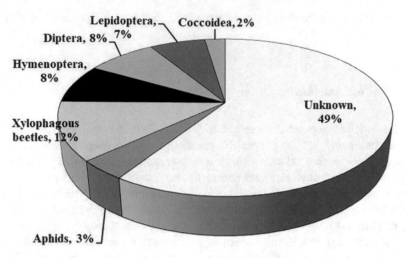

**Fig. 7.6** Biological association of the Pteromalidae family in the Iranian fauna

## 7.3.8    *Torymidae*

Some studies on Torymidae of Iran have recently been conducted by various local and foreign scientists (Modarres-Awal 1997; Ebrahimi and Ahmadian 2002; Rakhshani et al. 2003a, 2003b; Azizkhani et al. 2005; Delvare 2005; Lotfalizadeh and Gharali 2005; Askew et al. 2006; Fallahzadeh et al. 2008; Hesami et al. 2008; Lotfalizadeh and Khalghani 2008). The known Torymidae species of Iran belong to 14 genera and 45 species (Lotfalizadeh and Gharali 2005; Fallahzadeh et al. 2009).

Most of the collected species of Iranian torymids (76%) were reared on gall wasps (Hym.: Cynipidae) (Fig. 7.7). Mantodea and Cecidomyiidae (Diptera) take the second rank. The genus *Megastigmus* was reported as a phytophagous genus in Iran that this genus was treated as an independent family, Megastigmidae (Jansta et al. 2017). Rajabi et al. (2011) mentioned that a torymid species was the most important parasitoid of the Rosaceous branch borer in Esfahan.

## 7.4    The Superfamily Ichneumonoidea

The superfamily Ichneumonoidea is the largest and the most species-rich superfamily of the order Hymenoptera or perhaps of the class Insecta. Until 2016, 46,506 species of the superfamily were listed in the world catalogue of this superfamily prepared by Yu et al. (2012). It has been estimated that the number of species in this superfamily will exceed more than 100,000 species in the world. Almost all species of this superfamily are parasitoids of holometabolous insects and thus regulates the populations of their hosts in both agricultural and natural (forests and pastures)

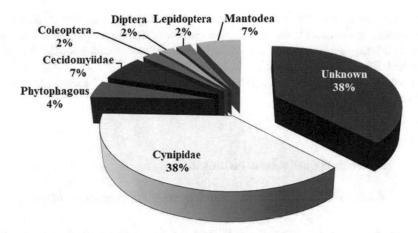

**Fig. 7.7** Biological association of the Torymidae family in the Iranian fauna

ecosystems. This large superfamily consists of two rather well-known families, Braconidae and Ichneumonidae.

The family Braconidae comprises more than 21,221 valid species classified into 45 subfamilies (Yu et al. 2012). Braconids parasitize larvae of holometabolous and hemimetabolous insects and develop on or inside the body of their hosts as ectoparasitoids and endoparasitoids, respectively. Ectoparasitoids can be found in the subfamilies Braconinae and Doryctinae and endoparasitoids in the rest of the subfamilies. Adults of ectoparasitoids search for larvae of mainly Lepidoptera or Coleoptera in the cryptic places and after finding a larva, paralyze the larva by injecting venoms from the poison glands located at the base of ovipositor and then lay their egg(s) near or on the body of the host larva. Ectoparasitoids are not highly hosted specific and can be developed on different hosts (polyphagous). Endoparasitoids start their developing growth from egg, larva or even adult stages of their hosts. They are known as egg-larval, egg-pupal or larva-pupal (Opiinae and Alysiinae, parasitoids of Diptera), larval or adult parasitoids. Endoparasitoids use both exposed and concealed hosts mainly in the orders Lepidoptera and Coleoptera. Although parasitism of Diptera is very rare in the ectoparasitoids of Braconidae, species of the two subfamilies, Opiinae and Alyssinae are mainly specialized in dipterans of the families Agromyzidae and Tephritidae. Endoparasitoid braconids have evolutionarily a narrow, taxonomic related host range. They are not polyphagous as seen in ectoparasitoids and many of them represent oligophagous or monophagous life history. Members of the subfamilies Agathidinae, Cheloninae and Microgasterinae are specialized in the Lepidoptera, Acaeliinae on the lepidopterous miners, Brachistinae on coleopterous weevils, Helconinae on cerambycid beetles and Ichneutinae on sawflies (Tobias 1995).

The family Ichneumonidae contains more than 25,285 extant species classified into 6 higher groups and 44 subfamilies (Yu et al. 2012). Ichneumon wasps comprise about 1.4% and 2.3% of all described organisms and insects, respectively. Almost all ichneumonids are parasitoids. Ichneumonids parasitize holometabolous insects and less commonly egg sacs and adults of spiders. Ichneumonidae wasps represent different life strategies in their life history, endoparasitism and ectoparasitism, koinobiosis and idibiosis, solitary and gregarious, primary and secondary (Hyperparasitism) (Quicke et al. 2009).

## 7.5 The Superfamily Ichneumonoidea

### 7.5.1 Basic Research on the Superfamily Ichneumonoidea

Ichneumonoidea distributed universally and can be found in all ecosystems. The greatest diversity of most subfamilies of Ichneumonoidea is in the northern hemisphere where Iran is located. So far, 780 species in 141 genera and 26 subfamilies of Braconidae and over 502 species of 189 genera and 26 subfamilies of Ichneumonidae have been recorded from Iran (Barahoei et al. 2012; Farahani et al. 2016). Comparing the ichneumonoid fauna of Iran with some of its well-studied neighbour countries like Russia and Turkey shows that the actual number of species of these two large families in Iran will certainly increase by future studies. Also, most species of ichneumonoids have been reported from the northern part of Iran where most research centers are located. During the recent decade, extensive and long-term samplings were carried out in different regions of Iran to identify regional assemblages of Braconidae and Ichneumonidae (Ameri et al. 2014; Farahani and Talebi 2012; Mohammadi-Khoramabadi et al. 2016a, 2016b; Mohammadi-Khoramabadi et al. 2013). These studies substantially increased our knowledge of the diversity, phenology and flight period of the parasitoid wasps of the two large families in Iran.

The first step toward understanding the role of ichneumonoid parasitoids in agricultural ecosystems is collecting and identifying the parasitoid assemblages within and around crop fields. In alfalfa fields of Khorasan in the northeast of Iran, 2 years of sampling conducted by Barahoei et al. (2014) showed that 26 species of 25 genera and 12 subfamilies of Ichneumonidae were present.

Rearing is an accurate method for determining the host-parasitoid relationships and assessment of the rate of parasitism of Ichneumonoidea. Many studies have been conducted in Iran to the rear and evaluate the parasitism rate of ichneumonoids on their hosts. Such studies have been focused on major pests of crops, fruit trees and forest pests. Three lists of known ichneumonoid (Braconidae and Ichneumonidae) parasitoids of major pests in Iran are presented in Tables 7.5, 7.6 and 7.7.

**Table 7.5** The known species of Ichneumonoidea associated with major pests of the order Coleoptera in Iran

| Family | Pest | Host plant | Parasitoid | Family | Reference (s) |
|---|---|---|---|---|---|
| Buprestidae | Sphenoptera davatchii Desc. And S. kambyses (Obenb) | Rosaceous fruit trees | Spathius polonicus (Niezabitowski) | Braconidae | Rajabi et al. (2011) |
| | | | Atanycolus ivanov-i (Kokujev) | Braconidae | Rajabi et al. (2011) |
| | Chrysobothris affinis (Fabricius) | Rosaceous fruit trees | Spathius polonicus (Niezabitowski) | Braconidae | Rajabi et al. (2011) |
| | | | Atanycolus ivanovi (Kokujev) | Braconidae | Rajabi et al. (2011) |
| Cerambycidae | Osphranteria coerulescens (Redtenbacher) | Rosaceous fruit trees | Xorides corcyrensis (Kriechbaumer) | Ichneumonidae | Rajabi et al. (2011) |
| | Calchaenesthes pistacivora (Holzschuh) | Pistachio | Megalommum pisacivorae (van Achterberg & Mehrnejad) | Braconidae | Van Achterberg & Mehrnejad (2001) |
| Chrysomelidae | Bruchus rufimanus (Boheman) | Stored products | Triaspis thoracica (Curtis) | Braconidae | Shahhosseini and Kamali (1989) |
| Curculionidae | Anthonomus pomorum (L.) | Apple | Apanteles sp. | Braconidae | Rajabi et al. (2011) |
| | Hypera postica (Gyllenhal) | Alfalfa | Microctonus aethiops (Nees) | Braconidae | Mirabzadeh (1968) |
| | | | Microctonus aethiopoides loan | Braconidae | Arbab (2011) |
| | | | Microctonus coles. Drea | Braconidae | Bartlet et al. (1978) |
| | | | Bathyplectes anurus (Thomson) | Ichneumonidae | Gonzalez et al. (1980); Rowshandel (2000) |
| | | | Bathyplectes curculionis (Thomson) | Ichneumonidae | Gonzalez et al. (1980) |
| | Scolytus rugulosus (Muller) | Rosaceous fruit trees | Atanycolus ivanowi (Kokujev) | Braconidae | Rajabi et al. (2011) |
| | | | Ecphylus silesiacus (Ratzeburg) | Braconidae | Rajabi et al. (2011); Shojaei (1996) |
| | | | Leluthia ragulosco?yti (Fischer) | Braconidae | Rajabi et al. (2011); Shojaei (1996) |
| | | | Hecabalodes xylophagi (Fischer) | Braconidae | Shojaei (1996) |
| | | | Dendrosoter middendorffii (Ratzeburg) | Braconidae | Basiri et al. (2013) |
| | Scolytus multistriatus (Marsham) | Elm | Leluthia ragulosco?yti (Fischer) | Braconidae | Shojaei (1996) |
| | Phloeosinus bicolor (Brull) | | Leluthia ragulosco?yti (Fischer) | Braconidae | Shojaei (1996) |

**Table 7.6** The known species of Ichneumonoidea associated with major pests of the order Lepidoptera in Iran

| Family | Pest | Host plant | Parasitoid | Family | Reference (s) |
|---|---|---|---|---|---|
| Arctiidae | *Hyphantria cunea* (Drury) | Polyphagous (forest and fruit trees, crops) | *Cotesia hyphantriae* (Riley) | Braconidae | Abai (2011) |
| | | | *Cotesia glomerata* (L.) | Braconidae | |
| | | | *Protapanteles liparidis* (Bouché) | Braconidae | |
| | | | *Netelia testacea* (Gravenhorst)? | Ichneumonidae | |
| | | | *Campoletis fasciata* (Bridgman) | Ichneumonidae | |
| | | | *Theronia atalantae* (Poda) | Ichneumonidae | |
| | | | *Theroscopus esenbeckii* (Gravenhorst) | Ichneumonidae | |
| Crambidae | *Chilo suppressalis* Walker | Rice | *Cotesia chilonis* (Munakata) | Braconidae | Rassipour (1993) |
| Gelechidae | *Achroia grisella* (Fabricius) | Honey, bee wax | *Apanteles galleriae* Wilkinson | Braconidae | Goldansaz et al. (1996) |
| | *Recurvaria nanella* (Denis & Schiffermuller) | Rosaceous fruit trees | *Orgilus obscurator* (Nees) | Braconidae | Sabzevari (1968) |
| | *Batrachedra amydraula* Meyrick | Palm | *Habrobracon hebetor* (say) | Braconidae | Karampour and Fasihi (2004) |
| | | | *Phanerotoma leucobasis* Kriechbaumer | Braconidae | Gharib (1968) |
| | *Scrobipalpa ocellatella* Boyd | Sugar beet | *Bracon intercessor* Nees | Braconidae | Mahmudi et al. (2013) |

| | | | *Chelonus contractus* (Nees) | Braconidae | Farahbakhsh (1961) |
|---|---|---|---|---|---|
| | *Phthorimaea operculella* (Zeller) | Potato | *Chelonus subcontractus* Abdinbekova | Braconidae | Abbasipour et al. (2012) |
| | | | *Habrobracon hebetor* (say) | Braconidae | Dezianian and Jalali (2004) |
| | | | *Habrobracon radialis* Telenga | Braconidae | Dezianian and Quicke (2006) |
| Geometridae | *Ennomos quercinaria* (Hufnagel) | Oak | *Pimpla turionellae* (L.) | Ichneumonidae | Babaei et al. (2012) |
| | | | *Theronia atalantae* (Poda) | Ichneumonidae | |
| Gracillariidae | *Phyllonorycter corylifoliella* (Hubner) | Rosaceous fruit trees | *Pholetesor bicolor* (Nees) | Braconidae | Rajabi et al. (2011) |
| | *Phyllonorycter platani* (Staudinger) | | *Pholetesor circumscriptus* (Nees) | Braconidae | Shojaei (1996) |
| | *Phyllonorycter populifoliella* Treitschke | | *Protapanteles liparidis* (Bouché) | Braconidae | Žikić et al. (2014) |
| | *Leucoptera malifoliella* (Costa) | | *Apanteles corvinus* Reinhard | Braconidae | Žikić et al. (2014) |
| Lymantriidae | *Euproctis chrysorrhoea* (L.) | Rosaceous fruit trees and forest | *Meteorus obsoletus* (Wesmael) | Braconidae | Nikdel et al. (2004) |
| | | | *Meteorus versicolor* (Wesmael) | Braconidae | Nikdel et al. (2004); Shojai (1998) |

(continued)

**Table 7.6** (continued)

| Family | Pest | Host plant | Parasitoid | Family | Reference (s) |
|---|---|---|---|---|---|
| | *Lymantria dispar* (L.) | Forest trees | *Meteorus pulchricornis* (Wesmael) | Braconidae | Abai (2011); Herard et al. (1979) |
| | | | *Apanteles lacteicolor* Viereck | Braconidae | |
| | | | *Cotesia melanoscela* (Ratzeburg) | Braconidae | |
| | | | *Protapanteles liparidis* (bouche) | Braconidae | Abai (2011) |
| | | | *Casinaria tenuiventris* (Gravenhorst) | Ichneumonidae | |
| | | | *Pimpla instigator* (Fabricius) | Ichneumonidae | |
| | | | *Theronia atalantae* (Poda) | Ichneumonidae | |
| | | | *Lymantrichneumon disparis* (Poda) | Ichneumonidae | |
| Noctuidae | *Agrotis segetum* (Denis & Schiffermuller) | Many crops and vegetables | *Chelonus inanitus* (L.) | Braconidae | Alizadeh and Javan Moghaddam (2004) |
| | | | *Macrocentrus collaris* (Spinola) | Braconidae | Shojai (1998) |
| | | | *Cotesia glomerata* (L.) | Braconidae | Alizadeh and Javan Moghaddam (2004) |
| | | | *Cotesia tibialis* (Curtis) | Braconidae | Shojai (1998) |
| | *Helicoverpa armigera* (Hubner) | Crops and vegetables | *Cotesia kazak* (Telenga) | Braconidae | Ghadiri Rad and Ebrahimi (2010) |
| | | | *Habrobracon hebetor* (say) | Braconidae | Rafiee Dastjerdi et al. (2008a, b) |
| | | | *Barylypa amabilis* (Tosquint) | Ichneumonidae | Mojeni and Sedivy (2001) |

| Host | Crops | Parasitoid | Family | Reference |
|---|---|---|---|---|
| *Heliothis viriplaca* (Hufnagel) | Many crops | *Barylypa pallida* (Gravenhorst) | Ichneumonidae | Mojeni and Sedivy (2001) |
| | | *Ichneumon sarcitorius* L. | Ichneumonidae | Mojeni and Sedivy (2001) |
| | | *Hyposoter didymator* (Gravenhorst) | Ichneumonidae | Ghadiri Rad et al. (2007) |
| | | *Habrobracon hebetor* (Say) | Braconidae | Adldoost (2010) |
| | | *Habrobracon iranicus* Fischer | Braconidae | Shojai (1998) |
| *Leucania loreyi* (Duponchel) | Corn, rice, sorghum, sugarcane, wheat and barley | *Habrobracon hebetor* (say) | Braconidae | Shayesteh Siahpoush et al. (1993) |
| | | *Cotesia rubecula* (Marshall) | Braconidae | Shayesteh Siahpoush et al. (1993) |
| | | *Cotesia ruficrus* (Haliday) | Braconidae | Shayesteh Siahpoush et al. (1993) |
| *Pseudaletia unipuncta* (Haworth) | Corn | *Meteorus pendulus* (Müller) | Braconidae | Abbasipour (2001); Abbasipour et al. (2004) |
| | | *Cotesia ruficrus* (Haliday) | Braconidae | Abbasipour et al. (2004) |
| *Ostrinia nubilalis* (Hubner) | Corn, wheat | *Habrobracon hebetor* (say) | Braconidae | Najafi Navaei et al. (2002); Taghizadeh & Basiri et al. (2013) |
| *Spodoptera exigua* (Hubner) | Crops and vegetables | *Chelonus inanitus* (L.) | Braconidae | Kheyri (1977) |
| | | *Meteorus pendulus* (Müller) | Braconidae | Farahani and Talebi (2012) |
| | | *Meteorus rubens* (Nees) | Braconidae | Farahani and Talebi (2012); Shojai (1998) |
| | | *Microplitis fulvicornis* (Wesmael) | Braconidae | Karimi-Malati et al. (2014) |

(continued)

**Table 7.6** (continued)

| Family | Pest | Host plant | Parasitoid | Family | Reference (s) |
|---|---|---|---|---|---|
| | *Simyra dentinosa* Freyer | -- | *Cotesia ofella* (Nixon) | Braconidae | Karimpour et al. (2001) |
| | | | *Cotesia vanessae* (Reinhard) | Braconidae | Karimpour et al. (2001) |
| | | | *Cotesia vestalis* (Haliday) | Braconidae | Karimpour et al. (2005) |
| Nolidae | *Earias insulana* (Boisduval) | Cotton | *Bracon lefroyi* (Dudgeon & Gough) | Braconidae | Hussain et al. (1976) |
| | | | *Habrobracon hebetor* (Say) | Braconidae | Gadallah and Ghahari (2015); Modarres-Awal (1997) |
| Oecophoridae | *Syringopais temperatella* (Lederer) | Wheat | *Apanteles longipalpis* Reinhard | Braconidae | Pirhadi et al. (2008) |
| | | | *Compoletis* sp.? (as *Anilastus trictintus*) | Ichneumonidae | |
| Pieridae | *Aporia crataegi* (L.) | Rosaceous fruit trees | *Cotesia glomerata* (L.) | Braconidae | Rajabi et al. (2011) |
| | *Pieris brassicae* (L.) and *P. rapae* (L.) | Crucifers | *Cotesia glomerata* (L.) | Braconidae | Hasanshahi et al. (2014); Modarres-Awal (1997) |
| | | | *Cotesia vestalis* (Haliday) | Braconidae | Hasanshahi et al. (2015a) |
| | | | *Diadegma anurum* (Thomson) | Ichneumonidae | Hasanshahi et al. (2015b) |
| | | | *Hyposoter ebeninus* (Gravenhorst) | Ichneumonidae | Hasanshahi et al. (2015b) |
| | | | *Hyposoter leucomerus* (Thomson) | Ichneumonidae | Hasanshahi et al. (2015b) |
| Plutellidae | *Plutella xylostella* (L.) | Cruciferes | *Habrobracon hebetor* (Say) | Braconidae | Afunizadeh et al. (2010) |
| | | | *Apanteles appellator* Telenga | Braconidae | Kazemzadeh et al. (2014) |

| | | | | | |
|---|---|---|---|---|---|
| Pyralidae | *Apomyelois ceratoniae* (Zeller) | Pomegranate | *Cotesia vestalis* (Haliday) | Braconidae | Afunizadeh et al. (2010); Golizadeh et al. (2008a, b, c); Karimzadeh Esfahani et al. (2013) |
| | | | *Habrobracon hebetor* (Say) | Braconidae | Norouzi et al. (2009) |
| | | | *Phanerotoma leucobasis* Kriechbaumer | Braconidae | Sobhani et al. (2012) |
| | | | *Apanteles laspeyresiella* Papp | Braconidae | Norouzi et al. (2009) |
| | | | *Apanteles myeloenta* Wilkinson | Braconidae | Kishani-Farahani et al. (2012a); Kishani-Farahani and Goldansaz (2013) |
| | | | *Temelucha decorata* (Gravenhorst) | Ichneumonidae | Kishani Farahani, et al. (2010) |
| | | | *Campoplex tumidulus* Gravenhorst | Ichneumonidae | Kishani Farahani et al. (2010a, b) |
| | | | *Venturia canescens* (Gravenhorst) | Ichneumonidae | Kishani Farahani et al. (2008) |
| | | | *Exeristes roborator* (Fabricius) | Ichneumonidae | |
| | | | *Cryptus* sp. | Ichneumonidae | |
| | *Arimania komaroffi* (Ragonot) | Pistachio | *Apanteles myeloenta* Wilkinson | Braconidae | Mehrnejad (2010) |
| | | Saxaul, *Haloxylon* sp. | *Habrobracon telengai* Mulyarskaya | Braconidae | Mehrnejad (2010) |
| | *Proceratia caesariella* Hampson | | *Cardiochiles shestakovi* Telenga | Braconidae | Shamszadeh (1998) |
| | | | *Pharetrophora iranica* Narolsky & Schonitzer | Ichneumonidae | Narolsky and Schonitzer (2003); Shamszadeh et al. (2015) |

(continued)

**Table 7.6** (continued)

| Family | Pest | Host plant | Parasitoid | Family | Reference (s) |
|---|---|---|---|---|---|
| | *Etiella zinckenella* (Treitschke) | Bean | *Cardiochiles fallax* Kokujev | Braconidae | Taghizadeh et al. (2013) |
| Tineidae | *Kermania pistaciella* Amsel | Pistachio | *Centistidea pistaciella* van Achterberg & Mehrnejad | Braconidae | Van Achterberg and Mehrnejad (2002) |
| | | | *Chelonus kermakiae* (Tobias) | Braconidae | Van Achterberg and Mehrnejad (2002) |
| | | | *Gelis exareolatus* (Forster) | Ichneumonidae | Van Achterberg and Mehrnejad (2002) |
| | | | *Gelis liparae* (Giraud) | Ichneumonidae | Van Achterberg and Mehrnejad (2002) |
| | | | *Gelis kermaniae* Schwarz | Ichneumonidae | Schwarz (2009) |
| Tortricidae | *Cydia pomonella* (L.) | Mainly apple, rosaceous fruit trees | *Lytopylus rufipes* (Nees) | Braconidae | Ranjbar-Aghdam and Fathipour (2010) |
| | | | *Habrobracon iranicus* Fischer | Braconidae | Shojai (1998) |
| | | | *Ascogaster quadridentata* Wesmael | Braconidae | Ranjbar-Aghdam and Fathipour (2010) |
| | | | *Liotryphon caudatus* (Ratzeburg) | Ichneumonidae | Rajabi et al. (2011) |
| | | | *Pimpla turionellae* (L.) | Ichneumonidae | Rajabi et al. (2011) |
| | | | *Pristomerus vulnerator* (panzer) | Ichneumonidae | Ranjbar-Aghdam and Fathipour (2010) |
| | *Lobesia (Polychrosis) botrana* (Denis & Schiffermuller) | Grape | *Lytopylus rufipes* (Nees) | Braconidae | Akbarzadeh-Shoukat et al. (2015) |
| | | | *Habrobracon hebetor* (Say) | Braconidae | Akbarzadeh-Shoukat (2012) |

| Family | Host species | Plant | Parasitoid | | Reference |
|---|---|---|---|---|---|
| | | | *Ascogaster quadridentata* Wesmael | Braconidae | Akbarzadeh-Shoukat et al. (2015) |
| | | | *Enytus apostata* Gravenhorst | Ichneumonidae | Akbarzadeh-Shoukat (2012) |
| | | | *Pristomerus vulnerator* (Panzer) | Ichneumonidae | Akbarzadeh-Shoukat (2012) |
| | *Spilonota ocellana* (Denis & Schiffermuller) | Apple | *Lytopylus rufipes* (Nees) | Braconidae | Shojai (1998) |
| | *Tortrix viridana* (L.) | Oak | *Lissonota palpalis* Thomson | Ichneumonidae | Mohammadi-Khoramabadi, et al. (2016) |
| | | | *Diadegma longicaudatum* Horstmann | Ichneumonidae | |
| | | | *Temelucha lucida* (Szepligeti) | Ichneumonidae | |
| | | | *Scambus elegans* (Woldstedt) | Ichneumonidae | |
| Yponomeutidae | *Yponomeuta malinella* Zeller | Apple | *Habrobracon iranicus* Fischer | Braconidae | Shojai (1998) |
| | | | *Campoletis ensator* (Gravenhorst) | Ichneumonidae | Rajabi et al. (2011) |
| | | | *Campoplex tumidulus* Gravenhorst | Ichneumonidae | Talebi et al. (2006) |
| | | | *Itoplectis tunetana* (Schmiedeknech) | Ichneumonidae | Rajabi et al. (2011); Talebi et al. (2006) |
| | | | *Itoplectis maculator* (Fabricius) | Ichneumonidae | Mohammadi-Khoramabadi et al. (2013) |
| | | | *Pimpla turionellae* (L.) | Ichneumonidae | Rajabi et al. (2011) |

Table 7.7 The known species of Ichneumonoidea associated with major pests of the order Diptera in Iran

| Family | Pest | Host plant | Parasitoid | Family | Reference (s) |
|---|---|---|---|---|---|
| Agromyzidae | Chromatomyia horticola (Goureau) | Vegetables | Chorebus aphantus (Marshall) | Braconidae | Hazini et al. (2015) |
| | | | Chorebus uliginosus (Haliday) | | |
| | | | Dacnusa hospita (Förster) | | |
| | | | Dacnusa sibirica (Telenga) | | Fathi (2011) |
| | Agromyza sp. | | Opius basalis (Fischer) | Braconidae | Sakenin et al. (2012) |
| | Liriomyza cicerina (Rondani) | | Opius monilicornis (Fischer) | Braconidae | Adldoost (1995) |
| Tephritidae | Acanthiophilus helianthi (Rossi), Chaetorellia carthami (Stackelberg), Terellia luteola (Wiedemann) and Urophora mauritanica (Macquart) | Safflower | Bracon luteator (Spinola) and Habrobracon hebetor (say) | Braconidae | Bagheri and Nematollahi (2006); Gharali (2004); Lotfalizadeh and Gharali (2014) |
| | Carpomya vesuviana (Costa) | Ber | Fopius carpomyiae (Silvestri) | Braconidae | Farrar et al. (2009) |
| | Urophora terebrans (Loew) | -- | Cremastus lineatus (Gravenhorst) | Ichneumonidae | Pourhaji et al. (2016) |
| | Hypenidium roborowskii (Becker) | | Diadegma maculatum (Gravenhorst) | Ichneumonidae | |
| | Sphenella marginata (fallen) | | Scambus brevicornis (Gravenhorst) | Ichneumonidae | |
| | Terellia gynaecochroma (Hering) | | Scambus rufator (Aubert) | | |

## 7.6   The Known Hosts of Ichneumonoidea in Iran

Excluding the hosts recorded for the species of the subfamily Aphidiinae which are specialized on Aphids, there have been yet reported 75 host-parasitoid associations between parasitoid wasps of the family Braconidae and species from Lepidoptera, Coleoptera and Diptera (Tables 7.5, 7.6 and 7.7) (Farahani et al. 2016).

From about 502 recorded species of Ichneumonidae from Iran, there have been yet reported only 53 host-parasitoid associations for the parasitoid wasps of Ichneumonidae and species of Coleoptera, Lepidoptera and Diptera (Tables 7.5, 7.6 and 7.7). Following the major pests in Iran and their known parasitoids in the Braconidae and Ichneumonidae families are presented.

### 7.6.1   The Major Coleopteran Pests and their Ichneumonoidea Parasitoids

Several families of the order Coleoptera are xylophagous and may appear as economically important pests of fruit trees or forests. Rajabi et al. (2011) represented 98 coleopterans as pests of rosaceous fruit trees in Iran. From the families Buprestidae and Cerambycidae there have been yet recorded 369 and 396 species from Iran, respectively (Barimani Varandi et al. 2009; Tavakilian and Chevillotte 2012). The larva of the jewel beetle or metallic wood-boring beetles, Buprestidae (Coleoptera) bores into roots, stems and branches of main trees and causes severe damage to its host plant. Rajabi et al. (2011) provided bioecological and integrated management of 30 important buprestid pest species in Iran. The long-horned beetles, Cerambycidae (Coleoptera) is another diverse group of beetles distributed all over the world. Their larvae bore into woods of living trees or unprotected lumbers. Some of them are among the cosmopolitan and destructive pests of fruit trees in Iran. The Rosaceous branch borer, *Osphranteria coerulescens* Redtenbacher is the most economically important pest of many tree species of the family Rosaceae like apple, pear, apricot, almond, plum and peach (Rajabi 2011). Several hymenopterous parasitoids from the superfamilies Chalcidoidea and Ichneumonoidea were collected on this pest in different regions of Iran (Steffan 1968; Sharifi and Javadi 1971a, 1971b; Rajabi 2011; Mohammadi-Khoramabadi and Lotfalizadeh 2011). The parasitism rate of *Xorides corcyrensis* Kriech. (Hymenoptera: Ichneumonidae) on larvae of *O. coerulescens* fluctuated between 5% and 22% in different regions of Iran (Rajabi 2011). Apparently, the larvae of *X. corcyrensis* may be attacked by a hyperparasitoid, *Leucospis dorsigera* Fabricius (Hymenoptera, Leucospidae) (Hesami et al. 2005). On pistachio, the longhorn beetle *Calchaenesthes pistacivora* (Holzschuh), is a destructive xylophage of the weakened pistachio trees in Rafsanjan, Kerman province. A project was conducted to find a biological control agent for the pest and finally resulted in finding a solitary parasitoid belonging to the family Braconidae, *Megalommum pistacivorae* (van Achterberg & Mehrnejad) (Van Achterberg and Mehrnejad 2011).

The bark beetles, Scolytinae Latreille (Coleoptera: Curculionidae), are an important group of beetles in orchards and forests. Their larvae feed on barks and the adults from the buds of their hosts. So far, 79 species of Scolytinae have been recorded from Iran (Beaver et al. 2016). Table 7.5 shows that members of the family Braconidae are the main parasitoids of the bark beetles in the superfamily Ichneumonoidea in Iran. This is in congruence with other parts of the world. For example, there have been globally recorded 27 parasitoid species of the superfamily Ichneumonoidea on *Scolytus rugulosus* (Muller) just from the family Braconidae (Yu et al. 2012).

Apple blossom weevil, *Anthonomus pomorum* (L.) (Col.: Curculionidae) is a small beetle that distributed in all apple areas of Iran. The beetle mainly feeds on apple but it can attack the pear. In Iran, the beetle has one generation a year and overwinters as the adult stage. Adults feed on leaves, flowers and buds of apple trees but the main damage caused by the larvae which feed on the inner tissues of the apple flowers (Rajabi 2011). There have been established 66 species of Ichneumonoidea associated with *A. pomorum* in the world (Yu et al. 2012) but only one species of the family Braconidae, *Apanteles* sp. have been yet reported as a parasitoid of this beetle from Iran (Rajabi 2011). There is no any documented information on the parasitism rate of hymenopterous natural enemies of Apple blossom weevil in Iran.

The alfalfa weevil, *Hypera postica* (Gyllenhal) (Col.: Curculionidae) is the key pest of alfalfa widely distributed in Iran. Adults and larvae of the alfalfa weevil feed on leaves and apical meristem of alfalfa plants but the main damage caused by the larval stage (Khanjani 2004). In order to find natural enemies of the pest, several investigations have been carried on within Alfalfa fields of Iran which dates back to about 1960 (Gonzalez et al. 1980; Mirabzadeh 1968). Some of the collected parasitoids were subject to classical biological of the alfalfa weevil (Barttlet et al. 1978). So far, there have been identified four parasitoids on this pest in Iran (Table 7.5). The alfalfa weevil has 29 parasitoids from the superfamily Ichneumonoidea and 26 parasitoids from the superfamily Chalcidoidea worldwide (Yu et al. 2012). Most of the studies conducted on the Alfalfa weevil parasitoids have not gone beyond estimating the natural parasitism rate of different species or reporting the adverse effects of sublethal doses on survival and parasitism rate of its parasitoids. The parasitism rate of *Bathyplectes anurus* (Hym.: Ichneumonidae) as a dominant species of alfalfa parasitoids has been estimated as 56% in Maku and Chalderan region (West Azerbaijan province, Northwest of Iran) (Alizadeh et al. 2000). Similarly, the highest and lowest parasitism rate by *B. anurus* was estimated at 63% and 1.25% in Chahar-Mahal va Bakhtiari province (Rowshandel 2000). The adverse effect of phosalone spraying on *B. cucrculionis* showed that the residue level of phosalone 1 week after treatment is still harmful to *B. curculionis* and thus the treatment should be used at least 7 days earlier (Sabahi and Talebi, 2006). Moreover, in a similar study, the undesirable effects of etrimfos on the saurvival of *B. anurus* larvae and life-history parameters of *Oomyzus incertus* (Ratzeburg) (Hym.: Eulophidae) had been confirmed (Sabahi et al. 2002; Sabahi and Talebi 2004). Alioghli et al. (2012)

reported higher value for natural parasitism rate and correlation with the host population for *B. annurus* while the one of *B. curculionis* was negligible. Unlikely, it has been reported that *B. curculionis* was the dominant species (ca 64.4%) of alfalfa weevil parasitoid in Kurdistan province, in the west of Iran (Kamangar and Ghazi 2006) They reported that the highest parasitism rate of *B. curculionis* and *B. anurus* reached to 33.3% maximally, which seems could not keep the Alfalfa weevil population build-up. The parasitism rate of near species, *O. incertus* as one of the alfalfa weevil parasitoids in East Azerbaijan province was estimated as 18% (Aligholi et al. 2012). As a rare case, the study of the spatial distribution of alfalfa natural enemies showed a random distribution pattern of *B. anurus* at alfalfa fields of Hamedan province (Zahiri et al. 2006).

The biology of a Persian race of *Microctonus aethiopoides* Loan (Hym.: Braconidae) as an alfalfa weevil parasitoid has been investigated under laboratory conditions which suggest that the low threshold temperature for egg and larvae was 9.8 °C and this parasitoid has two generations per year (Arbab 2011).

## 7.6.2 The Major Lepidopterous Pests and their Ichneumonoidea Parasitoids

The order Lepidoptera comprises destructive and key pests of crops, fruit trees and forests in Iran (Abai 2011; Khanjani 2004; Rajabi 2011). Table 7.6 shows the key pests from the family Lepidoptera and their parasitoids from the families Braconidae and Ichneumonidae in Iran.

The fall webworm, *Hyphantria cunea* (Drury) (Lep.: Arctiidae) distributed in the Holarctic region and Iran in the northern slops of the Alborz Mountains. It is a very polyphagous moth that has been reported on 636 plant species. The caterpillars of these pests spin a tent web on the tree limbs and feed on leaves (Abai 2011). Association of three species of the family Braconidae and four species of the family Ichneumonidae with the pest has been yet established in Iran (Table 7.6).

The rice stem borer, *Chilo suppressalis* (Walker) (Lep.: Crambidae) is the key pest of rice in Iran. A huge amount of pesticides was annually used for chemical control of this pest. Biological control of this pest using natural enemies is an environmentally safe method for decreasing of chemical applications and sustainable control of the pest (Niyaki 2010). There have been globally reported 47 species of Ichneumonoidea as a parasitoid of *C. suppressalis* (Yu et al. 2012) but in Iran, only four species of Ichneumonoidea have been yet reported which are listed in Table 7.6.

The lesser bud moth, *Recurvaria nanella* (Denis & Schiffermuller) (Lep.: Gelechidae) is a univoltine species feed on apple, pear, almond, apricot and many other rosaceous fruits. It is widely distributed in the western Palaearctic and Nearctic where it is probably introduced. In Iran, the lesser bud moth usually occurs in the higher altitudes (Rajabi 2011). The larva of the pest feeds on the buds, flowers and young leaves of its host at the beginning of the growing season (at early spring) and then mine the leaves. Damage of the pest in Iran reaches to 35–40% of the leaves of its host (Rajabi 2011). Although there were reported 27 parasitoid species of the superfamily Ichneumonoidea and six species of the superfamily Chalcidoidea on this pest all over the world (Yu et al. 2012), there is not enough information on the biological control agents on the lesser bud moth in Iran. The only reported parasitoid of the lesser bud moth in Iran is *Orgilus obscurator* (Nees) (Hym.: Braconidae) (Sabzevari 1968).

The peach twig borer, *Anarsia lineatella* (Zeller) (Lep.: Gelechiidae) is widely distributed in Europe and North America where is believed to be introduced. Apricot, peach, almond and plum have been recorded as main hosts of this pest in Iran. It has two generations annually and three to four generations in North America. The primary damage of the pest is that the larva burrows to tender shoots and kill them. The larva also feeds on the fruits of its hosts (Rajabi 2011). There have been recorded 25 species of the superfamily Ichneumonoidea and 16 species of the superfamily Chalcidoidea as a parasitoid of the peach twig borer in the world (Yu et al. 2012) among them one species of Chalcidoidea, *Copidosoma varicorne* (Nees) (Hym.: Encyrtidae) (Rajabi 2011) and one species of the family Braconidae have been collected in Iran, *Apanteles xanthostigma* (Haliday) (Hym.: Braconidae) (Esmaili 1983).

Several species of lepidopterous leaf miners from the family Gracillariidae occur in the apple orchards of Iran. They also feed on pear, quince, cherry, sour cherry, plum and hawthorn. The leaf miners have important hymenopterous natural enemies which may effectively regulate and suppress their host populations (Esmaili 1983; Rajabi 2011). Four species of Braconidae have been reported as parasitoids of leaf miners of the family Gracillariidae from Iran (Table 7.6).

The cutworm, *Agrotis segetum* (Denis & Schiffermuller) (Lep.: Noctuidae) is a serious pest of many agricultural crops. The larva cut the young plants off near the ground level and causes economic damage at the early stages of plant growth. Yu et al. (2012) listed 78 species of Ichneumonoidea associated with *A. segetum* in the world but in Iran, there have been reported just four species of Braconidae as parasitoids of this pest (Table 7.6).

The cotton bollworm, *Helicoverpa armigera* (Hubner) (Lep.: Noctuidae) is a polyphagous and worldwide crop pest that attacks to cotton, soybean, tomato, sorghum and groundnut. The hatched larvae feed on leaves and in the later instars enter to boll, square or fruits of their hosts and cause economic damage (Rafiee Dastjerdi et al. 2008a). Yu et al. (2012) listed 182 host-parasitoid associations of Ichneumonoidea and Chalcidoidea with the cotton bollworm. In Iran, there have been five species of the superfamily Ichneumonoidea as a parasitoid of *H. armigera* (Table 7.6). It has been suggested that natural enemies especially *Cotesia kazak* (Telenga) reduce the population and thus being responsible of suppresing *H. armigera* in tomato fields (Ghadiri Rad and Ebrahimi 2010). Besides, the mass rearing of *H. hebetor* for control of *H. armigera* and *Ostrinia nubilalis* was initiated from 2002 in Iran (Najafi Navaei et al. 2002). The functional response of *H. hebetor* is also investigated in different aspects. The type III responses have been indicated for females of *H. hebetor* to Indian meal moth (*Plodia interpunctella* Hubner), flour moth (*Ephestia kuehniella* Zeller) (Mostaghimi et al. 2010) and greater wax moth (Vaez and Pourgoli 2016). According to b constant, *H. hebetor* females preferred the flour moth. Although, type II response has been recorded for *H. hebetor* under different cold storage duration and it has been recorded that cold storage at 4 °C did not reduce the efficiency of *H. hebetor* significantly (Abedi et al. 2016). The type of host larvae affects on type of functional response and *H. hebetor* exhibits type II functional response when cotton bollworm offered (Vaez and Pourgoli 2016). Besides, the functional response parameters of *H. hebetor* could be influenced by the type of flour used for the rearing of larvae and accordingly, the best flour type for rearing *H. hebetor* provides from corn (Mostafazadeh and Mehrkhou 2016b). The temperature has been considered as another influential factor in the functional response type and parameters of *H. hebetor* (Alikhani et al. 2010). Additionally, different insecticides and chemicals may change the type of functional response or its parameters (Faal-MohamadAli et al. 2010a). Sublethal doses of some insecticides can affect on functional response parameters of *H. hebetor* adversely (Rafiee Dastjerdi et al. 2008b). According to functional response parameters, it has been reported that pyriproxyfen was less toxic rather than phozalone and it might be possible to use it with this biocontrol agent jointly (Rashidi et al. 2016).

The effect of host larvae density on the parasitism rate of *H. hebetor* showed that the parasitism rate varies according to host density and quality and this issue could be more considered in the mass rearing of *H. hebetor* (Mostaghimi et al. 2012; Saadat et al. 2012a). It has been shown that the parasitism rate of *H. hebetor* significantly influenced by the parasitoid population (Saadat et al. 2012d). Scarcely, the parasitism rate of *H. hebetor* under real filed conditions has been estimated, but in one rare case, the parasitism rate of this species on *Helicoverpa armigera* and *Spodoptera exigua* on different tomato cultivars was not significantly different and ranged from 25% to 49% maximally (Salehipour et al. 2017). Moreover, it has been estimated that maximally 48% of late instars larvae of legume pod borer *Heliothis viriplaca* parasitized by *H. hebetor* (Adldoost 2010). One of the other gaps in *Habrobracon*- related studies in Iran is the researches on optimization mass rearing. There are few documents about how it is possible to improve the efficiency of rearing through modification and amendment of the rearing system according to Iran's conditions (Musavi et al. 2018). The results about the number of eggs, dispersion ability and economic of rearing showed that the Aquarium method was superior to other methods used to produce *H. hebetor* (Musavi et al. 2018). Mousapour et al. (2015) showed that under short-term cold storage of this parasitoid the biological parameters did not significantly influence, therefore, it is possible to store females under temporary cold temperatures. The same is true and it has been shown that 48 hours of cold storage of adults had a negative effect on the mean number of egg production and sex ratio (Moody et al. 2004). Although, cold storage of *H. hebetor*'s pupae and adults up to 1 and 2 weeks respectively did not have a negative effect on the reproductive performance of *H. hebetor* (Dastmalchi et al. 2018a, 2018b).

Life table studies of *H. hebetor* are fairly common and the effect of different diets, supplements, physical conditions, different populations and even rearing duration on demographic parameters have been studied in different cases (Aleosfoor et al. 2004; Shojaei et al. 2006a; Abdi Bastami et al. 2011; Alikhani et al. 2013b; Baradaran et al. 2016a). Quite expectedly, honey and honey solution had a very positive effect on the fecundity of *H. hebetor* (Shojaei et al. 2006a). Additionally, an interesting issue about the rearing of *H. hebetor* is that the mass rearing efficiency of *H. hebetor* was greatest under the white light than natural light (Aleosfoor et al. 2004). Tritrophic studies of *H. hebetor* and its hosts have also been considered commonly (Dehghani et al. 2016). For instance, the effect of susceptible and resistant tomato cultivars on the life history of *H. hebetor* showed the possible compatibility of using resistant tomato to *Tuta absoulta* and *H. hebetor* (Heidari et al. 2016). As another tritrophic case, the effect of different flour composition and types as a diet for Mediterranean flour moth on the life history of *H. hebetor* considered (Mehdi Nasab et al. 2014; Mostafazadeh and Mehrkhou 2016a). Karimzadeh et al. (2016a, b) also studied the influence of rearing on flour moth different instars and consecutive rearing on some biological parameters of *H. hebetor*. They showed those fifth instar larvae were the best stages in terms of parasitism rate, adult longevity and fecundity (Karimzadeh et al. 2016a, b). The effect of pollen was investigated and it has been cleared that almond pollen increases the finite rate of increase ($\lambda$) and birth rate (b) of *H. hebetor* (Hadidi et al. 2018). Recently, some studies about the suitability of different host larvae e.g. greater wax moth (*Galleria melonella* (L.)) and *Plodia interpunctella* (Hubner) for rearing of *H. hebetor* through demographic parameters have been carried on at laboratory scale (Forouzan et al. 2008, 2009; Dastmalchi et al. 2017). Furthermore, the temperature range 25–32 °C and more specifically 30 °C was the best rearing temperature in terms of the intrinsic rate of increase and other demographical parameters (Forouzan et al. 2008, 2009; Alikhani et al. 2013b). Those results are in line with other similar studies which stated that 100% of eggs hatched at 30 °C (Shojaei et al. 2006b). In a related study, 25 and 30 °C were reported as suitable temperatures for *H. hebetor* reproduction (Alikhani et al. 2013a, 2013b); however, in earlier documents, the 25–28 °C and 60–70% relative humidity have been reported as the optimum condition for *H. hebetor* reproduction and rearing (Daniali 1993).

Host density influences on life table parameters of *H. hebetor* and investigated that intrinsic rate of the increase was not significantly influenced by different host densities. However, the highest parasitism rate was recorded at the lowest host density (Atrchian et al. 2018). The effect of temperature on the life table showed that the highest fecundity of *H. hebetor* has been recorded at 20 °C (Saadat et al. 2012b). Besides, the presence of males increased female sex ratio and the number of eggs produced (Saadat et al. 2012c). Also, rearing duration influenced life table parameters and fitness of *H. hebetor* and this issue should be considered in insectarium where long-term rearing of this species has been carried on (Baradaran et al. 2016a, b, 2017).

Moreover, the importance of sugar properties as a supplement on longevity and fecundity of *H. hebetor* has been demonstrated (Yazdanian and Khabbaz Saber 2013; Yazdanian et al. 2014; Khabbaz Saber et al. 2010), e.g. it has been proved that higher glucose concentration significantly affected on mean longevity of *H. hebetor* females (Khabbaz Saber et al. 2010).

Similar to other biocontrol agents, *H. hebetor* could be influenced by bottom-up forces, which might be influence on parasitism rate and other biological characters. For instance, it has been shown that resistant cultivars to tomato leafminer could be used with *H. hebetor* simultaneously in terms of parasitism rate (Heidari et al. 2018). The effect of host plant quality on life history and efficiency of natural enemies has been documented in many cases. In a tritrophic study, it has been demonstrated that there is a positive correlation between fecundity and longevity of *H. hebetor* with increasing nitrogen fertilizers used in greenhouse tomato infested by tomato fruit worm (Gharekhani 2018). The effect of ectosymbiont *Wolbachia* sp. demonstrated that occurrence of this bacterium in *H. hebetor* contributed to sexual reproduction and increases sex ratio in favour of females (Bagheri et al. 2017b; Nasehi et al. 2018), but crosses which females voided this symbiont therein, produced male progenies (Bagheri et al. 2017a). This effects and consequently increases the transferring probability of this bacterium to the next generation (Bagheri et al. 2017b).

Furthermore, many demographic toxicological studies have been conducted to evaluate the adverse effects of different classes of insecticides on *H. hebetor* life history. The interesting point is that there are great variations in obtained results e.g. some researchers addressed imidaclopride as a safe compound quietly compatible in IPM programs (Mahdavi et al. 2010) while at the same time, others indicated that it has a serious adverse effect on age parameters or functional response of *H. hebetor* (Sarmadi et al. 2010). In one of those bioassays, it has been shown that some insecticides e.g. indoxacarb, fenpropathrin and hexaflumuron sound to be less toxic to *H. hebetor* and thus could be used in an IPM program (Rafiee Dastjerdi et al. 2008a; Sarmadi et al. 2010; Faal-MohamadAli et al. 2010b). Additionally, it has been demonstrated that sublethal effect of flufenoxuron for *H. hebetor* was less toxic than lufenuron in terms of tintrinsic rate of increase (Shishcbor and Faal Mohammadali 2012), Although Rezaei et al. (2014a) showed that flufenoxuhe ron and diflubenzuron significantly influenced on life table parameters of *H. hebetor* comparing to control. The same authors noted that the least intrinsic rate of increase for *H. hebetor* has been gained with sublethal doses of Thiaclopride (Rezaei et al. 2014b). A bioassay study proved that recommended concentrations of pyriproxyfen did not have a significant harmful influence on *H. hebetor* and this insecticide could be used as a compatible compound with *H. hebetor* in IPM projects (Jarrahi, et al., 2013). Another study indicated that sublethal doses of pyridalil did not have any adverse effect on the adult emergence rate however the reproduction parameters and mortality rate have been influenced by (Abedi et al. 2013). The demographic parameters of *H. hebetor* were reduced by tomato methanolic extract (Deljavan Anvari et al. 2016). In another related study, it has been indicated that Asafoetida (*Ferula assa-foetida* L.) essential oil reduced stable population growth parameters of *H. hebetor* and thus they could not be used simultaneously to control carob moth in pomegranate (Hashemi et al. 2013). Among the insecticides used against tomato pinworm, it has been proved that sublethal concentrations of thiocyclam (Evisect®) significantly reduced life-history parameters of *H. hebetor* and care should be made to use this compound (Rezaei et al. 2018a, b, c). Similarly, the negative effect of sublethal concentrations of flubendiamide (Takumi) on intrinsic rate of increase and other life-history parameters of *H. hebetor* have been indicated (Fariborzi et al. 2018), while, absolutely opposed, it has been demonstrated that sublethal concentration of Takumi did not have any adverse effect on sex ratio and fecundity of *H. hebetor* (Rostami et al. 2016). Although, the sublethal effect of thiacloprid, azadirachtin and red pepper extract reduces life-history parameters of *H. hebetor* but the last two compounds were less toxic and they could be used within IPM programs (Rezaei et al. 2018a). The safety of BioNeem® and Neem Guard®, two commercial formulations of azadirachtin on fecundity and sex ratio of *H. hebetor* also have been proved (Abedi et al. 2012). The bioassay study of some IGRs (Match, Calypso and Lufox) on *H. hebetor* adults implied that among them Match is the least toxic for this beneficial (Hooshmandi et al. 2012). A similar study by those authors showed that botanical insecticide, sirinol® has the least toxicity to *H. hebetor* among other insecticides tested (Rezaei et al. 2018b). Moreover, these botanical insecticides did not have a significant negative effect on functional response parameters of *H. hebetor* (Asadi et al. 2018). However, the toxic effect of galbanum, *Ferrula gummosa* (Umbelliferae) essential oil on *H. hebetor* survival has been reported (Seyedi et al. 2012). Another study showed that sublethal concentrations of *Bacillus thuringiensis* significantly reduced the intrinsic rate of increase (Sedaratian et al. 2012).

There are not few behavioural studies of *H. hebetor* in Iran but among others, more attention has been paid to olfactory studies. In one of them, it has been indicated that *H. hebetor* used kairomones of fifth instar larvae of *H. armigera* to locate hosts (Davoudi Dehkordi et al. 2018). Additionally, fifth instar larvae and its feces were more attractive to *H. hebetor* in olfactometer assay and their components were analyzed by GC-MS (Shafaghi et al. 2018a, 2018b).

Most of the above studies about *H. hebetor* bounds to artificial and sophisticated experiments in laboratories while the number of field trials about *H. hebetor* efficiency and parasitism rate is scarce. For instance, it has been estimated that *H. hebetor* parasitized 64.44% of bollworm larvae (*Helicoverpa* sp.) in corn fields of Moghan region (Northwest of Iran) after seven times of releases (Najafi Navai et al. 2002). Accordingly, the natural parasitism rate of *H. hebetor* on *Tuta absoluta* has been estimated as 1.2% which is too low to suppress tomato pinworm population substantially and should be complemented with artificial releasing (Afshari et al. 2014).

The European corn borer (ECB), *Ostrinia nubilalis* (Hubner) (Lep.: Noctuidae) is an important pest of corn in Iran as well as other corn cultivating regions of the world. The insect is polyphagous and attacks many grains such as wheat, sugar cane, cotton and soybean (Taghizadeh and Basiri 2013). There have been reported 104 and 41 species of the superfamilies Ichneumonoidea and Chalcidoidea, respectively associated with *O. nubilalis* in the world (Yu et al. 2012). Mass rearing and releasing of egg parasitoids of *Trichogramma brassicae*, *T. pintoi* (Hym.: Trichogrammatidae) and larval parasitoid of *H. hebetor* (Hym.: Braconidae) have been recommended for controlling the ECB as a part of IPM projects against it (Najafi Navaei et al. 2002; Taghizadeh and Basiri 2013).

The spiny bollworm, *Earias insulana* (Boisduval) (Lep.: Nolidae) is one of the most economically important pests of cotton in Iran. It is widely distributed in the world, but not in America (Hajatmand et al. 2014). The larva of the spiny bollworm enters the tender shoots at the beginning of the growing season and causes the shoots to dry. At the later generations, the larva enters the bolls and squares resulted in the dropping of small squares or bad opening of ripening squares. There have been recorded 23 species of the superfamily Ichneumonoidea and 25 species of the superfamily Chalcidoidea associates with the pest all over the world (Yu et al. 2012). In Iran, *Bracon lefroyi* (Dudgeon & Gough) was the first known parasitoid of the pest (Hussain et al. 1976). *Habrobracon hebetor* was another reported parasitoid of the pest (Gadallah and Ghahari 2015; Modarres-Awal 1997).

Diamondback Moth (DBM), *Plutella xylostella* L. (Lep.: Plutellidae) is one of the most important pests of crucifer crops throughout the world. This pest shows a high resistance to different classes of insecticides which makes that chemical control of diamondback moth difficult in many regions. Thus, many attempts have been conducted to control the pest using biological control agents. There have been reported about 120 species of Ichneumonoidea associated with *P. xylostella* in the world (Yu et al. 2012). So far, six species of the families Braconidae and Ichneumonidae have been reported as parasitoids of this pest in Iran (Table 7.6). Cauliflower fields of the south of Tehran were searched in a faunistic survey of DBM parasitoids and their parasitism percentages were estimated based on different regions and sampling dates (Hasanshahi et al. 2012a). In a comprehensive study, nine parasitoid species of DBM larvae and pupae have been reported in four regions which among them *Diadegma semiclausum* (Hellen) reported from Alborz, Isfahan and Khuzestan provinces was dominant species and could be used as a potential candidate against DBM in biocontrol projects (Pourian et al. 2015). *Diadegma anurum* (Thomson) is also collected from DBM larvae (Golizadeh et al. 2008a, b, c) which is a misidentification of *D. semiclausum* (Karimzadeh and Broad 2013).

Bioecology, demography and potential of the most important braconid and ichneumonid parasitoid of *P. xylostella* have been studied for biological control program and integrated management of the pest in Iran (Afunizadeh et al. 2010; Alizadeh et al. 2011; Ebrahimi et al. 2013; Fathi et al. 2012; Heidary et al. 2012; Karimzadeh Esfahani et al. 2013; Kazemzadeh et al. 2014). Fathi et al. (2012) showed that using resistant cultivars integrating with its ichneumonid parasitoid *Diadegma majale* (Hym.: Ichneumonidae) would result in effective and sustainable control of DBM in canola fields. *D. semiclausum* is another important parasitoid of DBM in Iran. Several researches have been carried on about the ecology and efficiency of this parasitoid to control DBM. In one of them, it has been shown that *D. semiclausum* preferred third instar larvae of *P. xylostella* than other larval stages (Pourian et al. 2013a). This study provides hints for the mass rearing of *D. semiclausum* on the laboratory scale. Moreover, the reproductive biology of *D. semiclausum* showed that this parasitoid could be considered as a synovigenic species (Pourian et al. 2013b). The crucifer-DBM-*Cotesia* tritrophic system has been studied deeply both in the greenhouse and under real field conditions and it has been shown that host plant susceptibility and releasing rate significantly affected on DBM and *C. vestalis* population size and parasitism rate of DBM by *Cotesia vestalis* (Karimzadeh et al. 2006, 2012; Heidari and Karimzadeh 2012a, 2012b; Karimzadeh and Heidary 2012; Soufbaf et al. 2012). Unfortunately, some olfactory studies confirmed that *C. plutella* was not been compatible with host plant resistance and thus they could not be used simultaneously (Karimzadeh et al. 2006). Accordingly, it has been suggested that using *C. vestalis* (Haliday) and *Bacillus thuringiensis* in an IPM program framework could be may successfully suppress the diamondback moth population and reduce unnecessary sprayings (Karimzadeh Esfahani and Besharat Nejad 2017). Moreover, the insect growth regulator precocene has been proved to interfere in the immune system of *P. xylostella* thus it might be possible to use this compound with *C. vestalis* simultaneously (Alizadeh et al. 2012). The effect of three mass rearing procedures on life-history parameters of *C. vestalis* has been also investigated (Rabiei et al. 2017). Additionally, the effect of different mass rearing procedures on the fertility of *C. vestalis* and its quality control has been examined (Karimzadeh et al. 2016a, b).

The carob moth (pomegranate neck worm), *Apomyelois ceratoniae* (Zeller) (Lep.: Pyralidae) is a polyphagous moth widely distributed in tropical and subtropical regions of the world. It attacks to many fruit trees such as citrus, almond, date, pistachio and fig. The carob moth is the key pest of pomegranate in Iran, which can cause serious damage up to 80% of pomegranate fruits in Iran (Shakeri 2004). Pesticide applications did not effectively control the pest due to the special feeding habits of larvae, which feed on the internal tissues of the pomegranate fruit and protected from exposure to insecticides. Therefore, mechanical control by collecting and destroying infested fruits and biological control mainly using the egg parasitoids of the family Trichogrammatidae are recommended methods for controlling the pest. In the biological control program of the pomegranate neck worm, several studies were carried out to collect, identify and evaluate the parasitism rate and the conservation of native parasitoids of the superfamily Ichneumonoidea in different regions of Iran (Kishani Farahani et al. 2010a, b, 2012a, b; Kishani-Farahani and Goldansaz 2013; Norouzi et al. 2009). In a survey carried on three regions of pomegranate production, it has been reported that The braconid, *Apanteles myeloenta* Wilkinson (Hym.: Braconidae) was the dominant species attacking larvae and preferred second instars. They reported that the parasitism rate reached its peak value in September at all three selected regions (Kishani Farahani et al. 2008, 2010a, b). In the world, Yu et al. (2012) listed 15 species of the superfamily Ichneumonoidea and 10 species of the superfamily Chalcidoidea as a parasitoid of the pomegranate neck worm. Delvare et al. (2011) reported three new species of the family Chalcididae from Iran: *Brachymeria ceratoniae* Delvare, *Proconura persica* Delvare and *Psilochalcis ceratoniae* Delvare, as parasitoids of *A. ceratoniae*. The reported Braconidae and Ichneumonidae parasitoids of the pomegranate neck worm in Iran are presented in Table 7.6.

The pistachio fruit hull borer moth (PFHBM), *Arimania komaroffi* (Ragonot) (Lep.: Pyralidae) was not a pest until early 2000 but it changed gradually to an important pest of pistachio trees in Rafsanjan county, Kerman province, the largest area of pistachio plantations in Iran. The larva of this moth feeds on pistachio fruit hull and causes the dropping of young fruits. The moth has 3–4 generations annually (Basirat et al. 2015). The main control method of this pest currently based on the application of insecticides. However, some studies have been recently conducted to find natural enemies of the pest and to evaluate the potential of natural enemies in natural biological control of PFHBM (Mehrnejad and Speidel 2011; Mehrnejad 2010, 2012). Table 7.6 presented the known parasitoids of PFHBM in Iran.

The pistachio twig borer moth (PTBM), *Kermania pistaciella* (Amsel) (Lep.: Tineidae) is an important pest of pistachio in Iran (Mehrnejad 2001). The pest is univoltine. The adult moth lays their eggs in flower and fruit clusters in early spring and then the newly hatched larva bores into the flower tissue and continue its penetration to twigs. Chemical control cannot effectively control the pest. Biological control agents on the PTBM were studied which are listed in Table 7.6 to evaluate their role in controlling the pest. Van Achterberg and Mehrnejad (2002) studied the parasitism rate of the PTBM in different localities of Rafsanjan and showed that the most common parasitoid of this pest was *Chelonus kermakiae* (Tobias) which was responsible for 37–50% parasitism of the cocoons of the pest. Mehrnejad and Basirat (2009) in an extensive cocoon sampling of the PTBM showed that *Chelonus kermaniae* was the most abundant parasitoids in all studied sites.

The codling moth, *Cydia pomonella* (L.) (Lep.: Tortricidae) is the key pest of apple orchards throughout Iran as well as other apple cultivated areas of the world. The codling moth also damages economically the Asian pear, walnut fruits and less apricot, plum and peach. It passes one to four generations per year and overwinters as a fully grown larva within thick pupa under the bark of apple trees (Rajabi 2011). So far, an association of 151 species of Braconidae and Ichneumonidae has been established with the codling moth in the world (Yu et al. 2012). Many studies have been done in order to collect and identify the hymenopterous natural enemies of the codling moth in Iran and to determine the parasitism rate and their potential for biological control or integrated management of the pest. In Table 7.6, a list of ichneumonoid parasitoids of the codling moth in Iran is provided. The bioecology and parasitism rate of *Dibrachys boarmiae* (Walker) (Hym.: Pteromalidae) showed that this ectoparasitoid of codling moth overwinters as fully grown larvae (Mashhadi Jafarloo and Talebi Chaichi 2002). Different values have been reported as parasitism rates of this species which ranges from 7.2% in Karaj region (Habibi 1977) to a peak value at 21.4% in Tabriz, East Azarbaidjan (Mashhadi Jafarloo and Talebi Chaichi 2002). Rajabi et al. (2011) reported three ichneumonid parasitoids on the codling moth and mentioned that the parasitism rate of *Pimpla turionella* (L., 1758) was higher among them. Ranjbar-Aghdam and Fathipour (2010) collected three ichneumonoid parasitoids on the codling moth in the East-Azarbaijan province and showed that the parasitism rate of them varied between 0.88% and 9.47% depends on the studied regions.

The grape berry moth, *Lobesia botrana* (Denis & Schiffermuller) (Lep.: Tortricidae) is a key pest of vineyards in Iran. It is also an economically important pest of grape in the Palaearctic and northern Africa. Feeding of the larvae on flowers and unripped and ripening berries of grape cause severe damage and favored the growth of rot fungi. This moth passes three generations annually in Iran (Akbarzadeh-Shoukat 2012). In the world, more than 80 species have been reported of the superfamily Ichneumonoidea and 31 species of the superfamily Chalicoidea associated with Grape berry moth (Yu et al. 2012). Biological control of Grape berry moth by natural enemies is based on the mass rearing and releasing of egg parasitoids of the family Trichogrammatidae but there have been reported several larval or pupal ichneumonoid parasitoids in Iran which are listed in Table 7.6. In the studied areas of Urmia, a parasitism rate of hymenopterous larval parasitoids varied from 1% to 16.8% during 2004–2006 (Akbarzadeh-Shoukat 2012).

The eye-spotted bud moth, *Spilonota ocellana* (Denis & Schiffermüller) (Lep.: Tortricidae) is a univoltine moth, feed on the lower surface of leaves of apple trees. This moth is not an important pest of apple in Iran (Rajabi 2011). The eye-spotted bud moth has a large complex of parasitoids in the superfamily Ichneumonoidea (101 species) in the world (Yu et al. 2012). Although, just one species, named *Lytopylus rufipes* (Nees) on this pest has been reported from Iran (Shojai 1998).

The oak leaf roller moth, *Tortrix viridana* (L.) (Lep.: Tortricidae) is the most serious pest of the oak trees in the western Palaearctic. The moth takes one generation a year and passes most of the year from July to March in the egg developmental stage. The larvae of the moth feed on buds and leaves of the oak trees and in high populations can completely defoliate their tree host. This pest distributed in the north and west parts of Iran. Populations of *T. viridana* has been affected by different species of Braconidae and Ichneumonidae. There have been yet identified 142 species of Ichneumonoidea associated with the oak leaf roller moth in the world (Yu et al. 2012). A recent study in the west of Iran showed that four species of Ichneumonidae parasitized the larval and pupal stages of the pest (Table 7.6) (Mohammadi-Khoramabadi et al. 2016a).

The apple ermine, *Yponomeuta malinella* Zeller (Lep.: Yponomeutidae) is a univoltine species distributed in Europe, Asia and North America. The larvae of the apple ermine feed gregariously on leaves and can severely defoliate apple trees (Rajabi 2011). Rajabi et al. (2011) reared 13 parasitoid species of Hymenoptera and 2 parasitoid species of Diptera on the larvae and pupae of the apple ermine. There have been globally reported 32 species of Chalicidoidea and 86 species of Ichneumonidae associated with the apple ermine (Yu et al. 2012). The known parasitoids of the apple ermine in Iran belonging to the superfamily Ichneumonoidea are listed in Table 7.6. The rate of parasitism of these ichneumonoid parasitoids was not documented.

*Etiella zinckenella* (Treitschke) parasitizes with six chalcidoids species including three Eulophidae (*Aprostocetus arrabonicus* (Erdös), *Elasmus biroi* Erdös and *E. platyedrae* Ferrière), two Eurytomidae (*Aximopsis augasmae* (Zerova) and *A. near ghazvini* (Zerova)) and one Pteromalidae species (*Cyrtoptyx lichtensteini* (Masi)) (Lotfalizadeh and Hosseini 2014). Although this pest has nine Braconidae and eight Ichneumonidae species in Iran (see Table 7.6).

### 7.6.3 The Major Dipterous Pests and Their Ichneumonoidea Parasitoids

Within the order Diptera, two families Agromyzidae and Tephritidae comprise the most important and key pests of crops and fruit trees in Iran (Esmaili 1983; Khanjani 2004; Rajabi 2011).

The family Agromyzidae is a species-rich family of Diptera commonly known as leafminers. In Iran, there have been known 26 species of the family Agromyzidae and most of them have agricultural importance and known as pests (Dousti 2010). *Chromatomyia horticola* (Goureau) is a polyphagous pest of vegetable crops in the families Asteraceae, Brassicaceae and Fabaceae like tomato and cabbage and distributed in temperate and tropical regions of the world. So far, four species of Braconidae have been reported as parasitoids of *C. horticola* from Iran (Table 7.7) (Hazini et al. 2015). Globally, 35 species of the family Braconidae and 40 species of Chalicidoidea have been associated with *C. horticola* (Yu et al. 2012). *Liriomyza* is another genus of the family Agraomyzidae contains more than 300 species worldwide. Species of the genus injures to a wide range of vegetables and ornamental plants. One Braconid parasitoid has been reported on *Liriomyza cicerina* from Iran (Table 7.7).

Tephritid fruit flies (Dip.: Tephritidae) comprise more than 4400 phytophagous species in the world feeding on the reproductive organs of plants. Some species such as *Ceratitis capitata* (Wiedemann), *Bacterocera oleae* (Rossi) and *Rhagoletis cerasi* (L.) are among well-known and economically the most important pests of fruit crops around the world as well as Iran (Rajabi 2011). A wide effort has been made on the implementation and releasing the biological control agents as an ecologically safe and long term integrated pest management programs of tephritid fruit flies. For example, *Phygadeuon wiesmanni* (Sachtleben) (Hym.: Ichneumonidae) is a major parasitoid of European cherry fruit fly, *R. cerasi* and released against *R. pomonella* (Walsh) from 1985 in Ontario (Hagley et al. 1993). Recently, Pourhaji et al. (2016) studied the Ichneumonidae parasitoids of tephritid fruit flies on the plants of the family Asteraceae in the west of Iran and found five ichneumonid species on four species of tephritid fruit flies (Table 7.7). The biology and parasitism rate of *Fopius carpomyiae* as a parasitoid of the fruit fly, *Carpomyia vesuviana* Costa (Dip.: Tephritidae) showed that naturally 20–26% of larvae were parasitized by this species (Farar et al. 2002).

## 7.7 Conclusion and Future Directions

Research on the collecting, rearing, identification and possibility of application of parasitoid wasps of the superfamilies Chalcidoidea and Ichneumonoidea have been accelerated during two recent decades in Iran. Future studies on different agricultural and forest ecosystems especially intact ones will introduce valuable native parasitoids that are adapted to local environmental conditions. We need to investigate more intensively the parasitoid assemblages of pests and their trophic relationships for a successful biological control program.

Demography and biological attributes of some parasitoid wasps of these two superfamilies have been studied recently under laboratory conditions. Our next programs are to assess, to establish, to colonize, to evaluate and to increase the efficacy of the biocontrol candidates at natural field conditions.

Conservation of hymenopterous biological control agents is another issue that should be more considered in Iran. Several studies in Iran showed that a rich complex of Chalcidoidea and Ichneumonoidea presented in agricultural and natural ecosystems. Although we do not have in many cases an accurate assessment of the parasitism rate of them, we certainly protect them by decreasing the application of chemical and nonselective pesticides.

# References

Abai M (2011) An introduction to biological control in forest, vol 1. Iranian Research Institute of Plant Protection, Tehran

Abbasipour H (2001) Report of endoparasitoid wasp, *Meteorus gyrator* (Thunberg) (Hym.: Braconidae) on rice armyworm, *Mythimna unipuncta* (Howarth) from Iran. J Ent Soc Iran 20 (2):101–102

Abbasipour H, Amini Dehaghi M, Taghavi A (2004) Evalution of parasitoids efficiency to control the cereal armyworm, *Mythimna unipuncta* (Howarth) (Lep.: Noctuidae) in the rice fields of Iran. 16th Iranian Plant Protection Congress, 28 August–1 September 2004, Tehran, p 86

Abbasipour H, Mahmoudvand M, Basij M, Lozan A (2012) First report of the parasitoid wasps, *Microchelonus subcontracts* and *Bracon intercessor* (Hym.: Braconidae), from Iran. J Ent Soc Iran 32(1):89–92

Abdi Bastami F, Fathipour Y, Talebi AA (2011) Comparison of life table parameters of three populations of braconid wasp, *Habrobracon hebetor* (Hym.: Braconidae) on *Ephestia kuehniella* (Lep.: Pyralidae) in laboratory conditions. Ent Phytopat 78(2):153–176

Abedi Z, Saber M, Gharekhani G, Mehrvar A (2012) Lethal and sublethal effects of two formulations of azadirachtin on *Habrobracon hebetor* Say (Hym.: Braconidae). 20th Iranian Plant Protection Congress, 25–28 August 2012, Shiraz, p 244

Abedi Z, Saber M, Gharekhani G, Mehrvar A (2013) The effects of sublethal does of pyridalil on life table parameters of the parasitoid *Habrobracon hebetor* say (Hym., Braconidae). J Ent Res 5 (3):271–282

Abedi Z, Golizadeh A, Hassanpour M (2016) The effect of cold storage of *Habrobracon hebetor* in different periods on age-specific functional response to *Anagasta kuehniella*. Paper presented at the 3rd National Meeting on Biological Control in Agriculture and Natural Resources, Iran, 2–3 February 2016, p 37

Abolhassanzadeh F, Lotfalizadeh H, Madjdzadeh SM (2017) Updated checklist of Pteromalidae (Hymenoptera: Chalcidoidea) of Iran, with some new records. J Ins Biod System 3(2):119–140

Adldoost H (1995) Study of population dynamics of rainfed chickpea leafminer in West Azarbaijan. 12th Iranian Plant Protection Congress, 2–7 September 1995, Tehran, p 140

Adldoost H (2010) Study on the larval parasitism of *Heliothis viriplaca* Huf. in the western Azarbaijan province of Iran. 19th Iranian Plant Protection Congress, 31 July- 3 August 2010, Tehran, p 127

Afshari A, Yazdanian M, Shabanipour M, Ghadiri-Rad S (2014) Natural parasitism of tomato fruitworm (*Helicoverpa armigera* Hübner) in tomato fields of Golestan province, northern Iran. Paper presented at the 3rd Integrated Pest Management Conference (IPMC), Kerman, Iran. 21–22 January, p 503

Afunizadeh M, Karimzadeh J, Broad G, Shojai M, Emami M, Lotfalizadeh H, Papp J, La Salle J, Whitefield JB, van Achterberg K, Shaw MR (2010) Larval and pupal parasitoids of *Plutella xylostella* in Isfahan province. 19th Iranian Plant Protection Congress, 31 July- 3 August 2010, Tehran, p 115

Akbarzadeh-Shoukat G (1998) The first report on the occurrence of the egg parasitoid of pear lace bug in Iran. Applied Ent Phytopat 66:44–45

Akbarzadeh-Shoukat G (2012) Larval Parasitoids of *Lobesia botrana* (Denis and Schiffermiiller, 1775) (Lepidoptera: Tortricidae) in Orumieh Vineyards. J Agric Sci Technol 14:267–274

Akbarzadeh-Shoukat G, Ebrahimi E (2008) Egg parasitoids and their role in biological control of grape berry moth *Lobesia botrana* (Denis & Schiff.) (Lep.: Tortricidae) in Orumieh vineyards. 18th Iranian Plant Protection Congress, 24–27 September 2008, Hamedan, p 19

Akbarzadeh-Shoukat G, Ebrahimi E, Masnadi-Yazdinejad A (2008a) Larval parasitoids of *Lobesia botrana* (Denis & Schiff.) (Lep.: Tortricidae) on grape in Orumieh, Iran. 18th Iranian Plant Protection Congress, 24–27 September 2008, Hamedan, p 20

Akbarzadeh-Shoukat G, Horstmann K, Ebrahimi E (2008b) Study on the pupal parasitoids of grape berry moth *Lobesia botrana* (Denis & Schiff.) (Lep.: Tortricidae) and their role in an IPM program in vineyard. P18th Iranian Plant Protection Congress, 24–27 September 2008, Hamedan, p 21

Akbarzadeh-Shoukat G, Safaralizadeh M, Ranjbar Aghdam H, Aramideh S (2015) Study on the parasitoid wasps belonging to the superfamily, Ichneumonoidea on grape berry moth, *Lobesia botrana* (Lep.: Tortricidae), in Urmia vineyards. Paper presented at the proceedings of 1st Iranian international congress of entomology, Iran

Aleosfoor M, Soleyman Nejadian E, Savary A, Moody S (2004) The effect of thee different types of light, white, yellow and natural, on some biological factors of *Habrobracon hebetor* say (Hym.: Braconidae). Paper presented at the 16 th Iranian plant protection congress, Tabriz, Iran. 28 August-1 September, p 1

Alikhani M, Hassanpour M, Golizadeh A, Rafiee-Dstjerdi H, Razmjou J (2010) Temperature-dependent functional response of *Habrobracon hebetor* Say (Hym.: Braconidae) to larvae of *Anagasta kuehniella* Zeller (Lep.: Pyralidae). 19th Iranian Plant Protection Congress, 31 July–3 August 2010, Tehran, p 48

Alikhani M, Golizadeh A, Hassanpour M, Rafiee-Dstjerdi H, Razmjou J (2013a) Some life table and reproductive parameters of *Habrobracon hebetor* at five constant temperatures under laboratory conditions. Paper presented at the Conference of Biological Control in Agriculture and Natural Resources, Tehran, Iran, 26–27 August 2013, p 20

Alikhani M, Golizadeh A, Hassanpour M, Rafiee-Dstjerdi H, Razmjou J (2013b) Stable population growth parameters of *Habrobracon hebetor* on *Anagasta kuehniella* larva at five constant temperatures under laboratory conditions. Paper presented at the Conference of Biological Control in in Agriculture and Natural Resources, Tehran, Iran, 26–27 August 2013, p 21

Alioghli N, Mollazadeh Agdam M, Ghane Jahromi M (2012) Survey on the efficiency and population dynamics of larval parasitoids of Hypera postica Gyll. (Coleoptera: Curculionidae) in Sarab region. 20th Iranian Plant Protection Congress, 25–28 August 2012, Shiraz, p 10

Alizadeh S. Javan Moghaddam H (2004) Introduction of some natural enemies of common cutworm (*Agrotis segetum* Schiff.) in Miyandoab. Paper presented at the Proceedings of the 16 th Plant Protection Congress of Iran, Iran

Alizadeh I, Safaralizadeh MH, Shayesteh N (2000) Biology of alfalfa weevil (*Hypera postica* Gyll.) and identification of it's parasitoides and dominant species in Maku and Chalderan regions. 14 th Iranian Plant Protection Congress, 5–8 September, Isfahan, p 27

Alizadeh M, Rassoulian G, Karimzadeh J, Hosseini-Naveh V, Farazmand H (2011) Biological study of *Plutella xylostella* (L.) (Lep: Plutellidae) and it's solitary endoparasitoid, *Cotesia vestalis* (Haliday) (Hym. Braconidae) under laboratory conditions. Pak J Biol Sci 14 (24):1090–1099

Alizadeh M, Rassoulian G, Karimzadeh J, Farazmand H, Hosseini-Naveh V, Pourian HR (2012) The effects of three insect growth regulators on immune responses of *Plutella xylostella* (L.) larvae to *Cotesia vestalis* (Haliday). 20th Iranian Plant Protection Congress, 25–28 August 2012, Shiraz, p 576

Ameri A, Talebi AA, Rakhshani E, Beyarslan A, Kamali K (2014) Study of the genus *Opius* Wesmael (Hymenoptera: Braconidae: Opiinae) in southern Iran, with eleven new records. Zootaxa 3884(1):1–26

Amir-Maafi M, Chi H (2006) Demography of *Habrobracon hebetor* (Hymenoptera: Braconidae) on two pyralid hosts (Lepidoptera: Pyralidae). Ann Ent Soc Am 99(1):84–90

Annecke DP, Doutt RL (1961) The genera of the Mymaridae. Hymenoptera: Chalcidoidea Entomol Mem Dept Agr Un S Africa 5:1–71

Arbab A (2006) Spatial distribution pattern of immature stages of alfalfa seed weevil, Tychius aureolus (Keiswetter) (Col., Curculionidae), and alfalfa seed wasp, Bruchophagus roddi (Gussakovski) (Hym.: Eurytomidae) in alfalfa seed fields. J Agric Sci 12(2):263–268

Arbab A (2011) Introducing *Microctonus aethiopoides* loan (Hym., Braconidae) parasitoid of the alfalfa root and leaf weevils and some of its behavioral and biological characters. Paper presented at the biological control development congress in Iran, Tehran, Iran, 27–28 July, p 425

Asadi R, Talebi AA, Fathipour Y, Moharamipour S (2006) Life table, reproduction and growth population parameters of *Diglyphus isaea* (Walker) (Hym.: Eulophidae), a larval parasitoid of *Liriomyza sativae* (Blanchard) (Dip.: Agromyzidae) on bean bushes in laboratory conditions. 17th Iranian Plant Protection Congress, 2–5 September 2006, Karaj, p 26

Asadi M, Alizadeh M, Rafiee-Dastjerdi H (2018) Sublethal effects of Sirinol and Cyromazine insecticides on the functional response of ectoparasitoid wasp *Habrobracon hebetor* say (Hym.: Braconidae) on *Ephestia kuehniella* Zeller (Lep.: Pyralidae). 23th Iranian Plant Protection Congress, 27–30 August 2018, Gorgan, pp 1358–1359

Askew RR, Sadeghi SE, Tavakoli M (2006) Chalcidoidea (Hym.) in galls of *Diplolepis mayri* (Schlechtendal) (Hym., Cynipidae) in Iran, with the description of a new species of *Pseudotorymus* Masi (Hym., Torymidae). Entomologist's Monthly Magazine 142:1–6

Atrchian H, Mahdian K, Jalali MA (2018) Effect of host density on life table parameters of *Habrobracon hebetor* (Hym.: Braconidae) on *Tuta absoluta* (Lep.: Gelechiidae). 23th Iranian Plant Protection Congress, 27–30 August 2018, Gorgan, pp 1219–1220

Azizkhani E, Rasulian G, Kharazi-Pakdel A, Sadeghi S, Tavakoli M, Melika G (2005) Report of eight species of parasitoid wasps belonging to Chalcidoidea from cynipid galls on oak trees. J Ent Soc Iran 25(1):79–80

Babaei MR, Barimani H, Mafi S (2012) Investigation of parasitism on pupae of *Ennomos quercinaria* (Huf.) (Lep.: Geometridae) in forest park of Dashte-Naz, sari-Mazandaran. 20th Iranian Plant Protection Congress, 25–28 August 2012, Shiraz, p 18

Bagheri MR, Nematollahi MR (2006) Biology and damage rate of safflower shoot fly *Acanthiophilus helianthi* Rossi in Isfahan Province. 17th Iranian Plant Protection Congress, 2–5 September 2006, Karaj, p 162

Bagheri MR, Farrokhi S, Almasi H, Ahmadi F (2013) Investigating the possibility of biological control of cucumber leafminer *Liriomyza* spp. (Dip.: Agromyzidae) using parasitoid wasp, *Diglyphus isaea* (Hym.: Eulophidae). Paper presented at the Conference of Biological Control in in Agriculture and Natural Resources, Tehran, Iran, 26–27 August 2013, p 4

Bagheri Z, Mehrabadi M, Talebi AA (2017a) *Wolbachia*-induced cytoplasmic incompatibility (CI) in the parasitoid wasp, *Habrobracon hebetor*. Paper presented at the 2nd Iranian International Congress of Entomology, Karaj, 2–4 September, p 81

Bagheri Z, Mehrabadi M, Talebi AA (2017b) *Wolbachia* infection suppresses inbreeding-avoidance in *Habrobracon hebetor* to increase the likelihood of female offspring and its transmission to the next generation. Paper presented at the 2nd Iranian International Congress of Entomology, Karaj, 2–4 September, p 82

Baradaran F, Fathipour Y, Attaran MR (2016a) Generation-dependent life table parameters of the parasitoid wasp *Habrobracon hebetor* (Hymenoptera: Braconidae) in laboratory condition. Paper presented at the 22nd Iranian Plant Protection Congress, Karaj, Iran. 27–30 August, p 611

Baradaran F, Fathipour Y, Attaran MR (2016b) Comparative performance of the wild and reared populations of the parasitoid wasp *Habrobracon hebetor* (Hymenoptera: Braconidae). Paper presented at the 22nd Iranian Plant Protection Congress, Karaj, Iran. 27–30 August, p 666

Baradaran F, Fathipour Y, Attaran MR (2017) Generation-dependent life table parameters of the parasitoid wasp *Habrobracon hebetor* (Hymenoptera: Braconidae). Paper presented at the 2nd Iranian International Congress of Entomology, Karaj, 2–4 September, p 143

Barahoei H, Rakhshani E, Riedel M (2012) A checklist of Ichneumonidae (Hymenoptera: Ichneumonidae) from Iran. Iran J Anim Biosystem 8(2):83–132

Barahoei H, Rakhshani E, Fathabadi K, Moradpour H (2014) A survey on the fauna of Ichneumonidae (Hymenoptera) associating with alfalfa fields of Khorasan Razavi province. Iran J Anim Biosystem 10(2):145–160

Barimani Varandi H, Kalashian MY, Barari H (2009) Contribution to the knowledge of the jewel beetles (Coleoptera: Buprestidae) fauna of Mazandaran province of Iran. Caucas Ent Bul 5 (1):63

Barttlet BR, Clausen CP, DeBach P, Goeden RD, Legner EF, McMurtry JA, Oatman ER (1978) Introduced parasites and predators of arthropod pests and weeds: a world review. USA Agricultural Research Service, United States Department of Agriculture

Basirat M, Golizadeh A, Fathi SAA, Hassanpour M (2015) Demography of pistachio fruit hull borer moth, *Arimania komaroffi* (Lepidoptera: Pyralidae) under different constant temperatures. J Asia-Pacific Ent 18(3):501–505

Basiri N, Lotfalizadeh H, Kazemi M-H (2013) *Dendrosoter middendorffii* (Ratzeburg, 1848) (Hymenoptera: Braconidae) a parasitoid of the fruit bark beetles in Iran. Bihar Biol 7 (2):104–105

Beaver RA, Ghahari H, Sanguansub S (2016) An annotated checklist of Platypodinae and Scolytinae (Coleoptera: Curculionidae) from Iran. Zootaxa 4098(3):401–441

Behbahani AA, Tirgari S, Ghasemi MJ (1995) Effects of temperature on ovipositional behavior and developmental time of housefly pupal parasite *Spalangia endius* Walker (Hymenoptera: Pteromalidae). 12th Iranian Plant Protection Congress, 2–7 September 1995, Tehran, p 291

Bouček Z (1952) The first revision of the European species of the family Chalcididae (Hymenoptera). Acta Ent Mus Nat Pragae 27(supl 1): 1–108

Bouček Z (1970) Contribution to the knowledge of Italian Chalcidoidea based mainly on a study at the Institute of Entomology in Turin, with descriptions of some new European species (Hymenoptera). Mem Soc Ent Ital 49:35–102

Bouček Z (1977) A faunistic review of the Yugoslavian Chalcidoidea (parasitic Hymenoptera). Acta Ent Jugos, Supl 13:1–145

Bouček Z, Rasplus J-Y (1991) Illustrated key to west-Palearctic genera of Pteromalidae (Hymenoptera: Chalcidoidea). Institut National de la Recherche Agronomique (INRA), Paris

Burks B (1971) A synopsis of the genera of the family Eurytomidae (Hymenoptera: Chalcidoidea). Trans Am Ent Soc 97(1):1–89

Daniali M (1993) Biology and methods of mass production of *Bracon hebetor* ectoparasite of cotton boll worm (Heliothis armigera) in Gorgan and Gonabad. Paper presented at the 11 th Iranian Plant Protection Congress, Rasht, Iran. 28 August- 2 September, p 109

Dashtbani KH, Baniameri V, Shojai M, Rajabi M (2012) Evaluation of an exogenous strain of *Diglyphus isaea* (Hymenoptera: Eulophidae) for biological control of *Liriomyza sativae* (Diptera: Agromyzidae) on Greenhouse cucumber in Varamin. 20th Iranian Plant Protection Congress, 25–28 August 2012, Shiraz, p 39

Dashtbani K, Baniameri V, Shojai M, Rajabi M (2013) Evaluation of *Diglyphus isaea* (Hymenoptera: Eulophidae) (Miglyphus®) for biological control of *Liriomyza sativae* (dip.: Agromyzidae) on greenhouse cucumber. Ent Phytopat 81(1):31–42

Dastmalchi P, Vaez N, Mehrvar A (2017) Effect of short time storage on biological parameters of *Habrobracon hebetor* (Say) adults. Paper presented at the 2nd Iranian International Congress of entomology, Karaj, 2–4 September, p 187

Dastmalchi P, Vaez N, Mehrvar A (2018a) Effect of cold storage period of *Habrobracon hebetor* (Say) pupa on adult's quality. 23th Iranian Plant Protection Congress, 27–30 August 2018, Gorgan, pp 1102–1103

Dastmalchi P, Vaez N, Mehrvar A (2018b) Effect of low temperature on the biological parameters of *Habrobracon hebetor* (Say) parasitoid wasp on *Galleria mellonella* L. larvae. 23th Iranian Plant Protection Congress, 27–30 August 2018, Gorgan, pp 1104–1105

Davatchi A, Chodjai M (1968) Les Hymenopteres entomophages de l'Iran. Publication, Université de Teheran, Faculté d'Agronomie, Publication No 107:1–88

Davoudi Dehkordi S, Seraj AA, Avand-Faghih A (2018) Olfactory response of the parasitoid *Habrobracon hebetor* say to kiromonal volatile from cotton bollworm *Helicoverpa armigera* Hunber. 23th Iranian Plant Protection Congress, 27–30 August 2018, Gorgan, pp 1064–1065

Dehghani S, Talebi AA, Hajiqanbar HR (2016) Life table of *Habrobracon hebetor* (Hymenoptera: Braconidae) on tomato leafminer, *Tuta absoluta* (Lepidoptera: Gelechiidae) in laboratory conditions. 23th Iranian Plant Protection Congress, 27–30 August 2018, Gorgan, p 684

Deljavan Anvari S, Jamshidi M, Jafarlou M (2016) Lethal and sublethal effects of tomato Methanolic extract on *Bracon hebetor* (Hym.: Braconidae) in vitro. Paper presented at the third National Meeting on Biological Control in Agriculture and Natural Resources, Iran, 2–3 February 2016, p 29

Delvare G (2005) A revision of the West-Palearctic *Podagrion* (Hymenoptera: Torymidae), with the description of *Podagrion bouceki* sp. nov. Acta Soc Zool Bohem 69:65–88

Delvare G, Bouček Z (1992) On the new world Chalcididae (Hymenoptera). Mem Am Ent Inst 53:1–466

Delvare G, Talaee L, Goldansaz SH (2011) New chalcididae (Hymenoptera: Chalcidoidea) of economic importance from Iran. Ann Zool 61(4):789–801

Dezianian A, Jalali A (2004) Investigation on the distribution and seasonal fluctuation and important natural enemies of potato tuber moth *Phthorimaea operculella* (Zeller) in Shahrood region. 16th Iranian plant protection congress, 28 August-1 September 2004, Tehran

Dezianian A, Quicke D (2006) Introduction of potato tuber moth parasite wasp, *Bracon* (*Habrabracon*) aff. *radialis* Telenga from Iran. 17th Iranian Plant Protection Congress, 2–5 September 2006, Karaj, p 17

Dousti AF (2008) First record of Diglyphus poppoea (Hym.: Eulophidae) from Iran. J Ent Soc Iran 27(2):25–26

Dousti A (2010) Annotated list of Agromyzidae (Diptera) from Iran, with four new records. J Ent Res Soc 12(3):1–6

Ebrahimi E, Ahmadian HA (2002) Report of *Podagrion pachymerum* (Hym., Torymidae) from Iran. 15th Iranian Plant Protection Congress, 7–11 September 2002, Kermanshah, p 166

Ebrahimi E, Akbarzadeh-Shoukat G (2008) Report of *Trichogramma ingricum* (Hym.: Trichogrammatidae) from Iran. J Ent Soc Iran 27(2):43–45

Ebrahimi M, Sahragard A, Talaei-Hassanloui R, Kavousi A, Chi H (2013) The life table and parasitism rate of *Diadegma insulare* (Hymenoptera: Ichneumonidae) reared on larvae of *Plutella xylostella* (Lepidoptera: Plutellidae), with special reference to the variable sex ratio of the offspring and comparison of jackknife and bootstrap techniques. Ann Ent Soc Am 106 (3):279–287

Ebrahimi E, Tavakoli Korghond G, Mianbandi K, Mahmoodi H, Mohammadipour K, Noyes J (2014) Ooencyrtus ferdowsii sp. n. (Hymenoptera: Encyrtidae), an egg parasitoid of Osphranteria coerulescens (Coleoptera: Cerambycidae) in Iran. Zool Mid East 61:45–49

Esmaili M (1983) The important pests of fruit trees. Sepehr Publication, Tehran

Eslamizadeh R, Barzkar M, Shooshidezfuli AA, Karami Nejad M (2008) Investigation on the effect of last cutting time on damage rate of Bruchophagus gibbus on Baghdadi alfalfa seed yield in Khuzestan. Proceedings of 18th Iranian plant protection congress. Pests 1:388

Faal-MohamadAli H, Seraj AA, Talebi-Jahromi K, Shishebor P, Mosadegh MS (2010a) The effect of sublethal concentration on functional response of *Habrobracon hebetor* Say (Hymenoptera: Braconidae) in larval and pupal stages. 19th Iranian Plant Protection Congress, 31 July–3 August 2010, Tehran, p 236

Faal-MohamadAli H, Seraj AA, Talebi-Jahromi K, Shishebor P, Mosadegh MS(2010b) Investigation of the sublethal effects of chlorpyrifos and fenpropathrin on stable population parameters of parasitoid wasp *Habrobracon hebetor* Say (Hymenoptera: Braconidae) in larval stage. 19th Iranian Plant Protection Congress, 31 July–3 August 2010, Tehran, p 265

Falahatpisheh A, Fallahzadeh M, Dousti A, Delvare G (2018) Review of Iranian Chalcididae (Hymenoptera, Chalcidoidea) with nomenclatural notes. Zootaxa 4394(2):251–269

Fallahzadeh M, Japoshvili G (2010) Checklist of Iranian Encyrtids (Hymenoptera: Chalcidoidea) with descriptions of new species. J Ins Sci 10:1–24

Fallahzadeh M, Japoshvili G (2013) Corrections to the list of Encyrtidae (Hymenoptera: Chalcidoidea) from Iran. J Ent Res Soc 15(2):117–121

Fallahzadeh M, Japoshvili G (2017) An updated checklist of Iranian Encyrtidae (Hymenoptera, Chalcidoidea). Zootaxa 4344(1):1–46

Fallahzadeh M, Shojaei M, Ostovan H, Kamali K (2008) The first report of *Pseudotorymus stachidis* (Hym.: Torymidae) from Iran. J Ent Soc Iran 27(2):17–18

Fallahzadeh M, Narendran TC, Saghaei N (2009) Insecta, Hymenoptera, Chalcidoidea, Eurytomidae and Torymidae in Iran. Check List 5(4):830–839

Farahani S, Talebi A (2012) A review of the tribe Meteorini (Cresson, 1887) (Hymenoptera: Braconidae, Euphorinae) in northern Iran, with eight new records. Iran J Anim Biosystem 8 (2):133–153

Farahani S, Talebi AA, Rakhshani E (2016) Iranian Braconidae (Insecta: Hymenoptera: Ichneumonoidea): diversity, distribution and host association. J Ins Biod System 2(1):1–92

Farahbakhsh G (1961) Checklist of important insects and other enemies of plants and agricultural products in Iran (Vol. 1). The Ministry of Agriculture, Department Plant Protection

Farar N, Minaii K, Askari H (2002) Some biological aspects of *Fopius carpomyie* Silvestri (Hymenoptera: Braconidac), the parasitoid of Carpomyia vesuviana Costa (Dip.: Tephritidae). 15th Iranian Plant Protection Congress, 7–11 September 2002, Kermanshah, p 115

Fariborzi E, Ghane-Jahromi M, Sedaratian-Jahromi M, Sahraeian H, Rassaei A (2018) Sub-lethal effects of Flubendiamide (Takumi) on reproductive and population growth parameters of *Habrobracon hebetor* (Hym.: Braconidac). 23th Iranian Plant Protection Congress, 27–30 August 2018, Gorgan, pp 1396–1397

Farooqi SI, Menon MGR (1972) A new phytophagous species of *Systasis* Walker (Hymenoptera: Pteromalidae) infesting seeds of *Cenchrus* species. Mushi 46(9):111–114

Farrar N, Golestaneh R, Askari H, Assareh MH (2009) Studies on parasitism of *Fopius carpomyie* (Silvestri) (Hymenoptera: Braconidae), an egg-pupal parasitoid of ber (Konar) fruit fly, *Carpomyia vesuviana* Costa (Diptera: Techritidae), in Bushehr-Iran. Acta Hort 84:431–438

Fathi SAA (2011) Tritrophic interactions of nineteen canola cultivars-*Chromatomyia horticola*-parasitoids in Ardabil region. Munis Ent Zool 6(1):449–454

Fathi S, Bozorg-Amirkalaee M, Sarfraz R, Rafiee-Dastjerdi H (2012) Parasitism and developmental parameters of the parasitoid *Diadegma majale*(Gravenhorst) in control of *Plutella xylostella* (L.) on selected cultivars of canola. BioControl 1(57):49–59

Forouzan M, Sahragard A, Amir-Maafi M (2008) Temperature-dependent development of *Habrobracon hebetor* (Hym.: Braconidae) reared on larvae of *Galleria mellonella* (Lep.: Pyralidae). J Ent Soc Iran 28(1):67–78

Forouzan M, Sahragard A, Amir-Maafi M (2009) Demography of *Habrobracon hebetor* (Hym.: Braconidae) on *Galleria mellonella* (Lep.: Pyralidae) at different temperatures. J Ent Soc Iran 28 (2):27–44

Fry JM (1989) Natural enemy databank: a catalogue of natural enemies of arthropods derived from records in the CIBC natural enemy databank (pp. 79). Wallingford, Oxford, UK: CAB international, commonwealth Institute of Biological Control

Gadallah NS, Ghahari H (2015) An annotated catalogue of the Iranian Braconinae (Hymenoptera: Braconidae). Entomofauna 36:121–176

Ghadiri Rad S, Ebrahimi E (2010) Analysis regression of the effects pf *Cotesia kazak* Telenga (Hym.: Braconidae) on population densities of *Helicoverpa armigera* Hubner (Lep.: Noctuidae) and crop injury indices. 19th Iranian Plant Protection Congress, 31 July- 3 August 2010, Tehran, p 67

Ghadiri Rad S, Ebrahimi E, Akbarpour A (2007) Report of two parasitoid wasps on *Helicoverpa armigera* (Lep.: Noctuidae) from Iran. J Ent Soc Iran 26(2):93–94

Gharali B (2004) Study of natural enemies of safflower shoot flies in Ilam Province. Proceedings of 16th Iranian Plant Protection Congress, Tabriz, 28 August-1 September 2004, Tabriz University, p 54

Gharekhani G (2018) Nutrition interactions on different levels of nitrogen fertilizer, between tomato fruit worm, *Helicoverpa armigera* (Lepidoptera: Noctuidae) and ectoparasitoid *Habrobracon hebetor* (Hymenoptera: Braconidae) in different tomato cultivars. 23th Iranian Plant Protection Congress, 27–30 August 2018, Gorgan, pp 980–981

Gharib A (1968) *Batrachedra amydraula* Meyr (Superfamily: Gelechoidea)-(Momphidae (Cosmopterygidae)). Ent Phytop Appl 27:63–67

Ghasemzadeh M, Shakhsi Zare F, Shojaei M (2012) Comparison of population fluctuations of *Liriomyza trifolii* (Dip.: Agromyzidae) and parasitoid wasp *Diglyphus iasea* (Hym.: Eulophidae) in field conditions. 20th Iranian Plant Protection Congress, 25–28 August 2012, Shiraz, p 635

Gibson GA (1995) Parasitic wasps of the subfamily Eupelminae: classification and revision of the world genera (Hymenoptera: Chalcidoidea: Eupelmidae). Mem Ent Inter 5:1–421

Goldansaz SH, Esmaili M, Ebadi R (1996) Lesser wax moth, *Achroia grisella* F., and its parasitic wasps. Paper presented at the XX International Congress of Entomology, Firenze, Italy

Golizadeh A, Kamali K, Fathipour Y, Abbasipour H, Baur H (2008a) Report of the parasitoid wasp, *Oomyzus sokolowskii* (Hym.: Eulophidae), from Iran. J Ent Soc Iran 27(2):29–30

Golizadeh A, Kamali K, Fathipour Y, Abbasipour H, Lozan A (2008b) Report of the parasitoid wasp, *Cotesia plutellae* (Hym.: Braconidae), from Iran. J Ent Soc Iran 27(2):19–20

Golizadeh A, Kamali K, Fathipour Y, Abbasipour H, Jussila R (2008c) Report of the parasitoid wasp, *Diadegma anurum* (Hym.: Ichneumonidae), from Iran. J Ent Soc Iran 27(2):15–16

Gonzalez D, Etzal L, Esmaili M, El-Heneidy AH, Kaddou I (1980) Distribution of *Bathyplectes curculionis* and *Bathyplectes anurus* (Hym.: Ichneumonidae) from *Hypera postica* (Col.: Curculionidae) on alfalfa in Eagypt, Iraq and Iran. Entomophaga 25(2):111–121

Habibi J (1977) Dynamic and importance of *Dibrachys boarmiae* as parasite of *Laspeyresia pomonella* and its role in biological control. Paper presented at the 6 th Iranian Plant Protection Congress, Karaj, Iran. 12–14 September, 1

Habibian A (1981) Some studies on *Pseudaulacaspis pentagona* Targ. and its newly imported parasite (*Prospaltella berlesi*) in Guilan Province. Ent Phytop Appl 49(1):65–72

Habibpour B, Kamali K, Meidani J (2002) Insects and mites associated with stored products and their arthropod parasites and predators in Khuzestan province (Iran). Bull Sec Rég Oue Paléarctique 25(3):89–92

Habu A (1960) A revision of the Chalcididae (Hymenoptera) of Japan, with descriptions of sixteen new species. Bul Nat Inst Agric Sci Ser C (Phytopat Ent) 11:131–357

Hadidi MS, Abbasipour H, Askarianzadeh A (2018) Effect of feeding diets containing pollen of different plants on the biological parameters of parasitoid wasp, *Habrobracon hebetor* Say in laboratory conditions. 23th Iranian Plant Protection Congress, 27–30 August 2018, Gorgan, pp 1136–1137

Hagley EAC, Biggs AR, Timbers GE, Coutu-Sundy J (1993) Effect of age of the puparium of the apple maggot, *Rhagoletis pomonella* (Walsh) (Diptera: Tephritidae), on parasitism by *Phygadeuon wiesmanni* Sachtl. (Hymenoptera: Ichneumonidae). Can Ent 125:721–724

Hajatmand F, Abbasipour H, Amin G, Askarianzadeh A, Karimi J (2014) Evaluation of infestation percentage of cotton fields to the spiny bollworm, *Earias insulana* Boisduval (Lep.: Noctuidae), and its relationship with pheromone traps. Arch Phytopat Plant Prot 47(12):1523–1529

Hasanshahi G, Askarianzadeh A, Abbasipour H, Karimi J (2012a) IIdentification of parasitoids of diamondback moth, *Plutella xylostella* (Lep.: Plutellidae) and their parasitism rate in cauliflower fields of south of Tehran. 20th Iranian Plant Protection Congress, 25–28 August 2012, Shiraz, p 116

Hasanshahi G, Askarianzadeh A, Abbasipour H, Karimi J (2012b) Natural parasitism of diamondback moth, *Plutella xylostella* (L.) (Lep.: Plutellidae) on different cultivars of cauliflower. 20th Iranian Plant Protection Congress, 25–28 August 2012, Shiraz, p 11

Hasanshahi G, Abbasipour H, Jussila R, Jahan F, Dosti Z (2014) Host report of *Hyposoter clausus* (Brischke, 1880) (Ichneumonidae: Campopleginae), a larval parasitoid of the cabbage white butterfly, *Pieris rapae* from cauliflower fields in Tehran. Bioc Plant Prot 2(1):95–97

Hasanshahi G, Abbasipour H, Gharaei AM, Jussila R, Mohammadi-Khoramabadi A (2015a) First report of *Hyposoter ebeninus* a larval parasitoid of small cabbage butterfly, *Pieris rapae* from Tehran province. Applied Ent Phytopat 82(2):185–186

Hasanshahi G, Abbasipour H, Gharaei AM, Jussila R, Mohammadi-Khoramabadi A (2015b) First report of the parasitoid wasp, *Hyposoter leucomerus* Thomson (Hym.: Ichneumonidae, Campopleginae) from Iran. J Biol Con 29(1):47–48

Hashemi Z, Goldansaz H, Hosseini Naveh V (2013) Study parameter life table of *Habrobracon hebetor* say, larval parasitoid of carob moth in *Ferula asafoetida* essential oil. Paper presented at the Conference of Biological Control in Agriculture and Natural Resources, Tehran, Iran, 26–27 August 2013, p 37

Hazini F, Zamani AA, Peris-Felipo FJ, Yari Z, Rakhshani E (2015) Alysiinae (Hymenoptera: Braconidae) parasitoids of the pea leaf miner, *Chromatomyia horticola* (Goureau, 1851) (Diptera: Agromyzidae) in Kermanshah, Iran. J Crop Prot 4(1).97–108

Hedicke H (1921) Beiträge zu einer Monographie der paläarktischen Isosominen (Hym., Chalc.). Arch Nat 86:1–167

Heidari SM, Karimzadeh J (2012a) The effects of host-plant nutrition and availability of *Cotesia vestalis* (Haliday) on the population abundance of *Plutella xylostella* (L.). 20th Iranian Plant Protection Congress, 25–28 August 2012, Shiraz, p 25

Heidari SM, Karimzadeh J (2012b) The influences of host-plant resistance and initial parasitoid release on *Cotesia vestalis* (Haliday) population size in field. 20th Iranian Plant Protection Congress, 25–28 August 2012, Shiraz, p 50

Heidari N, Sedaratian-Jahromi A, Ghane-Jahromi M (2016) Effects of susceptible and resistant tomato cultivars to *Tuta absoluta* (Lep.: Gelechiidae) on the life table parameters of parasitoid wasp *Habrobracon hebetor* (Hym.: Braconidae). Paper presented at the 22nd Iranian Plant Protection Congress, Karaj, Iran. 27–30 August, p 666

Heidari N, Sedaratian-Jahromi A, Ghane-Jahromi M, Asadi P (2018) Investigating effects of different tomato cultivars on parasitism parameters of *Habrobracon hebetor* (Hym.: Braconidae) in management of *Tuta absoluta* (Lep.: Gelechiidae). 23th Iranian Plant Protection Congress, 27–30 August 2018, Gorgan, pp 1263–1264

Heidary SM, Karimzadeh J, Ravan S, Khani A, Besharatnejad MH (2012) The effects of host-plant type and initial density of *Cotesia vestalis* (Haliday) on the population abundance of *Plutella xylostella* (L.) in field. 20th Iranian Plant Protection Congress, 25–28 August 2012, Shiraz, p 28

Herard F, Mercadier G, Abai M (1979) Situation de *Lymantria dispar* (Lep.: Lymantriidae) et son complexe parasitaire en Iran, en. Entomophaga 24(4):371–384

Hesami S, Seyedaleslami H, Ebadi R (2001) Morphological notes on *Anagrus atomus* (Hym.: Mymaridae), an egg parasitoid of grape leafhopper, *Arboridia kermanshah* (Hom.: Cicadellidae) in Isfahan. J Ent Soc Iran 21:51–67

Hesami S, Seyedoleslami H, Hatami B (2002) Impact of *Anagrus atomus* (Hym.: Mymaridae) overwintering refuge on its abundance and parasitism of grape leafhopper *Arboridia kermanshah* (Hom.: Cicadellidae) eggs in Isfahan. 15th Iranian Plant Protection Congress, 7–11 September 2002, Kermanshah, p 113

Hesami S, Seyedoleslami H, Ebadi R (2004) Biology of *Anagrus atomus* (Hymenoptera: Mymaridae), an egg parasitoid of the grape leafhopper *Arboridia kermanshah* (Homoptera: Cicadellidae). Ent Sci 7(3):271–276

Hesami S, Akrami M, Baur H (2005) *Leucospis dorsigera* Fabricius (Hymenoptera, Leucospidae) as a hyperparasitoid of Cerambycidae (Coleoptera) through Xoridinae (Hymenoptera: Ichneumonidae) in Iran. J Hym Res 14:66–68

Hesami S, Behzadi M, Ebrahimi E, Miresmaili S, Doganlar M (2008) Report of *Torymus lapsanae* (Hym.: Torymidae), a parasitoid of *Diplolepis rosae* (Hym.: Cynipidae) from Iran. J Ent Soc Iran 27(2):17–18

Hesami S, Seyedoleslami H, Hatami B (2009) Impact of overwintering refugia of *Anagrus atomus* (Hym.: Mymaridae) on egg parasitism of grape leafhopper *Arboridia kermanshah* (Hem.: Cicadellidae). Plant Prot 1(1):95

Hesami S, Ebrahimi E, Ostovan H, Shojaei M, Kamali K, Yefremova Z, Yegorenkova E (2010) Contribution to the study of Eulophidae (Hymenoptera: Chalcidoidea) of Fars Province of Iran: I-subfamilies Entedoninae and Tetrastichinae. Mun Ent Zool 5(1):148–157

Hooshmandi M, Alichi M, Minaei K, Aleosfoor E (2012) TToxicity estimation of insect growth regulators insecticides (IGR); match, lufox and calypso on adult parasitoid wasp (*Habrobracon hebetor*) (Hym. Braconidae) under laboratory condition. 20th Iranian Plant Protection Congress, 25–28 August 2012, Shiraz, p 381

Huber JT (1986) Systematics, biology, and hosts of the Mymaridae and Mymarommatidae (Insecta: Hymenoptera): 1758–1984. Entomography 4:185–243

Hussain M, Askari A, Asadi G (1976) A study of *Bracon lefroyi* (Hymenoptera: Braconidae) from Iran [Parasite of *Earias insulana*, insect pest of cotton]. Ent News (USA) 87:299–302

Jalilvand N, Gholipour Y (2002) Pistachio production in Iran: II Main Iranian pistachio pests. NUCIS Newsletter 11:23–25

Jansta P, Cruaud A, Delvare G, Genson G, Heraty J, Krizkova B, Rasplus J-Y (2017) Torymidae (Hymenoptera, Chalcidoidea) revised: molecular phylogeny, circumscription and reclassification of the family with discussion of its biogeography and evolution of life-history traits. Cladistics 34:627–651

Japoshvili G (2007a) New records of Encyrtidae (Hymenoptera, Chalcidoidea) with the description of three new species from Georgia. Caucas Ent Bull 3(1):81–84

Japoshvili G (2007b) New Data on Species of *Syrphophagus* (Hymenoptera: Encyrtidae) from Transcaucasia and Turkey. Ann Entomol Soc Am 100(5):683–687

Japoshvili G, Abrantes IM (2006) New records of encyrtids (Chalcidoidea: Encyrtidae) from Portugal, including descriptions of two new species. Ent News 117(4):423–431

Japoshvili G, Noyes JS (2005) New record of Encyrtidae (Hymenoptera: Chalcidoidea). Caucas Entomol Bul 1(2):159–160

Japoshvili G, Noyes J (2006) New records of encyrtids (Hymenoptera: Chalcidoidea) from Europe. Entomol Rev 81(1):218–225

Kamangar S, Ghazi MM (2006) Investigation on population changes and efficiency of *B. anurus* Thom. and *Bathyplectes curculionis* Thom. on population of alfalfa weevil (*Hypera postica* Gyll.). 17th Iranian Plant Protection Congress, 2–5 September 2006, Karaj, p 31

Karampour F, Fasihi M (2004) Collection, identification and study of the natural enemies of *Batrachedra amydraula* Meyr. Paper presented at the Proceeding of 16 th Iranian Plant Protection Congress, Iran

Karimi-Malati A, Fathipour Y, Talebi A, Lozan A (2014) The first report of *Microplitis fulvicornis* (Hym.: Braconidae: Microgastrinae) as a parasitoid of *Spodoptera exigua* (Lep.: Noctuidae) from Iran. J Ent Soc Iran 33(4):71–72

Karimpour Y, Fathipour Y, Talebi A, Moharramipour S (2001) Report of two endoparasitoid wasps, *Cotesia ofella* (Nixon) & *C. vanessae* (Reinhard)(Hym.: Braconidae) on larvae of *Simyra dentinosa* Freyer (Lep.: Noctuidae) from Iran. J Ent Soc Iran 21(2):105–106

Karimpour Y, Fathipour Y, Talebi A, Moharramipour S, Horstmann K, Papp J (2005) New records of two parasitoid wasps of *Simyra dentinosa* Freyer (Lep., Noctuidae) larvae from Iran. Appl Ent Phytopat (1):73, 133

Karimzadeh J, Broad G (2013) Amendment to "report of the parasitoid wasp, *Diadegma anurum* (Hym.: Ichneumonidae), from Iran". J Ent Soc Iran 33:91–92

Karimzadeh Esfahani A, Besharat Nejad MH (2017) Integrated management of diamondback moth (*Plutella xylostella* (L.)) using *Cotesia vestalis* (Haliday) and *Bacillus thuringiensis* Berliner. J Plant Prot 40(1): 81–96

Karimzadeh Esfahani A, Rezapanah M, Emami M, Rahimi H, Mohammadi M (2013) Integrated management of diamondback moth using native parasitoids and commercial Bt products. FAO

Karimzadeh J, Heidary SM (2012) The role of bottom-up and top-down forces on host-parasitoid population variability in field. 20th Iranian Plant Protection Congress, 25–28 August 2012, Shiraz, p 51

Karimzadeh F, Hardie J, Wright DJ (2006) The effect of host plant resistance on foraging behaviour and parasitism success of *C. plutellae*. 17th Iranian Plant Protection Congress, 2–5 September 2006, Karaj, p 307

Karimzadeh J, Heidary SM, Ravan S, Khani A, Besharatnejad MH (2012) The effects of host-plant type and initial parasitoid abundance on parasitism of *Plutella xylostella* (L.) by *Cotesia vestalis* (Haliday) in field. 20th Iranian Plant Protection Congress, 25–28 August 2012, Shiraz, p 26

Karimzadeh J, Rabiei A, Shakarami J, Jafary S (2016a) Promotion of mass-rearing procedure of *Cotesia vestalis* (Haliday) and its quality control measured by fertility. 22th Iranian Plant Protection Congress, 27–30 August 2016, Karaj, p 660

Karimzadeh F, Shakarami J, Goldasteh S (2016b) Study on some biological parameters of *Habrobracon hebetor* (Hym.: Braconidae) on *Ephestia kuehniella* (Lep.: Pyralidae) *in vitro*. J Ent Res 8(2):125–134

Kazemi F, Talebi AA, Fathipour Y, Moharramipour S (2004) Host stage preference and functional response of *Anisopteromalus calandrae* (Hym.; Pteromalidae), a larval parasitoid of *Callosobruchus maculatus* (Col.: Bruchidae) on chickpea in laboratory conditions. 16th Iranian Plant Protection Congress, 28 August–1 September 2004, Tehran, p 29

Kazemzadeh Z, Shaw M, Karimzadeh J (2014) A new record for Iran of *Dolichogenidea appellator* (Hym.: Braconidae: Microgastrinae), a larval endoparasitoid of diamondback moth, *Plutella xylostella* (Lep.: Plutellidae). J Ent Soc Iran 33(4):81–82

Khabbaz Saber H, Yazdanian M, Afshari A (2010) EEffect of different glucose concentrations on longevity of the parasitoid wasp, *Habrobracon hebetor* (Hym., Braconidae) under laboratory conditions. 19th Iranian Plant Protection Congress, 31 July–3 August 2010, Tehran, p 41

Khanjani M (2004) Field crop pests in Iran. Bu-Ali Sina University, Hamedan

Kheyri M (1977) The necessary of integrated control application against beet armyworm. Entomol Phytopathol Appl 45:5–28

Kishani Farahani H, Goldansaz SH, Sabahi G, Ziaadini M, Haghani S (2008) Biology of *Apanteles myeloenta* (Hym.: Braconidae) a parasitoids of carob moth *Ectomyelois ceratoniae* Zeller (Lep.: Pyralidae). 18th Iranian Plant Protection Congress, 24–27 September 2008, Hamedan, p 474

Kishani Farahani H, Goldansaz SH, Sabahi G, Shakeri M (2010a) Larval parasitoids of *Ectomeylois ceratoniae* Zeller (Lep.: Pyralidae) in three regions of Iran: Varamin, Qom, and Saveh. Iran J Plant Prot Sci 41(2):337–344

Kishani Farahani H, Goldansaz SH, Sabahi G, Hortsmann K (2010b) First report of two ichneumonid wasp species from Iran. J Ent Soc Iran 2(30):59–60

Kishani-Farahani H, Goldansaz SH (2013) Is host age an important factor in the bionomics of *Apanteles myeloenta* (Hymenoptera: Braconidae)? Eur J Ent 110(2):277

Kishani-Farahani H, Bell H, Goldansaz SH (2012a) Biology of *Apanteles myeloenta* (Hymenoptera: Braconidae), a larval parasitoid of carob moth *Ectomyelais ceratoniae* (Lepidoptera: Pyralidae). J Asia Pac Ent 15(4):607–610

Kishani-Farahani H, Goldansaz SH, Sabahi Q (2012b) A survey on the overwintering larval parasitoids of *Ectomyelois ceratoniae* in three regions in Iran. Crop Prot 36:52–57

Latifian M, Soleyman-Nejadian E (2009) Study on the effects of spatial distribution and density of the parasitoid *Anagrus atomus* L. (Hym., Mymaridae) on its searching efficiency on garpe leafhopper eggs *Arboridia kermanshah* Delabola (Hem., Cicadellidae). J Ent Res 1:239–248

Lin N-Q, Huber JT, LaSalle J (2007) The Australian genera of Mymaridae (Hymenoptera: Chalcidoidea). Zootaxa 1596:1–111

Lotfalizadeh H (2002a) Natural enemies of cotton aphids in Moghan region, northwest of Iran. 15th Iranian Plant Protection Congress, 7–11 September 2002, Kermanshah

Lotfalizadeh H (2002b) Parasitoids of cabbage aphid, *Brevicoryne brassicae* (L.) (Hom.: Aphididae) in Moghan region. Agric Sci (Tabriz) 12(1):15–25

Lotfalizadeh H (2004) Introduction of two species of the genus *Spalangia* Lat.(Hym.: Pteromalidae) from Iran. 16th Iranian Plant Protection Congress, 28 August-1 September 2004, Tehran

Lotfalizadeh H (2010a) The genus *Metaphycus* Mercet, 1917 (Hym.: Encyrtidae) of the Iranian fauna with description of a new species. North-West J Zool 6:255–261

Lotfalizadeh H (2010b) Some new data and corrections on Iranian encyrtid wasps (Hymenoptera: Chalcidoidea, Encyrtidae) fauna. Bihar Biol 4(2):173–178

Lotfalizadeh H (2015) Review of the Iranian Pteromalinae with spiculated antennae, and description of a new species of *Norbanus* Walker (Hymenoptera: Chalcidoidea, Pteromalidae). Zootaxa 4013(3):428–434

Lotfalizadeh H, Ahmadi A (1998) New record of *Schizonotus sieboldi* Rarzburg (Hym.: Pteromalidae), pupal parasitoid of poplar leaf beetle, *Chrysomela populi* L. (Col.: Chrysomelidae) from Iran. Appl Ent Phytopat 66(1/2):45–46

Lotfalizadeh H, Ahmadi A (2000) Natural enemies of cypress tree mealybug, *Planococcus vovae* (Nasonov), and their parasitoids in shiraz, Iran. Iran Agric Res 19(2):145–154

Lotfalizadeh H, Ghadirzadeh L (2016) Review of Iranian Eupelmidae (Hymenoptera: Chalcidoidea), with five new records. J Ins Biod System 2(2):181–192

Lotfalizadeh H, Gharali B (2005) Introduction to the Torymidae fauna (Hymenoptera: Chalcidoidea) of Iran. Zool Midd East 36(1):67–72

Lotfalizadeh H, Gharali B (2008) Pteromalidae (Hymenoptera: Chalcidoidea) of Iran: new records and a preliminary checklist. Entomofauna 29(6):93–120

Lotfalizadeh H, Gharali B (2014) Hymenopterous parasitoids of safflower seed pests in Iran. Appl Ent Phytopat 82:1–11

Lotfalizadeh H, Hosseini F (2013) A survey of storage pests parasitoids (Hymenoptera) in Iran. Ege Üniv Zira Faki Derg 1(Spec Iss):113–120

Lotfalizadeh H, Hosseini F (2014) Chalcidoid parasitoids (Hymenoptera) of *Etiella zinckenella* (Treitschke) (Lep.: Pyralidae) on *Sophora alopecuroides* L. in Iran. North West J Zool 10 (2):251–258

Lotfalizadeh H, Jafari-Nadushan A (2015) New records of two rare species of the family Chalcididae (Hymenoptera: Chalcidoidea) in Iran, with data on their associations. Acta Zool Bul 67(2):297–298

Lotfalizadeh H, Khalghani J (2008) Hymenopterous parasitoids (Hym.: Chalcidoidea) of xylophagous beetles in Iran. Entomofauna 29(19):249–264

Lotfalizadeh H, Delvare G, Rasplus J-Y (2007a) Phylogenetic analysis of Eurytominae (Chalcidoidea: Eurytomidae) based on morphological characters. Zool J Linnean Soc 151 (3):441–510

Lotfalizadeh H, Delvare G, Rasplus J-Y (2007b) *Eurytoma caninae* sp. n. (Hymenoptera, Eurytomidae), a common species previously overlooked with *E. rosae*. Zootaxa 1640:55–68

Lotfalizadeh H, Ebrahim E, Delvare G (2012) A contribution to the knowledge of the family Chalcididae (Hym.: Chalcidoidea) in Iran. J Ent Soc Iran 30(2):67–100

Lotfalizadeh H, Talaei-Hassanloui R, Pourian HR (2013) Review of diamondback moth, *Plutella xylostella* (L.) parasitoids in Iran. Paper presented at the Conference of Biological Control in Agriculturale and Natural Resources, Karaj, p 27

Lotfalizadeh H, Pourhaji AR, Zargaran MR (2015) Hymenopterous parasitoids (Hymenoptera: Braconidae, Eulophidae, Pteromalidae) of the alfalfa leafminers in Iran and their diversity. Far East Ent 288:1–24

Mahdavi V, Saber M, Rafiee Dastjerdi H, Mehrvar A (2010) Susceptibility of *Habrobracon hebetor* Say (Hym.: Braconidae) to carbaryl, deltamethrin and imidacloprid. 19th Iranian Plant Protection Congress, 31 July- 3 August 2010, Tehran, p 194

Mahmudi J, Askarianzadeh A, Karimi J, Abbasipour H (2013) Introduction of two parasitoids of braconid wasps on the sugar beet moth, *Scrobipalpa ocellatella* Boyd.(Lep.: Gelechidae) from Khorasan-e-Razavi province. J Sugar Beet 28(2):189–197

Maroof A (2002) Assessment of damage caused by pests of stored wheat and barley in Tehran province. Paper presented at the 15th Iranian Plant Protection Congress, Kermanshah, Iran, 7-11 September, p 85

Mashhadi Jafarloo M, Talebi Chaichi P (2002) Bioecology of *Dibrachys boarmiae* (Walker) (Hym.: Pteromalidae) in East Azarbaidjan. 15th Iranian Plant Protection Congress, 7—11 September 2002, Kermanshah, p 86

Masi L (1924) Nuove species di Chalcis raccolte nella Persia da March. G. Doria. Ann Mus Civ Stor Nat Giac Doria, Genova 10(50):187–192

Mehdi Nasab Z, Shishchbor P, Faal Mohammad Ali H (2014) Effect of different diet regimes of Mediterranean flour moth *Ephestia kuehniella* (Zeller) on biological characteristics and life table parameters of *Habrobracon hebetor* Say (Hymenoptera: Braconidae) under laboratory conditions. Plant Prot 37(3):814–896

Mehrnejad MR (2001) The current status of pistachio pests in Iran. Paper presented at the XI GREMPA Seminar on Pistachios and Almonds, Zaragoza

Mehrnejad MR (2002) The natural parasitism ratio of the pistachio twig borer moth, *Kermania pistaciella*, in Iran. Acta Hort 591:541–544

Mehrnejad MR (2003) The influence of host species on some biological and behavioural aspects of *Dibrachys boarmiae* (Hymenoptera: Pteromalidae), parasitoid of *Kermania pistaciella* (Lepidoptera: Tineidae). Biocontrol Sci Tech 13(2):219–229

Mehrnejad MR (2010) The parasitoids of the pistachio fruit hull borer moth, *Arimania komaroffi*. Appl Ent Phytopat 78(1):129–130

Mehrnejad MR (2012) Biological parameters of *Elasmus nudus* (Hymenoptera, Eulophidae), a parasitoid of the pistachio fruit hull borer moth, *Arimania komaroffi* (Lepidoptera, Pyralidae). Biocontrol Sci Tech 22(6):659–670

Mehrnejad MR, Basirat M (2009) Parasitoid complex of the pistachio twig borer moth, *Kermania pistaciella*, in Iran. Biocontrol Sci Tech 19(5–6):499–510

Mehrnejad MR, Speidel W (2011) The pistachio fruit hull borer moth, *Arimania komaroffi* Ragonot, 1888 (Lepidoptera, Pyralidae). Entomofauna 32:5–16

Mirabzadeh A (1968) Investigation of integrated control against alfalfa weevil (*Hypera postica*). Paper presented at the Proceedings of the 8th Iranian Plant Protection Congress, Iran

Modarres-Awal M (1997) List of agricultural pests and their natural enemies in Iran. Ferdowsi University Press, Iran

Mohammadi-Khoramabadi A, Lotfalizadeh H (2011) *Eurytoma iranicola* (Hym.: Eurytomidae) as a gregarious ectoparasitoid of the rosaceous branch borer *Osphranteria coerulescens* (Col.: Cerambycidae) in Iran. Plant Prot J 3:275–279

Mohammadi-Khoramabadi A, Talebi AA, Zwakhals K (2013) A study of the subfamily Pimplinae (Hymenoptera: Ichneumonidae) in the north of Iran, with eleven new species records. Entomofauna 34(2):29–56

Mohammadi-Khoramabadi A, Hesami S, Shafiei S (2016a) A contribution to the knowledge of the fauna of Ichneumonidae in Rafsanjan county of Kerman province, Iran. Entomofauna 37 (29):453–468

Mohammadi-Khoramabadi A, Kamangar S, Lotfalizadeh H (2016b) Ichneumonid parasitoids of *Tortrix viridana* (Lepidoptera, Tortricidae) in the west of Iran. Linz Biol Beit 48(1):681–691

Mojeni T, Sedivy J (2001) New report of parasitoid ichneumonid wasps of cotton bollworm *Helicoverpa armigera* (Hub.) (Lep. Noctuidae) in Iran. J Ent Soc Iran 21(1):107–108

Moody S, Soleyman Nejadian E, Savary A, Aleosfoor M (2004) Storage duration effect on the oviposition rate and sex ratio in parasitoid *Habrobracon hebetor* Say. in the insectarium. 16th Iranian Plant Protection Congress, 28 August–1 September 2004, Tehran, p 4

Mostafazadeh N, Mehrkhou F (2016a) Effect of different cereal on demographic parameters of *Habrobracon hebetor* (Say) (Hym.: Braconidae). 22th Iranian Plant Protection Congress, 27–30 August 2016, Karaj, p 635

Mostafazadeh N, Mehrkhou F (2016b) Functional response and biological characteristics of *Habrobracon hebetor* (Say) (Hym: Braconidae) on fifth instar larvae of Mediterranean flour moth, *Ephestia kuehniella* Zeller, reared on of different cereal. 22th Iranian Plant Protection Congress, 27–30 August 2016, Karaj, p 637

Mostaghimi N, Fathi SAA, Nouri Ganbalani G (2010) Functional response of *Habrobracon hebetor* (Say) (Hym.: Braconidae) to various densities of two hosts, *Ephestia kuehniella* Zeller and *Plodia interpunctella* Hubner (Lepidoptera: Pyralidae). Iran J Plant Prot Sci 41(1):1–8

Mostaghimi N, Fathi SAA, Nouri Ganbalani G, Razmjoo G, Rafiei Dastjerdi H (2012) The effect of different larvae densities of *Ephestia kuehniella* and *Plodia interpunctella* on the parasitism efficiency of *Habrobracon hebetor*. Iran J Plant Prot Sci 43(2):243–250

Mousapour Z, Askarianzade A, Abbasipour H (2015) Cold storage of adult parasitoid wasp, *Habrobracon hebetor* (Say) (Hymenoptera: Braconidae) and the flour moth larvae, *Anagasta kuehniella* (Zeller) at 12°C. Plant Pest Res 5(3):17–29

Musavi SA, Rezapanah MR, Avalin Chahrsoughi K, Attaran MR (2018) Domestication technology of producing *Habrobracon hebetor* Say parasitoid of *Helicoverpa armigera*. 23th Iranian Plant Protection Congress, 27–30 August 2018, Gorgan, pp 1172–1173

Najafi Navaei I, Taghizadeh M, Javan Moghaddam H, Oskoo T, Attaran MR (2002) Efficiency of parasitoid wasps *Trichogramma pintoi* and *Habrobracon hebetor* against *Ostrinia nubilalis* and *Helicoverpa* sp. on Maize in Moghan. 15th Iranian Plant Protection Congress, 7–11 September 2002, Kermanshah, p 193

Narendran TC (1986) Family Chalcididae. In: Subba Rao BR, Hayat M (eds) The Chalcidoidea (Insecta: Hymenoptera) of India and the adjacent countries, Oriental Insec, vol 20, pp 11–41

Narendran TC, Lotfalizadeh H (1999) New species of *Eurytoma* Illiger (Hym.: Eurytomidae) parasitic on *Eulecanium rugulosum* arch.(Hom.: Coccidae) from Iran. Iran Agric Res 18:197–204

Narolsky N, Schonitzer K (2003) Eine neue Art der Gattung *Pharetrophora* Narolsky aus dem Iran (Hymenoptera, Ichneumonidae, Cremastinae). Spixiana 26(2):155–158

Nasehi SF, Mehrabadi M, Bagheri Z, Talebi AA (2018) *Wolbachia* enhance sex-ratio rate in the inbred and outbred lines of *Habrobracon Hebetor*. 23th Iranian Plant Protection Congress, 27–30 August 2018, Gorgan, pp 1090–1091

Naseri M, Safaralizadeh MH, Yazdanshenas OA (1998) Introduction of *Euplecterus bicolor* Swederus ectoparasite of Mythimna loreyi Dup in Iran. Paper presented at the 13th Iranian Plant Protection Congress, Karaj, Iran. 23–27 August, p 58

Nikdel M, Sadeghian B, Dordaei M (2004) Collection and identification of brown-tail moth's natural enemies in Arasbaran forest. Paper presented at the The Joint Agriculture and Natural Resources Symposium, Tabriz, Iran

Nikol'skaya M (1952) Chalcids of the fauna of the USSR (Chalcidoidea). Moscow and Leningrad: Opredeliteli po Faune SSSR: Zoologicheskim Institutom Akademii Nauk SSSR

Nikol'skaya MN (1978) Fam. Serphitidae. In: Medvedeva GS (ed) Handbook of the insects of the European part of the U.S.S.R, vol III. Zoological Institute, Academy of Science U.S.S.R, Leningard, pp 646–647

Niyaki S (2010) Decline of pesticides application by using biological control: the case study in North of Iran. Middle East J Sci Res 6:166–169

Norouzi A, Talebi AA, Fathipour Y, Lozan AI (2009) *Apanteles laspeyresiellus* (Hymenoptera: Braconidae), a new record for Iran insect fauna. J Ent Soc Iran 28(2):79–80

Noyes JS (2018) Universal Chalcidoidea Database: World Wide Web electronic publication. Retrieved 20 August 2016, from The Natural Historu Museum http://www.nhm.ac.uk/entomology/chalcidoids/index.html

Pirhadi A, Rajabi G, Ebrahimi E, Ostovan H, Shekarian Moghadam B, Mohiseni AA, Mozaffarian F, Ghavami S (2008) Natural enemies of cereal leaf miner *Syringopais temperatella* Led. (Lep.: Elachistidae) in Lorestan province. 18th Iranian Plant Protection Congress, 24–27 September 2008, Hamedan

Pourhaji A, Lotfalizadeh H, Farshbaf-Pourabad R, Gharali B, Mohammadi-Khoramabadi A (2016) Ichneumonid parasitoids (Hymenoptera: Ichneumonidae) of fruit flies (Diptera: Tephritidae) in the northwest of Iran. J Ins Biod System 2(2):193–202

Pourian HR, Talaei-Hassanloui R, Ashouri A, Lotfalizadeh H, Nozari J (2013a) Host stages and suitability in parasitoid wasp *Diadegma semiclausum* (Hym.: Ichneumonidae): effect on parasitism, developmental time and sex ratio. Paper presented at the Conference of Biological Control in Agriculture and Natural Resources, Tehran, Iran, 26–27 August 2013, p 7

Pourian HR, Talaei-Hassanloui R, Ashouri A, Lotfalizadeh H, Nozari J (2013b) Ontogenic study and reproductive biology of *Diadegma semiclausum* (Hym.: Ichneumonidae), larval endoparasitoid of Diamondback moth, *Plutella xylostella* (Lepidoptera: Plutellidae). Paper presented at the Conference of Biological Control in in Agriculture and Natural Resources, Tehran, Iran, 26–27 August 2013, p 8

Pourian HR, Talaei-Hassanloui R, Ashouri A, Lotfalizadeh H, Nozari J (2015) Abundance and parasitism rate of larval and pupal parasitoids of Diamondback Moth, *Plutella xylostella* L. (Lepidoptera: Plutellidae) in four regions of Iran. Iran J Plant Protect Sci 45(2):265–278

Quicke DLJ (2015) The braconid and ichneumonid parasitoid wasps: biology, systematics, evolution and ecology. Chichester, UK, John Wiley & Sons, Ltd.

Quicke DLJ, Laurenne NM, Fitton MG, Broad GR (2009) A thousand and one wasps: a 28S rDNA and morphological phylogeny of the Ichneumonidae (Insecta: Hymenoptera) with an investigation into alignment parameter space and elision. J Nat Hist 43(23–24):1305–1421

Rabiei A, Shakarami J, Karimzadeh Esfahani J, Jafari S (2017) Study on some biological characteristics of parasitoid *Cotesia vestalis* in different mass rearing conditions. J Bioc Plant Prot 4(2):99–108

Rafiee Dastjerdi H, Hejazi M, Ganbalani GN, Saber M (2008a) Toxicity of some biorational and conventional insecticides to cotton bollworm, *Helicoverpa armigera* (Lepidoptera: Noctuidae) and its ectoparasitoid, *Habrobracon hebetor* (Hymenoptera: Braconidae). J Ent Soc Iran 28(1):27–37

Rafiee Dastjerdi H, Hejazi M, Ganbulani GN, Saber M (2008b) Effect of LC$_{25}$ of profenofos, thiodicarb, hexaflumuron and spinosad on functional response of *Habrobracon hebetor* say. (Hym.: Braconidae). 18th Iranian Plant Protection Congress, 24–27 September 2008, Hamedan, p 171

Rajabi G (2011) Insect pests of rosaceous fruit trees in Iran, management based on ecological priciples, 2nd edn. Iranian Research Institute of Plant Protection, Tehran

Rajabi M, Lotfalizadeh H, Madjdzadeh SM (2011) The family Chalcididae (Hym.: Chalcidoidea) from Kerman province, southeastern Iran with some new records. Acta Zool Bulg 63(3):263–268

Rakhshani E, Talebi AA, Fathipour Y, Moharramipour S (2003a) The first report of rose seed gall wasp, *Megastigmus aculeatus* Swederus (Hymenoptera: Torymidae) from Iran. Paper presented at the Applied-Scientific Seminar on Flowers and Ornamental Plants, Semnan, Iran

Rakhshani E, Talebi A, Sadeghi S, Ebrahimi E, Thuroczy C (2003b) Report of five wasps species associated with dog rose galls in Iran. J Ent Soc Iran 23(1):107–108

Ranjbar-Aghdam H, Fathipour Y (2010) Fist report of parasitoid wasps, *Ascogaster quadridentata* and *Bassus rufipes* (Hym.: Braconidae) on codling moth (Lep.: Tortricidae) larvae from Iran. J Ent Soc Iran 30(1):55–58

Ranji H, Karimpour Y, Dousti A, Lotfalizadeh H, Dursun O (2015) Report of *Diglyphus sensilis* parasitoid of Chromatomyia horticola from Iran. Bioc Plant Prot 3(1):113–115

Rashidi F, Nouri-Ganbalani G, Imani S, Mahdavi V (2016) Sublethal effects of pyriproxyfen and phozalone on the functional response of *Habrobracon hebetor* Say (Hymenoptera: Braconidae) to *Helicoverpa viriplaca* Hufn larvae. Paper presented at the 22nd Iranian Plant Protection Congress, Karaj, Iran. 27–30 August, p 767

Rassipour A (1993) Etude biologique d'*Apanteles chilonis* Mun. (Hym.: Braconidae) en vue de la lutte biolgique contre la pyrrale du riz, *Chilo suppressalis* Walk. (Lep.: Pyralidae). Bull Prot Organ Iran 29:1–24

Rezaei V, Moharramipour S, Talebi A (2003) The first report of *Psychophagus omnivorus* (Walker) and *Chouioia cunea* (Yang) parasitoid wasps of American white webworm *Hyphantria cunea* Drury (Lep.: Arctiidae) from Iran. Appl Entomol Phytopathol 70(2):33

Rezaei M, Fallahzadeh M, Zohdi H, Hasani MR (2014a) Sublethal effects of diflubenzuron and flufenoxuron on the biological parameters of *Habrobracon hebetor* (Hym., Braconidae). Paper presented at the 3rd Integrated Pest Management Conference (IPMC), Kerman, Iran. 21–22 January, p 591

Rezaei M, Fallahzadeh M, Zohdi H, Hasani MR (2014b) Assessment of sublethal effects of acetamiprid and thiacloprid on life-table parameters in *Habrobracon hebetor* (Hym., Braconidae). Paper presented at the 3rd Integrated Pest Management Conference (IPMC), Kerman, Iran. 21–22 January, p 592

Rezaei M, Gheibi M, Hesami S, Zohdi H (2018a) Sub-lethal effect of thiacloprid, azadirachtin, and red pepper extract on biological parameters of *Habrobracon hebetor* Say (Hym.,: Braconidae). 23th Iranian Plant Protection Congress, 27–30 August 2018, Gorgan, pp 1010–1011

Rezaei M, Gheibi M, Hesami S, Zohdi H (2018b) Sub-lethal effects of Vertimec®, Proteus® and Sirinol® on biological parameters of *Habrobracon hebetor* Say (Hym.,: Braconidae) in laboratory conditions. 23th Iranian Plant Protection Congress, 27–30 August 2018, Gorgan, pp 1012–1013

Rezaei Z, Ghane-Jahromi M, Sedaratian-Jahromi A, Sahraeian H, Azizinesar R (2018c) Evalaution of sub-lethal effects of Thiocyclam (Evisect) on demographic parameters of *Habrobracon hebetor* (Hym.: Braconidae). 23th Iranian Plant Protection Congress, 27–30 August 2018, Gorgan, p 1393

Rostami F, Zandi-Sohani N, Yarahmadi F, Ramezani L, Avalin Chaharsoghi K (2016) Effects of Azadirakhtin and Takomi on some biological parameters of *Habrobracon hebetor* (Hym.: Braconidae). 22th Iranian Plant Protection Congress, 27–30 August 2016, Karaj, p 25

Rowshandel S (2000) Biology and efficiency of *Bathyplectes anurus* in biological control of alfalfa weevel *Hypera postica* Gyll. 14th Iranian Plant Protection Congress, 27–30 August 2000, Isfahan, p 28

Saadat D, Rasekh A, Seraj AA (2012a) Effect of host density and host quality on oviposition strategy of *Habrobracon hebetor* (Hymenoptera: Braconidae). Mod Tech Agric 5(3):37–54

Saadat D, Seraj AA, Goldansaz H (2012b) Investigation of biological parameters of *Habrobracon hebetor* Say (Hym.: Braconidae) parasitoid of larva carob moth in different tempreture. 20th Iranian Plant Protection Congress, 25–28 August 2012, Shiraz, p 45

Saadat D, Seraj AA, Goldansaz H (2012c) Investigation on effect male presence of *Habrobracon hebetor* (Hym.: Braconidae) fallow female on some of biological parameters. 20th Iranian Plant Protection Congress, 25–28 August 2012, Shiraz, p 46

Saadat D, Seraj AA, Goldansaz H (2012d) Host finding behavior in different populations of *Habrobracon hebetor* Say (Hym.: Braconidae) parasitoid of carob moth larvae. 20th Iranian Plant Protection Congress, 25–28 August 2012, Shiraz, p 57

Sabahi Q, Talebi K (2004) Demographic studies on *Oomyzus incertus* (Ratzburg) (Hym.: Eulophidae), the larval parasitoid of *Hypera postica* (Gyllenhal) (Col.: Curculionidae), treated with four organophosphorous insecticides. 16th Iranian Plant Protection Congress, 28 August–1 September 2004, Tehran, p 40

Sabahi Q, Talebi K (2006) Effects of phosalone residues on alfalfa weevil larval parasitoid, Bathyplectes curculionis (Hym.: Ichneumonidae). Lett Ent Soc 26(2):11–22

Sabahi Q, Talebi K, Radjabi GH (2002) Effects of four organophosphorus insecticides on mortality and fecundity of *Bathyplectes anurus* (Thomson) (Hymenoptera: Ichneumonidae) larval parasitoid of alfalfa weevil, *Hypera postica* (Coleoptera: Curculionidae). 15th Iranian Plant Protection Congress, 7–11 September 2002, Kermanshah, p 27

Sabzevari A (1968) Lepidopterous pest on apricot. Paper presented at 1th Iranian Plant Protection Congress, Tehran, Iran

Sadeghi S, Askary H (2001) Parasitism rate of *Schizonotus sieboldii* Ratzeburg (Hymenoptera: Pteromalidae) a parasitoid of poplar leaf beetle pupa, on different poplar species. Paper presented at the International symposium: Parasitic Hymenoptera: taxonomy and biological control

Sadeghi S, Ebrahimi E (2001) New report of *Pachyneuron grande* Thomson (Hym. Pteromalidae) from Iran. J Ent Soc Iran 21(1):113–114

Saghaei N, Fallahzadeh M, Lotfalizadeh H (2018) Annotated catalog of Eurytomidae (Hymenoptera: Chacidoidea) from Iran. Trans Am Ent Soc 144:263–293

Sakenin H, Naderian H, Samin N, Rastegar J, Tabari M, Papp J (2012) On a collection of Braconidae (Hymenoptera) from northern Iran. Linz Biol Beit 44(2):1319–1330

Salehipour H, Vahedi HA, Moeini Naghadeh N, Zamani AA (2017) Population density and parasitism evaluation of *Helicoverpa armigera* (Hubner) and *Spodoptera exigua* (Hubner) on twelve varieties of tomato in the field. J Plant Prot 31(3):527–539

Sarmadi S, Nouri-Ganbalani G, Hassanpour M, Rafiee Dastjerdi H (2010) Effect of imidacloprid, indoxacarb, and deltamethrin on stable population growth parameters of *Habrobracon hebetor* Say. (Hym.: Braconidae) in pupal stage treatment. 19th Iranian Plant Protection Congress, 31 July–3 August 2010, Tehran, p 235

Schwarz M (2009) Ostpaläarktische und orientalische *Gelis*-Arten (Hymenoptera, Ichneumonidae, Cryptinae) mit macropteren weibchen. Linz Biol Beit 41:1103–1146

Sedaratian A, Fathipour Y, Talaei-Hassanlouei R (2012) Antagonism effects of *Bacillus thuringiensis* on life table parameters of H*abrobracon hebetor* (Hym.: Braconidae) in integrated management of *Helicoverpa armigera* (Lep.: Noctuidae). 20th Iranian Plant Protection Congress, 25–28 August 2012, Shiraz, p 86

Seyedi A, Abbasipour H, Moharramipour S (2012) Effect of *Ferula gummosa* Boiss. resin essential oil on *Ephestia kuehniella* Zeller (Lep.: Pyralidae) and its parasitoid *Habrobracon hebetor* (Say) (Hym.: Braconidae). 20th Iranian Plant Protection Congress, 25–28 August 2012, Shiraz, p 258

Shafaghi F, Goldansaz SH, Avand-Faghih A (2018a) Olfactory responses of parasitoid wasp *Habrobracon hebetor* to volatile compounds of host insect and pomegranate fruit under laboratory conditions. Ent Phytopat 86(1):91–102

Shafaghi F, Goldansaz SH, Avand-Faghih A (2018b) Identification of volatile components of *Habrobracon hebetor, a* parasitoids of Carob Moth. 23th Iranian Plant Protection Congress, 27–30 August 2018, Gorgan, pp 1215–1216

Shahhosseini MJ, Kamali K (1989) A checklist of insects mites and rodents affecting stored products in Iran. J Ent Soc Iran 5:1–47

Shakeri M (2004) A review on investigations on pomegranate neck worm in Iran. A proceeding on evaluation of finding and current problems associated with Spectrobates ceratoniae Management in Pomegranate. Tehran, Iran Ministry of Jihad-e-Agriculture, Organization of Research and Education, 33pp

Shamszadeh M (1998) Introducing of *Cardiochiles shestakovi* (Hym.: Braconidae) parasitoid of *Proceratia caesariella* larva in Yazd province. Pajohesh & Sazandegi 39:47–49

Shamszadeh M, Jafary A, Bishe G (2015) Natural parasitism of *Pharetrophora iranica* (Hym.: Ichneumonidae) on saxaul seed moths larvae in Yazd province. Iran J Forest Range Prot Res 12 (2):160–162

Sharifi S, Javadi I (1971a) Control of Rosaceae branch borer in Iran. J Econ Ent 64(2):484–486

Sharifi S, Javadi I (1971b) Biology of *Xorides corcyrensis* Kriech.(Hymenoptera: Ichneumonidae), a parasite of the Rosaceae branch borer *Osphranteria coerulescens* Redt.(Coleoptera: Cerambycidae) 1. Zeitsch Ang Ent 68:25–31

Shayesteh Siahpoush A, Azimi A, Rabee, Mozaffari M (1993) Introduction of three species of *Mythimna* (Lep.: Noctuidae) at Khuzestan. 11th Iranian Plant Protection Congress, 28 Aug. – 1 Sept. 1993, Rasht

Shishebor P, Faal Mohammadali H (2012) Sublethal effects of flufenoxuron and lufenuron on life table parameters of *Habrobracon hebetor* (Hymenoptera: Braconidae). Iranian J Plant Prot Sci 43(2):233–242

Shojaei M (1996) Entomology, Entomophages, 2nd edn. Tehran University Publication, Tehran

Shojaei Sh, Safaralizadeh M, Shayesteh N, Nikpay A (2006a) Effect of different nutrition diets on longevity and fecundity of *Habrobracon hebetor* Say (Hym.: Braconidae). 17th Iranian Plant Protection Congress, 2–5 September 2006, Karaj, 294

Shojaei Sh, Safaralizadeh M, Shayesteh N (2006b) Development and Survival *Habrobracon hebetor* Say (Hym.: Braconidae) in relation with different temperature on larvae of *Plodia interpunctella* Hubner (Lep.: Pyralidae). 17th Iranian Plant Protection Congress, 2–5 September 2006, Karaj, p 30

Shojai M (1998) Entomology (ethology, social life and natural enemies), Biological control, vol III. Tehran University Publications, Tehran

Sobhani M, Goldansaz SH, Hatami B (2012) Study of larval parasitoids of carob moth *Ectomyelois ceratoniae* (Lep.: Pyralidae) in Kashan region. 20th Iranian Plant Protection Congress, Iran, Shiraz, 25–28 August 2012, p 83

Soufbaf M, Fathipour Y, Karimzadeh Isfahani J (2012) Concurrent effect of plant availability and competition of parasitoids *Diadegma semiclausum* (Hym.: Ichneumonidae) and *Cotesia vestalis* (Hym.: Braconidae) on population dynamics of a tritrophic system. 20th Iranian Plant Protection Congress, 25–28 August 2012, Shiraz, p 656

Steffan J (1968) Observation sur *Chalcedectus sinaiticus* (Ms.) et descriptions de *C. balachowskyi* sp. n. [Hym. Chalcedectidae] et d'*Oopristus safavii* gen. n., sp. n. [Hym. Torymidae], deux parasites d'importance économique en Iran. BioControl 13(3):209–216

Taghizadeh M, Basiri G (2013) Technical guide for European corn borer *Ostrinia nubilalis* Hbn. and its integrated management. Iranian Reseach Institute of Plant Protection, Tehran

Taghizadeh R, Talebi AA, Fathipour Y, Lozan AI (2013) *Cardiochiles fallax* (Hym.: Braconidae), a new species record for Iran. J Ent Soc Iran 33(2):83–85

Talebi AA, Rakhshani E, Daneshvar S, Fathipour Y, Moharramipour S, Horstman K (2006) Report of *Campoplex tumidulus* and *Itoplectis tunetana* (Hym.: Ichneumonidae), parasitoids of *Yponomeuta malinellus* Zell. (Lep.: Yponomeutidae) from Iran. Appl Ent Phytopat 73(1):134

Talebi AA, Mohammadi-Khoramabadi A, Rakhshani E (2011) Checklist of eulophid wasps (Insecta: Hymenoptera: Eulophidae) of Iran. Check List 7(6):708–719

Tavakilian G, Chevillotte H (2012) Titan: base de données internationales sur les Cerambycidae ou Longicornes. Version 3.0 Retrieved 16 August 2016 http://lully.snv.jussieu.fr/titan/

Tavakoli Roodi T, Fallahzadeh M, Lotfalizadeh H (2016) Fauna of chalcid wasps (Hymenoptera: Chalcidoidea, Chalcididae) in Hormozgan province, southern Iran. J Ins Biod System 2 (1):155–166

Tobias VI (1995) Keys to the insects of the European Part of the USSR. Amerind Publishing Co. Pvt. Ltd., New Delhi

Triapitsyn SV (2013) Review of *Gonatocerus* (Hymenoptera: Mymaridae) in the Palaearctic region, with notes on extralimital distributions. Zootaxa 3644(1): 1–178

Vaez N, Pourgoli Z (2016) Functional response of *Habrobracon hebetor* (Say) (Hym.; Braconidae) on *H. armigera*, *Anagasta kuehniella* and *Galleria mellonella* larvae. Paper presented at the 22nd Iranian Plant Protection Congress, Karaj, Iran. 27–30 August, p 531

Van Achterberg C, Mehrnejad M (2002) The braconid parasitoids (Hymenoptera: Braconidae) of *Kermania pistaciella* Amsel (Lepidoptera: Tineidae: Hieroxestinae) in Iran. Zool Med 76:27–39

Van Achterberg C, Mehrnejad M (2011) A new species of *Megalommum Szépligeti* (Hymenoptera, Braconidae, Braconinae); a parasitoid of the pistachio longhorn beetle (*Calchaenesthes pistacivora* Holzschuh; Coleoptera, Cerambycidae) in Iran. ZooKeys 112:21–38

Yasnosh V, Japoshvili G (1999) Parasitoids of the genus *Psyllaephagus* Ashmead (Hymenoptera: Chalcidoidea: Encyrtidae) in Georgia with the description of *P. georgicus* sp. nov. Bul Georg AcadSci 159(3):516–519

Yazdanian M, Khabbaz Saber H (2013) Feeding of adults of the parasitoid wasp, *Habrobracon hebetor* (Hym., Braconidae) from different concentrations of glucose, fructose, and sucrose in the absence of host larvae and its effect on their longevity under laboratoryconditions. J Ent Res 5(1):67–82

Yazdanian M, Khabbaz Saber H, Afshari A (2014) Effect of sugar concentration and feeding frequency on adult's longevity and progeny production of the parasitoid wasp, *Habrobracon hebetor* (Hymenoptera: Braconidae). Plant Pest Res 3(4):1–16

Yu DS, Van Achterberg K, Horstmann K (2012) World Ichneumonoidea 2011. Taxonomy, biology, morphology and distribution. Retrieved 25 May 2016, from Taxapad.com http://www.taxapad.com/

Zahiri B, Kiabi BH, Madjnoonian H (2002) The natural areas and landscape of Iran: an overview. Zool Midd East 26(1):7–10

Zahiri B, Fathipour Y, Khanjani M, Moharramipour S (2006) Spatialdistribution of alfalfa weevil and its some natural enemies in Hamedan. 17th Iranian Plant Protection Congress, 2–5 September 2006, Karaj, p 52

Zehzad B, Kiabi B, Madjnoonian H (2002) The natural areas and landscape of Iran: an overview. Zool Midd East 26(1):7–10

Zerova M, Mehrnejad M, Gharali B, Seryogina LY (2004) Two new species of the genus Eurytoma (Hymenoptera, Eurytomidae) from Iran. Vest Zool 38:81–84

Zerova M, Seryogina L, Karimpour Y (2008) New species of the chalcidoid wasps of the families Eurytomidae and Torymidae (Hymenoptera, Chalcidoidea) from Iran. Vest Zool 42(6):101–108

Žikić V, Lotfalizadeh H, Sadeghi S, Petrović A, Janković M, Tomanović Ž (2014) New record and new associations of two leaf miner parasitoids (Hymenoptera: Braconidae: Microgastrinae) from Iran. Arch Biol Sci Belg 66(4):1591–1594

# Chapter 8
# Superfamily Platygastroidea: Natural Enemies of True Bugs, Moths, Other Insects, and Spiders

Shahzad Iranipour

## 8.1 Introduction

In previous classifications, both families of Scelionidae and Platygastridae included in superfamily Proctotrupoidea (Kozlov 1988). Later these families separated from Proctotrupoidea and appeared as superfamily Platygastroidea (Triplehorn and Johnson 2005). However, more recently, both families merged as a single family of Platygastridae and hence subfamilies of Scelioninae, Teleasinae and Telenominae account as subfamilies of platygastridae (Aguiar et al. 2013). These are minute (at most a few millimetres), and often blackish wasps with reduced venation particularly in platygastrins. Maximum four longitudinal veins namely submarginal, marginal, post-marginal and radial or stigmal veins appear in forewings. Such venation resembles those of the chalcids (superfamily Chalcidoidea), but platygastrids can easily be separated with the structure of thorax. The triangular shape of mesonotum in lateral view, an absence of post-spiracular sclerite (prepectus) and tegulae touching mesonotum are characteristics separating the members of Platygastridae from those of the Chalcidoidea. Former Scelionidae is endoparasitoids of insect eggs (mainly Hemiptera, Lepidoptera and Orthoptera in Iran) as well as spiders. Former Platygastridae is also mainly parasitoids of larvae of gall midges (Cecidomyiidae) (Triplehorn and Johnson 2005) that are not the subject of this chapter.

The most important target pests of these egg parasitoids in Iran are true bugs of cereals and pistachio; serious pests of their host plants. Pistachio is exported to different Asian and European countries. On the other hand, wheat is a strategic crop in Iran that constitutes the highest item in the food basket of Iranian people. Self-sufficiency in wheat production is the first priority of Iran's Ministry of Agriculture,

S. Iranipour (✉)
University of Tabriz, Tabriz, Iran
e-mail: shiranipour@tabrizu.ac.ir

© The Author(s), under exclusive license to Springer Nature Switzerland AG 2021
J. Karimi, H. Madadi (eds.), *Biological Control of Insect and Mite Pests in Iran*,
Progress in Biological Control 18, https://doi.org/10.1007/978-3-030-63990-7_8

but unfortunately, it depends mostly on environmental factors including rainfall and also pest loss. Iran located at one of the driest areas of the world, so wheat importation is unavoidable some years. Common sunn pest *Eurygaster integriceps* Puton is the most serious insect pest of cereals and control measurements mainly focused on this key pest. However, the economic importance of the pest is not the same everywhere. It seems that the central plateau of Iran is the main hotspot of the pest throughout the world. The most serious invasions occur in central provinces such as Tehran (Varamin is a typical infested area), Markazi, Isfahan, Fars and rain-fed fields of western Iran such as Kermanshah, Hamedan, Kordestan and Lorestan (Radjabi 2000). Unfortunately, both groups are direct pests of grain and kernels. Considering the strategic and economic importance of these crops in Iran, the role of any control agent of key pests (such as the egg parasitoids) to suppress those pest populations will be obvious. Furthermore, the role of the egg parasitoids is outstanding among other natural enemies because pest suppression occurs prior to damage stage.

## 8.2    Iranian Platygastroidea

More than 70 species of this family were reported by researchers in Iran (Table 8.1) which *Telenomus* spp. and *Trissolcus* spp. are prevalent genera. Nevertheless, a comprehensive revision is needed. Some references in Table 8.1 (for example, Modarres-Awal 1997 and Ghahari et al. 2015) are checklists that just reflect reports of the other researchers, so they may repeat some incorrect data. As an early work, Safavi (1973), published an identification key to distinguishing six species of *Trissolcus* using some simple characters such as the presence of parapsidal grooves (notauli) and leg coloration. This identification key probably was used by subsequent users and lead to some mistakes in identifications because the key was very simplified. For example, all the species with yellowish legs in *semistriatus* species-group were distinguished as *T. basalis*. Consequently, some reports of *T. basalis* may be belonging to relative species such as *T. djadetshkoe*, *T. pseudoturesis*, etc. Just following to translation of Kozlov's (1988) key to English and further to Persian by Radjabi (2000) a more comprehensive key to Palearctic Platygastroidea became available. But this did not meet requirements of an intact identification because often access to holotypes or other type specimens was impossible to most Iranian researchers. As another example, the distinction of *T. semistriatus* from *T. grandis* commonly has been based on yellowish hind tibiae in former species. But one should not forget that *T. grandis* specimens tend to present lighter hind tibiae in warmer seasons (particularly in males) and separation of these two species with highly emphasize on leg coloration may be misleading. Finally, the absence of enough familiarity with key characters and different interpretations about one character by different users may be another source of the mistake. The present identification keys are ancient and relatively confusing for separating characters such as a relative ratio of the post marginal vein (pm) to the stigmal vein (st). For example in Kozlov's key

**Table 8.1** Iranian Platygastroidea and their hosts and distribution in Iran

| Taxon | Host | Area in Iran (Province) | References |
|---|---|---|---|
| Family Platygastridae (Haliday, 1840) | | | |
| 1. Subfamily Scelioninae (Forster, 1856) | | | |
| 1.1. Genus *Anteris* (Forster, 1856) | | | |
| 1.1.1. *Anteris simulans* (Kieffer, 1908) | – | 1. Kerman<br>2. West Azarbaijan | Samin et al. (2011a, b) and Ghahari et al. (2015) |
| 1.2. *Baeus* (Haliday, 1833) | | | |
| 1.2.1. *Baeus seminulum* (Haliday, 1833) | – | East Azarbaijan | Shamsi et al. (2014a) |
| 1.3. *Baryconus* (Foerster, 1856) | | | |
| 1.3.1. *Baryconus europaeus* (Kieffer, 1908) | – | East Azarbaijan | Shamsi (2014) |
| 1.4. *Calliscelio* (Ashmead, 1893) | | | |
| 1.4.1. *Calliscelio ruficollis* (Szeleyi, 1941) | – | East Azarbaijan | Shamsi (2014) |
| 1.5. Genus *Gryon* (Haliday, 1833) | | | |
| 1.5.1. *Gryon fasciatum* (Priesner, 1951) | *Eurygaster integriceps* Puton (Hem. Scutelleridae) | Golestan | Sakenin et al. (2008a), Samin et al. (2010c) and Ghahari et al. (2015) |
| 1.5.2. *Gryon monspeliense* (Picard, 1924) | 1. *Eurygaster integriceps* Puton | 1. Alborz | Radjabi and Amir-Nazari (1989), Iranipour et al. (1998a), Samin et al. (2010c), Samin and Asgari (2012b) and Ghahari et al. (2015) |
| | 2. *Dolycoris baccarum* L. (Hem. Pentatomidae) | 2. Hamadan<br>3. Isfahan<br>4. Lorestan<br>5. Markazi<br>6. Tehran | |
| 1.5.3. *Gryon muscaeformis* (Nees, 1834) | – | East Azarbaijan | Shamsi (2014) |

(continued)

**Table 8.1** (continued)

| Taxon | Host | Area in Iran (Province) | References |
|---|---|---|---|
| 1.5.4. *Gryon pedestre* (Nees, 1834) | 1. *Dolycoris penicillatus* Horváth (Hem. Pentatomidae) | 1. Isfahan | Samin et al. (2010c), Samin and Asgari (2012b) and Ghahari et al. (2015) |
|  | 2. *Eurygaster integriceps* Puton (Hem. Scutelleridae) | 2. Tehran |  |
| 1.5.5. *Gryon solutus* (Kononova & Petrov, 2001) | – | East Azarbaijan | Shamsi (2014) |
| 1.6. *Idris* (Foerster, 1856) |  |  |  |
| 1.6.1. *Idris aureonitens* (Szabo, 1965) | – | East Azarbaijan | Shamsi et al. (2014b) |
| 1.6.2. *Idris clypealis* (Huggert, 1979) | – | East Azarbaijan | Shamsi et al. (2014b) |
| 1.6.3. *Idris desertorum* (Priesner, 1951) | – | East Azarbaijan | Shamsi et al. (2014b) |
| 1.6.4. *Idris diversus* (Wollaston, 1858) | – | East Azarbaijan | Shamsi et al. (2014b) |
| 1.6.5. *Idris rufescens* (Kieffer, 1908) | – | East Azarbaijan | Shamsi et al. (2014b) |
| 1.7. Genus *Scelio* (Latreille, 1805) |  |  |  |
| 1.7.1. *Scelio flavibarbis* (Marshall, 1874) | *Locusta migratoria* (L.) (Orth. Acrididae) | 1. Golestan<br>2. Isfahan<br>3. Khuzestan | Khajehzadeh and Ghazavi (2000), Khajehzadeh (2002, 2004), Sakenin et al. (2008b), Ghahari et al. (2009, 2015), Samin et al. (2011a, 2012) and Samin and Asgari (2012a, b) |
| 1.7.2. *Scelio nitens* (Brues, 1906) | *Locusta migratoria* (L.) (Orth. Acrididae) | Isfahan | Sakenin et al. (2008b) and Ghahari et al. (2009, 2015) |
| 1.7.3. *Scelio poecilopterus* (Priesner, 1951) | – | Sistan-and-Baluchestan | Samin et al. (2011b) and Ghahari et al. (2015) |
| 1.7.4. *Scelio remaudierei* (Ferrière, 1952) | *Locusta migratoria* (L.) (Orth. Acrididae) | 1. Golestan<br>2. Isfahan<br>3. Kermanshah<br>4. Khuzestan | Ghahari et al. (2009, 2015), Samin et al. (2011a, 2012) and Samin and Asgari (2012b) |
| 1.7.5. *Scelio rugosulus*(Latreille, 1805) | – | East Azarbaijan | Shamsi et al. (2015a) |

(continued)

**Table 8.1** (continued)

| Taxon | Host | Area in Iran (Province) | References |
|---|---|---|---|
| 1.7.6. *Scelio zolotarevskyi* (Ferrière, 1930) | *Locusta migratoria* (L.) (Orth. Acrididae) | Mazandaran | Sakenin et al. (2008b) and Ghahari et al. (2009, 2015) |
| 1.8. Genus *Sparasion* (Latreille, 1802) | | | |
| 1.8.1. *Sparasion emarginatum* (Kieffer, 1906) | – | Khorasan | Samin et al. (2011b) and Ghahari et al. (2015) |
| 1.8.2. *Sparasion punctatissimum* (Kieffer, 1906) | – | Khorasan | Samin et al. (2011b) and Ghahari et al. (2015) |
| 1.8.3. *Sparasion subleve* (Kieffer, 1906) | – | Khorasan | Samin et al. (2011b) and Ghahari et al. (2015) |
| 2. Subfamily Telenominae (Thomson, 1860) | | | |
| 2.1. Genus *Paratelenomus* (Dodd, 1914) | | | |
| 2.1.1. *Paratelenomus saccharalis* (Dodd, 1914) | – | Khuzestan | Samin et al. (2012) and Ghahari et al. (2015) |
| 2.1.2. *Paratelenomus striativentris* (Risbec, 1950) | – | Khuzestan | Samin et al. (2012) and Ghahari et al. (2015) |
| 2.2. Genus *Psix* (Kozlov & Lê, 1976 | | | |
| 2.2.1. *Psix abnormis* (Kozlov & Lê, 1976) | – | 1. Golestan 2. Khorasan | Samin et al. (2011b), Samin and Asgari (2012a) and Ghahari et al. (2015) |
| 2.2.2. *Psix lacunatus* (Johnson & Masner, 1985) | – | 1. Isfahan 2. Kerman | Samin et al. (2011b), Samin and Asgari (2012b) and Ghahari et al. (2015) |
| 2.2.3. *Psix saccharicola* (Mani, 1941) | *Acrosternum arabicum* Wagner, (Hem. Pentatomidae) *Brachynema germari* (Kolenati) (Hem. Pentatomidae) | Kerman | Mehrnejad (2013) and Ghahari et al. (2015) |

(continued)

**Table 8.1** (continued)

| Taxon | Host | Area in Iran (Province) | References |
|---|---|---|---|
| 2.2.4. *Psix striaticeps* (Dodd, 1920) | 1. *Acrosternum heegeri* Fieber (Hem. Pentatomidae) | 1. Kerman | Hashemi-Rad et al. (2002), Hashemi-Rad (2008), Samin et al. (2011a) and Ghahari et al. (2015) |
| | 2. *Brachynema signatum* Jakovlev, (Hem. Pentatomidae) | 2. Kordestan | |
| | 3. *Croantha ornatula* (Herrich-Schaeffer) (Hem. Pentatomidae) | | |
| | 4. *Graphosoma lineatum* (L.) (Hem. Pentatomidae) | | |
| 2.3. Genus *Telenomus* (Haliday, 1833) | | | |
| 2.3.1. *Telenomus acrobates* (Giard, 1895) | *Chrysoperla carnea* (Stephens) (Neur. Chrysopidae) | 1. Chaharmahal–o-Bakhtiari | Talebi et al. (2005), Shahpouri-Arani et al. (2005), Ghahari et al. (2010, 2015), Samin et al. (2010a, d) and Samin and Asgari (2012b) |
| | | 2. East Azarbaijan | |
| | | 3. Isfahan | |
| | | 4. Tehran | |
| 2.3.2. *Telenomus angustatus* (Thomson, 1860) | *Chrysops* (*Petersenychrysops*) *hamatus* Loew (Dip. Tabanidae) | 1. East Azarbaijan | Samin et al. (2011b) and Shamsi (2014) |
| | | 2. Sistan–and-Baluchestan | |
| 2.3.3. *Telenomus benefactor* (Crawford, 1911) | – | Khorasan | Samin et al. (2011b) and Ghahari et al. (2015) |
| 2.3.4. *Telenomus beneficiens* (Zehntner, 1896) | *Scirpophaga novella* (F.) (Lep. Pyralidae) | Kerman | Samin et al. (2010d, 2011b) and Ghahari et al. (2015) |
| 2.3.5. *Telenomus busseolae* (Gahan, 1922) | 1. *Sesamia nonagrioides* Lefevbre (Lep. Noctuidae) | 1. East Azarbaijan | Abbasipour et al. (1991) and Abdul Razzagh (1995), Sharififar (2000), Abbasipour (2004), Narehi et al. (2004), Jamshidnia et al. (2009), Samin et al. (2010a, d, 2011a, 2012), Samin and Asgari (2012a, b), Sayad Mansour et al. (2011) and Ghahari et al. (2015) |
| | 2. *Sesamia cretica* Lederer (Lep. Noctuidae) | 2. Golestan | |
| | | 3. Isfahan | |
| | | 4. Khuzestan | |

(continued)

**Table 8.1** (continued)

| Taxon | Host | Area in Iran (Province) | References |
|---|---|---|---|
| 2.3.6. *Telenomus chloropus* (Thomson, 1861) | 1. *Eurygaster integriceps* Puton (Hem. Scutelleridae) | 1. Alborz | Shojai (1968), Mohaghegh-Neyshabouri (1993), Modarres-Awal (1997), Samin (2010), Samin et al. (2010a, b, c, d, 2011b, c), Samin and Asgari (2012b) and Ghahari et al. (2015) |
| | 2. *Eurygaster testudinaria* (Geoffroy) (Hem. Scutelleridae) | 2. East Azarbaijan | |
| | 3. *Dolycoris baccarum* L. (Hem. Pentatomidae) | 3. Isfahan | |
| | | 4. Khorasan | |
| | | 5. Mazandaran | |
| | | 6. Tehran | |
| 2.3.7. *Telenomus chrysopae* (Ashmead, 1893) | 1. *Chrysoperla* sp. (Neuroptera: Chrysopidae) | 1. East Azarbaijan | Rakhshani et al. (2008), Samin et al. (2010a, d) and Ghahari et al. (2015) |
| | 2. *Catolaccus* sp. (Hymenoptera: Pteromalidae) | 2. Golestan | |
| | | 3. Isfahan | |
| 2.3.8. *Telenomus dignus* (Gahan, 1925) | *Scirpophaga* sp. (Lepidoptera: Pyralidae) | Kerman | Samin et al. (2010d, 2011b) and Ghahari et al. (2015) |
| 2.3.9. *Telenomus harpyiae* (Mayr, 1879) | – | East Azarbaijan | Shamsi (2014) |
| 2.3.10. *Telenomus heydeni* (Mayr, 1879) | *Lixus incanescens* Boheman (Coleoptera: Curculionidae) | 1. West Azarbaijan | Parvizi and Javan Moghaddam (1988), Samin et al. (2010d, 2011a), Samin and Asgari (2012b) and Ghahari et al. (2015) |
| | | 2. Isfahan | |
| 2.3.11. *Telenomus hofmani* (Matr, 1879) | – | East Azarbaijan | Shamsi (2014) |
| 2.3.12. *Telenomus minimus* (Ashmead, 1893) | – | 1. Guilan | Modarres-Awal (1997), Samin et al. (2010d) and Ghahari et al. (2015) |
| | | 2. Mazandaran | |
| 2.3.13. *Telenomus pentopherae* (Mayr, 1879) | – | East Azarbaijan | Shamsi (2014) |
| 2.3.14. *Telenomus phalaenarum* (Nees & Esenbeck, 1834) | – | Kermanshah | Samin et al. (2010d, 2011a) and Ghahari et al. (2015 |
| 2.3.15. *Telenomus politus* (Thomson, 1861) | *Eurygaster integriceps* Puton | 1. Isfahan | Modarres-Awal (1997), Samin et al. (2010c, d), Samin and Asgari (2012b) and Ghahari et al. (2015) |
| | | 2. Tehran | |

(continued)

**Table 8.1** (continued)

| Taxon | Host | Area in Iran (Province) | References |
|---|---|---|---|
| 2.3.16. *Telenomus punctatissimus* (Ratzeburg, 1844) | – | East Azarbaijan | Shamsi (2014) |
| 2.3.17. *Telenomus remus* (Nixon, 1937) | – | Sistan-and-Baluchestan | Samin et al. (2011b) and Ghahari et al. (2015) |
| 2.3.18. *Telenomus sechellensis* (Kieffer, 1910) | – | Khuzestan | Samin et al. (2010d, 2011a, 2012) and Ghahari et al. (2015) |
| 2.4. Genus *Trissolcus* (Ashmead, 1893) | | | |
| 2.4.1. *Trissolcus agriope* (Kozlov & Lê, 1976) | Hemiptera: Pentatomidae: | Kerman | Hashemi-Rad et al. (2000a, b, 2002), Hashemi-Rad (2008), Ghahari et al. (2011, 2015) and Mehrnejad (2013) |
| | 1. *Brachynema* sp. | | |
| | 2. *Acrosternum arabicum* Wagner | | |
| | 3. *Brachynema germari* (Kolenati) | | |
| 2.4.2. *Trissolcus basalis* (Wollaston, 1858) | Hemiptera: Pentatomidae: | 1. East Azarbaijan | Shojai (1968), Martin et al. (1969), Radjabi and Amir-Nazari (1989), Iranipour et al. (1998a), Taghaddosi and Radjabi (1998), Mehravar et al. (2000), Samin (2010), Samin et al. (2010a, c, 2011a, b, c, 2012), Ghahari et al. (2011, 2015) and Samin and Asgari (2012a, b) |
| | 1. *Aelia acuminata* (L.) | 2. Golestan | |
| | 2. *Apodiphus amygdali* (Germar) | 3. Hamadan | |
| | 3. *Carpocoris fuscipinus* (Boheman) | 4. Lorestan | |
| | 4. *Dolycoris baccarum* (L.) | 5. Markazi | |
| | 5. *Eurygaster integriceps* Puton | 6. Isfahan | |
| | 6. *Eurygaster maura* (L.) | 7. Kermanshah | |
| | 7. *Nezara viridula* (L.) | 8. Khorasan | |
| | | 9. Khuzestan | |
| | | 10. Tehran | |
| | | 11. Zanjan | |
| 2.4.3. *Trissolcus cantus* (Kozlov & Le, 1977) | – | East Azarbaijan | Shamsi (2014) |
| 2.4.4. *Trissolcus cephalotes* (Kozlov & Le, 1977) | – | East Azarbaijan | Shamsi (2014) |

(continued)

**Table 8.1** (continued)

| Taxon | Host | Area in Iran (Province) | References |
|---|---|---|---|
| 2.4.5. *Trissolcus circus* (Kozlov & Lê, 1976) | – | Tehran | Samin et al. (2010b) and Ghahari et al. (2015) |
| 2.4.6. *Trissolcus crypticus* (Clarke, 1993) | *Nezara viridula* (L.) | Sistan–and-Baluchestan | Samin et al. (2011b) and Ghahari et al. (2011, 2015) |
| 2.4.7. *Trissolcus delucchii* (Kozlov, 1968) | 1. *Sesamia nonagrioides* Lefevbre (Lepidop-tera: Noctuidae) | 1. Alborz | Modarres-Awal (1997), Iranipour et al. (1998b), Samin et al. (2010c) and Ghahari et al. (2011, 2015) |
| | 2. *Eurygaster integriceps* Puton | 2. Khuzestan | |
| | 3. *Eurygaster maura* (L.) (Hemiptera: Pentatomidae) | 3. Tehran | |
| 2.4.8. *Trissolcus djadetshko* (Rjachovsky, 1959) | 1. *Eurydema ornatum* (L.) | 1. Ardabil | Sakeniń et al. (2008a), Nozad-Bonab and Iranipour (2010), Samin (2010), Samin et al. (2010a, b, c, 2011a, b, 2012), Hejazi et al. (2011), Shafaei et al. (2011), Ghahari et al. (2011, 2015) and Samin and Asgari (2012a, b) |
| | 2. *Eurygaster integriceps* Puton (Hem: Pentatomidae) | 2. East Azarbaijan | |
| | | 3. Golestan | |
| | | 4. Isfahan | |
| | | 5. Kerman | |
| | | 6. Khuzestan | |
| | | 7. Mazandaran | |
| | | 8. Tehran | |
| | | 9. West Azarbaijan | |
| 2.4.9. *Trissolcus dryope* (Kozlov & Lê, 1976) | Hemiptera: Pentatomidae: 1. *Acrosternum* sp. | 1. Isfahan | Hashemi-Rad et al. (2000b, 2002), Ghahari et al. (2011, 2015) and Samin and Asgari (2012b) |
| | 2. *Brachynema* sp. | 2. Kerman | |
| 2.4.10. *Trissolcus esmailii* (Radjabi, 2001) | Hemiptera: Pentatomidae: 1. *Dolycoris baccarum* L. | 1. Fars | Radjabi (2001), Samin et al. (2010c), Ghahari et al. (2011, 2015) and Samin and Asgari (2012b) |
| | 2. *Eurygaster integriceps* Puton | 2. Isfahan | |
| 2.4.11. *Trissolcus festivae* (Viktorov, 1964) | Hemiptera: Pentatomidae: 1. *Eurydema ornatum* L. | 1. Alborz | Radjabi (1994), Modarres-Awal (1997), Iranipour et al. (1998a), Samin (2010), Samin et al. (2010a, b, c, 2011b, c), Ghahari et al. (2011, 2015) and Samin and Asgari (2012a, b) |
| | 2. *Eurygaster integriceps* Puton | 2. Ardabil | |
| | | 3. Golestan | |
| | | 4. Isfahan | |
| | | 5. Kerman | |

(continued)

**Table 8.1** (continued)

| Taxon | Host | Area in Iran (Province) | References |
|---|---|---|---|
| | | 6. Markazi | |
| | | 7. Qazvin | |
| | | 8. Zanjan | |
| | | 9. Tehran | |
| 2.4.12. *Trissolcus grandis* (Thomson, 1861) | Hemiptera: Pentatomidae: 1. *Aelia acuminata* (L.) | 1. Alborz | Zomorrodi (1962), Shojai (1968) and Martin et al. (1969), Radjabi and Amir-Nazari (1989), Asgari et al. (1995, Noorbakhsh and Razavi (1995), Iranipour et al. (1998b), Allahyari and Azmayeshfard (2002), Haghshenas (2004), Nozad-Bonab and Iranipour (2010), Samin (2010), Samin et al. (2010a, b, c, 2011a, b, c, 2012), Fathi et al. (2011), Ghahari et al. (2011, 2015), Hejazi et al. (2011), Noori et al. (2011), Shafaei et al. (2011) and Samin and Asgari (2012a, b) |
| | 2. *Apodiphus amygdali* (Germar) | 2. Ardabil | |
| | 3. *Eurygaster integriceps* Puton | 3. Chaharmahal-and-Bakhtiari | |
| | 4. *Graphosoma lineatum* (L.) | 4. East Azarbaijan | |
| | 5. *Carpocoris fuscipinus* (Boheman) | 5. Golestan | |
| | 6. *Dolycoris baccarum* (L.) | 6. Ilam | |
| | 7. *Eurygaster maura* (L.) | 7. Kermanshah | |
| | 8. *Podisus maculiventris* (Say) | 8. Kordestan | |
| | 9. *Andrallus spinidens* (F.) | 9. Isfahan | |
| | 10. *Perillus bioculatus* (F.) | 10. Kerman | |
| | | 11. Khorasan | |
| | | 12. Sistan-and-Baluchestan | |
| | | 13. Khuzestan | |
| | | 14. Mazandaran | |
| | | 15. Qazvin | |
| | | 16. Tehran | |
| | | 17. West Azarbaijan | |
| 2.4.13. *Trissolcus manteroi* (Kieffer, 1909) | Hemiptera: Pentatomidae: | 1. Mazandaran | Sakenin et al. (2008a), Samin et al. (2010c, 2011a) and Ghahari et al. (2011, 2015) |
| | 1. *Carpocoris coreanus iranus* (Tamanini) | 2. West Azarbaijan | |
| | 2. *Dolycoris penicillatus* Horváth | | |

(continued)

**Table 8.1**  (continued)

| Taxon | Host | Area in Iran (Province) | References |
|---|---|---|---|
| 2.4.14. *Trissolcus mentha* (Kozlov & Lê, 1977) | Hemiptera: Pentatomidae: | 1. Alborz | Iranipour et al. (1998b), Hashemi-Rad et al. (2002) and Samin et al. (2010c) |
| | 1. *Eurygaster integriceps* Puton | 2. Kerman | |
| | 2. *Apodyphus amygdali* (Germar) | 3. Tehran | |
| 2.4.15. *Trissolcus* sp. near *mitsukurii* (Ashmead, 1904) | Hemiptera: Pentatomidae: *Brachynema* spp | Kerman | Hashemi-Rad et al. (2000b) |
| 2.4.16. *Trissolcus pseudoturesis* (Rjachovsky, 1959) | Hemiptera: Pentatomidae: | 1. Khuzestan | Sakenin et al. (2008a), Samin et al. (2010c, 2011a, 2012) and Ghahari et al. (2011, 2015) |
| | 1. *Eurygaster integriceps* Puton, | 2. Kordestan | |
| | 2. *Eurygaster testudinaria* (Geoffroy) | 3. Mazandaran | |
| 2.4.17. *Trissolcus radjabii* (Iranipour, 2010) | Hemiptera: Pentatomidae: *Apodiphus amygdali* Germar | 1. Alborz | Iranipour and Johnson (2010) and Ghahari et al. (2011, 2015) |
| | | 2. East Azarbaijan | |
| | | 3. Kerman | |
| | | 4. Tehran | |
| 2.4.18. *Trissolcus rufiventris* (Mayr, 1908) | Hemiptera: Pentatomidae: | 1. Alborz | Shojai (1968), Martin et al. (1969), Radjabi and Amir-Nazari (1989), Modarres-Awal (1997), Iranipour et al. (1998b), Mehravar et al. (2000), Samin (2010), Samin et al. (2010a, b, c, 2011a, b, c, 2012), Hejazi et al. (2011), Ghahari et al. (2011, 2015), Shafaei et al. (2011), Samin and Asgari (2012b) and Shamsi (2014) |
| | 1. *Aelia furcula* Fieber | 2. East Azarbaijan | |
| | 2. *Dolycoris penicillatus* Horváth | 3. Hamadan | |
| | 3. *Eurygaster integriceps* Puton | 4. Lorestan | |
| | | 5. Markazi | |
| | | 6. Mazandaran | |
| | | 7. Isfahan | |
| | | 8. Kerman | |
| | | 9. Kermanshah | |
| | | 10. Khuzestan | |
| | | 11. Tehran | |
| | | 12. West Azarbaijan | |
| 2.4.19. *Trissolcus saakowi* (Mayr, 1903) | *Apodiphus* sp. (Hemiptera: Pentatomidae) | Alborz | Iranipour et al. (1998a) and Ghahari et al. (2011) |

(continued)

**Table 8.1** (continued)

| Taxon | Host | Area in Iran (Province) | References |
|---|---|---|---|
| 2.4.20. *Trissolcus semistriatus* (Nees, 1834) | Hemiptera: Pentatomidae: | 1. Alborz | Alexandrov (1948a, b), Vaezi (1950), Zomorrodi (1962), Martin et al. (1969), Safavi (1973), Radjabi and Amir-Nazari (1989), Modarres-Awal (1997), Iranipour et al. (1998b), Taghaddosi and Radjabi (1998), Noori and Asgari (2000), Asgari and Sahragard (2002), Haghshenas (2004), Sakenin et al. (2008a), Nozad-Bonab and Iranipour (2010), Samin (2010), Samin et al. (2010a, b, c, 2011a, b, c, 2012), Ghahari et al. (2011, 2015), Hejazi et al. (2011), Noori et al. (2011), Shafaei et al. (2011) and Samin and Asgari (2012a, b) |
| | 1. *Aelia acuminata* (L.) | 2. Ardabil | |
| | 2. *Apodiphus amygdali* (Germar) | 3. Chaharmahal–and-Bakhtiari | |
| | 3. *Dolycoris baccarum* L. | 4. East Azarbaijan | |
| | 4. *Eurygaster integriceps* Puton | 5. Fars | |
| | 5. *Carpocoris fuscipinus* (Boheman) | 6. Hamadan | |
| | 6. *Eurygaster maura* (L.) | 7. Lorestan | |
| | 7. *Graphosoma lineatum* (L.) | 8. Markazi | |
| | 8. *Carpocoris pudicus* (Poda) | 9. Golestan | |
| | 9. *Holcostethus sphacelatus* (F.) | 10. Ilam | |
| | 10. *Carpocoris pudicus* (Poda) | 11. Kermanshah | |
| | 11. *Holcostethus sphacelatus* (F.) | 12. Kordestan | |
| | | 13. Isfahan | |
| | | 14. Kerman | |
| | | 15. Sistan–and-Baluchestan | |
| | | 16. Khorasan | |
| | | 17. Khuzestan | |
| | | 18. Mazandaran | |
| | | 19. Qazvin | |
| | | 20. Tehran | |
| | | 21. West Azarbaijan | |
| | | 22. Zanjan | |
| 2.4.21. *Trissolcus simoni* (Mayr, 1879) | Hemiptera: Pentatomidae: | 1. Alborz | Shojai (1968), Modarres-Awal (1997), Iranipour et al. (1998b), Sakenin et al. (2008a) and Samin (2010), Samin et al. (2010a, b, c, 2011a, b, c, 2012), Ghahari et al. (2011, 2015) and Samin and Asgari (2012b) |
| | 1. *Aelia acuminata* (L.) | 2. Ardabil | |
| | 2. *Dolycoris baccarum* L. | 3. Golestan | |

(continued)

**Table 8.1** (continued)

| Taxon | Host | Area in Iran (Province) | References |
|---|---|---|---|
| | 3. *Aelia melanota* (Fieber) | 4. Isfahan | |
| | 4. *Apodiphus amygdali* (Germar) | 5. Kerman | |
| | 5. *Carpocoris fuscipinus* (Boheman) | 6. Khuzestan | |
| | 6. *Eurydema ornatum* L. | 7. Mazandaran | |
| | 7. *Eurygaster integriceps* Puton | 8. Tehran | |
| 2.4.22. *Trissolcus tumidus* (Mayr, 1879) | Hemiptera: Pentatomidae: | 1. Fars | Zomorrodi (1962), Shojai (1968), Modarres-Awal (1997), Samin et al. (2010c, 2011a) and Ghahari et al. (2011, 2015) |
| | 1. *Aelia acuminata* (L.) | 2. Isfahan | |
| | 2. *Apodiphus amygdali* (Germar) | 3. Tehran | |
| | 3. *Carpocoris fuscipinus* (Boheman) | 4. Zanjan | |
| | 4. *Dolycoris baccarum* (L.) | 5. West Azarbaijan | |
| | 5. *Eurygaster integriceps* Puton | | |
| 2.4.23. *Trissolcus vassilievi* (Mayr, 1903) | Hemiptera: Pentatomidae: | 1. Alborz | Alexandrov (1948a, b), Vaezi (1950), Zomorrodi (1962), Shojai (1968), Martin et al. (1969), Safavi (1973), Farid (1985), Radjabi and Amir-Nazari (1989), Asgari et al. (1995), Modarres-Awal (1997), Iranipour et al. (1998b), Taghaddosi and Radjabi (1998), Mansour-Ghazi and Radjabi (2000), Noori and Asgari (2000), Haghshenas (2004), Sakenin et al. (2008a), Samin (2010), Samin et al. (2010a, b, c, 2011a, b, c, 2012), Ghahari et al. (2011, 2015), Hejazi et al. (2011), Shafaei et al. (2011) and Samin and Asgari (2012a, b) |
| | 1. *Aelia acuminata* (L.) | 2. Chaharmahal-and-Bakhtiari | |
| | 2. *Apodiphus amygdali* (Germar) | 3. East Azarbaijan | |
| | 3. *Carpocoris fuscispinus* (Boheman) | 4. Fars | |
| | 4. *Dolycoris baccarum* L. | 5. Hamadan | |
| | 5. *Eurygaster integriceps* Puton | 6. Kermanshah | |
| | 6. *Graphosoma lineatum* (L.) | 7. Lorestan | |
| | 7. *Eurygaster maura* (L.) | 8. Markazi | |
| | 8. *Carpocoris mediterraneus* (Tamanini), | 9. Golestan | |

(continued)

**Table 8.1** (continued)

| Taxon | Host | Area in Iran (Province) | References |
|---|---|---|---|
| | 9. *Graphosoma semipunctatum* (F.) | 10. Isfahan | |
| | | 11. Kerman | |
| | | 12. Khorasan | |
| | | 13. Khuzestan | |
| | | 14. Kordestan | |
| | | 15. Mazandaran | |
| | | 16. Qazvin | |
| | | 17. Tehran | |
| | | 18. West Azarbaijan | |
| | | 19. Zanjan | |
| 2.4.24. *Trissolcus volgensis* (Viktorov, 1964) | Hemiptera: Pentatomidae: | Kerman | Mehrnejad (2013) and Ghahari et al. (2015) |
| | 1. *Acrosternum arabicum* (Wagner) | | |
| | 2. *Brachynema germari* (Kolenati) | | |
| 3. Subfamily Teleasinae (Ashmead, 1902) | | | Rahnemaye-Shahsavari et al. (2011) |
| 3.1. Genus *Proteleas* (Kozlov, 1961) | | | |
| 3.1.1. *Proteleas rugosus* (Kozlov, 1961) | – | East Azarbaijan | Shamsi (2014) |
| 3.2. Genus *Teleas* (Latreille, 1805) | | | |
| 3.2.1. *Teleas rugosus* (Kieffer, 1908) | – | East Azarbaijan | Shamsi et al. (2015b) |

one may decide if pm is twice as long as st or 1.8 times longer than it. A small error in measurement may lead to an erroneous decision and misidentification. We need a more precise key and might be necessary to add molecular traits and more detailed morphological characters in future work. This may necessitate a comprehensive study and revision of previous works.

Among platygastroid species, just some species, of *Trissolcus* were used for control of sunn-pests' complex (*Eurygaster* spp. and on a smaller scale *Aelia* spp., *Dolycoris* spp., *Carpocoris* spp. etc) in cereal fields (Alexandrov 1948a, b; Martin et al. 1969; Safavi 1973; Radjabi 2000). A potential target pest complex also may be true bugs related to pistachio, with a rich fauna of these parasitoids upon them reported in Kerman province (Hashemi-Rad et al. 2000a, b, 2002). Although *Nezara viridula* L.-*Trissolcus basalis* (Wollaston) system is also a successful biological control system in some parts of the world (Jones 1988), and records of both species are present in Iran (see Modarres-Awal 1997), but no attempt was done by Iranian researchers for studying this system, maybe due to lower importance in comparison to bollworm *Helicoverpa* spp.

## 8.3   Egg Parasitoids of Sunn Pests

Several species of true bugs belonging to *Eurygaster* (Hem. Scutelleridae), *Aelia*, *Dolycoris*, and *Carpocoris* (Hem. Pentatomidae) attack cereals in Iran (Salavatian 1991; Radjabi 2000). Undoubtedly, *Eurygaster* species are the most important among them. Three species of *Eurygaster* exist in Iran including *E. integriceps* Puton, *E. maura* L. and *E. testudinaria* Geoffroy. The latter species so far reported from Tabriz (north-west of the country; Brown & Eralp, 1962) and Mazandaran (Northern Iran; Mohaghegh-Neyshabouri, 1993) which is a rare species without economic importance. Just a single species of egg parasitoids namely *Telenomus chloropus* (Thomson) was reported upon it (Mohaghegh-Neyshabouri 1993). The European sunn pest *E. maura* L. is distributed in Northern provinces (predominantly Golestan, followed by Mazandaran, Guilan, East and West Azarbaijan; Radjabi 2000), although it seems that it has displaced by *E. integriceps* at recent years in north-western provinces (East and West Azarbaijan). The economic importance of this species is lower than the more virulent species of *E. integriceps* that known as common sunn pest or sunn pest and is prevalent in the central plateau of Iran. At least four species of *Aelia* (*A. furcula* Fieb., *A. melanota* Fieb, *A. virgata* Klug, and *A. accuminata* L.), two species of *Dolycoris* (*D. baccarum* (L.) and *D. penisilatus* (Horvath)), and *Carpocoris fuscispinus* (Boh.) also attack cereal fields in Iran.

Two groups of natural enemies are important in the sunn-pest complex. The first one egg parasitoids belonging mainly to Platygastroidea and a few species distributed in some families of Chalcidoidea mainly Encyrtidae. The second one nymph-adult parasitoids belonging to Phasiinae flies (Dip. Tachinidae) (Radjabi 2000). However, our focus is on the most important of them platygastroid egg parasitoids in this chapter. A rich fauna of Platygastroidea, as well as, encyrtids are active on sunn pest eggs. Regional diversity of sunn pests' egg parasitoids has been studied by many researchers (Table 8.1). These studies show that in different parts of the country, miscellaneous species established and there is spatial and temporal heterogeneity among regions, and even among adjacent localities (Radjabi and Amir-Nazari 1989; Shafaei et al. 2011). For example in 20 places out of 37 places of five provinces (Tehran, Alborz, Hamedan, Markazi and Lorestan) studied by Radjabi and Amir-Nazari (1989), *Trissolcus grandis* was dominant in number followed by *T. semistriatus* (13 places), *T. vassilievi* (3 places) and *T. basalis* just at one place. These four species, as well as *T. rufiventris* and *Ooencyrtus telenomicida* (Hym. Encyrtidae) are common species that are encountered in major parts of Iran.

However, just three species of *T. grandis*, *T. semistriatus* and *T. vassilievi* are considered as main candidates for biological control of the sunn-pest complex.

## 8.3.1 Factors Influencing Species Richness and Distribution

Weather condition is the most important factor that influences geographical distribution and the parasitism rate. For example, the parasitism rate is very low in dry regions and increases around water resources like irrigation channels. In addition, parasitism rate is always lower in rain-fed fields (Martin et al. 1969; Radjabi 2000). On the other hand, lower parasitism was recorded in colder regions like Chaharmahal-o-Bakhtiari and East Azarbaijan provinces (Haghshenas 2004; Nozad-Bonab and Iranipour 2010). In a laboratory study, an equal number of three species of *T. grandis*, *T. semistriatus* and *T. vassilievi* was released on *E. integriceps* eggs at 25, 30 and 34 °C and 10, 33, 55 and 85% RH regimes. *T. semistriatus* preceded other species just at 30 °C and 85% RH, while it declined to zero at 34 °C. In other RH regimes at 30 °C *T. vassilievi* preceded the others and *T. grandis* preceded at all RH regimes at extreme temperatures (Radjabi 2000). This may imply a wider thermal range of the recent species. In fact, *T. grandis* is a prevalent species in many regions in Iran. It can found equally in cold and warm conditions or dry and humid ones. This species recorded as prevalent species in Golestan province on *E. maura* (Iranipour et al. 1998a), which is one of the most humid regions of the distribution of sunn pest in Iran. It is also dominant species in cold regions such as northwestern provinces (Shafaei et al. 2011; Nozad-Bonab and Iranipour 2013) as well as warmer and drier region such as Varamin (Iranipour et al. 2011). In recent place, however, *T. vassilievi* and *T. grandis* precede the other species at different years. This is mainly due to temporal density dependence of *T. vassilievi* that led this species to exceed *T. grandis* in those years that outbreak occurs. *T. grandis* caused density independently 19.5–29.2% egg mortality, while *T. vassilievi* killed density dependently 0–59.7% at different place-years. Although Iranipour et al. (2011) observed no temporal reaction of *T. grandis* in this study, spatial density dependence was recorded in another study (Amir-Maafi and Parker 2002). *T. vassilievi* was the second key factor with a very weaker impact on host population compared to adult mortality in diapause phase, which was the first key factor in population dynamics of the sunn pest in Varamin fields (Iranipour et al. 2011).

Safavi (1973) believes that *T. vassilievi* is prevalent species of southern provinces of Iran and has a tendency toward the dry and warm condition. However, In Isfahan with a similar climate to Varamin, *T. vassilievi* has not reported yet. On the other hand, in colder provinces such as East and West Azarbaijan, this species is an occasional parasitoid of sunn pest (Shafaei et al. 2011; Nozad-Bonab and Iranipour 2010).

Hence, different combinations of species are active in different regions. For example, in Alborz wheat fields, *T. grandis* reported as prevalent species and seven other species (*T. semistriatus*, *T. vassilievi*, *T. basalis*, *T. rufiventris*,

*T. festivae*, *Gryon monspeliensis* and *Ooencyrtus telenomicida*) also parasitized sunn pest eggs. *T. grandis* emerges prior to the remnant species, but often leaves wheat field late spring and before harvesting. Except for *T. grandis*, the other species had no determined trend and some were rare (Such as *T. festivae* and particularly *G. monspelinsis*). *T. basalis*, *T. rufiventris* and *O. telenomicida* often were seen late spring and predominated in number when *T. grandis* left the cereal fields toward other crops, adjacent orchards, and shade trees at June. Harsh condition of June (dry and warm) lead *T. grandis* to leave cereal fields and inferior species become prevalent. The majority of these species also trapped in other field crops later. Three species also were added in orchards and shade trees including *T. delucchii*, *T. radjabii* and *Anastatus bifasciatus* Forcroy (Hym.: Eupelmidae). The latter species included only males. Although, these species can successfully develop within sunn pest eggs, they never have seen in cereal fields. Their target hosts are other bugs and moths (Iranipour et al. 1998b).

The richness of parasitoids in East Azarbaijan province is poorer. In this province, just five species collected in fields; *T. grandis* is prevalent again, followed by *T. djadetshkoe*, with late emergence of *O. telenomicida* and *O. fecundus*. *T. vassilievi* also is an occasional parasitoid (Nozad-Bonab and Iranipour 2010).

In Golestan province during spring, *T. grandis* (60%) followed by *T. simoni* (40%) was observed by sweeping in a single sampling date (Iranipour et al. 1998a). This record is important because *T. simoni* seems has a very restricted distribution in Iran (northern province of Golestan with relatively high precipitation and relative humidity). The previous record from Varamin with a hot and dry climate (Shojai 1989) is not confirmed by a subsequent study in a four-year interval (Iranipour et al. 2011). The mentioned record may be related to a relative species, *T. festive* with an elegant difference in scutellar sculptures. This species is a specific parasitoid of *Eurydema* spp. and seldom attack sunn pest eggs and two records of it are present in other parts of Tehran province and adjacent Alborz province (Radjabi 1994; Iranipour 1996). Another important record is *T. agriopae* by Mehravar et al. (2000). This was the first record of a species from *oobius* group of Kozlov and Kononova (1983) in Iran. One probable explanation for observed difference among different regions is thermal requirements of different species (see Sect. 8.5.1.3).

## 8.3.2  How Competing Parasitoids Co-exist?

Different thermal requirements may be also a possible mechanism of co-existence and avoiding competitive displacement, although other separation mechanisms such as host range, density dependence, etc. also can be involved. Shafaei et al. (2011) showed that *T. grandis* and *T. djadetshkoe* are only species that continuously present in West Azarbaijan fields during the growing season. Three overlapping generations of the former embedded two mainly separate generations of the latter. Peaks of the *T. djadetshkoe* population coincides *T. grandis* declines. Thus, *T. djadetshkoe* appears later and terminates its activity sooner or simultaneously and therefore have a shorter duration of parasitism. This sequence of generations let the two species co-exist.

Co-existence mechanisms need further be studied. Salavatian (1991) believed that 5–6 generation a year spent by wasps which 2–3 of them spent on other bugs. Two to three generations in cereal fields accepted by authors (Safavi 1973; Zatyamina and Kletchkovsky 1974; Radjabi 1994, 2000).

### 8.3.3   Aestivation

The parasitic wasps leave cereal fields and emigrate toward adjacent habitats at the end of growing season and even prior to it while cereals are going to ripe and are often semi-green. Other crops, green weeds around the channels, orchards and shade trees are the places where parasitoids inhabit (Safavi 1959; Viktorov 1967; Martin et al. 1969; Zatyamina et al. 1976; Radjabi and Amir-Nazari 1989; Salavatian 1991; Asgari 1995; Iranipour 1996). More temperate and shaded climate, presence of nectar, honeydew and water sources as well as alternative hosts such as *Apodiphus*, *Brachynema*, *Carpocoris*, *Chroanta*, *Dolycoris*, *Eurydema*, *Graphosoma*, *Holcostethus*, *Piezodorus*, *Stollia*, . . . are crucial for survival during aestivation. Alfalfa fields are among the most desirable habitats within which parasitoids spend considerable time even during spring (Zatyamina and Burakova 1980). Parasitism rates higher than cereal fields were recorded by egg traps tied to plants (Iranipour 1996). Many species of parasitoids are common among cereal fields and aestivation sites but they are observed in fewer number and discontinueously in resting sites. In addition, some species only inhabit trees and often emerge late-spring or early summer coincides to the arrival of sunn pest parasitoids. Alternative hosts have an important role in the enrichment of overwintering populations. No reproductive diapause was recorded in these wasps. However, in some regions such as Isfahan, parasitism was not observed in aestivation sites at all (Safavi 1973).

### 8.3.4   Hibernation

The parasitoids hibernate inside bark clefts of different trees close to ground level situated in vicinity of cereal fields. What kind of tree is not important, the only physical trait of bark is crucial. Elder trees with thicker barks are preferable. In such places, parasitoids hibernate alone or in small groups including 10–30 virgin or inseminated females of different species (Safavi 1973). Romanova (1953) and Radjabi and Amir-Nazari (1989) showed that the presence of trees is not a necessity in overwintering and parasitoids can survive in perennial weeds, soil crevices etc. A small reservoir of parasitoid passes into the next year and often emerges prior to host (Martin et al. 1969; Popov and Paulian 1971; Safavi 1973; Iranipour 1996).

## 8.3.5 Host-Parasitoid Synchrony Problem

The wasps emerge when the temperature reaches to 13 °C in spring, begin to feed on wild flowers around fields, and then enter cereal fields. Up to three weeks later, their hosts arrive (Şimşek 1986; Şimşek and Yaşarakinci 1990). Such a delay has a negative impact on parasitoids. Effect of such a delay studied by Bazavar et al. (2015; see Sect. 8.5.2). Discontinuous access may occur when the density of host is scarce and female wasp has to spend some time to find an egg mass. Delayed access followed by subsequent unlimited access is possible when the delay in host arrival occurs. Salavatian (1991) also believed that pioneer eggs are not attacked due to colder condition of early spring. In fact, early-season parasitism or parasitism in the colder regions is always low (Amir-Maafi 2000; Haghshenas 2004; Nozad-Bonab and Iranipour 2010; Iranipour and Kharrazi-Pakdel 2011). Unfortunately, pioneer eggs have a more important role in crop damage compared to subsequent ones (Radjabi 1995). Some other reasons also are present to explain the low rate of parasitism early season. For example, a longer pre-oviposition period of host and density dependence of parasitoid may decrease the attack rate. In a multi-patch experiment, it was shown that a female *T. vassilievi* spent 4.3, 13.0, 19.1, 23.9, and 18.4% of her time in patches with 1, 2, 4, 8 and 15 clutches and 21.3% out of patches. Corresponding parasitism rate in those patches was 4.0, 13.3, 22.5, 34.2 and 25.9% respectively. When parasitoid was alone no one-clutch patch was met. Other patches were met equally. This may suggest a threshold density of two clutches, which parasitoid can detect chemical cues. Lower amount of signals are neglected and thus parasitoid avoids exploitation in low-density patches except in random encounters. Inclusion of additional foraging females in this experiment caused the parasitoids disperse more severely among the patches and hence, the one-clutch patch also was met. This is clearly due to mutual interference among foraging parasitoids. Pseudo-interference had a stronger influence on per capita attack rate. Actual interference had only a minor role in per capita exploitation. These results suggest that interference is an unimportant force in sunn pest-egg parasitoid systems (Iranipour et al. 2020). The parasitoids encounter other kinds of problems late season. They will encounter low quality hosts (parasitized eggs or developed host-embryogenesis).

## 8.3.6 Indirect Effects of Parasitism

In addition to direct mortality, parasitic wasps can disturb synchrony between most vulnerable stage of the crop (ripening seeds) and most injurious stage of the pest (adults and higher instar nymphs). Radjabi (1995) showed that higher parasitism rate disturbs this synchrony. In those years or places where parasitism is high, stage-structure of the pest tend to consist of younger stages at ripening. In this regards, early-season parasitism is very crucial. So in the future, our attempts should be focused on enhancemant of early-season parasitism. In this regards selection or breeding parasitoids with the lower thermal threshold for reproduction and development may be useful. Iranipour et al. (2010, 2015) compared two populations of *T. grandis* and *T. vassilievi* respectively in terms of their thermal phenotypes. The more detectable difference was observed in *T. grandis*. Reciprocal crosses between two populations of the latter species revealed that thermal phenotypes are inheritable and a maternal effect was observed. This may show that we can find, select or breed parasitoids with superior traits among different populations and enhance early season parasitism.


## 8.4 Augmentation Efforts

Applied biological control against sunn pest started in Varamin in 1941. An Iranian entomologist (Kowsari) harvested parasitized eggs from an infested field in Garmsar and then released them in an uninfested field in Varamin. The first augmentative effort was carried out at 1947 in Varamin. However, due to inappropriate conditions in Varamin (drought and absence of wide spread orchards and/or shade trees), the control measurements was ceased and re-established in Isfahan. Reports showed successful control in Isfahan in a time interval as long as nine years (1951–1959). During this interval, pest density declined from 8 to $0.25/m^2$ by releasing $\approx$779 million *T. semistriatus* (Safavi 1960). No artificial diet introduced for mass culture of *Trissolcus* spp. Therefore, natural rearing method has been an obvious choice (Alexandrov 1948b; Vaezi 1950). However, there was not a straightforward method for the rearing of the host. Initial efforts were concentrated on rearing on grown wheat plants. Sunn pest adults were collecting from their resting sites in altitudes (2300 m ASL, Qara-aghach Mountains, Tehran province, 30 km distant to Varamin wheat fields) which flew to find their resting sites during June in Varamin. Formerly it was believed that such a flight is necessary for maturation of ovarioles. Hence, scientists waited for migration toward mountains and then arranged a trip for collecting aestivating adults in autumn. The collected individuals were kept in boxes among debris to spend the remaining of their diapause. Finally, they were reared on grown wheat and the obtained eggs were offered to parasitoids. Parasitized eggs, as well as, adult wasps themselves were released in wheat fields at a rate of 10,000 wasps/4–5 adult bugs per $m^2$. Furthermore, close field cages were used for rearing both the host and parasitoids in wheat fields and then parasitoids were released via small windows with fine mesh to permit wasps only to leave the cage. Approximately 500,000 parasitoids per a cage ($2 \times 1.2 \times 0.3$ m) can be reared and this quantity adequate pest control in 50–100 ha. Next efforts showed that flight is not a necessity in terminating reproductive diapause and starting oogenesis. Collecting from resting sites was difficult and this small modification accounted as a large success (Safavi 1960). Adult bugs are collected in larger number and by lower labour in wheat fields prior to emigration. Another improvement was rearing on kernels instead grown plants. Higher reproductive output, lower costs and smaller-scale facilities are advantages (Zomorrodi 1962). However, the main problem in natural rearing method has been obligatory reproductive diapause of sunn pest and annual life cycle of the pest. This means that the host is available for a short period (at most two months). This is also a restriction in laboratory studies on parasitoids. One can partially overcome such a restriction by some foresees. For example, one can search for adults in the resting site during January–March. In this period, reproductive diapause terminates and bugs begin to oviposit if they are maintained in warm condition. They need only spend a short pre-oviposition period as long as one week (Martin et al. 1969; Abdollahy 1989; Iranipour 2008). Thus, eggs will be available three months before arrival of immigrant bugs. Another way for overcoming the problem is collecting from different regions with different thermal patterns. A three-month difference is present among different regions in terms of seasonal activity of sunn pest (Radjabi 2000). For example, one can sweep bugs upon cereal ears during February in southern regions of country such as Jiroft (Kerman province), Darab (Fars province) or Saravan (Sistan-and-Baluchestan province). On the other hand in colder regions such as Ardabil and Meshkinshahr (Ardabil province) and Faridan (Isfahan province), bugs only leave their resting sites during May. In both areas, the duration of reproductive phase takes two months. A third way to overcome the problem has been manipulating of diapause by JH mimics such as pyriproxyfen (Zarnegar and Noori 2006). Unfortunately, this method has not been entirely flawless. Parasitoids can attack and successfully develop in the majority of eggs deposited by JH-treated bugs (Iranipour 1996), but bug embryo cannot develop (Zarnegar and Noori 2006). In some cases, using factitious hosts may be a solution. Safavi (1973) had limited studies on the possibility of parasitoid rearing upon *Aelia*, *Dolycoris* and *Brachynema* species, but unfortunately, no report of success degree or probable obstacles has published. Some attempts also were done for the rearing of egg parasitoids on *Graphosoma lineatum* (Asgari 1995; Shahrokhi et al. 1998), *Podisus maculiventris* (Allahyari et al. 2002) and *Andralus spinidens* (F.) (Mohaghegh-Neyshabouri and Amir-Maafi 2000; Najafi-Navaei et al. 2000). None of these hosts has been comparable to a target host. *P. maculiventris* is not a native species of Iran fauna and no effort was done for an establishment in Iran. It just introduced from Canada and investigated in the laboratory for a short time. Among other species, *G. lineatum* has been a promising factitious host. An easy way for rearing, the absence of diapause, the suitable rate of development and reproduction, and high acceptance by parasitoids are positive properties that make this species a good candidate for biological control programs. Nevertheless, inundative release programs of late 1940s and 1950s finally were abandoned for unknown reasons. No published article explained the reasons for such avoidance of inundation. Afterward, only some restricted efforts in research scale were done for using these parasitoids (*T. vassilievi* in Asgari's 2011 study with acceptable results) in biocontrol programs. Nowadays chemical control is the main control measurement used against sunn pests. In the last 10 years, governmental incorporation in applications using aircraft were replaced by farmer applications.

## 8.5    Biostatistics of the Egg Parasitoids of Sunn Pest

Many criteria are used for preliminary screening of natural enemies. Searching efficiency, attack rate, reproductive potential, population growth rate, host specificity, and synchrony to host are among the most important of them. Synchronization is not important in inundation programs (van Driesche and Bellows 1996). Many of these criteria are examined by functional and numerical response, life table, and host preference studies in laboratory. These parameters and effect of different environmental factors such as host quality, temperature, host density, and other competitors as well as the intrinsic difference among geographical populations and species were evaluated on egg parasitoids in numerous studies.

### 8.5.1    Effect of Internal Factors

The intrinsic difference among species, populations, individuals, subspecies, biotypes etc. was studied in different works. Comprehensive comparison among species hardly is feasible. This is because there are few studies in which two species or more were studied simultaneously. On the other hand, comparison among different studies is not an intact and flawless work. Different scientists obtained very different results even for a single species because parasitoids are very sensitive to physical conditions. A small difference in experiment condition or regional populations reflects a large discrepancy in biological parameters. For example, total fecundity of *T. grandis* showed three-fold difference between Amir-Maafi (2000) and Bazavar et al. (2015) researches. Physical conditions, as well as, the intrinsic difference between populations may be responsible for these variations. In order to separate these sources of variation, comparisons should be done only under the same physical conditions and at the same time. Unfortunately, few studies provide this requirement. Hence, we can do only some comparisons among species by referring to different works.

#### 8.5.1.1    Effect on Life History

In a moderate temperature of 24–26 °C, by providing unlimited resources, least value for the intrinsic rate of increase ($r_m$) among egg parasitoids of sunn pest was recorded as 0.14–0.21 $d^{-1}$ for *T. djadetshkoe* (Abdi 2014), raised to 0.366 $d^{-1}$ in *T. grandis* (Amir-Maafi 2000). The value of this statistic for the other species is 0.3–0.314 for *T. vassilievi* (BenaMolaei 2014), 0.266–0.272 for *T. semistriatus* (Asgari et al. 2001; Kivan and Kiliç 2006a), 0.205 to 0.252 for *Ooencyrtus fecundus* (Ahmadpour et al. 2013), 0.224–0.234 for *O. telenomicida* (Rafat et al. 2013).

Reproductive parameters such as total fecundity, gross and net reproduction rates (GRR and $R_0$ respectively) showed the highest variability among the other life history parameters. It seems that reproduction is more sensitive than the other fitness components to environmental variables. For example, 136.4, 85.4 and 36.2 daughters per a female *T. grandis* were recorded during a generation in researches of Amir-Maafi (2000), Bazavar et al. (2015) and Nozad-Bonab et al. (2014) respectively. The same parameter was estimated as 198 and 63 for *O. fecundus* in studies of Ahmadpour et al. (2013) and Bazavar et al. (2015) respectively, 76–82 for *O. telenomicida* (Rafat et al. 2013), 130 for *T. semistriatus* (Asgari et al. 2001), 167–216 for *T. vassilievi* (BenaMolaei 2014) and 20–40 for *T.djadetshkoe* (Abdi 2014). The highest recorded value for total fecundity is 355.5 eggs for *O. fecundus* (Ahmadpour et al. 2013), but a low value for sex ratio (0.55) caused $R_0$ fall within the range of *T. vassilievi* in BenaMolaei's (2014) study. Total fecundity of the recent species was 180–280 progenies per female with a strong female biased sex ratio of 0.8–0.95. The sex ratio of sunn pest's egg parasitoids generally is female biased and often exceeds 0.7, but the least value was recorded for *T. djadetshkoe* and *O. fecundus* with 0.57 and 0.55 respectively. The GRR value is often closely related or even is equal to $R_0$ because survivorship curves of these parasitoids generally are convex (Type I) and mortality occurs mainly at senescence.

Development and therefore generation time are very homoscedastic variables among life-history parameters. The difference between individuals and hence variance of these variables is negligible. Some external factors like temperature, however (see Sect. 8.5.3) can create large discrepancies among treatments. Nevertheless, within-group variation is always small. Hence, these are good indicators for quality control and comparing physical, nutritional and other rearing conditions. Development time is sensitive to temperature and always shorter in males (protandry). Therefore, a comparison among species is possible only at the same temperature and optimum temperature of 25–26 °C will be recommendable due to the convergence of developmental rate in different treatments. It means that effect of external factors such as host quality etc. minimizes (Nozad-Bonab and Iranipour 2013). Development time was 9.6 ± 0.07 for *T. grandis* disregarding gender (Nozad-Bonab et al. 2014), 13.5 ± 0.1 for males and 14.6–15.0 ± 0.04 for females of *T. vassilievi* (Iranipour et al. 2015), 15–16 days for *T. djadetshkoe* (Abdi 2014), 14.4 ± 0.17 for *O. telenomicida* (Rafat et al. 2013) and 13.4 ± 0.05 for males and 14.2 ± 0.06 d for females of *O. fecundus* (Ahmadpour et al. 2013). Small value of SE in all studies is very interesting. Generation time differs between 10–21 days in different species at similar temperature regims.

### 8.5.1.2  Effect on Foraging Behaviour

Searching efficiency, handling time and maximum attack rate are functional response parameters that may represent the efficiency of a parasitoid, predict density dependence and also can use for calculating number of parasitoids required for release. Numerical and aggregation studies also may enhance our knowledge about the results of release.

There is a clear relationship between maximum attack rate of a parasitoid in a functional response study and daily parasitism of a parasitoid in life table studies. Daily parasitism in an unlimited environment may reflect maximum daily attack rate of a parasitoid of age x. Unfortunately, in many functional response studies, age-specific attack rates were neglected. Amir-Maafi's (2000) study is an exception. In most studies, the functional response of newborn females or those of a specific age only included. However, life table studies reveal that parasitism often varies age-dependently. A general pattern is not obvious in different studies, but except *T. grandis* in Amir-Maafi's (2000) study in which a gradual decrease occurred, in other studies, a sudden decline in parasitism occurred (Nozad-Bonab et al. 2014; BenaMolaei 2014; Abdi 2014). Hence, parasitism of the first day is often the highest. Sometimes parasitism reaches the maximum at the second or third day because all females do not emerge on the same day. Maximum of oviposition occurs at the first week of oviposition (62.8% in *T. grandis* and 63–84% in *T. vassilievi* (Amir-Maafi 2000; BenaMolaei 2014). After the 10th day, only 10–20% of total fecundity realizes. Oviposition often terminates at the end of the third week of oviposition and seldom continues at 4th week. Therefore, one week old or younger females are the best candidates not only in functional response studies but also in releasing programs. Oviposition pattern is somewhat different in *Ooencyrtus* spp. It takes longer and with the lower daily rate. It is high at initial 10 days with limited fluctuations, but often reaches a maximum at the 3rd–7th day of oviposition and then suffers a sharp decline (Rafat et al. 2013; Ahmadpour et al. 2013).

A comparison among different species is not a straightforward task. At first physical conditions, particularly temperature and host quality must be similar. Secondly, parameter estimation based on classic fixed-time single-patch experiments bears heavy biases. Direct observation on handling time, comparisons among experiments with equal total times or comparison among daily attack rates (24 h-based estimates of $T/T_h$ i.e. $24/T_h$) is a solution.

Type III functional response seems to be the most common among sunn pest's egg parasitoids (Abdi et al. 2015). This is true for *T. vassilievi* (BenaMolaei et al. 2018b) and *O. fecundus* (Iranipour et al. 2013a) on *E. integriceps* eggs, and four species of *Trissolcus* spp. on *Echistus hero* F. (Laumann et al. 2008). The response of *O. telenomicida* to sunn pest eggs was type II (Iranipour et al. 2013b), and that of the *T. grandis* and *T. semistriatus* was host-dependent. The response of the *T. semistriatus* to *E. integriceps* and *Graphosoma lineatum* L. was type II and III respectively (Asgari et al. 2001). The response of *T. grandis* to both *E. integriceps* and *Podissus maculiventris* Say was type III (Allahyari et al. 2004), while the functional response of 2–11 day old females of *T. grandis* was type II (Amir-Maafi 2000).

In a few studies, direct observations were done upon parasitoid's behaviour, but whenever such observations are made, a comprehensive conclusion is possible. For example, Amir-Maafi (2000) showed that the actual handling time of *T. grandis* is three mins, whereas model estimate was 17 mins. Furthermore, the maximum attack rate estimated to be 85 d$^{-1}$ where the actual rate was 28. This is because parasitoid is egg-limited rather than time-limited. Similar handling time was recorded for

*T. vasslievi* (BenaMolaei et al. 2018c), while that of the *O. fecundus* was very longer (11.4–14.3 with an average of 12.7 min; Iranipour et al. 2013a). In addition, the maximum attack rate was 42–48 in *T. vassilievi* and < 21 in *O. fecundus*. This may imply that total time of 2.5 and 4.5 h is adequate for a complete daily clutch by those species respectively. Considering further time will increase handling time estimates unrealistically. In similar 6 h lasted experiments, the higher slope of searching efficiency (parameter *b* of the type III functional response) in *T. vassilievi* compared to *T. djadetshkoe* (Abdi et al. 2015) shows a stronger density-dependent response of the former. In addition, the constant value of searching efficiency in type II response of *T. grandis* and *O. telenomicida* imply that these species find host eggs more rapidly than *T. djadetshkoe* in densities below 4–8 host eggs whereas in higher densities the status is the reverse. This suggests an advantage of type III response in higher densities.

### 8.5.1.3   Effect on Thermal Requirements

Thermal requirements may determine distribution range of a species, can use for predicting the result of competition in different climates and estimate species evenness in a guild. In a few studies, the thermal requirements of *Trissolcus* spp. were studied. Safavi (1973) suggests that species with parapsidal grooves have overall longer development time than those one lacking them. This statement confirmed by later works (see discussion by Iranipour et al. 2015). As an instant, *T. brochymenae* has the longest development among *Trissolcus* species (214.7 DD; Torres et al. 1997; Cividanes et al. 1998). Furthermore, the shortest one recorded for *T. rufiventris* (125.0 and 111.1 DD for females and males respectively; Kivan and Kiliç 2006b). Comparison between *T. vassilievi* (Iranipour et al. 2015) and *T. grandis* (Iranipour et al. 2010) also show 10–40% lower DD for the latter. In both cases, two populations were studied simultaneously. Despite the closer distance, *T. grandis* populations showed larger differences. Two degrees Celsius difference in thermal threshold (12.5 vs. 14.5 °C) and 27–38 DD in thermal constant (116.9 vs. 143.8 in males and 124.6 vs. 162.9 DD in females) was observed in *T. grandis*. Similar statistics for *T. vassilievi* was 13.0–13.8 °C and 12.2–12.6 °C for the threshold and 192.2–204.2 and 164.0–173.9 for thermal constant of females and males respectively. The authors concluded that higher phenotypic plasticity in *T. grandis* enables it to develop in wider ranges and adapt to colder regions.

## 8.5.2   Effect of External Factors – I. Biotic Factors

### Host Plant
Host plant mainly affects sunn pest itself, but some indirect effects may be expected. For example, often barley is more developed when parasitoids and hosts come in fields. It may attract a higher number of them and cause density-dependence effects.

Furthermore, a change in functional response from type II to III was observed depending on wheat cultivar in *T. grandis* (Fathipour et al. 2000).

**Host Access** Among external factors, host availability and quality are among the most important of them. There are many reports of host escape early season due to lack of synchrony between host eggs and egg parasitoids (Kaitazov 1968; Popov and Paulian 1971; Safavi 1973; Kartavtsev et al. 1975; Şimşek 1986; Şimşek and Yaşarakinci 1990; Asgari 1995; Iranipour 1996; Radjabi 2000). Egg parasitoids come sooner and encounter to lack of the host. This event causes a negative effect on parasitoid. Such an effect was investigated by Bazavar (2013) and Bazavar et al. (2015) on *O. fecundus* and *T. grandis*. Five treatments of the experiment were (1) unlimited access to host after one-week deprivation (2) access every other day, (3) access every third days (4) access once a week, and (5) control (unlimited access). Interrupted access had worse effect while effects were enhanced by access intervals. A partial compensation occurred in access dates. The *O. fecundus* had a slower and longer rhythm of oviposition. Oviposition rate was slow at the first and second day, increased at the third day of oviposition, remained unchanged between third and eighth days, and continued by a lower rate untill 23rd day. In *T. grandis* oviposition rate was highest on the first day, continued by a high rate for a few days, and then declined to zero at 19th day. Overall fecundity also was higher. Cumulatively 75 and 95% of oviposition occurred at first and second weeks of oviposition respectively in *T. grandis* and 44 and 84% in *O. fecundus*. Compensation in *T. grandis* was more considerable but did not extend to older ages. In *O. fecundus* however, it was slower and extend to older ages. Final reproductive output in first treatment (one-week deprivation followed by full access) was 78 and 92% of control in *O. fecundus* and *T. grandis* respectively. It was 2/3 and 90% of control in every other day treatment. In two other treatments, it was 54 and 38% of control in *O. fecundus*, 68, and 47% in *T. grandis*.

**Host Quality** Many factors contributed to host quality. Host species and therefor biochemistry of egg (factors like total protein and contents of aminoacids and other compounds), the size of host egg, the degree of host embryogenesis, previous parasitism, duration of cold storage etc. are among the most important of them.

**Host Species** Earlier in this chapter, I mentioned species examined as factitious hosts. Among them, *Graphosoma lineatum* has a unique position. Different works, however, show advantages of target pest *E. integriceps* to factitious hosts. For example, Asgari (2004a) showed that *T. semistriatus* prefer *E. integriceps* eggs to *G. lineatum*. Handling time on the latter was three times as much as the former. Maximum daily attack rate was 63.5 and 20.3 on the two hosts respectively (Asgari et al. 2001). The wasps emerged from *E. integriceps* eggs showed higher parasitism and population growth rate (Asgari. 2004b). Furthermore, temperature-dependent developmental rate of *T. grandis* has a steeper line slope when developed on *G. lineatum*, *G. semipunctatum*, and *E. integriceps* respectively. Such a pattern led to higher estimates of thermal threshold and smaller thermal constant in above-mentioned hosts respectively. Those lines intersect around 26.5 °C that may be considered as an optimal temperature in which development time is independent of host quality. Total protein content could properly explain variations (Nozad-Bonab and Iranipour 2013).

**Host Size** It is considered as an effective factor in host quality. Often larger hosts are preferable although this statement is not always true. In gregarious parasitoids such as *Ooencyrtus* spp. the size of the wasp depends on number of broods within a single host egg (Iranipour et al. 2013b; Ahmadpour et al. 2013). In Solitary parasitoids, however, it depends on other factors. BenaMolaei et al. (2018a) showed that between two populations of *E. integriceps*, Varamin population has slightly heavier eggs than Tabriz population (0.81 vs. 0.75 μg), although the diameter of eggs was very close (1.07 vs. 1.06 mm). Such a small difference led to slightly larger parasitoids with corresponding detectable although often insignificant advantages. Females affected slightly more than males but the size difference was significant in both. Allahyari et al. (2002) obtained 7 and 11% smaller *T. vassilievi* and *T. grandis* respectively on *Podissus maculiventris* compared to *E. integriceps* eggs. In contrast, Asgari et al. (2001) observed no significant difference in *T. semistriatus* obtained from either *E. integriceps* or *G. lineatum* eggs.

**Host Age** In some cases, parasitoid size is a function of other factors like host age. Host age determines embryogenesis level and it determines in turn host quality. In an experiment, 0–5 day old eggs of *E. integriceps* were offered to two populations of *T. vassilievi* (BenaMolaei et al. 2015a). No parasitoid, but six males developed within 5-day old eggs. Number of the parasitized eggs, emergence rate, sex ratio, total fecundity and wasp size was similar in 0–2 day old eggs but declined in older eggs. This allows harvesting done every third day that may decrease costs in mass cultures. Tabriz population was smaller and therefore suffered more seriously from host age than Varamin population. For example, the sex ratio was 44–47% in 3–4 day old eggs in the former population vs. 61–68% in the latter one. However, the difference was only in females and males were of the same size. This suggests that more vigor parasitoids should be selected for mass culture initiation. Development time differed as small as 0.6 days or less, although such minor differences were significant. Fecundity also was affected by host age. Fecundity was 40–60% higher at commence in control compared to wasps emerged from 2–4 d old eggs. However, a partial compensation in subsequent days occurred. Nasiri et al. (2020) studied functional response and host preference of *O. fecundus* among three kinds of hosts. Fresh host eggs, 6 d old developed host eggs and 8 d old paerasitized eggs by *T. grandis*. Effect of development of host embryo was very worse than previous parasitism. Females of *O. fecundus* preferred parasitized eggs to developed ones and fresh eggs to both. No swithing was observed in lower rates of preferred hosts. It was concluded that *O. fecundus* can play a minor negative role on host/parasitoid ratio, because it often parasitizes late season eggs that often are parasitized or developed and therefore kill more *Trissolcus* than *E. integriceps* offspring. Similar results by different Platygastroid egg parasitoids and different host species was obtained. For example *T.semistriatus* on *E. integriceps*, *G. lineatum*, *Dolycoris baccarum*, *Eurydema ornata* (L.), and *Holcostethus vernalis* (Wolff) (Kivan and Kiliç 2004),

*T. vassilievi* and *T. grandis* on *E. integriceps* (Sheikhmoss 2009), *T. megallocephalus* on *Nezara viridula* (Awadalla 1996), *Telenomus isis* (Polaszek) on *Sesamia calamistis* Hampson (Chabi-Olaye et al. 2001).

**Host Storage** Cold-stored host eggs also may decrease parasitoid efficiency in a similar manner. In a similar study 0, 1, 2, 4, 6 and 8 week stored eggs were offered to *T. vassilievi* females (BenaMolaei et al. 2015b). Many statistics remained unchanged up to two weeks, but storing for longer times caused undesirable effects. Parasitism decreased somewhat linearly at a rate of one less host egg parasitized per 8.5 days storing at 4 °C. Such trend predicts ceasing parasitism after 4 months. This prediction was true in Kivan and Kiliç's (2005) studies on *T. semistriatus*. Cold-storage had a stronger effect than fetal development on the development time of the parasitoid. It caused four days (32%) delay in emergence following eight weeks storing at the cold condition. Effect on emergence rate was more moderate (maximum 12% lower emergence). This trend also is true in *T. semistriatus* (Kivan and Kiliç 2005), *T. grandis* and *Telenomus chloropus* (Asgari 1995). Sex ratio was high (80–90%) prior to the first month of storage but beyond a month, it declined to 50–60%. Higher allocation of low-quality hosts to male progeny is a known phenomenon among parasitoids (Wage 1986), particularly in *T. grandis* and *T. semistriatus* (Gusev and Shmettser 1975). *Anastatus bifasciatus* Forcroy lay only male eggs within *E. integriceps* eggs (Iranipour et al. 1998b). This is due to the small size of these eggs. Total fecundity of *T. vassilievi* reached to 2/3 of control (182 vs. 118) in eggs stored for 1.5 months at refrigerator. The detectible difference was observed in the second week of storage. A continuous reduction in age-specific oviposition was observed in fresh eggs and one week stored ones, whereas in remaining treatments oviposition reached to a maximum at the end of the first week (BenaMolaei et al. 2015b). Such pattern resembles those of the developed host eggs (BenaMolaei et al. 2015a). The wasp size reduced -steeper in females- as the duration of storage increased. The results of this research showed that host eggs can store and used for two months with undesirable effects on parasitoids. The higher rate of release (up to twice as many as parasitoids developed on fresh eggs) may overcome the problem in inundation programs, but in research programs, it is recommended to avoid more than two week stored eggs. Two week stored eggs may cause biases <10%. Gusev and Shmettser (1975) obtained similar results with *T. grandis* and *T. semistriatus* in response to cold-stored eggs. After 1.5 months, sex ratio was affected and after 2 months emergence rate declined suddenly. In addition, egg stored at 6 and − 20 °C were suitable for two and four months respectively (Kivan and Kiliç 2005). Previous works on telenomin wasps showed that cold storage of wasps themselves had poor results (Asgari 1995; Foerster et al. 2004; Kodan and Gürkan 2004; Foerster and Doetzer 2006).

## Competition

The behaviour of the parasitoid in a single-patch experiment cannot show all behavioural components of a host-parasitoid system. A better understanding achieves when experiments are conducted in conditions nearer to the real world. Natural enemies often are foraging in a multi-patch environment while other competitors of the same species or relative or homologue species that are foraging simultaneously. Hence, the presence of choice among patches of different densities as well as competitors is near to what occurs in nature. In a multi-patch experiment, Iranipour et al. (2020) released 1–16 *T. vassilievi* females into an arena with five patches of 1 to 15 egg clutches (14 eggs/clutch). They observed that one-clutch patch was avoided by a single wasp, while other patches were visited equally. The motivation threshold was concluded to be two clutches i. e. 28 eggs. In no-choice experiments however, the wasp handled majority of eggs in tubes containing 4–28 eggs. This may suggest that the parasitoid can avoid poorer patches when it accesses better ones. However, their results also revealed that the parasitoid seldom leaves a poorer patch of its first choice. This led to lower exploitation in some cases in the experimental arena. Although such behaviour may seem maladaptive in the experimental arena, it may be adaptive in a natural situation with scarce host and high risk of leaving a worse situation hoping to find a better one. As soon as they added number of foraging females within an arena, the restlessness of the parasitoids increased. They left their situations repeatedly and even one-clutch patch also was visited. This is due to interference between them. Time allocation to different patches, however, was not equal. Both parasitism and time spent increased from 1 to 8 clutch patches, but then decreased at the highest density of 15 clutches. This may be due to kairomone saturation and therefore habituation and confusion of the parasitoid around the highest host density. The high correlation between patch-time and parasitism suggests unimportance of interference because aggregation of wasps in a patch did not lead to lower exploitation. Nevertheless, a significant negative correlation was observed between searching efficiency and parasitoid density. This may suggest the presence of mutual interference that divided into two components of actual interference and pseudo-interference. The latter component was more important than the former one. Pseudo-interference arises from aggregation in some patches (often denser ones) and avoiding the others that final result will be lower exploitation. Therefore, parasitoid's fidelity to a discovered patch benefits its host. In this case, per capita parasitism declined as the number of foraging wasps increased from 1 to 16 by a rate of 0.4. Inundation may be a solution. Results also showed that the parasitoid responds to the presence of other individuals by accelerating its handling and more efficient utilization of time. The total response (functional plus numerical) of the parasitoid was type III and parasitism rate reached a maximum at 28 eggs. A similar study by Amir-Maafi (2000) had somewhat similar results by a larger slope of decrease (coefficient *m* of Hassell and Varley, 1969) in searching efficiency by wasp density (0.37 compared with 0.18–0.24). Iranipour et al. (2018a) studied reaction of two species of *T. vassilievi* and *T. grandis* to each other and to presence of conspecifics in a multipatch environment. Densities of 1, 8 and 16 parasitoids of single species with combinations of the same number of two species (either 4 *vassilievi* + 4 *grandis* or 8 *vassilievi* + 8 *grandis*) were released on the same arrange of host clutches. Results revealed that competition among *T. grandis* females was more burdensome as it caused a strong decline in searching rate. Effect of *T. vassilievi* females on conspecifics was moderate, but interspecific confrontation of the two species had interesting results. *T. grandis* reduced intra-specific interference and searching rate of this species enhanced compared to equal numbers of conspecifics in a single species arena. In contrast, *T. vassilievi* could not handle more than ¼ of the total parasitized eggs. This may imply that *T. grandis* is more aggressive than *T. vassilievi*, however, in muti-species situations, it neglecs presence of conspecific more frequently than the other species. Larval competition between the two species was studied by Najafipour et al. (2018). Overall, first invador will be win the contest disregarding species and interval between attacks (3–48 h), however success rate of *T. vassilivi* decreases as second attack delays.

## 8.5.3   Effect of External Factors – II. Abiotic Factors

**Temperature**

Among abiotic factors, the temperature is the most influencing one. Insects are poikilothermic animals and their body temperature is the function of their environment. Body temperature determines the speed of biochemical reactions and activity of enzymes. Temperatures in which development occur can be wider than temperatures within which reproduction realizes. There is a thermal threshold for development above which development and biochemical reactions accelerate up to an optimal temperature. In colder temperatures, all activities stop. Beyond optimal temperature, reactions slow down. This range of temperature may determine the fundamental niche and distribution pattern of an insect. In optimum temperature, speed of reactions is the maximum and hence handling time is the shortest, the searching rate is highest, development is most rapidly and reproduction occurs in the shortest interval. Both in colder and warmer temperatures speed of these events decreases. Parasitoids often experience colder than optimum temperatures in fields. In this condition, daily attack rate reduces. Total fecundity is lower and parasitism occurs with a lower daily rate but in a longer period. Lower parasitism rates in colder regions were reported (Haghshenas 2004; Nozad-Bonab and Iranipour 2010). On the other hand, cold springs can limit the activity of some species with higher thermal threshold and heat requirements such as *T. vassilievi* in colder provinces (Iranipour et al. 2015). The main problem in natural biological control of sunn pest complex is early season low parasitism due to higher reproductive requirements of parasitoids compare to hosts (Salavatian 1991). In previous sections, detailed effects of temperature were discussed.

Effect of low temperatures on survival of *T. grandis* and *T. vassilievi* was studied by Iranipour et al. (2018b). Females of both species were introduced to 4 °C after a 10 day acclimatization in 10 °C. They were confined individually or in batches of 3–30 individuals among shelters constructed from the paper balls. Their mortality then was monitored weekly in room air. In all duration of the experiment, mortality of *T. vassilievi* was higher than *T. grandis*. On the other hand, aggregation in higher numbers protected females from higher mortalities observed in lower batches. An advantage of 1.35, 3.04 and 6.24 times more *T. grandis* survivors was predicted by dose-response curves after 1, 2 and 3 months storing in cold condition respectively. This may explain the reason of scarceness of *T. vassilievi* in colder provinces.

Both *T. grandis* and *O. fecundus* showed tendency to temperature of 26 °C in a confined container in which a gradient between 15 and 50 °C was created (Nozad-Bonab et al. 2014; Ahmadpour et al. 2014). More than 85% of the former crowded in spaces with temperatures between 20 and 32 °C and 90% of the latter between 20 and 30 °C.

**Other Factors**
Effect of the other factors like light intensity, photoperiod and relative humidity was not extensively studied. A single observation shows that parasitism by *T. grandis* has limited to days longer than 13: 11 h (L: D) photoperiod (Iranipour 1996). Quantitative measurements, showed limited effects of both photoperiod and light intensity on *T. grandis* (Teimouri et al. 2019). Females reproduced and larvae developed even in darkness. Foerster et al. (2004) found that cold temperatures in short days may induce overwintering in *T. basalis* and *Telenomus podisi*.

Some species like *T. simoni* and *Telenomus chloropus* are distributed only in more humid Northern provinces. This may show their dependence on moisture. Relation to humidity must be studied in close connection to temperature.

## 8.6   Future Prospect

It seems that the future of sunn pest IPM must be constituted of integration of different tactics. If other measurements do properly, it may lead to a reduction in spraying area and also the complete omission of insecticides in some regions. Biological control measurements will constitute an inseparable component of future sunn pest management strategy. Both conservation and early season inundation will be central in this program. The major actions in order to conserve natural enemies and primarily egg parasitoids are exact timing and SSIPM (Karimzadeh et al. 2011a, b). This measurement avoids insecticide application in throughout a field and spraying was done only in those places where the population exceeds ET. Other places remain untreated as refuges for natural enemies. Choosing low-persistent selective insecticides that benefit parasitoid/host ratio also is very important. An early-season release program upon pioneer eggs will strengthen the crop protection. In order to enhance parasitoids, habitat modification also will be very useful. Among other measurements, vicinity of cereal fields to orchards, shade trees, other crops and especially alfalfa fields seems to be very effective. This vicinity help parasitoids to reproduce, feed, overwinter and find alternative hosts. Also, the vicinity of barley to wheat cause sunn pest to attract to barley and do not attend in wheat. This has some advantages. First barley is cheaper crop and has no tolerance level because only is used as feed for livestock. On the other hand, infested crops can consume by livestock. Second, barley always is in a more developed stage when pest comes in and then the parasitoid-pest system can colonize sooner in barley and then spread in wheat fields. Finally, leaving weeds around fields where is possible, and avoiding unnecessary applications are the other actions.

# References

Abbasipour H (2004) Biological characteristics of *Platytelenomus hylas* (Hym.: Scelionidae) on egg parasitoid of corn and sugarcane stalk borer *Sesamia nonagrioides* (Lep.: Noctuidae) in Khuzestan province. J Entomol Soc Iran 23(2):116. (in Persian)

Abbasipour H, Shojai M, Nasrollahi AA (1991) A survey on the effectiveness of *Sesamia* egg parasitoid, *Platytelenomus hylas* Nixon in corn fields of Khuzestan province. Paper presented at the 10th Iranian Plant Protection Congress, Bahonar University, Kerman, p 49

Abdi F (2014) Age-specific fertility life table and functional response for *Trissolcus djadetshko* Rjachovsky (Hym: Scelionidae). M.Sc. thesis, Faculty of Agriculture, University of Tabriz, Tabriz, Iran, 59 pp

Abdi F, Iranipour S, Hejazi MJ (2015) Effect of mating and previous parasitism on functional response of *Trissolcus djadetshkoe* (Hym.: Scelionidae) an egg parasitoid of *Eurygaster integriceps* (Hem.: Scutelleridae). Iran J Plant Prot Sci 46(1):131–139. (in Persian)

Abdollahy A (1989) An investigation on diapausing period of *E. integriceps* Put (Hem. Scutelleridae) in overwintering sites. Paper presented at the 9th Iranian Plant Protection Congress, Ferdowsi University of Mashhad, Mashhad, 8–13 September 1989, p 33

Abdul Razzagh ZA (1995) Successful transporting, rearing and releasing of *Sesamia nonagrioides* egg parasitoid on *Sesamia cretica* in corn fields in Esfahan. Paper presented at the 12th Iranian Plant Protection Congress, University of Tehran, Tehran, 7 September 1995, p 91

Aguiar AP, Deans AR, Engel MS, Forshage M, Huber JT, Jennings JT, Johnson NF, Lelej AS, Longino JT, Lohrmann V, Miko I, Ohl M, Rasmussen C, Taeger A, Ki Yu DS (2013) Order Hymenoptera. Zootaxa 3703(1):51–62

Ahmadpour S, Iranipour S, Asgari S (2013) Effects of superparasitism on reproductive fitness of *Ooencyrtus fecundus* Ferriere & Voegele (Hym. Encyrtidae), egg parasitoid of sunn pest, *Eurygaster integriceps* Puton (Hem. Scutelleridae). Biol Control Pests Plant Dis 2(2):97–105. (in Persian)

Ahmadpour S, Iranipour S, Asgari S (2014) Thermal tendencies in *Ooencyrtus fecundus* Ferriere & Voegele (Hymenoptera: Encyrtidae), egg parasitoid of sunn pest. Biol Control Pests Plant Dis 3 (1):1–6. (in Persian)

Alexandrov N (1948a) *Eurygaster integriceps* Put. a Varamine et ses parasites (1). Entomol Phytopathol Appl 6–7:28–47

Alexandrov N (1948b) *Eurygaster integriceps* Put. a Varamine et ses parasites (2). Entomol Phytopathol Appl 8:16–52

Allahyari H, Azmayeshfard P (2002) Comparison between functional responses of the first generation of *Trissolcus grandis*, emerged from egg of sunn pest and *Podisus maculiventris*. Paper presented at the 15th Iranian Plant Protection Congress, University of Kermanshah, Kermanshah, 7–10 September 2002, p 7

Allahyari H, Azmayeshfard P, Kharrazi-Pakdel A, Radjabi G (2002) Rearing of predatory bug *Podisus maculiventris* (Say) (Hem. Pentatomidae) on artificial diet. Iran J Agric Sci 33 (3):445–454

Allahyari H, Azmayeshfard P, Nozari J (2004) Effect of host on functional response of offspring in two populations of *Trissolcus grandis* on the sunn pest. J Appl Entomol 128:39–43

Amir-Maafi M (2000) An investigation on the host-parasitoid system between *Trissolcus grandis* Thomson (Hym.: Scelionidae) and sunn pest eggs. PhD thesis, Faculty of Agriculture, University of Tehran, Karaj, Iran, 220 pp

Amir-Maafi M, Parker BL (2002) Density dependence of *Trissolcus* spp. (Hymenoptera: Scelionidae) on eggs of *Eurygaster integriceps* Puton (Hemiptera: Scutelleridae). Arab J Plant Prot 2:62–64

Asgari S (1995) A study on possibility of mass rearing of sunn bug egg parasitoids on its alternative host, *Graphosoma lineatum* L. (Het., Pentatomidae). M.Sc. thesis, Faculty of Agriculture, University of Tehran, Tehran, Iran, 220 pp

Asgari S (2004a) Host preference and switching in the egg parasitoid *Trissolcus semistriatus* to the eggs of stripped pentatomid and sunn pest. Paper presented at the 16th Iranian Plant Protection Congress, University of Tabriz, Tabriz, 28 August–1 September 2004, p 15

Asgari S (2004b) Comparing population parameters of egg parasitoid, *Trissolcus semistriatus* on the host eggs, *Graphosoma lineatum* and *Eurygaster integriceps* for host fitness determination. Paper presented at the 16th Iranian Plant Protection Congress, University of Tabriz, Tabriz, 28 August–1 September 2004, p 38

Asgari S (2011) Inundative release of the sunn pest egg parasitoid and evaluation of its performance. Paper presented at the Biological Control Development Congress, University of Tehran, Tehran, 27–28 July 2011, pp 423–428

Asgari S, Sahragard A (2002) Comparing the biology of egg parasitoid, *Trissolcus semistriatus* reared on egg of sunn pest and stripped pentatomid. Paper presented at the 15th Iranian Plant Protection Congress, University of Kermanshah, Kermanshah, 7–10 September 2002, p 7

Asgari S, Kharazi-Pakdel A, Esmaili M (1995) *Graphosoma lineatum* L. (Het.: Pentatomidae) as an alternative host for mass rearing of egg parasitoids, *Trissolcus* spp. (Hym.: Scelionidae). Paper presented at the 12th Iranian Plant Protection Congress, University of Tehran, Tehran, 2–7 September 1995, p 17

Asgari S, Sahragard A, Kamali K, Soleymannezhadian E, Fathipour Y (2001) Functional and numerical responses of sunn pest egg parasitoid, *Trissolcus semistriatus*, reared on *Eurygaster integriceps* and *Graphosoma lineatum*. Appl Entomol Phytopathol 69(2):97–110

Awadalla SS (1996) Influence of temperature and age of *Nezara viridula* L. eggs on the Scelionid egg parasitoid, *Trissolcus megallocephalus* (Ashm.) (Hym., Scelionidae). J Appl Entomol 120:445–448

Bazavar A (2013) Effect of host unavailability durations on parasitism behavior of *Trissolcus grandis* (Hymenoptera: Scelionidae) and *Ooencyrtus fecundus* Ferriere &Voegele (Hym.: Encyrtidae) egg parasitoids of sunn pest. M.Sc. thesis of Agricultural Entomology, University of Tabriz, Faculty of Agriculture, 67 pp

Bazavar A, Iranipour S, Karimzadeh R (2015) Effect of host unavailability durations on parasitism behavior of *Trissolcus grandis* (Hymenoptera: Scelionidae) egg parasitoid of sunn pest. Appl Res Plant Prot 4(1):41–56

BenaMolaei P (2014) Comparison of biological, demographic and behavioral characteristics of two populations of *Trissolcus vassilievi* Mayr (Hym., Scelionidae), egg parasitoid of sunn pest on two populations of the host. Ph.D. dissertation in Agricultural Entomology, Faculty of Agriculture, The University of Tabriz, 222 pp

BenaMolaei P, Iranipour S, Asgari S (2015a) Effect of the host embryogenesis on efficiency of *Trissolcus vassilievi*. Biocontrol Plant Prot 3(1):83–100

BenaMolaei P, Iranipour S, Asgari S (2015b) Biostatistics of *Trissolcus vassilievi* (Hym., Scelionidae) developed on sunn pest eggs cold-stored for different durations. Munis Entomol Zool 10(1):259–271

BenaMolaei P, Iranipour S, Asgari S (2018a) Host preference and switching of two populations of *Trissolcus vassilievi* between two host populations. Biocontrol Plant Prot 5(2):27–42

BenaMolaei P, Iranipour S, Asgari S (2018b) Functional response of two population of *Trissolcus vassilievi* (Mayr) on sunn pest eggs (*Eurygaster integriceps* Puton). J Appl Res Plant Prot 6 (4):89–106

BenaMolaei P, Iranipour S, Asgari S (2018c) Time allocation by female parasitoid *Trissolcus vassilievi* (Mayr) encountering with sunn pest eggs (*Eurygaster integriceps* Puton). J Appl Res Plant Prot 7(2):1–11

Brown ES, Eralp M (1962) The distribution of the species of *Eurygaster* Lap. (Hemiptera, Scutelleridae) in Middle East countries. Ann Mag Nat Hist 5(50):65–81

Chabi-Olaye A, Schulthess F, Poehling HM, Borgemeister C (2001) Factors affecting the biology of *Telenomus isis* (Polaszek) (Hymenoptera: Scelionidae), an egg parasitoid of cereal stem borers in West Africa. Biol Control 21(1):44–54

Cividanes FJ, Figueiredo JG, Carvalho DR (1998) Prediction of the emergence of *Trissolcus brochymenae* (Ashmead) and *Telenomus podisi* Ashmead (Hym: Scelionidae) in the field. Sci Agric 55:43–47

Farid A (1985) Preliminary investigations on sunn pests (*Eurygaster intergriceps* Put.) In Jiroft (Kerman). Bull Plant Prot Org Iran 30:74–86

Fathi AA, Nouri-Ganbalani G, Honarmand P (2011) Parasitoids of sunn pest (*Eurygaster integriceps* Puton) and percent parasitism of egg in Quraim, Ardabil province. Sci J Agric 34 (1):47–57

Fathipour Y, Kamali K, Khalgani J, Abdollahi G (2000) Functional response of *Trissolcus grandis* (Hym., Scelionidae) to different egg densities of *Eurygaster integriceps* (Het., Scutelleridae) and effects of different wheat genotypes on it. Appl Entomol Phytopathol 68:123–136

Foerster LA, Doetzer AK (2006) Cold storage of the egg parasitoids *Trissolcus basalis* (Wollaston) and *Telenomus podisi* Ashmead (Hymenoptera: Scelionidae). Biol Control 36:232–237

Foerster LA, Doetzer AK, Castro LCF (2004) Emergence, longevity and fecundity of *Trissolcus basalis* and *Telenomus podisi* after cold storage in the pupal stage. Pesqui Agropecu Bras 39:841–845

Ghahari H, Havaskary M, Tabari M, Ostovan H, Sakenin H, Satar A (2009) An annotated catalogue of Orthoptera (Insecta) and their natural enemies from Iranian rice fields and surrounding grasslands. Linz Biol Beitr 41(1):639–672

Ghahari H, Satar A, Anderle F, Tabari M, Havaskary M, Ostovan H (2010) Lacewing (Insecta: Neuroptera) of Iranian rice fields and surrounding grasslands. Munis Entomol Zool 5(1):65–72

Ghahari H, Buhl PN, Kocak E (2011) Checklist of Iranian *Trissolcus* Ashmead (Hymenoptera: Platygastroidea: Scelionidae: Telenominae). Int J Environ Stud:1–9

Ghahari H, Neerup B, Kocak E, Iranipour S (2015) An annotated catalogue of the Iranian Scelionidae (Hymenoptera: Platygastroidea). Entomofauna 36(28):349–376

Gusev GV, Shmettser NV (1975) Effect of ecological factors on the rearing of telenomines in artificial conditions. Trudy Vsesoyuznogo Nauchno Issledo Vatelskogo Instituta Zashchity Rastenii 44:70–82

Haghshenas A (2004) Protection and conservation of sunn pest tachinid and scelionid parasitoids in Chahar Mahal va Bakhtiari province. Paper presented at the 16th Iranian Plant Protection Congress, University of Tabriz, Tabriz, 28 August–1 September 2004, p 5

Hashemi-Rad H (2008) Study on egg parasitoids *Trissolcus agriope* and *Psix* sp. using egg of five bug species in the laboratory condition. Paper presented at the 18th Iranian Plant Protection Congress, University of Bu-Ali Sina, Hamedan, 2–5 September 2004, p 47

Hashemi-Rad H, Soleymannejadian E, Radjabi G, Rajabi A (2000a) Biology of the parasitoid, *Trissolcus agriope* (Hym.: Scelionidae), the predominant species of genus *Trissolcus* in the Rafsanjan area. Paper presented at the 14th Iranian Plant Protection Congress, Isfahan University of Technology, Isfahan, 5–8 September 2000, p 105

Hashemi-Rad H, Soleymannejadian E, Radjabi G, Iranipour S (2000b) Introduction of egg-parasitoids of pentatomid pistachio bugs in Kerman province. Paper presented at the 14th Iranian Plant Protection Congress, Isfahan University of Technology, Isfahan, 5–8 September 2000, p 269

Hashemi-Rad H, Soleymannejadian E, Radjabi G (2002) Population fluctuation of the pistachio green stink bugs (*Brachynema* spp. & *Acrosternum* spp.) and their egg parasitoids in Rafsanjan. Paper presented at the 14th Iranian Plant Protection Congress, Isfahan University of Technology, Isfahan, 5–8 September 2000, p 107

Hassell MP, Varley GG (1969) New inductive population model for insect parasites and its bearing on biological control. Nature 223:1133–1136

Hejazi MJ, Karimzadeh R, Iranipour S (2011) *Eurygaster integriceps* egg parasitoids in Gharamalek. Paper presented at the global conference on Entomology, University of Chiang Mai Thailand, Chiang Mai, 5–9 March 2011, p 45

Iranipour S (1996) A study on population fluctuation of the egg parasitoids of *Eurygaster integriceps* Put. (Heteroptera: Scutelleridae) in Karaj, Kamalabad, and Fashand. M.Sc. thesis on Agricultural Entomology, University of Tehran, Karaj, Iran, 187 pp

Iranipour S (2008) Relationship between fecundity, weight and body dimentions in *Eurygaster integriceps* Puton (Hem., Scutelleridae). Paper presented at the 18th Iranian Plant Protection Congress, University of Bu-Ali Sina, Hamadan, 24–27 August 2008, p 454

Iranipour S, Johnson NF (2010) *Trissolcus radjabii* n.sp. (Hymenoptera: Platygastridae), an egg parasitoid of the shield bug, *Apodiphus amygdali* (Heteroptera: Pentatomidae) and the sunn pest, *Eurygaster integriceps* (Heteroptera: Scutelleridae). Zootaxa 2515:65–68

Iranipour S, Kharrazi-Pakdel A (2011) Relationship between parasitism rates in egg traps and natural egg populations of sunn-pest *Eurygaster integriceps* Put. J Sustain Agric Prod 22 (4):45–55

Iranipour S, Kharrazi-Pakdel A, Esmaili M, Radjabi G (1998a) Introduction of two species of egg parasitoids of pentatomid bugs from genus *Trissolcus* (Hym.: Scelionidae). Paper presented at the 13th Iranian Plant Protection Congress, Campus of Agriculture and Natural Resources, University of Tehran, Karaj, 1998, p 4

Iranipour S, Kharrazi-Pakdel A, Esmaili M, Radjabi G (1998b) Introduction of a Chalcid species from Eupelmidae for Iran. Paper presented at the 13th Iranian Plant Protection Congress, University of Tehran, Karaj, 23–27 August 1998, p 5

Iranipour S, Nozad Bonab Z, Michaud JP (2010) Thermal requirements of *Trissolcus grandis* (Hymenoptera: Scelionidae), an egg parasitoid of sunn pest. Eur J Entomol 107(1):47–53

Iranipour S, Kharrazi Pakdel A, Radjabi G, Michaud JP (2011) Life tables for sunn pest, *Eurygaster integriceps* (Heteroptera: Scutelleridae) in Northern Iran. Bull Entomol Res 101:33–44

Iranipour S, Ahmadpour S, Asgari S (2013a) Effects of superparasitism on searching ability of *Ooencyrtus fecundus* (Hym.: Encyrtide), egg parasitoid of sunn pest, *Eurygaster integriceps* (Hem., Scutelleridae). Paper presented at the 2nd global conference on Entomology, Kuching Malaysia, 8–12 November 2013, p 125

Iranipour S, Rafat A, Safavi SA (2013b) Functional and numerical response of *Ooencyrtus telenomicida* (Hym.: Encyrtidae) against sunn pest *Eurygaster integriceps* (Hem.: Scutelleridae) eggs. Paper presented at the 2nd global conference on Entomology, Kuching Malaysia, 8–12 November 2013, p 127

Iranipour S, BenaMolaei P, Asgari S, Michaud JP (2015) Reciprocal crosses between two populations of *Trissolcus vassilievi* (Mayr) (Hymenoptera: Scelionidae) reveal maternal effects on thermal phenotypes. Bull Entomol Res 105(3):355–363

Iranipour S, Najafipour M, Khaghaninia S (2018a) Intraspecific and interspecific competition between *Trissolcus grandis* and *T. vassilievi* (Hym., Platygastridae) females in multi-patch environment. XI European Congress of Entomology, Naples, Italy, 2–6 July 2018, p 24

Iranipour S, Najafipour M, Khaghaninia S (2018b) Cold hardiness comparison between *Trissolcus grandis* and *T. vassilievi* in cold-storage condition. Paper presented at the 23rd Iranian Plant Protection Congress, Gorgan University of Agricultural Sciences and Natural Resources, Gorgan, 27–30 Augest 2018, p 971

Iranipour S, BenaMoleai P, Asgari S, Michaud JP (2020) Foraging egg parasitoids, *Trissolcus vassilievi* (Hymenoptera: Platygastridae), respond to host density and conspecific competitors in a patchy laboratory environment. J Econ Entomol 13(2):760–769

Jamshidnia A, Kharazi-Pakdel A, Allahyari H, Soleymannejadian E (2009) Natural parasitism of *Telenomus busseolae* (Hym.: Scelionidae) an egg parasitoid of sugarcane stem borers, *Sesamia* spp. (Lep.: Noctuidae), on sugarcane commercial varieties in Khuzestan. J Entomol Soc Iran 29 (2):99–112

Jones WA (1988) World review of the Parasitoids of the Southern Green Stink Bug *Nezara viridula* (L.) (Heteroptera: Pentatomidae). Ann Entomol Soc Am 81(2):262–273

Kaitazov A (1968) The role of egg parasites in the reduction of the population of cereal bugs and possibilities of integration with chemical for control. Zashchita Rastenii 61:9–12

Karimzadeh R, Hejazi MJ, Helali H, Iranipour S, Mohammadi SA (2011a) Assessing the impact of site-specific spraying on control of Eurygaster integriceps (Hemiptera: Scutelleridae) damage and natural enemies. Precis Agric 12:576–593

Karimzadeh R, Hejazi MJ, Helali H, Iranipour S, Mohammadi SA (2011b) Analysis of the spatio-temporal distribution of Eurygaster integriceps (Hemiptera: Scutelleridae) by using spatial analysis by distance indices and geostatistics. Environ Entomol 40(5):1253–1265

Kartavtsev NI, Voronin KE, Sumarokh AF, Dzyuba ZA, Pukinskaya GA (1975) Investigations over many years on the seasonal colonization of Telenomines in the control of the noxious pentatomid in the Krasnodar region. Trudy Vsesoyuznogo Nauchno Issledovatel'Skogo Instituta Zashchity Rastenii 44:83–90

Khajehzadeh Y (2002) Investigation of parasitoid wasps Scelio flavibarbis M. and its effect on the population dynamics of Locusta migratoria L. in sugarcane fields in Khuzestan. Paper presented at the 14th Iranian Plant Protection Congress, Isfahan Uiversity of Technology, Isfahan, 5–8 September 2000, p 52

Khajehzadeh Y (2004) Investigation on the grasshoppers and their natural enemies in Khuzestan province. Paper presented at the 16th Iranian Plant Protection Congress, University of Tabriz, Tabriz, 28 Augest–1 September 2004, p 93

Khajehzadeh Y, Ghazavi M (2000) Some investigations on injurious grasshoppers of rice fields in Khuzestan province. Paper presented at the 14th Iranian Plant Protection Congress, Isfahan University of Technology, Isfahan, 5–8 September, p 230

Kivan M, Kiliç N (2004) Influence of host species and age on host preference of Trissolcus semistriatus. BioControl 49:553–562

Kivan M, Kiliç N (2005) Effects of storage at low-temperature of various heteropteran host eggs on the egg parasitoid, Trissolcus semistriatus. BioControl 50:589–600

Kivan M, Kiliç N (2006a) Age-specific fecundity and life table of Trissolcus semistriatus, an egg parasitoid of the sunn pest Eurygaster integriceps. Entomol Sci 9:39–46

Kivan M, Kiliç N (2006b) A comparison of the development time of Trissolcus rufiventris (Mayr) and Trissolcus simoni (Mayr) (Hym.: Scelionidae) at three constant temperatures. Turk J Agric For 30:383–386

Kodan M, Gürkan MO (2004) Investigations on the mass production and storage possibilities of the egg parasitoid, Trissolcus grandis (Thomson) (Hymenoptera: Scelionidae). Paper presented at the second international conference on Sunn Pest, ICARDA, Aleppo, Syria, 19–22 July 2004, p 55

Kozlov MA (1988) Superfamily Proctotrupoidea. In: Medvedev GS (ed) Key to insects of the European portion of the USSR, Part 2, vol 3. Nauka, Leningrad, pp 983–1212

Kozlov MA, Kononova SV (1983) Telenominae of the fauna of the USSR (Hymenoptera, Scelionidae, Telenominae). Leningrad Nauka Publisher 136:336 pp

Laumann RA, Moraes MCB, Pareja M, Alarcao GC, Botelho AC, AHN M, Leonardecz E, Borges M (2008) Comparative biology and functional response of Trissolcus spp. (Hymenoptera: Scelionidae) and implications for stink bugs (Hemiptera: Pentatomidae) biological control. Biol Control 44:32–41

Mansour-Ghazi M, Radjabi GR (2000) Sunn pest tachinid and scelionid parasitoids in Kurdistan. Paper presented at the 14th Iranian Plant Protection Congress, Isfahan University of Technology, Isfahan, 5–8 September 2000, p 219

Martin H, Javaheri M, Radjabi G (1969) Note la punaise des cereals, Eurygaster intergriceps Put. et de ses parasites du genre Asolcus en Iran. Entomol Phytopathol App 28:56–65

Mehravar M, Radjabi G, Shojai M (2000) Introduction of species of egg parasitoids of Eurygaster intergriceps Put. in the region of Isfahan. Paper presented at the 14th Iranian Plant Protection Congress, Isfahan University of Technology, Isfahan, 5–8 September 2000, p 220

Mehrnejad MR (2013) Abundance of parasitoids associated with two major stink bugs on pistachio trees. Appl Entomol Phytopathol 81(1):83–84

Modarres-Awal M (1997) Family Scelionidae (Hymenoptera). In: Modarres Awal M (ed) List of agricultural pests and their natural enemies in Iran. Ferdowsi University Press, pp 279, 429 pp–280

Mohaghegh-Neyshabouri J (1993) A report on *Eurygaster testudinaria* Geoffr. and some of its biological properties. Paper presented at the 11th Iranian Plant Protection Congress, University of Guilan, Guilan, 28 Augest–2 September 1993, p 16

Mohaghegh-Neyshabouri J, Amir-Maafi M (2000) A method on laboratory rearing of predatory bug *Andrallus spinidens* (F.) (Het: Pentatomidae). Paper presented at the second national conference on Optimum Utilization of Chemical Fertilizers & Pesticides in Agriculture, Karaj, February 2000, pp 15–16

Najafi-Navaei I, Bayat-Asadi H, Oskou T (2000) Predatory bug *Andrallus spinidens* F. as laboratory host for mass production of *Trissolcus* spp. important egg parasitoids of sunn pest. Paper presented at the second national conference on Optimum Utilization of Chemical Fertilizers & Pesticides in Agriculture, Karaj, February 2000, p 28

Najafipour M, Iranipour S, Khaghaninia S (2018) Larval competition and multiparasitism between *Trissolcus grandis* and *T. vassilievi* (Hym., Platygastridae), egg parasitoids of sunn pest. Paper presented at the the 23rd Iranian Plant Protection Congress, Gorgan University of Agricultural Sciences and Natural Resources, Gorgan, 27–30 August, p 969

Narehi A, Askarianzadeh A, Taherkhani K (2004) Biological control of sugar cane stem borers with *Platytelenomus hylas* Nixon (Hym.: Scelionidae) in Khuzestan. Paper presented at the 3rd national conference on the Development in the Application of Biological Products & Optimum Utilization of Chemical Fertilizers & Pesticides in Agriculture. University of Tehran, Karaj, February 2004, p 409

Nasiri R, Iranipour S, Karimzadeh R (2020) Host preference of *Ooencyrtus fecundus* Ferriere & Voegele (Hym. Encyrtidae) egg parasitoid of sunn pest and hyperparasitoid of *Trissolcus* spp. Nato Adv Sci I A-Lif (in press)

Noorbakhsh SH, Razavi S (1995) Distribution of sunn pest (*Eurygaster integriceps* Put.) and its natural enemies in Chaharmahal Bakhtiari. Paper presented at the 12th Iranian Plant Protection Congress, University of Tehran, Karaj, 1–6 September 1995, p 15

Noori H, Asgari S (2000) Study and identification of sunn pest egg parasitoid in Qazvin Province. Paper presented at the 14th Iranian Plant Protection Congress, Isfahan University of Technology, Isfahan, p 218

Noori H, Amir-Maafi M, Frouzan M (2011) Introduction of sunn pest egg pentatomids in Qazvin, Iran. Paper presented at proceedings of the Biological Control Development Congress in Iran, Tehran, 27–28 July 2011, pp 417–422

Nozad-Bonab Z, Iranipour S (2010) Seasonal fluctuation in egg parasitoid fauna of sunn-pest *Eurygaster integriceps* Puton in wheat fields on New Bonab County, East Azarbaijan province, Iran. J Sustain Agric Prod Sci 20(3):71–83

Nozad-Bonab Z, Iranipour S (2013) Development of *Trissolcus grandis* (Thomson) (Hymenoptera: Scelionidae) on two factitious hosts *Graphosoma lineatum* (L.) and *G. semipunctatum* (F.) (Hemiptera: Scutelleridae) at three constant temperatures. J Appl Res Plant Prot 1(1):41–54

Nozad-Bonab Z, Iranipour S, Farshbaf-Pourabad R (2014) Demographic parameters of two populations of *Trissolcus grandis* (Thomson) (Hymenoptera: Scelionidae) at five constant temperatures. J Agric Sci Technol 16:969–979

Parvizi R, Javan Moghaddam H (1988) Investigation on some biological features of the sugar-beet weevil (*Lixus incanescens* Boh.) in Azarbayjan. Entomol Phytopathol Appl 55(1–2):1–8

Popov C, Paulian F (1971) Present possibilities of using parasites in the control of cereal bugs. Probleme Agricole 23:53–61

Radjabi G (1994) First report of the existence of sunn pest egg parasitoid, *Trissolcus festivae* Viktorov (Hym.: Scelionidae) in Iran and some preliminary studies on its biology. J Entomol Soc Iran 14:1–7

Radjabi G (1995) Investigations on the various aspects of Hymenopterous egg parasitoids in alleviating the outbreak occurrence of *Eurygaster integriceps* in Iran. Appl Entomol Phytopathol 62:13–14

Radjabi G (2000) Ecology of cereal's sunn pests in Iran. Agricultural Research, Education and Extension Organization, Tehran, Iran, 343 pp

Radjabi G (2001) *Trissolcus esmailii* sp.n. (Hym.: Scelionidae), an egg parasitoid of *Eurygaster intergriceps* Put. and *Dolycoris baccarum* L. in Iran. J Agric Sci 7(2):106–112

Radjabi G, Amir-Nazari M (1989) Egg parasites of sunn pest in the central part of Iranian plateau. Entomol Phytopathol Appl 56(1–2):1–12

Rafat A, Iranipour S, Safavi SA (2013) Fecundity-life tables of *Ooencyrtus telenomicida* (Hym., Encyrtidae), egg parasitoid of *Eurygaster integriceps* (Hem., Scutelleridae). Paper presented at the 2nd global conference on Entomology, Kuching, Malaysia 8–12 November 2013, p 119

Rahnemaye-Shahsavari M, Lotfalizadeh H, Iranipour S, Pourmohammadi S (2011) First report of the subfamily Teleasinae (Hym.: Scelionidae) from Iran. J Field Crop Entomol 1(1):77–79

Rakhshani H, Ebadi R, Rakhshani E (2008) Report of *Telenomus chrysopae* (Hym.: Scelionidae) from Iran. J Entomol Soc Iran 27(2 suppl):21–22

Romanova VP (1953) Oophages de la punaise des cereals d'Apres des observations faites dans la region de Rostov. Zool Zhurn 32(2):238–248

Safavi M (1959) Monographie des Hemipteres Heteropteres del'Iran. Entomol Phytopathol Appl 18:31–39

Safavi M (1960) Progress and experiments in the biology of the sunn pest *Eurygaster integriceps* Put. Entomol Phytopathol Appl 19:14–17

Safavi M (1973) Etude bio-ecologiue des hymenopteres parasites des oeufs des punaises des cereals en Iran. Plant Pests and Diseases Research Institute, Tehran, 159 pp

Sakenin H, Imani S, Shirdel F, Samin N, Havaskary M (2008a) Identification of Pentatomidae (Heteroptera) and their host plants in central and eastern Mazandaran province and introducing of many dominant natural enemies. J Plant Ecol 15:37–51

Sakenin H, Raheb J, Imani S, Havaskary M, Shirdel F, Mohseni H (2008b) A preliminary survey on dipteran predators and parasitoids and Odonata in Iranian rice fields. Paper presented at the national conference of agronomical rice breeding, Ghaemshahr Islamic Azad University, Ghaemshahr, 26–27 November 2008, p 79

Salavatian M (1991) The necessity of studying ecological and biological effective factors in controlling field crop pests. Agricultural Extension Organization Publication, Tehran, Iran

Samin N (2010) Scelionid wasps (Hymenoptera: Scelionidae) and dominant species in wheat fields of Varamin and vicinity. M.Sc thesis of Entomology, Islamic Azad University, Tehran Science and Research Branch, 161 pp

Samin N, Asgari S (2012a) A study on the Scelionidae (Hymenoptera: Platygastroidea) of Golestan province, Iran. CAL 200:1–5

Samin N, Asgari S (2012b) A study on the fauna of scelionid wasps (Hymenoptera: Platygastroidea: Scelionidae) in the Isfahan province, Iran. Arch Biol Sci 64(3):1073–1077

Samin N, Shojai M, Koçak E, Havaskary M (2010a) An annotated list of the Platygastroidea (Hymenoptera) from the Arasbaran biosphere and vicinity, northwestern Iran. Far East Entomol 210:1–8

Samin N, Shojai M, Ghahari H, Koçak E (2010b) A contribution to the scelionid wasps (Hymenoptera: Platygastroidea: Scelionidae), egg parasitoids of Pentatomidae (Heteroptera) in Tehran province, Iran. Efflatounia 10:7–14

Samin N, Shojai M, Asgari S, Ghahari H, Koçak E (2010c) Sunn pest (*Eurygaster integriceps* Puton, Hemiptera: Scutelleridae) and its scelionid (Hymenoptera: Scelionidae) and tachinid (Diptera: Tachinidae) parasitoids in Iran. Linz Biol Beit 42(2):1421–1435

Samin N, Koçak E, Ghahari H, Shojai M (2010d) A checklist of Iranian *Telenomus* Haliday (Hymenoptera: Platygastroidae: Scelionidae: Telenominae). Linz Biol Beit 42(2):1437–1444

Samin N, Shojai M, Koçak E, Ghahari H (2011a) Distribution of scelionid wasps (Hymenoptera: Platygastroidea: Scelionidae) in Western Iran. Klapalekiana 47:75–82

Samin N, Ghahari H, Koçak E, Radjabi G (2011b) A contribution to the scelionid wasps (Hymenoptera: Scelionidae) from some regions of Eastern Iran. Zoosyst Ross 20(2):299–304

Samin N, Shojai M, Asgari S, Ghahari H, Khoddam H (2011c) Population flactuations of the sunn pest, *Eurygaster integriceps* (Hemiptera: Scutelleridae) in the wheat and barley fields, and introducing of its important parasitoids in aestivation and hibernation shelters in Varamin and Shahre Rey, Iran. J Agron Sci 5(6):78–91

Samin N, Radjabi G, Asgari S (2012) Contribution to the knowledge of Scelionidae (Hymenoptera: Platygastroidea) from Khuzestan province, Southwestern Iran. Entomofauna 33(3):17–24

Sayad Mansour A, Latifian M, Soleymannejadeian E, Askarianzadeh AR (2011) Phenology study of parasitoid *Telenomus busseola* Gahan (Hym., Scelionidae) based on egg population of sugar cane stem borer's egg *Sesamia nonagrioides* Lefebvre (Lep., Noctuidae) and comparison of parasitism percent by recruitment and common method. J Entomol Res (Islamic Azad University, Arak Branch) 3(1):29–41

Shafaei F, Iranipour S, Kazemi MH, Alizadeh E (2011) Diversity and seasonal fluctuations of sunn pest's egg parasitoids (Hymenoptera: Scelionidae) in central regions of West-Azarbaijan province, Iran. J Field Crop Entomol 1(1):39–54

Shahpouri-Arani S, Talebi AA, Fathipour Y, Moharramipour S (2005) The comparison of population paramcters in green lacewing, *Chrysoperla carnea* (Steph.) (Neur.: Chrysopidae) and its egg parasitoid wasp, *Telenomus acrobates* Giard (Hym.: Scelionidae). J Agric Sci 11 (1):107–115

Shahrokhi S, Esmaili M, Kharrazi-Pakdel A, Radjabi G (1998) Rearing of *Graphosoma lineatum* (Hem., Pentatomidae) on Various Diets. Paper presented at the 13th Iranian Plant Protection Congress, University of Tehran, Karaj, 1998, p 24

Shamsi M (2014) Fauna of the subfamily Scelioninae (Hym: Platygastroidea) in central parts of East-Azarbaijan province. M.Sc. thesis at Agricultural Entomology, Islamic Azad University, Tabriz branch, Tabriz, Iran, 122 pp

Shamsi M, Lotfalizadeh H, Iranipour S (2014a) Report of *Baeus seminulum* Haliday (Hym.: Platygasteridae, Scelioninae) from Iran. Paper presented at the first National Congress of Biology and Natural Sciences of Iran, 18 December 2014, p 7

Shamsi M, Lotfalizadeh H, Iranipour S (2014b) A review of species of the genus *Idris* (Hym.: Platygasteridae, Scelioninae) in Iran. The first conference on Traditional Medicin, Medicinal Plants and Organic Agriculture, Shahid Mofatteh University, Hamadan 29 November 2014, pp 1–8

Shamsi M, Lotfalizadeh H, Iranipour S (2015a) New finding of *Scelio rugosulus* Latreille (Hymenoptera, Scelioninae) in Iran. Biharean Biol 9(2):162–163

Shamsi M, Lotfalizadeh H, Iranipour S (2015b) Report of *Teleas rugosus* Kieffer, 1908 (Hym.: Platygasteridae, Teleasinae) from Iran. Biocontrol Plant Prot 2(2):79–81

Sharififar S (2000) Mass rearing and release of wasp *Platytelenomus hylas* in Haft tappeh sugarcane fields. Paper presented at the 14th Iranian Plant Protection Congress, Isfahan University of Technology, 5–8 September 2000, p 52

Sheikhmoss S (2009) Ecological and biological study of two species of genus *Trissolcus* spp. The egg parasitoids on sunn pest eggs in Syria. General Commission for Scientific Agricultural Research, Kameshli Research Center. From: http://www.gcsar.gov.sy/gcsarEN/spip.php?article228. Accessed on 10 Feb 2013

Shojai M (1968) Resultats de l'etude faunestiques des hymenopteres parasites (Terebrants) en Iran et l'importance de leur utilization dans la lutte biologique. Paper presented at the 1st Iranian Plant Protection Congress, University of Tehran, Karaj, 1968, pp 25–35

Shojai M (1989) Entomology Vol. III. Ethology, social life and natural enemies "Biological control". University of Tehran Publications, 406 pp

Şimşek N (1986) Investigations on some interactions between Sunn-pest (*Eurygaster integriceps* Put.) and its egg parasite (*Trissolcus semistriatus* Nees) in southern Anatolia. Paper presented at the first Turkish National Congress of Biological Control, University of Adana, Adana, 12–14 February 1986

Şimşek N, Yaşarakinci N (1990) The bioecology of the egg parasites (*Trissolcus* spp.) of sunn pest (*Eurygaster integriceps* Put.) in South-East Anatolia Region, Turkey, pp 81–88. Türkiye Π. Biolojik Mucadele Kongresi, 26–29 Eylül 1990, Ankara, Türkiye

Taghaddosi MV, Rajabi G (1998) Sunn pest egg parasitoids in Zanjan province. Paper presented at the 13th Iranian Plant Protection Congress, Campus of Agriculture and Natural Resources, University of Tehran, Karaj, p 10

Talebi AA, Shahpouri S, Fathipour Y, Moharramipour S (2005) Study on the some biological characteristics of *Telenomus acrobates* Girard (Hym.: Scelionidae), egg parasitoid of *Chrysoperla carnea* (Steph.) (Neur.: Chrysopidae). Appl Entomol Phytopathol 74(1):65–79

Teimouri N, Iranipour S, Benamolaei P (2019) Effect of light intensity and photoperiod on development, fecundity and longevity of *Trissolcus grandis* (Hym.: Platygastridae), egg parasitoid of sunn pest, *Eurygaster integriceps* Puton (Hem.: Scutelleridae). J Appl Res Plant Prot 8 (3):77–93

Torres JB, Pratissoli D, Zanuncio JC (1997) Thermal requirements and development of the egg parasitoids *Telenomus podisi* Ashmead and *Trissolcus brochymenae* (Ashmead) on the predator stink bug *Podisus nigrispinus* (Dallas). Ann Entomol Soc Bras 26:445–453

Triplehorn CA, Johnson NF (2005) Borror and DeLong's introduction to the study of insects. Thomson Books/Cole, CA, USA

Vaezi M (1950) Rapport du laboratoire d'elevage des parasites d' *Eurygaster intergriceps* Put. Entomol Phytopathol Appl 11:27–41

van Driesche RG, Bellows JTS (1996) Biological control. Chapman and Hall Pub, New York

Viktorov GA (1967) Probleme de la dynamique des populations de puniases des cereal. Akad. Nauk. SSSR. Ins Morf. Zhivotn Moscow, 272 pp

Zarnegar A, Noori H (2006) The effect of JHM (pyriproxyfen) on the nymph of sunn bug (*E. integriceps* Put.) in laboratory and field conditions. J Agric Sci 12(1):215–221

Zatyamina VV, Burakova VI (1980) Supplementary feeding of Telenomini. Zashchita Rastenii 10:24

Zatyamina VV, Kletchkovskii ER (1974) Telenomines of the Voronezh region. Zashchita Rastenii 4:32

Zatyamina VV, Kletchkovskii ER, Burakova VI (1976) Ecology of the egg parasites pentatomid bugs in the Voronezh region. Zool Zhurnal 55(7):1001–1005

Zomorrodi A (1962) Les experiences et observation sur la lutte biologique d' *Eurygaster intergriceps* Put. Entomol Phytopathol Appl 20:16–23

# Chapter 9
# Aphid Parasitoids: Aphidiinae (Hym., Braconidae)

**Ehsan Rakhshani and Petr Starý**

## 9.1 Introduction

As a valuable ecological model, the aphid parasitoids (Hymenoptera, Braconidae, Aphidiinae) are among the well known group of insects, exclusively associating with the aphids (Hemiptera, Aphididae). They are a small group of parasitic wasps comprises a number 420–505 species worldwide, mainly distributed in the Holarctic region (Žikić et al. 2017). Among the aphid natural enemies, the Aphidiinae have an important position as biological control agents, which are targeted for both conservation programs and introduction into the new areas for control of invasive pest aphids. Some species are mass reared as commercial products and successfully used for control of important aphids on protected crops. The aphid parasitoids can be found both in natural and agricultural ecosystems, in association with their preferred host aphids. It is believed the aphid parasitoids co-evolved by the divergence of their respective host aphids (Starý 1981), as strictly specific to generalist parasitoids. Species in both confidences were used as successful biological control agents. A complex trophic web is suggested for community structure of insects associated with aphids including parasitoids, predators and pathogens may have in direct and indirect interactions (van Veen et al. 2008).

In biological view, Aphidiine are solitary koinobiont endoparasitoids. The foraging female lays a single egg inside the body of host aphid, which manifest various ovipositing behavior and structure adaptations on its ovipositor depending on the route of co-evolution. The ovarian fluid, as well as the venom are also injected to

E. Rakhshani (✉)
Department of Plant Protection, Faculty of Agriculture, University of Zabol, Zabol, Iran
e-mail: rakhshani@uoz.ac.ir

P. Starý
Institute of Entomology, Biology Center AVCR, České Budejovice, Czech Republic
e-mail: stary@entu.cas.cz

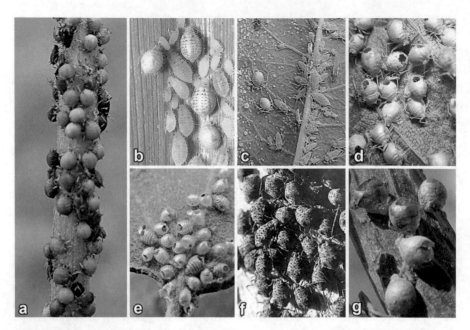

**Fig. 9.1** Various parasitized aphids, which their empty bodies turned into the mummies: (**a**) *Aphis craccivora* Koch – *Lysiphlebus fabarum* (Marshall); (**b**) *Sipha flava* (Forbes) – *Adialytus ambiguus* (Haliday); (**c**) *Hyalopterus amygdali* (Blanchard) – *Aphidius transcaspicus* Telenga; (**d**) *Aphidius matricariae* Haliday – *Dysaphis plantaginea* (Passerini); (**e**) *Chaitophorus populialbae* (Boyer de Fonscolombe) – *Adialytus salicaphis* (Fitch); (**f**) *Pterochloroides persicae* (Cholodkovsky) – *Pauesia antennata* (Mukerji); (**g**) *Cinara tujafilina* (Del Guercio) – *Pauesia hazratbalensis* Bhagat

aphid haemolymph for subsequent regulation of the host development (Falabella et al. 2000). The larva starts feeding the internal body of its host and the aphid will then being killed upon final feeding steps by the mature mandibulate larva. At the end, only the cuticle of host remains which is known as "mummified aphid" (Fig. 9.1). Shape and color of the mummies may even differ based on the species of the parasitoid within or under which spins a cocoon and pupates. The aphid mummy is usually attached to the plant substrate from which the adult parasitoid emerges in few days. Small round exit hole is cut by the emerging parasitoid usually at the posterior part of the mummy, which is different from those cutted by the aphid hyperaparasitoids, the latter manifesting irregular margins.

Various aphids have been considered as pest species in Iran, as well as in other parts of the world. Majority of these aphids are naturally controlled by the native aphid parasitoids in various field crops, fruit orchards and urban areas. The common aphids in greenhouses, *Aphis gossypii* Glover and *Myzus persicae* (Sulzer), need further attention, since they have selected protected habitats and can easily escape from their natural enemies, including parasitoids. Some aphid parasitoids including *Aphidius colemani* Viereck, *Aphidius matricariae* Haliday, *Aphidius ervi* Haliday and *Lysiphlebus testaceipes* (Cresson), are commercially produced mainly for biological control of the pest aphids on protected crops (Boivin et al. 2012). The recent attempts in reducing the amount of pesticides in the protected crops in Iran led to limited application of these biological agents, which are essentially not produced in Iran. Few attempts have been done for evaluation of the efficiency of endemic species that needs to be organized within a more robust organization for subsequent steps until commercial production of efficient aphid parasitoids.

In this chapter, the background information on the aphid parasitoids in Iran, including basic and applied aspects are provided. Taxonomic studies both on primary and secondary aphid parasitoids in Iran have been reviewed that provided a rich database of trophic associations. Such an approach is necessary for the subsequent biological studies, which provide the framework for selection of potential candidates in various biological control programs. Similarly, the significance of conservation programs on the basis of adequate knowledge about habitat manipulation is discussed. Many aphids in this respect are economically unimportant, but may have a critical role, when positioned inside the host range of the parasitoid species, those parasitized the economically important aphids in the course of growing season. Finally, the preliminary attempts on the potential mass rearing of efficient parasitoids are reviewed and compared.

## 9.2  Recent Advances in Taxonomy of Aphid Parasitoids

### 9.2.1  Knowledge About Taxonomy of Aphid Parasitoids in Iran

Research on the Aphidiines has been fortunate due to several reasons in Iran. Whereas Iran has prevailingly been focused as an area of "developing world" yielding promising biocontrol agents within a framework of biocontrol programmes in the abroad (namely for USA aphids on alfalfa, walnut aphids), later fundamental studies have been authorized by sound research fellows in Europe, with more or less participation of the Iranian specialists, and this phenomenon has increased in the subsequent years. The co-work has been realized between specialists on the hosts-the aphids, together with the parasitologists. Both the local and foreign workers closely participated in the research. Additionally, many foreign specialists have had personal experience in the research on the targets. The basal approach of field studies has been initiated from the sampling of tritrophic associations (plant-aphid-parasitoid) in various ecosystems and agroecosystems. Such an approach has allowed later elaboration of databases which have been useful in faunal studies, determinations of the associations including those on the pest species and ecosystem relation. At last, increase of knowledge of molecular methods not only in the world but also in the local research institutions has allowed up-dated use resulting in DNA studies and their application in the taxonomy, population studies compatible with the world levels and interactions.

The early evidences on fauna of Aphidiinae in Iran have been summarized in Starý et al. (2000) which includes 49 species belonging to 13 genera. Since then, many other attempts have been done on biosystematics of the aphidiine parasitoids, including tritrophic associations in different parts of Iran (Starý et al. 2005; Rakhshani et al. 2005a, b, 2006a, b, 2007a, b, 2008a, b, c, d, 2012a, b, 2013; Nazari et al. 2012; Tomanović et al. 2007; Kazemzadeh et al. 2009; Barahoei et al. 2012, 2013; Jafari and Modarres Awal 2012; Mossadegh et al. 2011; Rakhshani 2012; Alikhani et al. 2013; Farahani et al. 2015; Hadadian et al. 2016). Iran includes a diverse array of habitats and seems to be the most explored country in the Middle East, concerning the aphid parasitoids. Barahoei et al. (2014) listed 78 species of 17 genera from Iran which includes some suspicious records/and unidentified species. The last up-dated list of Aphidiinae of Iran (Farahani et al. 2016) includes 73 species and 15 genera, in which the genus *Toxares* Haliday is excluded. Species of the genus *Euaphidius* Mackauer (*Euaphidius cingulatus* and *Euaphidius setiger*) are also classified within *Aphidius* Nees. By including the missing record of *Lysiphlebus cardui* (Alikhani et al. 2013), as well as the newly recorded species, *Monoctonia pistaciaecola* (Kargarian et al. 2016), number of Aphidiinae in Iran reached to 75. Yet, it is expected many more species are occurring in several unexplored areas of the country. Many aphids have not yet the recorded parasitoid in Iran and no aphid parasitoid is known in association with root aphids (Rakhshani et al. 2019).

## 9.2.2   Importance of Accurate Identification

For a successful biological control program, it is critical to search for effective natural enemies, including a very careful and accurate identification, which sometimes led to complete failure. This is important both on the host aphid and its parasitoid, especially at classical biological control, where a huge amount of resources are spending for laboratory and field experiences (Rosen 1986). The situation is more complicated about closely related taxa. Concerning the aphids, it is critically important to find the identity of the host plants, too. It will give a very useful perspective about the aphids living on these plants based on the literature. In his comprehensive handbooks, Rezwani (2001, 2004, 2010) provided reliable data on the host range of aphids, as well as the key for identification at the species level. In general, there is considerable amount of available data for the aphids of economic importance, but in the course of biological control programs, it is needed to have an expanded knowledge about the host range of the parasitoids, to be identified as much as possible. It is sometimes necessary to have a medium to large scale rearing for the parasitoid species on the alternative aphid host which should be very carefully selected and identified. In every case, an expert aphidologist should confirm the identifications.

Like many other braconids, identification of Aphidiinae species are very difficult. Beside the occurrence of intraspecific variabilities on morphological characters, lacking the reliable characters in some genera made several inconsistencies. The identification is mainly based on the female specimens and it is necessary to separate them very carefully. The male specimens generally have a stout and apically truncated gaster, and their body is darker than females. They have more number of antennal segments that bearing more pubescence, comparing the females. In the genus *Ephedrus*, both male and females have 11-segmented antennae. There is a series of identification keys on the aphidiinae of Iran, segregating species of various genera (Rakhshani et al. 2007a, 2008a, 2012a), specific guilds (Rakhshani et al. 2006a, b, 2007b, 2011) or regional fauna (Rakhshani et al. 2012b; Barahoei et al. 2012, 2013). Identification at the generic level is better to be done based on the keys covering a broader area (Starý 1976, 1979). There are many morphological characters used for identification at generic/species level in Aphidiinae. The characters in the head (antennae, maxillary and labial palpomeres) as well as on mesosoma (forewings, propodeum) and metasoma (petiole, female genitalia) bearing the main diagnostic criteria (Fig. 9.2).

Normally the female genitalia acquired the adaptive evolution for best handling of the specific host aphids. The generalist parasitoids have also their own adaptations to cover a more general range of host aphids (Völkl and Mackauer 2000). Therefore, it has the most value for the expert taxonomists on Aphidiinae, as well as a few other characters combined. It is needed in many cases to make slide at least from the most important characters. Usually the coloration has a least value for identification of Aphidiinae because of intraspecific variability and also great dependence to the environmental factors, including temperature (Shu-Sheng and Carver 1982).

According to the moderately well-defined host range pattern of the aphid parasitoid, it is possible to do the identification in an easier way. In the case of some genera including *Aphidius* and *Praon*, it is almost impossible to have a trustful identification without host data. Information about host plants and habitat give also very useful insights helping the identification. For example, a well-known assemblage of the aphid parasitoids inhabiting alfalfa fields and it is also possible to predict the occurring species based on the regions they were collected in Iran (Rakhshani et al. 2006a, 2007b). Species-specific molecular markers have also been used for identification of complicated species like *Lysiphlebus fabarum* (Marshall) (Rahimi et al. 2012; Farrokhzadeh et al. 2014), *Aphidius transcaspicus* Telenga (Jafari et al. 2011) and *Ephedrus persicae* Froggatt (Farrokhzadeh et al. 2014). The complicated species, *Lysiphlebus fabarum* is believed to be a young and under evolution lineage (Rakhshani et al. 2013) including many cryptic species (Starý et al. 2014). The recent surveys on the taxonomy of the genus *Lysiphlebus* resulted in identification of some new species as well as clarifications on the taxonomy of some cryptic species (Starý et al. 2010; Tomanović et al. 2018). *Aphidius colemani* Viereck was known in Iran for several years (Starý et al. 2000; Rakhshani et al. 2008a, b, c, d), as a common and efficient aphid parasitoid. Great biological variations in ecology and host range of *A. colemani* led to confusion on its taxonomy including a long list of synonyms (Starý 1972a, b, 1973; Takada 1998; Kavallieratos and Lykouressis 1999). Recent molecular and morphological analyses (Tomanović et al. 2014) have concluded the Iranian specimens all belong to *Aphidius platensis* Brethes and true *Aphidius*

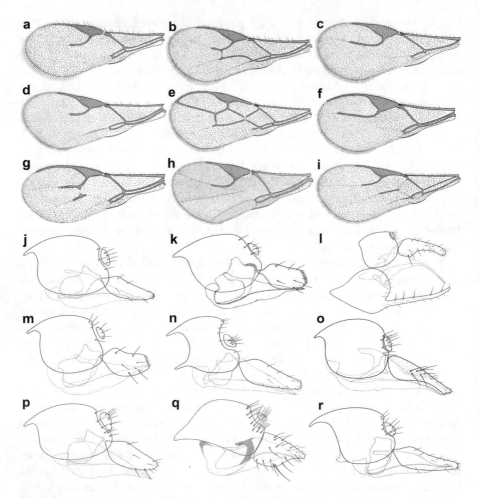

**Fig. 9.2** A schematic view of the two main diagnostic characters in Aphidiinae: Fore wing venation and female genitalia. (**a, j**) *Adialytus ambiguus* (Haliday); (**b, k**) *Aphidius platensis* Brethes; (**c, l**) *Binodoxys acalephae* (Marshall); (**d, m**) *Diaeretiella rapae* (M'Intosh); (**e, n**) *Ephedrus persicae* Froggatt; (**f, o**) *Lipolexis gracilis* (Förster); (**g, p**) *Lysiphlebus fabarum* (Marshall); (**h, q**) *Monoctonia vesicarii* Tremblay; (**i, r**) *Praon barbatum* Mackauer

*colemani* is distributed in the Mediterranean area (Kavallieratos et al. 2004; Rakhshani et al. 2015b). Re-examination of the voucher specimens used in studies on the biology of *Aphidius colemani* in Iran (Zamani et al. 2006, 2007, 2012; Talebi et al. 2006) is necessary to confirm the recorded data, as they most probably belong to *Aphidius platensis* Brethes.

## 9.2.3   Host Associations, Parasitoid Assemblages

In many cases, exact identification of the aphidiinae parasitoids is not possible without host data, both aphid and its host plant. There are many aphidiinae species described without host data (Rakhshani et al. 2012c; Davidian 2016), but further identification would need comparison with the type specimens. Data about host associations greatly helps by search through a narrow assemblage of species and generally gives more reliable identification. Based on these requirements, many researchers tried to work on the reared material, which their host aphids and plants are already identified. Aphids are group of insects with mostly very simple rearing methods and their colonies consisting parasitized individuals can be easily kept inside the plastic boxes at room temperature. In order to increase the rate of parasitoid emergence, it is recommended to remove all other insects, which are associated with the same plant material. It is critically important with the predators like coccinellids, chrysopids, anthocorids, etc. Ignoring other insects can also have led to emergence of the parasitoid species, which were incorrectly assigned as aphid parasitoids (see Samanta et al. 1983). Many aphid species occur on the same host plant and at same time with mixed or overlapped colonies. Rearing the mixed aphid colonies result to incorrect host association records (see Jafari and Modarres Awal 2012).

The pattern of host specificity was preliminarily classified based on their host aphid phylogeny (Starý 1981), by which five subgroups were defined. It ranges from the parasitoids associated with single host aphid species to those parasitizing the species from several genera in two or more aphid subfamilies. In the recent most comprehensive analysis, Žikić et al. (2017) suggested three groups including specialists, generalist and oligophagous species. The latter group was also subdivided into three subgroups of narrow, moderate and broad oligophagous species.

The aphids occur on various host plants, both in natural and agro-ecosystems manifesting a heterogeneous parasitoid complex of different genera (Rakhshani et al. 2005a, 2007b, 2011). On the other hand, there are aphids with only a single parasitoid species or even without parasitoids, at least in Iran and adjacent countries. Some parasitoid species can be found in the colonies of many common aphids in ago-ecosystems and their nearby areas, which have a great potential to shift their host aphid, when the environments is changing. In general, it is important to have the parasitoid species with a narrow and preferably single host specific behavior in biological control programs. It is expected that the specific parasitoids are more successful in control of their host aphid, since they will not disperse around on the other non-target host aphids. In such a way, the environmental risk of ecological interference can also be minimized. Surprisingly, the important parasitoid species are seldom host specific, especially those are commercially produced. Among them, *Lysiphlebus testaceipes* is known as a parasitoid with unbelievably wide host range pattern (Starý et al. 2004). It was found occasionally in Iran, reared from *Aphis craccivora* (Rakhshani et al. 2005a), but never recovered again. It could be an occasional application of commercially reared specimens by the local peoples or an invasion through the neighbouring countries (Iraq, Turkey). Further exploration is needed to find the source and the reasons, which disclosed the expansion of this opportunistic species into Iran. More than 500 parasitoid-aphid associations are recorded in various regions of Iran (Barahoei et al. 2014; Farahani et al. 2016) including the data about parasitoid complex of many pest aphids. Further investigations with emphasis on the natural habitat and especially on the area surrounding the field crops and orchards is needed to realize the host range pattern of the known species throughout the season, as well as to determine the rare parasitoids, which strictly attack the aphids on their natural and intact environments.

### 9.2.3.1 Host Specific Aphid Parasitoids

There are many monophagous species of aphid parasitoids, which are mainly considered as rare species with unknown significance in biological control. Host specificity in the aphid parasitoids is believed to be a result of parallel evolution, apart of their phylogeny. The host plant distribution has a signification attribution in this respect (Žikić et al. 2017). Strict host specificity increased to cost of ecological interaction for the parasitoid species and it needs to acquire additional adaptations, which are not necessary for other species, in general. The well-known species is *Monoctonia vesicarii* Tremblay which is specific parasitoid of *Pemphigus* species on poplars (Rakhshani et al. 2015a). It has only one generation per year and spending most of the year as diapausing pupa inside the mummified aphid (Tremblay 1991). The term "monophagous" should be re-defined in the case of aphid parasitoids, referring the species which are associated with aphids in less diverse genera or even monotypic. So that the parasitoids have no chance to find an appropriate host among the related aphid taxa. Beside this, there are groups of specialized parasitoids evolved within the more diverse aphid taxa, which can parasitize the closely related aphids within the small genera. The border of their host range seems to be defined ecologically by the habitat properties and host plant diversity. It means the parasitoid can find its specific hosts only on the host plants of the same family level. According to this explanation, this group of parasitoids can be categorized in a group named as "narrow oligophagous" species.

There are some species that are very specific or have a very narrow host range. They are mainly targeted in classical biological control programs of the invasive aphids of economically important. The list of species exported from Iran are shown in Table 9.1, among them, the specific parasitoids are dominant. *Trioxys pallidus* (Haliday) is a specific parasitoid of *Chromaphis juglandicola* (Kaltenbach), but it is also recorded in association with closely related aphids on different host plants, including *Eucalipterus tiliae* L., *Myzocallis coryli* (Goetze), *Tuberculoides annulatus* (Hartig) (Starý 1978), *Pterocallis alni* (De Geer) (Babaee et al. 2000) and some other species. So, how it can be a specific parasitoid? and why many other similar cases have been considered as generalist?. *Trioxys pallidus* is native to Iran and can be found with its host aphid throughout the country. The association with others, but closely related aphids has been confirmed and discussed for this species

**Table 9.1** The representative parasitoid species with narrow host range, in Iran

| Parasitoid species | Host aphid | Host plants |
|---|---|---|
| *Adialytus ambiguus* (Haliday) | *Sipha* spp | Cereals |
| *Adialytus salicaphis* (Fitch) | *Chaitophorus* spp. | Poplars and willows |
| *Adialytus thelaxis* (Starý) | *Thelaxes* spp. | Oaks (*Quercus* spp.) |
| *Aphidius absinthii* Marshall | *Macrosiphoniella* spp. | Wormwoods (*Artemisia* spp.) |
| *Aphidius eadyi* Starý, González & Hall | *Acyrthosiphon pisum* | Alfalfa and clover |
| *Aphidius hieraciorum* Starý | *Nasonovia* spp. | Asteraceae, etc.. |
| *Aphidius funebris* Mackauer | *Uuroleucon* spp. | Asteraceae |
| *Aphidius popovi* Starý | *Amphorophora catharinae* | *Rosa* spp. |
| *Aphidius persicus* Rakhshani & Starý | *Uroleucon* spp. *Macrosiphoniella* spp. | Asteraceae |
| *Aphidius rosae* Haliday | *Macrosiphum rosae* | *Rosa* spp. |
| *Aphidius smithi* Sharma & Subba Rao | *Acyrthosiphon pisum* | Alfalfa and clover |
| *Aphidius salicis* Haliday | *Cavariella* spp. | Salix and crucifers |
| *Aphidius transcaspicus* Telenga | *Hyalopterus* spp. | *Prunus* spp. |
| *Areopraon lepelleyi* (Waterson) | *Eriosoma lanuginosum* | *Ulmus* spp. |
| *Betuloxys hortorum* (Starý) | *Tinocallis* spp., *Myzocalis* spp. | *Ulmus, Castanea, Corylus* |
| *Euaphidius cingulatus* Ruthe | *Pterocomma* spp. | Poplar and willows |
| *Euaphidius setiger* Mackauer | *Periphyllus* spp. | Maples (*Acer* spp.) |
| *Pauesia antennata* (Mukcrji) | *Pterochloroides persicae* | *Prunus* sp. |
| *Praon absinthii* Bignell | *Macrosiphoniella* spp. | Wormwoods (*Artemisia* spp.) |
| *Praon barbatum* Mackauer | *Acyrthosiphon pisum* | Alfalfa, clover |
| *Praon exsoletum* (Nees) | *Therioaphis trifolii* | Alfalfa, clover |
| *Praon pubescens* Starý | *Nasonovia* spp. | Asteraceae |
| *Praon yomenae* Takada | *Urolecuon* spp. | Asteraceae |
| *Trioxys asiaticus* Telenga | *Acyrthosiphon* spp. | *Sophora* spp. *Glycyrrhiza* spp. |
| *Trioxys cirsii* (Curtis) | *Drepanosiphum platanoidis* | Maples (*Acer* spp.) |
| *Trioxys complanatus* Quilis | *Therioaphis trifolii* | Alfalfa, clover |
| *Trioxys pallidus* (Haliday) | *Chromaphis juglandicola* *Hoplocallis pictus* *Pterocallis alni* *Eucalipterus tiliae* *Tinocallis* spp. | Walnut (*Juglans* spp.) Oak (*Quercus* spp.) Alder (*Alnus* spp.) Linden (*Tilia* spp.) Elms (*Ulmus* spp.) |

(Starý 1978). While, it is clearly evidenced that the host range/host specificity of the aphid parasitoid can be changed by the different races (Antolin et al. 2006; Henry et al. 2008), they exhibit a distinct preference for some aphid hosts (Starý 1978).

Therefore, *Trioxys pallidus* has been considered as the most important biological control of the walnut aphid, *Chromaphis juglandicola*, but it can be found rarely on other species. The same pattern is recorded from many other *Trioxys* species (Starý 1978), while *Trioxys complanatus* can be an exceptional case, which showed a great preference both on its specific host aphids, *Therioaphis trifolii* and the host plants, alfalfa and clover (Rakhshani et al. 2006a – alfalfa aphids). It is a good representative of successful classical biological control program, when introduced into USA at 1956 and it is now distributed in the North America (van den Bosch 1957; Flint 1980) and in Australia (Hughes et al. 1987).

The number of the narrow oligophagous species therefore covers the majority of Aphidiinae species both in Iran and other parts of the world. The representative species with their confirmed specific host range is presented in Table 9.1.

### 9.2.3.2   Broadly Oligophagous Aphid Parasitoids

Assigning the term "polyphagous" to the aphid parasitoids is essentially incorrect. Almost all parasitoids indicated some patterns of host specificity. Aphidiines are exclusively parasitoids of the aphids, which means a natural definition opposite to polyphagy. There is no absolute criterion to separate "narrow" and "broadly" oligophagous groups, based on the dynamic and poorly known biology and evolutionary history of the aphids and their parasitoids. The main aspect of this categorization is the rate of diversification in the aphid lineages. This statements might be more simplified by some examples, but clearly referring to association of a known parasitoid with aphid species of single highly diverse genus or several related genera. On the other hand, there are some parasitoids species, like *Praon volucre* (Haliday) acquired a more expanded host range. It was found in association with aphids of the genera, *Acyrthosiphon* Mordvilko (2 species), *Amphorophora* Buckton (1), *Aphis* L. (8), *Brachycaudus* van der Goot (4), *Diuraphis* Aizenberg (1), *Dysaphis* Börner (1), *Hyalopterus* Koch (2), *Hyperomyzus* Börner (1), *Macrosiphum* Passerini (1), *Macrosiphoniella* del Guercio (1), *Metopolophium* Mordvilko (1), *Myzus* Passerini (2), *Phorodon* Passerini (1), *Rhopalosiphum* Koch (2), *Schizaphis* Börner (1), *Sitobion* Mordvilko (1) and *Uroleucon* Mordvilko (4) in Iran. *Lysiphlebus fabarum* (Marshall) has been known as a parasitoid with a vast number of recorded host aphids, at least 65 species in Iran (Farahani et al. 2016). But surprisingly, its real host range is limited to few closely related aphid genera, *Aphis, Brachycaudus* and *Myzus* (Rakhshani et al. 2013). Other host records contributed to occasional associations or possibly separated but cryptic species inside the *Lysiphlebus fabarum* complex. There are many important aphid parasitoids including *Aphidius ervi* Haliday, *Aphidius platensis* Brethes, *Aphidius matricariae* Haliday, *Diaeretiella rapae* (M'Intosh) and *Ephedrus persicae* Froggatt, of which some of them are now the biological control products, commercially sold.

An important aspect on the biology of these species is that their host range covers many aphid pests on the field crops, fruit trees and greenhouses. Near to hundred aphid species infesting 180 plant species (43 families) are listed as hosts of

*Diaeretiella rapae* throughout 87 countries (Singh and Singh 2015). The cabbage aphid, *Brevicoryne brassicae* (L.) as its preferred and frequently encountered host aphid is considered as ancestral association for *Diaeretiella rapae* (Némec and Starý 1994; Pike et al. 1999). This aphid has no other parasitoid, indeed and other records seem to be the result of the observations on the mixed colonies (especially with *Myzus persicae*). *Diaeretiella rapae* has also a strong affinity to the Russian wheat aphid, *Diuraphis noxia* (Mordvilko) in cereal fields (Rakhshani et al. 2008b). It is yet unclear how this parasitoid species is able to overcome the host range barriers, while the genetic differences bearing not enough insights and correlations with its diversification and host range pattern (Baer et al. 2004).

## 9.3   Biology of Aphid Parasitoids

The followed fundamental research approach requires basically a good level of identification of all the food-web participants. A useful respective identification co-work net has to be organized as a background interacting with the database development. Many studies are contributed to various aspects of ecology and biology of the important aphid parasitoids including their seasonal occurrence (Monajemi and Esmaili 1981; Rakhshani et al. 2000, 2004b, 2009; Ghotbi Ravandi et al. 2017), life history and demography (Talebi et al. 2002, Tahriri et al. 2010; Tazerouni et al. 2012a, b, 2013; Zamani et al. 2007, 2012; Pourtaghi et al. 2016; Ameri et al. 2015), functional response and foraging behavior (Takalloozadeh et al. 2004b; Fathipour et al. 2006; Zamani et al. 2006; Tahriri et al. 2007; Rasekh et al. 2010a, b, c, d; Bazyar et al. 2012; Farhad et al. 2011; Tazerouni et al. 2011, 2012c, 2016, 2017; Ameri et al. 2013; Pasandideh et al. 2015) and host preference (Rakhshani et al. 2004a; Takalloozadeh et al. 2004a; Talebi et al. 2006; Tahriri et al. 2007; Pasandideh et al. 2015; Najafpour et al. 2016). These surveys are generally focused on the economically important aphids and their parasitoids in laboratory conditions, and in the field to a lesser extent. There are yet many important aphid parasitoids for which almost no biological data are available. Valuable data were achieved on the basis of laboratory experiments that should be expanded in the natural conditions, too.

### 9.3.1   Seasonal Occurrence

Aphid communities are characterized by their complicated trophic interactions and the associated natural enemies, among them the parasitoids have usually more dependence and adaptations to the host population. There is a moderately wide spectrum of the aphid parasitoids in the field crops, represented by the wheat (Rakhshani et al. 2008b) and alfalfa (Rakhshani et al. 2006a). Very few studies on the population dynamics and seasonal occurrence of the aphid parasitoids were realized in Iran (Rakhshani et al. 2000, 2004b, 2009; Monajemi and Esmaili 1981; Amini et al. 2012). Usually the population of both narrow and broadly oligophagous parasitoids fluctuate with their host aphid, where they are living in. Specific parasitoids are generally more adapted to the seasonal dynamics in the population of their host to reduce the risk of starvation. They need also the same preferred host for summer hibernating and overwintering within its empty body, as mature larva or pupa. The local survey on the seasonal occurrence of *Trioxys pallidus* on walnut orchards (Rakhshani et al. 2004b) is a good representative for the adaptation of the specific aphid parasitoids. The general scheme of seasonal activity for this species is an early emergence in the spring together with its host aphid, *Chromaphis juglandicola*, which sharply declined in early summer, so that no aphids and parasitoids were observed during the summer. Rearing the mummies collected in early summer did not give adults in subsequent weeks, indicating existence of an aestival diapause, which is expected to prolong until next spring. On the other hand, the weak resurgence of aphid population in early autumn led to appearance of few parasitoids, which successfully parasitized even the sexual form of *C. juglandicola*. There is no evidence explained, what was happened for the mummies went into the aestival diapause, and led to emergence in autumn, partly. Preservation of the mummies collected during autumn in natural condition confirmed existence of an adaptive diapause, which synchronize the emergence of the parasitoids with its specific host in walnut orchards. The mid-season crash in the host aphid populations is believed to be the consequence of some natural phenomena including decline in plant nutritional quality, increased natural enemy pressure and extreme temperature (Karley et al. 2004) directly affecting the population of the specific parasitoids.

Similar adaptation can be found in the complex environments as in alfalfa field, where each parasitoid allocated a specified niche including the preferred ecological or physiological host aphids. There is a well-known group of aphids associated with alfalfa including, *Acyrthosiphon kondoi* Shinji, *A. pisum* (Harris), *Therioaphis trifolii* (Buckton) and *Aphis craccivora* in many parts of Iran. Each aphid species manifests a specific parasitoid complex, which seems not to have interference on each other. Among the parasitoid of the alfalfa aphids, two species, *Aphidius ervi* and *Lysiphlebus fabarum* are known as broadly oligophagous species, but they act as locally specific parasitoids in alfalfa ecosystem, associating only with their preferred hosts, *Acyrthosiphon pisum* (*A. kondoi*, as well) and *Aphis craccivora*, exclusively. *Praon volucre* was also found commonly in association with *Acyrthosiphon* aphids and very rarely on *Aphis craccivora* (Rakhshani et al. 2006a). The general trend of the seasonal dynamics of the parasitoid species is well representative of their host aphid, while the maximum rate of parasitism and occurrence of summer decline differ according to the climatic adaptation of the aphid/parasitoid species (Rakhshani et al. 2010). In the case of *Therioaphis trifolii*, two parasitoid species, *Praon exsoletum* Nees and *Trioxys complanatus* Quilis occurred simultaneously in major parts of the country, except few desertic areas in early summer. High populations of both species are generally active during spring, but *Trioxys complanatus* showed more activities in early summer, when *Praon exsoletum* disappeared. These observations need supplementary investigation to be confirmed and clarified. It is also necessary to investigate the capacity of aphid parasitoids in suppressing aphid populations under field conditions (Khajehzadeh et al. 2010), as well as surveys targeting to find how to manipulate the habitats for achieving the highest rate of parasitism and survival of the parasitoids.

## 9.3.2  Reproduction and Life Table Statistics

Among the natural enemies of a given pest aphid, those with greater specific association have higher priority in biological control programs. In the basic theory, the parasitoids have to combat against continuous impacts of the adverse environments even in their native area. So, they should acquire enough natural capacity to compensate mortalities and to establish the next generation. Rate of reproduction is the key element, which is closely connected to the survival of the populations through long term adaptations. Two non-equal terms fertility and fecundity are used for assessing the potential of growth rate in population of both pests and their natural enemies (Jervis and Kidd 1996). The reproduction of natural enemies generally defined with a statistical parameter known as "Intrinsic rate of increase $= r$" which is assessing in a set of laboratory conditions. Wide range of laboratory setup and analytical method are commonly used (Carey 1993; Chi and Su 2006) for the life table construction of the natural enemies including the aphid parasitoids. The various parameters within the life table statistics are the simple translation of a complicated reproductive strategy, including host selection, competition, rate of oviposition, survival, superparasitism, developmental time, sex ratio, mutual interference etc., which affects the fitness and dynamics of the host–parasitoid interactions (Roitberg et al. 2001). The intrinsic rate of increase of a parasitoid related to its hosts play an important role in biological control programs (Khatri et al. 2017). Parasitoids with the greatest value for the intrinsic rate of natural increase are obviously better candidates in biological control program. At the same way, it is very important that the selected parasitoid has a greater value comparing its host aphid (Bigler 1989).

The progeny sex ratio as main factor among the life table statistics, also affects the stability of the host parasitoid interactions and rate of controlling efficiency for the aphid parasitoids, and other natural enemies, as well. Various parameters including host aphid (Mackauer and Kambhampati 1988a; Pandey and Singh 1999; Jarosik et al. 2003; Kairo and Murphy 2005; Sidney et al. 2011), temperature (Pandey and Singh 1998; Singh et al. 2000; Zamani et al. 2007, 2012; Tazerouni et al. 2012b; Silva et al. 2015), intraspecific competition (Mackauer and Völkl 2002; West, 2009), and maternal effects (Hofsvang and Hägvar 1975; Godfray 1994), significantly affect the progeny sex ratio and life history parameters of the aphid parasitoids. Virgin female produce only male offspring, a phenomenon that frequently happened in the mass rearing and also in the aged females (Talebi et al. 2002; Sequeira and Mackauer 1993; Chi and Su 2006). Reproduction is a high temperature dependent process among insects inevitably alters their population dynamics, which has broad evolutionary and ecological consequences (Frazier et al. 2006).

Few studies were done on the reproductive life table of the aphid parasitoids in Iran, mainly focusing on the species associated with most common and economically important aphid species. Talebi et al. (2002) compared the reproductive rate of the walnut aphid, *Chromaphis juglandicola* (Kaltenbach) with its specific parasitoid, *Trioxys pallidus* (Haliday). The parasitoid showed a greater intrinsic rate of increase comparing its host aphid and can be an efficient biological control agent in its native area, which was already confirmed by introduction of this species in the in North America (van den Bosch et al. 1962, 1979), too. The subsequent studies were done on the parasitoids that have potential for mass rearing and an augmentation biological control (Zamani et al. 2007, 2012; Bagheri-matin et al. 2009; Tahriri et al. 2010; Tazerouni et al. 2012a, b, 2013; Pourtaghi et al. 2016; Mottaghinia et al. 2017). As a tritrophic study, the effect of soil media containing different vermicompost compositions on life history parameters of *A. matricariae* was also investigated (Mottaghinia et al. 2017). Those results confirmed the significant effect of soil composition on population parameters of *A. matricariae*. Both *Aphidius colemani* Viereck (=*Aphidius platensis* Brethes in Iran) and *Aphidius matricariae* Haliday are well-known parasitoids of the pest aphids on protected crops (*Aphis gossypii* Glover and *Myzus persicae* (Sulzer)) and their life tables were evaluated with more details (Zamani et al. 2007, 2012; Pourtaghi et al. 2016; Tazerouni et al. 2017), but yet many other aspects including the effects of host aphid size, inter and intra-specific competition (host aphid density), food plant species, semiochemicals, host aphid resistance, microbial symbiosis and genetics need further attention. Accordingly, it has been revealed that the interspecific interaction between *A. matricariae* and *P. volucre* on *M. persicae* influenced on biological traits of both parasitoids and their population size (Tazerouni et al. 2016, 2017). It is recently reported (unpublished) that commercially produced race of *Aphidius matricariae* (Koppert B.V.) did not accept *Aphis gossypii*, while this is a serious pest in the greenhouses. It might be a local drift of the reared population, sourced in Europe, but the Iranian population successfully parasitized this aphid both in natural (Rakhshani et al. 2008a) and laboratory condition (Zamani et al. 2012). It has worth to re-emphasize that the second commercially produced species, *Aphidius colemani* Viereck did not occur in Iran and the previous records (Rakhshani et al. 2008a, b, c, d; Zamani et al. 2007, 2012) all refer to *Aphidius platensis* Brethes (Tomanović et al. 2014), which seems not to be distributed in Europe. The specimens which were used in the studies by Hofsvang and Hägvar (1975) might also being originated from the area outside Europe, i.e. South America.

In environmental view, life table studies providing important data for the ecosystem management in the course of conservation biological control programs. There is no space to ignore the importance of reproduction rate and sex allocation strategies at the time of mass rearing for the aphid parasitoids. Considering the general problems exist in the practical mass rearing of the aphid parasitoids, the greater value of reproduction under controllable sex determination mechanisms (Heimpel and de Boer 2008; Singh et al. 2014) is rather critical. Beside the crucial parameters affecting the sex ratio, the food plant on which the host aphids are propagating, is also important (Prasad et al. 2005).

### 9.3.3   Functional Response and Foraging Behavior

The behavioral characteristics of the aphid parasitoids have certainly a great impact on their efficiency for the successful aphid biocontrol (Berryman 1999). Assessing the foraging behavior and functional response of the aphid parasitoids in one of the important criteria for selecting the effective species. It generally indicated how the parasitoid influence the population dynamics of its host species (Jervis and Kidd 1996). In the simplest definition, functional response is the number of hosts that successfully attacked per each parasitoid, depending to the host density (Solomon 1949). Various models are developed for clarification of the functional response of the parasitoids. The host exploitation by the aphid parasitoids may fall into Holling's type II or the type III with a sigmoid response (van Steenis and El-Khawass 1995). Parasitoids with type II functional response are not density dependent and their parasitism rate does not increase with host density (Hassell 1978). The parasitoids with a type III functional response, can regulate the population of their host to a definite density, since they can increase the rate of parasitism with increases in host density increase (Holling 1959). This behavior maximizes the exploitation of dense patches (Stilmant 1996). It was suggested that the parasitoids with a type II functional response are able to control the pest aphid more effectively on the protected crops, where the parasitoid can reach its hosts even at low densities (Lopes et al. 2009). The parasitoids with type III functional response are likely more successful in the field conditions, especially in classical biological control program, where such a density dependent regulation provide a stable and long term control (Fernandez-Arhex and Corley 2003).

In many cases, a type II functional response was recorded (Table 9.2), but few species of the aphid parasitoids showed the type III functional response (Rakhshani et al. 2004a; Jokar et al. 2012; Tazerouni et al. 2016, 2017). Various factors directly affect the functional response or its attributed parameters (Jervis and Kidd 1996). It has been shown that temperature has no effect on the type of functional response but may change some of its parameters (Zamani et al. 2006; Farhad et al. 2011; Tazerouni et al. 2012c; Moayeri et al. 2013; Hajrahmatollahi et al. 2015). Handling time is defined as the time used for handling and parasitizing the host, including the time for resting, feeding and cleaning (Jervis and Kidd 1996). Significantly higher handling time was recorded at the lowest temperature, which is likely a result of non-searching activities including resting (Moayeri et al. 2013; Hajrahmatollahi et al. 2015). The higher temperatures were also found to negatively affect the handling time (Farhad et al. 2011). The host plant can also change the foraging behavior and functional response of the parasitoids (Price 1986). It was shown that the host plant variety has also significant effect on the type of functional response in aphid parasitoids (Bazyar et al. 2012).

**Table 9.2** The functional response of aphid parasitoids surveyed in Iran

| Parasitoid | Host aphid | Host plant | Type | References |
|---|---|---|---|---|
| Aphidius colemani [=platensis Brethes] | Aphis gossypii | Cucumis sativus | II | Mottaghinia et al. (2017) |
| Aphidius ervi Haliday | Sitobion avenae | Triticum aestivum | II, III | Bazyar et al. (2012) |
| Aphidius matricariae Haliday | Aphis gossypii | Cucumis sativus | II | Zamani et al. (2006) |
| | | | II, III | Tazerouni et al. (2017) |
| | Aphis fabae | Beta vulgaris | II | Tahriri et al. (2007) |
| | Myzus persicae | Capsicum annuum | II, III | Tazerouni et al. (2016) |
| | | Solanum melongena | III | Rashki et al. (2013) |
| | Myzus persicae nicotianae Blackman | Nicotiana tabacum | II | Rezaei et al. (2019) |
| | Schizaphis graminum | Triticum aestivum | II | Hajrahmatollahi et al. (2015) |
| Aphidius platensis Brethes | Aphis gossypii | Cucumis sativus | II | Zamani et al. (2006) |
| Diaeretiella rapae (M'Intosh) | Brevicoryne brassicae | Brassicae olearcea | II | Fathipour et al. (2006) |
| | | Brassica oleracea | II | Moayeri et al. (2013) |
| | Diuraphis noxia | Triticum aestivum | II | Tazerouni et al. (2011, 2012c) |
| | Schizaphis graminum | Triticum aestivum | III | Jokar et al. (2012) |
| | Lipaphis erysimi | Brassica napus | II | Rezaei et al. (2014) |
| Praon volucre (Haliday) | Aphis gossypii | Cucumis sativus | II | Tazerouni et al. (2017) |
| | Myzus persicae | Capsicum annuum | II | Tazerouni et al. (2016) |
| | Sitobion avenae | Triticum aestivum | II | Farhad et al. (2011) |
| Trioxys pallidus (Haliday) | Chromaphis juglandicola | Juglans regia | III | Rakhshani et al. (2004a) |

Few attempts were made to clarify the effect of aging in foraging behavior of the aphid parasitoids. Age specific functional response has received little attention (Pasandideh et al. 2015; Tazerouni et al. 2016). Functional response of *Aphidius matricariae* showed a fluctuated pattern between II and III by aging the foraging female (Tazerouni et al. 2016, 2017). The similar experiences by the same authors indicated no changes in type of functional response for *Praon volucre*.

Learning in foraging female can change the parameters in the functional response of aphid parasitoids (Byeon et al. 2011). Within a series of experimental assays, Rasekh et al. (2010a, b, c, d) surveyed the effects of various parameters on the foraging behavior of *Lysiphlebus fabarum* (Marshall), as parasitoid of black bean aphid, *Aphis fabae* Scopoli. They found no significant differences between young and old females, suggesting that age had also no effect on proportional time allocation to various activities during patch exploitation (Rasekh et al. 2010b). On the same way, encounters with con-specific females had no effect on foraging behavior of this species, which was interpreted as a consequence of cuticular camouflage interferes with conspecific recognition (Rasekh et al. 2010d).

Several different factors, e.g. interactions with other natural enemies, soil media composition, etc. have also been used to study the functional response of aphid parasitoids (Rashki et al. 2013; Mottaghinia et al. 2017). Considering the complication and considerable number of the factors affecting the functional responses, there are yet some ambiguities in the relevance of the experiments dealing in this respect (Hassell et al. 1977). It should be point out that these kinds of experiments are conducting under laboratory condition, where parasitoids are forced to remain in the patch (van Steenis and El-Khawass 1995), that is not consistent with the real conditions. Therefore, the functional response may simply have altered in natural condition, but the basic consequences can be predicted, at least (Houck and Strauss 1985). Recently, the effect of interspecific interactions on parasitism rate and efficiency of different species have been investigated. Accordingly, it has been revealed that interspecific interaction negatively affecting the parasitism rate of *A. matricariae* and *P. volucre* against second and third instars of green peach aphid (Tazerouni et al. 2016).

## 9.3.4   Host Stage Preference

An important aspect of behavioral and physiological ecology of the aphid parasitoids is the host preference which guarantees the survival of the next generation. In a broad sense, it includes various definitions ranging from habitat preference to host instar selection. The host range pattern of the aphid parasiotids discussed in a separate section, is a matter of their evolutionary history connected with their host aphid's evolution and host plant distribution (Žikić et al. 2017). Ignoring the expanded definition, the behavioral host preference including the host nymphal instar selection, has been considered as important criteria in selecting the efficient aphid parasitoids, both at the time of mass rearing and in their practical application for achieving a successful biological control. In general, the larvae of aphid parasitoids can successfully develop in all nymphal instars, while their size is highly variable, which means different value of the food resources. The early nymphal instars may even be smaller than the ovipositing female parasitoid (Mackauer 1986; Mackauer and Kambhampati 1988b). However, the female parasitoid mainly foraging within

**Table 9.3** The preferred developmental stages of the host aphids for the aphid parasitoids in laboratory conditions

| Parasitoid | Host aphid | Host plant | Preferred instars | References |
|---|---|---|---|---|
| Aphidius matricariae (Haliday) | Aphis gossypii Glover | Cucumis sativus | 3rd, 4th | Talebi et al. (2006) |
| | Aphis fabae Scopoli | Beta vulgaris | 3rd | Tahriri et al. (2007) |
| | Myzus persicae nicotianae Blackman | Nicotiana tabacum | 3rd, 4th | Rezaei et al. (2019) |
| | Schizaphis graminum | Triticum aestivum | 3rd | Hajrahmatollahi et al. (2015) |
| Aphidius colemani Viereck | Aphis gossypii Glover | Cucumis sativus | 3rd | Talebi et al. (2006) |
| Diaeretiella rapae (M'Intosh) | Diuraphis noxia (Mordvilko) | Triticum aestivum | 3rd, 4th | Tazerouni et al. (2012c) |
| | Schizaphis graminum (Rondani) | Triticum aestivum | 2nd | Jokar et al. (2012) |
| Lysiphlebus fabarum (Marshall) | Aphis fabae Scopoli | Vicia faba | 3rd | Ameri et al. (2013) |
| | | | 2nd | Najafpour et al. (2016) |
| | Aphis craccivora Koch | Medicago sativa | 3rd | Takalloozadeh et al. (2004a) |
| Praon volucre (Haliday) | Acyrthosiphon pisum (Harris) | Vicia faba | 1st, 2nd | Pasandideh et al. (2015) |
| Praon volucre (Haliday) | Sitobion avenae (Fabricius) | Triticum aestivum | 2nd | Farhad et al. (2011) |
| Trioxys pallidus (Haliday) | Chromaphis juglandicola (Kaltenbach) | Juglans regia | 3rd, 4th | Rakhshani et al. (2004a) |

the mixed-aged aphid colonies of its adapted host and do what the natural selection already defined.

The female parasitoids generally try to decrease the risks for their progenies by selecting the most appropriate host stage (Pyke 1984), while it may be affected by the various factors (Jervis and Kidd 1996). The size of the host aphids, depending to the nymphal instar influences directly biological characteristics of the adult emerging parasitoids, including size (Sequeira and Mackauer 1992b), developmental time (Sequeira and Mackauer 1992a), fecundity (Hägvar and Hofsvang 1990) and sex ratio (Cloutier et al. 2000) that can be defined by the behavior of foraging female at the time of oviposition. The aphid parasitoids showed a moderately different pattern of host stage preference in laboratory conditions (Rakhshani et al. 2004a; Takalloozadeh et al. 2004a; Talebi et al. 2006; Tahriri et al. 2007; Tazerouni et al. 2011; Farhad et al. 2011; Jokar et al. 2012; Pasandideh et al. 2015; Najafpour et al. 2016), which may be altered in natural condition (Sequeira and Mackauer 1993). Both preference for younger and older nymphal instars are reported (Table 9.3), that may have its own advantages in the variable environment with different host aphid inhabited therein.

The surveys on the behavior of *Lysiphlebus fabarum* (Marshall) on the black bean aphid, *Aphis fabae* Scopoli indicated different strategies under choice and no-choice host accessibility (Najafpour et al. 2016). No preference for specific nymphal instar was documented, when the female parasitoids were offered different host instars separately, while host instar preference progressively decreased with host age. On the other hand, the host stage preference in the aphid parasitoids are known as a constitute in the network of host aphid – parasitoid interaction (Henry et al. 2005). It means the generalist parasitoids may show a variable host stage preference on different host aphid species. Preference for the older nymphal instars can normally led to a greater reduction in the population growth rate of the host aphid (Lin and Ives 2003), then the population will be stabilized with the rest non-preferred instars. On the other hand, preference on the younger nymphal instars (Perdikis et al. 2004) is also suggested as an advantage for the aphid parasitoids enabling them to control the more numerous fraction (Kouame and Mackauer 1991) of the aphid populations in early stages of their increase (Perdikis et al. 2004). Defensive behavior of the aphid in late instars (third and fourth) may also result in the lower preference for some parasitoids (Kouamé and Mackauer 1991; Kairo and Murphy 1999; Wyckhuys et al. 2008), which increases the handling time of the foraging female (Kant et al. 2008). The parasitoids develop at a slower rate in the smaller aphid hosts (Henry et al. 2005; Najafpour et al. 2016). This prolonged developmental time is necessary in order to acquire additional resources while the host is developing. Theoretically, a preference for the younger nymphal instars may increase the yield of mass rearing efforts, since it is easier to provide a large number of first and second nymphal instars. Najafpour et al. (2016) recommend a synchronous population of second nymphal instar of *Aphis fabae* for mass rearing of *Lysiphlebus fabarum*. Rearing on the younger nymphal instars may result to a significantly male –biased population, since the female parasitoids have preferentially a sex allocation behavior through which they lay the unfertilized eggs into the smallest aphids (Pandey and Singh 1999; Cloutier et al. 2000).

## 9.3.5   Ant Associations

The aphid colonies are frequently attracting the ants (Fig. 9.3), which may range from an occasional encounter to highly evolved coexistence, as bilateral mutualism (Starý 1966). Interactions between ants and aphids as common honeydew-producing insects are widespread and abundant, but their ecological consequences are very poorly known. One of the more advanced relationships is the transportation of the aphid nymphs by ants into the suitable parts of the host plant, then achieving more honeydew secretions. Some aphid species alter their feeding behavior and the composition of their honeydew in the presence of the attending ants (Yao et al. 2000). The ants are known also protecting the aphid colonies from the natural

**Fig. 9.3** Ant associated aphid colony: (**a**) *Myrmica* sp. attended to *Aphis fabae* Scopoli; (**b**) Behavior of *Lysiphlebus cardui* (Marshall) female for soliciting honeydew by antennation on *Aphis fabae*. (Courtesy of Jan Havelka)

enemies by different level of adaptations (Addicott 1979). In the case of some aphid parasitoids, higher rate of oviposition was observed in ant attended colonies (Völkl and Novak 1997) indicating they could overcome this protection by some adaptations. The root aphid parasitoid, *Paralipsis enervis* (not recorded from Iran) is a highly evolved species co-existing in the ant attended aphid colonies. The ants may mutilate large portions of its wings, but leave it alive, dispersing flightless (Starý 1970a, b). The parasitoid is believed to acquire some of the ants cuticular hydrocarbons by rubbing her legs and body on them, and become readily accepted to the ants (Takada and Hashimoto 1985).

*Lysiphlebus fabarum* has taken more stages of adaptation. It might be unique in soliciting honeydew secretion from the aphid host by the ant mimicry. Consequently, aphids serve both as hosts and as a source of necessary hydrocarbon for the adult females. Experienced female parasitoid also performing an ant-like antennation to reduce the defensive behavior of the host aphid (*Aphis fabae*), which is mainly the frequent kicking by the hind legs. In such a peculiar case, the mutualisms seem to evolve between *L. fabarum* and the ants instead of the aphids, so that the progeny of *L. fabarum* also receive ant protection against the predators and secondary parasitoids (Rasekh et al. 2010e). Other parasitoids of the same aphids greatly suffered from the hyperparasitism caused by lack of the ants in the colonies that the female parasitoids preferred for oviposition (Völkl 1992). The same behavior has also been recorded in some other *Lysiphlebus* species (Kaneko 2002), but ant attendance had a negative effect on the efficiency of *Lysiphlebus testaceipes*

(Cresson) (Vinson and Scarborough 1991). Many aphid species are recorded as attended by ants, which are also parasitized by *Lysiphlebus fabarum* in Iran (Rakhshani et al. 2013).

## 9.4   Aphid Parasitoids in Various Ecosystems

### 9.4.1   Aphid Parasitoids in Farmlands and Orchards, Host Range and Economic Importance

The best model of aphid parasitoid association with economic importance can be found in the farmlands and fruit orchards as the main source of food production for the human society. The agro-ecosystem is generally referring to a simplified eco-system, in which the core lies on the agricultural activity by human. It is not an appropriate situation to open the big problems occurred in the ecosystems by these modifications, but an escape to the changes in natural aphid-parasitoid association would be useful enough. Cereals are the main crop that has occupied the vast majority of the agricultural fields. There are many other field crops, which more or less suffering from the aphid infestation, but no concern was paid on the cereal aphids in Iran, comparing other parts of the world, especially in Europe and North America. A series of researches arbitrarily targeted the aphid parasitoids in farm-lands and fruit orchards, but a valuable amount of data from the surrounding areas were also compiled (Rakhshani et al. 2012b; Nazari et al. 2012; Alikhani et al. 2013; Barahoei et al. 2013). A moderately common and well known aphids are associated with cereals throughout the world, with some relevant differences on their biology and economic importance (Dixon 1987). They may occur irregularly in various regions or with some gradient patterns (Rakhshani et al. 2008b). Depending on the host plants, different aphids may be more frequent than others, i.e. *Rhopalosiphum maidis* (Fitch) is rather common in corn fields comparing other aphids. Ignoring the host plant associations, an efficient group of aphid parasitoids can be found in cereal fields, each of them have its own preference, when encountering various host aphids.

The rare and non-economic *Sipha* aphids are associated with *Adialytus ambiguus* (Haliday), which has no other host aphid in such environment, was occurred. Other aphids including *Diuraphis noxia* (Mordvilko), *Sitobion avenae* (Fabricius), *Schizaphis graminum* (Rondani), *Metopolophium dirhodum* (Walker), *Rhopalosiphum padi* (L.) and *Rhopalosiphum maidis* (Fitch), all can be parasitized by *Praon volucre* (Haliday). The cabbage aphid parasitoid, *Diaeretiella rapae* (M'Intosh) has a significant affinity to *Diuraphis noxia* (Mordvilko) (Rakhshani et al. 2008b), but the records (Pike et al. 1999; Singh and Singh 2015) indicating it does not have limit for the host association at least in the cereal fields. At the same way, *Aphidius uzbekistanicus* Luzhetzki is mainly in association with *Sitobion avenae* (Rakhshani et al. 2008a), the preferred host for *Aphidius ervi* Haliday, which is not so common in cereal fields in Iran.

There is a good example of local adaptation for the generalist parasitoid, *Ephedrus persicae* Froggatt in Sistan plain with extreme temperature in mid spring. The most common cereal aphid in this area is *Schizaphis graminum*, which never find opportunity to escape from the highly efficient parasitoid. No live aphids were found in field observations during 10 years at the time of ripening, while high population of the emerged parasitoids swarming around the fields. It is widely distributed in the Palaearctic region and other parts of the world and has a wide range of the host aphids (Žikić et al. 2009). *Diuraphis noxia* is one of the most important aphid pest of cereals in many parts of the world, but no serious damage was recorded in Iran, as the native area for this aphid (Liu et al. 2010). The occasional outbreak of the other aphids is most probably because of the chemical treatments done especially against sun pest, *Eurygaster integiceps* Puton, that disrupt the efficiency of aphid parasitoids. Excluding the occasional, but important member of cereal aphid parasitoids in Iran, *Aphidius platensis* Brethes and *Aphidius matricariae* Haliday, other *Aphidius* species (*Aphidius rhopalosiphi* De Stefani-Perez and *Aphidius uzbekistanicus* Luzhetzki) seem having a permanent occurrence in the cereal fields throughout the country (Rakhshani et al. 2012b; Nazari et al. 2012; Alikhani et al. 2013; Barahoei et al. 2013). *Praon gallicum* Starý and *Praon necans* Mackauer have also recently been recorded as members of cereal aphid parasitoid guilds in western part of Iran (Bagheri-Matin et al. 2010; Nazari et al. 2012). In total, existence of a complete assemblage of the efficient parasitoids in cereal fields of Iran never allowed the dangerous pest aphids outbreaks. All of these parasitoids lose their habitat by the time of harvesting and have to migrate to the surrounding weeds. Therefore, a simple conservation program including the avoidance of the complete harvesting of the plant material, irrigation after harvesting, ignoring the unimportant graminaceous weeds around the fields and most importantly avoiding the chemical treatments, as much as possible can greatly help protection of the aphid parasitoids.

The aphid parasitoid complex in alfalfa fields of Iran has been very well studied throughout several years (Van den Bosch 1957; Monajemi and Esmaili 1981; Rakhshani et al. 2006a, b, 2009, 2010). The associated aphids are discussed in other sections. Similar to cereals, there is no problem of aphid population in alfalfa fields of Iran, because of the efficient native aphid parasitoids. The large populations of *Acyrthosiphon pisum* (Harris) in humid regions was observed after unnecessary treatments for control of the alfalfa weevil. A rich complex of parasitoids including *Aphidius ervi* Haliday, *Aphidius smithi* Sharma & Subba Rao, *Aphidius eadyi* Starý, Gonzales & Hall, *Praon barbatum* Mackauer and *Praon volucre* Haliday are strictly associated with this aphid in alfalfa fields across the various regions of Iran (Rakhshani et al. 2006a). Additional species, *Aphidius urticae* Haliday was also recorded from this aphid, which comprises a complex species including *Aphidius banksae* Kittel, that occurred in Turkey, but its existence has not been confirmed in Iran, yet. Majority of these species, excluding *Praon volucre* are specific and permanent members of alfalfa fields and related legume crops. *Aphidius ervi* is considered as one of the most generalist species in this group, but it showed the highest rate of parasitism of *Acyrthosiphon* aphids (Rakhshani et al. 2010; Alikhani et al. 2013). Other important aphid on alfalfa is *Therioaphis trifolii* (Monel), which has two specific parasitoids, *Trioxys complanatus* Quilis and *Praon exsoletum* (Nees). Other parasitoids are host specific with slightly different population fluctuation along the season, whole the country. Both species can be found together in many parts of the country, but *Trioxys complanatus* is more common in late spring, when the temperature starts the increasing. *Aphis craccivora* Koch is a permanent member of alfalfa aphids in early season, has its own parasitoid complex (Rakhshani et al. 2005a), which some of them, especially *Lysiphlebus fabarum* are rather common in alfalfa fields. Other parasitoids including, *Aphidius platensis* Brethes and *Diaeretiella rapae* (M'Intosh) can be found in association with this aphid, but as occasional member with less value of regulatory effect. *Ephedrus persicae* Froggatt was also found parasitizing this aphid in Sistan plain (Rakhshani et al. 2006a), sourcing from the nearby cereals fields.

Sporadically researches have been done on the aphid parasitoids in fruit orchards in various regions of Iran (Rakhshani et al. 2005b, 2008a; Jafari et al. 2011; Jafari and Modarres Awal 2012; Nazari et al. 2012; Barahoei et al. 2013), supplemented by an additional compilation on the aphid parasitoids, which occurred in pome and stone fruit trees (Rakhshani 2012). The latter contribution supplied the major list of the known parasitoid species attacking common aphids on various host plants categorized as economically important species grown in commercial orchards. The pesticide application is impartible element of the agricultural operations in fruit orchards, but the aphid parasitoid could survive in very small areas, escaped from the treatments. Parasitoids of the leaf curling aphids may also be protected from the contact insecticides. Numerous species of the aphids belonging to various genera are feeding on the leaves, branches and even trunk of the fruit trees, which differ in biology and associated parasitoids. Many of them have also shared host plants inside the same orchard or within the neighboring areas. In the ecological view, it should be considered as important colonizing patches for the aphid parasitoids, too. *Pauesia antennata* (Mukerji) is a strictly specific parasitoids of the giant brown peach aphid, *Pterochloroides persicae* (Cholodkovsky) on its Prunaceae host plants (Rakhshani et al. 2005b). The aphid appeared in the high populations with economically injurious situation, which is a consequence of chemical treatment with wide spectrum insecticides. According to field observation, both parasitoid and predatory coccinellids are rather common in the orchards without chemical treatments. List of the parasitoids and their common host aphids in pome and stone fruit trees of Iran is presented in Table 9.4. There are yet some unknown parasitoid species in pome and stone fruit orchards of Iran, which seems having an obligative diapause in main part of season. Attempts for rearing adult specimens was failed and only existence of peculiar mummies is documented (Rakhshani 2012). Many other aphids are also known with other fruit trees (Rezwani 2001, 2004), with or without known parasitoids. The pomegranate aphid, *Aphis punicae* Passerini has also a well-known group of the common parasitoids, including *Lysiphlebus fabarum*, *Aphidius platensis*, *Binodoxys angelicae*, *Aphidius matricariae* and *Ephedrus persicae* (Rakhshani et al. 2008a; Talebi et al. 2009; Farrokhzadeh et al. 2014).

**Table 9.4** Aphid parasitoids and their hosts in pome and stone fruit trees in Iran

| Parasitoids species | Host aphids |
|---|---|
| *Aphidius matricariae* Haliday | *Aphis pomi* de Geer; *Aphis spiraecola* Patch; *Brachycaudus amygdalinus* (Schouteden); *Brachycaudus cardui* (L.); *Brachycaudus helichrysi* (Kaltenbach); *Brachycaudus persicae* (Passerini); *Myzus cerasi* (Fabricius); *Myzus persicae* (Sulzer); *Ovatus insitus* (Walker); *Rhopalosiphum padi* (L.) |
| *Aphidius platensis* Brethes | *Brachycaudus amygdalinus* (Schouteden); *Myzus persicae* (Sulzer); *Phorodon humuli* (Schrank) |
| *Aphidius transcaspicus* Telenga | *Hyalopterus amygdali* (Blanchard); *Hyalopterus pruni* (Geoffrey) |
| *Binodoxys acalephae* (Marshall) | *Aphis pomi* de Geer; *Aphis spiraecola* Patch |
| *Binodoxys angelicae* (Haliday) | *Aphis pomi* de Geer; *Aphis spiraecola* Patch |
| *Diaeretiella rapae* (M'Intosh) | *Myzus persicae* (Sulzer) |
| *Ephedrus cerasicola* Starý | *Dysaphis plantaginea* (Passerini); *Myzus cerasi* (Fabricius); *Myzus persicae* (Sulzer) |
| *Ephedrus persicae* Froggatt | *Brachycaudus amygdalinus* (Schouteden); *Brachycaudus helichrysi* (Kaltenbach); *Dysaphis plantaginea* (Passerini); *Dysaphis pyri* (Boyer de Fonscolombe); *Dysaphis reaumuri* (Mordvilko); *Hyalopterus amygdali* (Blanchard); *Myzus cerasi* (Fabricius); *Myzus persicae* (Sulzer); *Phorodon humuli* (Schrank) |
| *Ephedrus plagiator* (Nees) | *Brachycaudus helichrysi* (Kaltenbach); *Dysaphis pyri* (Boyer de Fonscolombe) |
| *Lysiphlebus fabarum* (Marshall) | *Aphis pomi* de Geer; *Aphis spiraecola* Patch; *Brachycaudus amygdalinus* (Schouteden); *Brachycaudus helichrysi* (Kaltenbach); *Brachycaudus persicae* (Passerini); *Dysaphis plantaginea* (Passerini) |
| *Pauesia antennata* (Mukerji) | *Pterochloroides persicae* (Cholodkovsky) |
| *Praon abjectum* (Haliday) | *Brachycaudus cardui* (L.) |
| *Praon volucre* (Haliday) | *Aphis pomi* de Geer; *Brachycaudus amygdalinus* (Schouteden); *Brachycaudus cardui* (L.); *Brachycaudus helichrysi* (Kaltenbach); *Brachycaudus persicae* (Passerini); *Dysaphis pyri* (Boyer de Fonscolombe); *Hyalopterus amygdali* (Blanchard); *Hyalopterus pruni* (Geoffrey); *Myzus persicae* (Sulzer) |

## 9.4.2 Aphid Parasitoids in Urban Areas

Urban ecosystems are significantly influenced by an area and settlement intensity, and green area size. In a milder zone, city parks, arboretums, road line shading groves, and small private gardens yield a lot of plant mixes composed from ornamentals, shady species, fruit trees, and vegetable crops. More or less, various kinds of weeds are present, everywhere yield a usually rich aphid and parasitoid fauna, to be considered even what their interactions and possible management concerns

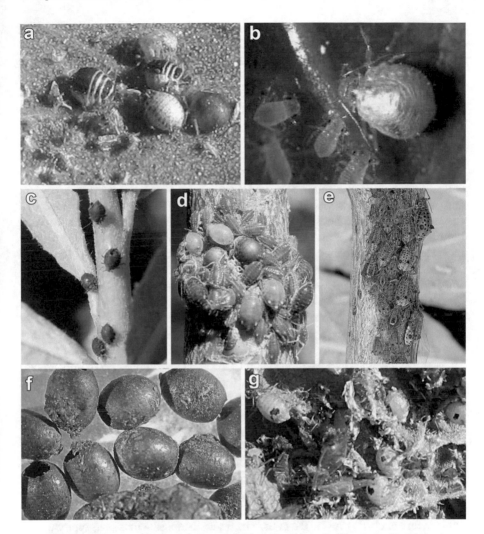

**Fig. 9.4** The common aphids and their parasitoids on poplar and willows: (**a, b**) *Chaitophorus salijaponicus niger* Mordvilko and *Chaitophorus leucomelas* Koch on *Salix* sp. and *Populus nigra*, respectively parasitized by *Adialytus salicaphis* (Fitch); (**c**) *Aphis farinosa* Gmelin on *Salix* sp. parasitized by *Lysiphlebus confusus* Tremblay & Eady; (**d**) *Pterocomma* sp. on *Salix* sp. parasitized by *Euaphidius cingulatus* (Ruthe); (**e**) colony of *Tuberolachnus salignus* (Gmelin) on *Salix* sp.; (**f**) *Pemphigus spirothecae* Passerini parasitized by *Monoctonia vesicarii* Tremblay (and including hyperparasitoid) inside the leaf galls; (**g**) Colony of *Eriosoma lanuginosum* (Hartig) on leaves of *Ulmus carpinifolia* parasitized by *Areopraon lepelleyi* (Waterson)

(Fig. 9.4). In the subtropical zone and up to the semi desert and desert zones the plant diversity may even be very rich up to respectively reduced, the same with the aphid and parasitoid associations (Pons and Lumbierres 2004; Mossadegh et al. 2011, 2016). The composition of plants, aphids and parasitoids is usually reduced to some mostly common pests on the cultivated plants and weeds.

Poplar and willows are the major ornamental trees both in urban areas and natural bio-corridors in Iran. They are plants of the same family, and share the pest aphids of at least same genera, *Chaitophorus* Koch and *Pterocomma* Buckton. Many other aphids belonging to the genera *Aphis* Linnaeus, *Cavariella* del Guercio, *Tuberolachnus* Mordvilko and *Plocamaphis* Oestlund are exclusively associated with willows. Two latter genera have not known parasitoids in Iran, while the rest encompass a moderately rich parasitoid assemblage (Rakhshani et al. 2007b). *Chaitophorus* aphids both on poplars and willows have known to be parasitized in early season by a well-adapted parasitoid, *Adialytus salicaphis* (Fitch). The parasitoid indicated the ability to handle whole the colonies on a single patch. There are no studies on the biology of this species and the existence of separate species on poplars and willows is not accepted. *Euaphidius cingulatus* (Ruthe) is the parasitoid of *Pterocomma* aphids on both group of host plants and occasionally has been observed in urban areas. It is a moderately large species that can parasitized the large individuals of *Pterocomma* aphids, but never found in association with *Tuberolachnus* aphids, even where it was colonized on the nearby branches.

The poplars host also other aphids of the genera *Pemphigus* Hartig and *Phelomyzus* de Horváth in Iran. No parasitoid species is recorded from the latter species in the world. *Pemphigus* species were found heavily parasitized by *Protaphelinus nikolskayae* (Yasnosh) (Hymenoptera: Aphelinidae) in north central part of Iran with parasitism rate of 50–98% (Rakhshani et al. 2007b). It is a specific parasitoid of *Pemphigus* species inside the galls as a closed micro-ecosystem. A rare parasitoid, *Monoctonia vesicarii* Tremblay was found in association with *Pemphigus spirothecae* Passeriini on *Populous nigra* in the area from north central to north western parts of the country (Ghafouri-Moghaddam et al. 2012; Rakhshani et al. 2015a). It is a parasitoid of many *Pemphigus* species in Europe and is known having only one generation per year (Starý 1968; Tremblay 1991; Rakhshani et al. 2015a). Despite lacking the information about efficacy and parasitism rate of *Monoctonia vesicarii*, it can be expected as main bio-control agent of *Pemphigus* species both in natural and urban area. It is a very specific and ecologically adapted parasitoid, emerging from the diapaused mummies early in the season to find the fundatrix aphid before enclosing inside the galls. The fundatrix aphid will lose its reproduction capacity or possibly produce very few progenies, then it becomes mummified. The parasitoid remains inside the mummy until the next spring as mature diapaused larva or pupa.

Various species of coniferous plants are the main elements of the evergreen areas in urban ecosystems. Pines (*Pinus* spp.), cypress (*Cedrus, Cupressus* spp., *Juniperus, Thuja*), and spruce (*Picea*) are the well-known conifers planted as ornamental trees in the cities. The Lachninae aphids are associated with these plants and can be the serious pests in the absence of their natural enemies, especially the parasitoids. Aphid species of the genera *Cinara* Curtis are frequently active on these plants, and *Eulachnus* spp. are living on pines, as well. Majority of these aphids have been considered as invasive species imported along with the seedlings of their host plants. Therefore, except a few cases, no parasitoid species was detected (Starý et al.

**Table 9.5**  The common aphids and their parasitoids associated with Rosa spp. in Iran

| Aphids | Parasitoids |
|---|---|
| *Metopolophium dirhodum* (Walker) | *Aphidius platensis* Brethes; *Praon volucre* (Haliday) |
| *Wahlgreniella nervata* (Gillette) | *Aphidius ervi* Haliday; *Aphidius matricariae* Haliday |
| *Amphorophora catharinae* (Nevsky) | *Aphidius popovi* Starý; *Praon volucre* (Haliday) |
| *Macrosiphum rosae* (L.) | *Aphidius rosae* Haliday; *Praon rosaecola* Starý; *Praon volucre* (Haliday) |
| *Aphis gossypii* Glover | *Lysiphlebus fabarum* (Marshall) |
| *Aphis craccivora* Koch | *Lysiphlebus fabarum* (Marshall); *Binodoxys acalephae* (Marshall); *Aphidius platensis* Brethes |

2005). The large populations of *Cinara cedri* Mimeur make a considerable amount of honeydew around the tree and greatly disrupt its ornamental values.

The situation is worse, in the case of *Eulachnus* species on pines, where the aphid honeydew makes a sticky surface on glass of the cars stopped under the trees. All of these aphids have their own specific parasitoids at their area of origin. Species of the genus *Pauesia* Quilis are the most common and well known parasitoids of *Cinara* aphids on conifers in (Europe, Far East, Oriental region, North America). A single species *Pauesia hazratbalensis* Bhagat is detected on *Thuja orientalis* in association with its common aphid, *Cinara thujafilina* (Del Guercio) in north central (Starý et al. 2005) and north eastern (Rakhshani et al. 2012b) parts of Iran. It is necessary to make program for introduction of some efficient parasitoid species for the rest of coniferous aphids, especially *Eulachnus* species. *Diaeretus leucopterus* (Haliday) is a taxonomically closely related species to *Pauesia* and known as the specific parasitoid of *Eulachnus* aphids. It is widely distributed in Europe (Kavallieratos et al. 2004) and even in North Africa (Ben Halima et al. 2020). Rose (*Rosa* spp.), including a number of famous and popular flowering shrubs, which categorized as the most common ornamental plants in gardens and outdoors. Many aphids are known attacking various commercial rose (Rezwani 2001) causing defoliation, reducing the ornamental value of the flowers and transmitting the viral diseases. Among them, some species including *Macrosiphum rosae* (L.), *Amphorophora catharinae* (Nevsky), *Aphis craccivora* Koch, *Aphis gossypii* Glover, *Wahlgreniella nervata* (Gillette) are frequently encountered in the urban areas. Fortunately, there are active parasitoids associated with all of this aphid species in their local area of plantations (Table 9.5). The application of chemical pesticides is restricted in both urban areas and where commercial varieties for the scent extraction (Golab, Rose-water) are planted. However, no attempt is done for application of the aphid parasitoids, even at higher level of infestations, which most probably may have an economical basis. *Amphorophora catharinae* (Nevsky) and its specific parasitoid, *Aphidius popovi* Starý seems not endemic to Iran, and accidentally arrived together with host plant material (Rakhshani et al. 2008a). *Betuloxys hortorum* (Starý) and *Praon flavinode* (Haliday) are known as parasitoids of *Tinocallis* aphids on elms

(*Ulmus* spp.) (Starý et al. 2000; Barahoei et al. 2013). *Trioxys pallidus* was also reared from *Tinocallis* species on elms in urban areas.

Furthermore, there are many gall making aphids of the genera *Eriosoma* Leach, *Colopha* Monell, *Kaltenbachiella* Schouteden, *Tetraneura* Hartig are living on elms (Rezwani 2001, 2004), free of parasitoid. *Areopraon lepelleyi* (Waterson), is recently recorded as a parasitoid of *Eriosoma lanuginosum* (Hartig) on *Ulmus carpinifolia* (Kazemzadeh et al. 2009; Rakhshani et al. 2012b). Because of various reasons, especially activity of the predatory bugs, these aphids have no economic importance.

## 9.4.3   Aphid Parasitoids in Natural Ecosystems, Economically Indifferent Species

It has already been confirmed that the majority of species can be found in natural ecosystems with considerably higher rate of diversity both in flora and fauna (Alikhani et al. 2013; Nazari et al. 2012; Barahoei et al. 2013; Taheri and Rakhshani 2013). It is critically important to understand the species diversity and the trophic networks for both aphids and their parasitoids in the simple or complex natural environments. Many aphid species are strictly associated with the host plant, which never considered for the economic purposes by the human. From the same group, there are even very rare species, which has not even a single recorded parasitoid. It is clear that the phrase of economically indifferent species directly refers to the importance of the host plant. The point is on the other hand, aphids which are the host of parasitoid species manifesting a broader host range including the economically important aphids. These group of parasitoids can colonize on these alternative hosts out of growing season, then they can recover their population on the aphids on crops. A list of economically indifferent aphid species with their parasitoids is presented in Table 9.6. There are two groups of parasitoid species: a) the species that are associated with aphids living solely on few non-economic host plants; b) the parasitoids species associated with the aphids that encountered on various host plant including economic species (Agricultural, ornamental, etc..). The second group are generally the parasitoids with broader host range that can occasionally be considered as economically important species. In some cases, the aphid may have different parasitoids on different host plants. The similar consequences can be simply happened for the holocyclic aphids living on two different host plant groups. Generally, parasitoids are dispersing, where the alate aphids migrate in the late season. Depending on preference of the parasitoids, they may share the same host aphid in two different habitats or even move to the colony of second aphid species, living on the secondary host plant. As it is shown in Table 9.6, *Macrosiphoniella* aphids are not economically important species. They are generally associated with *Artemisia* and do not have any direct connection with crops, as well as their aphidiine parasitoids (Kavallieratos et al. 2004; Starý 2006).

**Table 9.6** Aphid parasitoids associated with economically indifferent aphids and the representative host plants in Iran

| Parasitoid | Host aphids | Host plant | Importance |
|---|---|---|---|
| *Aphidius absinthii* Marshall | *Macrosiphoniella* spp. | *Artemisia* spp. | Indifferent |
| *Aphidius asteris* Haliday | *Macrosiphoniella* spp. | *Artemisia* spp. | Indifferent |
| *Aphidius funebris* Mackauer | *Uroleucon* spp. | *Acroptilon, Centaurea, Chondrilla, Conyza, Sonchus, Tragopogon* | Important |
| *Aphidius iranicus* Rakhshani & Starý | *Coloradoa* spp. | *Artemisia* spp. | Indifferent |
| *Aphidius persicus* Rakhshani & Starý | *Uroleucon* spp. | *Artemisia, Chondrilla, Lactuca, Launaea, Picnomon, Sonchus* | Indifferent |
| *Aphidius stigmaticus* Rakhshani & Tomanović | *Macrosiphoniella* spp. | *Tanacetum polycephalum* | Indifferent |
| *Aphidius urticae* Haliday | *Microlophium carnosum* | *Urtica dioica* | Indifferent |
| *Ephedrus niger* Gautier, Bonnamour & Gaumont | *Macrosiphoniella* spp. *Uroleucon* spp. | *Chysanthemum, Chondrilla, Acroptilon, Sonchus, Conyza* | Important |
| *Lysiphlebus desertorum* Starý | *Protaphis* spp. | *Artemisia* spp. | Important |
| *Praon absinthii* Bignell | *Macrosiphoniella* spp. | *Artemisia* spp. | Indifferent |
| *Praon orpheusi* Kavallieratos, Athanassiou & Tomanović | *Hyperomyzus lactucae* (L.) | *Sonchus* spp. | Indifferent |
| *Praon unitum* Mescheloff & Rosen | *Uroleucon* spp. | *Acroptilon, Sonchus* | Indifferent |
| *Praon yomenae* Takada | *Uroleucon* spp. | *Acroptilon, Centaurea, Cichorium, Launaea, Picnomon, Sonchus* | Important |
| *Trioxys metacarpalis* Rakhshani & Starý | *Chaitaphis tenuicaudata* Nevsky | *Kochia scoparia* | Indifferent |
| *Trioxys panonnicus* Starý | *Titanosiphon* sp. | *Artemisia* spp. | Indifferent |
| *Trioxys tanaceticola* Starý | *Coloradoa* spp. | *Artemisia* spp. | Indifferent |

The same situation is also known for the *Uroleucon* aphids, which manifest a similar parasitoid assemblage, in part (Rakhshani et al. 2006a, b). *Ephedrus niger* and *Binodoxys centaureae* (European species) are the most common parasitoids attacking aphids of both genera. *Praon unitum* Mescheloff & Rosen is also an uncommon parasitoid of the same group (Fig. 9.5).

There are also some generalist parasitoids, especially *Praon volucre* (Haliday) which have a considerably wider host range (Rakhshani et al. 2007a, b). Except a few cases, the parasitoid complex of *Macrosiphoniella* aphids is a good

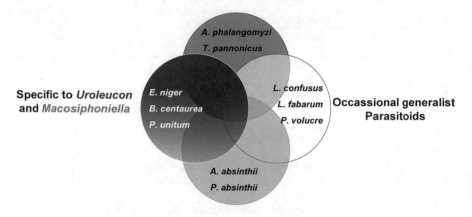

**Fig. 9.5** Categorization of parasitoids of *Macrosiphoniella* aphids based on their host range. (Data from Rakhshani et al. 2011)

representative of the Western Palaearctic region. *Aphidius absinthii* Marshall (including *Aphidius asteris* Haliday) can be considered as a the most adapted parasitoid species attacking about 30 species of *Macrosiphoniella* del Guercio, followed by *Ephedrus niger* Gautier, Bonnamour & Gaumont (10 species) in the Palaearctic region (Fig. 9.6). *Aphidius phalangomyzi* Starý and *Praon absinthii* Bignell are also the strictly specific parasitoids of *Macrosiphoniella* aphids on *Artemisia*, but *Trioxys pannonicus* Starý is a parasitoid of *Titanosiphon* Nevsky aphids on the same host plant and its association with *Macrosiphoniella* seems erroneous and needs further investigations. The same problem is also existing for *Aphidius arvensis* (Starý) – a parasitoid of *Coloradoa* Wilson as well. Since their Asteraceae host plants represent an important part of environmental structure, both *Macrosiphoniella* and *Uroleucon* Mordvilko aphids are very common around agro-ecosystems and can play important role in association with broadly oligophagous parasitoids, as well as the predatory insects.

## 9.5   Negative Agents on the Efficieny of Aphid Parasitoids

### 9.5.1   *Hyperparasitoids*

Hyperparasitoids, which are commonly known as "secondary parasitoids" are group of insects which their progeny develops on a primary parasitoid (Sullivan 1987). Many studies have been conducted on the various aspect of the biology (Höller et al. 1993; Brodeur and McNeil 1994), ecology (Mackauer and Völkl 1993; Völkl et al.

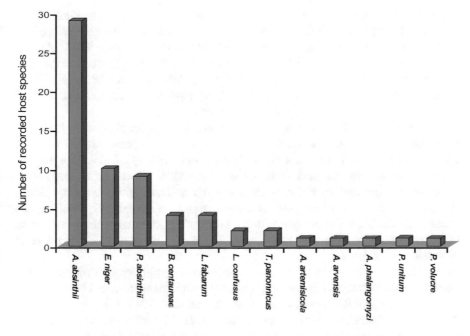

**Fig. 9.6** Comparison of host ranges for the member of *Macrosiphoniella*-parasitoid complex in the Palaearctic region. (Data from Rakhshani et al. 2011)

1994), behavior (Völkl and Kranz 1995; Hübner and Völkl 1996; Völkl and Sullivan 2000) and taxonomy (Ferrer-Suay et al. 2013a, b, c, 2014a, b, 2015) of the aphid hyperparasitoids, worldwide. The relevant classification of the aphid hyperparasitoids, as well as their general ecology can be found in Sullivan (1987) and in Sullivan and Völkl (1999). Aphid hyperparasitism is evolved within three superfamily of hymenoptera including Chalcidoidea, Ceraphronoidea and Cynipoidea. Few genera of the large families within Chalcidoidea are the aphid hyperparasitoids including *Asaphes* Walker, *Coruna* Walker, *Euneura* Walker and *Pachyneuron* Walker (Pteromalidae), among them the *Pachyneuron aphidis* (Bouché) is rather common in association with many aphid parasitoid complexes (González et al. 1978; Talebi et al. 2009; Rakhshani et al. 2004b, 2005b, 2009; Farsi et al. 2010; Darsoei et al. 2011; Amini et al. 2012; Jahan et al. 2013; Nematollahi et al. 2014). *Pachyneuron solitarium* (Hartig) has recently been recorded as hyperparasitoid of *Monoctonia vesicarii* Tremblay, a parasitoid of the gall making aphid, *Pemphigus spirothecae* Passerini on *Populus nigra* (Ghafouri-Moghaddam et al. 2014). *Asaphes suspensus* (Nees) is another common aphid hyperparasitoid in Iran (Lotfalizadeh and Gharali 2008; Mitroiu et al. 2011), known to occurs across the entire Holarctic region (Noyes 2019). There is no record of the genus *Conura* in Iran and no host data is also documented for *Euneura lachni* (Ashmead), except in association with *Pauesia antennata/Pterochloroides persicae* (Rakhshani et al. 2005b). A single record of aphid hyperparasitism within the genus *Tetrastichus*

Walker (Eulophidae) has also presented in the aforementioned citation. Many Pteromalidae hyperparasitoids are considered in general association with various host aphids and their parasitoids (Kamijo and Takada 1973; Sullivan 1987). Among the chalcidoids, an aphelinid, *Marietta picta* (André) has generally been known as a hyperparasitoid of the mealy bugs (Talebi et al. 2010; Fallahzadeh et al. 2011), coccids (Arnaoudov et al. 2006) and even the psyllids (Mehrnejad and Emami 2005), but it is also found as aphid hyperparasitoid.

One of the most common aphid hyperparasitoids is *Syrphophagous aphidivorus* (Mayr) belonging to the large family Encyrtidae. The female parasitoid has a distinct preference on the mummified aphid, in which the host aphid is killed and the primary parasitoid is already pupated (Buitenhuis et al. 2004). It can be a competitive behavior to other hyperparasitoids (Matejko and Sullivan 1984) and may reduce the risk of later attack of the host by other female hyperparasitoids (Roitberg and Mangel 1988). On the other hand, it can be rather destructive in the mass reared colonies of aphid parasitoids, where a single foraging female can destroy several mummies in few hours.

*Dendrocerus carpenteri* (Curtis) is also a common aphid hyperparasitoid that has been recorded in various regions of the country both in field crops and fruit orchards (Rezaei et al. 2006; Rakhshani et al. 2009; Darsoei et al. 2011; Jafari and Modarres Awal 2012; Rakhshani 2012; Farsi et al. 2014). Members of the superfamily Cynipoidea are known as phytophagous group, but surprisingly majority of them are parasitoids of other insects. The genera *Alloxysta* and *Phaenoglyphis* from the subfamily Charipinae (Figitidae) are obligatory endo-hyperparasitoids of aphids (Carver 1992). While some patterns of host specificity are recorded for the Charipinae aphid hyperparasitoids (Rakhshani et al. 2001; Lotfalizadeh and van Veen 2004), it seems as only a local isolation and greatly depending on the host plant community and ecological niches. Qualitative analysis of the trophic association for aphid, primary parasitoids and their *Alloxysta* hyperparasitoids were indicated to manifest no pattern of host specificity, but some species have few or only one host record (Ferrer-Suay et al. 2014a).

The recent revision of Charipinae aphid hyperparasitoids in Iran revealed existence of many species occurred within a wide range of natural habitats and agro-ecosystems (Khayrandish-Koshkooei et al. 2013; Ferrer-Suay et al. 2013a, b, 2014b, 2015). A summary of the recorded Charipinae from Iran as well as their host associations is presented in Table 9.7. As a general framework for the future studies both on host associations and biological control programs, it is necessary to pay enough attention to the associated hyperparasitoids.

In the native area, the hyperparasitoids are believed to have a regulatory role of the population dynamic of both aphids and their parasitoids (Bennett 1981; Mackauer and Völkl 1993), but long-term, multi-generation experiments are needed to test the hypothesis of stabilizing effects by the hyperparasitoids on the herbivore-parasitoid interaction (Rosenheim 1998).

**Table 9.7**  Charipinae aphid hyperparasitoids and their host associations in Iran

| Hypeparasitoids | Primary parasitoids | Host aphids |
|---|---|---|
| *Alloxysta arcuata* (Kieffer) | *Adialytus ambiguus* | *Sipha maydis* |
| | *Aphidius ervi* | *Sitobion avenae* |
| | *Aphidius matricariae* | *Aphis craccivora*; *Brachycaudus helichrysi*; *Capitophorus similis* |
| | *Aphidius rhopalosiphi* | *Schizaphis graminum* |
| | *Aphidius rosae* | *Macrosiphum rosae* |
| | *Binodoxys acalephae* | *Aphis craccivora*; *Aphis fabae* |
| | *Diaeretiella rapae* | *Aphis fabae*; *Brevicoryne brassicae*; *Rhopalosiphum padi* |
| | *Ephedrus persicae* | *Aphis fabae* |
| | *Lysiphlebus fabarum* | *Aphis craccivora*; *Aphis fabae*; *Aphis gossypii*; *Aphis idaei*; *Aphis nerii*; *Aphis rubiae*; *Aphis urticata*; *Brachycaudus tragopogonis* |
| | *Praon exsoletum* | *Therioaphis trifolii* |
| | *Praon volucre* | *Aphis fabae*; *Uroleucon sonchi* |
| *Alloxysta brevis* (Thomson) | *Adialytus ambiguus* | *Sipha elegans* |
| | *Adialytus salicaphis* | *Chaitophorus* sp. |
| | *Aphidius ervi* | *Sitobion avenae* |
| | *Aphidius matricariae* | *Aphis craccivora, Aphis fabae, Aphis umbrella, Aphis solanella* |
| | *Aphidius salicis* | *Cavariella* sp. |
| | *Binodoxys acalephae* | *Aphis fabae* |
| | *Diaeretiella rapae* | *Aphis fabae, Aphis gossypii* |
| | *Ephedrus persicae* | *Aphis fabae* |
| | *Lysiphlebus fabarum* | *Aphis craccivora*; *Aphis punicae*; *Aphis fabae, Aphis umbrella*; *Aphis origani*; *Aphis gossypii, Aphis solanella*; *Aphis terricola* |
| | *Praon volucre* | *Aphis fabae* |
| *Alloxysta castanea* (Hartig) | *Trioxys asiaticus* | *Acyrthosiphon gossypii* |
| | *Aphidius persicus* | *Uroleucon sonchi* |

(continued)

**Table 9.7** (continued)

| Hypeparasitoids | Primary parasitoids | Host aphids |
|---|---|---|
| Alloxysta circumscripta (Hartig) | Ephedrus niger | Uroleucon sp. |
| | Praon yomenae | Uroleucon sp.; Uroleucon sonchi |
| Alloxysta citripes (Thomson) | Trioxys pallidus | Chromaphis juglandicola |
| Alloxysta darci (Girault) | No host record | No host record |
| Alloxysta erythrothorax (Hartig) | Praon volucre | Aphis fabae |
| Alloxysta fuscicornis (Hartig) | Aphidius platensis | Aphis fabae |
| | Lysiphlebus fabarum Diaeretiella rapae | Aphis fabae Brevicoryne brassicae |
| Alloxysta macrophadna (Hartig) | Aphidius uzbekistanicus | Sitobion avenae |
| Alloxysta melanogaster (Hartig) | Aphidius ervi | Sitobion avenae |
| Alloxysta mullensis (Cameron) | Aphidius ervi | Sitobion avenae |
| | Aphidius matricariae | Aphis craccivora; Aphis umbrella |
| | Binodoxys acalephae | Aphis umbrella |
| | Diaeretiella rapae | Aphis sp. |
| | Lysiphlebus fabarum | Aphis craccivora; Aphis fabae; Aphis gossypii; Aphis umbrella |
| | Trioxys pallidus | Chromaphis juglandicola |
| Alloxysta pleuralis (Cameron) | Unknown | Aphis gossypii |
| Alloxysta pusilla (Kieffer | Lysiphlebus fabarum | Aphis nasturtii Kaltenbach; Aphis fabae |
| Alloxysta ramulifera (Thomson) | Adialytus ambiguus | Sipha maydis |
| Alloxysta ruficollis (Cameron) | Aphidius platensis | Aphis nerii |
| Alloxysta tscheki (Giraud) | Aphidius ervi | Sitobion avenae |
| | Aphidius uzbekistanicus | Sitobion avenae |

(continued)

**Table 9.7** (continued)

| Hypeparasitoids | Primary parasitoids | Host aphids |
|---|---|---|
| *Alloxysta ullrichi* (Giraud) | *Aphidius funebris* | *Uroleucon sonchi* |
| *Alloxysta victrix* (Westwood) | *Aphidius absinthii* | *Macrosiphoniella abrotani* |
| | *Aphidius salicis* | *Cavariella* sp. |
| | *Lysiphlebus fabarum* | *Aphis craccivora*; *Aphis fabae*; *Aphis urticata* |
| *Phaenoglyphis villosa* (Hartig) | *Adialytus ambiguus* | *Sipha maydis* |
| | *Adialytus salicaphis* | *Chaitophorus pakistanicus* |
| | *Adialytus veronicaecola* | *Aphis gossypii* |
| | *Aphidius persicus* | *Uroleucon sonchi* |
| | *Aphidius platensis* | *Aphis fabae*; *Schizaphis graminum*; |
| | *Aphidius rosae* | *Macrosiphum rosae* |
| | *Aphidius smithi* | *Acyrthosiphon pisum* |
| | *Aphidius ervi* | *Sitobion avenae* |
| | *Aphidius matricariae* | *Aphis craccivora*; *Aphis solanella*; *Capitophorus similis*; *Myzus persicae*; *Schizaphis graminum* |
| | *Aphidius transcaspicus* | *Hyalopterus pruni* |
| | *Binodoxys acalephae* | *Aphis fabae* |
| | *Binodoxys angelicae* | *Aphis fabae* |
| | *Diaeretiella rapae* | *Aphis craccivora*; *Aphis fabae*; *Brevicoryne brassicae*; *Rhopalosiphum padi*; *Lipaphis pseudobrassicae*; *Myzus persicae*; *Schizaphis graminum* |
| | *Ephedrus niger* | *Uroleucon* sp. |
| | *Ephedrus persicae* | *Aphis fabae* |
| | *Lysiphlebus fabarum* | *Aphis affinis*; *Aphis craccivora*; *Aphis fabae*; *Aphis nerii*; *Aphis origani* |
| | *Praon abjectum* | *Aphis solanella* |
| | *Praon barbatum* | *Acyrthosiphon pisum* |

(continued)

**Table 9.7** (continued)

| Hypeparasitoids | Primary parasitoids | Host aphids |
|---|---|---|
| | *Praon rosaecola* | *Macrosiphum rosae* |
| | *Praon volucre* | *Aphis fabae*; *Schizaphis graminum*; *Uroleucon cichori*; *Uroleucon sonchi* |
| | *Praon yomenae* | *Uroleucon sonchi* |
| | *Trioxys pallidus* | *Chromaphis juglandicola* |

Data compiled from Ferrer-Suay et al. (2013a, b, 2014b, 2015)

## 9.5.2 Predators

Despite the relative trophic simplicity in the agroecosystems, the associated aphids generally have a series of multiple natural enemies (Dixon 1998), interacting with each other in various ways including predation (Snyder and Wise 1999; Symondson et al. 2002). Direct and indirect interactions may occur through shared natural enemies (predators and parasitoids) of the aphids, which may have various consequences both on community structure and biocontrol impact (van Veen et al. 2008). The natural sequence of the parasitism on an aphid species starts usually prior to the peak and increases in the population peak-decrease state. At this state, also the effect and interactions with the other natural enemies becomes important. Immature stages of the aphid parasitoids may be attacked by hyperparasitoids and the predators, as well (Al-Rawy et al. 1969). This is a common trophic relation within the insect communities (Rosenheim et al. 1995). The predatory insects normally kill the other members of the same guild, therefore they reduce the potential competition and also benefited by them as a food source. It is generally known as intra-guild predation (Brodeur and Rosenheim 2000). The generalist predators like carabids interrupting aphid population control by the specialist parasitoids. They are able to climb into plants and prey upon the mummies, actively (Snyder and Ives 2001). The effects of intra-guild predation cannot be defined in a simple way, since it is capable to produce diverse impacts on biological control of herbivorous insects (Diehl 1993).

The aphids attacked by the parasitoids, usually remain active on the host plants and continue feeding and even producing few number of progenies. The parasitized aphids becoming less active with growth of the parasitoids, while staying among the colonies or leaving for the nearby branches/leaves, being mummified. The dead aphids containing the mature larva or pupa of parasitoid may be more vulnerable comparing to the healthy aphids. Larvae and adults of various species of coccinellids, larvae of the lacewings, adults and nymphs of the anthocorids and nabids, as well as the larvae of syrphid flies and chamaemyiids are the major predatory guild members associated with aphid colonies, many of them are generalist and are not strictly dependent to their host. The predatory insects like ceccidomyiid larvae and coccinellids showed no preference for healthy or parasitized (mummified) aphids (Colfer and Rosenheim 2001; Brodeur and Rosenheim 2000). The mummy provides not enough protection for the parasitoid, since it can easily be torn by the mandibles of the coccinellids or pierced by the stylets of predatory bugs and lacewings. The small bite can also result in fatal effects of the pupa or led to emergence of a defective adult. It is not always easy to recognize the attacked mummies of the second group, indicated only by small holes with darkened margins. Dense aggregations of aphidiine mummies may be preyed upon through sucking by the mandibles of the chrysopid larvae, which results in the presence of two very small apertures in a mummy which is but actually dead and empty (Al-Rawy et al. 1969). The larvae of syrphid flies are likely have less preference on the parasitized aphids, since they are either unable to open the mummy with their mouthparts or they could not recognize the mummies as suitable food source/pray (Meyhöfer and Klug 2002).

It is evident that the parasitized aphids produce more honeydew, that is likely attracting the aphid predators (Carter and Dixon 1984), therefore the parasitized aphids are more vulnerable to the attacks by predators which use honeydew as contact kairomone. Predation on the mummies by the coccinellids can be very heavy in the field (Colfer and Rosenheim 2001). The heavy predation may disrupt regulation of host aphid populations by the parasitoids. On the other hand, the successful results of the combined effects by the aphid parasitoids and predators (Rosenheim et al. 1997) cannot be overlooked. The effects of natural enemies in this system are non-additive (Snyder and Ives 2003). Ecologically, the competitive interaction of the predator-parasitoid can suppress efficiency of parasitoid, but the predator can regulate the population of the pest aphid, solely (Costamagna et al. 2007) or within a diverse assemblage of the predators (Snyder et al. 2006). In the positive view, the predators may have a regulatory effect on the parasitoids on the same way that suggested for the hyperparasitoids. The negative effect of predation has a dynamic nature and can be affected by the habitat complexity (Janssen et al. 2006), migrations (Briggs and Borer 2005), or the availability of alternative resources (Daugherty et al. 2007).

The adult parasitoids can also be preyed by the predatory insects. It is not a rare phenomenon, when the anthocorids feed on the adult aphid parasitoids among the aphid colony. The carabids, lady beetles and staphylinid beetles as well as the spiders are known as predators of the adult aphid parasitoids (Traugott et al. 2012). There are some evidences indicated the existence of some behavioral or chemical defensive behavior in the adult parasitoids against the generalist predators (Godfray 1994; Völkl 1997; Wells et al. 2001), but it needs to be further investigated in the case of aphid parasitoids. The situation is more complicated in the case of some parasitoids, like *Lysiphlebus fabarum* (Marshall), which adapted in the ant-attending aphid colonies. The ants act as effective guards for the tended aphids in warding off the predators (Jiggins et al. 1993). In the absence of protecting ants, all developmental stages of *L. fabarum* suffered from a high risk of predation. Foraging females of *L. fabarum* did not show an effective defence behavior even in direct confrontations with the predator species, while they are able to kill the adult parasitoid, readily (Meyhöfer and Klug 2002). Foraging female of *Aphidius ervi* make patch leaving decisions according to both the relative requirements for oviposition and avoidance of intraguild predation, which may be adjusted by adult experiences (Nakashima and Senoo 2003). The parasitoid is able to utilize indirect cues of the predator, which has recently been removed (Taylor et al. 1998). This might have a chemical basis that left either by the predator or even the alarm pheromones of the reacting aphids. The spiders are group of generalist predators, the adult parasitoids caught within their web, and even have been eaten as prey. These interactions were studied in the case of some aphidiine, partly (Völkl and Kraus 1996), but may have a large impact on the efficiency of parasitism, where the spiders are rather common.

## 9.5.3   Agricultural Activities

The naturally heterogeneous and complex environments including grasslands, forests, hedgerows, and other semi-natural habitats are well-balanced in the proportions. The agricultural activities, transforming these habitats into homogeneous and simple landscapes dominated with preferred field crops. Agro-technical activities greatly change the insect community structure and may significantly interrupt the population-relations between the aphid and parasitoids. The simplified agro-ecosystems cannot maintain high populations of the natural enemies (Perović et al. 2010; Jonsson et al. 2012), while common agricultural practices including application of pesticides, tillage, habitat isolation, and harvesting, require arthropods to frequently recolonize crops (Wissinger 1997) and make severe reductions on their population density. Disturbance caused by the agricultural activities by the human and the habitat isolation resulting the significant decreases in biodiversity and disrupt the structure of resident communities (Andrade et al. 2015). Aphid parasitoids are very susceptible to environmental changes, and simply lose their natural habitat or any transformation in the farmlands (Brewer et al. 2008; Lohaus et al. 2013). Population of the aphid parasitoids sharply disappear after each harvesting step and they can start with delays, after the successful establishment of their hosts. Periodic harvesting of alfalfa affects adversely the aphid population as well as the number of their parasitoids. Strip-cutting program might be adapted in respective time.

The effects of fertilizers, as a common agricultural element on efficiency of the aphid parasitoids surveyed partially (Garratt et al. 2010). Changes on the host plant volatiles can affect the heterogeneity and size of the associated insect community including aphid parasitoids (Krauss et al. 2007). Suitability of the host can also be affected through fertilizer treatment and subsequently influences on fitness and abundance of the aphid parasitoids. Results of the experiments conducted by Garratt et al. (2010) indicated fertilizer treatments that improve fitness of the cereal-aphid (*Metopolophium dirhodum* Walker) will also improve the fitness of parasitoid (*Aphidius ervi* Haliday), measured by parasitoid size, but may not influence percentage of parasitism. Applications of both fertilizers and pesticides is believed to result in losses in the biodiversity and habitat degradation.

There is yet great debate about how the agricultural activities affect the species diversity (Tylianakis et al. 2007; Vollhardt et al. 2008; Gagic et al. 2012). Currently the definition of "Agricultural landscape" is largely accepted in the pest management strategies. In such system, there are habitats that provide alternative food sources, shelter and overwintering sites for the aphid parasitoids according to their composition (Landis et al. 2000; Roschewitz et al. 2005; Bianchi et al. 2006; Liu et al. 2013). In some cases, the positive effects of managed agricultural activity on the species diversity have been evidenced (Hole et al. 2005). In the local survey by Alikhani et al. (2013), greater values of species diversity for the aphid parasitoids were found comparing the non-cultivated areas. In the small scale it was interpreted as negative effects of the desertic climate in the non-cultivated areas (in Markazi province), which is not favourable for the development of both plants and aphids. The agricultural activities, on the other hand provide enough irrigated areas favouring the environments for growth of many host plants the aphids and their parasitoids.

Application of pesticides has also significant effect on the efficiency of the aphid parasitoids, and other natural enemies, as well (Bacci et al. 2009). Conventional insecticides readily kill majority of the foraging female parasitoids and reduce their survival rate. Sub-lethal effects of the pesticides led to further changes in the life history parameter of the aphid parasitoids including developmental time, fecundity, longevity, egg viability, sex ratio, host selection, foraging behavior and habitat preference (Desneux et al. 2004, 2006; Rezaei et al. 2014). Indirect exposure to spray droplets, or through contact with residue of insecticides on the host plant foliage, as well as feeding on contaminated honcydews are the common routes, negatively affecting the foraging female aphid parasitoids. Feeding on the extra-floral nectars of the plants, which their seed was treated with insecticide was found inducing some changes in sex ratios of the offspring (Moscardini et al. 2014). Adult parasitoids migrate into the treated area or emerge from the mummies and they can subsequently be exposed to the insecticides, because the parasitoids spend much of their life time foraging for food and host aphids for oviposition on the contaminated plants.

The residue of chemical insecticides may alter behavior of the parasitoids through disrupting activity of the central nervous system even at very small doses (Haynes 1988; Elzen 1989). The low doses of lambda-cyhalothrin was found to impair the orientation and oviposition behaviors of *Aphidius ervi* (Haliday), when allocating its host aphid, *Myzus persicae* (Sulzer) (Desneux et al. 2004). A single application of insecticide is enough to destroy the whole population of the adult parasitoids; however, the immature stages are more protected inside the body of their host aphid. It was believed that the late developmental stages of the aphid parasitoids inside the mummified bodies of their aphid hosts have the least susceptibility to the conventional insecticides (Starý 1970a, b). Considering the importance of ecological selectivity (Ripper et al. 1951) for the broad–spectrum insecticides, this period can be an opportunity of the pesticide applications, which should be timed so that the major parts of parasitoid populations are in pupal stage within the mummified aphids (Newsom et al. 1976). Even at this situation, the sub-lethal effects on the emerging adults can be detrimental (Desneux et al. 2006; Sabahi et al. 2011; Purhematy et al. 2013; Mardani et al. 2016). Mortality of the emerging parasitoids was observed in the case of treatment by dimethoate or deltamethrin and has assumed caused by ingestion of insecticide residues, when the adult parasitoid was trying to cut a hole in the mummy with its mandibles (Polgar and Sagi 1983). Aphids in alfalfa fields of Iran have a complex of efficient parasitoids (Rakhshani et al. 2006a), which occasionally help in biocontrol of the aphids in neighbouring fields, too. Normally no chemical treatments are needed for alfalfa even at the period of outbreak in population of its key pest, *Hypera postica* (Gyllenhal). It has also been strongly prohibited because of foraging value for the livestock. Occasional application of chemical insecticide directly reflecting the emergence of large aphid populations represented by *Acyrthosiphon pisum* (Harris).

The adult parasitoid leaves the treated patches and avoid oviposition in these areas (Longley and Jepson 1996). The catastrophic aphid mortality following insecticide treatment can also result migration of the parasitoids to the nearby areas, where they may fail to find suitable food resources. This phenomena is quite common in isolated monocultures and cause a longer delay in establishing the new generation of the parasitoids on the growing population of the survived aphids. Organic farming is an alternative to the conventionally managed farms, in which a higher proportion of semi-natural habitat are exist (Langer 2001). Inter-cropping as a model that enhances the plants diversity (Gibson et al. 2007) and the pesticide application is prohibited (Lampkin 1999) to improve colonization of aphid parasitoids. Within such environment, pest control process is naturally occurring by the same predators and parasitoids, but very efficiently in absence of the insecticides. Presence of alternate aphid hosts, as well as food resources for the adult parasitoids enhance their efficiency in a complex organic farming system (Langer 2001; Macfadyen et al. 2009).

## 9.6    Conservation of the Aphid Parasitoids

### 9.6.1    Concept of Refugiums for the Aphid Parasitoids

The term "refugia" has been originally used for a site or area where some rare/endangered species have succeeded to survive. It has been used for the endemic species, "living relics", and potentially also for rare species which required conservation and environmental protection. In this respect, the meaning still has been respectively used. However, diversification of the landscape and namely, the development of agroecosystems has determined some situations, where the pests as well as useful natural enemies are concentrated and/or surviving after unfavourable periods of the season, and may again re-expand in the environment. In the natural plant community conservation programmes, it is easy to join it with the aphid and parasitoids associations, as each aphid and a parasitoid is generally capable to find a certain trophic association in an environment, even at very low population level. Thus, aphid-parasitoid conservation reflects the conservations of both the insect groups. There are explicit, but small-scale evidences indicating that parasitoids benefit from the resources in non-crop habitats and then spill over into adjacent agricultural areas to parasitize their host insects (Starý 1975; Rand et al. 2006). Such situations have become common today in the cultivated landscape, from the fields, meadows, orchards, up to the urban ecosystems. A special attention also represents so-called non-cultivated habitats such as wastelands, hedges and fallows as well as their interactions with the nearby occurring forests. However, some crops and their respective management may themselves also represent refugia of natural enemies (parasitoids). The occurrence of a particular refugium can be short- or long-termed, respectively, as well as its environmental interactions. There are different refugiums derivable from the target enemies. For the aphidiine parasitoids, the host spectrum and associations are important.

A target aphid-parasitoid association works also in the refugium, but sometimes there are differences between the aphids and parasitoids. The aphids are generally followed by the parasitoids in all the environments, but, on the other hand, the parasitoids manifest also more or less other interactions derivable from their host range, i.e. attacking other aphid/plant associations, apart from the target pest species. These situations are perhaps best to be exemplified, as follows: Alfalfa is attacked by *Acyrthosiphon pisum* throughout the season, with respective seasonal population fluctuation. There is an almost obligate association with *Aphidius ervi* (Rakhshani et al. 2006a), thus alfalfa works as a semi-perennial seasonal refugium of this parasitoid, which can disperse through the other preferred host in favourable condition. However, alfalfa may be cut off- and the population association are temporarily more or less broken until the autumn. *Acyrthosiphon pisum* looks for the legumes (other alfalfa fields, peas), the parasitoid (*Aphidius ervi*) may find also other hosts but both in other legumes fields or in the cereals (cereal aphids). In the annual cereals, which are temporary refugia, the population of cereal aphids and their parasitoids colonize on the crop (Rakhshani et al. 2008b; Tomanović et al. 2009), but seasonally affected by crop ripening causing subsequent leave of the parasitoids and aphids into the surrounding environment including alfalfa fields, where merely *Aphidius ervi* can survive on *A. pisum*.

However, the situation is different for *Therioaphis trifolii*, another pest aphid on alfalfa. The aphid is holocyclic on alfalfa, manifesting a different seasonal population peak than *A. pisum*, both the parasitoids, *P. exsoletum* and *T. complanatus* are specific to *Therioaphis*, and cannot find other host species on nearby cereals. The only refugium are alfalfa fields in nearby areas or possible associated with same host aphid the weeds. Strip-harvesting system is a conservation method that provided two separate stages of alfalfa in the same field throughout the growing season (Stern et al. 1964). The more stable environment prevents emigration of the natural enemies (Summers 1976) and provide a local refugium as expected (Hossain et al. 2000; Rakhshani et al. 2010). In a refugium, there may be a set of other plant-aphid associations which may be but economically indifferent if related to a target pest aphid., i.e. wild grasses related to alfalfa like *Melilotus* spp.. The effect of a refugium is doubtlessly also definable by the number of finding prey availability, hence more diversified communities may also increase the role of refugium for target parasitoids.

Another aspect of a refugium is dealing with the aphids migrating to their secondary host plant in late season. The parasitized alate aphids may also transfer the parasitoids into the colonies in the refugium, where they can multiply for the rest of season. *Hyalopterus* aphids are good example, which migrates from the weeds (*Phragmites*) to plums, peaches, apricots, where it is a pest. The parasitoid, *A. transcaspicus* is associated with the aphid in both these systems. So, it is not far expected to find this parasitoid in the wetlands, where it can attack the secondary host aphid, *Melanaphis donacis* (Passerini) on *Arundo donax* (Tomanović et al. 2012). Inter-cropping is a new agro-technical approach, which believe to increase the diversity and population of the natural enemies and should be considered in insect pest management programs. The alternative host plant for both pest aphid and its parasitoids known as "reservoir" can be provided both within and outside of the field. Preservation of the naturally growing plants (not the persistent weeds) provide shelter and food source for the aphid parasitoids during and after the growing season. The parasitoids can also multiply their populations on the innocuous aphid species living on the surrounding plants (Starý 1972b; Tylianakis et al. 2004). According to the database of the host records, it is possible to select the alternative host plants, which can act as the best reservoirs at each agro-ecosystem. Such reservoirs of the beneficial aphid parasitoids are investigated by various authors (Perrin 1975; Starý 1982, 1983, 1986a, b; Kavallieratos et al. 2002, 2008; Havelka et al. 2012).

## 9.6.2   Importance of Conservation Programs

Aphid parasitoids play an important role in natural control of the aphid population in various habitats (Starý 1970a, b). Fauna of Iran includes a number of endemic plants and associated food webs composed also from aphids and their parasitoids. Primarily, such food webs are associated with plants and, therefore, all the conservation programmes targeting the plants simultaneously conserve the associated webs of aphids and parasitoids. This concerns namely the important, often endemic, but possibly less known natural ecosystems of Iran, from the mountains to the semi-deserts (Barahoei et al. 2013; Nazari et al. 2012; Alikhani et al. 2013; Taheri and Rakhshani 2013). These evidences are important in the cases, where the aphid species have their native home and are in association with their native parasitoids.

Conservation is one of the major aspect of biological control strategies that has received the least attention, at least in the case of aphid parasitoids. It has been defined as increase the effectiveness of natural enemies through manipulation of the environment and agricultural landscapes, which led to enhancing the longevity, fecundity and survival rate of the natural enemies (Cortesero et al. 2000; Landis et al. 2000). Many farming practices can be categorized as conservation methods, even those can relieve the unfavourable factors. Providing the supplementary food resources, shelters and refugium within or outside the arable crop are the most common conservation efforts. The plant habitat manipulations (Fiedler et al. 2008) including production of nectars (Zhao et al. 1992), attractiveness to natural enemies (Patt et al. 1997) and phenology of the flowering period (Rebek et al. 2005) have direct impact on the parasitoids. The objective of conservation programs is to ensure about proper occurrence of the parasitoid with its resource requisites including the preferred host, so maximizing plant species diversity may accomplish the matter in its own way. The complexity of the plant community, as an element of the sustainable agriculture may has benefits for both specialist and generalist aphid parasitoids, providing the shelter and protect them against hyperparasitoids and extreme weather conditions. The situation is different in the case of broadly oligophagous aphid parasitoids, which has different host aphids in various habitats (Barbosa and Benrey 1998). The result of an experimental survey on the rate of complexity and its effect on conservation biological control (Jonsson et al. 2015) suggested that conservation programs are most effective in moderately simple agricultural landscapes, and less effective in either very simple (no capacity for response), or in highly complex landscapes (already saturated in responses). In general, the effect of plant diversity on the efficiency of conservation biological control is complicated, since the host aphid may also benefit from the same modifications. Inter-cropping and other practices that increase the plant species diversity bring further modification both on intraguild predations, as well as hyperparasitism that is hard to find a clear assessment. The hydrocarbon resources including nectars can be supplied by the various flowering plants in surrounding area. Effect of hyperparasitism has generally been ignored in the context of conservation biological control (Polis and Winemiller 1996). The surveys by Araj et al. (2011) indicated the searching efficiency of primary and secondary parasitoids of *Acyrthosiphon pisum* was enhanced, when accessing floral nectar. They concluded that nectar provision can potentially have positive or negative effects, depending on the relative proportion of each species. Various plants of the same or different family supporting the host aphids of the parasitoid, which primarily attacking the pest aphid. Further details about reservoir host plants in the field conditions, and in similar way to the "banker plants" in the greenhouse are presented in previous section. Surprisingly that is an important aspect of conservation program for biological control of the pest aphids (Brown and Mathews 2007) with the same definition of increase in plant diversity. No need for additional efforts in plantation of these plants in many cases, since they are growing naturally, around or inside the arable crops and especially in fruit orchards.

The continuous application of the chemical insecticides in large scale caused many environmental and ecological problems including development of resistance in serious pest aphids (Ghadamyari et al. 2008) and subsequent resurgence of their populations in the absence of their natural enemies which destroyed by the direct and indirect effects of the pesticides (Hardin et al. 1995). The lethal and sub-lethal effects of insecticides on the aphid parasitoids can be reduced in different ways, mainly by decreasing the doses and frequency of applications. Achieving the ecological selectivity by timing treatments is likely to be impractical within the current situation of agriculture in many parts of Iran. Simultaneously, problems with the other key pests are the major reason prohibiting the scheduled treatments for the aphids, when their parasitoids are inside the mummies. Nonetheless, the frequency and efficiency of the native aphid parasitoids in the field condition, make enough sense for eliminating the unnecessary chemical treatments, and performing local patch applications, where the aphids and other pests are colonized. The selective aphicides (Sabahi et al. 2011; Mardani et al. 2016) with least negative effects on the natural enemies, especially aphid parasitoids are preferred over the broad spectrum insecticides. These practices will certainly give enough chance to the aphid parasitoid to recover their population and provide enough control on the aphid population below the economic injury level. Conducting the supplementary field surveys in the various agro-ecosystems with plant suffering from aphid infestations are necessary to clarify the status of current natural biological control in Iran.

## 9.7   The Aphid Parasitoids Imported from Iran for Biological Control of Pest Aphids

The area of Iran, because of the rich faunal and floral composition, as well as of the respective associations of insects including aphids and their parasitoids become doubtlessly as an outstanding area, where a number of aphids have their native home. For this reason, many researches have been centred to search and find potentially useful parasitoids to be used in the biocontrol of some pest aphids in other countries, especially in the North America. Often, Iran has been covered within a framework of search over the Central Asian and the Mediterranean area. The definition of "Classical Biological Control - CBC" can be explained as purposeful introduction of an exotic natural enemy of an invasive pest from its area of origin, in order to suppress the abundance of the pest in the newly invaded region (DeBach 1964). Many aphids have known as alien species, invaded the new area together with plant material transported for various purposes. *Aphis illinoisensis* Shimer is a grapevine originally distributed in North, central and South America (Blackman and Eastop 2006) is now widely distributed in the Mediterranean countries including North Africa (Havelka et al. 2011). The native parasitoids (*Aphidius colemani*, *Aphidius matricariae* and *Lysiphlebus testaceipes*) indicated that they are able to successfully parasitize the aphid (Havelka et al. 2011). There are some non-confirmed field evidences of this aphid in Saudi Arabia, too (Zubair Ahmad, pers. com.). Many Lachninae aphids of the genera *Cinara* and *Eulachnus* achieved the pest status in Iran in the absence of their parasitoids. It can be expected a long history of invasion for this aphids, but the frequent chemical treatments seem to have more disrupting effect of their endemic predatory insects and led to the occasional outbreaks in the urban areas. There is no attempt for introduction of the aphid parasitoids into Iran, whereas many successful species in CBC originated from Iran. The best representative attempts for aphid biological control programs are the search for parasitoids of alfalfa and walnut aphids in Iran (Mediterranean countries, as well).

*Trioxys pallidus* was imported into California, where is an area with similar climate to Iran and it became quickly established in all walnut-growing areas, and led to a very successful biological control of the walnut aphid (van den Bosch et al. 1962, 1979). Simultaneous explorations were done for the parasitoid of the spotted alfalfa aphid, *Therioaphis trifolii* in Iran and some other countries in the Middle East (van den Bosch 1957). Among the collected parasitoids, *Trioxys complanatus* showed the greatest activity during hot periods and successfully established in California (Schlinger and Hall 1959). Further investigations indicated that Iranian strain of *Trioxys complanatus* is highly tolerant to the extreme temperatures (Flint 1980). *Praon exsoletum* was also among the imported species from Iran to California which was subsequently produced in very large numbers and colonized at many localities throughout the alfalfa growing areas and successfully established (van den Bosch 1957; van den Bosch et al. 1959).

Recently, because of unknown reasons the ability *T. pallidus* significantly decreased to provide sufficient pest suppression (Hougardy and Mills 2009; Walton et al. 2009). The evidences for hybridization of various strains of *Trioxys pallidus* in California (Messing and Aliniazee 1988) has considered as potential reason of this failure, but further investigations indicated that the "inbreeding depression" may be contributing to the breakdown of biological control by *T. pallidus* (Andersen and Mills 2016). Through the connection between the aforementioned researchers in California (Andersen and Mills), Ilania Astorga, a specialist from Division of Agricultural Protection and Forestal del SAG, Chile, tried for a fresh introduction of *Trioxys pallidus* from Iran to the southern hemisphere, in 2013 & 2014. It was expected that through program, the walnut producers will no longer need to make exclusive applications of pesticides. The introduction is believed to be successful (unpublished data). *Aphidius transcaspicus* Telenga is specialized parasitoid of *Hyalopterus* aphids on *Prunus* spp. and widely distributed in central and western Asia (Iran), as well as Mediterranean basin (Starý 1976, 1979). It has recently been targeted for use as a biological control agent against the mealy plum aphid *Hyalopterus pruni* (Geoffroy) in prune orchards in California (Latham and Mills 2010).

González et al. (1978) reviewed the geographical distribution of the imported parasitoid to California for biological control of *Acyrthosiphon* aphids on alfalfa. The area of explorations were some countries in the southeastern Europe and the Middle East. Four species including *Aphidius ervi* Haliday, *Aphidius smithi* Sharma & Subba Rao, *Aphidius urticae* Haliday and *Praon barbatum* Mackauer were collected from Iran (and Afghanistan) that showed a different rate of parasitism in the field condition. A considerable effort was also done to find the exotic parasitoids of the Russian Wheat aphid (RWA) in Iran and some other countries (González et al. 1990, 1992). The most recent activity is introduction of *Pauesia antennata* (Mukerji) for the biological control of the giant brown peach aphid, *Pterochloroides persicae* (Cholodkovsky) from Iran to Tunisia (Mdellel et al. 2015), that is yet in evaluation process.

It should be emphasized that search for potentially useful aphid parasitoid species has reflected the urgent need to obtain some parasitoid targets, but often with a lack of the broader ecological data on the respective guild members and their ecosystem interactions (such as host range). A clear centring to a target-aphid parasitoid associations did not include description of a more or less complete host species spectrum and corresponding interactions with the other (agro)- ecosystems. This is a feature which might affect adversely the predicted/realized interactions of the target species in an area of destination. In some cases, a parasitoid biocontrol agent might affect even populations of a non-target (pest) aphid in the area. A more or less successive respective research on such targets has been also obtained in a framework of an up-dated and on-going research derived basically from the tri-trophic associations realized in Iran (Nazari et al. 2012; Barahoei et al. 2013; Rakhshani et al. 2012b; Mossadegh et al. 2011, 2016). Last but not least, exportation-successive information has contributed also to the detection and elucidation of some species-complexes earlier classified as "species exported" in the abroad (*L. testaceipes, A. colemani*, etc.).

## 9.8   Potential Aphid Parasitoids for Mass Rearing

The sufficiently elaborated and supplemented databases allow an easy derivation of all the required tritrophic combinations such as lists of parasitoid-host aphids, aphid-plant combinations up to the determination of the individual assemblages and their habitats. Successful mass rearing of the aphid parasitoids is a crucial element of the aphid biological control especially in glasshouses. Various aphids are known to have the pest status in the glasshouses, of which two species, *Myzus persicae* (Sulzer) and *Aphis gossypii* glover are more common on the vegetable and ornamental plants in Iran. The augmentation (inundative) application of the aphid parasitoid requires mass production of thousands adults. It is also necessary to have small scale mass rearing, when the aphid parasitoids are introduced into the new environment in the course of classical biological control programs (Persad et al. 2007). The procedures are the same in general, but the major difference is the designation of economically affordable practices. The long term benefits from introduction of a successful aphid parasitoid justify the costs for mass rearing, but it should be very carefully balanced, when the parasitoids are commercially produced. In this way, the simple procedures should be changed into rearing on the cost-effective methods, which include rearing on the alternative host aphid/plants, appropriate storage and packing the biological products.

Among several potentially effective aphid parasitoids, few species are commercially produced, including *Aphidius matricariae* Haliday, *Aphidius ervi* Haliday and *Aphidius colemani* Viereck (Leppla 2013). Some other species have also been successfully produced in smaller scale, like *Lysiphlebus testaceipes* (Cresson) that may have the same or greater potential. In the simplest definition, the selected parasitoids must be easy to rear both on the main and/or on alternative host and applicable for control of a wide range of aphids. *Myzus persicae* (Sulzer) is known as a generalist aphid that can feed on over 400 plant species (Weber 1985), of which

**Table 9.8** Potential species for mass rearing and area of applicability in Iran

| Parasitoid species | *Hosts records | Preferred Target host aphids | Target plant | Score |
|---|---|---|---|---|
| Adialytus salicaphis | 9 | Chaitophorus spp. | Poplars willows | Medium |
| Aphidius ervi | 9 | Acyrthosiphon pisum; Myzus persicae; Sitobion avenae | Field crops Greenhouses | High |
| Aphidius matricariae | 51 | Aphis craccivora; Brachycaudus helichrysi; Myzus persicae | Field crops Greenhouses | High |
| Aphidius platensis | 28 | Aphis fabae; Aphis gossypii; Brachycaudus helichrysi; Myzus persicae | Field crops Greenhouses | High |
| Aphidius smithi | 1 | Acyrthosiphon pisum | Alfalfa, Clover | High |
| Aphidius transcaspicus | 2 | Hyalopterus spp. | Prunus spp. | Medium |
| Binodoxys acalephae | 13 | Aphis craccivora; Aphis fabae; Aphis gossypii | Field crops Greenhouses | Medium |
| Binodoxys angelicae | 16 | Aphis craccivora; Aphis fabae; Aphis gossypii | Field crops Greenhouses | Medium |
| Diaeretiella rapae | 24 | Brevicoryne brassicae; Diuraphis noxia; Myzus persicae | Rapeseed Cereals Greenhouses | High |
| Ephedrus persicae | 30 | Brachycaudus spp.; Schizaphis graminum | Cereals Prunus spp. | High |
| Lysiphlebus fabarum | 65 | Aphis craccivora; Aphis fabae | Field crops Greenhouses | High |
| Praon volucre | 35 | Acyrthosiphon spp.; Aphis fabae; Hyalopterus spp. | Legum crops Prunus spp. | Medium |

*Total number of the host aphids in Iran including pest and non-pest species

some plants species like peppers can be grown throughout the year in greenhouses. It is a very good direct and alternative host for mass rearing of the various generalist aphid parasitoids (Wei et al. 2005; Silva et al. 2011; Vafaie et al. 2013).

Ignoring the type of application, i.e. glasshouses or open field, further aphid parasitoids species can be listed for mass rearing on the basis of the importance of their preferred host aphid and also their efficiency (Table 9.8). A good knowledge about biology, host preference and behavior of the candidate aphid parasitoids is necessary for assembling the mass-rearing techniques. An excellent review for mass rearing of the aphid parasitoid is provided by Boivin et al. (2012). There are many important factors affecting the efficiency of the mass reared products (Leppla 2013). The host plants directly affect the host aphids, which can be considered as unit of the products. The host aphid quality and its size are two factors affecting the efficiency of the produced parasitoids, through its survival and reproductive rate (Henry et al. 2008). The genetic composition of the maternal populations is rather important and should be kept far from the artificial disorders like drifts and inbreeding which are common in laboratory reared populations (Steiner and Teig 1989). The adult

**Fig. 9.7** General procedure for mass production of *Aphidius colemani/platensis* (**a, b**) colony of parasitized host aphids on their food plants – (**a**) *Phaseolus vulgaris*, (**b**) *Triticum vulgare*; (**c, d**), separating the detached mummies containing live pupae of the aphid parasitoids; (**e**) cold storage; (**f**) packing the boxes for transportation to the field and/or to the costumers

experiences and learning opportunity are also important factors, which contributed to the host preference and behavior of the produced adult parasitoids (Pungerl 1984; Powell and Wright 1992; Vafaie et al. 2013). Several technical methods are suggested (Halfhill 1967; Halfhill and Featherstone 1967; Starý 1970b; Hägvar and Hofsvang 1990) for the mass rearing of the common aphid parasitoids. A general scheme is presented in Fig. 9.7.

Depending on ability of the aphid parasitoid species, cold storage can be applied to enhance the durability of product by keeping the mummies at low temperatures (Hofsvang and Hägvar 1977). Like many other commercial activity, it is necessary for a mass propagation program to have an effective storage protocol, which include even the pre-storage practices. The sold products also need to be used in appropriate time, to avoid unfavorable weather conditions and perfect synchrony with the host aphid outbreaks (Frère et al. 2011).

The mummy is known as best stage for cold storage (Starý 1970a, b). In optimal condition the mummies can be keep alive for over 1 month (Archer et al. 1973; Polgar 1986; Hofsvang and Hägvar 1977; Rabasse and Ibrahim 1987; Frère et al. 2011; Mahi et al. 2014; Al-Antary and Abdel-Wali 2016). The mummies of *Acyrthosiphon pisum* (Harris), containing mature larvae and pupae of *Aphidius ervi* Haliday were kept under 3–10 °C for 196–202 days without significant effect on fecundity and viability (He et al. 1983). On the other hand, the low temperatures can have a detrimental effect on the emergence rate, sex ratio and adult behavior (Ismail et al. 2010; Frère et al. 2011; Bourdais et al. 2012). The quality control is a necessary step in mass rearing (van Lenteren, 2003) of the aphid parasitoids, which may use various indicators including sex ratio (Silva et al. 2011), Adult size (Pavlik 1993; Ameri et al. 2014) and morphometrics (Ameri et al. 2013; Mohammadi et al. 2015), rate of emergence (Frère et al. 2011), fecundity (Portilla et al. 2013), adult longevity and ovipostion period (Rezaei et al. 2020). Thorough inspecting of the colonies for possible contamination with the hyperparasitoids both at the time of rearing and in products is necessary, since they can destroy all the colonies.

# References

Addicott JF (1979) A multispecies aphid–ant association: density dependence and species-specific effects. Can J Zool 57:558–569

Al-Antary TM, Abdel-Wali MI (2016) Response of the parasitoid, *Aphidius matricariae* Haliday (Hymenoptera: Aphidiidae) mummy to cold storage. Adv Environ Biol 10:124–130

Alikhani M, Rezwani A, Starý P, Kavallieratos NG, Rakhshani E (2013) Aphid parasitoids (Hymenoptera: Braconidae: Aphidiinae) in cultivated and non-cultivated areas of Markazi Province, Iran. Biologia 68:966–973

Al-Rawy MA, Kaddou IK, Starý P (1969) Predation of *Chrysopa carnea* Steph. On mummified aphids and its possible significance in population growth of pea aphid (Homoptera: Aphididae). Bull Biol Res Centre, Baghdad 4:30–40

Ameri M, Rasekh A, Michaud JP, Allahyari H (2013) Morphometric indicators for quality assessment in the aphid parasitoid, *Lysiphlebus fabarum* (Braconidae: Aphidiinae). Eur J Entomol 110:519–525

Ameri M, Rasekh A, Michaud JP (2014) Body size affects host defensive behavior and progeny fitness in a parasitoid wasp, *Lysiphlebus fabarum*. Entomol Exp Appl 150:259–268

Ameri M, Rasekh A, Mohammadi Z (2015) A comparison of life history traits of sexual and asexual strains of the parasitoid wasp, *Lysiphlebus fabarum* (Braconidae: Aphidiinae). Ecol Entomol 40:50–56

Amini B, Madadi H, Desneux N, Lotfalizadeh HA (2012) Impact of irrigation systems on seasonal occurrence of *Brevicoryne brassicae* and its parasitism by *Diaeretiella rapae* on canola. J Entomol Res Soc 14:15–26

Andersen JC, Mills NJ (2016) Geographic origins and post-introduction hybridization between strains of *Trioxys pallidus* introduced to western North America for the biological control of walnut and filbert aphids. Biol Control 103:218–229

Andrade TO, Outreman Y, Krespi L, Plantegenest M, Vialatte A, Gauffre B, van Baaren J (2015) Spatiotemporal variations in aphid-parasitoid relative abundance patterns and food webs in agricultural ecosystems. Ecosphere 6(7):113

Antolin MF, Bjorksten TA, Vaughn TT (2006) Host-related fitness trade-offs in a presumed generalist parasitoid, *Diaeretiella rapae* (Hymenoptera: Aphidiidae). Ecol Entomol 31:242–254

Araj SE, Wratten S, Lister A, Buckley H, Ghabeish I (2011) Searching behavior of an aphid parasitoid and its hyperparasitoid with and without floral nectar. Biol Control 57:79–84

Archer TL, Murray CL, Eikenbary RD, Starks KJ, Morrison RD (1973) Cold storage of *Lysiphlebus testaceipes* mummies. Environ Entomol 2(6):1104–1108

Arnaoudov V, Olszak R, Kutinkova H (2006) Natural enemies of plum brown scale *Parthenolecanium corni* Bouche (Homoptera: Coccidae) in plum orchards in the region of Plovdiv. IOBC/WPRS Bull 29(10):105–109

Babaee MR, Sahragard A, Rezwani A (2000) Three species of parasitoids (Aphidiidae) on forest trees aphids in Mazandaran and a new method for determining percent parasitism. Paper presented at the 14th Iranian Plant Protection Congress, Isfahan, Iran 5–8 September 2000

Bacci L, Picanço MC, Rosado JF, Silva GA, Crespo ALB, Pereira EJG, Martins JC (2009) Conservation of natural enemies in brassica crops: comparative selectivity of insecticides in the management of *Brevicoryne brassicae* (Hemiptera: Sternorrhyncha: Aphididae). Appl Entomol Zool 44:103–113

Baer CF, Tripp DW, Bjorksten TA, Antolin MF (2004) Phylogeography of a parasitoid wasp (*Diaeretiella rapae*): no evidence of host-associated lineages. Mol Ecol 13:1859–1869

Bagheri-Matin S, Sahragard A, Rasoolian G (2009) Some biological parameters of *Lysiphlebus fabarum* (Hymenoptera: Aphidiidae), a parasitoid of *Aphis fabae* (Homoptera: Aphididae) under labaratory conditions. Munis Entomol Zool 4:193–200

Bagheri-Matin S, Shahrokhi S, Starý P (2010) Report of *Praon gallicum* (Hym.: Braconidae, Aphidiinae) from Iran. Appl Entomol Phytopathol 78:33–34

Barahoei H, Madjdzadeh SM, Mehrparvar M (2012) Aphid parasitoids (Hymenoptera: Braconidae: Aphidiinae) and their tritrophic relationships in Kerman province, southeastern Iran. Iran J Anim Biosyst 8:1–14

Barahoei H, Rakhshani E, Madjdzadeh SM, Alipour A, Taheri S, Nader E, Mitrovski Bogdanović A, Petrović-Obradović O, Starý P, Kavallieratos NG, Tomanović Ž (2013) Aphid parasitoid species (Hymenoptera: Braconidae: Aphidiinae) of central submountains of Iran. North West J Zool 9:70–93

Barahoei H, Rakhshani E, Nader E, Starý P, Kavallieratos NG, Tomanović Ž, Mehrparvar M (2014) Checklist of Aphidiinae parasitoids (Hymenoptera: Braconidae) and their host aphid associations in Iran. J Crop Prot 3:199–232

Barbosa P, Benrey B (1998) The influence of plants on insect parasitoids: implications for conservation biological control. In: Barbosa P (ed) Conservation biological control. Academic, San Diego, pp 55–82

Bazyar M, Hodjat M, Alichi M (2012) The functional response of *Aphidius ervi* (Haliday)(Hym.: Braconidae, Aphidiinae) to different densities of *Sitobion avenae* (Fabricius)(Hom.: Aphididae) on two wheat cultivars. Iran Agric Res 30:61–72

Blackman R, Eastop VF (2006) Aphids on the world´s herbaceous plants and shrubs. John Wiley & Sons, London

Ben Halima M, Kavallieratos NG, Starý P, Rakhshani E (2020) First record of *Diaeretus leucopterus* (Haliday) (Hymenoptera, Braconidae, Aphidiinae), the parasitoid of the aphid species, *Eulachnus agilis* (Kaltenbach)(Hemiptera, Aphididae) in North Africa. Egypt J Biol Pest Control. In press

Bennett FD (1981) Hyperparasitism in the practice of biological control. In: Rosen D (ed) The role of hyperparasitism in biological control, University of California, publication no 4103. Division of Agricultural Sciences, Oakland, pp 43–49

Berryman AA (1999) The theoretical foundations of biological control. In: Hawkins BA, Cornell HV (eds) Theoretical approaches to biological control. Cambridge University Press, Cambridge, pp 3–21

Bianchi FJ, Booij CJH, Tscharntke T (2006) Sustainable pest regulation in agricultural landscapes: a review on landscape composition, biodiversity and natural pest control. Proc R Soc Lond Biol Sci 273:1715–1727

Bigler F (1989) Quality assessment and control in entomophagous insects used for biological control. J Appl Entomol 108:390–400

Boivin G, Hance T, Brodeur J (2012) Aphid parasitoids in biological control. Can J Plant Sci 92:1–12

Bourdais D, Vernon P, Krespi L, van Baaren J (2012) Behavioral consequences of cold exposure on males and females of *Aphidius rhopalosiphi* De Stephani Perez (Hymenoptera: Braconidae). BioControl 57:349–360

Brewer MJ, Noma T, Elliott NC, Kravchenko AN, Hild AL (2008) A landscape view of cereal aphid parasitoid dynamics reveals sensitivity to farm-and region-scale vegetation structure. Eur J Entomol 105(3):503–511

Briggs CJ, Borer ET (2005) Why short-term experiments may not allow long-term predictions about intraguild predation. Ecol Appl 15:1111–1117

Brodeur J, McNeil JN (1994) Life history of the aphid hyperparasitoid *Asaphes vulgaris* Walker (Pteromalidae): possible consequences on the efficacy of the primary parasitoid *Aphidius nigripes* Ashmead (Aphidiidae). Can Entomol 126:1493–1497

Brodeur J, Rosenheim JA (2000) Intraguild interactions in aphid parasitoids. Entomol Exp Appl 97:93–108

Brown MW, Mathews CR (2007) Conservation biological control of rosy apple aphid, *Dysaphis plantaginea* (Passerini), in eastern North America. Environ Entomol 36:1131–1139

Buitenhuis R, Boivin G, Vet LEM, Brodeur J (2004) Preference and performance of the hyperparasitoid *Syrphophagus aphidivorus* (Hymenoptera: Encyrtidae): fitness consequences of selecting hosts in live aphids or aphid mummies. Ecol Entomol 29:648–656

Byeon YW, Tuda M, Kim JH, Choi MY (2011) Functional responses of aphid parasitoids, *Aphidius colemani* (Hymenoptera: Braconidae) and *Aphelinus asychis* (Hymenoptera: Aphelinidae). Biocontrol Sci Tech 21:57–70

Carey JR (1993) Applied demography for biologists with special emphasis on insects. Oxford University Press, New York

Carter MC, Dixon AFG (1984) Honeydew, an arrestment stimulus for coccinellids. Ecol Entomol 9:383–387

Carver M (1992) Alloxystinae (Hymenoptera: Cynipoidea: Charipidae) in Australia. Invert Taxon 6:769 785

Chi H, Su HY (2006) Age-stage, two-sex life tables of *Aphidius gifuensis* (Ashmead) (Hymenoptera: Braconidae) and its host *Myzus persicae* (Sulzer) (Homoptera: Aphididae) with mathematical proof of the relationship between female fecundity and the net reproductive rate. Environ Entomol 35:10–21

Cloutier C, Duperron J, Tertuliano M, McNeil JN (2000) Host instar, body size and fitness in the koinobiotic parasitoid *Aphidius nigripes*. Entomol Exp Appl 97:29–40

Colfer RG, Rosenheim JA (2001) Predation on immature parasitoids and its impact on aphid suppression. Oecologia 126:292–304

Cortesero AM, Stapel JO, Lewis WJ (2000) Understanding and manipulating plant attributes to enhance biological control. Biol Control 17:35–49

Costamagna AC, Landis DA, Difonzo CD (2007) Suppression of soybean aphid by generalist predators results in a trophic cascade in soybeans. Ecol Appl 17:441–451

Darsoei R, Karimi J, Modarres Awal M (2011) Parasitic wasps as natural enemies of aphid population in the Mashhad region of Iran: new data from DNA sequences and SEM. Arc Biol Sci 63:1225–1234

Daugherty MP, Harmon JP, Briggs CJ (2007) Trophic supplements to intraguild predation. Oikos 116:662–677
Davidian EM (2016) A new genus and new species of subfamily Trioxinae (Hymenoptera: Aphidiidae) from the Far East of Russia. Zootaxa 4205:475–479
DeBach P (1964) Biological control of insect pests and weeds. Reihhold Publishing Corporation, New York
Desneux N, Pham-Delègue MH, Kaiser L (2004) Effects of sub-lethal and lethal doses of lambda-cyhalothrin on oviposition experience and host-searching behavior of a parasitic wasp, Aphidius ervi. Pest Manag Sci 60:381–389
Desneux N, Ramirez-Romero R, Kaiser L (2006) Multistep bioassay to predict recolonization potential of emerging parasitoids after a pesticide treatment. Environ Toxicol Chem 25:2675–2682
Diehl S (1993) Relative consumer sizes and the strengths of direct and indirect interactions in omnivorous feeding relationships. Oikos 68:151–157
Dixon AFG (1987) Cereal aphids as an applied problem. Agric Zool Rev 2:1–57
Dixon AFG (1998) Aphid ecology, 2nd edn. Chapman and Hall, London
Elzen GW (1989) Sublethal effects of pesticides on beneficial parasitoids. In: Jepson PC (ed) Pesticides and non target invertebrates. Intercept Ltd., Wimborne, pp 129–150
Falabella P, Tremblay E, Pennacchio F (2000) Host regulation by the aphid parasitoid Aphidius ervi: the role of teratocytes. Entomol Exp Appl 97:1–9
Fallahzadeh M, Japoshvili G, Saghaei N, Daane KM (2011) Natural enemies of Planococcus ficus (Hemiptera: Pseudococcidae) in Fars province vineyards, Iran. Biocontrol Sci Tech 21:427–433
Farahani S, Talebi AA, Barahoei H (2015) Occurrence of the rare aphid parasitoid Praon bicolor Mackauer, 1959 (Hymenoptera, Braconidae, Aphidiinae) in Central Asia. J Insect Biodiver Syst 1:11–15
Farahani S, Talebi AA, Rakhshani E (2016) Iranian Braconidae (Insecta: Hymenoptera: Ichneumonoidea): diversity, distribution and host association. J Insect Biodiver Syst 2:1–92
Farhad A, Talebi AA, Fathipour Y (2011) Foraging behavior of Praon volucre (Hymenoptera: Braconidae) a parasitoid of Sitobion avenae (Hemiptera: Aphididae) on wheat. Psyche J Entomol 868546:1–7
Farrokhzadeh H, Moravvej G, Awal MM, Karimi J (2014) Molecular and morphological identification of hymenoptran parasitoids from the pomegranate aphid, Aphis punicae in Razavi Khorasan province, Iran. Turk J Entomol 38:291–306
Farsi A, Kocheili F, Soleymannezhadian E, Khajehzadeh Y (2010) Population dynamics of canola aphids and their dominant natural enemies in Ahvaz. J Plant Prot (Ahvaz Univ) 32(2):55–65
Farsi A, Kocheili F, Mossadegh MS, Rasekh A, Tavoosi M (2014) Natural enemies of the currant lettuce aphid, Nasonovia ribisnigri (Mosely) (Hemiptera: Aphididae) and their population fluctuations in Ahvaz, Iran. J Crop Prot 3:487–497
Fathipour Y, Hosseini A, Talebi AA, Moharramipour S (2006) Functional response and mutual interference of Diaeretiella rapae (Hymenoptera: Aphidiidae) on Brevicoryne brassicae (Homoptera: Aphididae). Entomol Fenn 17:90–97
Fernandez-Arhex V, Corley J (2003) The functional response of parasitoids and its implications for biological control. Biocontrol Sci Tech 13:403–413
Ferrer-Suay M, Selfa J, Seco-Fernández MV, Melika G, Alipour A, Rakhshani E, Talebi AA, Pujade-Villar J (2013a) Contribution to the knowledge of Charipinae from Iran (Hymenoptera: Cynipoidea: Figitidae) associating with aphids (Hemiptera: Aphididae), including new records. North West J Zool 9:30–44
Ferrer-Suay M, Selfa J, Safoora F, Karimi J, Pujade-Villar J (2013b) First records of Alloxysta ramulifera (Thomson, 1862) and Asaphes vulgaris Walker, 1834 from Iran. Linzer Biol Beitr 45:671–672
Ferrer-Suay M, Selfa J, Tomanović Ž, Janković M, Kos K, Rakhshani E, Pujade-Villar J (2013c) Revision of Alloxysta from the North-Western Balkan Peninsula with description of two new species (Hymenoptera: Figitidae: Charipinae). Acta Entomol Mus Natl Pragae 53:347–368

Ferrer-Suay M, Janković M, Selfa J, van Veen FF, Tomanović Ž, Kos K, Rakhshani E, Pujade-Villar J (2014a) Qualitative analysis of aphid and primary parasitoid trophic relations of genus *Alloxysta* (Hymenoptera: Cynipoidea: Figitidae: Charipinae). Environ Entomol 43:1485–1495

Ferrer-Suay M, Selfa J, Feli-Kohikheili Z, Damavandian MR, Pujade-Villar J (2014b) First record of *Alloxysta pleuralis* from Iran (Hym.: Cynipoidea: Figitidae: Charipinae). J Entomol Soc Iran 34(2):67–68

Ferrer-Suay M, Selfa J, Rakhshani E, Nader E, Pujade-Villar J (2015) New host and new records of Charipinae (Hymenoptera: Cynipoidea: Figitidae) from Iran. Turk J Zool 39:1121–1131

Fiedler AK, Landis DA, Wratten SD (2008) Maximizing ecosystem services from conservation biological control: the role of habitat management. Biol Control 45:254–271

Flint ML (1980) Climatic ecotypes in *Trioxys complanatus*, a parasite of the spotted alfalfa aphid. Environ Entomol 9:501–507

Frazier MR, Huey RB, Berrigan D (2006) Thermodynamics constrains the evolution of insect population growth rates: "warmer is better". Am Natl 168:512–520

Frère I, Balthazar C, Sabri A, Hance T (2011) Improvement in the cold storage of *Aphidius ervi* (Hymenoptera: Aphidiinae). Eur J Environ Sci 1:33–40

Gagic V, Hänke S, Thies C, Scherber C, Tomanović Ž, Tscharntke T (2012) Agricultural intensification and cereal aphid–parasitoid–hyperparasitoid food webs: network complexity, temporal variability and parasitism rates. Oecologia 170:1099–1109

Garratt MP, Leather SR, Wright DJ (2010) Tritrophic effects of organic and conventional fertilisers on a cereal-aphid-parasitoid system. Entomol Exp Appl 134:211–219

Ghadamyari M, Mizuno H, Oh S, Talebi K, Kono Y (2008) Studies on pirimicarb resistance mechanisms in Iranian populations of the peach-potato aphid, *Myzus persicae*. Appl Entomol Zool 43:149–157

Ghafouri-Moghaddam M, Rakhshani E, Starý P, Tomanović Ž, Kavallieratos NG (2012) Occurrence of *Monoctonia vesicarii* Tremblay (Hym., Braconidae, Aphidiinae), a very rare parasitoid of the gall forming aphids, *Pemphigus* spp. (Hemi., Eriosomatidae) in Iran. Paper presented at 20th Iranian Plant Protection Congress, Shiraz, Iran, 22–28 August 2012

Ghafouri-Moghaddam M, Lotfalizadeh H, Rakhshani E (2014) A survey on hyperparasitoids of the poplar spiral gall aphid, *Pemphigus spyrothecae* Passerini (Hemiptera: Aphididae) in Northwest Iran. J Crop Prot 3:369–376

Ghotbi Ravandi S, Askari Hesni M, Madjdzadeh SM (2017) Species diversity and distribution pattern of Aphidiinae (Hym.: Braconidae) in Kerman province, Iran. J Crop Prot 6:245–257

Gibson RH, Pearce S, Morris RJ, Symondson WOC, Memmott J (2007) Plant diversity and land use under organic and conventional agriculture: a whole farm approach. J Appl Ecol 44:792–803

Godfray HCJ (1994) Parasitoids: behavioral and evolutionary ecology. Princeton University Press, Princeton

González D, White W, Hall J, Dickson RC (1978) Geographical distribution of Aphidiidae [Hym.] imported to California for biological control of *Acyrthosiphon kondoi* and *Acyrthosiphon pisum* [Hom.: Aphididae]. Entomophaga 23:239–248

González D, Gilstrap F, Zhang G, Zhang J, Zareh N, Wang R, Dijkstra E, McKinnon L, Starý P, Woolley J (1990) Foreign exploration for natural enemies of Russian wheat aphid in China, Iran Turkey, and the Netherlands. Paper presented at the fourth Russian wheat aphid conference, Bozeman, Montana, 10–12 October 1990

González D, Gilstrap F, McKinnon L, Zhang J, Zareh N, Zhang G, Starý P, Wolley J, Wang R (1992) Foreign exploration for natural enemies of Russian wheat aphid in Iran and in the Kunlun, Tian Shjan and Altai mountain valleys of the People's republic of China. Paper presented at the fifth Russian wheat aphid conference, Forth Worth, Texas, 26–28 January 1992

Hadadian M, Zamani AA, Marefat A, Rakhshani E (2016) Can morphological analysis reveal the existence of subspecies of *Praon exsoletum* (Nees, 1811) (Hymenoptera: Braconidae, Aphidiinae) in various geographical regions? J Insect Biodivers Syst 2:339–354

Hägvar EB, Hofsvang T (1990) The aphid parasitoid *Ephedrus cerasicola*, a possible candidate for biological control in glasshouses. SROP/WPRS Bull I5:87–90

Hajrahmatollahi F, Rashki M, Shirvani A (2015) Host stage preference and effect of temperature on functional response of *Aphidius matricariae* (Hym.: Aphididae) on common wheat aphid. Biol Control Pest Plant Dis 4:65–72

Halfhill JE (1967) Mass propagation of pea aphids. J Econ Entomol 60:298–299

Halfhill JE, Featherstone PE (1967) Propogation of braconid parasites of the pea aphid. J Econ Entomol 60:17–56

Hardin MR, Benrey B, Coll M, Lamp WO, Roderick GK, Barbosa P (1995) Arthropod pest resurgence: an overview of potential mechanisms. Crop Prot 14:3–18

Hassell MP (1978) The dynamics of arthropod predator-prey systems. Princeton University Press, Princeton

Hassell MP, Lawton JH, Beddington JR (1977) Sigmoid functional responses by invertebrate predators and parasitoids. J Anim Ecol 46:249–262

Havelka J, Shukshuk AH, Ghaliow ME, Laamari M, Kavallieratos NG, Tomanović Ž, Rakhshani E, Pons X, Starý P (2011) Review of invasive grapevine aphid, *Aphis illinoisensis* Shimer, and native parasitoids in the Mediterranean (Hemiptera, Aphididae; Hymenoptera, Braconidae, Aphidiinae). Arch Biol Sci 63:269–274

Havelka J, Tomanović Ž, Kavallieratos NG, Rakhshani E, Pons X, Petrović A, Pike KS, Starý P (2012) Review and key to the world parasitoids (Hymenoptera: Braconidae: Aphidiinae) of *Aphis ruborum* (Hemiptera: Aphididae) and its role as a host reservoir. Ann Entomol Soc Am 105:386–394

Haynes KF (1988) Sublethal effects of neurotoxic insecticides on insect behavior. Annu Rev Entomol 33:149–168

He W, Li XF, Zhao QH, Huang XY, Hou MX (1983) Studies on the mass rearing and utilization of *Aphidius ervi* Haliday, a natural enemy of *Acyrthosiphon pisum* Harris. Acta Phytophyl Sin 10:167–170

Heimpel GE, de Boer JG (2008) Sex determination in the Hymenoptera. Ann Rev Entomol 53:209–230

Henry LM, Gillespie DR, Roitberg BD (2005) Does mother really know best? Oviposition preference reduces reproductive performance in the generalist parasitoid *Aphidius ervi*. Entomol Exp Appl 116:167–174

Henry LM, Roitberg BD, Gillespie DR (2008) Host-range evolution in *Aphidius* parasitoids: fidelity, virulence and fitness tradeoffs on an ancestral host. Evolution 62:689–699

Hofsvang T, Hägvar EB (1975) Fecundity and oviposition period of *Aphidius platensis* Brethes (Hymenoptera, Aphidiidae) parasitizing *Myzus persicae* Sulz. (Homoptera, Aphididae) on paprika. Nor J Entomol 22:113–116

Hofsvang T, Hägvar EB (1977) Cold storage tolerance and supercooling points of mummies of *Ephedrus cerasicola* Starý and *Aphidius colemani* Viereck (Hymenoptera: Aphidiidae). Nor J Entomol 24:1–6

Hole DG, Perkins AJ, Wilson JD, Alexander IH, Grice PV, Evans AD (2005) Does organic farming benefit biodiversity? Biol Conserv 122:113–130

Höller C, Borgemeister C, Haardt H, Powell W (1993) The relationship between primary parasitoids and hyperparasitoids of cereal aphids: an analysis of field data. J Anim Ecol 62:12–21

Holling CS (1959) The components of predation as revealed by a study of small-mammal predation of the European pine sawfly. Can Entomol 91(5):293–320

Hossain Z, Gurr GM, Wratten SD (2000) The potential to manipulate lucerne insects by strip cutting. Aust J Entomol 39:39–41

Houck MA, Strauss RE (1985) The comparative study of functional responses: experimental design and statistical interpretation. Can Entomol 115:617–629

Hougardy E, Mills NJ (2009) Factors influencing the abundance of *Trioxys pallidus*, a successful introduced biological control agent of walnut aphid in California. Biol Control 48:22–29

Hübner G, Völkl W (1996) Behavioral strategies of aphid hyperparasitoids to escape aggression by honeydew-collecting ants. J Insect Behav 9:143–157

Hughes RD, Woolcock LT, Roberts JA, Hughes MA (1987) Biological control of the spotted alfalfa aphid, *Therioaphis trifolii* f. *maculata*, on lucerne crops in Australia, by the introduced parasitic hymenopteran *Trioxys complanatus*. J Appl Ecol 24:515–537

Ismail M, Vernon P, Hance T, van Baaren J (2010) Physiological costs of cold exposure on the parasitoid *Aphidius ervi*, without selection pressure and under constant or fluctuating temperatures. BioControl 55:729–740

Jafari N, Modarres Awal M (2012) Aphid parasitoids associations on stone fruit trees in Khorasan Razavi Province (Iran) (Hymenoptera: Braconidae: Aphidiinae). Munis Entomol Zool 7:418–423

Jafari N, Karimi J, Modarres Awal M, Rakhshani E (2011) Morphological and molecular methods in identification of *Aphidius transcaspicus* Telenga (Hym: Braconidae: Aphidiinae) endoparasitoid of *Hyalopterus* spp. (Hom: Aphididae) with additional data on Aphidiinae phylogeny. J Entomol Res Soc 13:91–103

Jahan F, Askarianzadeh A, Abbasipour H, Hasanshahi G, Saeedizadeh A (2013) Effect of various cauliflower cultivars on population density fluctuations of the cabbage aphid, *Brevicoryne brassicae* (L.)(Hom.: Aphididae) and its parasitoid *Diaeretiella rapae* (M'Intosh)(Hymenoptera: Braconidae). Arch Phytopathol Plant Prot 46:2208–2215

Janssen A, Montserrat M, Hille Ris Lambers R, de Roos AM, Pallini A, Sabelis MW (2006) Intraguild predation usually does not disrupt biological control. In: Brodeur J, Boivin G (eds) Trophic and guild interactions in biological control, vol 3. Springer SBS, Dordrecht, pp 21–44

Jarosik V, Holý I, Lapchin L, Havelka J (2003) Sex ratio in the aphid parasitoid *Aphidius colemani* (Hymenoptera: Braconidae) in relation to host size. Bull Entomol Res 93:255–258

Jervis M, Kidd N (1996) Insect natural enemies: practical approaches to their study and evaluation. Chapman and Hall, London

Jiggins C, Majerus M, Gough U (1993) Ant defence of colonies of *Aphis fabae* Scopoli (Hemiptera: Aphididae), against predation by ladybirds. Br J Entomol Nat Hist 6(4):129–137

Jokar M, Zarabi M, Shahrokhi S, Rezapanah M (2012) Host-stage preference and functional response of aphid parasitoid *Diaeretiella rapae* (M'Intosh)(Hym.: Braconidae) on greenbug, *Schizaphis graminum* (Rondani) (Hem: Aphididae). Arch Phytopathol Plant Prot 45:2223 2235

Jonsson M, Buckley HL, Case BS, Wratten SD, Hale RJ, Didham RK (2012) Agricultural intensification drives landscape-context effects on host–parasitoid interactions in agroecosystems. J Appl Ecol 49:706–714

Jonsson M, Straub CS, Didham RK, Buckley HL, Case BS, Hale RJ, Gratton C, Wratten SD (2015) Experimental evidence that the effectiveness of conservation biological control depends on landscape complexity. J Appl Ecol 52:1274–1282

Kairo MT, Murphy ST (1999) Host age choice for oviposition in *Pauesia juniperorum* (Hymenoptera: Braconidae: Aphidiinae) and its effect on the parasitoid's biology and host population growth. Biocontrol Sci Tech 9:475–486

Kairo MT, Murphy ST (2005) Comparative studies on populations of *Pauesia juniperorum* (Hymenoptera: Braconidae), a biological control agent for *Cinara cupressivora* (Hemiptera: Aphididae). Bull Entomol Res 95:597–603

Kamijo K, Takada H (1973) Studies on aphid hyperparasites of Japan. II. Aphid hyperparasites of the Pteromalidae occurring in Japan (Hymenoptera). Insecta Matsummurana 2:39–76

Kaneko S (2002) Aphid-attending ants increase the number of emerging adults of the aphid's primary parasitoid and hyperparasitoids by repelling intraguild predators. Entomol Sci 5:131–146

Kant R, He XZ, Wang Q (2008) Effect of host age on searching and oviposition behavior of *Diaeretiella rapae* (M'Intosh) (Hymenoptera: Aphidiidae). N Z Plant Prot 61:355–361

Kargarian F, Hesami S, Rakhshani E (2016) First report of *Monoctonia pistaciaecola* (Hymenoptera: Braconidae) from Iran. J Entomol Res 8:263–267

Karley AJ, Parker WE, Pitchford JW, Douglas AE (2004) The mid-season crash in aphid populations: why and how does it occur? Ecol Entomol 29:383–388

Kavallieratos NG, Lykouressis D (1999) Redescription of *Aphidius transcaspicus* Telenga and its distinction from *Aphidius colemani* Viereck (Hymenoptera Braconidae). Bol Lab Entomol Agrar F S 55:105–112

Kavallieratos NG, Stathas GJ, Athanassiou CG, Papadoulis GT (2002) *Dittrichia viscosa* and *Rubus ulmifolius* as reservoirs of aphid parasitoids (Hymenoptera: Braconidae: Aphidiinae) and the role of certain coccinellid species. Phytoparasitica 30:231–242

Kavallieratos NG, Tomanović Ž, Starý P, Athanassiou CG, Sarlis GP, Petrović-Obradović O, Niketić M, Veroniki MA (2004) A survey of aphid parasitoids (Hymenoptera: Braconidae: Aphidiinae) of Southeastern Europe and their aphid-plant associations. Appl Entomol Zool 39:527–563

Kavallieratos NG, Tomanović Ž, Starý P, Emmanouel NE (2008) *Vitex agnus-castus* and *Euphorbia characias* ssp. *wulfenii* as reservoirs of aphid parasitoids (Hymenoptera: Braconidae: Aphidiinae). Fla Entomol 91:179–191

Kazemzadeh S, Rakhshani E, Tomanović Ž, Starý P, Petrović A (2009) *Areopraon lepelleyi* (Waterston) (Hymenoptera: Braconidae: Aphidiinae), a parasitoid of Eriosomatinae (Hemiptera: Aphidoidea: Pemphigidae) new to Iran. Acta Entomol Serb 14:55–63

Khajehzadeh Y, Malkeshi SH, Keyhanian AA (2010) The population dynamism of canola aphids, biology of *Lipaphis erysimi* Kalt. and its natural enemies efficiency in the rapeseed fields of Khuzestan province, Iran. Iran J Plant Prot Sci 41:165–178

Khatri D, He XZ, Wang Q (2017) Effective biological control depends on life history strategies of both parasitoid and its host: evidence from *Aphidius colemani–Myzus persicae* system. J Econ Entomol 110:400–406

Khayrandish-Koshkooei MK, Talebi AA, Rakhshani E, Pujade-Villar J (2013) Two new records of aphid hyperparasitoids (Hym.: Figitidae) from Iran. J Entomol Soc Iran 32:137–140

Kouame KI, Mackauer M (1991) Influence of aphid size, age and behavior on host choice by the parasitoid wasp *Ephedrus californicus*: a test of host-size models. Oecologia 88:197–203

Krauss J, Härri SA, Bush L, Husi R, Bigler L, Power SA, Müller CB (2007) Effects of fertilizer, fungal endophytes and plant cultivar on the performance of insect herbivores and their natural enemies. Funct Ecol 21:107–116

Lampkin N (1999) Organic farming. Farming Press, Miller Freeman, Tonbridge

Landis DA, Wratten SD, Gurr GM (2000) Habitat management to conserve natural enemies of arthropod pests in agriculture. Annu Rev Entomol 45:175–201

Langer V (2001) The potential of leys and short rotation coppice hedges as reservoirs for parasitoids of cereal aphids in organic agriculture. Agric Ecosyst Environ 87:81–92

Latham DR, Mills NJ (2010) Life history characteristics of *Aphidius transcaspicus*, a parasitoid of mealy aphids (*Hyalopterus* species). Biol Control 54:147–152

Leppla NC (2013) Concepts and methods of quality assurance for mass-reared parasitoids and predators. In: Morales-Ramos JA, Rojas MG, Shapiro-Ilan DI (eds) Mass production of beneficial organisms: invertebrates and entomopathogens. Academic, Amsterdam, pp 277–320

Lin LA, Ives AR (2003) The effect of parasitoid host-size preference on host population growth rates: an example of *Aphidius colemani* and *Aphis glycines*. Ecol Entomol 28:542–550

Liu X, Marshall JL, Starý P, Edwards O, Puterka G, Dolatti L, El Bouhssini M, Malinga J, Lage J, Smith CM (2010) Global phylogenetics of *Diuraphis noxia* (Hemiptera: Aphididae), an invasive aphid species: evidence for multiple invasions into North America. J Econ Entomol 103:958–965

Liu JH, Yu MF, Cui WY, Song L, Zhao ZH, Ali A (2013) Association between patterns in agricultural landscapes and the abundance of wheat aphids and their natural enemies. Eur J Environ Sci 3:101–108

Lohaus K, Vidal S, Thies C (2013) Farming practices change food web structures in cereal aphid–parasitoid–hyperparasitoid communities. Oecologia 171:249–259

Longley M, Jepson PC (1996) Effects of honeydew and insecticide residues on the distribution of foraging aphid parasitoids under glasshouse and field conditions. Entomol Exp Appl 81:189–198

Lopes C, Spataro T, Lapchin L, Arditi R (2009) Optimal release strategies for the biological control of aphids in melon greenhouses. Biol Control 48:12–21

Lotfalizadeh H, Gharali B (2008) Pteromalidae (Hymenoptera: Chalcidoidea) of Iran: new records and a preliminary checklist. Entomofauna 29:93–120

Lotfalizadeh H, van Veen F (2004) Report of *Alloxysta fuscicornis* (Hym.: Cynipidae), a hyperparasitoid of aphids in Iran. J Entomol Soc Iran 23:119–120

Macfadyen S, Gibson R, Raso L, Sint D, Traugott M, Memmott J (2009) Parasitoid control of aphids in organic and conventional farming systems. Agric Ecosyst Environ 133:14–18

Mackauer M (1986) Growth and developmental interactions in some aphids and their hymenopterous parasites. J Insect Physiol 32:275–280

Mackauer M, Kambhampati S (1988a) Parasitism of aphid embryos by *Aphidius smithi*: some effects of extremely small host size. Entomol Exp Appl 49:167–173

Mackauer M, Kambhampati S (1988b) Sampling and rearing of aphids parasites. In: Minks AK, Harrewijn P (eds) Aphids, their biology, natural enemies and control. Elsevier, Amsterdam, pp 205–216

Mackauer M, Völkl W (1993) Regulation of aphid populations by aphidiid wasps: does parasitoid foraging behavior or hyperparasitism limit impact? Oecologia 94:339–350

Mackauer M, Völkl W (2002) Brood-size and sex-ratio variation in field populations of three species of solitary aphid parasitoids (Hymenoptera: Braconidae, Aphidiinae). Oecologia 131:296–305

Mahi H, Rasekh A, Michaud JP, Shishehbor P (2014) Biology of *Lysiphlebus fabarum* following cold storage of larvae and pupae. Entomol Exp Appl 153:10–19

Mardani A, Sabahi Q, Rasekh A, Almasi A (2016) Lethal and sublethal effects of three insecticides on the aphid parasitoid, *Lysiphlebus fabarum* (Marshall) (Hymenoptera: Aphidiidae). Phytoparasitica 44:91–98

Matejko I, Sullivan DJ (1984) Interspecific tertiary parasitoidism between two aphid hyperparasitoids: *Dendrocerus carpenteri* and *Alloxysta megourae* (Hymenoptera: Megaspilidae and Cynipidae). J Wash Acad Sci 74:31–38

Mdellel L, Ben Halima M, Rakhshani E (2015) Laboratory evaluation of *Pauesia antennata* (Hymenoptera: Braconidae), specific parasitoid of *Pterochloroides persicae* (Hemiptera: Aphididae). J Crop Prot 4:385–393

Mehrnejad MR, Emami SY (2005) Parasitoids associated with the common pistachio psylla, *Agonoscena pistaciae*, in Iran. Biol Control 32:385–390

Messing RH, Aliniazee MT (1988) Hybridization and host suitability of two biotypes of *Trioxys pallidus* (Hymenoptera: Aphidiidae). Ann Entomol Soc Am 81:6–9

Meyhöfer R, Klug T (2002) Intraguild predation on the aphid parasitoid *Lysiphlebus fabarum* (Marshall)(Hymenoptera: Aphidiidae): mortality risks and behavioral decisions made under the threats of predation. Biol Control 25:239–248

Mitroiu MD, Abolhassanzadeh F, Madjdzadeh SM (2011) New records of Pteromalidae (Hymenoptera: Chalcidoidea) from Iran, with description of a new species. North West J Zool 7:243–249

Moayeri HR, Madadi H, Pouraskari H, Enkegaard A (2013) Temperature dependent functional response of *Diaeretiella rapae* (Hymenoptera: Aphidiidae) to the cabbage aphid, *Brevicoryne brassicae* (Hemiptera: Aphididae). Eur J Entomol 110:109–113

Mohammadi Z, Rasekh A, Kocheli F, Habibpour B (2015) Determining the best morphometric indices for quality control in a sexual population of *Lysiphlebus fabarum* (Braconidae: Aphidiinae). Plant Pest Res 5:37–48

Monajemi N, Esmaili M (1981) Population dynamics of lucerne aphids and their natural controlling factors, in Karadj. J Entomol Soc Iran 6:41–63

Moscardini VF, Gontijo PC, Michaud JP, Carvalho GA (2014) Sublethal effects of chlorantraniliprole and thiamethoxam seed treatments when *Lysiphlebus testaceipes* feed on sunflower extrafloral nectar. BioControl 59:503–511

Mossadegh MS, Starý P, Salehipour H (2011) Aphid parasitoids in a dry lowland area of Khuzestan, Iran (Hymenoptera, Braconidae, Aphidiinae). Asian J Biol Sci 4:175–181

Mossadegh MS, Starý P, Sharaf M, Mohammadi S, Aldawood AS, Tamoli-Torfi E, Abolfärsi R, Bahrami R, Mohseni L, Shanini A, Seifollahi F, Soheilyfar P, Ravan B, Alaghemand A (2016) Aphid-ant-parasitoid and host plant associations in drylands of Khuzestan, Iran (Hemiptera: Aphidae; Hymenoptera: Formicidae; Hymenoptera: Braconidae, Aphidiinae). Entomol Mag 152:289–294

Mottaghinia L, Hassanpour M, Razmjou J, Chamani E, Hosseini M (2017) Effect of vermicompost on functional response of the parasitoid wasp Aphidius colemani (Hym., Braconidae) to the melon aphid, Aphis gossypii (Hem., Aphididae). J Entomol Soc Iran 37:81–93

Najafpour P, Rasekh A, Esfandiari M (2016) The effect of choice and no-choice access on host (Aphis fabae) instars preference, in different ages and sizes of Lysiphlebus fabarum (Hymenoptera: Braconidae: Aphidiinae). J Entomol Soc Iran 36:101–111

Nakashima Y, Senoo N (2003) Avoidance of ladybird trails by an aphid parasitoid Aphidius ervi: active period and effects of prior oviposition experience. Entomol Exp Appl 109:163–166

Nazari Y, Zamani AA, Masoumi SM, Rakhshani E, Petrović-Obradović O, Tomanović S, Starý P, Tomanović Ž (2012) Diversity and host associations of aphid parasitoids (Hymenoptera: Braconidae: Aphidiinae) in the farmlands of western Iran. Acta Entomol Mus Natl Pragae 52:559–584

Nematollahi MR, Fathipour Y, Talebi AA, Karimzadeh J, Zalucki MP (2014) Parasitoid-and hyperparasitoid-mediated seasonal dynamics of the cabbage aphid (Hemiptera: Aphididae). Environ Entomol 43:1542–1551

Némec V, Starý P (1994) Population diversity of Diaeretiella rapae (M'Int) (Hym, Aphidiidae), an aphid parasitoid in agroecosystems. Entomol Z 97:223–233

Newsom LD, Smith RF, Whitcomb WH (1976) Selective pesticides and selective use of pesticides. In: Huffaker CB, Messenger PS (eds) Theory and practice of biological control. Academic, New York, pp 565–591

Noyes JS (2019) Universal Chalcidoidea Database. The Natural History Museum. Available via: http://www.nhm.ac.uk/our-science/data/chalcidoids. Accessed 15 Aug 2019

Pandey S, Singh R (1998) Effect of temperature on the development and reproduction of a cereal aphid parasitoid, Lysiphlebia mirzai Shuja-Uddin (Hymenoptera: Braconidae). Biol Agric Hort 16:239–250

Pandey S, Singh R (1999) Host size induced variation in progeny sex ratio of an aphid parasitoid Lysiphlebia mirzai. Entomol Exp Appl 90:61–67

Pasandideh A, Talebi AA, Hajiqanbar H, Tazerouni Z (2015) Host stage preference and age-specific functional response of Praon volucre (Hymenoptera: Braconidae, Aphidiinae) a parasitoid of Acyrthosiphon pisum (Hemiptera: Aphididae). J Crop Prot 4:563–575

Patt JM, Hamilton G, Lashomb J (1997) Foraging success of parasitoid wasps on flowers: the interplay of insect morphology, floral architecture and searching behavior. Entomol Exp Appl 83:21–30

Pavlik J (1993) The size of the female and quality assessment of mass-reared Trichogramma spp. Entomol Exp Appl 66:171–177

Perdikis DC, Lykouressis DP, Garantonakis NG, Iatrou SA (2004) Instar preference and parasitization of Aphis gossypii and Myzus persicae (Hemiptera: Aphididae) by the parasitoid Aphidius colemani (Hymenoptera: Aphidiidae). Eur J Entomol 101:333–336

Perović DJ, Gurr GM, Raman A, Nicol HI (2010) Effect of landscape composition and arrangement on biological control agents in a simplified agricultural system: a cost–distance approach. Biol Control 52:263–270

Perrin RM (1975) The role of the perennial stinging nettle, Urtica dioica, as a reservoir of beneficial natural enemies. Ann Appl Biol 81:289–297

Persad AB, Hoy MA, Nguyen R (2007) Establishment of Lipolexis oregmae (Hymenoptera: Aphidiidae) in a classical biological control program directed against the brown citrus aphid (Homoptera: Aphididae) in Florida. Fla Entomol 90:204–213

Pike KS, Starý P, Miller T, Allison D, Graf G, Boydston L, Miller R, Gillespie R (1999) Host range and habitats of the aphid parasitoid *Diaeretiella rapae* (Hymenoptera: Aphidiidae) in Washington state. Environ Entomol 28:61–71

Polgar L (1986) Effect of cold storage on the emergence, sex ratio and fecundity of *Aphidus matricariae*. In: Hodek I (ed) Ecology of Aphidophaga. Academia, Prague, pp 255–260

Polgar L, Sagi K (1983) The effect of pesticides on a beneficial hymenopterous parasite: *Aphidius matricariae* Hal. Bull Int Conf Integr Plant Prot 4:65–69

Polis G, Winemiller K (1996) Food webs: integration of patterns and dynamics. Chapman and Hall, London

Pons X, Lumbierres B (2004) Aphids on ornamental shrubs and trees in an urban area of the Catalan coast: bases for an IPM programme. In: Simon JC, Dedryver CA, Rispe C, Hullé M (eds) Aphids in a new millennium. Institut National de la Recherche Agronomique, Paris, pp 359–364

Portilla M, Morales-Ramos JA, Guadalupe Rojas M, Blanco CA (2013) Life tables as tools of evaluation and quality control for arthropod mass production. In: Morales-Ramos JA, Rojas MG, Shapiro-Ilan DI (eds) Mass production of beneficial organisms: invertebrates and entomopathogens. Academic, Amsterdam, pp 214–276

Pourtaghi E, Shirvani A, Rashki M (2016) Effect of temperature on biological parameters of *Aphidius matricariae*, the *Aphis fabae* parasitoid. Anim Biol 66:335–345

Powell W, Wright AF (1992) The influence of host food plants on host recognition by four aphidiine parasitoids (Hymenoptera: Braconidae). Bull Entomol Res 81:449–453

Prasad S, Singh D, Singh R (2005) Effect of host plant resistance on the agespecific life table of the aphid parasitoid, *Diaeretiella rapae* (M'Intosh)(Hymenoptera: Braconidae, Aphidiinae) at variable host densities. J Aphidol 19:113–119

Price PW (1986) Ecological aspects of host plant resistance and biological control: interactions among three trophic levels. In: Boethel DJ, Eikenbary RD (eds) Interactions of plant resistance and parasitoids and predators of insects. Ellis Horwood Ltd, Chichester, pp 11–30

Pungerl NB (1984) Host preferences of *Aphidius* (Hymenoptera: Aphidiidae) populations parasitising pea and cereal aphids (Hemiptera: Aphididae). Bull Entomol Res 74:153–162

Purhematy A, Ahmadia K, Moshrefi M (2013) Toxicity of Thiacloprid and Fenvalerate on the black bean aphid, *Aphis fabae*, and biosafety against its parasitoid, *Lysiphlebus fabarum*. J Biopest 6:207–210

Pyke GH (1984) Optimal foraging theory: a critical review. Ann Rev Ecol Evol Syst 15:523–575

Rabasse JM, Ibrahim AMA (1987) Conservation of *Aphidius uzbekistanicus* (Luz.) (Hymenoptera: Aphidiidae), parasite on *Sitobion avenae* F. (Homoiptera: Aphididae). SROP/WPRS Bull 12:54–56

Rahimi S, Hosseini R, Hajizadeh J, Sohani MM (2012) Molecular identification and detection of *Lysiphlebus fabarum* (Hym.: Braconidae): a key parasitoid of aphids, by using polymerase chain reaction. J Agric Sci Technol 14:1453–1463

Rakhshani E, Talebi AA, Sadeghi SE (2000) Some biological characteristics of walnut aphid, *Chromaphis juglandicola* (Kaltenbach)(Homoptera: Aphididae) in Karaj. J Entomol Soc Iran 20:25–41

Rakhshani E (2012) Aphid parasitoids (Hym., Braconidae, Aphidiinae) associated with pome and stone fruit trees in Iran. Journal of Crop Protection 1:81–95

Rakhshani E, Talebi AA, Sadeghi SE (2001) The first record of aphid hyperparasitoid, *Alloxysta* (*Alloxysta*) *citripes* (Thomson) (Hymenoptera: Cynipidae) from Iran. Appl Entomol Phytopathol 69:184–185

Rakhshani E, Talebi AA, Kavallieratos NG, Fathipour Y (2004a) Host stage preference, juvenile mortality and functional response of *Trioxys pallidus* (Haliday) (Hymenoptera: Braconidae: Aphidiinae). Biologia 59:197–203

Rakhshani E, Talebi AA, Sadeghi SE, Kavallieratos NG, Rashed A (2004b) Seasonal parasitism and hyperparasitism of walnut aphid, *Chromaphis juglandicola* (Hom.: Aphididae) in Tehran province. J Entomol Soc Iran 23(2):1–11

Rakhshani E, Talebi AA, Kavallieratos NG, Rezwani A, Manzari S, Tomanović Ž (2005a) Parasitoid complex (Hymenoptera, Braconidae, Aphidiinae) of *Aphis craccivora* Koch (Hemiptera: Aphidoidea) in Iran. J Pest Sci 78:193–198

Rakhshani E, Talebi AA, Starý P, Manzari S, Rezwani A (2005b) Re-description and biocontrol information of *Pauesia antennata* (Mukerji) (Hym., Braconidae, Aphidiinae), parasitoid of *Pterochloroides persicae* (Chol.) (Hom., Aphidoidea, Lachnidae). J Entomol Res Soc 7 (3):59–69

Rakhshani E, Talebi AA, Manzari S, Rezwani A, Rakhshani H (2006a) An investigation on alfalfa aphids and their parasitoids in different parts of Iran, with a key to the parasitoids (Hemiptera: Aphididae; Hymenoptera: Braconidae: Aphidiinae). J Entomol Soc Iran 25(2):1–14

Rakhshani E, Talebi AA, Starý P, Tomanović Ž, Manzari S, Kavallieratos NG, Ćetković A (2006b) A new species of *Aphidius* Nees, 1818 (Hymenoptera, Braconidae, Aphidiinae) attacking *Uroleucon* aphids (Homoptera, Aphididae) from Iran and Iraq. J Nat Hist 40:1923–1929

Rakhshani E, Talebi AA, Manzari S, Tomanović Ž, Starý P, Rezwani A (2007a) Preliminary taxonomic study of the genus *Praon* (Hymenoptera: Braconidae: Aphidiinae) and its host associations in Iran. J Entomol Soc Iran 26(2):19–34

Rakhshani E, Talebi AA, Starý P, Tomanović Ž, Manzari S (2007b) Aphid-parasitoid (Hymenoptera: Braconidae: Aphidiinae) associations on willows and poplars in Iran. Acta Zool Acad Sci Hung 53:281–292

Rakhshani E, Talebi AA, Starý P, Tomanović Ž, Manzari S, Kavallieratos NG (2008a) A review of *Aphidius* Nees (Hymenoptera, Braconidae, Aphidiinae) in Iran: host associations, distribution and taxonomic notes. Zootaxa 1767:37–54

Rakhshani E, Tomanović Ž, Starý P, Talebi AA, Kavallieratos NG, Zamani AA, Stanković S (2008b) Distribution and diversity of wheat aphid parasitoids (Hymenoptera: Braconidae: Aphidiinae) in Iran. Eur J Entomol 105:863–870

Rakhshani E, Kazemzadeh S, Talebi AA, Starý P, Manzari S, Rezwani A, Asadi G (2008c) Preliminary taxonomic study of genus *Trioxys* Haliday (Hym., Braconidae, Aphidiinae) and its host associations in Iran. Paper presented at the 18th Iranian Plant Protection Congress, Bu Ali Sina University, Hamadan, Iran, 7–11 September 2008

Rakhshani E, Talebi AA, Manzari S, Starý P, Tomanović Ž, Arjmandi AA (2008d) *Lipolexis gracilis* Förster (Hymenoptera, Braconidae, Aphidiinae), an aphid parasitoid newly determined in Iran and Turkey. Paper presented at the 18th Iranian Plant Protection Congress, Bu Ali Sina University, Hamadan, Iran, 7–11 September 2008

Rakhshani H, Ebadi R, Mohammadi AA (2009) Population dynamics of alfalfa aphids and their natural enemies, Isfahan, Iran. J Agric Sci Technol 11:505–520

Rakhshani H, Ebadi R, Hatami B, Rakhshani E, Gharali B (2010) A survey of alfalfa aphids and their natural enemies in Isfahan, Iran, and the effect of alfalfa strip-harvesting on their populations. J Entomol Soc Iran 30(1):13–28

Rakhshani E, Tomanović Ž, Starý P, Kavallieratos NG, Ilić M, Stanković SS, Rajabi-Mazhar N (2011) Aphidiinae parasitoids (Hymenoptera: Braconidae) of *Macrosiphoniella* aphids (Hemiptera: Aphididae) in the western Palaearctic region. J Nat Hist 45:2559–2575

Rakhshani E, Starý P, Tomanović Ž (2012a) Species of *Adialytus* Förster, 1862 (Hymenoptera, Braconidae, Aphidiinae) in Iran: taxonomic notes and tritrophic associations. ZooKeys 221:81–95

Rakhshani E, Kazemzadeh S, Starý P, Barahoei H, Kavallieratos NG, Ćetković A, Popović A, Bodlah I, Tomanović Ž (2012b) Parasitoids (Hymenoptera: Braconidae: Aphidiinae) of north-eastern Iran: Aphidiine-aphid-plant associations, key and description of a new species. J Insect Sci 12(1):143

Rakhshani E, Barahoei H, Ahmed Z, Tomanović Ž, Janković M, Petrović A, Bodlah I, Starý P (2012c) New species and additional evidence of aphid parasitoids (Hymenoptera: Braconidae: Aphidiinae) from India. Zootaxa 3397:45–54

Rakhshani E, Starý P, Tomanović Ž (2013) Tritrophic associations and taxonomic notes on *Lysiphlebus fabarum* (Marshall) (Hymenoptera: Braconidae: Aphidiinae), a keystone aphid parasitoid in Iran. Arch Biol Sci 65:667–680

Rakhshani E, Starý P, Perez-Hidalgo N, Čkrkić J, Ghafouri-Moghaddam M, Tomanović S, Petrović A, Tomanović Ž (2015a) Revision of the world *Monoctonia* Starý, parasitoids of gall aphids: taxonomy, distribution, host range, and phylogeny (Hymenoptera, Braconidae: Aphidiinae). Zootaxa 3905:474–488

Rakhshani E, Starý P, Tomanović Ž, Mifsud D (2015b) Aphidiinae (Hymenoptera, Braconidae) aphid parasitoids of Malta: review and key to species. Bull Entomol Soc Malta 7:121–137

Rakhshani E, Barahoei H, Ahmad Z, Starý P, Ghafouri-Moghaddam M, Mehrparvar M, Kavallieratos NG, Čkrkić J, Tomanović Ž (2019) Review of Aphidiinae parasitoids (Hymenoptera: Braconidae) of the Middle East and North Africa: key to species and host associations. Eur J Taxon 552:1–132

Rand TA, Tylianakis JM, Tscharntke T (2006) Spillover edge effects: the dispersal of agriculturally subsidized insect natural enemies into adjacent natural habitats. Ecol Lett 9:603–614

Rasekh A, Allahyari H, Michaud JP (2010a) The effect of competition on foraging behavior of a thelytokous parasitoid, *Lysiphlebus fabarum* (Marshall) on *Aphis fabae* Scopoli. J Plant Prot 24:342–333

Rasekh A, Kharazi Pakdel A, Allahyari H, Michaud JP (2010b) The effect of experience and age on foraging behavior of a thelytokous parasitoid, *Lysiphlebus fabarum* (Marshall) on *Aphis fabae* Scopoli. J Plant Prot 24:187–195

Rasekh A, Kharazi Pakdel A, Allahyari H, Michaud JP, Farhadi R (2010c) The effect of hunger on foraging behavior of an aphid parasitoid, *Lysiphlebus fabarum* (Marshall) on *Aphis fabae* Scopoli. Iran J Plant Prot Sci 41:261–270

Rasekh A, Michaud JP, Allahyari H, Sabahi Q (2010d) The foraging behavior of *Lysiphlebus fabarum* (Marshall), a thelytokous parasitoid of the black bean aphid in Iran. J Insect Behav 23:165–179

Rasekh A, Michaud JP, Kharazi-Pakdel A, Allahyari H (2010e) Ant mimicry by an aphid parasitoid, *Lysiphlebus fabarum*. J Insect Sci 10(1):126

Rashki M, Kharazi Pakdel A, Allahyari H, Shirvani A, van Alphen JJM (2013) Effect of entomopathogenic fungus, *Beauveria bassiana*, on functional response and reproduction of parasitoid wasp, *Aphidius matricariae* Haliday (Hym.: Braconidae). Biol Control Pest Plant Dis 2:43–52

Rebek EJ, Sadof CS, Hanks LM (2005) Manipulating the abundance of natural enemies in ornamental landscapes with floral resource plants. Biol Control 33:203–216

Rezaei N, Mossadegh MS, Hojat SH (2006) Aphids and their natural enemies in wheat and barley fields in Khuzestan. Sci J Agric 29:127–137

Rezaei N, Kocheyli F, Mossadegh MS, Talebi Jahromi K, Kavousi A (2014) Effect of sublethal doses of thiamethoxam and pirimicarb on functional response of *Diaeretiella rapae* (Hymenoptera: Braconidae), parasitoid of *Lipaphis erysimi* (Hemiptera: Aphididae). J Crop Prot 3:467–477

Rezaei M, Talebi AA, Fathipour Y, Karimzadeh J, Mehrabadi M (2019) Foraging behavior of *Aphidius matricariae* (Hymenoptera: Braconidae) on tobacco aphid, *Myzus persicae nicotianae* (Hemiptera: Aphididae). Bull Entomol Res 109:840–848

Rezaei M, Talebi AA, Fathipour Y, Karimzadeh J, Mehrabadi M, Reddy GV (2020) Effects of cold storage on life-history traits of *Aphidius matricariae*. Entomol Exp Appl 168:800–807

Rezwani A (2001) Identification key of aphids in Iran. Agricultural Research, Education and Extension Organization, Tehran

Rezwani A (2004) Aphids on trees and shrubs in Iran. Agricultural Research, Education and Extension Organization, Tehran

Rezwani A (2010) Aphids (Hemiptera: Aphidoidea) of herbaceous plants in Iran. Entomological Society of Iran, Tehran

Ripper WE, Greenslade RM, Hartley GS (1951) Selective insecticides and biological control. J Econ Entomol 44:448–459

Roitberg BD, Mangel M (1988) On the evolutionary ecology of marking pheromones. Evol Ecol 2:289–315

Roitberg BD, Boivin G, Vet LEM (2001) Fitness, parasitoids, and biological control: an opinion. Can Entomol 133:429–438

Roschewitz I, Hücker M, Tscharntke T, Thies C (2005) The influence of landscape context and farming practices on parasitism of cereal aphids. Agric Ecosyst Environ 108:218–227

Rosen D (1986) The role of taxonomy in effective biological control programs. Agric Ecosyst Environ 15:121–129

Rosenheim JA (1998) Higher-order predators and the regulation of insect herbivore populations. Annu Rev Entomol 43:421–447

Rosenheim JA, Kaya HK, Ehler LE, Marois JJ, Jaffee BA (1995) Intraguild predation among biological-control agents: theory and evidence. Biol Control 5:303–335

Rosenheim JA, Wilhoit LR, Goodell PB, Grafton-Cardwell EE, Leigh TF (1997) Plant compensation, natural biological control, and herbivory by *Aphis gossypii* on pre-reproductive cotton: the anatomy of a non-pest. Entomol Exp Appl 85:45–63

Sabahi Q, Rasekh A, Michaud JP (2011) Toxicity of three insecticides to *Lysiphlebus fabarum*, a parasitoid of the black bean aphid, *Aphis fabae*. J Insect Sci 11(1):104

Samanta AK, Pramanik DR, Raychaudhuri D (1983) Some new aphid parasitoids (Hymenoptera: Aphidiidae) from Meghalaya, North East India. Akitu 54(8):1–8

Schlinger EI, Hall JC (1959) A synopsis of the biologies of three imported parasites of the spotted alfalfa aphid. J Econ Entomol 52:154–157

Sequeira R, Mackauer M (1992a) Covariance of adult size and development time in the parasitoid wasp *Aphidius ervi* in relation to the size of its host, *Acyrthosiphon pisum*. Evol Ecol 6:34–44

Sequeira R, Mackauer M (1992b) Nutritional ecology of an insect host-parasitoid association: the pea aphid-*Aphidius ervi* system. Ecology 73:183–189

Sequeira R, Mackauer M (1993) Seasonal variation in body size and offspring sex ratio in field populations of the parasitoid wasp, *Aphidius ervi* (Hymenoptera: Aphidiidae). Oikos 68:340–346

Shu-Sheng L, Carver M (1982) The effect of temperature on the adult integumental coloration of *Aphidius smithi*. Entomol Exp Appl 32:54–60

Sidney LA, Bueno VH, Pereira LH, Silva DB, Lins JC, van Lenteren JC (2011) Quality of *Myzus persicae* (Hem.: Aphididae) as host for *Praon volucre* (Hym.: Braconidae: Aphidiinae). IOBC/WPRS Bull 68:177–180

Silva RJ, Cividanes FJ, Pedroso EC, Sala SRD (2011) Host quality of different aphid species for rearing *Diaeretiella rapae* (M'Intosh)(Hymenoptera: Braconidae). Neotrop Entomol 40:477–482

Silva DB, Bueno VHP, Sampaio MV, van Lenteren JC (2015) Performance of the parasitoid *Praon volucre* in *Aulacorthum solani* at five temperatures. Bull Insectol 68:119–125

Singh R, Singh G (2015) Systematics, distribution and host range of *Diaeretiella rapae* (M'Intosh) (Hymenoptera: Braconidae, Aphidiinae). Int J Res Stud Biosci 3:1–36

Singh R, Pandey S, Singh A (2000) Effect of temperature and photoperiod on development, fecundity, progeny sex ratio and life-table of an aphid parasitoid *Binodoxys indicus*. Malaysian Appl Biol 29:79–94

Singh G, Singh NP, Singh R (2014) Maternal manipulation of progeny sex ratio in parasitic wasps with reference to Aphidiinae (Hymenoptera: Braconidae): a review. Indo-Am J Life Sci Biotechnol 2:1–23

Snyder WE, Ives AR (2001) Generalist predators disrupt biological control by a specialist parasitoid. Ecology 82:705–716

Snyder WE, Ives AR (2003) Interactions between specialist and generalist natural enemies: parasitoids, predators, and pea aphid biocontrol. Ecology 84:91–107

Snyder WE, Wise DH (1999) Predator interference and the establishment of generalist predator populations for biocontrol. Biol Control 15:283–292

Snyder WE, Snyder GB, Finke DL, Straub CS (2006) Predator biodiversity strengthens herbivore suppression. Ecol Lett 9:789–796

Solomon ME (1949) The natural control of animal populations. J Anim Ecol 18:11–35

Starý P (1966) Aphid parasites (Hym., Aphidiidae) and their relationship to aphid attending ants, with respect to biological control. Insect Soc 13:185–202

Starý P (1968) Diapause in *Monoctonia pistaciaecola* Starý, a parasite of gall aphids (Hymenoptera: Aphidiidae; Homoptera: Aphidoidea). Boll Lab Entomol Agrar F S 25:241–250

Starý P (1970a) Biology of aphid parasities (Hymenoptera: Aphidiidae) with respect to integrated control. Dr. W Junk, The Hague

Starý P (1970b) Methods of mass-rearing, collection and release of *Aphidius smithi* (Hymenoptera, Aphidiidae) in Czechoslovakia. Acta Entomol Bohem 67:339–346

Starý P (1972a) *Aphidius platensis* Brethes, its distribution and host range (Hymenoptera, Aphidiidae). Orient Insects 6:359–370

Starý P (1972b) Host range of parasites and ecosystem relations, a new viewpoint in multilateral control concept (Horn., Aphididae; Hym., Aphidiidae). Ann Entomol Soc Fr 8:351–358

Starý P (1973) A review of the *Aphidius*-species (Hymenoptera, Aphidiidae) of Europe. Annot Zool Bot 84:1–85

Starý P (1975) *Aphidius colemani* Viereck: its taxonomy, distribution and host range (Hymenoptera, Aphidiidae). Acta Entomol Bohem 72:156–163

Starý P (1976) Aphid parasites (Hymenoptera, Aphidiidae) of the Mediterranean area. Dr W Junk, The Hague

Starý P (1978) Parasitoid spectrum of the arboricolous callaphidid aphids in Europe (Hymenoptera: Aphidiidae; Homoptera: Aphidoidea, Callaphididae). Acta Entomol Bohem 75:164–177

Starý P (1979) Aphid parasites (Hymenoptera, Aphidiidae) of the central Asian area. Academia, Praha

Starý P (1981) On the strategy, tactics and trends of host specificity evolution in aphid parasitoids (Hymenoptera: Aphidiidae). Acta Entomol Bohem 78:65–75

Starý P (1982) The role of ash (*Fraxinus*) as a reservoir of aphid parasitoids with description of a new species in Central Europe (Hymenoptera: Aphidiidae). Acta Entomol Bohem 79:97–107

Starý P (1983) The perennial stinging nettle (*Urtica dioica* L.) as a reservoir of aphid parasitoids (Hymenoptera: Aphidiidae). Acta Entomol Bohem 80:81–86

Starý P (1986a) Creeping thistle *Cirsium arvense* L. as a reservoir of aphid parasitoids (Hymenoptera: Aphidiidae) in agroecosystems. Acta Entomol Bohem 83:425–431

Starý P (1986b) Reservoirs of beneficial insects and their classication in integrated pest control. Ziva 34:22–24

Starý P (2006) Aphid parasitoids of the Czech Republic (Hymenoptera: Braconidae: Aphidiinae). Academia, Praha

Starý P, Lumbierres B, Pons X (2004) Opportunistic changes in the host range of *Lysiphlebus testaceipes* (Cr.), an exotic aphid parasitoid expanding in the Iberian Peninsula. J Pest Sci 77:139–144

Starý P, Remaudiere G, Gonzalez D, Shahrokhi S (2000) A review and host a ssociations of aphid parasitoids (Hym., Braconidae, Aphidiinae) of Iran. Parasitica 56:15–41

Starý P, Rakhshani E, Talebi AA (2005) Parasitoids of aphid pests on conifers and their state as biocontrol agents in the Middle East to Central Asia on the world background (Hym., Braconidae, Aphidiinae; Hom., Aphididae). Egypt J Biol Pest Control 15:147–151

Starý P, Rakhshani E, Tomanović Ž, Hoelmer K, Kavallieratos NG, Yu J, Wang M, Heimpel GE (2010) A new species of *Lysiphlebus* Förster 1862 (Hymenoptera: Braconidae: Aphidiinae) attacking soybean aphid, *Aphis glycines* Matsumura (Hemiptera: Aphididae) from China. J Hym Res 19:179–186

Starý P, Kavallieratos NG, Petrović A, Žikić V, Rakhshani E, Tomanović S, Tomanović Ž, Havelka J (2014) Interference of field evidence, morphology, and DNA analyses of three related *Lysiphlebus* aphid parasitoids (Hymenoptera: Braconidae: Aphidiinae). J Insect Sci 14:1–6

Steiner WWM, Teig DA (1989) *Microplitis croceipes* (Cresson): genetic characterization and development insecticide resistant biotypes. Southwest Entomol 12:71–87

Stern V, Bosch RVD, Leigh T (1964) Strip cutting alfalfa for *Lygus* bug control. Calif Agric 18 (4):4–6

Stilmant D (1996) The functional response of three major parasitoids of *Sitobion avenae*: *Aphidius rhopalosiphi*, *Aphidius ervi* and *Praon volucre*; how could different behavior conduct to similar results? IOBC/WPRS Bull 19(3):17–29

Sullivan DJ (1987) Insect hyperparasitism. Annu Rev Entomol 32:49–70

Sullivan DJ, Völkl W (1999) Hyperparasitism: multitrophic ecology and behavior. Annu Rev Entomol 44:291–315

Summers CG (1976) Population fluctuations of selected arthropods in alfalfa: influence of two harvesting practices. Environ Entomol 5:103–110

Symondson WOC, Sunderland KD, Greenstone MH (2002) Can generalist predators be effective biocontrol agents? Annu Rev Entomol 47:561–594

Taheri S, Rakhshani E (2013) Identification of aphid parasitoids (Hym., Braconidae, Aphidiinae) and determination of their host relationships in Southern Zagros. J Plant Prot 27:85–95

Tahriri S, Talebi AA, Fathipour Y, Zamani AA (2007) Host stage preference, functional response and mutual interference of *Aphidius matricariae* (Hym.: Braconidae: Aphidiinae) on *Aphis fabae* (Hom.: Aphididae). Entomol Sci 10:323–331

Tahriri S, Talebi AA, Fathipour Y, Zamani AA (2010) Life history and demographic parameters of *Aphis fabae* (Hemiptera: Aphididae) and its parasitoid, *Aphidius matricariae* (Hymenoptera: Aphidiidae) on four sugar beet cultivars. Acta Entomol Serb 15:61–73

Takada H (1998) A review of the *Aphidius colemani* (Hymenoptera: Braconidae; Aphidiinae) and closely related species indigenous to Japan. Appl Entomol Zool 33:59–66

Takada H, Hashimoto Y (1985) Association of the root aphid parasitoids *Aclitus sappaphis* and *Paralipsis eikoae* (Hymenoptera, Aphidiidae) with the aphid-attending ants *Pheidole fervida* and *Lasius niger* (Hymenoptera, Formicidae). Kontyû 53:150–160

Takalloozadeh HM, Kamali K, Talebi AA, Fathipour Y (2004a) Alfalfa black aphid, *Aphis craccivora* Koch (Hom.: Aphididae) stage preferences by *Lysiphlebus fabarum* (Marshall) (Hym.: Aphidiidae). J Sci Technol Agric Nat Res 7:225–234

Takalloozadeh HM, Kamali K, Talebi AA, Fathipour Y (2004b) Functional response of *Lysiphlebus fabarum* (Marshall) to different densities of Alfalfa black aphid, *Aphis craccivora* Koch (Hom. Aphididae) on two host plants at two different temperatures. Agric Sci 15:33–43

Talebi AA, Rakhshani E, Sadeghi SE, Fathipour Y (2002) Comparative study on the fecundity, developmental time and adult longevity of walnut aphid, *Chromaphis juglandicola* (Kalt.) and its parasitoid wasp, *Trioxys pallidus* (Hal.). J Sci Technol Agric Nat Res 6:241–254

Talebi AA, Zamani A, Fathipour Y, Baniameri V, Kheradmand K, Haghani M (2006) Host stage preference by *Aphidius colemani* and *Aphidius matricariae* (Hymenoptera: Aphidiidae) as parasitoids of *Aphis gosoypii* (Hemiptera: Aphididae) on greenhouse cucumber. IOBC/WPRS Bull 29(4):181–185

Talebi AA, Rakhshani E, Fathipour Y, Starý P, Tomanović Ž, Rajabi-Mazhar N (2009) Aphids and their parasitoids (Hym., Braconidae: Aphidiinae) associated with medicinal plants in Iran. Am Eurasian J Sustain Agric 3:205–219

Talebi AA, Amiri A, Fathipour Y, Rakhshani E (2010) Natural enemies of Cypress tree mealybug, *Planococcus vovae* (Nasonov)(Hem., Pseudococcidae), and their parasitoids in Tehran, Iran. J Agric Sci Technol 10:123–133

Taylor AJ, Müller CB, Godfray HCJ (1998) Effect of aphid predators on oviposition behavior of aphid parasitoids. J Insect Behav 11:297–302

Tazerouni Z, Talebi AA, Rakhshani E (2011) The foraging behavior of *Diaeretiella rapae* (Hymenoptera: Braconidae) on *Diuraphis noxia* (Hemiptera: Aphididae). Arc Biol Sci 63:225–234

Tazerouni Z, Talebi AA, Rakhshani E (2012a) Comparison of development and demographic parameters of *Diuraphis noxia* (Hem., Aphididae) and its parasitoid, *Diaeretiella rapae* (Hym., Braconidae: Aphidiinae). Arch Phytopathol Plant Prot 45:886–897

Tazerouni Z, Talebi AA, Rakhshani E (2012b) Effect of temperature on biological characteristics and population growth parameters of *Diaeretiella rapae*, parasitoid of Russian wheat aphid, *Diuraphis noxia*. Iran J Plant Prot Sci 43:83–95

Tazerouni Z, Talebi AA, Rakhshani E (2012c) Temperature-dependent functional response of *Diaeretiella rapae* (Hymenoptera: Braconidae), a parasitoid of *Diuraphis noxia* (Hemiptera: Aphididae). J Entomol Res Soc 14:31–40

Tazerouni Z, Talebi AA, Rakhshani E, Zamani AA (2013) Comparison of life table parameters of Russian wheat aphid, *Diuraphis noxia*, and its parasitoid, *Diaeretiella rapae* under constant temperatures. Appl Entomol Phytopathol 81:1–10

Tazerouni Z, Talebi AA, Fathipour Y, Soufbaf M (2016) Age-specific functional response of *Aphidius matricariae* and *Praon volucre* (Hymenoptera: Braconidae) on *Myzus persicae* (Hemiptera: Aphididae). Neotrop Entomol 45:642–651

Tazerouni Z, Talebi AA, Fathipour Y, Soufbaf M (2017) Age-specific functional response of *Aphidius matricariae* and *Praon volucre* (Hym.: Braconidae) on *Aphis gossypii* (Hem.: Aphididae). J Entomol Soc Iran 36:239–248

Tomanović Ž, Rakhshani E, Starý P, Kavallieratos NG, Stanisavljević LŽ, Žikić V, Athanassiou CG (2007) Phylogenetic relationships between the genera *Aphidius* Nees and *Lysphidus* (Hymenoptera: Braconidae: Aphidiinae) with description of *Aphidius iranicus* sp. nov. Can Entomol 139:297–307

Tomanović Ž, Kavallieratos NG, Starý P, Ćetković A, Stamenković S, Jovanović S, Athanassiou CG (2009) Regional tritrophic relationship patterns of five aphid parasitoid species (Hymenoptera: Braconidae: Aphidiinae) in agroecosystem-dominated landscapes of southeastern Europe. J Econ Entomol 102:836–854

Tomanović Ž, Starý P, Kavallieratos NG, Gagić V, Plećas M, Janković M, Rakhshani E, Ćetković A, Petrović A (2012) Aphid parasitoids (Hymenoptera: Braconidae: Aphidiinae) in wetland habitats in western Palaearctic: key and associated aphid parasitoid guilds. Ann Entomol Soc Fr 48:189–198

Tomanović Ž, Petrović A, Mitrović M, Kavallieratos NG, Starý P, Rakhshani E, Rakhshanipour M, Popovic A, Shukshuk AH, Ivanović A (2014) Molecular and morphological variability within the *Aphidius colemani* group with redescription of *Aphidius platensis* Brethes (Hymenoptera: Braconidae: Aphidiinae). Bull Entomol Res 104:552–565

Tomanović Ž, Mitrović M, Petrović A, Kavallieratos NG, Žikić V, Ivanović A, Rakhshani E, Starý P, Vorburger C (2018) Revision of the European *Lysiphlebus* species (Hymenoptera: Braconidae: Aphidiinae) on the basis of COI and 28SD2 molecular markers and morphology. Arth Syst Phyl 76:179–213

Traugott M, Bell JR, Raso L, Sint D, Symondson WO (2012) Generalist predators disrupt parasitoid aphid control by direct and coincidental intraguild predation. Bull Entomol Res 102:239–247

Tremblay E (1991) On a new species of *Monoctonia* Starý (Hymenoptera, Braconidae) from *Pemphigus vesicarius* Pass. galls (Homoptera Pemphigidae). Boll Lab Entomol Agrar F S 48:137–142

Tylianakis JM, Didham RK, Wratten SD (2004) Improved fitness of aphid parasitoids receiving resource subsidies. Ecology 85:658–666

Tylianakis JM, Tscharntke T, Lewis OT (2007) Habitat modification alters the structure of tropical host-parasitoid food webs. Nature 445:202–205

Vafaie EK, Fitzpatrick SM, Cory JS (2013) Does rearing an aphid parasitoid on one host affect its ability to parasitize another species? Agric For Entomol 15:366–374

van den Bosch R (1957) The spotted alfalfa aphid and its parasites in the Mediterranean region, Middle East, and East Africa. J Econ Entomol 50:352–356 van den Bosch R, Schlinger EI, Dietrick EJ, Hagen KS, Holloway JK (1959) The colonization and establishment of imported parasites of the spotted alfalfa aphid in California. J Econ Entomol 52:136–141

van den Bosch R, Schlinger EI, Hagen KS (1962) Initial field observations in California on *Trioxys pallidus* (Haliday) a recently introduced parasite of the walnut aphid. J Econ Entomol 55:857–862

van den Bosch R, Hom R, Matteson P, Frazer B, Messenger P, Davis C (1979) Biological control of the walnut aphid in California: impact of the parasite, *Trioxys pallidus*. Calif Agric 47:1–13

van Lenteren J (2003) Need for quality control of massproduced biological control agents. In: van Lenteren J (ed) Quality control and production of biological control agents: theory and testing procedures. CABI, London, pp 1–18

van Steenis MJ, El-Khawass KAMH (1995) Behavior of *Aphidius colemani* searching for *Aphis gossypii*: functional response and reaction to previously searched aphid colonies. Biocontrol Sci Tech 5:339–348

van Veen FJF, Müller CB, Pell JK, HCJ G (2008) Food web structure of three guilds of natural enemies: predators, parasitoids and pathogens of aphids. J Anim Ecol 77:191–200

Vinson SB, Scarborough TA (1991) Interactions between *Solenopsis invicta* (Hymenoptera: Formicidae), *Rhopalosiphum maidis* (Homoptera: Aphididae), and the parasitoid *Lysiphlebus testaceipes* Cresson (Hymenoptera: Aphidiidae). Ann Entomol Soc Am 84:158–164

Völkl W (1992) Aphids or their parasitoids: who actually benefits from ant-attendance? J Anim Ecol 61:273–281

Völkl W (1997) Interactions between ants and aphid parasitoids: patterns and consequences for resource utilization. In: Dettner K, Bauer G, Völkl W (eds) Vertical food web interactions, ecological studies (analysis and synthesis), vol 130. Springer, Berlin/Heidelberg, pp 225–240

Völkl W, Kranz P (1995) Nocturnal activity and resource utilization in the aphid hyperparasitoid, *Dendrocerus carpenteri*. Ecol Entomol 20:293–297

Völkl W, Kraus W (1996) Foraging behavior and resource utilization of the aphid parasitoid *Pauesia unilachni*: adaptation to host distribution and mortality risks. Entomol Exp Appl 79:101–109

Völkl W, Mackauer M (2000) Oviposition behavior of aphidiine wasps (Hymenoptera: Braconidae, Aphidiinae): morphological adaptations and evolutionary trends. Can Entomol 132:197–212

Völkl W, Novak H (1997) Foraging behavior and resource utilization of the aphid parasitoid, *Pauesia pini* (Hymenoptera: Aphidiidae) on spruce: influence of host species and ant attendance. Eur J Entomol 94:211–220

Völkl W, Sullivan DJ (2000) Foraging behavior, host plant and host location in the aphid hyperparasitoid *Euneura augarus*. Entomol Exp Appl 97:47–56

Völkl W, Hübner G, Dettner K (1994) Interactions between *Alloxysta brevis* (Hymenoptera, Cynipoidea, Alloxystidae) and honeydew-collecting ants: how an aphid hyperparasitoid overcomes ant aggression by chemical defense. J Chem Ecol 20:2901–2915

Vollhardt IM, Tscharntke T, Wäckers FL, Bianchi FJ, Thies C (2008) Diversity of cereal aphid parasitoids in simple and complex landscapes. Agric Ecosyst Environ 126:289–292

Walton VM, Chambers U, Olsen JL (2009) The current status of the newly invasive hazelnut aphid in Oregon hazelnut orchards. Acta Hortic 845:479–485

Weber G (1985) Genetic variability in host plant adaptation of the green peach aphid, *Myzus persicae*. Entomol Exp Appl 38:49–56

Wei JN, Bai BB, Yin TS, Wang Y, Yang Y, Zhao LH, Kuang RP, Xiang RJ (2005) Development and use of parasitoids (Hymenoptera: Aphidiidae & Aphelinidae) for biological control of aphids in China. Biocontrol Sci Tech 15:533–551

Wells ML, Mcpherson RM, Ruberson JR (2001) Predation of parasitized and unparasitized cotton aphids (Homoptera: Aphididae) by larvae of two coccinellids. J Entomol Sci 36:93–96

West SA (2009) Sex allocation. Princeton University Press, Princeton

Wissinger SA (1997) Cyclic colonization in predictably ephemeral habitats: a template for biological control in annual crop systems. Biol Control 10:4–15

Wyckhuys KAG, Stone L, Desneux N, Hoelmer KA, Hopper KR, Heimpel GE (2008) Parasitism of the soybean aphid, *Aphis glycines* by *Binodoxys communis*: the role of aphid defensive behavior and parasitoid reproductive performance. Bull Entomol Res 98:361–370

Yao I, Shibao H, Akimoto SI (2000) Costs and benefits of ant attendance to the drepanosiphid aphid *Tuberculatus quercicola*. Oikos 89:3–10

Zamani AA, Talebi AA, Fathipour Y, Baniameri V (2006) Temperature-dependent functional response of two aphid parasitoids, *Aphidius colemani* and *Aphidius matricariae* (Hymenoptera: Aphidiidae), on the cotton aphid. J Pest Sci 79(4):183–188

Zamani AA, Talebi AA, Fathipour Y, Baniameri V (2007) Effect of temperature on life history of *Aphidius colemani* and *Aphidius matricariae* (Hymenoptera: Braconidae), two parasitoids of *Aphis gossypii* and *Myzus persicae* (Homoptera: Aphididae). Environ Entomol 36:263–271

Zamani AA, Haghani M, Kheradmand K (2012) Effect of temperature on reproductive parameters of *Aphidius colemani* and *Aphidius matricariae* (Hymenoptera: Braconidae) on *Aphis gossypii* (Hemiptera: Aphididae) in laboratory conditions. J Crop Prot 1:35–40

Zhao JZ, Ayers GS, Grafius EJ, Stehr FW (1992) Effects of neighboring nectar-producing plants on populations of pest Lepidoptera and their parasitoids in broccoli plantings. Great Lakes Entomol 25:253–258

Žikić V, Tomanović Ž, Ivanović A, Kavallieratos NG, Starý P, Stanisavljević LŽ, Rakhshani E (2009) Morphological characterization of *Ephedrus persicae* biotypes (Hymenoptera: Braconidae: Aphidiinae) in the Palaearctic. Ann Entomol Soc Am 102:1–11

Žikić V, Lazarević M, Milošević D (2017) Host range patterning of parasitoid wasps Aphidiinae (Hymenoptera: Braconidae). Zool Anz 268:75–83

# Part III
# Insect Pathogens

# Chapter 10
# Progress on the Bacterium *Bacillus thuringiensis* and Its Application Within the Biological Control Program in Iran

Ayda Khorramnejad, Javad Karimi, and Gholamreza Salehi Jouzani

## 10.1 Introduction

Microbial pest control has been considered as an appropriate substitution for chemical insecticides because of its potential for controlling agricultural and household pests, vectors of human and animal diseases without introducing non-biodegradable materials to the environment. Among the microbial control agents, entomopathogenic bacteria are of crucial importance due to the ease of their application, the low-cost of production, environmental persistence, specificity and their speed of action. Insects and bacteria are associated based on the wide variety of relationships from symbiosis to pathogenesis (Feldhaar 2011; Ruiu 2015). Insecticidal bacteria have an invasive power to colonize the insect gut. From the standpoint of the utilization of bacteria in biological control of pests, families of Bacillaccae, Paenibacillaceae and Streptococcaceae belonging to the group of Gram-positive bacteria, Enterobacteriaceae and Neisseriaceae belonging to Gram-negative bacteria, have been considered substantially (Jurat-Fuentes and Jackson 2012). However, the microbial control market has been dominated by the bacteria in the genus *Bacillus*, notably *Bacillus thuringiensis* (Bt). The application of insecticidal proteins produced by *B. thuringiensis* is highly desirable, due to its high specificity,

A. Khorramnejad (✉)
Department of Plant Protection, College of Agriculture and Natural Resources, University of Tehran, Karaj, Iran

Instituto BioTecMed, Departamento de Genética, Universitat de València, València, Spain
e-mail: ayda.khoramnejad@uv.es

J. Karimi
Department of Plant Protection, Ferdowsi University of Mashhad, Mashhad, Iran

G. S. Jouzani
Agricultural Biotechnology Research Institute of Iran (ABRII), Agricultural Research, Education and Extension Organization (AREEO), Karaj, Iran

© The Author(s), under exclusive license to Springer Nature Switzerland AG 2021          403
J. Karimi, H. Madadi (eds.), *Biological Control of Insect and Mite Pests in Iran*,
Progress in Biological Control 18, https://doi.org/10.1007/978-3-030-63990-7_10

great efficiency, environmental safety, and lack of harmful effects on mammals and non-target organisms. In Iran, likewise in other parts of the world, due to the success of Bt in biological control, research on this bacterium is being continued to discover novel Bt strains and toxins, find a broader spectrum of activity and eventually develop efficient bio-pesticide based on Bt.

## 10.2   Insect Pathogenic Bacteria

Based on the industrial activities and/or scientific interests' point of view, the most important entomopathogenic bacteria in the biological control of pests are mainly in the genera of the family Bacillaceae (*B. thuringiensis* and *Lysinibacillus sphaericus*), Paenibacillaceae (*Paenibacillus* spp. and *Brevibacillus lateropporous*), Entrobacteriaceae (*Photorhabdus* spp., *Xenorhabdus* spp., *Serratia* spp. and *Yersinia entomophaga*), Pseudomonaceae (*Psedomonas entomophila*) and Neisseriaceae (Jurat-Fuentes and Jackson 2012; Ruiu 2015). Among the mentioned entomopathogenic bacteria, the most widely used and well-studied is the *Bacillus* genus.

## 10.3   The Genus *Bacillus*

Bacterial species within the genus *Bacillus* represent the most successful microbial control agents to date. *B. thuringiensis* with other bacteria such as *B. mycoides*, *B. pseudomycoides*, *B. weihensteohanensis*, *B. cereus* and *B. anthracis* belong to the *B. cereus* group (Helgason et al. 2000). Due to the high degree of genetic similarity, *B. cereus* sensu stricto, Bt, *B. anthracis* are allocated into the *B. cereus* sensu *lato* group (Helgason et al. 2000; Bavykin et al. 2004). The presence of the insecticidal genes in the Bt plasmids differentiates this bacterium from the other members of the *B. cereus* sensu *lato* group, *B. anthracis* (the causative agent of anthrax) and *B. cereus* (an opportunistic human pathogen causing food poisoning) (Thorne 1993). Although the insecticidal activity of some *B. cereus* isolates has been previously reported (Strongman et al. 1997; Selvakumar et al. 2007; Chatterjee et al. 2010), due to the presence of some toxins toxic to vertebrates, the commercialization of this bacterium in biological control of pests has not yet been developed. Different strains of *L. sphaericus* produce mosquitocidal toxins toxic to Dipterans (Baumann et al. 1991). Commercial products based on the *L. sphaericus* target mosquitos and black files (Jurat-Fuentes and Jackson 2012). In the *Bacillus* genus, different strains of *B. thuringiensis* showed great potential for control of agricultural, forestry, medical and veterinary pests. Thus, Bt is the most commonly commercialized and used bio-insecticide worldwide.

## 10.4   *Bacillus thuringiensis* (Bt): Brief History, Ecology, and Biology

The discovery of *B. thuringiensis* dates back to over a century ago, in 1901, which had been isolated by a Japanese biologist, Shigetane Ishiwata, from the diseased larvae of *Bombyx mori* (Lep.: Bombycidae) (Ishiwata 1901) and named *B. sotto*. Afterward, in 1911, Bt was isolated from the flour moth, *Ephestia kuehniella* (Lep.: Pyralidae) infected larvae in the Thuringia region of Germany, described and denominated *B. thuringiensis* by Ernest Berliner (Berliner 1915). In 1928, Bt crystals for the first time were used to control the European corn borer, *Ostrinia nubilalis* Hüber (Lep.: Crambidae). Sporine was the first commercial Bt-based product produced in France in 1938 and used to control the flour moth. The identification and characterization of Bt toxins led to a revolution in genetically modified crops expressing *cry* genes and subsequently the introduction of transgenic plant resistant to different insect pests (Lambert and Peferoen 1992; Letourneau et al. 2003). Since 1996, the cultivation of Bt crops has been expanded worldwide (Kleter et al. 2007). Since the discovery of Bt, intensive research has been conducted to study profoundly different aspects of this bacterium. Amongst these, a long history of studies has been performed in Iran to isolate, characterize and mass-produce local Iranian Bt strains.

   *B. thuringiensis* is a ubiquitous bacterium widely distributed in diverse environments and can be isolated from soil, water, plant surface, insect feces, dust, and storage products (Federici et al. 2006). When the food is sufficient for its growth, the spores germinate, begin the growth stage and reproduce through duplicate reproduction. The bacteria continue to proliferate while food sources are sufficient to sustain vegetative growth. Under adverse conditions, Bt sporulates and produces one or more insecticidal crystal proteins (de Maagd et al. 2003).

   The endogenous crystalline inclusions produced during the sporulation phase, also called δ-endotoxins, are predominantly comprised of one or more Crystal (Cry) and Cytolytic (Cyt) proteins. These proteins have a specific insecticidal activity to certain groups of insects mostly within Lepidoptera, Diptera and Coleoptera, but also from other orders; Hymenoptera, Hemiptera, Orthoptera and Mallophaga. Beyond the insect worlds, Bt toxins are also found to be toxic to mites, nematodes, protozoa and human cancer cells (Schnepf et al. 1998; van Frankenhuyzen 2009; Ohba et al. 2009). The insecticidal activity of Bt strains also relies on the other toxins and virulence factors such as β-exotoxin, Vip (vegetative insecticidal protein), Sip (secreted insecticidal protein) and enzymes (phospholipase, protease, chitinase, AHL-degrading enzyme) (Regev et al. 1996; Jurat-Fuentes and Jackson 2012; Salehi Jouzani et al. 2017).

## 10.5   Mode of Action of Bt

Different models have been proposed for the mode of action of insecticidal Cry
toxins (Vachon et al. 2012). The commonly accepted mode of action of Bt toxin
(Fig. 10.1) relies on the ingestion of Bt crystal proteins by a susceptible host and
solubilization of the crystals. Bt crystals are solubilized in the alkaline environment
of the insect gut and converted to protoxins. The digestive proteases in the insect gut
cleave the protoxins to the activated toxin as a stable core. The activated toxin
interacts with the receptors on the midgut epithelial cells. This binding leads to toxin
insertion, pore formation, cell lysis and eventually insect death (Bravo et al. 2007).

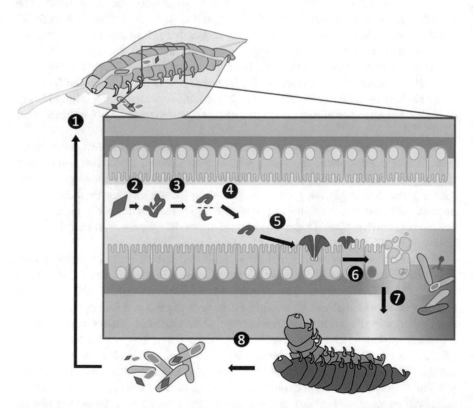

**Fig. 10.1**  Representation of the commonly accepted mode of action of Bt Cry toxins (adapted from
Pinos and Hernández-Martínez 2019). Spores and crystals of Bt are ingested by a susceptible larva
(1). Bt crystals are solubilized in the midgut of insect and convert to protoxin (2). The protoxins are
processed by the proteases to yield the stable core (3). The activated toxins traverse the peritrophic
membrane (4) and bind to the specific receptors on the brush border epithelial cells of insect midgut
(5). The insertion of toxin to the membrane leads to pore formation and cell lysis (6). Due to the
disruption of midgut epithelial cells, Bt spores enter the insect haemolymph, germinate and
replicate, eventually leading to septicaemia (7). The insect cadaver is the main reservoir of the
infection transmission and the dispersion of Bt spores (8)

## 10.6   Research Studies in Iran

Due to the importance of Bt in biological control of agricultural and medical pests, many countries have attempted to investigate novel Bt strains. The purpose of this research is to discover new toxins with a different mode of action, increased potency against pests, and a wider spectrum of activity. In Iran, researchers over the past decades have attempted to study different aspects of this entomopathogenic bacterium. An overview of studies for the application of local Bt strains in the biological control programs of Iran is summarized in Fig. 10.2. These studies are mainly concentrated on the isolation of Bt from diverse environmental sources (Keshavarzi 2008; Salehi Jouzani et al. 2008b, c; Seifinejad et al. 2008; Nazarian et al. 2009; Aramideh et al. 2010; Senfi et al. 2012; Shojaaddini et al. 2012; Moazamian et al. 2016; Khorramnejad et al. 2018), characterization of Bt strains and crystal morphology (Keshavarzi 2008; Senfi et al. 2012), protein profile (Salehi Jouzani et al. 2008b), gene content (Salehi Jouzani et al. 2008b, c), insecticidal activity (Hanafi-Bojd et al. 2006; Amiri-Besheli 2008; Seifinejad et al. 2008; Deilamy and Abbasipour 2013), cytocidal activity (Moazamian et al. 2018; Khorramnejad et al. 2018), nematicidal activity (Salehi Jouzani et al. 2008c; Ramezani Moghaddam et al. 2014; Baghaee and Moghaddam 2015), insect resistance mechanism to Bt toxins (Talaei-Hassanloui et al. 2014), Bt toxin mode of action (Khorramnejad et al. 2020), Bt crops (Ghareyazie et al. 1997; Tohidfar et al. 2005; Salehi Jouzani et al. 2005; Mousavi et al. 2007; Kiani et al. 2009a, b, 2012), optimization of Bt mass production

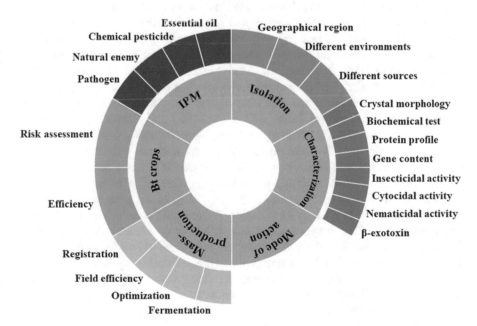

**Fig. 10.2** General outline of research studies performed on application of local Bt strains in biological control programs in Iran

(Keshavarzi et al. 2005; Shojaaddini et al. 2010; Marzban 2012a; Sarrafzadeh 2012; Salehi Jouzani et al. 2015b; Marzban et al. 2014, 2016; Maghsoudi and Jalali 2017), and utilization of Bt in IPM programs (Moazami 1997, 2004, 2005; Marzban et al. 2009; Magholli et al. 2013; Kalantari et al. 2014; Abedi et al. 2014; Nazarpour et al. 2014). These studies led to the development and application of Bt-based products in Iran. Based on the national Plant Protection Organization (PPO) report, the annual use of Bt-based bio-insecticides in Iran was about 200,000 ha (Personal communication). In the following sections, the significant research studies on Bt in biological control in Iran are listed and discussed.

## 10.7   Isolation, Natural Occurrence, and Geographical Diversity

Regarding the importance of Bt in biological control, several studies have focused on the isolation of Bt from diverse sources in different geographical regions worldwide (Bravo et al. 1998; Bel et al. 1997; Uribe et al. 2003; Djenane et al. 2017; Boonmee et al. 2019). In Iran also, researchers have attempted to collect native Bt strains from different habitats (Keshavarzi 2008; Salehi Jouzani et al. 2008b, c; Seifinejad et al. 2008; Nazarian et al. 2009; Aramideh et al. 2010; Senfi et al. 2012; Shojaaddini et al. 2012; Khojand et al. 2013; Moazamian et al. 2016; Khorramnejad et al. 2018). It has been clearly demonstrated that Bt is a ubiquitous bacterium. In this context, several Iranian authors attempted to correlate the natural occurrence of Iranian Bt strains isolated from diverse environments with ecological distribution.

The extensive screening program performed in different parts of Iran led to the establishment of many Bt collections at universities or research institutes. In 2008, Keshavarzi isolated 127 Bt strains from 514 samples collected from soil and dead larvae in diverse environments including high altitude mountains, non-cultivated lands, beaches, agricultural lands, urban locations, and forests, in Khorassan, Lorestan, Tehran, Ghazvin, East Azarbaijan, West Azarbaijan, Mazandaran and Hamedan Provinces. The relationship of Bt occurrence with vegetation was addressed. It was found that a high level of vegetation is not a requisite for high Bt occurrence. *B. thuringiensis* subsp. *kurstaki* was reported as the most frequent Bt strain in the studied Bt collection (Keshavarzi 2008). In the same year, a massive screening program funded by the Iranian Agricultural Research and Education Organization (AREO) was carried out as a collaboration between different Iranian institutes (Agricultural Biotechnology Research Institute of Iran (ABRII) and Plant Protection Institute of Iran) and universities (Zanjan University and University of Tehran). In this screening, 2292 soil samples from agricultural fields in 28 provinces located in the different geographic and climatic regions were collected (Salehi Jouzani et al. 2008b). This is one of the most important countrywide screenings carried out in Iran in terms of the number of provinces, samples and more importantly, the identification of different *cry* genes. Based on the isolation results, 128 Bt isolates have been isolated from the collected samples (Salehi Jouzani et al. 2008b). Most of the Bt isolates were from the humid region of Iran, particularly the Caspian zone, and fewer isolates were found in the dry and the semidry zones. The Bt distribution based on the plant flora showed that the largest number of Bt isolates was collected from cotton and sunflower fields. In 2010, samples were collected from the forest, grassland, and agricultural field soils, stored products, and insect cadavers in West Azerbaijan province. Out of 740 samples, 48 Bt strains were isolated (Aramideh et al. 2010). These isolates were characterized based on the morphology of crystal proteins, the biochemical types (aesculin utilization, lecithin-ase production and acid formation from salicin and sucrose), and the insecticidal activity against *Pieris brassica* (Lep.: Pieridae) and *Culex pipiens*. This study introduced the potent Iranian Bt strains for controlling *P. brassica* and *C. pipiens*.

In another screening, the natural occurrence of Bt strains in apple orchards was investigated collecting 180 soil samples from apple orchards in West Azerbaijan Province (Iran) (Senfi et al. 2012). Based on the sodium acetate selective method and the observation of parasporal crystals in bacterial cells, 40 Bt isolates were isolated and characterized. Investigations on the crystal morphology and insecticidal activity of Bt strains against *E. kuehniella* and *P. brassicae* showed that the majority of Bt strains produced bipyramidal crystals and a few had toxicity greater than *B. thuringiensis* subsp. *kurstaki* (Senfi et al. 2012). The isolation of Bt from infected larvae has resulted in the discovery of highly potent local Bt strains. A new *B. thuringiensis* subsp. *aizawai* strain EF495116 isolated from an infected *Plodia interpunctella* (Lep.: Pyralidae) larva was thoroughly characterized based on sero-logical identification, protein pattern, PCR-based analysis, sequencing of 16S rDNA and *gyrB* genes, and insecticidal activity against *P. interpunctella* and *Plutella xylostella* (Shojaaddini et al. 2012). Bt strain EF495116 was the cause of several epizootic outbreaks in the insect colonies, *P. interpunctella* and *E. kuehniella*, reared at Tabriz University, Iran. Different *cry* genes and virulence factors, namely *cry1Aa*, *cry1Ab*, *cry1Ac*, *cry1Ad*, *cry1B*, *cry1C*, *cry1D*, *cry1E*, *cry1F*, *cry1I*, *cry2A*, *cry9*, *cry39*, *cry40*, *plcR*, *chit*, *vip1*, *vip2*, *vip3*, and *inhA2* were traced in this Bt strain using general and specific primers. *P. interpunctella* and *P. xylostella* were found to be highly susceptible to Bt strain EF495116 in laboratory bioassays. This suggests that this strain has great potential to be used as an active ingredient in a bio-insecticide (Shojaaddini et al. 2012).

By focusing on a specific ecological niche, a total of 37 Bt strains were isolated from dead and infected *Helicoverpa armigera* Hüber (Lep.: Noctuidae) larvae in the cotton fields without a history of Bt application in Khorasan Razavi province, Iran (Keshavarzi 2008; Khojand et al. 2013). A high infestation rate of *H. armigera* and also considerable biodiversity of Bt strains in the cotton field were found (Khojand et al. 2013). A total of 100 soil samples from seven provinces, Guilan, Tehran, Isfahan, Yazd, Kerman, Fars, and Hormozgan, in Iran were collected and 60 Bt strains were isolated from these samples (Moazamian et al. 2016). The Bt isolates were characterized based on the crystal morphology of parasporal crystals, biochem-ical experiments, and SDS-PAGE analysis. The majority of Bt strains produced bipyramidal crystals and a relationship between crystal protein production and the geographical origin of Bt strains was found (Moazamian et al. 2016).

Altogether, the results of the screening program, whether on a vast or small scale, showed that Bt was predominantly isolated from diverse ecological habitats in different geographical regions of Iran. The high abundance of Bt isolates in the samples points to the richness of the selected habitats, which consequently led to the building of several local Bt collections.

## 10.8 Characterization

The characterization of novel Bt strains may lead to the identification of effective and highly potential Bt strains for application in the biological control of pests. The characterization of Bt strains is performed by different methods. The morphological characterization is initially carried out to somehow reflect the possible insecticidal activity of the Bt strains (Lecadet et al. 1999). Biochemical experiments (Martin et al. 1985) are used to identify the biochemical types of Bt strains. PCR-based characterization is utilized as a rapid means of Bt gene identification and prediction of insecticidal activity (Carozzi et al. 1991; Juárez-Pérez et al. 1997). Afterward, the expression of the detected genes in PCR is initially investigated by SDS-PAGE analysis, and subsequently, the protein composition can be detected by LC-MS/MS analysis. As the most important step of characterization, the biological activity of Bt strains against susceptible targets, from insect orders, mites, nematodes or even plant pathogens, has to be evaluated. Recently, the characterization of the insecticidal activity spectrum of Bt strains is performed following the in vitro toxicity assays using the insect cultured cell lines (Smagghe et al. 2009; Soberón et al. 2018). Indeed, insect cell lines have been considered as an easy, rapid and accurate technique for studying the spectra activity of Bt toxins and their mode of action.

The Iranian Bt collections have undergone different means of characterization in order to describe the local Bt strains and determine their biological activity. The most commonly used methods for characterization of Bt strains in Iran are mentioned here.

### 10.8.1   Crystal Morphology

B. thuringiensis isolates are characterized based on the presence of parasporal inclusions in the sporangium of the sporulating cell (de Maagd et al. 2001). Microscopic observations (Fig. 10.3) were performed in several Iranian laboratories to detect the Bt isolates based on the presence of parasporal inclusions and to study the Bt crystal morphology as the initial description of newly isolated Bt strains (Salehi Jouzani et al. 2008b; Keshavarzi 2008; Seifinejad et al. 2008; Nazarian et al. 2009). The majority of Bt strains isolated by Salehi Jouzani et al. (2008b) produced the

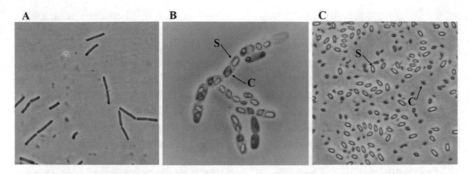

**Fig. 10.3** Different stages of an Iranian Bt strain growth cycle (original images, taken by phase contrast microscopy, by Khorramnejad). Panel (**a**) shows the vegetative growth stage, panel (**b**) indicates the stationary (sporulation of Bt cells) phase and panel (**c**) points out to the death phase, bacterial cell lysis and releasing of Bt spores and crystals. The crystal morphology of Bt strains are studied during the stationary phase in the sporulating cells (Panel **b**). Letters "S" and "C" point to spore and crystal, respectively

bipyramidal crystal and very few produced spherical parasporal inclusions. The shape of Bt crystals was correlated with the gene content, and based on the results of this correlation, the Bt strains harboring dipteran-active genes (*cry* and/or *cyt*) had cuboidal, rhomboidal and spherical parasporal inclusions (Salehi Jouzani et al. 2008b). The morphology of parasporal crystals was studied both based on shape and size (Keshavarzi 2008). Based on the crystal morphology studies, the bipyramidal crystals are the most frequent crystals in most of the Iranian Bt collections.

## 10.8.2   Biochemical Tests

Different biochemical experiments were carried out for the determination of *B. thuringiensis* sub-species. The esculin utilization, acid formation from salicin and sucrose, lecithinase production were performed to characterize the biochemical types of Bt strains (Keshavarzi 2008). Different biochemical characteristics of 37 Bt strains were examined for classification of Bt isolates such as hydrolysis of starch and gelatin, fermentation of glucose, sucrose, salicin, cellobiose, mannose, ADH and urease, and production of indole and $H_2S$. The reaction of Bt strains to biochemical discriminative tests differed and classified in the diverse groups (Khojand et al. 2013). Based on the biochemical experiments, *B. thuringiensis* subsp. *kurstaki* has been reported as the most abundant sub-species in Iran.

## 10.8.3   PCR-Based Screening

The identification of Bt insecticidal genes by PCR-based methods is a quick and useful method, with some limitations, for characterization and somehow prediction of the Bt isolates toxicity spectrum (Juárez-Pérez et al. 1997; Ferrandis et al. 1999). To date, different *cry*, *cyt*, *sip, vip* and *ps* genes were screened in the Iranian Bt collections. Correspondingly, the toxicity spectrum of Bt strains has been predicted based on the different δ-endotoxin genes. The presence of different lepidopteran-, coleopteran-, and dipteran-active genes, as well as Bt genes active against nematodes, was investigated in an Iranian Bt collection (Salehi Jouzani et al. 2008b, c; Seifinejad et al. 2008; Nazarian et al. 2009). Iranian Bt strains were characterized based on their gene content and the presence of lepidopteran-active genes, namely *cry1Aa, cry1Ab, cry1Ac, cry1Ad, cry1B, cry1C, cry1D, cry1E, cry1F, cry1G, cry1H, cry1I, cry1J, cry1K, cry2Aa, cry2Ab, cry2Ac, cry9A, cry9B, cry9C, cry9D* and *vip3Aa* (Seifinejad et al. 2008). Based on the PCR analysis, the Bt strains harboring lepidopteran-active genes were selected for the bioassays (Seifinejad et al. 2008). The presence of *cry1B, cry1I, cry3A, cry3B, cry3C, cry7A, cry8A, cry8B, cry8C, cry14, cry18, cry26, cry28, cry34* and *cry35* genes, coleopteran specific *cry* genes were determined (Nazarian et al. 2009). Afterward, the Bt strains containing coleopteran active *cry* genes were bioassayed against the elm leaf beetle, *Xanthogaleruca luteola* Muller (Col.: Chrysomellidae), first instar larvae (Nazarian et al. 2009). The number of 22 dipteran-active *cry* and *cyt* genes, including; *cry2Aa1, cry1Ab2, cry2Ac, cry4Aa, cry4B, cry10, cry11Aa, cry11Ba, cry11Bb, cry17Aa1, cry19Aa1, cry19Ba1, cry21Aa1, cry21Aa2, cry21Ba1, cry24Ba1, cry29, cry30, cry32Aa, cry32Ba, cry32Ca, cry32D, cry39, cry49Aa, cyt1Aa, cyt1Ab1, cyt1Ba, cyt2Aa, cyt2Ba, cyt2Bb* and *cyt2Ca*, were traced in the 128 Iranian local Bt strains (Salehi Jouzani et al. 2008b). The *cry2* gene family was the most frequent in this Bt collection. According to the PCR analysis and the gene content of the studied Bt strains, high potent Bt strains active against mosquitoes can be introduced from this collection. Besides the Bt genes targeting insects, the presence of *cry* genes active against parasitic or free-living nematodes, has been assessed (Salehi Jouzani et al. 2008c). The Bt strains were screened for the presence of *cry5, cry6, cry12, cry13, cry14* and *cry21* as the nematode-active genes. The presence of the nematode-active genes were found and these strains were bioassayed against nematodes. Although, there is a correlation between the gene profile and the geographical origin of Bt strains, the most frequent and common Bt genes found in different Bt screenings all over the country are, *cry2, cry1* and *vip3Aa* genes families.

## 10.8.4   Protein Profile

The PCR-based technique rapidly determines the presence or absence of genes, while it does not necessarily identify the expression of the identified gene. Therefore, protein electrophoresis of Bt crystal proteins is helpful for characterization and also for comparison of the protein profile. Based on the molecular masses of the observed band in SDS-PAGE analysis, the insecticidal activity of Bt strains is predicted. The protein profile of Bt isolates with dipteran-active *cry* and *cyt* genes was assessed by Salehi Jouzani et al. (2008b). Regardless of the origin of Bt strains, diverse protein profiles have been observed in this study. The molecular weight of the observed bands in SDS-PAGE was correlated with the molecular weight of dipteran active proteins. Based on this comparison, several protein profiles corresponded to the specific Cry or Cyt proteins toxic for dipterans (molecular masses of 17–28 kDa correspond to Cry17 and Cyt, 50 kDa to Cry2Ab, 65–75 kDa to Cry2Aa, Cry2Ac, Cry4D, Cry17A, Cry19 and Cry24, 80 kDa to Cry4C and Cry11B, and 128–140 kDa correspond to Cry4A, Cry4B, Cry32). The protein profile analysis revealed a great diversity of protein patterns within the Iranian Bt collections.

## 10.8.5 Biological Activity

All the mentioned means of characterization; parasporal crystal morphology, biochemical experiments, PCR-screening and SDS-PAGE analysis, are being used to predict the insecticidal activity of newly isolated Bt strains. Therefore, performing bioassays provides complementary information on the biological activity of the isolated Bt strains. Moreover, the final goal of these studies is to introduce a highly potent and effective Bt strain for biological pest control. Therefore, many studies were carried out to evaluate the insecticidal activity of Iranian Bt strains or Iranian commercial Bt-based products.

The bioassays with the Bt strains harboring genes encoding proteins active for lepidopterans were carried out against the cotton bollworm, *Helicoverpa armigera* Hüber (Lep.: Noctuidae). Several strains with high insecticidal activity have been introduced as good candidates for biocontrol of this pest (Seifinejad et al. 2008). The Bt strains containing coleopteran-active *cry* genes, were bioassayed against the elm leaf beetle, *X. luteola* first instar larvae (Nazarian et al. 2009), leading to the identification of novel *cry* genes based on the PCR analysis and also the introduction of potent Iranian Bt strains for biological control of coleopteran insect pests, particularly *X. luteola*. With the aim of finding the nematicidal Bt strain, 70 Bt isolates available in the Bt collection of Agricultural Biotechnology Research Institute of Iran (ABRII) were characterized. Based on the presence of the *cry* genes active against nematodes, the nematicidal activity of these Bt strains against *Meliodogyne incognita*, *Chiloplacus tenuis* and *Acrobeloides enoplus* were evaluated (Salehi Jouzani et al. 2008c). Based on the bioassays, two strains YD5 and KON4 were toxic to *M. incognita*, and two studied free-living nematodes, *C. tenuis* and *A. enoplus*, were susceptible to strains SN1 and KON4 (Salehi Jouzani et al. 2008c).

In several cases, the screening and characterization studies led to the introduction of potent Bt strains to industrial sections for mass production. The Iranian strains isolated from soil and infected larvae were thoroughly characterized (Khorramnejad et al. 2018). Bt strains with diverse larvicidal activity against Indian meal moth, *P. interpunctella*, were selected and characterized by employing different techniques. The insecticidal activity of selected Bt strains was investigated against different lepidopteran pest species; *Spodoptera exigua*, *Mamestra brassicae*,

*Grapholita molesta* and *Ostrinia nubilalis*. The cytotoxicity of the strains was evaluated against four lepidopteran cell lines from *Trichoplusia ni* (Hi5), *Helicoverpa zea* (HzGUT), *S. exigua* (UCR-SE) and *S. frugiperda* (Sf21). The results of the mentioned study (Khorramnejad et al. 2018) led to the introduction of three Iranian Bt strains (AzLp, IE-2 and IP-2), highly toxic for lepidopterans, as appropriate candidates for developing a Bt-based bio-insecticide. Indeed, based on the result of Khorramnejad et al. (2018) and complementary studies, AzLp Bt strain is being employed in the production of Rouin-2 as a Bt-based insecticide by Green Biotechnology Company of Iran (Karimi et al. 2019).

## 10.8.6   Determination of β-exotoxin Production

β-exotoxin, a non-proteinaceous and thermostable toxin, toxic to insects, nematodes, mites and mammals (Levinson et al. 1990; Liu et al. 2014), is secreted by some Bt strains, which has to be determined prior to the introduction of a Bt strain for mass-production. The β-exotoxin is secreted during the vegetative growth phase of some of the Bt strains. This non-proteinaceous toxin is an analog of ATP and inhibits DNA-dependent RNA polymerase activity (Farkas et al. 1976). Due to the toxicity of β-exotoxin to vertebrates, the production of Bt-based products with the ability of producing β-exotoxin is prohibited by law in many countries (McClintock et al. 1995; Meadows 1993). Therefore, Khorramnejad et al. (2018) determined the production of β-exotoxin in Iranian Bt strains using liquid chromatography and tandem mass spectrometry (LC-MS/MS) analysis. Based on their findings, none of the studied Bt strains produced β-exotoxin, thus these strains can be considered safe for the development of Bt-based insecticides.

## 10.9   Laboratory, Greenhouse and Field Experiments

Prior to the application of a microbial control agent in the greenhouse or field, the efficiency of the desired agent has to be evaluated under laboratory conditions. Therefore, numerous studies have been carried out to assess the biological activity and efficiency of local Bt strains or commercial formulations. Here, some examples of the extensive studies performed in the laboratories of different Iranian universities and institutes, greenhouse studies and field trials, are given.

It is truly worthy to mention the significant research studies performed by Dr. Nasrin Moazami the founder of the Biotechnology Research Centre at the Iranian Research Organization for Science and Technology in Tehran, and also Persian Gulf Biotechnology Research Center in Qeshm Island. Moazami (1997, 2004, 2005) has studied the potential and the feasibility of application of *B. thuringiensis* MH-14 to control malaria vectors.

The susceptibility of the citrus leafminer, *Phyllocnistis citrella* Stainton (Lep.: Gracillariidae) to the commercial *B. thuringiensis* formulation (Bithiran®, Mehr Asia Biotechnology Company, Iran) was determined under laboratory conditions (Amiri-Besheli 2008). The studied Bt formulation demonstrated great insecticidal activity against this pest and can be used in combination with bio-rational insecticides to increase the efficiency of leafminer management. The insecticidal activity of certain Bt strains available in the Iranian Bt collection with lepidopteran-active genes were assessed against the cotton bollworm, *H. armigera*. The toxicity of spore and crystal mixtures of Bt strains was evaluated against the first instar larvae of *H. armigera*. Two Bt strains (YD5 and KON4) were detected with great insecticidal activity against the cotton bollworm (Seifinejad et al. 2008). Under laboratory conditions, the insecticidal activity of spore and crystal suspensions of Iranian Bt strains was assessed against *P. brassica* and *C. pipiens*. Some of the studied Bt strains caused mortality higher than 75% on *P. brassica* and *C. pipiens* (Aramideh et al. 2010). So, these Bt strains, can be considered for biological control of insect pests. The insecticidal activity of four Iranian Bt strains isolated from soil were assessed against *P. xylostella*, one the major pests of cruciferous crops, especially cabbage fields in Tehran, and compared to that of Dipel as a commercial Bt based insecticide (Deilamy and Abbasipour 2013). The pathogenicity of the spore and crystal suspensions of Bt strains were investigated against third instar larvae of *P. xylostella*, with Bt strain 87 found to be highly toxic for diamondback moth (Deilamy and Abbasipour 2013).

The sub-lethal effects of *B. thuringiensis* subsp. *kurstaki* were evaluated on the fitness and the biological parameters of *H. armigera* (Sedaratian et al. 2013). Developmental time, fecundity, egg hatching percentage, net production rate ($R_0$), intrinsic ($r_m$) and finite ($\lambda$) rates of increase, mean generation time ($T$), mean generation doubling time ($DT$) were adversely affected in the *H. armigera* treated with sub-lethal concentrations of Bt. Sedaratian et al. (2013) concluded that biological performance and fitness of *H. armigera* were affected by exposing to the sub-lethal concentrations of Bt.

Beyond the insect world, the biological activity of Iranian Bt strains has also been assessed for nematodes (Salehi Jouzani et al. 2008c). The nematicidal activity of the mixture of spores and crystals of Bt strains harboring nematode-active *cry* genes was determined against three nematode species, *M. incognita* (plant parasite), *C. tenuis* and *A. enoplus* (free-living nematodes). The bioassays showed that two strains, YD5 and KON4, previously shown great insecticidal activity against lepidopteran pests (Seifinejad et al. 2008), also have a great potential to control *M. incognita*. Therefore, these two Iranian Bt strains can be considered as appropriate candidates for the development of bio-insecticides based on Bt. In another study, the nematicidal activity of Iranian Bt strains isolated from soil in the tomato fields was evaluated against *M. javanica*, root-knot nematode under in vivo and greenhouse conditions (Ramezani Moghaddam et al. 2014). The Bt strains prohibited nematode egg hatching and were toxic for *M. javanica*. The toxicity of two Bt strains containing Cry14 protein was also evaluated against *M. javanica* under laboratory and greenhouse conditions (Baghaee and Moghaddam 2015). These Bt strains could reduce the nematode population, kill the juveniles, prohibit the nematode egg hatching and also decrease the number of galls (Baghaee and Moghaddam 2015).

These results highlight the importance and the success of Bt in the biological control of pests under the laboratory conditions. Although the successful application of Bt is dependent on various factors such as applied dose, weather conditions and time of application, the efficacy of the desired Bt strain, Bt formulation or Bt crop has to be clearly determined under greenhouse and field conditions.

*B. thuringiensis* subsp. *israelensis* (serotype H14) is commonly used in the southern part of Iran to control the vectors of malaria, *A. dthali*, *A. fluviatilis* and *A. stephensi* (Hanafi-Bojd et al. 2006; Shahi et al. 2013). *B. thuringiensis* subsp. *israelensis* H14 effectively controls the dipterans and this success is due to the expression of insecticidal proteins, Cry4A, Cry4B, Cry11A and Cyt1A, in this serotype (Höfte and Whiteley 1989; Dulmage et al. 1990). The larvicidal activity of *B. thuringiensis* H14 has been studied and indeed demonstrated against *Anopheles dthali* and *A. fluviatilis*. Besides the application of chemical insecticides, these malaria vectors are being controlled by the utilization of Bt in the mosquitoes breeding sites in the south part of Iran (Hanafi-Bojd et al. 2006).

The effect of Bt and Azadirachtin individually and in combination against tomato leaf miner, *Tuta absoluta* Meyrick (Lep.: Gelechiidae) was evaluated under the field conditions. Due to the significant reduction in the population of *T. absoluta* in the tomato field after treating with Bt and Azadirachtin, the application of these two insecticides whether alone or in combination, has been strongly recommended by Nazarpour et al. (2014).

The possibility of simultaneous application of *Trichogramma* spp. and a commercial Bt formulation (Biolep®) based on the *B. thuringiensis* subsp. *kurstaki* for the control of tomato leafminer, *T. absoluta*, has been examined under greenhouse conditions (Alsaedi et al. 2017). Moreover, the effect of this commercialized Bt-based product was assessed against the non-target organism, *Trichogramma* spp., the egg parasitoids of *T. absoluta*. Based on the greenhouse observations, the number of mines detected in the treatment of the combination of Bt and *Trichogramma*, were significantly lower than the individual application of these two biological control agents. More importantly, following the application of Bt, no negative effect was observed on the biological parameters of *Trichogramma* spp. (Alsaedi et al. 2017).

## 10.10    Fermentation Optimization

Due to the importance of Bt in microbial pest control, the appropriate and economic production of Bt-based bio-insecticides is critical. Therefore, many studies have investigated optimization of the fermentation conditions. *B. thuringiensis* strains have special requirements during the fermentation process. Sporulation and germination of Bt are dependent on the available nutrition. Therefore, investigation of the nutritional requirements of Bt is of importance in Bt production. A summary of

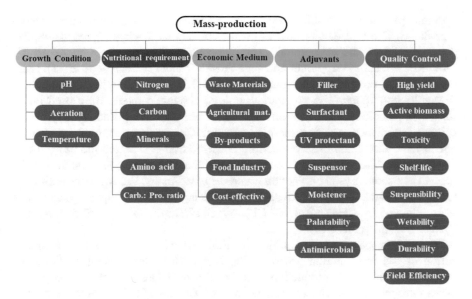

**Fig. 10.4** An overview of the studies on mass-production of Iranian Bt strains

research studies on the appropriate and economic production of Bt-based bio-insecticides and the parameters assessed to determine Bt product quality in Iran is presented in Fig. 10.4.

The necessity of nutrition, minerals and oxygen rate has been questioned in different studies. Several studies have focused on different aspects of Bt mass production. Keshavarzi et al. (2005) investigated the cultural conditions and the nutritional requirements of *B. thuringiensis* subsp. *kurstaki* with the aim of optimizing growth conditions (pH and aeration), active biomass yield and designing the economic culture media, by using different agricultural materials and by-products. The effect of total carbohydrate, total protein, carbohydrate/protein ratio and amino acid content were also investigated. The results indicated that the highest biomass was obtained with carbohydrate/protein ratio of 0.4–0.5, 13.9% glutamic acid and pH of 7–8 (Keshavarzi et al. 2005). The oxygen transfer rate is an important parameter in Bt fermentation and the effect of the oxygen supply on the sporulation phase, δ-endotoxin synthesis and toxicity was also investigated by Sarrafzadeh and Navarro 2006.

Salehi Jouzani et al. (2011) tried to optimize the growth conditions for two native Bt strains KH4 and YD5 using economic media. At the first step, different kinds of available agricultural and food industry wastes such as carbon, nitrogen and mineral sources for Bt strains mass production, were selected. The selection was based on the cost, chemical composition, and physiological characteristics. Finally, molasses (lactose) and starch liquor (polysaccharide) were used as carbon sources. Corn steep liquor and sea minerals were used as nitrogen and mineral sources, respectively. Based on the different designed treatments as economic media, the growth kinetic and the spore-crystal production were evaluated. The results showed that the

optimum pH for YD5 & KH4 strains were 6.5 and 7 and the optimum temperature
was 30 °C. The best medium for YD5 strain contained 3% hydrolyzed starch, 3%
corn steep liquor and sea minerals. The best medium for strain KH4 contained 2%
molasses, 3% corn steep liquor and sea minerals (Salehi Jouzani et al. 2011). Later
on, the mass production of strains YD5 and KH4 was optimized under batch
fermenter conditions. The previous optimized economic media were used under
fermenter conditions, and the effects of different factors, including quantity of
primary inoculation, aeration and pH were evaluated on the growth rate, and spore
and crystal production. The pre-culture concentration of 2% and the oxygen rate
ranging from 50 to 90%, were found to be optimal. Moreover, aeration during the
growth and spore-crystal production was necessary. Finally, following the optimized
growth condition, the maximum growth and the highest yield of spores and crystals
were achieved for KH4 ($6.2 \times 10^9$ CFU/ml) and YD5 ($5.5 \times 10^9$ CFU/ml) isolates
(Salehi Jouzani et al. 2011).

The success of a commercial bio-insecticide effectively depends on the optimization of the fermentation process (Fig. 10.5). Therefore, several studies have
focused to improve the production of Bt based insecticides. The nutritional parameters, nitrogen carbon and mineral sources of Bt strain H14 medium have been
modified to study their effects on Bt growth phase, sporulation and germination
(Sarrafzadeh 2012). The effect of different nutritional components such as glucose,
glycerol, sodium acetate, corn steep liquor, yeast extract, hydrolyzed casein,
$(NH_4)_2SO_4$, $Ca^{2+}$, $Mg^{2+}$ and $Mn^{2+}$ were assessed. The effect of corn steep liquor
on bacterial growth, spore production and spore germination were more important
than other elements (Sarrafzadeh 2012). Marzban (2012b) optimized the solid
fermentation of three local Iranian Bt strains, KON3, KN3, KD2 using four solid
culture media: rice bran, wheat bran, rice flour and barley supplemented with mineral
salts, $MnSO_4$, $MgSO_4$, $ZnSO_4$ and $FeSO_4$. The number of Bt spores and the
insecticidal activity of produced bacteria against *H. armigera* were investigated
and compared with that of HD-1 as a reference. Wheat bran was the best media
for Bt production (Marzban 2012b).

The cost of raw materials used in the mass production of a bio-insecticide is one
of the main expenditures in the production process. Therefore, there is a growing

**Fig. 10.5** The fermenters of the Nature Biotechnology Company, Biorun, for microbial control
agents' production

interest in the economic production of bio-insecticides with low-cost materials, industrial by-products and even waste materials (Shojaaddini et al. 2010). Shojaaddini et al. (2010) focused on the economic production of Bt based insecticide by developing fermentation media based on food barley as a substitution of commercial starch. They investigated different carbon and nitrogen sources; hull-less barley grain, commercial soybean cake, and groundnut, for fermentation of newly isolated Iranian Bt strain and *B. thuringiensis* subsp. *kurstaki* HD-1 as a reference. To determine the best substitution, different parameters such as biomass, δ-endotoxin concentration and the number of Bt spores (CFU/ml) were estimated. The use of barley in Bt fermentation media led to faster sporulation, similar final biomass, shorter fermentation time and more importantly economic production.

Previously it was reported that KH4 Bt strain from the Bt collection of Agricultural Biotechnology Research Institute of Iran (ABRII) is highly toxic for coleopteran species namely *X. luteola* and *Leptinotarsa decemlineata* (Nazarian et al. 2009; Salehi Jouzani et al. 2015a). Due to the great potential of this Bt strain Salehi Jouzani et al. (2015a) attempted to optimize the growth condition and develop a cost-effective medium for its mass production. Different temperatures and pH conditions were examined to determine the appropriate growth conditions. A pH of 6.5 and temperature of 30 °C were found to be the optimum. Available agricultural waste sources were used as carbon and nitrogen sources, and sea salt was utilized as a cheap source of minerals. Molasses and corn steep liquor were the appropriate sources of carbon and nitrogen. These findings (Salehi Jouzani et al. 2015a) could be used to successfully develop an appropriate fermentation condition and design a cost-effective medium for mass production of a local high potent Bt strain. An economic fermentation medium and an appropriate condition for mass production of YD5 Bt strain toxic for lepidopterans was designed and examined by Salehi Jouzani et al. (2015b). The optimum reported pH and temperature for mass production of YD5 Bt strain were estimated at 7 and 30 °C. The utilization of different cost-effective materials from wastes of agriculture and industry led to discovery of an economical medium for mass production of YD5 Bt strain. The highest number of spores and crystals were obtained in the medium containing 3% hydrolyzed starch, 3% corn steep liquor and 0.003% sea minerals (Salehi Jouzani et al. 2015b).

It is important to recover Bt spores and crystals from the fermentation medium as the last step of Bt mass production. Several methods such as precipitation, absorption, evaporation and centrifugation (Brar et al. 2006a, b; Adjalle et al. 2007; Naseri Rad et al. 2016), have been suggested and indeed used for recovery of the Bt spores and crystals. Amongst them, Marzban et al. (2014, 2016) investigated the membrane-based filtration method. Based on their observation, microfiltration and ultrafiltration were found to be effective for the separation of Bt spores and crystals from the medium.

## 10.11   Enhancing the Efficiency and the Durability of Bt Products

Ultra Violet (UV) radiation affects the survival and activity of Bt spores and δ-endotoxins (Frye et al. 1973; Liu et al. 1993; McGuire et al. 2000) which leads to instability of Bt-based products in the environment and inconsistency of the pest control. Therefore, protectants need to be added to the formulations. The efficiency of different materials in the protection of Bt spores and crystals from UV radiation have been evaluated (Zareie et al. 2003; Moxtarnejad et al. 2014; Maghsoudi and Jalali 2017; Jalali et al. 2020). In 2003, the efficiency of 16 formulations of Bt after mixing with molasses and henna as the protectants from sunlight, particularly ultraviolet radiation, were determined (Zareie et al. 2003). In the group I, Bt formulations were not exposed to UV light, while group II and III were exposed to UV for 1 and 2 days. The insecticidal activity of all these Bt formulations were assessed against5th instar larvae of *Galleria mellonella* (Lep.: Pyralidae). The highest mortality was observed in the formulations of group I. In group II, the highest insecticidal activity was recorded in the molasses and henna encapsulated formulations, as in group III. The efficiency of the Bt formulations decreased with increasing exposure to UV. Interestingly, the addition of sugar increased the stability of the tested Bt formulation (Zareie et al. 2003). The role of UV-protection of different compounds such as henna, Congo red, rhodamine B, active carbon, molasses, conventional coal, sodium alginate, gelatin, starch, methyl green and gelatin were evaluated by Moxtarnejad et al. (2014) under laboratory and field conditions. The insecticidal activity of *B. thuringiensis* subsp. *kurstaki* strain ABTS-315 was assessed alone and in combination with protectants against *P. brassicae*. Henna and starch powder protected the Bt strain from UV radiation and can be suggested for commercial use in mass production. In another study (Maghsoudi and Jalali 2017), the protective effect of graphene oxide (GO) and olive oil from UV radiation was examined. The spore viability and the toxicity were considered as the indicators for evaluating the efficiency of the selected protectants. The mixture of GO and olive oil was found to effectively protect Bt spores and retain the insecticidal activity of Bt against *E. kuehniella* (Maghsoudi and Jalali 2017). In the next research, Jalali et al. (2020) studied the effect different polymorphs of titanium dioxide ($TiO_2$) nanoparticles on the viability of Bt after exposure to UV radiation. Afterward the biological activity of UV-exposed Bt was evaluated against *E. kuehniella* second instar larvae. Based on their results, some polymorphs of titanium dioxide could enhance the stability of Bt against UV radiation (Jalali et al. 2020). Salehi Jouzani et al. (2011) prepared an optimized wettable powder formulation for two Bt isolates (YD5 and KH4) in order to enhance their efficiency against lepidopteran and coleopteran pests, and increase the durability of the bio-products. Mass production of these strains was done in the batch fermenter based on the optimized conditions. To improve the formulation for two Bt strains, different adjuvants including fillers, surfactants, materials enhancing durability, UV protectants, suspensors, moisteners, palatability materials and antimicrobials were mixed with 25 grams of spores and crystals mixture of Bt strains (containing $6 \times 10^9$ CFU/mg). After drying and milling by jet mil, the quality indexes including suspensability, wettability, durability and the $LC_{50}$ values against pests were measured for different treatments. The best treatment for YD5 and KH4 strains had 73 and 71% suspender, and 25 and 24% wettable agents, respectively. The $LC_{50}$ values of the selected formulations for YD5 and KH4 strains against cotton bollworm larvae were 550 and 510 ng/cm$^2$, indicating the efficiency of the selected formulations (Salehi Jouzani et al. 2011).

## 10.12  Bt Based Bio-insecticides

The most widely produced microbial control agents in Iran are based on different strains of *B. thuringiensis* (Moosavi and Zare 2016). Several Bt-based products have been registered and developed in Iran with the aim of controlling agricultural pests and also mosquitoes as the vectors of human diseases. Three Iranian companies; Nature Biotechnology Company (Biorun), Mehr Asia Biotechnology Company, and Green Biotechnology Company, produce Bt-based bio-insecticides in Iran (Karimi et al. 2019). The registered products are mainly based on the spores and crystals of *B. thuringiensis* subsp. *kurstaki*, lepidopteran active bio-insecticide. Other Bt-based insecticides composed of *B. thuringiensis* subsp. *israelensis* active against dipterans, *B. thuringiensis* subsp. *tenebrionis* and *B. thuringiensis* subsp. *morrisoni* toxic for coleopterans have also been developed in Iran (Karimi et al. 2019). The commercial bio-pesticides based on Bt are reviewed here.

Bactospein is the first local bio-insecticide, registered in Iran based on Bt and formulated as 90% wettable powder. This bio-insecticide preliminarily used against the forest insect pests (Marzban 2012a). Later, the insecticidal activity of Bactospein was evaluated against different lepidopterans including, *Tortrix viridana* (Lep.: Tortricidae), *H. armigera*, *Ostrinia nubilalis* (Lep.: Crambidae), *Lymantria dispar* (Lep.: Erebidae) and *P. interpunctella* (Marzban 2012a).

Bithiran® is based on the *B. thuringiensis* subsp. *tenebrionis*, produced and registered by Mehr Asia Biotechnology Company. The effect of this commercial Bt-based product has been evaluated against the elm leaf beetle, *X. luteola*, as one of the most important pests of elm trees in Iran. The bioassay results showed that Bithiran could effectively control third instar larvae and adults of *X. luteola* under the laboratory conditions (Hajialiloo et al. 2016). Bithurin® is another bio-insecticide based on the *B. thuringiensis* subsp. *kurstaki*, produced by Mehr Asia Biotechnology Company. This bio-insecticide is active against lepidopteran larvae (Karimi et al. 2019).

The Nature Biotechnology Company with the Biorun brand has been registered and indeed produces two important Bt-based bio-insecticides; Biolep® suspension concentrate and Biolep® wettable powder formulations. Both of these products are a mixture of spores and crystals of *B. thuringiensis* subsp. *kurstaki*, but formulated in two different formulations. Biolep is active against different lepidopteran pests including agricultural and forest pests. BioBeet® is also produced by the same company, based on the *B. thuringiensis* subsp. *tenebrionis*, active against coleopteran larvae. This product has not yet been registered (Karimi et al. 2019). Different studies have evaluated the efficiency of Biolep under laboratory, greenhouse and field conditions (Alsaedi et al. 2017). Bioflash® is a local bio-insecticide based on the *B. thuringiensis* subsp. *israelensis* (MH14) also produced by Nature Biotechnology Company (Biorun) in Iran. It is composed of a mixture of spores and crystals of *B. thuringiensis* subsp. *israelensis* (MH14) and manufactured as wettable powder and granular formulations. The *B. thuringiensis* subsp. *israelensis* (MH14) was isolated from an *A. stepehensis* larva by a research group led by Nasrin Moazami. The larvicidal activity of *B. thuringiensis* subsp. *israelensis* (MH14) has been evaluated in the Qeshm Island (Moazami 1997). The efficiency of Bioflash® was assessed against the immature stages of *Anopheles* spp. and *Culex* spp. under laboratory, semi-field and field conditions (Gezelbash et al. 2014). The laboratory assays resulted in 100% mortality of *A. stephensi* larvae at the concentration of 512 ppm after 24 h. While, the results of, semi-field and field tests indicated a reduction in efficiency of Bioflash® granule and wettable powder formulations. Therefore, further investigations are needed to improve the efficiency of this local bio-insecticide (Gezelbash et al. 2014).

Rouin-2 is another Bt-based insecticide produced as a wettable powder by Green Biotechnology Company. This formulation is based on spores and crystals of *B. thuringiensis* subsp. *kurstaki* (AzLp Bt strain) isolated from infected lepidopteran larvae by Khorramnejad et al. (2018).

The commercially available Bt insecticides are claimed by the respective companies to control many lepidopteran insect species including; *H. armigera*, *H. viriplaca*, *O. nubilalis*, *Spodoptera* spp., *Pieris* spp., *Plutella xylostella*, *T. absoluta*, *Chilo suppressalis*, *Yponomeuta malinellus*, *Lobesia botrana*, *Porthesia melania*, *Leucoma wiltshirei*, *Lymantria dispar*, *Batrachedra amydraula*, *Hyphanteria cunea* and *Cydalima perspectalis*.

## 10.13    Bt Crops

Management of agricultural pests solely relying on the sprayable products based on Bt is insufficient due to several constraints. Bt formulations are generally sensitive to UV radiation and have limited activity against boring and sucking insects. These are the main restrictions of Bt sprays. The incorporation of Bt genes into agricultural crops has resulted in the production of insect-resistant crops providing an effective strategy to control the agricultural pests. In Iran, as shown in Fig. 10.6, different Bt synthetic genes have been transferred into the different plants, mainly rice (Ghareyazie et al. 1997; Mousavi et al. 2007; Kiani et al. 2009a, b, 2012), cotton (Tohidfar et al. 2005), potato (Salehi Jouzani et al. 2008a), sugar beet (Jafari et al. 2007, 2009a; Norouzi et al. 2011), alfalfa (Zare et al. 2008), maize and date (Tohidfar et al. 2008) by the aim of developing the transgenic plants with enhanced

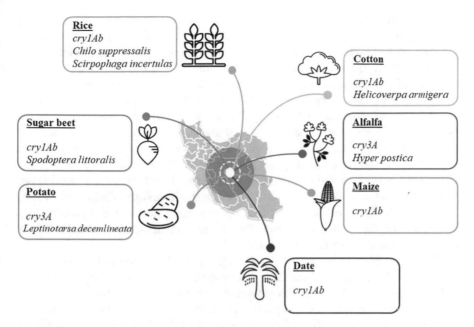

**Rice**
*cry1Ab*
*Chilo suppressalis*
*Scirpophaga incertulas*

**Cotton**
*cry1Ab*
*Helicoverpa armigera*

**Sugar beet**
*cry1Ab*
*Spodoptera littoralis*

**Alfalfa**
*cry3A*
*Hyper postica*

**Potato**
*cry3A*
*Leptinotarsa decemlineata*

**Maize**
*cry1Ab*

**Date**
*cry1Ab*

**Fig. 10.6** Bt crops with insect resistance traits, under study in Iran. The transformed plants, the corresponding transgene and the target pest are indicated

resistance to insects. In 2004, the first filed experiments for evaluating the efficiency of Bt rice was carried out in Iran (Tohidfar and Salehi Jouzani 2008).

Rice as one of the most important food crops is widely produced and indeed consumed in Iran. Iran is also one of the pioneers in the development of transgenic lines of rice (James 2007). Many studies have focused on the development of transgenic rice lines with enhanced resistance to lepidopterans (Ghareyazie et al. 1997; Alinia et al. 2000). Bt-transformed Tarom Molaii rice, a high quality aromatic rice, expressing *cry1Ab* Bt gene, showed resistance to the major rice pests in Iran, striped stem borer, *C. suppressalis* Walker (Lep.: Crambidae) and yellow stem borer, *Scirpophaga incertulas* Walker (Lep.: Pyralidea) (Ghareyazic et al. 1997). Different field and greenhouse studies indicated the efficiency of this transgenic rice line. Bt rice was the first genetically modified (GM) crop grown in Iran and the first GM rice released worldwide was generated by one of the most important Iranian agricultural biotechnology institutes, the Agricultural Biotechnology Research Institute of Iran (ABRII) with collaboration with the International Rice Research Institute (Mousavi et al. 2007). The Bt rice line expressing *cry1Ab* gene has undergone different risk assessments and safety analyses. Moreover, different physiological characteristics have been studied in Bt rice and compared with non-transgenic rice. The incorporation of Bt *cry1Ab* gene in rice has not adversely affected the plant features. Many studies have focused on the efficiency of Bt crops under field conditions.

Field trials were conducted to evaluate the efficiency and safety of genetically modified rice lines in Iran (Mousavi et al. 2007). Resistance of Bt rice lines was assessed against *C. suppressalis* (Kiani et al. 2009a). Symptoms of stem borer damage, like dead hearts and white heads, were recorded. The Iranian Bt rice resistant to lepidopterans, "Tarom Molaii expressing *cry1Ab*" was developed by the Agricultural Biotech Research Institute (Iran) and submitted in GM approval database of International Service for the Acquisition of Agri-biotech Applications (ISAAA) (www.isaaa.org). Three Iranian lines of Bt rice, Khazar, Neda and Nemat, expressing *cry1Ab* gene were developed and showed significant resistance to rice pests, particularly *C. suppressalis*. Different physiological properties of Bt rice such as; plant height, number of tillers, panicle length, the number of panicles per plant, maturity time, weight of 1000-grain and tone of yield per hectare, were recorded under field conditions and compared with non-transgenic rice. Prior to field experiments, the quality of rice grain was assessed in the laboratory (Kiani et al. 2009b). Transgenic rice lines, especially Neda and Nemat lines, showed better physiological properties compare to non-transgenic ones. Moreover, all three Bt rice lines showed significant resistance to *C. suppressalis*.

The stable inheritance and the level of transgene expression in transgenic crops are of crucial importance. Therefore, the inheritance of *cry1Ab* in Bt rice lines was studied under field conditions, moreover, the inheritance of *cry1Ab* gene in crosses between transgenic and non-transgenic rice varieties has been evaluated (Kiani et al. 2009b). The stable integration of *cry1Ab* gene was shown in the three transgenic rice lines (Tarom Molaii, Neda and Nemat). These transgenic lines showed resistance to stem borer infestations and also could be employed as a donor source of the transgene in recombination breeding (Kiani et al. 2009b). Three Bt rice lines, Neda, Nemat and Khazar, expressing *cry1Ab* gene were generated by Rice Research Institute (Amol, Iran). After evaluating the quality of transgenic rice grains in the laboratory, under the field conditions at Sari University of Agricultural Sciences and Natural Resources (Sari, Iran), the physiological properties of transgenic rice expressing synthetic *cry1Ab* gene and field resistance to insect were recorded and compared to that of non-transgenic plants (Kiani et al. 2009b). Neda and Nemat transgenic lines showed better physiological properties based on height and maturity time, and all three transgenic lines showed a high degree of resistance to *C. suppressalis*. The development of a second-generation transgenic plant pyramided with different genes relating to resistance to insects and rice quality was carried out as a part of the cultivar development program in Iran (Kiani et al. 2012). Two Iranian Br rice lines (Neda and Nemat) expressing *cry1Ab* gene were crossed with an aromatic variety to develop insect-resistant (*cry1Ab* gene) and aromatic rice (*fgr* gene) in $F_2$ plants. Gene pyramiding was successfully carried out and resistant and aromatic rice with favorable agricultural characteristics were detected (Kiani et al. 2012).

Cotton is one of the most important crops in the northeast of Iran deals with a high level of yield loss due to insect damage. The major cotton pest that threatens cotton production in Iran is the cotton bollworm, *H. armigera*. Developing transgenic cotton expressing Bt gene is one of the strategies employed to enhance the plant resistance to insect pests. The cotton var. Coker has been transformed with the synthetic *cry1Ab* gene to enhance resistance towards the cotton bollworm (Tohidfar et al. 2005, 2008). The presence of the transgene and the Cry1Ab protein were determined in the transgenic cotton by PCR and western immunoblot analysis. Cotton transgenic line, namely line 61, was developed and showed a high level of resistance to *H. armigera* (Tohidfar et al. 2008). Afterward, the effect of Iranian Bt cotton was assessed on the cotton secondary pest, *Bemiscia tabaci*, and also *Encarsia formosa*, its parasitoid (Azimi et al. 2012). The Bt cotton negatively affected *E. formosa* in terms of parasitoid developmental time and weight.

The collaboration between the Vavilov Institute of General Genetics, Russian Academy of Sciences, and Agricultural Biotechnology Research Institute of Iran, resulted in the development of transgenic potato plants expressing the modified *cry3A* gene (Salehi Jouzani et al. 2008a). The hybrid *cry3a*M-*lic*BM2 gene was synthesized by Salehi Jouzani et al. (2005). The transgenic potatoes showed a high level of resistance to Colorado potato beetles, *L. decemlineata* (Col.: Chrysomelidae).

The alfalfa weevil, *Hyper postica* (Col.: Curculionidae) is one of the most destructive pests of alfalfa in Iran. The transgenic alfalfa expressing *cry3A* gene was developed and the efficiency of these plants was evaluated in the greenhouse (Zare et al. 2008).

The transgenic line of sugar beet plants resistant to lepidopteran pests was developed and generated by the Agricultural Biotechnology Research Institute of Iran (ABRII) and Sugar Beet Institute of Iran (SBSI). The sugar beet, *Beta vulgaris* L., transformed with the synthetic *cry1Ab* gene from Bt showed enhanced resistance to *S. littoralis* (Jafari et al. 2007). Probably due to the low expression level of Cry1Ab, the full protection of sugar beet against *S. littoralis* was not achieved and only the enhancement of resistance was observed, ranging from 37 to 70% mortality within the first week of infestation (Jafari et al. 2009b). The inheritance of synthetic *cry1Ab* gene to the next generation of sugar beet was successfully carried out (Jafari et al. 2009a). After demonstrating the successful inheritance of *cry1Ab* gene to $F_1$ progeny, the morphological properties and the resistance of transgenic sugar beet to *S. littoralis* were evaluated (Norouzi et al. 2011). As a result, the morphological traits of transgenic sugar beet did not differ compared to controls, while, the transgenic plants showed enhanced resistance to *S. littoralis*.

Although GM crops expressing *cry* genes is considered as the successful substitute for chemicals, evaluation of potential hazards or in general, risk assessment is required before deployment of this technology. A significant study has been devoted to the investigation of the environmental and human health effects of Bt rice production vs. non-transgenic rice in Iran (Dastan et al. 2019). This study was performed by the Agricultural Biotechnology Research Institute of Iran in the paddy fields of three northern cities, Amol, Sari and Rasht (Mazandaran and Guilan provinces), under the control of Rice Research Institute of Iran (RRII). Four transgenic rice lines were employed. Different parameters such as climate change, ozone layer depletion, terrestrial acidification, freshwater eutrophication, marine eutrophication, human toxicity, photochemical oxidant formation, particulate matter formation, terrestrial eco-toxicity, freshwater eco-toxicity, marine eco-toxicity, ionizing radiation, agricultural land occupation, urban land occupation, natural land transformation, water depletion, soil depletion, metal depletion, carcinogens and non-carcinogens, water footprint, ecological footprint, greenhouse gases emission, cumulative energy demand, and global warming, were measured as life cycle assessment. Briefly, some parameters, particularly the level of energy consumption and greenhouse gas emission, the non-transgenic rice affected the ecosystem adversely two times more than the transgenic rice. Since, in the Bt rice fields no chemical insecticides were used, the ecological impact of chemical application had to be taken into consideration. Moreover, the emission of heavy metals in air, water and soil in transgenic cultivars was less than in non-transgenic lines. Therefore, due to the lower consumption of agricultural machinery and management practices in the Bt rice fields, fewer pollutants were emitted into the environment (Dastan et al. 2019).

## 10.14  Implementation of Bt in IPM Programs

To increase the activity spectrum of a microbial insecticide, effective management of insect pests, application of lower numbers of microbial agents, and more importantly, overcoming or at least delaying the emergence of resistance, integration of microbial insecticides with other means of insect pests management have been suggested. Therefore, many studies focused on the compatibility of Bt formulation with other control practices. Indeed, the interaction and the possible combination of Bt toxins with other insect pathogens (Marzban et al. 2009; Marzban 2012b; Magholli et al. 2013; Kalantari et al. 2014; Allahyari et al. 2019), parasitoids (Sedaratian et al. 2014; Alsaedi et al. 2017; Allahyari et al. 2019), other bio-insecticides such as Azadirachtin and Spinosad (Abedi et al. 2014; Nazarpour et al. 2014; Hosseinizadeh and Aramideh 2014; Nouri-Ghnbalani et al. 2016), and also plant extracts or essential oils (Ghassemi-Kahrizeh et al. 2014), have been studied. Based on the findings, the combination of Bt with biological control agents, bio-insecticides and plant materials may increase the efficiency of the insect pest management.

Due to the emergence of resistance in *P. xylostella* to several chemical insecticides (Sarfraz et al. 2005) and also in some cases to Bt (Tabashnik et al. 1990), the application of two different microbial control agents might improve the efficiency of *P. xylostella* management. Therefore, the feasibility of using *B. thuringiensis* subsp. *kurstaki* in combination with *H. armigera* single polyhedrosis virus, HaSNPV, under laboratory conditions was investigated (Magholli et al. 2013). A synergistic effect was observed when the lowest concentrations of Bt ($2.4 \times 10^1$ spores/ml) and HaSNPV ($2.3 \times 10^3$ OB/ml) were used together. Moreover, the biological parameters of *P. xylostella* were negatively affected due to the simultaneous application of Bt and a baculovirus (Magholli et al. 2013). Further experiments are needed to show the applicability of this combination in the field.

The simultaneous application of Bt strain and nucleopolyhedrovirus (NPV) in integrated pest management programs of cotton bollworm, *H. armigera*, was questioned. Both microbial control agents, the bacterium and the virus, have to be ingested to exert the insecticidal activity. The interaction between Bt and HaSNPV in biological control of *H. armigera* was assessed with the aim of solving the insect resistance problems, using a low concentration of bio-insecticides and more importantly, increasing the efficiency of cotton bollworm management (Kalantari et al. 2014). Three Iranian Bt strains, KD-2, 20 and 6R, and *H. armigera* single polyhedrosis virus (HaSNPV) provided from the national collection of Iranian Research Institute of Plant Protection, were employed. The virulence of Bt strains and HaSNPV were investigated individually and in combination at different concentrations of bacterium and virus. KD-2 Bt strain and HaSNPV were highly toxic to *H. armigera*. The joint use of Bt and HaSNPV led to significant differences in *H. armigera* larval mortality. Depending on the concentration, all three possible interactions, antagonism, additive, and synergism, were observed when two microbial control agents were used jointly. The application of Bt at a concentration lower than $LC_{50}$ value in combination with HaSNPV at the concentration equal to $LC_{50}$, resulted in a synergistic effect. Kalantari et al. (2014) showed that the simultaneous application of Bt and HaSNPV adversely affected the growth and development of *H. armigera*, therefore, the appropriate combination of these two microbial control agents may increase the efficiency in *H. armigera* control.

The interaction between Bt and Azadirachtin, a tetranortriterpenoid compound derived from the neem tree, *Azadirachta indica*, in biological control of pests have been investigated against different insect species under laboratory and field conditions (Abedi et al. 2014; Nazarpour et al. 2014; Nouri-Ghanbalani et al. 2016). The combination of *B. thuringiensis* subsp. *kurstaki* strain ABTS-351 and Azadirachtin for the control of *H. armigera* under laboratory conditions has been assessed (Abedi et al. 2014). The combination of Bt and Azadirachtin had a synergistic effect on the mortality of third instar larvae of *H. armigera*, therefore, after field trials, the joint use of these two insecticides can be suggested (Abedi et al. 2014).). The same results were obtained by Nouri-Ghanbalani et al. (2016) showing synergistic interaction between Bt and Azadirachtin on the mortality of *P. interpunctella* third instar larvae.

Under laboratory conditions, the combined effect of Bt and spinosad against beet armyworm larvae, *S. exigua*, was studied (Hosseinizadeh and Aramideh 2014). Combination of Bt and spinosad resulted in a high *S. exigua* neonate mortality, therefore, the simultaneous application of Bt and spinosad is recommended in the biological control of beet armyworm (Hosseinizadeh and Aramideh 2014).

The susceptibility of Colorado potato beetle larvae, *L. decemlineata* (Col.: Chrysomelidae), to *B. thuringiensis* subsp. *tenebrionis* alone and in combination with henna powder and cinnamon extract were investigated (Ghassemi-Kahrizeh et al. 2014). The synergistic effect on *L. decemlineata* mortality was observed following combination of both Bt with henna powder and also Bt with cinnamon extract.

The compatibility of Bt application simultaneously with chemical insecticides has been also studied. The interaction between *B. thuringiensis* subsp. *kurstaki* with abamectin, azadirachtin, indoxacarb, chlorantraniliprole, dichlorvos and metaflumizone, in the control of *T. absoluta* were investigated (Amizadeh et al.

2015). The colonization of Bt was affected by metaflumizone. Moreover, both antagonism and synergism effects were observed based on the time of application of Bt treatment and concentration of chemical insecticides. When Bt was applied immediately after chemical insecticides, an antagonistic effect was observed. Whereas, synergistic activity was detected when Bt was used 12 and 24 h after treatment at a concentration equal to $LC_{25}$ value of chlorantraniliprole, dichlorvos, and abamectin. The simultaneous use of Bt and chemical insecticides was not recommended by Amizadeh et al. (2015).

The possibility of simultaneous application of parasitoid wasps and Bt has been questioned (Sedaratian et al. 2014; Alsaedi et al. 2017). The braconid, *Habrobracon hebetor* Say (Hym.: Braconidae) and *B. thuringiensis* subsp. *kurstaki* were selected to control the larval stages of *H. armigera* (Sedaratian et al. 2014). Prior to the integration of *H. hebetor* with Bt, the effect of Bt on the wasp or in general, natural enemies has to be precisely assessed. Sedaratian et al. (2014) studied some of biological parameters of *H. hebetor* parasitizing the Bt-infected *H. armigera* larvae. Bt negatively affected the development of *H. hebetor* wasps, suggesting that risk assessment is required to evaluate the side effects of Bt on natural enemies. In another study, the possibility of application of the egg parasitoid, *Trichogramma* spp., with a commercial Bt formulation (Biolep®) based on the *B. thuringiensis* subsp. *kurstaki* for the control of tomato leafminer, *T. absoluta*, was examined under greenhouse conditions (Alsaedi et al. 2017). Moreover, the effect of this commercialized Bt-based product was assessed against the non-target organism, *Trichogramma* spp. Based on greenhouse observations, the infestation rate was significantly lower in the treatment of the combination of Bt and *Trichogramma*, compared to the individual application of these two biological control agents. Interestingly, Bt had no negative effect on the biological parameters of *Trichogramma* spp. (Alsaedi et al. 2017), suggesting the simultaneous application of these two biological control agents. The combined use of two insect pathogens, Bt and HaNPV, targeting the first instars of *H. armigera* larvae with the larval parasitoid, *H. hebetor*, was evaluated under field conditions (Allahyari et al. 2019). Synergistic interaction between the applications of insect pathogens with releasing the larval parasitoid in the control of *H. armigera* in chickpea field was found. Therefore, the simultaneous applications of these biological control agents in the IPM program of *H. armigera* in chickpea field is suggested.

## 10.15   Challenges

About five microbial pesticides have been registered in Iran (Karimi et al. 2019). Biological control by the means of insect pathogens faces different challenges in Iran. The development of the bio-insecticide mass production, finding a wider spectrum of activity, improvements in Bt formulation, application strategies, persistence in the environment, preserving shelf-life, effective delivery system to all agricultural sections in Iran, cost-effective production, registration of microbial pesticides, and optimizing the quality of Bt formulations are the main concerns regarding microbial control in Iran. Although Bt-based insecticides effectively control the target insect species, they still have not significantly substituted the application of chemical insecticides. Moosavi and Zare (2016) have comprehensively discussed the limitations of bio-insecticide application or applying microbial control. The major restrictions are that the available commercial Bt products do not target all major pests of strategic Iranian crops. These Bt products are not available in all agricultural regions, the cost is high, farmers lack knowledge, and more importantly, chemical pesticides have a raid speed of action and in some cases chemicals have greater effectiveness. Therefore, in the market, bio-insecticides cannot compete with chemical insecticides. Moreover, the efficiency of microbial pesticides is not stable, limited shelf-life and specific requirements for storage are other constraints that restrict their application.

On the other hand, the proper application of Bt definitely leads to successful pest management. Several parameters affect the efficiency of Bt-based insecticides and subsequently change the result of pest control. As it has been previously mentioned, the spores and crystals of Bt have to be ingested by the susceptible larvae to exert toxicity. Therefore, the coincidence of applying Bt in the field and the emergence of susceptible larvae is a critical factor directly affecting the efficiency of a Bt product. The instability of Bt formulations under field conditions threatens their effectiveness. Additives such as protectants, dispersants, surfactants, and attractants, are required in Bt formulation for their success in biological control. The limited shelf-life is another constraint that has been considered for Bt formulation. All of the mentioned parameters affect the efficiency of biological control based on Bt.

Further, the difficulties of microbial pesticide registration has to be taken into consideration. In general, the registration of microbial bio-pesticides is a time and money consuming process, although attempts have been made in Iran to simplify the process (Moosavi and Zare 2016). Development and mass production of bio-insecticides receive less attention and inadequate financial support compared to what is dedicated to chemical insecticides.

The development and application of Bt crops is another controversial issue. The possible adverse effects of transgenic crops on human and other non-target organisms, and also the environmental potential risks of these genetically modified crops have been the matter of debate for at least the last decade. It seems that it is necessary to repeatedly shed light on the side effects of chemical insecticides and to raise the question: does the application of chemicals as an alternative have less environmental and health hazards compare to transgenic crops? Based on these challenges, GM crops have not been commercially released in Iran yet. Therefore, more scientific-based studies including biosafety regulation, risk assessment, field experiments, and regulatory considerations are needed (Salehi Jouzani 2012).

## 10.16    Future Research Direction

Biological control by means of Bt has proven to be an effective and valuable strategy for insect control and also for diminishing the application of chemical insecticides. Moreover, due to the multifunctional role of Bt, as it has been comprehensively discussed in Salehi Jozanai et al. (2017) review, this bacterium has a great potential to be applied in different areas. Therefore, the search for discovering novel Bt strains and toxins with improved efficiency, wider insecticidal toxicity, antagonistic activity against plant pathogens and plant growth-promoting function, has to be continued. After discovering novel Bt toxins, in the next step, the appropriate formulation of bacterial products that enhance both survival and efficiency of the organisms is needed. Bt products are not cost-effective, though the effectiveness of Bt based insecticides has to be improved and the cost of production has to be lowered. Evidently, optimization of the growth conditions leads to the highest yields, while utilization of locally available materials, low-cost materials, agricultural and industrial by-products, improves the local cost-effective production. Undoubtedly, this issue needs more sufficient techniques in the fermentation, quality control, formulation and delivery system of Bt. Finally, the current registration protocols of the Plant Protection Organization (PPO) needs to be revised to facilitate the procedures. Decreasing the costs related to the Bt-based mass-production and supporting native Bt products could be helpful in the development of local production and the application of these products in the country.

Understanding the mechanism of action of Bt toxins would definitely lead to the development of high potent and more efficient Bt-based products and Bt crops. On the other hand, pyramiding two or more Bt toxins with different modes of action has been considered as an effective strategy for delaying resistance (Ferré et al. 2008). Therefore, future studies can be focused on the mode of action and binding behavior of Bt toxins in the midgut of susceptible larvae. Genetic manipulation of the Bt toxins with enhanced and broadened insecticidal activity must be taken into consideration.

## 10.17    Conclusion

*B. thuringiensis* as one the most successful microbial control agents for controlling agricultural, forestry, medical and veterinary pests, is being used as part of integrated pest management strategies. Commercial Bt products are used worldwide, as well as in Iran. Based on the several extensive Bt screening projects carried out in Iran, the high rate of Bt recovery from different ecosystems, depending on the geographical region, demonstrates the richness and variability of biological control agent resources. In these screenings, Iranian Bt isolates were found to be highly toxic to different species within Lepidoptera, Coleoptera, and Diptera, pointing to the fact that the Iranian Bt strains have a huge potential to control different agricultural pests.

Based on the literature, many studies as master thesis, PhD dissertation and projects have been performed in Iran and focused on the discovery of new Bt strains with potent pesticidal activity. The final aim of all these studies is to introduce an appropriate candidate for the development of future Bt-based products. While so far, only a few of these candidates have been developed as Bt-based bio-insecticides (Karimi et al. 2019). Indeed, more considerations have to be given to bio-insecticide production in order to develop the microbial pest control. Definitely, innovative production techniques, optimization of mass production protocols, assessment of product quality and quantity, preserved shelf-life and high biological activity are required for the commercialization of bio-insecticides based on Bt. Undoubtedly, financial support and simplification of microbial insecticide registration lead to the development of bio-insecticide production. Finally, the consistency of microbial agents' effectiveness and low prices of commercialized bio-insecticides encourage the farmers to substitute Bt-based products for chemicals.

Bt has been predominantly isolated from soils of different regions of Iran. In different studies, *B. thuringiensis* subsp. *kurstaki* was reported as the most abundant Bt serovar, although it highly depended on the geographical region and the environment where sampling was carried out. The bipyramidal crystalline inclusions were commonly observed in different Bt screening projects. Due to the economic and environmental benefits, several companies produce Bt-based insecticides. The bacterial formulations registered as bio-insecticides in Iran are mainly based on *B. thuringiensis* subsp. *kurstaki* following by *B. thuringiensis* subsp. *israelensis*. Different kinds of formulations are used in Iran for bacterial production, including suspension concentrates, wettable powders and granular formulations.

*B. thuringiensis* synthetic genes were transferred into different plants, mainly rice, cotton, potato, sugar beet, alfalfa, maize and date with the aim of developing transgenic plants with enhanced resistance to insects. Although GM crops expressing *cry* genes have been considered as the successful substitute for chemicals, risk assessments are performed before their deployment.

Overall, the screening and characterization studies performed in Iran have introduced native Bt strains with high potential for controlling agricultural, household and medical insect pests. These Bt strains are appropriate candidates for application as commercialized biocontrol agents. The complementary research is certainly necessary under laboratory, greenhouse, semi-field and field conditions for evaluating the efficiency of Bt products and Bt crops. The effect of Bt products and Bt crops has to be assessed on non-target organisms (i.e. parasitoids, predators and pollinators). Optimization of mass production, fermentation, and formulation is critical for development of successful commercial products for us in biological control and providing a healthier environment.

**Acknowledgement**  We are deeply grateful to Dr. Mark Goettel for his valuable comments and improvement of the text. Ayda Khorramnejad acknowledges Dr. Yolanda Bel, Dr. Baltasar Escriche and Dr. Patricia Hernández-Martínez for their supports and thoughtful comments on the manuscript.

# References

Abedi Z, Saber M, Vojoudi S, Mahdavi V, Parsaeyan E (2014) Acute, sublethal, and ombination effects of azadirachtin and *Bacillus thuringiensis* on the cotton bollworm, *Helicoverpa armigera*. J Insect Sci 14:30–39

Adjalle KD, Brar SK, Verma M, Tyag RD, Valero JR, Surampalli RY (2007) Ultrafiltration recovery of entomotoxicity from supernatant of *Bacillus thuringiensis* fermented wastewater and wastewater sludge. Process Biochem 42:1302–1311

Alinia F, Ghareyazie B, Rubia L, Bennett J, Cohen MB (2000) Effect of plant age, larval stage, and fertilizer treatment on resistance of a *cry1Ab*-trasformed aromatic rice to Lepidopterous stem borers and foliage feeders. J Econ Entomol 93:484–493

Allahyari R, Aramideh S, Safaralizadeh MS, Rezapanah M, Michaud JP (2019) Synergy between parasitoids and pathogens for biological control of *Helicoverpa armigera* in chickpea. Entomol Exp Appl 168:1–6

Alsaedi G, Ashouri A, Talaei-Hassanloui R (2017) Assessment of two *Trichogramma* species with *Bacillus thuringiensis* var. *kurstaki* for the control of the tomato leafminer *Tuta absoluta* Meyrick (Lepidoptera: Gelechiidae) in Iran. Open J Ecol 7:112–124

Amiri-Besheli B (2008) Efficacy of *Bacillus thuringiensis*, mineral oil, insecticidal emulsion and insecticidal gel against *Phyllocnistis citrella* Stainton (Lepidoptera: Gracillariidae). Plant Prot Sci 44:68–73

Amizadeh M, Hejazi MJ, Niknam G, Azanlou M (2015) Compatibility and interaction between *Bacillus thuringiensis* and certain insecticides: perspective in management of *Tuta absoluta* (Lepidoptera: Gelechiidae). Biocontrol Sci Tech 25:671–684

Aramideh S, Safaralizadeh MH, Pourmirza AA, Rezazadeh Bari M, Keshavarzi M, Mohseniazar M (2010) Characterization and pathogenic evaluation of *Bacillus thuringiensis* isolates from West Azerbaijan province, Iran. Afr J Microbiol Res 12:1224–1231

Azimi S, Ashouri A, Tohidfar M, Talaei-Hassanlouei R (2012) Effect of Iranian Bt cotton on *Encarsia formosa*, parasitoid of *Bemisia tabaci*. Int Res J Basic Appl Sci 3:2248–2251

Baghaee RS, Moghaddam EM (2015) Efficacy of *Bacillus thuringiensis* Cry14 toxin against root knot nematode, *Meloidogyne javanica*. Plant Prot Sci 51:46–51

Baumann P, Clark MA, Baumann L, Broadwell AH (1991) *Bacillus sphaericus* as a mosquito pathogen: properties of the organism and its toxins. Microbiol Rev 55:425–436

Bavykin SG, Lysov YP, Zakhariev V, Kelly JJ, Jackman J, Stahl DA, Cherni A (2004) Use of 16S rRNA, 23S rRNA, and *gyrB* gene sequence analysis to determine phylogenetic relationships of *Bacillus cereus* group microorganisms. J Clin Microbiol 42:3711–3730

Bel Y, Granero F, Alberola TM, Sebastian MJM, Ferre J (1997) Distribution, frequency and diversity of *Bacillus thuringiensis* in olive tree environments in Spain. Syst Appl Microbiol 20:652–658

Berliner E (1915) Uber die schlaffsucht der mehlmottenraupe (*Ephestia kuhniella*, Zell.) und ihren erreger *B. thuringiensis* n. sp. Z. Angew Entomol 2:29–56

Boonmee K, Thammasittirong SN, Thammasittirong A (2019) Molecular characterization of lepidopteran-specific genes in *Bacillus thuringiensis* strains from Thailand. 3. Biotech 9:117–128

Brar SK, Verma M, Tyagi RD, Valero JR, Surampalli RY (2006a) Efficient centrifugation recovery of *Bacillus thuringiensis* biopesticides from fermented wastewater and wastewater sludge. Water Res 40:1310–1320

Brar SK, Verma M, Tyagi RD, Valero JR (2006b) Recent advances in downstream processing and formulations of *Bacillus thuringiensis* based biopesticides. Process Biochem 41:323–342

Bravo A, Sarabia S, Lopez L, Ontiveros H, Abarca C, Ortiz A, Ortiz M, Lina L, Villalobos FJ, Peña G, Nuñez-Valdez ME, Soberón M, Quintero R (1998) Characterization of *cry* genes in a Mexican *Bacillus thuringiensis* strain collection. Appl Environ Microbiol 64:4965–4972

Bravo A, Gill SS, Soberón M (2007) Mode of action of *Bacillus thuringiensis* Cry and Cyt toxins and their potential for insect control. Toxicon 49:423–435

Carozzi NB, Kramer VC, Warren GW, Evola S, Koziel MG (1991) Prediction of insecticidal activity of *Bacillus thuringiensis* strains by polymerase chain reaction product profiles. Appl Environ Microbiol 57:3057–3061

Chatterjee S, Ghosh TS, Das S (2010) Virulence of *Bacillus cereus* as natural facultative pathogen of *Anopheles subpictus* Grassi (Diptera: Culicidae) larvae in submerged rice-fields and shallow ponds. Afr J Biotechnol 9:6983–6987

Dastan S, Ghareyazie B, Pishgar AH (2019) Environmental impacts of transgenic Bt rice and non-Bt rice cultivars in northern Iran. Biocatal Agric Biotechnol 20:101160

de Maagd RA, Bravo A, Crickmore N (2001) How *Bacillus thuringiensis* has evolved specific toxins to colonize the insect world. Trends Genet 17:193–199

de Maagd RA, Bravo A, Berry C, Crickmore N, Schnepf HE (2003) Structure, diversity, and evolution of protein toxins from spore-forming entomopathogenic bacteria. Annu Rev Genet 37:409–433

Deilamy A, Abbasipour H (2013) Comparative bioassay of different isolates of *Bacillus thuringiensis* subsp. *kurstaki* on the third larval instars of diamondback moth, *Plutella xylostella* (L.) (Lep.: Plutellidae). Arch Phytopathol Plant Protect 46:1480–1487

Djenane Z, Nateche F, Amziane M, Gomis-Cebolla J, El-Aichar F, Khorf H, Ferré J (2017) Assessment of the antimicrobial activity and the entomocidal potential of *Bacillus thuringiensis* isolates from Algeria. Toxins 9:139

Dulmage HT, Correa JA, Gallegos-Morales G (1990) Potential for improved formulations of *Bacillus thuringiensis* var *israelensis* through standardization and fermentation development. In: De Barjac H, Sutherland DJ (eds) Bacterial control of mosquitoes and blackflies: biochemistry, genetics and applications of *Bacillus thuringiensis israelensis* and *Bacillus sphaericus*. Rugers University Press, New Burnswick, pp 110–133

Farkas J, Sebesta K, Horská K, Samek Z, Dolejs L, Sorm F (1976) Structure of thuringiensin, the thermostable exotoxin from *Bacillus thuringiensis*. Collect Czechoslov Chem Commun 42:909–929

Federici BA, Park HW, Sakano Y (2006) Insecticidal protein crystals of *Bacillus thuringiensis*. In: Shively JM (ed) Inclusions in prokaryotes, Microbiology monographs, vol 1. Springer, Berlin/Heidelberg, pp 195–123

Feldhaar H (2011) Bacterial symbionts as mediators of ecologically important traits of insect hosts. Ecol Entomol 36:533–543

Ferrandis MD, Juárez-Pérez VM, Frutos R, Bel Y, Ferré J (1999) Distribution of *cryI*, *cryII* and *cryV* genes within *Bacillus thuringiensis* isolates from Spain. Syst Appl Microbiol 22:179–185

Ferré J, Van Rie J, MacIntoch SC (2008) Insecticidal genetically modified crops and insect resistance management (IRM). In: Romeis J, Shelton AM, Kennedy GG (eds) Integration of insect-resistant geneticaly modified crops within IPM programs. Springer Science, NewYork, pp 41–85

Frye RD, Scholl CG, Scholz EW, Funke BR (1973) Effect of weather on a microbial insecticide. J Invertebr Pathol 22:50–54

Gezelbash Z, Vatandoost H, Abai MR, Raeisi A, Rassi Y, Hanafi-bojd AA, Jabbari H, Nikpoor F (2014) Laboratory and field evalustion of two formulations of *Bacillus thuringiensis* M-H-14 against mosquito larvae in the Islamic Republic of Iran, 2012. East Mediterr Health J 4:229–235

Ghareyazie B, Alinia F, Menguito CA, Rubia LG, De Palma JM, Liwanag EA, Cohen MB, Khush GS, Bennett J (1997) Enhanced resistance to two stem borers in an aromatic rice containing a synthetic *cryIAb* gene. Mol Breed 3:401–414

Ghassemi-Kahrizeh A, Mohammadzadeh S, Miandoab MP (2014) The effect of *Bacillus thuringiensis* var. *tenebrionis* on Colorado Potato Beetle larvae, *Leptinotarsa decemlineata* (Say) (Coleoptera: Chrysomelidae) and synergistic role of henna and cinnamon in increasing its efficiency. Paper presented at the international conference on biopesticides, Antalya, Turkey, 19–25 Octuber 2014

Hajialiloo S, Moravvej G, Sadeghi H (2016) Comparative study on the efficeincy of *Bacillus thuuringiensisi* subsp. *tenebrionis* and a neem based insecticide on adults and larvae of *Xanthogaleruca luteola* (Mull) (Col.: Chrysomelidae) in laboratory conditions. J Entomol Zool Stud 4:1122–1125

Hanafi-Bojd AA, Vatandoost H, Jafari R (2006) Susceptibility status of *Anopheles dthali* and *An. fluviatilis* to commonly used larvicides in an endemic focus of malaria, southern Iran. J Vect Borne Dis 43:34–38

Helgason E, Økstad OA, Caugant DA, Johansen HA, Fouet A, Mock M, Hegna I, Kolstø AB (2000) *Bacillus anthracis*, *Bacillus cereus*, and *Bacillus thuringiensis* – one species on the basis of genetic evidence. Appl Environ Microbiol 66:2627–2630

Höfte H, Whiteley HR (1989) Insecticidal crystal proteins of *Bacillus thuringiensis*. Microbiol Rev 53:242–255

Hosseinizadeh A, Aramideh A (2014) Effect of *Bacillus thuringiensis* var. *kurstaki* and Spinosad on beet armyworm, *Spodoptera exigua* (Hübner) (Lepidoptera: Noctuidae) neonates in laboratory conditions. Paper presented at the international conference on biopesticides, Antalya, Turkey, 19–25 Octuber 2014

Ishiwata S (1901) On a kind of flacherie (sotto disease). Dainihon Sanshi Keiho 114:1–5

Jafari M, Valizadeh M, Malboobi MA, Ghareyazie B, Mohammadi SA, Mosavi M, Norouzi P (2007) Agrobacterium-mediated transformation of suger beet (*Beta vulgaris* L.) with *cry1Ab* gene and development of resistant sugar beet plants against lepidopteran pests. Paper presented at the 5th National Biotechnology Congress of Iran, Tehran, 24–26 November 2007

Jafari M, Norouzi P, Malboobi MA, Ghareyazie B, Valizadeh M, Mohammadi SA, Mosavi M (2009a) Enhanced resistance to a lepidopteran pest in transgenic sugar beet plants expressing synthetic *cry1Ab* gene. Euphytica 165:333–344

Jafari M, Norouzi P, Malboobi MA, Ghareyazie B, Valizadeh M, Mohammadi SA (2009b) Transformation of *cry1Ab* gene to sugar beet (*Beta vulgaris* L.) by *Agrobaterium* and development of resistant plants against *Spodoptera littoralis*. J Sugar Beet Res 24:37–55. (In Persian)

Jalali E, Maghsoudi S, Noroozian E (2020) A novel method for biosynthesis of different polymorphs of TiO$_2$ nanoparticles as a protector for *Bacillus thuringiensis* from Ultra Violet. Sci Rep 10:426

James C (2007) Global status of commercialized Biotech/GM Crops: 2007. ISAAA Brief No. 37. ISAAA, Ithaca, NY

Juárez-Pérez VM, Ferrandis MD, Frutos R (1997) PCR-based approach for detection of novel *Bacillus thuringiensis cry* genes. Appl Environ Microbiol 63:2997–3002

Jurat-Fuentes JL, Jackson TA (2012) Bacterial entomopathogens. In: Vega FE, Harry KK (eds) Insect pathology, 2nd edn. Academic Press, pp 265–349

Kalantari M, Marzban R, Imani S, Askari H (2014) Effects of Bacillus thuringiensis isolates and single nuclear polyhedrosis virus in combination and alone on Helicoverpa armigera. Arch Phytopathol Plant Protect 47:42–50

Karimi J, Dara SK, Arthurs S (2019) Microbial insecticides in Iran: history, current status, challenges and perspective. J Invertebr Pathol 165:67–73

Keshavarzi M (2008) Isolation, identification and differentiation of local *B. thuringiensis* strains. J Agric Sci Technol 10:493–499

Keshavarzi M, Salimi H, Mirzanamadi F (2005) Biochemical and physical requirements of *Bacillus thuringiensisi* subsp. *kurstaki* for high biomass yield production. J Agric Sci Technol 7:41–47

Khojand S, Keshavarzi M, Zargari K, Abdolahi H, Rouzbeh F (2013) Presence of multiple *cry* genes in *Bacillus thuringiensis* isolated from dead cotton bollworm *Heliothis armigera*. J Agric Sci Technol 15:1285–1292

Khorramnejad A, Talaei-Hassanloui R, Hosseininaveh V, Bel Y, Escriche B (2018) Characterization of new *Bacillus thuringiensis* strains from Iran, based on cytocidal and insecticidal activity, proteomic analysis and gene content. BioControl 63:807–818

Khorramnejad A, Domínguez-Arrizabalaga M, Caballero P, Escriche B, Bel Y (2020) Study of the *Bacillus thuringiensis* Cry1Ia protein oligomerization promoted by midgut brush border membrane vesicles of Lepidopteran and Coleopteran insects, or cultured insect cells. Toxins 12:133

Kiani G, Nematzadeh GA, Ghareyazie B, Sattari M (2009a) Comparing the agronomic and grain quality characteristics of transgenic rice lines expressing *cry1Ab*vs. Non-transgenic controls. Asian J Plant Sci 8:64–68

Kiani G, Nematzadeh GA, Ghareyazie B, Sattari M (2009b) Genetic analysis of *cry1Ab* gene in segregating populations of rice. Afr J Biotechnol 8:3703–3707

Kiani G, Nematzadeh GA, Ghareyazie B, Sattari M (2012) Pyramiding of *cry1Ab* and *fgr* genes in two Iranian rice cultivars Neda and Nemat. J Agric Sci Technol 14:1087–1092

Kleter GA, Bhula R, Bodnaruk K, Carazo E, Felsot AS, Harris CA, Katayama A, Kuiper HA, Racke KD, Rubin B, Shevah Y, Stephenson GR, Tanaka K, Unsworth J, Wauchope RD, Wong SS (2007) Altered pesticide use on transgenic crops and the associated general impact from an environmental perspective. Pest Manag Sci 63:1107–1115

Lambert B, Peferoen M (1992) Insecticidal promise of *Bacillus thuringiensis*: facts and mysteries about a successful biopesticide. Bioscience 42:112–122

Lecadet MM, Frachn E, Casmao V, Ripouteau H, Hamon S, Laurent P, Thiery I (1999) Updating the H-antigen classification of *Bacillus thuringiensis*. J Appl Microbiol 86:660–672

Letourneau DK, Robinson GS, Hagen JA (2003) Bt crops: predicting effects of escaped transgenes on the fitness of wild plants and their herbivores. Environ Biosaf Res 2:219–246

Levinson BL, Kaysan KJ, Chiu SS, Currier TC, Gonzalez JM (1990) Identification of beta-exotoxin production, plasmids encoding beta-exotoxin, and a new exotoxin in *Bacillus thuringiensis* by using high-performance liquid chromatography. J Bacteriol 172:3172–3179

Liu YT, Sui MJ, Ji DD, Wu IH, Chou CC, Chen CC (1993) Protection from ultraviolet irradiation by melanin of mosquitocidal activity of *Bacillus thuringiensis* var. *israelensis*. J Invertebr Pathol 62:131–136

Liu X, Ruan L, Peng D, Li L, Sun M, Yu Z (2014) Thuringiensin: a thermostable secondary metabolite from *Bacillus thuringiensis* with insecticidal activity against a wide range of insets. Toxins 6:2229–2238

Magholli Z, Marzban R, Abbasipour H, Shikhi A, Karimi J (2013) Interaction effects of *Bacillus thuringiensis* subsp. *kurstaki* and single nuclear polyhedrosis virus on *Plutella xylostella*. J Plant Dis Protect 120:173–178

Maghsoudi S, Jalali E (2017) Noble UV protective agent for *Bacillus thuringiensis* based on a combination of graphene oxide and olive oil. Sci Rep 7:11019

Martin PAW, Haransky EB, Travers RS, Reichelderfer CF (1985) Rapid biochemical testing of large numbers of *Bacillus thuringiensis* isolates using agar dots. BioTechniques 3:386–392

Marzban R (2012a) Investigation on the suitable isolate and medium for production of *Bacillus thuringiensis*. J Biopest 5:144–147

Marzban R (2012b) Midgut pH profile and energy differences in lipid, protein and glycogen metabolism of *Bacillus thuringiensis* Cry1Ac toxin and cypovirus-infected *Helicoverpa armigera* (Hübner) (Lepidoptera: Noctuidae). J Entomol Res Soc 14:45–53

Marzban R, He Q, Liu X, Zhang Q (2009) Effects of *Bacillus thuringiensis* toxin Cry1Ac and cytoplasmic polyhedrosis virus of *Helicoverpa armigera* (Hübner) (HaCPV) on cotton bollworm (Lepidoptera: Noctuidae). J Invertebr Pathol 101:71–76

Marzban R, Saberi F, Shirazi MMA (2014) Separation of *Bacillus thuringiensis* from fermentation broth using microfiltration: optimization approach. Res J Biotechnol 9:33–37

Marzban R, Saberi F, Shirazi MMA (2016) Microfiltration and ultrafiltration of *Bacillus thuringiensis* fermentation broth: membrane performance and sporecrystal recovery approaches. Braz J Chem Eng 33:783–791

McClintock J, Schaffer CR, Sjoblad RD (1995) A comparative review of the mammalian toxicity of *Bacillus thuringiensis*-based pesticides. J Pest Sci 45:95–105

McGuire MR, Behle RW, Goebel HN, Fry TC (2000) Calibration of a sunlight simulator for determining solar stability of *Bacillus thuringiensis* and *Anagrapha falcifera* nuclear polyhedrovirus. Biol Control 29:1070–1074

Meadows MP (1993) *Bacillus thuringiensis* in the environment: ecology and risk assessment. In: Entwistle PF, Cory JS, Bailey MJ, Higgs S (eds) *Bacillus thuringiensis*, an environmental biopesticide: theory and practice. Wiley & Sons Ltd, West Sussex, pp 193–220

Moazami N (1997) Large scale production of slow release formation of *Bacillus thuringiensis* M-H-4 in Qeshm Island. Proceeding of Second Technical Meeting and The First Regional Conference on Combating Malaria IROST, UNDP/UNESCO, 4–8 December 1995

Moazami N (2004) The role of *Bacillus thuringiensis* H-14 in malaria control. Paper presented at the 4th intercountry meeting of national malaria programme manages. Isfahan, Iran, 22–25 May 2004

Moazami N (2005) Controlling malaria, the vampire of the technological age. A World Sci 3:16–19

Moazamian E, Bahador N, Rasouli M, Azarpira N (2016) Diversity, identification and biotyping of *Bacillus thuringiensis* strains from soil samples in Iran. Nat Environ Pollut Technol 15:947–950

Moazamian E, Bahador N, Azarpira N, Rasouli M (2018) Anti-cancer parasporin toxins of new *Bacillus thuringiensis* against human colon (HCT-116) and blood (CCRF-CEM) cancer cell lines. Curr Microbiol 75:1090–1098

Moosavi MR, Zare R (2016) Present status and the future prospects of microbial biopesticides in Iran. In: Singh HB, Sarma B, Keswani C (eds) Agriculturally important microorganisms. Springer, Singapore, pp 293–305

Mousavi A, Malboobi MA, Esmailzadeh NS (2007) Development of agricultural biotechnology and biosafety regulations used to assess the safety of genetically modified crops in Iran. J AOAC Int 90:1513–1516

Moxtarnejad E, Safaralizade MH, Aramideh S (2014) The protective material effect in combination with *Bacillus thuringiensis* var. *kurstaki* (*Btk*) against UV for control *Pieris brassicae* L. (Lep.: Pieridae). Arch Phytopathol Plant Protect 47:2414–2420

Naseri Rad S, Shirazi MMA, Kargari A, Marzban R (2016) Application of membrane separation technology in downstream processing of *Bacillus thuringiensis* biopesticide: a review. J Membr Sci Res 2:66–77

Nazarian A, Jahangiri R, Salehi Jouzani G, Seifinejad A, Soheilivand S, Bagheri O, Keshavarzi M, Alamisaeid K (2009) Coleopteran-specific and putative novel *cry* genes in Iranian native *baciilus thuringiensis* collection. J Invertebr Pathol 102:101–109

Nazarpour L, Yarahmadi F, Rajabpour A, Saber M (2014) Effects of some bio-insectcides against *Tuta absoluta* Meyrick in tomato fields. Paper presented at the international conference on biopesticides Antalya, Turkey, 19–25 October 2014

Norouzi P, Jafari M, Malboobi MA, Ghareyazie B, Rajabi A (2011) Inheritance of transgene and resistance to a Lepidopteran pest, *Spodoptera littoralis*, in treansgenic sugar beet plants harboring a synthetic *cry1Ab* gene. Trangenic Plant J 5:62–66

Nouri-Ghnbalani G, Borzoui E, Abdolmaleki A, Abedi Z, Kamita SG (2016) Individual and combined effects of *Bacillus thuringiensis* and Azadirakhtin on *Plodia interpunctella* Hübner (Lepidoptera: Pyralidae). J Insect Sci 16:1–8

Ohba M, Mizuki E, Uemori A (2009) Parasporin, a new anticancer protein group from *Bacillus thuringiensis*. Anticancer Res 29:427–433

Pinos D, Hernández-Martínez P (2019) Modo de acción de las proteínas insecticidas de *Bacillus thuringiensis*. In: Tena A, Bielza P, Ferré J (eds) Boletín de la Sociedad Española de Entomología Aplicada. SEEA, Madrid, pp 26–30

Ramezani Moghaddam M, Moghaddam EM, Ravari SB, Rouhani H (2014) The nematicidal potential of local *Bacillus* species against the root-knot nematode infecting greenhouse tomatoes. Biocontrol Sci Tech 24:279–290

Regev A, Keller M, Strizhov N, Sneh B, Prudovsky E, Chet I, Ginzberg I, Koncz-Kalman Z, Koncz C, Schell J, Zilberstein A (1996) Synergistic activity of a *Bacillus thuringiensis* delta-endotoxin and a bacterial endochitinase against *Spodoptera littoralis* larvae. Appl Environ Microbiol 62:3581–3586

Ruiu L (2015) Insect pathogenic bacteria in integrated pest management. Insects 6:352–367

Salehi Jouzani G (2012) Risk assessment of GM crops; challenges in regulations and science. Biosafety 1:e113. https://doi.org/10.4172/2167-0331.1000e113

Salehi Jouzani G, Komakhin RA, Piruzian ES (2005) Comparative study of the expression of the native, modified, and hybrid *cry3a* genes of *Bacillus thuringiensis* in prokaryotic and eukaryotic cells. Russ J Genet 41:116–121

Salehi Jouzani G, Goldenkova IV, Piruzian ES (2008a) Expression of hybrid *cry3a*M-*lic*BM2 genes in transgenic potatoes (*Solanum tuberosum*). Plant Cell Tissue Organ Cult 92:321–325

Salehi Jouzani G, Pourjan Abad A, Seifinejad A, Marzban R, Kariman K, Maleki B (2008b) Distribution and diversity of dipteran-specific *cry* and *cyt* genes in native *Bacillus thuringiensis* strains obtained from different ecosystems of Iran. J Ind Microbiol Biotechnol 35:83–94

Salehi Jouzani G, Seifinejad A, Saeedizadeh A, Nazarian A, Yousefloo M, Soheilivand S, Mousivand M, Jahangiri R, Yazdani M, Maali Amiri R, Akbar S (2008c) Molecular detection of nematicidal crystalliferous *Bacillus thuringiensis* strains of Iran and evaluation of their toxicity on free-living and plant-parasitic nematodes. Can J Microbiol 54:812–822

Salehi Jouzani GH, Moradali MF, Morsali H, et al. (2011) Isolation and identification of native *Bacillus thuringiensis* isolates and production of biological pesticide based on effective strains, Final report of the megaproject NO: 1-013-140000-05-8512-00000, Agricultural Research, Education and Extension Organization (AREEO)

Salehi Jouzani G, Abbasalizadeh A, Moradali MF, Morsali H (2015a) Development of a cost effective bioprocess for production of an Iranian anti-coleopteran *Bacillus thuringiensis* strain. J Agric Sci Technol 17:1183–1196

Salehi Jouzani G, Moradali MF, Abbasalizadeh A (2015b) Optimization of economic medium and fermentation process of a lepidopteran active native *Bacillus thuringiensis* strain to enhance spore/crystal production. J Agric Biotechnol 7:93–113. (In Persian)

Salehi Jouzani G, Valijanian F, Sharafi R (2017) *Bacillus thuringiensis*: a successful insecticide with new environmental features and tidings. Appl Microbiol Biotechnol 101:2691–2711

Sarfraz M, Keddie AB, Dosdall LM (2005) Biological control of the diamondback moth, *Plutella xylostella*: a review. Biocon Sci Technol 15:763–789

Sarrafzadeh MH (2012) Nutritional requirements of *Bacillus thuringiensis* during different phases of growth, sporulation and germination evaluated by Plackett-Burman method. Iran J Chem Chem Eng 31:131–136

Sarrafzadeh MH, Navarro JM (2006) The effect of oxygen on the sporulation, δ-endotoxin synthesis and toxicity of *Bacillus thuringiensis* H14. World J Microbiol Biotechnol 22:305–310

Schnepf E, Crickmore N, Van Rie J, Lereclus D, Baum J, Feitelson J, Zeigler DR, Dean DH (1998) *Bacillus thuringiensis* and its pesticidal crystal proteins. Microbiol Mol Biol Rev 62:775–806

Sedaratian A, Fathipour Y, Talaei-Hassanloui R, Jurat-Fuentes JL (2013) Fitness costs of sublethal exposures to *Bacillus thuringiensis* in *Helicoverpa armigera*: a carryover study on offspring. J Appl Entomol 137:540–549

Sedaratian A, Fathipour Y, Talaei-Hassanloui R (2014) Deleterious effects of *Bacillus thuringiensis* on biological parameters of *Habrobracon hebetor* parasitizing *Helicoverpa armigera*. BioControl 59:89–98

Seifinejad A, Salehi Jouzani G, Hosseinzadeh A, Abdmishani C (2008) Characterization of Lepidopter-active *cry* and *vip* genes in Iranian *Bacillus thuringiensis* strain collection. Biol Control 44:216–226

Selvakumar G, Mohan M, Sushil SN, Kundu S, Bhatt JC, Gupta HS (2007) Characterization and phylogenetic analysis of an entomopathogenic *Bacillus cereus* strain WGPSB-2 (MTCC 7182) isolated from white grub, *Anomala dimidiata* (Coleoptera: Scarabaeidae). Biocontrol Sci Tech 17:525–534

Senfi F, Safaralizadeh MH, Safavi SA, Aramideh S (2012) Isolation and characterization of native *Bacillus thuringiensis* strains from apple orchards at Urmia, Iran and their toxicity to lepidopteran pests, *Ephestia kuehniella* Zeller and *Pieris brassicae* L. Egypt J Biol Pest Co 22:33–37

Shahi M, Hanafi-Bojd AA, Vatandoost H, Soleimani Ahmadi M (2013) Susceptibility status of *Anopheles stephensi* Liston the main malaria vector, to deltamethrin and *Bacillus thuringiensis* in the endemic malarious area of Hormozgan province, southern Iran. J Kerman Univ Med Sci 20:87–95

Shojaaddini M, Moharramipour S, Khodabandeh M, Talebi AA (2010) Development of a cost effective medium for production of *Bacillus thuringiensis* bioinsecticide using food barley. J Plant Prot Res 50:9–14

Shojaaddini M, López MJ, Moharramipour S, Khodabandeh M, Talebi AA, Vilanova C, Latorre A, Porcar M (2012) A *Bacillus thuringiensis* strain producing epizootics on *Plodia interpunctella*: a case study. J Stored Prod Res 48:52–60

Smagghe G, Goodman CL, Stanley D (2009) Insect cell culture and applications to research and management. In Vitro Cell Dev Biol-Anim 45:93–105

Soberón M, Portugal LC, Garcia-Gómez BI, Sánchez J, Onofre J, Gómez I, Pacheco S, Bravo A (2018) Cell lines as models for the study of Cry toxins from *Bacillus thuringiensis*. Insect Biochem Mol Biol 93:66–78

Strongman DB, Eveleigh ES, van Frankenhuyzen K, Royama T (1997) The occurrence of two types of entomopathogenic *bacilli* in natural populations of the spruce budworm, *Choristoneura fumiferana*. Can J For Res 27:1922–1927

Tabashnik BE, Cushing NL, Finson N, Johnson MW (1990) Field development of resistance to *Bacillus thuringiensis* in diamondback moth (Lepidoptera: Plutellidae). J Econ Entomol 83:1671–1676

Talaei-Hassanloui R, Bakhshaei R, Hosseininaveh V, Khorramnejad A (2014) Effect of midgut proteolytic activity on susceptibility of lepidopteran larvae to *Bacillus thuringiensis* subsp. *kurstaki*. Front Physiol 4:406

Thorne CB (1993) Bacillus anthracis. In: Sonenshein AL, Hoch JA, Losick R (ed) *Bacillus subtilis* and other gram-positive bacteria. American Society for Microbiology, Washington, DC p 113–124

Tohidfar M, Salehi Jouzani G (2008) Genetic engineering of crop plants for enhanced resistance to insects and diseases in Iran. Transgenic Plant J 2:151–156

Tohidfar M, Ghareyazie B, Mohammadi M (2005) *Agrobacterium*-mediated transformation of cotton using a chitinase gene. Plant Cell Tiss Org Cult 83:83–96

Tohidfar M, Ghareyazie B, Mosavi M, Yazdani S, Golabchian R (2008) *Agrobacterium*-mediated transformation of cotton (*Gossypium hirsutum*) using a synthetic *cry1Ab* gene for enhanced resistance against *Heliothis armigera*. Iran J Biotechnol 6:164–173

Uribe D, Martinez W, Cerón J (2003) Distribution and diversity of *cry* genes in native strains of *Bacillus thuringiensis* obtained from different ecosystems from Colombia. J Invertebr Pathol 82:119–127

Vachon V, Laprade R, Schwartz JL (2012) Current models of the mode of action of *Bacillus thuringiensis* insecticidal crystal proteins: a critical review. J Invertebr Pathol 111:1–12

van Frankenhuyzen K (2009) Insecticidal activity of *Bacillus thuringiensis* crystal proteins. J Invertebr Pathol 101:1–16

Zare N, Valizadeh M, Malboobi M, Tohifar M (2008) Regeneration of Iranian alfalfa by somatic embryo and sheep apex. In: Proceeding of the 2nd National Congress of Cellular and Molecular Biology, Kermanshah, Iran, 29–30 January 2008

Zareie R, Shayesteh N, Pourmirza AA (2003) Starch-encapsulating of *Bacillus thuringiensis* Berliner containning different additives and evaluation of their efficiency. Iranian J Agric Sci 34:854–862. (In Persian)

# Chapter 11
# Fungal Entomopathogens of Order Hypocreales

**Hassan Askary, Sepideh Ghaffari, Mina Asgari, and Javad Karimi**

## 11.1  Introduction

Insect pathogenic fungi are a great group of microorganisms that attack a wide range of insect and mite species. Some species and strains of the fungi are very specific. Generally, the fungi produce spores which infect their host by germinating on their body surface and entering into its hemocoel. Entomopathogenic fungi have the greatest potential for controlling sucking insect pests because they have the contact mode of action, infecting through the cuticle rather than requiring ingestion (Vega and Kaya 2012; Chandler 2017).

Early research on entomopathogenic agents started about 50 years ago in Iran, firstly by Dr. Aziz Kharazi Pakdel, emeritus professor of entomology, Tehran University. His experiences on *Nosema* sp. (Microsporidia: Microsporea) and *Beauveria bassiana* (Ascomycota: Hypocreales) were the basis of researches and thesis of his students. Since the 1990s, researchers have been paid more attention to insect fungal pathogen. Isolation of native fungal isolates and their bioassay on

H. Askary (✉)
Biological Control Department, Iranian Research Institute of Plant Protection, Agricultural Research, Education and Extension Organization, Tehran, Iran
e-mail: askary@areo.ac.ir

S. Ghaffari
Department of Plant Protection, Faculty of Agriculture, Ferdowsi University of Mashhad, Mashhad, Iran

M. Asgari
Department of Entomology, Science and Research Branch, Islamic Azad University, Tehran, Iran

J. Karimi
Department of Plant Protection, Ferdowsi University of Mashhad, Mashhad, Iran
e-mail: jkb@um.ac.ir

© The Author(s), under exclusive license to Springer Nature Switzerland AG 2021
J. Karimi, H. Madadi (eds.), *Biological Control of Insect and Mite Pests in Iran*, Progress in Biological Control 18, https://doi.org/10.1007/978-3-030-63990-7_11

different hosts were the basis of most research area. In recent years, more researches have been realized to develop suitable culture media, determine optimal condition for mass production and development an effective formulation (Karimi et al. 2018).

In last decades, few commercialized insect pathogenic fungi have been registered in Iran. These commercial mycopesticides are *B. bassiana* (Naturalis-L®) for *Bemisia tabaci* (Hemiptera: Aleyrodidae) in cotton, *Akanthomyces dipterigenus* (Petch) Spatafora, Kepler, Zare & B. Shrestha (= *Verticillium lecanii, Lecanicillium longisporum*) (Vertalec®) (Ascomycota: Hypocreales) for aphids and *Akanthomyces lecanii* (= *Verticillium lecanii, Lecanicillium lecanii*) (Mycotal®) (Ascomycota: Hypocreales) for whitefly larvae in glasshouses.

A number of researchers have tried collecting fungi related to insects and mites from the ecological regions of Iran (Ghazavi 2009; Nouri-Aiin et al. 2014; Asadalapour et al. 2011). The most important species of EPFs are stored in the Iranian fungal culture collection. This collection is founded in 1968, placed in Iranian Research Institute of Plant Protection and known as IRAN WDCM 939 in "World Federation of Culture Collections: WFCC". Totally, the ninety-one isolates of entomopathogenic species belonging to twenty-six species and fifteen genera are collected and conserved in this collection from 1995 (Table 11.1).

## 11.2   Mode of Action

The infectious propagules of insect pathogenic EPFs are generally aerial conidia and blastospores, both commonly referred to as "spores". When these fungal bodies attach to host cuticle, a cascade of recognition and germination events happen during a few hours. Charnley (1989); Boucias and Pendland (1991); Ortiz-Urquiza and Keyhani (2013) reviewed the infection process of the fungi on their hosts. In this process, Pr1 serine protease enzyme has a prominent role during infection by *B. bassiana* Safavi (2012). Talaei-Hassanloui et al. (2007) emphasized a significant role of nutrient media in germination of *B. bassiana*. Additional aspects for the fungal mode of action by ultrastructural and cytochemical characterization of the host invasion by the *V. lecanii* have been documented by Askary et al. (1997) and Askary et al. (1999).

The process of infection in *Macrosiphum euphorbiae* (Hemiptera: Aphididae) and *Sphaerotheca fuliginea* (Ascomycota: Erysiphales) by *V. lecanii* follows a sequence of events from the exposure of the host to conidia to the production of conidiophores on host cadavers. These events include initial adherence of conidia to the host, then germination of the conidia and production of either germ tubes or branched hyphae that colonize the surface of the cuticle or powdery mildew body. Generally, *V. lecanii* hyphae colonize host structures by tight adhesion, apparently mediated by a thin mucilaginous matrix. Subsequent penetration by germ tubes involves mechanical pressure and production of cuticle-degrading enzymes, such as chitinases, as confirmed by the pattern of labelling obtained with the WGA/ovomucoid-gold complex (Askary et al. 1997). The next phase involves extensive 'lateral' colonization of the procuticle in aphid and hyphaea cell wall of powdery

**Table 11.1** Iranian isolates of insect pathogenic fungi stored in national fungal culture collection

| Accession N. | Taxonomic Name | Substrate | Isolated By | Region and place |
|---|---|---|---|---|
| IRAN 1229 C | *Acremonium alternatum* (cf.) | *Ommatissus lybicus* | M. Aminaei | Kerman |
| IRAN 1168 C | *Acremonium egyptiacum* (F. H. Beyma) W. Gams | *Agrilus aurichalceus* (Coleoptera: Buprestidae) | M. M. Aminaei | Kerman |
| IRAN 1167 C | *Sarocladium kiliense* (Grütz) Summerbell (= *Acremonium kiliense*) | *Hartigia trimaculata* (Hymenoptera: Cephidae) | M. M. Aminaei | Kerman |
| IRAN 2213 | *Actinomucor elegans* (Eidam) C. R. Benj. & Hesselt. | *Acythopeus curvirostris* (pupa) | N. Sepasi | Birjand |
| IRAN 2234 | *Aspergillus flavus* Link | *Acythopeus curvirostris persicus* (adult) | N. Sepasi | Nehbandan |
| IRAN 187 C | *Beauveria bassiana s.l.* (Bals.-Criv.) Vuill. | *Leptinotarsa decemlineata* Say | R. Zare | Ardabil |
| IRAN 428 C | *Beauveria bassiana s.l.* | *Chilo suppressalis* Walker | M. Ghazavi | Guilan, Rasht |
| IRAN 429 C | *Beauveria bassiana s.l.* | *Chilo suppressalis* | M. Ghazavi | Hassan Rud |
| IRAN 1395 | *Beauveria bassiana s.l.* | *Zeuzera pyrina* on *Juglans regia* | M. M. Aminaee | Kerman |
| IRAN 441 C | *Beauveria bassiana s.l.* | *Rhynchophorus ferrugineus* Olivier | M. Ghazavi | Saravan |
| IRAN 789 C | *Beauveria bassiana s.l.* | *Melolontha melolontha* | R. Zare | Gorgan-Alang Darreh forest |
| IRAN 923 C | *Beauveria bassiana s.l.* | *Chilo suppressalis* | R. Zare | Babol – Golmahalleh |
| IRAN 2240 | *Beauveria bassiana s.l.* | *Agrilus coxalis* (adult) | ? | Ilam |
| IRAN 1217 C | *Beauveria bassiana s.l.* | *Leptinotarsa decemlineata* | M. Hadayegh & H. Asgari | Ardabil |
| IRAN 1228 C | *Beauveria bassiana s.l.* | *Ommatissus lybicus* | M. Aminaei | Kerman |
| IRAN 1025 C | *Beauveria brongniartii* (Sacc.) Petch | *Coccinella septempunctata* | M. H. Ghasemi | Tehran-Varamin |
| IRAN 2214 | *Cunninghamella echinulata* var. *nodosa* R. Y. Zheng | *Acythopeus curvirostris* (pupa) | N. Sepasi & M. R. Mirzaei | Birjand, Mohammadieh |
| IRAN 934 C | *Cylindrocarpon* sp. | *Chilo suppressalis* | R. Zare | Mazandaran – Babol |

(continued)

**Table 11.1** (continued)

| Accession N. | Taxonomic Name | Substrate | Isolated By | Region and place |
|---|---|---|---|---|
| IRAN 1218 C | *Fusarium* cf. *proliferatum* (Matsush.) Nirenberg ex Gerlach & Nirenberg | *Lixus* sp. on *Beta vulgaris* | M. Ghazavi | Ghazvin, Hezarjolfa farm |
| IRAN 1193 C | *Fusarium incarnatum* (Roberge) Sacc. (= *Fusarium semitectum*) | *Pulvinaria* sp. | R. Zare | Behshahr |
| IRAN 1220 | *Fusarium lateritium s.l.* | *Pulvinaria* sp. | R. Zare | Tonekabon (Khorramabad) |
| IRAN 1285 | *Neocosmospora solani* (Mart.) L. Lombard & Crous (= *Fusarium solani*) | *Zeuzera pyrina* on *Juglans regia* | M. M. Aminaee | Kerman |
| IRAN 1638 | *Neocosmospora solani* | *Zeuzera pyrina* on *Juglans regia* | M. M. Aminaee | Kerman |
| IRAN 2193 | *Neocosmospora solani* | Larvae of *Zeuzera pyrina* | M. R. Mirzaei | South Khorasan, Sarayan, Fathabad |
| IRAN 2040 | *Neocosmospora solani* | *Coccinella septempunctata* | Fakhredini | Tehran, Bustan Besat |
| IRAN 739 C | *Neocosmospora solani* | *Euzophera bigella* Zeller | ? | Mazandaran |
| IRAN 2028 | *Hirsutella* cf. *versicolor* Petch | Beetle larva (*Amphimalon* sp.) | M. Ghazavi | Kermanshah, Bilvar |
| IRAN 1216 C | *Cordyceps farinosa* (Holmsk.) Kepler, B. Shrestha & Spatafora (= *Isaria farinosa*) | *Euproctis chrysorrhoea* L. | H. Asgari & P. Molaei | Arasbaran forest |
| IRAN 1782 | *Cordyceps farinosa* | Overwintering pupae of *Hyphantria* sp. | M. Salehi | Guilan, Rezvanshahr |
| IRAN 2257 | *Cordyceps farinosa* | *Galleria mellonella* on *Astragalus* sp. | M. Nori | Ilam, Sarable, Karezan |
| IRAN 2260 | *Cordyceps farinosa* | *Galleria mellonella* on *Astragalus* sp. | M. Nori | Ilam, Sheshdar |
| IRAN 1823 | *Lecanicillium* sp. | Scale insect | S. Aghajanzadeh | Mazandaran, Ramsar |
| IRAN 1282 C | *Lecanicillium aphanocladii* Zare & W. Gams (cf.) | *Pulvinaria aurantii* Cockerell | M. Davudi & A. Khazaeipour | Noshahr |
| IRAN 1750 | *Lecanicillium aphanocladii* | *Pulvinaria aurantii* on *Citrus sinensis* (L.) Osbeck | F. Sahbanian | Mazandaran, Abbasabad |

(continued)

**Table 11.1** (continued)

| Accession N. | Taxonomic Name | Substrate | Isolated By | Region and place |
|---|---|---|---|---|
| IRAN 1751 | *Lecanicillium aphanocladii* | *Pulvinaria aurantii* on *Citrus sinensis* | F. Sahbanian | Guilan, Chaboksar |
| IRAN 1754 | *Lecanicillium aphanocladii* | *Pulvinaria aurantii* on *Citrus aurantium* | F. Sahbanian | Guilan, Chaboksar |
| IRAN 1756 | *Lecanicillium aphanocladii* | *Pulvinaria aurantii* on *Citrus sinensis* | F. Sahbanian | Mazandaran, Ramsar |
| IRAN 1757 | *Lecanicillium aphanocladii* | *Pulvinaria aurantii* on *Citrus sinensis* | F. Sahbanian | Mazandaran, Abbasabad |
| IRAN 1758 | *Lecanicillium aphanocladii* | *Pulvinaria aurantii* on *Citrus sinensis* | F. Sahbanian | Guilan, Langrud |
| IRAN 1759 | *Lecanicillium aphanocladii* | *Pulvinaria aurantii* on *Citrus sinensis* | F. Sahbanian | Mazandaran, Ramsar, Liasar |
| IRAN 1760 | *Lecanicillium aphanocladii* | *Pulvinaria aurantii* on *Citrus sinensis* | F. Sahbanian | Guilan, Langrud, Leilakuh |
| IRAN 1761 | *Lecanicillium aphanocladii* | *Pulvinaria aurantii* on *Citrus sinensis* | F. Sahbanian | Guilan, Chaboksar |
| IRAN 1762 | *Lecanicillium aphanocladii* | *Pulvinaria aurantii* on *Citrus nobilis* var. *unshiu* | F. Sahbanian | Guilan, Chaboksar |
| IRAN 1763 | *Lecanicillium aphanocladii* | *Pulvinaria aurantii* on *Citrus sinensis* | F. Sahbanian | Guilan, Chaboksar |
| IRAN 1773 | *Lecanicillium aphanocladii* | *Pulvinaria aurantii* on *Citrus grandis* | F. Sahbanian | Guilan, Chaboksar |
| IRAN 1775 | *Lecanicillium aphanocladii* | *Pulvinaria aurantii* on *Citrus sinensis* | F. Sahbanian | Guilan, Langrud, Leilakuh |
| IRAN 1776 | *Lecanicillium aphanocladii* | *Pulvinaria aurantii* on *Citrus sinensis* | F. Sahbanian | Guilan, Langrud, Komele |
| IRAN 1777 | *Lecanicillium aphanocladii* | *Pulvinaria aurantii* on *Citrus sinensis* | F. Sahbanian | Guilan, Langrud, Leilakuh |
| IRAN 1778 | *Lecanicillium aphanocladii* | *Pulvinaria aurantii* on *Citrus grandis* | F. Sahbanian | Guilan, Malat |
| IRAN 1779 | *Lecanicillium aphanocladii* | *Pulvinaria aurantii* on *Citrus sinensis* | F. Sahbanian | Mazandaran, Ramsar |
| IRAN 1770 | *Lecanicillium aphanocladii* | *Pulvinaria aurantii* on *Citrus sinensis* | F. Sahbanian | Mazandaran, Chalus |
| IRAN 1764 | *Lecanicillium aphanocladii* | *Pulvinaria aurantii* on *Citrus sinensis* | F. Sahbanian | Guilan, Chaboksar |
| IRAN 1765 | *Lecanicillium aphanocladii* | *Pulvinaria aurantii* on *Citrus aurantium* | F. Sahbanian | Guilan, Chaboksar |
| IRAN 1767 | *Lecanicillium aphanocladii* | *Pulvinaria aurantii* on *Citrus nobilis* var. *unshiu* | F. Sahbanian | Guilan, Chaboksar |

(continued)

**Table 11.1** (continued)

| Accession N. | Taxonomic Name | Substrate | Isolated By | Region and place |
|---|---|---|---|---|
| IRAN 1030 C | *Lecanicillium aphanocladii* | Aphids in glass-house, vectors of viruses | A.-H. Mohammadi | Shiraz-College of Agriculture |
| IRAN 1032 C | *Lecanicillium aphanocladii* | Aphids in glass-house, vectors of viruses | A.-H. Mohammadi | Shiraz-College of Agriculture |
| IRAN 1787 | *Lecanicillium* cf. *dimorphum* (J. D. Chen) Zare & W. Gams | *Bemisia tabaci* | T. Safari | Hamedan |
| IRAN 1649 | *Akanthomyces lecanii* (Zimm.) Spatafora, Kepler & B. Shrestha (= *Lecanicillium* cf. *lecanii*) | *Pulvinaria floccifera* | Naeim Amini | Mazandaran, Tonekabon, Balaband |
| IRAN 1822 | *Akanthomyces lecanii* | Scale insect | S. Aghajanzadeh | Mazandaran, Ramsar |
| IRAN 1825 | *Akanthomyces lecanii* | Scale insect | S. Aghajanzadeh | Mazandaran, Ramsar |
| IRAN 1826 | *Akanthomyces lecanii* | Scale insect | S. Aghajanzadeh | Mazandaran, Ramsar |
| IRAN 1827 | *Akanthomyces lecanii* | Scale insect | S. Aghajanzadeh | Mazandaran, Ramsar |
| IRAN 1828 | *Akanthomyces lecanii* | Scale insect | S. Aghajanzadeh | Mazandaran, Ramsar |
| IRAN 1650 | *Akanthomyces muscarius* (Petch) Spatafora, Kepler & B. Shrestha (= *Lecanicillium* cf. *muscarium*) | *Pulvinaria floccifera* | Naeim Amini | Mazandaran, Tonekabon, Balaband |
| IRAN 1766 | *Akanthomyces muscarius* | *Pulvinaria aurantii* on *Citrus aurantium* | F. Sahbanian | Guilan, Langrud |
| IRAN 1768 | *Akanthomyces muscarius* | *Pulvinaria aurantii* on *Citrus sinensis* | F. Sahbanian | Guilan, Katalem |
| IRAN 1769 | *Akanthomyces muscarius* | *Pulvinaria aurantii* on *Citrus sinensis* | F. Sahbanian | Guilan, Langrud |
| IRAN 1824 | *Akanthomyces muscarius* | Scale insect | S. Aghajanzadeh | Mazandaran, Ramsar |
| IRAN 1163 C | *Akanthomyces muscarius* | *Hartigia trimaculata* (Hymenoptera: Cephidae) | M. M. Aminaei | Kerman |
| IRAN 1352 | *Akanthomyces muscarius* | *Zeuzera pyrina* on *Juglans regia* | M. M. Aminaee | Kerman |

(continued)

**Table 11.1** (continued)

| Accession N. | Taxonomic Name | Substrate | Isolated By | Region and place |
|---|---|---|---|---|
| IRAN 1774 | *Akanthomyces muscarius* | *Pulvinaria aurantii* on *Citrus aurantium* | F. Sahbanian | Guilan, Kalachai |
| IRAN 1771 | *Akanthomyces muscarius* | *Pulvinaria aurantii* on *Citrus sinensis* | F. Sahbanian | Guilan, Rahimabad |
| IRAN 1772 | *Akanthomyces muscarius* | *Pulvinaria aurantii* on *Citrus sinensis* | F. Sahbanian | Guilan, Langrud, Leilakuh |
| IRAN 1752 | *Akanthomyces muscarius* | *Pulvinaria aurantii* on *Citrus aurantium* | F. Sahbanian | Guilan, Lisserud |
| IRAN 1753 | *Akanthomyces muscarius* | *Pulvinaria aurantii* on *Citrus aurantium* | F. Sahbanian | Guilan, Liasar |
| IRAN 1755 | *Akanthomyces muscarius* | *Pulvinaria aurantii* on *Citrus sinensis* | F. Sahbanian | Mazandaran, Abbasabad |
| IRAN 250 C | *Akanthomyces muscarius* | *Pulvinaria aurantii* | Mirabzadeh | – |
| IRAN 1157 C | *Lecanicillium psalliotae* (Treschow) Zare & W. Gams | *Eurygaster integriceps* | D. Zafari | Lorestan, Kuhdasht |
| IRAN 437 C | *Metarhizium anisopliae s.l.* (Metschn.) Sorokīn | *Chilo suppressalis* | – | Rasht |
| IRAN 715 C | *Metarhizium anisopliae s.l.* | Locust | A. Dehghan | Ahwaz |
| IRAN 1018 C | *Metarhizium anisopliae s.l.* | *Parandra caspia* (Coleoptera) | M. Ghazavi | Mazandaran – Nour Forest |
| IRAN 2252 | *Metarhizium anisopliae s.l.* | *Galleria mellonella* on *Quercus* sp. | M. Nori | Ilam, Sarable, Ghalaje |
| IRAN 1020 C | *Metarhizium rileyi* (Farl.) Kepler, S. A. Rehner & Humber (= *Nomuraea rileyi*) | Lasiocampidae | S. Zanganeh & M. Ghazavi | Mazandaran – Haraz road, 15 km to Amol |
| IRAN 2039 | *Paecilomyces* cf. *cinnamomeus* (Petch) Samson & W. Gams | Dead larvae of *Ospheranteria coerulescens* | H. Mahmoudi & M. R. Mirzaee | Iran |
| IRAN 1026 C | *Purpureocillium lilacinum* (Thom) Luangsa-ard, Houbraken, Hywel-Jones & Samson (= *Paecilomyces lilacinus*) | *Lymantria dispar* | A. Abaei | Iran |

(continued)

**Table 11.1** (continued)

| Accession N. | Taxonomic Name | Substrate | Isolated By | Region and place |
|---|---|---|---|---|
| IRAN 1019 C | *Cordyceps tenuipes* (Peck) Kepler, B. Shrestha & Spatafora (= *Paecilomyces tenuipes*) | *Lymantria dispar* | S. Zanganeh & M. Ghazavi | Kordkoy |
| IRAN 2237 | *Mucor circinelloides* Tiegh. (= *Rhizomucor regularior*) | *Acythopeus curvirostris persicus* (adult) | N. Sepasi | South Khorasan, Birjand, Mohammadieh |
| IRAN 1274 C | *Verticillium* sp. | Cuticle of *Leptoglossus occidentalis* Heidemann, hibernating adult specimen in a pine wood | Marta Mossenta | – |
| IRAN 2189 | *Verticillium epiphytum* Hansf. | Larvae of *Zeuzera pyrina* | Baradaran | Kerman, Rabar |
| IRAN 2191 | *Verticillium epiphytum* | *Thrips tabaci* | T. Safari | Mashhad |
| IRAN 2192 | *Verticillium epiphytum* | *Thrips tabaci* | T. Safari | Mashhad |

Source: Database of IRAN WDCM 939 in World Federation of Culture Collections: WFCC

mildew, followed by production of blastospores and massive invasion of the aphid hemolymph and other internal tissues, or suppressing growth and spore production of *S. fuliginea*. Pathogenesis is caused by direct assimilation of nutrients and accumulation of lipids. Finally, the fungus, when environmental conditions are permissive, produces conidiophores pushed through the host barrier and release of conidia from host cadavers or dead cells of the plant pathogen. One typical feature of the developmental process of *V. lecanii* is the overlap of certain stages of infection. For example, under favorable environmental conditions, the fungus may proliferate simultaneously both on the surface and/or within the host body. As a consequence, sporulation may originate from conidiophores emerging from the host barrier or from those produced by hyphae developing on the host.

Askary et al. (2008), Benhamou and Brodeur (2000, 2001) and Benhamou (2004) showed that *Akanthomyces muscarius* (= *Lecanicillium muscarium*) (Ascomycota: Hypocreales) is an opportunistic antagonist for some plant pathogenic fungi such as *Penicillium digitatum* (Ascomycota: Eurotiales) and *Pythium ultimatum* (Oomycota: Peronosporales). Results of their studies provide that antagonism, triggered *A. muscarius*, is a multifaceted process in which antibiosis, with alteration of the host hyphae prior to contact with the antagonist, appears to be the key process in the antagonism against the mentioned plant pathogenic fungi.

## 11.3   Development of Entomopathogenic Fungi in Iran

Previous research activity on entomopathogenic fungi in Iran occurred 50 years ago, mostly concerning bioassays of some insect fungi with different hosts. Since that early work, more researchers and students have conducted studies on insect pathogenic fungi. But to date, except the introduction of different fungal strains. We still cannot claim registration of any commercial product from native strains of entomopathogenic fungi.

There are many reasons that have prevented the development of entomopathogenic fungi products. Iran is a vast country with different types of climate: wet and mild on the coast of the Caspian Sea, continental and arid in the plateau, cold in high mountains, desert and hot on the southern coast and the southeast. However, this climate variation can be a cause for genetic variation in microorganisms, but because of lower humidity, the occurrence of naturally epidemic entomopathogenic fungi and their applied use is inhibited.

- Most regions of the country are sunny, especially from mid-spring, summer and until mid-autumn. Also, ultraviolet radiation is high in most areas which situation is destructive for fungal propagules.
- There is a perception among farmers that compared to conventional chemical products, biologicals are not as fast acting, lose their effectiveness more rapidly, have a narrower host spectrum and require more knowledge to use effectively (St. Leger and Screen 2001).
- In a general view, few joint-type researches have been attended by a group of entomologist, microbiologist and chemist to develop the entomopathogenic fungal products. In addition, the relationship between researchers and businesses has been poor for commercialization of products.

## 11.4   Application of Entomopathogenic Fungi with Other Natural Enemies

In most cases, successful biological control of sucking insect pests, especially in greenhouses, needs the combined use of parasitoids, predators and/or fungal pathogens (Mesquita and Lacey 2001). In this complex, both deleterious and beneficial aspects in host-natural enemies' relation should be taken into account to recognize the efficiency of biological control agents (Hochberg and Lawton 1990; Memmott et al. 2000). Therefore, studies on their relationships can provide basic knowledge in integrated pest management.

Overall, many studies have shown that EPFs with broad host ranges can interact antagonistically with arthropod natural enemies either under environmental conditions that are optimal for the fungus or field condition. Laboratory studies often accentuate the adverse effect of a fungus on a beneficial insect while in nature, ecological barriers can minimize the exposure to fungus (James and Lighthart 1994; Lacey et al. 1997; Askary and Brodeur 1999; Jaronski et al., 2004). Also, deleterious effects of insect pathogenic fungi have various effect on predators and parasitoids (Ashouri et al. 2004; Danfa and Van der Valk 1999; Hall 1981). For example, exposure of parasitic wasp, *Bracon hebetor* (Hymenoptera: Braconidae) and *Apoanagyrus lopezi* (Hymenoptera: Encyrtidae) to *Metarhizium anisopliae* (Ascomycota: Hypocreales) and *B. bassiana*, resulted in 100% mortality of the parasitoids. However, no infection was observed in the tenebrionid, *Pimelia senegalensis* (Coleoptera: Tenebrionidae) and *Trachyderma hispida* (Coleoptera: Tenebrionidae), exposed to high doses of the same isolates (Danfa and Van der Valk 1999). There are a number of papers since the mid-2000s that indicate behavioral barriers, whereby a parasitoid can detect mycosis and not oviposit. Deleterious effect of EPFs on natural enemies especially on parasitoids depends strongly to fungal isolate and its concentration, inoculation time of host and parasitoid, life stage of host and parasitoid and their behaviors (Askary and Brodeur 1999). Brooks (1993) listed parasitoids that are unable to complete their development due to premature deaths by fungus-infection.

The studies of Askary and Ajam Hassani (2009) on *A. muscarius* (= *V. lecanii* isolate DAOM 198499) and *Aphidius nigripes* (Hymenoptera: Aphidiidae) on potato aphid, *M. euphorbiae* clarified some of these host-parasitoid-pathogen relations. They showed that both *A. nigripes* males and females were infected by high concentrations of *A. muscarius*, however, females were three times less susceptible than males. Adult wasps were also infected by fungus spores of inoculated plants; however, the mortality of parasitoid was less than that of direct infection (sprayed on adults). However, development of fungus mycelium was observed on mummies, but mortality of parasitoid pupae was significantly lower than its larval stages (Askary et al. 2007; Askary and Ajam Hassani 2009).

Results demonstrated that there was no transmission of fungus from infected to healthy individuals during mating. Also, survival of healthy females, parasitism rate and sex ratio of progeny are same of control, when infected males mated with healthy females. Whereas, survival, the ability to mating and parasitism rate of infected females decrease when a heavy concentration of fungi infects them. The sex ratio and other biological traits of progeny were normal (Askary and Ajam Hassani 2009).

In order to determine the possible role of *Myzus persicae* (Homoptera: Aphididae) treated with *B. bassiana* on patch time allocation of the parasitoid wasp, *Aphidius matricariae* (Hymenoptera: Aphidiidae), Rashki et al. (2009) found that there was no significant difference in patch leaving time of the parasitoid, either in presence or absence of fungus. Interaction between presence of eggplant volatile and fungus did not have effect on patch time allocation. Moreover, they observed *A. matricariae* has significant effect on transmission of *B. bassiana* from *M. persicae* sporulating cadavers to non-treated aphids.

Fazeli-Dinan et al. (2016b) investigated virulence of three isolates of *A. dipterigenus* (LRC190, LRC216 and LRC229) on third-instar nymphs of greenhouse whitefly, *Trialeurodes vaporariorum* (Hemiptera: Aleyrodidae), as well the virulence of the most effective isolate, LRC216 on its parasitoid, *Encarsia formosa* (Hymenoptera: Aphelinidae). The results indicated that the larval stage of the parasitoid was more susceptible than the pupa to fungus application; the adult parasitoid had low susceptibility to fungus treatment. Also they found the parasitoid displayed a preference for the healthy whitefly nymphs. So a negative effect of simultaneous application of fungus and parasitoid within integrated pest management for *T. vaporariorum* could be reduced (Fazeli-Dinan et al. 2016a).

In another survey, Jarrahi and Safavi (2016) examined effect of two native isolates of *M. anisopliae* M14 and DEMIOO1 on biological parameters of parasitoid wasp, *Habrobracon hebetor* (Hymenoptera: Braconidae) attacking larval stage of cotton bollworm, *Helicoverpa armigera* (Lepidoptera: Noctuidae). Bioassay experiment revealed that M14 isolate caused higher mortality than DEMIOO1, but its sub-lethal effect on the intrinsic rate of population growth ($r_m$), finite rate of increase ($\lambda$) and net reproductive rate ($R_0$) weren't affected by this fungal isolate. Therefore, M14 isolate had no negative effect on biological parameters of *H. hebetor* and can be used within integrated pest management program.

These results could be useful for design the management programs for susceptible pests to fungi, especially for pests in greenhouses. Due to an increasing trend of greenhouse areas in most provinces of the country, which have reached to about 15,000 ha, the combined use of EPFs with other agents will be a promising manner for greenhouse management.

## 11.5  Researches on Entomopathogenic Fungi on Crops

### 11.5.1  Cereal (Wheat) Pests

#### 11.5.1.1  Sunn Pest

Sunn pest, *Eurygaster integriceps* (Hemiptera: Scutelleridae) is the key pest of cereals in Central and West Asia and East Europe (Javahery 1995). During spring and early summer, *E. integriceps* attack cereals especially wheat by feeding on leaves, stems and grains, reducing yield and injecting a toxin into the grains which reduce the quality of flour (Hariri et al. 2000; Brown 1965).

In the absence of control measures, infestations can lead to 100% crop loss (Javahery 1995; Amir-Maafi et al. 2007). During spring and early summer, the pest feeds on wheat and barley in the fields and then migrates to the hills and mountains for the rest of the year which is called overwintering period (Brown 1965). The Sunn pest spends a dormant period of some nine months at overwintering sites, under some plants such as *Astragalus* sp., *Artemisia* sp., *Acantholimon bracteatum*, *Quercus* sp., *Amygdalus* sp. and *Crataegus aronia*. After hibernation, adult migrates again to farms at early spring, feed, mate and lay eggs on cereal leaves and weeds (Amir-Maafi et al. 2007; Schuh and Slater 1995).

Control of Sunn pest is mainly based on chemical control in farms which is widely used in different countries. In recent years, chemical control of Sunn pest spanned more than 1,500,000 ha in Iran. Chemical insecticides have negatively affected many beneficial insects, wildlife, waters and causing even more harm to humans (Javahery 2004). In general, a few parasitoids, predators and pathogens are known as natural enemies of Sunn pest. However, the scelionid wasp, *Trissolcus grandis* (Hymenoptera: Scelionidae), has shown considerable promise due to its ability to regulate Sunn pest populations in farm (Amir-Maafi 2000; Amir-Maafi and Parker 2002, 2003). The entomopathogenic fungi have demonstrated a favorable potential to affect this pest when other natural enemies are not present especially during its diapause in overwintering sites (El-Bouhssini et al. 2004; Nouri-Aiin et al. 2014). Parker et al. (2000, 2003) collected EPFs on Sunn pest from some parts of Syria, Turkey, Iran, Uzbekistan, Kazakhstan, Kyrgyz Republic and Russia. The most common species was *B. bassiana*, followed by other species such as *Verticillium* and *Paecilomyces*. Skinner et al. (2007) obtained more than 220 isolates of fungi from the soil of hibernating sites of Sunn pest in Russia, East and central part of Asia. In addition, Kazemi Yazdi et al. (2011) and Nouri-Aiin et al. (2014) isolated more than 170 strains of different fungal species from overwintering sites of Sunn pest in various regions of Iran. Some of these isolates preserve in the fungal culture collection of IRIPP. In other studies, Parsi and Mohammadipour (2017) collected different isolates of *B. bassiana* from overwintering areas in the center of Iran. They collected 510 soil samples and cadavers from overwintering sites of Sunn pest in Semnan, Qom, Kerman and Markazi provinces in three elevation classes (1900, 2100 and 2300 m), in late summer and winter 2015 and 2016. Their results indicate that from 250 samples in winter, 50.4% and from 260 samples in summer, 17% had *B. bassiana* spores. The result showed more isolates were found at 2300 meters elevation although some were present at 1900 to 2100 meters.

Furthermore, *B. bassiana* and *M. anisopliae* have been subjected to bioassays on different developmental stages of Sunn pest. Rastegar et al. (2006) evaluated the virulence of eight Iranian isolates of *B. bassiana* and also the American GHA isolate on adult Sunn pest by immersion of insects in spore suspensions. The lowest LC50 was $3.78 \times 10^3$ spores ml$^{-1}$ and lowest LT50 was 8.55 days with isolate DEBI002 and the highest LC50 ($5.06 \times 10^5$ spore ml$^{-1}$) and LT50 (17.96 days) recorded for DEBI008, respectively. In another work, Bandani and Esmailpour (2006) proved that an oil formulation of *B. bassiana* against the Sunn pest increased fungal virulence. Oil formulation enhances the adhesion of spore to the insect cuticle, especially epicuticle, through hydrophobic interaction between the spore and cuticle surface.

A study was conducted by Tork et al. (2008) to determine the virulence and enzyme production of two isolates of *M. anisopliae* (4556 and M187) on Sunn pest. Comparison of the LC50s of these isolates showed that the 4556 isolate, with LC50 of $3.38 \times 10^5$ spores ml$^{-1}$, was more virulent than isolate M189, with an LC50 of $7.70 \times 10^5$ spores ml$^{-1}$. The assays also demonstrated that enzyme production in the 4556 isolate was more than M198. Therefore, there may be a direct relationship between enzyme production and virulence, at least with these two isolates. Also,

**Fig. 11.1** Infected Sunn pest by entomopathogenic fungi, *Beauveria bassiana* (Haji Allahverdi Pour et al. 2008)

Haji Allahverdi Pour et al. (2008) evaluated the virulence of four native isolates and two foreign isolates of *B. bassiana* on fifth instar nymphs and adults of *E. integriceps* using dipping and topical, micro-application techniques, respectively. Nymphs were highly susceptible to the isolates without a significant difference among the isolates. For adults, however, the native DEBI 002 isolate showed the lowest LD50 value among the others (Fig. 11.1).

Another aspect of entomopathogenic fungi is their ability for horizontal transmission after the initial treatment, epizootic spread. The first such study was designed by Talaei-Hassanloui et al. (2009). They reported no significant difference among the *E. integriceps* adults in different treatment (inoculated males + non-inoculated males and treatment with inoculated females + non-inoculated males). Due to the mating behavior, this kind of horizontal transmission did not evidently occur. Among the five body-part treatments, however, there was a significant difference among adult mortality, ranging from 16.6% to 48.9%, with the lowest mortality for pronotal application of spores and the highest mortality for total body treatment. This trait could be useful for possible auto-dissemination programs.

Another important issue in using fungi to control insect pests is their ready mass production on natural media. Roshandel et al. (2013) evaluated the effect of culture substrate on virulence of *M. anisopliae* conidia and blastospores against Sunn pest in 2011 and 2012. There were significant differences between virulence of both propagule types, conidia and blastospores produced in different media and also there was a significant difference between different developmental stages. For blastospores, the lowest LT50 recorded (2.33 days) on the wheat bran + yeast extract medium was on second instar nymphs and the highest LT50 recorded (10.18 days) on the wheat bran + rice bran extract on fifth instar nymph. For conidia, the lowest

LT50 recorded 4.68 days on the rice bran extract for second instar nymph and the top LT50 (13.13 days) was recorded on rice grain versus fourth instar nymphs. Susceptibility of overwintered adults to conidia and blastospores of *M. anisopliae* was greater than the aestivated Sunn pest adult.

Bioassays clearly indicate that *E. integriceps* is susceptible to infection by *M. anisopliae s.l.* and the nymphal stages more susceptible than the adults (Sedighi et al. 2014). Variable susceptibility of different nymphal stages could be attributed to cuticle structure and molting, but this relationship needs elucidation. Ghamari Zare et al. (2014) showed significant differences between diapause and summer adult susceptibility to fungi. Diapausing populations were more susceptible than summer populations, presumably because of reduced food storage and reduced fat body.

Large scale application of *B. bassiana* to overwintering sites in winter (before the immigration of adults) showed high fungal epizooty among Sunn pest populations especially in the year following treatment (Askary, unpublished data) (Fig. 11.2). On the basis of results, finding fungal isolates for biological control of Sunn pest in overwintering sites or farms is promising.

The entomopathogenic fungi have considerable potential for managing the Sunn pest in wheat fields, in a general sense, but the suitability of these agents for using in wheat crop lands is questionable. Most of the wheat fields are dry farming, which situation needs special formulations. An additional issue conservation of fungi in hibernating sites of the Sunn pest. Maintaining plant diversity in those sites and avoiding any disruption should increase the fungal prevalence, which can have a significant role in regulating the Sunn pest density.

**Fig. 11.2** Infected Sunn pest by entomopathogenic fungi in hibernate site (Askary, H)

## 11.5.1.2  Locusts and Grasshoppers

Locusts and grasshoppers are known to be one of the important phytophagous insects in different agricultural, forests and rangeland ecosystems of Iran. *Dericorys albidula* (Orthoptera: Dericorythidae) is an important monophagous pest of saxoul trees, *Haloxylon* sp. (Chenopodiaceae) (Adeli and Abaei 1989; Moniri 1998). *Haloxylons* spp. are shrubs found or planted in the desert regions of Iran. In the past two decades, most outbreaks of *D. albidula* grasshopper likely developed mainly due to the expansion of cultivated *Haloxylon* trees. The main control strategies for *D. albidula* are based on the use of chemical insecticides or mechanical methods. But *Metarhizium acridum* (as Green Muscle® which is now licensed by and produced by Elephant Vert in Morrocco, along with their own strain, as Novacrid) which applied at the rate of 50 g ha$^{-1}$ to control *D. albidula* nymphs on *Haloxylon* in different regions of Isfahan and Qom Province by Farsi et al. (2009). Experiments were conducted on *Haloxylon* plantation in which trial conidia were suspended in diesel fuel and sprayed on nymphs by ULV and electrostatic sprayers. Plants were subsequently enclosed in cages. Results showed that this product can control successfully *D. albidula* nymphs by spraying methods in low relative humidity (less than 30% RH). Field application of Green Muscle® on *D. albidula* in *Halocxylon* plantation was successful in controlling pest populations on a large scale (Fig. 11.3). The other important grasshopper is *Esfandiara obesa* (Orthoptera: Acrididae) that feeds on oak in the southwest, Khuzestan Province. Oak trees are the most important species in Zagros Mountains from North-West to South-West of the country. In recent years, *E. obesa* has caused heavy damage on oak by feeding on the leaves. Despite considerable damage by *E. obesa*, there was no effective method for

**Fig. 11.3**  *Dericorys albidula* infected by *Metarhizium anisopliae* (Askary, H)

its control. The results of field observations on *E. obesa* showed that Green Muscle®
was more effective than other treatments when it had been sprayed on oak trunks or
foliage, where the nymphs feed. Dimilin® with oil carrier was the second effective
product (with 70–83.8% mortality) whereas, *B. bassiana* did not cause any considerable mortality on the nymphal stage. It seems that Green Muscle® with oil carrier
can be applied within management program of this pest.

Furthermore, one of the important aspects of EPFs is their impact on immune
defenses of their host. Ghazavi et al. (2004) evaluated the cellular immune response
of *Locusta migratoria* (Orthoptera: Acrididae) to Fashand isolate of *B. bassiana*.
Their results indicated that the total hemocyte count (THC) and differential hemocyte count (DHC) increased during the first three days of infection but subsequently
decreased. The phagocytic potential of blood cells decreased with the development
of the fungus. Although the hemocytes were able to phagocytize the aerial conidia
injected to insects' hemocoel at the beginning of infection and blastospores produced
in the course of fungal proliferation, these fungal elements were not killed and after
germination or multiplication, they destroyed the cells, in which they were enclosed
and colonized the host hemocoel.

One of the important criteria in applying fungi is their persistence post application. Bagheri and Tajvand (2006) designed an experiment to control *E. obesa* with
*M. acridum* with application to foliage. There were significant differences among
different time intervals, including immediately after application, one day after
application, three days after application, seven days after application, ten days
after application, and with no application as a control treatment. They also addressed
the effect of Green Muscle® on survival and feeding behavior of praying mantis,
*Mantis religiosa* (Mantodea: Mantidae), the most important natural enemy for
*E. obesa* in Khuzestan province. Mortality of mantis was not observed in any
treatment but also mantid nymphs continued to feed and molt, however there was
a decrease in feeding behavior, from 78% to 54% (Bagheri and Tajvand 2008).

Finally, another important and harmful pest of field crops in Sistan and
Blouchistan province is the sugarcane grasshopper *Chrotogonus trachypterus*
(Orthoptera: Pyrgomorphidae). This insect has high populations in this province
and is one of the most important pest in early stages of wheat growth and other crops,
especially in seedling stage. Mirshekar et al. (2004) evaluated effects of two native
isolates of *B. bassiana* and *M. anisopliae* and also Green Muscle® on
*C. trachypterus*. Comparison between LD50s of the isolates indicated that native
isolate of *M. anisopliae* had the lowest LD50 and was more effective than others.
Also, native isolate of *B. bassiana* was more virulent than a commercialized
formulation of *M. anisopliae*, also Green Muscle®.

*Trichoderma harzianum* (Ascomycota: Hypocreales) is a fungus which also has
an important role in pest management. Hamzehei et al. (2018) reported the pathogenicity of native isolate of *T. harzianum* on the grasshopper, *C. trachypterus* under
laboratory conditions with a calculated LC50 of $1.01 \times 10^6$ conidia ml$^{-1}$.

### 11.5.1.3 Aphids

Several species of aphid, including *Schizaphis graminum* and *Rhopalosiphum padi* (Hemiptera: Aphididae), are the most abundant species in the wheat fields (Shayesteh et al. 2015; Karami et al. 2016). Also, Russian wheat aphid, *Diuraphis noxia* (Hemiptera: Aphididae), has been reported as an endemic pest in the majority of the wheat producing areas of Iran (Rezwani 2010). Aphids are harmful in wheat due to their effect on the grain protein by feeding on wheat and injecting toxic enzymes into plants during feeding (Shayesteh et al. 2015). Application of EPFs can be an ideal management tool to reduce the spread and incidence of these group of pests because of inadequate efficiency of the parasitoids on aphids in cereal fields. Also, exploring different aspects are recommended to evaluate the fungus application in field conditions.

Initially, under the laboratory conditions, efficacy of Iranian isolates of *M. anisopliae*, *A. muscarius*, *Lecanicillium aphanocladii* (Ascomycota: Hypocreales) and *B. bassiana* against *D. noxia* were demonstrated by Mohammadipour et al. (2009). Several surveys were also conducted to evaluate the effect of different isolates of EPF on bird cherry oat aphid, *R. padi* in different conditions. The first such study (Sedighi et al. 2012) examined the effects of seven Iranian isolates of *B. bassiana* (DEBI001, DEBI002, DEBI003, DEBI004, DEBI008, DEBI010, DEBI015,) on the adult *R. padi*, under laboratory conditions. The lowest and the highest LC50s were 0.059 and 162.28 spores ml$^{-1}$, respectively, and the shortest and longest LT50s were 2.08 and 4.57 days, respectively, with DEBI001 and DEBI002 isolates.

Fadayivatan et al. (2014) found significantly different susceptibilities between two cereal aphids, *Sipha maydis* and *Metopolophium dirhudum* (Heteroptera: Aphididae) to *A. dipterigenus* isolate LRC 190 under greenhouse conditions. The results indicated that both aphid populations were significantly decreased, and the estimated LC50 and other parameters indicated that *A. dipterigenus* was more virulent to *S. maydis* than to *M. dirhudum*.

Recently, Mossavi et al. (2016) evaluated the effect of different isolates of *B. bassiana* including IRAN429C, IRAN108 and LRC127 on the life table parameters of adult *S. graminum* under laboratory conditions. Adult death percentages had significant difference among all isolates; IRAN429C caused the highest death rate of adults. As expected, with increasing the fungal concentration, death rate increased. Life expectancy ($e_x$) in adults for control, IRAN429C and IRAN108 isolate were 28.25, 13.65 and 17.1 day, respectively. Net reproductive rate ($R_0$) among aphids in the IRAN429C and IRAN108 treatments were 19.9 and 25.5 (females/female/ generation), which decreased significantly compared to the control (61.6). There is no difference among the finite rate of increase ($\lambda$) for those two isolates of fungus.

Moreover, Ebadollahi et al. (2017) studied *A. muscarius* effects on the melon aphid, *Aphis gossypii* (Hemiptera: Aphididae) under laboratory conditions. The pathogenic fungus and both essential oils had useful toxicity against *A. gossypii*. Aphid mortality also increased when the essential oils were combined with

*A. muscarius*, although the phenomena was additive rather than synergistic. Mycelial growth inhibition of *A. muscarius* exposed to the essential oils was also very low.

The overall data on efficacy of fungi for aphids shows their potential, but aphid damage on wheat and other cereals rendeers them less important as major pests. Here, it is important to consider natural infection of aphids by Entomophthoromycota and also some hypocrealean fungi, and improve the environment for the conservation.

## 11.5.2   Rice Pests

Rice (*Oryza sativa* L.) is one of the most important agricultural crops in the north of Iran (Lotfalizadeh et al. 2016). Within four decades (from 1980 to 2010), Iran has changed from the leading rice importing country to the eleventh rice producer of the world (Poor Amiri et al. 2017). Some main important rice pests are *Chilo suppressalis* (Lepidoptera: Pyralidae) and *Naranga aenescens* (Lepidoptera: Noctuidae). These pests highly damage this strategic crop annually (Lotfalizadeh et al. 2016). Chemical insecticides have been the most common tools for management of these two rice pests. The integrated pest management (IPM) concept proposes the use of environmentally friendly strategies to decrease the rapid development of pesticide resistance by these pests and also combat with the environmental pollution, human health risks, and negative affect of pesticides beneficial organisms (Poor Amiri et al. 2017).

Several surveys have been conducted to determine the effect of EPFs on rice pests in Iran. Initially, laboratory evaluation of pathogenicity of six *B. bassiana* isolates on the striped stem borer, *C. suppressalis* had been done by Majidi-Shilsar et al. (2004). Among the isolates, there were significant differences and the Mcb18 isolate caused greater mortality than other isolates. Subsequently, Majidi-Shilsar et al. (2008) tested combined effect of this fungus isolate and diazinon on the pest in rice fields. The trials were carried out during 2003 and 2004 in Rasht, Guilan province. Not only did the *B. bassiana* cause mortality in the first generation of *C. suppressalis* but also in the second generation. Evidently, spores of *B. bassiana* could persist in rice and cause mortality of *C. suppressali* larvae. On the other hand, diazinon was effective on just the first generation.

The biochemical assessment and the virulence of *B. bassiana*, *M. anisopliae*, *Isaria fumosorosea* (Ascomycota: Hypocreales) and *A. lecanii* on the larvae of *C. suppressalis* were determined by Ramzi and Zibaee (2014). They recorded the highest amounts of total protein and hydrophobin for the isolate BB3 (*B. bassiana*) and *I. fumosorosea*. The *M anisopliae* isolates and BB3 showed the highest activities of chitinases and lipases. The fungi presumably had different virulence for the larvae by producing extracellular enzymes and adhering protein. Overall, they concluded that larvae of *C. suppressalis* was more susceptible to BB2 isolate. To investigate the effect of fungi on the immune system of the larvae, Zibaee and Malagoli (2014) determined fluctuation in cell populations, nodule formation and phenoloxidase activity at 1, 3, 6, 12, 24, 48 and 72 h after injection of different isolates of

*B. bassiana, M. anisopliae, I. fumosorosea* and *A. lecanii.* Within 6 h spores of *B. bassiana* had the highest effects on the number of circulating hemocytes, although the highest number of granulocytes and plasmatocytes were observed 3–6 h post injection. The *B. bassiana* induced the highest increase in phenoloxidase activity between 1 and 12 h post injection. Finally, they detected antimicrobial peptide activity in the hemolymph of larvae.

Virulence of ten fungal isolates consisting of eight Iranian isolates and two foreign isolates of *B. bassiana* on *Ostrinia nubilalis* (Lepidoptera: Pyralidae) larvae was determined by Safavi et al. (2010). Their BEH isolate had the greatest mortality, a mean of 57.7%. Other isolates, such as EVIN I and DEBI007, were scored lower, with 42.7% and 53.4% percent mortality, respectively.

More recently, Fazeli-Dinan et al. (2012) investigated the pathogenicity of four Iranian *B. bassiana* isolates (DEBI001, DEBI003, DEBI007 and DEBI008) on rice green semi looper (*N. aenescens*) using two methods, dipping and spraying. In dipping method, DEBI003 isolate at concentration of $1 \times 10^7$ (spores $ml^{-1}$) caused the highest rate of mortality, at 50.8%, compared to other treatments. In spraying, DEBI003 isolate had the most impact on *N. aenescens* larvae with a 41.3% mortality. So with both methods, this isolate showed the greatest efficacy. With older larval instars, mortality decreased; the LC50 of DEBI003 on second instar larvae was $8.1 \times 10^7$ spores $ml^{-1}$. Due to importance of *N. aenescens* on rice where its larval feeding causes the reduction of photosynthesis levels and thereby reducing plant efficiency, Aboutalebian et al. (2017) evaluated the mortality of *N. aenescens* at different larval instars with *Bacillus thuringiensis* var. *kurstaki, B. bassiana* and *M. anisopliae* and control (water) in field conditions on Tarom rice variety. Based on their result, *B. thuringiensis* and *B. bassiana* had high impact in controlling the pest in the field and can be used within pest management program.

Overall, fungal application in rice is promising due to suitable conditions namely humidity. Moreover, the rice farmers are familiar with biocontrol because most of them have experience in application of *Trichogramma* as an initial biocontrol. Based on recent national standards for organic farming, appropriate formulations and application methods could be potential research line for this area.

## 11.5.3   Beet Pests

The sugar beet armyworm, *Spodoptera exigua* (Lepidoptera: Noctuidae), is a major pest of many agricultural crops in tropical and semitropical areas of the world. The intensive feeding of larvae on leaves can cause significant yield loss and defoliating small plants and frequently causing economic damage (Talaee et al. 2017; Darsouei et al. 2018).

Ajamhassani (2014) investigated cellular immune response of fourth-instar larvae of *S. exigua* to two isolates of *B. bassiana* (Fashand and 566). Total number of plasmatocytes, granulocytes and vermicytes reached a maximum in 3–6 h post injection. Then total hemocyte decreased gradually. Prohemocytes decreased within

3 h post injection. Plasmatocytes, granulocytes and vermicytes engulf foreign particles to form phagocytosis and nodule. Phagocytosis occurred within less than 1 h after injection, but nodulation happened in 3–6 h post injection. There was a correlation between phenoloxidase activity and total hemocyte count. Phenoloxidase activity was the highest in 3–6 h post injection.

In another work, for the first time Darsouei et al. (2018) reported *Beauveria varroae* (Ascomycota: Hypocreales) isolated from *S. exigua* through an extensive sampling during 2014–2015 in the Mashhad region, Khorasan Razavi province (Fig. 11.4). More research is needed to exploit fungi against this pest in future.

### 11.5.4   Cotton Pests

Cotton bollworm, *H. armigera,* is the most destructive pest on economically important plants such as cotton, tomato and chickpea in Iran and throughout the world. Applying insecticides with different mode of action and from different classes causes resistance in cotton bollworm and leads to reduction in the efficacy of some conventional insecticide groups such as pyrethroids, organophosphates and carbamates. Using different management tools such as microbial control is necessary for solving this resistance problem (Naseri et al. 2011; Bagheri et al. 2019).

The first such study, Javar and Kharazi-Pakdel (2004) assessed the pathogenicity of some native isolates of *B. bassiana,* including IRAN 440C, IRAN403C, IRAN 441C and KCF105, against third instar *H. armigera.* Because the mortality rate of IRAN 441C isolate was significantly higher in both tests and had high spore production, it was identified as the most effective one among those tested.

Because of the importance of sublethal effects caused by EPFs on pest population dynamics, Kalvandi et al. (2018) evaluated sublethal impact of *B. bassiana* DC7 isolate on cotton bollworm. Life table analysis indicated that *B. bassiana* had a negative effect on total developmental time by elongation of larval development period and reducing reproduction and egg laying duration in females. Also, this fungus has a great potential to reduce the fitness of *H. armigera* populations, thus it could be considered as a good candidate for biological control program of *H. armigera.*

### 11.5.5   Subterranean Termites

Subterranean termites are one of the most important pests of buildings, historic monuments and agricultural crops in some parts of Iran. These termites are cryptic insects and reside underground, inside trees and remain hidden which make their control very difficult. In agriculture, termites' attacks occur at different development stages of crops, particularly at seedling and maturity stages. In general, damage is

**Fig. 11.4** Scanning
Electron Micrographs of
spore of *Beauveria varroae*.
(**a**) Arrangement of conidia
on conidiogenous cell, (**b**)
Conidia germination, (**c**)
Appressorium and germ
tube. (Courtesy of Reyhaneh
Darsouei, Ferdowsi
University of Mashhad)

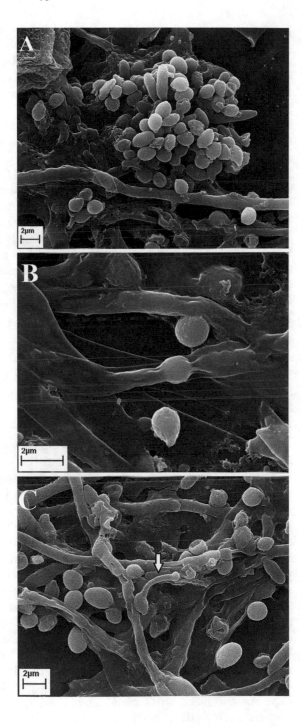

greater in rain-fed than irrigated crops and during dry periods than periods of regular rainfall so rainfall affect the termite's geographical distribution very well.

A number of physical, chemical and biological control measures are used to prevent termite attack. During a preliminary study, Rahimzadeh et al. (2006) confirmed the potency of four isolates of *M. anisopliae* to cause mortality on *Microcerotermes gabrielis* (Isoptera: Termitidae) under laboratory conditions. Later, Rahimzadeh et al. (2012) evaluated the pathogenicity of *M. anisopliae* (DEMI 001) isolated from *Rhynchophorus ferrugineus* (Coleoptera: Curculionidae) against two subterranean termites, *Amitermes vilis* (Isoptera: Termitidae) and *M. gabrielis* under laboratory conditions. The LC50 values for *A. vilis* and *M. gabrielis* were $8.5 \times 10^3$ and $0.2 \times 10^2$ spores ml$^{-1}$, respectively. According to the results of the bioassay, *M. anisopliae* was more effective for controlling *M. gabrielis* than that for *A. vilis*. Due to increasing concern of termites in urban habitats, there is increasing interest in environmentally friendly tools against termites including fungal insecticides. An information deficit concerns the behavior of most destructive Iranian termite species. The most important factor is the behavior of termite population when encountering a fungal pathogen. Based on this finding, an optimal application tactic could be designed, based on direct mortality and repellency of fungal pathogen.

## 11.6    Fruit Trees, Forests and Riparian Area Pests

Some main wood borer pests from Lepidoptera, (*Zeuzera pyrina*), Coleoptera (*Osphranteria coerulescens*, *Capnodis* spp., and species of Scolytinae) and Hymenoptera (*Synanthedon tabaniformis*) cause significant crop losses annually in the country, with increasing damage in the past two decades, mostly correlated to climate change issues and water crises. Application of chemical treatment for management of these pests has some restrictions, including overlapping of life stages, the issue of sufficient translocation of pesticides in infested trees and other known concerns which has led to a restriction of available tactics. Among biocontrol agents, entomopathogenic fungi are an accepted group for use against this group. Here we can recommend to design plans for incorporating EPFs within management program of each mentioned species of the wood borers. The main question is the best method for applying these agents, including injection a spore suspension, or a paste form by mixing the fungal conidia with waste molasses and sweet potato starch and smearing into excretion hold of larva (Huang et al. 1990). Alternatively the fungi could be applied using a fungus-impregnated cloth band around the trees (Hajek and Bauer 2007). These approaches could be effective in regulating population density of those pest groups along with pheromone trapping. Several studies have been conducted on efficiency of EPFs as biocontrol agents for the control of a number of important fruit tree pests (Table 11.2). In general, these studies have shown promising results, indicating that these fungi could developed for using within management program for some important pest in the country.

**Table 11.2** List of insect pest on fruit trees, forests and riparian area pests which considered as target of EPFs research in Iran

| Order | Family | Species | Cultivation | Susceptibility stage to EPF | EPN species (isolate) recommended | References |
|---|---|---|---|---|---|---|
| | Tortricidae | *Cydia pomonella* | Apple trees | Egg | *Metarhizium anisopliae* (M14) | Ghasemi et al. (Unpublished) |
| | Tortricidae | *Cydia pomonella* | Apple trees | Egg | *Beauveria bassiana* (Fashand isolate) | Ghasemi et al. (Unpublished) |
| | Cerambycidae | *Aeolesthes sarta* | Poplar (*Populous nigra*) | Larval stage | *Beauveria bassiana* (PTCC5197) | Farashiani et al. (2008) |
| | Arctiidae | *Hyphantria cunea* | Boxelder (*Acer negundo*), Mulberry (*Morus alba*) | – | *Beauveria bassiana* (ETU105) | Zibaee et al. (2013) |
| | Chrysomelidae | *Xanthogaleruca luteola* | *Ulmus* spp. | Adult | *Metarhizium anisopliae*, *Beauveria bassiana* | Ebrahimifar et al. (2016) |
| Coleoptera | Scarabaeidae | *Polyphylla adspersa* | – | Third instar | *Metarhizium reobertsii* (FUM04, FUM05) | Sargazi et al. (Unpublished) |
| Hemiptera | Coccidae | *Pulvinaria aurantii* | Citrus | First instar | *Verticillium lecanii* | Mirabzadeh et al. (2000b) |
| | Aphalaridae | *Agonoscena pistaciae* | Pistachio | Fifth instar | *Beauveria bassiana* (DEBI008) | Alizadeh et al. (2007) |
| | Coccidae | *Pulvinaria aurantii* | Citrus | First instar | *Acanthomyces muscarius* | Aghajanzadeh and Taheri (2018) |
| | Pseudococcidae | *Planococcus ficus* | Grape, Fig, Apple, Citrus | Adult, Second instar | *Beauveria bassiana* (FUM02) | Amiri Karizkani et al. (2017) |

One such study, conducted by Karimi et al. (2010), surveyed the effect of *Heterorhabditis bacteriophora* (isolate Iran3) (Rhabditida: Heterorhabditidae) and *M. anisopliae* applied simultaneously on second instar larva of white grub, *Polyphylla adspersa* (Coleoptera: Scarabaeidae). Results revealed that application of both pathogens had an additive effect.

Another pest targeted for field trials with fungal entomopathogens was the rosaceous longhorn beetle, *Osphranteria coerulescens* (Coleoptera: Cerambycidae). Mohammadyani et al. (2016) found high mortality in various larval stages of the beetle. Their results also implied a potential of fungi against the larvae within galleries. The challenge remains an application method for field usage to maintain the fungal viability in extreme environmental condition. Another work investigated the immunological response of the larvae to *B. bassiana* (Fashand and Ir-k40 isolates) and *Cordyceps* (*Isaria*) *farinosa* (1872c isolate) (Ascomycota: Hypocreales) (Ajamhassani 2019). There was a variable cellular and enzymatic response of larvae to different isolates of fungi.

## 11.6.1 Date Palm Pest

*Microcerotermes diversus* (Isoptera: Termitidae) is considered as a major pest of date palm (*Phoenix dactylifera* L.) trees, in Iran, Iraq and Saudi Arabia. This pest attacks root, trunk, stem, petiole and cluster of palm. Because *M. anisopliae* is considered to be a suitable agent for biological control of this pest Cheraghi et al. (2012b) investigated whether social behavior of *M. diversus*, such as grooming, can be effective in promoting epizootic outbreaks of *M. anisopliae* in a colony. The highest mortality of recipient workers was observed after 14 days after being treated with the concentration of $3.5 \times 10^8$ conidia ml$^{-1}$ with a donor: recipient ratio of 1:1. The mortality of recipient workers was less than 20% at all concentrations at a donor: recipient ratio of 1:9.

Cheraghi et al. (2012a) performed a field experiment in order to evaluate the efficiency of *M. anisopliae* (DEMI001) to control a *M. diversus* population. After introduction of wooden blocks treated with *M. anisopliae*, mean density of termites decreased from 1756 to 691 individuals compared to the control. Mean percentage of feeding decreased from 59.75 to 27.81 g. They concluded that treatment of wooden blocks with *M. anisopliae* is a suitable control method for *M. diversus*.

Proliferation and virulence of EPFs can be affected by exposure to pesticides. Shaabani et al. (2015) tested the compatibility of *M. anisopliae* (DEMI 001) with imidacloprid (IMI) for control of *M. diversus* under laboratory conditions. IMI had no negative effect on conidial germination or mycelial growth. However, sporulation was progressively reduced as IMI concentrations increased. According to these results, IMI is reasonably compatible with isolate of DEMI001. In addition, mortality of *M. diversus* in combined treatments was significantly higher than non-combination treatments, indicating possible synergistic effects. Dastbarjan et al. (2016) assessed viability of *M. anisopliae* conidia in vegetable oil formulations and their virulence on the termite, *M. diversus*. Olive oil formulations had the highest average conidial germination while the sesame oil formulation had the lowest germination. The highest mean termite mortality (91.25%) was obtained with the sesame oil formulation and the lowest (58.33%) with control formulation. The lowest LT50 and LC50 occurred with the sesame oil formulation in contrast to other treatments.

## 11.7 Vegetable and Greenhouse Pests

Vegetable and greenhouse production is very common in Iran, but allows for a rapid spread of different pests such as aphids. Chemical pesticides are the most commonly used methods in vegetable farms and greenhouses, but environmental concerns as well as adverse effects on human health and other organisms, while public desire for vegetables and fruits with minimal pesticide application has restricted pesticide usage in vegetable farms. To date, various isolates of *A. muscarius*, *A. lecanii*, *B. bassiana* and *M. anisopliae* have been tested to control aphids and other pests in Iran (Table 11.3). In general, these studies have shown promising results for some important pest in the country.

Rashki and Kharazi-Pakdel (2010) studied the behavior of the green peach aphid, *M. persicae* when it encountered *B. bassiana* conidia. Green peach aphid was able to recognize and avoid *B. bassiana* conidia. However, fungus transmission did occur. Numbers of recovered aphids on eggplants containing fungus between undamaged and damaged eggplants were significantly different. Fungus transmission occurred during aphid colonization. There was no significant difference in proportion of sporulating cadavers on damaged and undamaged eggplants. Mean number of sporulating cadavers on damaged and undamaged plants were 5.25 and 4, respectively.

Biological control of *Aphis fabae* (Hemiptera: Aphididae) and *B. tabaci* with *Ferula assa-foetida* extract and entomopathogenic fungus *A. muscarius* was performed by Zamani et al. (unpublished). After 5, 7 and 9 days, the percentage mortality in treatments with *A. muscarius* ($1 \times 10^8$ conidia ml$^{-1}$) were 17.3%, 25%, 36% for *A. fabae* and 24%, 28.8% and 40.7% for *B. tabaci*, respectively. Sprays with *F. assa-foetida* suspension caused 18.6%, 27.1% and 40.96% for *A. fabae*, and 32%, 35.7% and 38.1% for *B. tabaci* after 5, 7 and 9 days.

Biocontrol efficiency of *Planococcus citri* (Hemiptera: Pseudococcidae) by *A. dipterigenus* and *A. lecanii* under laboratory and greenhouse conditions was studied by Ghaffari et al. (2017). Nymphal stages of *P. citri* were more susceptible than adults to fungal infection. Susceptibility at all stages was dose dependent. This study indicated that the entomopathogenic fungi *A. dipterigenus* and *A. lecanii* are potentially useful biological control agents for the citrus mealybug.

Valizadeh et al. (2017) investigated the responses of a new cell line from hemocytes of rose sawfly *Arge ochropus* (Hymenoptera: Argidae) to spores of *B. bassiana*. The changes in cell number and morphology in response to treatment

**Table 11.3** The insect pests of vegetable and greenhouse crops which used for research on fungal entomopathogens in Iran

| Order | Family | Species | Cultivation | Susceptibility stage to EPF | EPN species (isolate) recommended | References |
|---|---|---|---|---|---|---|
| Hemiptera (Sternorrhyncha) | Aleyrodidae | *Trialeurodes vaporariorum* | Beans | Second instar nymph | *Verticillium lecanii* | Mirabzadeh et al. (2000a) |
| | | *Trialeurodes vaporariorum* | – | Third instars nymph | *Akanthomyces muscarius* (DAOM198499) | Tabadkani et al. (2010) |
| | | *Trialeurodes vaporariorum* | – | Third and fourth instars nymph | *Akanthomyces muscarius* | Malekan et al. (2015) |
| | | *Bemisia tabasi* | Eggplant | Second instar nymph | *Akanthomyces muscarius* (DAOM198499), *Beauveria bassiana* (DEBI001) | Kuhestani et al. (2010) |
| Hemiptera (Sternorrhyncha) | Aphididae | *Brevicoryne brassicae* | Beans | Third instar nymph | *Verticillium lecanii* | Mirabzadeh et al. (2000a) |
| | | *Brevicoryne brassicae* | Brassicaceae family | – | *Beauveria bassiana, Verticillium lecanii, Metarhizium anisopliae* | Derakhshan Shadmehri (2008) |
| | | *Myzus persicae* | Rape seed | Adult | *Akanthomyces muscarius* | Mossavi et al. (2010) |
| | | *Myzus persicae* | – | Third instar nymph | *Beauveria bassiana* (EUT116) | Emami et al. (2013) |

| | | Host | Plant | Stage | Fungus | Reference |
|---|---|---|---|---|---|---|
| | | *Aphis gossypii* | Cucumber | Adult | *Akanthomyces muscarius* | Mossavi et al. (2010) |
| | | *Aphis gossypii* | Cucumber | Adult | *Beauveria bassiana* | Asgarpour et al. (2010) |
| | | *Aphis fabae* | Rose leaves | Adult | *Metarhizium anisopliae* (DEMI001) | Nategh-Jashiran et al. (2013) |
| | | *Aphis craccivora* | Bean (Chiti and Red cultivars) | – | *Beauveria bassiana* | Ezzat Abadi Pour et al. (Unpublished) |
| Coleoptera | Chrysomelidae | *Leptinotarsa decemlineata* | Potato | Adult | *Beauveria bassiana* (IRAN 441C, IRAN 429C, IRAN 187C) | Mahdneshin et al. (Unpublished) |
| | | *Leptinotarsa decemlineata* | – | Second instar larvae | *Beauveria bassiana* (DEBI007) | Shafighi et al. (2012) |
| | | *Leptinotarsa decemlineata* | – | Second and third instar larvae | *Beauveria bassiana* (AKB, LRC107, IRAN429C, LRC137, IRAN441C, Z-1), *Metarhizium anisopliae* (DEMI001, IRAN437C) | Akbarian et al. (2012) |

with *B. bassiana* were repeatable and readily observed. The immune response of the fourth instar larvae of *Phthorimaea operculella* (Lepidoptera: Gelechiidae) to two isolates of *B. bassiana* (Fashand and 47) were investigated by Pourali and Ajam Hassani (2017). Total hemocyte count for plasmatocytes and granulocytes increased significantly after three hours post infection compared to control. This increase in the number of hemocysts in in response to Fashand was clearly seen. The total hemocyte count, Plasmatocytes and granulocytes number decreased gradually at six and ten hours after injection.

## 11.8   Stored Product Pests

Stored products are vulnerable to infestation by many species of insects and mites. The impact of pests is not only direct in terms of weight loss or expense of sanitation, but also indirect because of toxic effects, allergy or environmental problems from excessive use of chemical treatments. Effective protective pest management approaches should be frequently applied to prevent stored pest infestation. To be efficient and environmentally safe, stored product production should be accomplished by different pest management programs (Akbari Asl et al. 2009; Forghani and Marouf 2015). Several studies have been conducted on efficiency of EPF as biocontrol agents for control of stored product pests (Table 11.4). These studies have shown promising results indicating that these fungi could developed for integrated pest management of some important pest in the country.

Selection of the most virulent isolate from three isolates of *B. bassiana* (IRAN 441C, IRAN 403C and IRAN 440C) and the effect of temperature on germination and radial growth and virulence were studied on larvae and adult stages of *Oryzaephilus surinamensis* (Coleoptera: Silvanidae) (Latifian et al. 2009). There were significant differences in germination and radial growth among the isolates at different temperatures. IRAN 441C had the greatest germination and radial growth at wide range of temperatures. Khashaveh and Sakenin Chelav (2013) evaluated the pathogenicity of Iranian isolates of *M. anisopliae* against *Sitophilus granarius* (Coleoptera: Curculionidae), *Tribolium castaneum* (Coleoptera: Tenebrionidae) and *O. surinamensis*. All isolates were virulent for these species and mortality increased with increasing concentration and time. The cumulative mortality 10 days after treatment varied from 62.8% with IRAN715C for *O. surinamensis* to 90% with IRAN1081C for *T. castaneum*. In conclusion, *S. granarius* is the most susceptible of the species. Masoudi et al. (2013) surveyed the virulence of 66 isolates of *Beauveria* spp. and the role of conidial count on biocontrol efficiency of *Sitophilus oryzae* (Coleoptera: Curculionidae). Out of the 66 isolates tested, 65 isolates demonstrated a mean mortality rate greater than 22% in 12 days post inoculation. The number of produced spores on culture media had significant differences among isolates. There seemed to be direct correlation between virulence and conidial production; isolates with more conidial production were more virulent. In contrast,

**Table 11.4** The stored product pests as target of research on EPFs in Iran

| Order | Family | Species | Susceptibility stage to EPF | EPN species (isolate) recommended | References |
|---|---|---|---|---|---|
| Lepidoptera | Pyralidae | Ephestia kuehniella | Egg, larvae, Pupa | Beauveria bassiana | Bahmani et al. (2012) |
| | | Ephestia kuehniella | Third instar larvae | Trichoderma atroviridae | Azimiyan et al. (2013) |
| | | Ephestia kuehniella | Last instar larvae | Beauveria bassiana | Bahrampour et al. (2014) |
| | | Plodia interpunctella | Larvae | Beauveria bassiana | Arooni Hessari et al. (2013) |
| Coleoptera | Dermestidae | Trogoderma granarium | Larvae, Adult | Metarhizium anisopliae (IRAN1018C, DEMI001) | Khashaveh et al. (2011) |
| | Bostrichidae | Rhizopertha dominica | Adult | Beauveria bassiana (IRAN187C, IRAN429C) | Mahdneshin et al. (Unpublished) |
| | | Rhizopertha dominica | Adult | Metarhizium anisopliae (IRAN715C, DEMI001) | Mahdneshin et al. (Unpublished) |
| | Chrysomelidae | Callosobruchus maculatus | Adult | Beauveria bassiana (IRAN187C, IRAN429C) | Mahdneshin et al. (Unpublished) |
| | | Callosobruchus maculatus | Adult | Metarhizium anisopliae (IRAN715C, DEMI001) | Mahdneshin et al. (Unpublished) |
| | | Callosobruchus maculatus | Adult | Beauveria bassiana (IRAN187C) | Mahdneshin et al. (Unpublished) |
| | | Callosobruchus maculatus | Adult | Beauveria bassiana (Beauvarin®) | Shams et al. (2011) |
| | | Callosobruchus maculatus | – | Beauveria bassiana (B.IRAN 1217, B.74, MC6₆) | Esmaili and Saber (2016) |
| | Nitidulidae | Carpophilus hemipterus | Larvae, Adult | Metarhizium anisopliae (IRAN1018C) | Jamali et al. (2014) |
| | Curculionidae | Sitophilus granarius | Adult | Beauveria bassiana (Beauvarin®) | Shams et al. (2011) |
| | Tenebrionidae | Tribolium castaneum | Adult | Metarhizium anisopliae (DEMI001) | Jahanbazian et al. (Unpublished) |

Jokar et al. (2013) reported only a moderate association between pathogenicity and spore count for *T. castaneum* treated with different isolates of *M. anisopliae*.

In 2017, Chehri reported incidence of *Fusarium keratoplasticum* and *Fusarium proliferatum* by molecular method on *Tribolium* species and found high susceptibility of the *Tribolium confusum* to those fungi (Figs. 11.5 and 11.6).

The role of temperature on the pathogenicity of *B. bassiana* against *O. surinamensis* fed on stored date fruits had been investigated by Latifian et al. (2018). Results showed that the mean death rate under 15, 20, 25, 30 and 35 °C temperatures was 0.89, 1.15, 1.40, 1.21, and 1.11 larvae/day, respectively: while the values for adults were 0.99, 1.38, 1.47, 1.18 and 1.16 insects/day for adults, respectively.

Hoseinizadeh et al. (2018) tested the interaction of botanical insecticides, Tondexir (pepper extract) and Palizin (Eucalyptus extract), with *B. bassiana*, *M. anisopliae* and *A. lecanii* on *G. mellonella*. The *B. bassiana* germination was not affected by either insecticide. However, *A. lecanii* germination was negatively affected by both Tondexir and Palizin. This study showed that while Tondexir and Palizin as plant-based insecticides, were effective against *G. mellonella* larvae; the levels of control achieved did not increase by the mixture of these insecticides and fungi, and in fact, were usually lower than when fungi were used alone.

In an evaluation of the behavioral response and conidia transmission of *B. bassiana* by *T. castaneum* on different wheat cultivars including Chamran, Chamran2 and Pishtaz, Ghayeb et al. (2019) observed that the flour beetle could distinguish the presence of the fungus. However, the seed condition (damaged or

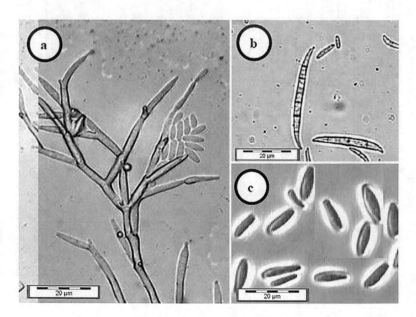

**Fig. 11.5** *Fusarium proliferatum* as natural pathogen of *Tribolium confusum*. (**a**) Polyphialides, (**b**) Macroconidia, (**c**) Microconidia. (Permission from Elsevier, License number 4802110578427)

**Fig. 11.6**
Entomopathogenic
*Fusarium keratoplasticum*
on cadaver of *Tribolium
confusum*. (Permission from
Elsevier, License number
4802110499874)

undamaged) did not have any influence on the interaction between the beetle and the fungus. The fungus was transmitted properly through the beetle colony. The authors declared the efficiency of the sporulating-cadavers can be a remarkable reservoir in grain stores.

The major concern about these research is an issue about target niche of stored product pests. EPF can kill the pests in these niches, but there are limitations to their use in storage facilities. In all cases, the insects were treated directly, not within the infested grains. So there is a gap among this aspect for usage of EPF in storage facilities and more work needs to be done elaborating practical applications.

## 11.9   Household Pests

High level of insecticide resistance in hygienic pests and public demand for reducing broad pesticide usage has developed an interest in biocontrol strategies of these pests.

Initially, Sharififard et al. (2011) found an interaction between *M. anisopliae* and sublethal doses of Spinosad and indicated a synergistic effect on mortality of house fly, *Musca domestica* (Diptera: Muscidae) as well as reduction in the speed of mortality. Afterward, effect of temperature (15, 20, 25, 30, 35 °C) and humidity (45 and 75%) on the pathogenicity of *B. bassiana* (isolate Iran 187C) and *M. anisopliae* (isolate Iran 437C) in controlling the house fly under laboratory condition was examined by Sharififard et al. (2012). The result indicated Bb 187C caused the highest mortality and sporulation on cadavers at 25–30 °C and 75% RH but there were no significant differences for Ma 437C under the same conditions. In addition, there was no significant difference in adult mortality between temperatures for both fungi.

In a study of the virulence of ten Iranian isolates of *B. bassiana* and *M. anisopliae* against *Anopheles stephensi* (Diptera: Culicidae) larvae, Fakor-ziba et al. (2014) showed that *B. bassiana* isolate 429C and *B. bassiana* isolate 796C were the most virulent isolates of *B. bassiana* causing 100% larval mortality. The most efficient isolate among *M. anisopliae* was Ma 1018C. Veys-Behbahani et al. (2014) subsequently recommended isolate Iran 429C of *B. bassiana* for the control of *Anophele* larvae.

Sharififard et al. (2014) tested a conidia-dust formulation of *M. anisopliae* prepared in proportions of 1%, 5%, 10%, 25%, 50% and 100% with the carrier of wheat flour to biocontrol the cockroach *Supella longipalpa* (Blattaria: Blattellidae). They found that with an increased proportion of conidia from 1% to 100%, the cockroach survival times (ST50) decreased from 1% to 25% but mortality and survival time at proportions of 25%, and above were not significantly different from each other. The conidia dust-formulation of *M. anisopliae* isolate IRAN 437C showed a promising alternative to control the brown-banded cockroach. In a similar work, Sharififard et al. (2016) founded that *M. anisopliae* IRAN 437C was the most virulent isolate against the brown-banded cockroach, *S. longipalpa*, causing 100% mortality in adults at seven days post-exposure. Moreover, they found treating surfaces with conidia as an aqueous suspension or oil-in- water formulation was more effective than the bait formulation against the cockroach, causing 39–97% mortality after two days. Spraying the conidia formulated with sunflower oil was an effective formulation causing 76% reduction in the cockroach density on the third day post treatment in the houses.

## 11.10    Conclusion

To date, numerous EPFs have been isolated and identified from various ecosystems of the country. It is evident that many of these fungi naturally regulate population density of a wide range of arthropod pests attacking human health or their agricultural products. Generally, EPFs are safe and can be specific for restricted virulence on a narrow range of arthropods. The application of fungi as bio-insecticides seems to have developed rapidly where chemical control is impractical or restricted.

There are yet many challenges which limit and restrict the EPFs usage in the field. By contrast, there are few successful practices in aphids, whiteflies, locusts and grasshoppers by these biological agents. Research developments that enhance the use of fungi and their efficacy in pest control of the country should address the following items.

- Addressing more diverse isolates of EPFs from various regions of the country, screening the isolates and selecting the most appropriate ones for specific uses.
- Extensive research to gain an in-depth understanding of the critical ecological factors affecting on virulence and persistence of EPFs in various crop situations.
- Increasing fungal stability by making advances in fungal nutrition, mass production and formulation techniques.
- Extending the role of fungal pesticides for farmers, especially those how to deal with greenhouses.
- Highlighting the role of agricultural practices in conservation of natural occurring fungi, like those of the Sunn pest.
- Developing education about EPF potential use strategies, their cost, and other factors which have a major influence on using EPF.
- Investigating the role of fungal entomopathogens as endophytes.
- Attempting to explore the role of EPFs as plant growth promoters and long term protectants against pest and diseases.

**Acknowledgment** The authors appreciate Stephan Jaronski for critical review and edit the last version of the manuscript and Surendra K Dara for his suggestion on early draft of chapter. The authors thank from Shokoofeh Kamali for assistance in the images processing.

# References

Aboutalebian A, Toorani AH, Abbasipour H, Heydari S, Amiri B (2017) Comparison of effects of microbial pesticides on the rice green leaf semi-looper, *Naranga aenescens* in field conditions. Paper presented at the 8th national conference on Biological Control in Agriculture and Natural Resources, University of Guilan, Rasht, 1–2 November 2017

Adeli A, Abaei M (1989) Study of harmful pests of saxaul plants of Iran. Annual report of research project. Tehran, Iran, Environmental Research Studies Center, University of Tehran

Aghajanzadeh S, Taheri H (2018) Study on efficiency of different isolates of *Lecanicillium muscarium* against *Pulvinaria aurantii*. Biol Control Pest Plant Dis 6(2):257–260

Ajamhassani M (2014) Cellular reactions of *Spodoptera littura* (Fabricus) (Lepidoptera: Noctuidae) against entomopathogenic fungi *Beauveria bassiana*. Plant Pest Res 4(2):59–68. (in Persian)

Ajamhassani M (2019) Study on morphology and frequency of hemocytes in *Osphranteria coerulescense* (Redt) (Coleoptera: Cerambycidae) and *Zeuzera pyrina* L. (Lepidoptera: Cossidae) larvae, two wood boring insects of Iran. Iran J For Range Prot Res 17(1):96–106

Akbari Asl MHA, Talebi AA, Kamali H, Kazemi S (2009) Stored product pests and their parasitoid wasps in Mashhad. Iran Adv Environ Biol 3(3):239–243

Akbarian J, Ghosta Y, Shayesteh N, Safavi SA (2012) Pathogenicity of some isolates of *Beauveria bassiana* (Bals.) Vuill. and *Metarhizium anisopliae* (Metsch.) Sorokin on 2nd and 4th larval instars of Colorado potato beetle, *Leptinotarsa decemlineata* (Say) (Col.: Chrysomelidae), under laboratory conditions. Afr J Microbiol Res 6(34):6407–6413

Alizadeh A, Kharrazi Pakdel A, Talebi K, Samih MA (2007) Effect of some *Beauveria bassiana* (Bals.) Viull. isolates on common Pistachio psyllam *Agonoscena pistaciae* Burck. and Laut. Int J Agric Biol 9(1):76–79

Amiri Karizkani SH, Moravvej G, Sadeghi Namaghi H (2017) Pathogenecity of *Beauveria bassiana* on adult and nymphal stage of vine mealybug, *Planococcus ficus* under laboratory conditions. Paper presented at the 8th national conference on Biological Control in Agriculture and Natural Resources, University of Guilan, Rasht, 1–2 November 2017

Amir-Maafi M (2000) An investigation on the host-parasitoid system between *Trissolcus grandis* Thomson (Hym.: Scelionidae) and Sunn pest eggs. Ph.D. thesis, University of Tehran, Iran, 220 pp

Amir-Maafi M, Parker BL (2002) Density dependence of *Trissolcus* spp. (Hym.: Scelionidae) on eggs of *Eurygaster integriceps* Puton [Het.: Scutelleridae]. Arab J Plant Prot 20:62–64

Amir-Maafi M, Parker BL (2003) Efficiency of *Trissolcus* spp. (Hym.: Scelionidae) as an egg parasitoids of *Eurygaster integriceps* Puton [Het.: Scutelleridae] in Iran. Arab J Plant Prot 21:69–72

Amir-Maafi M, Majdabadi M, Aghdasi F, Parker BL, Parsi F (2007) Degree days for forcasting migration of Sunn Pest. In: Parker BL, Skinner M, El-Bouhssini M, Kumari SG (eds) Sunn Pest management, a decade of progress 1994–2004. Arab Society for Plant Protection, pp 139–145

Arooni Hessari M, Talaei-Hasanloei R, Sabahi G, Hosseini Naveh V (2013) Combined use of Diatomaceous earth-*Beauveria bassiana* and Calcofluor-NPV in controlling *Plodia interpunctella* (Lepidoptera: Pyralidae). Paper presented at the conference of Biological Control in Agriculture and Natural Resources, University of Tehran, Karaj, 27–28 August 2013

Asadalapour M, Zafari D, Zare R (2011) Hyphomycetous fungi isolated from insects and their pathogenic effect on Colorado beetle in Hamedan Province. J Plant Prot 24(4):465–470

Asgarpour R, Soleyman-Nejadian E, Shafizadeh S, Ghazavi M (2010) Investigation on the effect of four isolates of *Beauveria bassiana* (Balsamo) on the melon aphid on greenhouse cucumber under laboratory condition. Paper presented at the 19th Iranian Plant Protection Congress, Iranian Institute of Plant Protection Research, Tehran, 31 July–3 August 2010

Ashouri A, Arzanian N, Askary H, Rasoulian GR (2004) Pathogenicity of the fungus, *Verticillium lecanii*, to the green peach aphid, *Myzus persicae* (Hom.: Aphididae). Commun Agric Appl Biol Sci 69(3):205–209

Askary H, Ajam Hassani M (2009) Effect of *Lecanicillium muscarium* (Deut.: Moniliales) on longevity, fecundity and mating behavior of *Aphidius nigripes* (Hym.: Aphididae). Appl Entomol Phytopathol 76(2):15–31

Askary H, Brodeur J (1999) Susceptibility of larval stages of the aphid parasitoid *Aphidius nigripes* to the entomopathogenic fungus *Verticillium lecanii*. J Invertebr Pathol 73:129–132

Askary H, Benhamou N, Brodeur J (1997) Ultrastructural and cytochemical investigations of the antagonistic effect of *Verticillium lecanii* on cucumber powdery mildew. Phytopathology 87:359–368

Askary H, Benhamou N, Brodeur J (1999) Ultrastructural and cytochemical characterization of aphid invasion by the hyphomycete *Verticillium lecanii*. J Invertebr Pathol 74(1):1–13

Askary H, Ajam Hassani M, Yarmand H (2007) Investigation on survival of *Aphidius nigripes* Ashmead (Hym.: Aphidiidae) reared on infected potato aphid by *Lecanicillium muscarium* (Deut.: Moniliaceae). Commun Agric Appl Biol Sci 71(2):375–385

Askary H, Morad Ali M, Ajam Hassani M, Zamani SM (2008) Preliminary laboratory investigation of *Lecanicillium muscarium* affecting some phytopathogenic fungi. Iran J For Range Prot Res 5 (2):151–159

Azimiyan S, Zafari D, Madadi H, Ghobadi Anvar F (2013) Survey on the control possibility of *Ephestia kuehniella* using *Trichoderma atroviridae* under laboratory condition. Paper presented at the conference of Biological Control in Agriculture and Natural Resources, University of Tehran, Karaj, 27–28 August 2013

Bagheri S, Tajvand B (2006) Determination of survival *Metarhizium anisopliae* var. *acridium* in foliage application for control of *Esfandiaria obesa* Popov. Paper presented at the 17th Iranian Congress of Plant Protection, Tehran University, Karaj, 2–5 September 2006

Bagheri S, Tajvand B (2008) The effect of application *Metarhizium anisopliae* var. *acridium* for controlling *Esfandiaria obesa* Popov (Orth.: Acrididae) on the *Mantis religiosa* L. (Dict.: Mantidae) in arid forests of Khuzestan. Paper presented at the 18th Iranian Plant Protection Congress, University of Bu Ali Sina, Hamedan, 26–27 August 2008

Bagheri A, Seyahooei MA, Fathipour Y, Famil M, Koohpayma F, Mohammadi-Rad A, Parichehreh S (2019) Ecofriendly managing of *Helicoverpa armigera* in tomato field by releasing *Trichogramma evanescence* and *Habrobracon hebetor*. J Crop Prot 8(1):11–19

Bahmani N, Latifian M, Rad B, Ostovan H, Haghani M (2012) Study the lethal and sublethal doses of suitable isolate of *Beauveria bassiana* for microbial control of *Ephestia kuehniella* on Sayer cultivar Date. Paper presented at the 20th Iranian Plant Protection Congress, Shiraz University, Shiraz 25–28 August 2012

Bahrampour N, Morravej G, Taheri P (2014) Pathogenicity of entomopathogenic fungus, *Beauveria bassiana* against last instar larvae of *Ephestia kuehniella* in laboratory condition. Paper presented at the 21th Iranian Plant Protection Congress, University of Urmia, 23–26 August 2014

Bandani A, Esmailpour N (2006) Oil formulation of entomopathogenic fungus, *Beauveria bassiana*, against Sunn pest, *Eurygaster integriceps* puton (Heteroptera: Scutelleridae). Commun Agric Appl Biol Sci 71(2 Pt B):443–448

Benhamou N (2004) Potential of the mycoparasite, *Verticillium lecanii*, to protect citrus fruit against *Penicillium digitatum*, the causal agent of green mold: a comparison with the effect of chitosan. Phytopathology 94:693–705

Benhamou N, Brodeur J (2000) Evidence for antibiosis and induced host defense reactions in the interaction between *Verticillium lecanii* and *Penicillium digitatum*, the causal agent of green mold. Phytopathology 90(9):932–943

Benhamou N, Brodeur J (2001) Pre-inoculation of Ri T-DNA transformed cucumber roots with the mycoparasite, *Verticillium lecanii*, induces host defense reactions against *Pythium ultimum* infection. Physiol Mol Plant Pathol 58:133–146

Boucias D, Pendland J (1991) Attachment of mycopathogens to cuticle. In: Cole GT, Hoch HC (eds) The fungal spore and disease initiation in plants and animals. Plenum Press, New York, pp 101–127

Brooks WM (1993) Host-parasitoid-pathogen interactions. In: Beckage NE, Thompson SN, Federici BA (eds) Parasites and pathogens of insects: pathogens. Academic, San Diego, pp 231–272

Brown ES (1965) Notes on the migration and directions of flight of *Eurygaster* and *Aelia* species (Hemiptera, Pentatomoidea) and their possible bearing of the invasion of cereal crops. J Anim Ecol 34(1):93–107

Chandler D (2017) Basic and applied research on entomopathogenic fungi. In: Lacey LA (ed) Microbial control of insect and mite pests. Academic, pp 69–89

Charnley AK (1989) Mechanisms of fungal pathogenesis in insects. In: Whipps JM, Lumsden RD (eds) The biotechnology of fungi for improving plant growth. Cambridge University, London, pp 85–125

Chehri K (2017) Molecular identification of entomopathogenic *Fusarium* species associated with *Tribolium* species in stored grains. J Invertebr Pathol 144:1–6

Cheraghi A, Habibpour B, Mossadegh MS, Mahinpour V (2012a) Impact of treated wood with the fungus *Metarhizium anisopliae* on controlling *Microcerotermes diversus* (Isoptera: Termitidae). J Plant Prot Res 26(2):209–216

Cheraghi A, Habibpour B, Mossadegh MS, Sharififard M (2012b) Horizontal transmission of the entomopathogen fungus *Metarhizium anisopliae* in *Microcerotermes diversus* groups. Insects 3 (3):709–718

Danfa A, Van der Valk HCHG (1999) Laboratory testing of *Metarhizium* spp. and *Beauveria bassiana* on Sahelian non-target arthropods. Biocontrol Sci Tech 9(2):187–198

Darsouei R, Karimi J, Ghadamyari M, Hosseini M (2018) Natural enemies of the sugar beet army worm, *Spodoptera exigua* (Lepidoptera: Noctuidae) in Northeast Iran. Entomol News 127 (5):446–464

Dastbarjan M, Habibpour B, Eslamizadeh R (2016) Viability of *Metarhizium anisopliae* (Metschnikoff) Sorokin conidia in vegetable oil formulations and their virulence on the termite *Microcerotermes diversus* Silvestri (Isoptera: Termitidae). Appl Res Plant Prot 5(1):35–48

Derakhshan Shadmehri A (2008) Evaluation on entomopathogenic fungi for biological control of cabbage aphid, *Brevicoryne brassicae*. Paper presented at the 18th Iranian Plant Protection Congress, University of Bu Ali Sina, Hamedan, 26–27 August 2008

Ebadollahi A, Davari M, Razmjou J, Naseri B (2017) Separate and combined effects of *Mentha piperata* and *Mentha pulegium* essential oils and a pathogenic fungus *Lecanicillium muscarium* against *Aphis gossypii* (Hemiptera: Aphididae). J Econ Entomol 110(3):1025–1030

Ebrahimifar J, Jamshidnia A, Sadeghi R, Babamir A (2016) Pathogenicity evaluation of *Beauveria bassiana* and *Metarhizium anisopliae* on elm leaf beetle, *Xanthogaleruca luteola* (Col.: Chrysomelidae). Paper presented at the 3th Iranian meeting on Biocontrol in Agriculture and Natural Resources Ferdowsi University, Mashhad, 2–3 February 2016

El-Bouhssini M, Parker BL, Skinner M, Reid B, Canhilal R, Aw-Hassan A, Nachit M, Valcun Y, Ketata H, Moore D, Amir-Maafi M, Kutuk H, Abdel Hay M, El-Haramein J (2004) Integrated pest management of Sunn Pest in west and central Asia. World Entomological Congress, Brisbane, Australia. Australian Entomological Society

Emami F, Alich M, Minaei K (2013) Interaction between the entomopathogenic fungus, *Beauveria bassiana* (Ascomycota: Hypocreales) and the parasitoid wasp, *Aphidius Colemani* Viereck (Hymenoptera: Braconidae). J Entomol Acarol Res 45(1):14–17

Esmaili M, Saber M (2016) Effect of *Beauveria bassiana* isolates on *Callosobruchus maculatus* (Col.: Bruchidae) under laboratory conditions. Paper presented at the 3th Iranian meeting on Biocontrol in Agriculture and Natural Resources Ferdowsi University, Mashhad, 2–3 February 2016

Fadayivatan S, Moravvej G, Karimi J (2014) Pathogenicity of the fungus *Lecanicillium longisporum* against *Sipha maydis* and *Metopolophium dirhodum* in laboratory conditions. J Plant Prot Res 54(1):67–73

Fakor-ziba MR, Veys-Behbahani R, Dinparast-Djadid N, Azizi K, Sharififad M (2014) Screening of the entomopathogenic fungi, *Metarhizium anisopliae* and *Beauveria bassiana* against early larval instars of *Anopheles stephensi* (Diptera: Culicidae). J Entomol 11(2):87–94

Farashiani ME, Askary H, Ehteshamhoseini M (2008) Laboratory investigation on virulence of three entomopathogenic fungi against the larvae of *Aeolesthes sarta* (Col.:Cerambycidae). JESI 28(142):19–34

Farsi MJ, Moniry VR; Askary H; Zamani M, Azizkhani E, Omid, R, Zeinaly S (2009) Investigation on biological control of *Dericorys albidula* with *Metarhizium anisopliae* var. *acridum*. Research project report, Research Institute of Forests and Rangelands, 60p

Fazeli-Dinan M, Kharazi-Pakdela A, Aliniab F, Tabaric MA-O, Fahimia A (2012) Field biology of the green semi-looper, *Naranga aenescens* Moore (Lepidoptera: Noctuidae) and efficiency determination of *Beauveria bassiana* isolates. SOAJ Entomol Stud 1:68–80

Fazeli-Dinan M, Talaei-Hassanloui R, Allahyari H (2016a) Host preference of *Encarsia formosa* (Hym.: Aphelinidae) towards untreated and *Lecanicillium longisporum*-treated *Trialeurodes vaporariorum* (Hem.: Aleyrodidae). J Asia Pac Entomol 19(4):1145–1150

Fazeli-Dinan M, Talaei-Hassanloui R, Goettel M (2016b) Virulence of the entomopathogenic fungus *Lecanicillium longisporum* against the greenhouse whitefly, *Trialeurodes vaporariorum* and its parasitoid *Encarsia formosa*. Int J Pest Manag 62(3):251–260

Forghani SH, Marouf A (2015) An introductory study of storage insect pests in Iran. Biharean Biol 9(1):59–62

Ghaffari S, Karimi J, Kamali S, Moghadam EM (2017) Biocontrol of *Planococcus citri* (Hemiptera: Pseudococcidae) by *Lecanicillium longisporum* and *Lecanicillium lecanii* under laboratory and greenhouse conditions. J Asia Pac Entomol 20(2):605–612

Ghamari Zare Z, Askary H, Abbasipour H, Sheikhi Gorjan A, SaeidiZadeh A (2014) Virulence of four *Beauveria bassiana* isolates on aestivating and overwintering populations of *Eurygaster integriceps*. Biocontrol Plant Prot 2(1):31–41

Ghayeb S, Rashki M, Shirvani A (2019) Evaluation of the behavioral response and conidium transmission by *Tribolium castaneum* in presence of the entomopathogenic fungus, *Beauveria bassiana* on different wheat cultivars. Anim Res 32(1):62–78

Ghazavi M (2009) Isolation and identification of the protist pathogens of the insect classes, Lepidoptera, Orthoptera, Coleoptera and Diptera, Project report, Iranian Research Institute of Plant Protetion

Ghazavi M, Ershad J, Kharazi Pakdel A (2004) The response of the cellular immune system of *Locusta migratoria* (Orth.: Acrididae) to Fashand isolate of *Beauveria bassiana* (Moniliales, Moniliaceae). JESI 23(2):77–101

Hajek AE, Bauer LS (2007) Microbial control of wood-boring insects attacking forest and shade trees. In: Field manual of techniques in invertebrate pathology. Springer, Dordrecht, The Netherlands, pp 505–525

Haji Allahverdi Pour H, Ghazavi M, Kharazi-Pakdel A (2008) Comparison of the virulence of some Iranian isolates of *Beauveria bassiana* to *Eurygaster integriceps* (Hem.: Scutelleridae) and production of the selected isolate. JESI 28(1):12–26

Hall RA (1981) The fungus *Verticillium lecanii* as a microbial insecticide against aphids and scales. In: Burges HD (ed) Microbial control of pests and plant diseases. Academic, London, pp 483–498

Hamzehei M, Mirshekar A, Salari M, Khani A, Sabbagh SK (2018) Pathogenicity assessment of an indigenous strain of *Trichoderma harzianum* Rifai (Ascomycota: Hypocreales) to the grasshopper, *Chrotogonus trachypterus* (Orth.: Pyrgomorphidae) (B.) in laboratory conditions. J Exp Anim Biol 2(26):71–79

Hariri G, Williams PC, El-Haramain FJ (2000) Influence of Pentatomid insect on the physical and dough properties and two-layered flat bread baking quality of Syrian wheat. J Cereal Sci 31 (2):111–118

Hochberg ME, Lawton JH (1990) Competition between kingdoms. Trends Ecol Evol 5:367–371

Hoseinizadeh ZS, Sohrabi F, Jamali F (2018) The effects of botanical insecticides tondexir and palizin on *Galleria mellonella* L. (Lepidoptera: Pyralidae) and their interaction with entomopathogenic fungi. Paper presented at the 23rd Iranian Plant Protection Congress, Gorgan University of Agricultural Sciences and Natural Resources, 27–30 August 2018

Huang J, He Y, Lin Q (1990) Use of a new paste preparation of *Beauveria bassiana* in the forest to control *Zeuzera multistrigata* (Lep.: Cossidae). Chin J Biol Control 6(3):121–123

Jamali F, Sohrabi F, KohanMoo MA (2014) Biocontrol activity of *Metarhizium anisopliae* (Metsch.) sorokinin against larvae and adults of date sap beetle *Carpophilus hemipterus* (Linnaeus). In: 21st Iranian Plant Protection Congress, University of Urmia, 23–26 August 2014, p 515

James RR, Lighthart B (1994) Susceptibility of the convergent lady beetle to four entomopathogenic fungi. Environ Entomol 23:188–190

Jaronski ST, Goettel MA, Lomer C (2004) Regulatory requirements for ecotoxicological assessments of microbial insecticides – how relevant are they? In: Hokkanen H, Hayek A (eds) Environmental impacts of microbial insecticides: need and methods for risk assessment. Kluwer Academic Publishers, Dordrecht, pp 237–260

Jarrahi A, Safavi SA (2016) Sublethal effects of *Metarhizium anisopliae* on life table parameters of *Habrobracon hebetor* parasitizing *Helicoverpa armigera* larvae at different time intervals. BioControl 61(2):167–175

Javahery M (1995) A technical review of Sunn Pests (Heteroptera: Pentatomidae) with special reference to *Eurygaster integriceps* Puton. FAO/RNE, 80 pp

Javahery M (2004) Sustainable management of cereal Sunn Pests in the 21st century. World Entomological Congress, Brisbane, Australia. Australian Entomological Society

Javar S, Kharazi-Pakdel A (2004) Laboratory investigation on pathogenicity of the fungus *Beauveria bassiana* (Bals.) Vuill. on cotton boll worm, *Helicoverpa armigera* (Hubner.). Paper presented at the 16th Iranian Plant Protection Congress, Tabriz University, Tabriz 27 August–1 September 2004

Jokar M, Farrokhi N, Derakhshan Shadmehri A, Masoudi A, Parsaeeian M, Matin S (2013) Evaluation of *Metarhizium* spp. isolates as a control of *Tribolium castaneum* (Col.: Tenebrionidae) and correlation with spore count. Paper presented at the conference of Biological Control in Agriculture and Natural Resources, University of Tehran, Karaj, 27–28 August 2013

Kalvandi E, Mirmoayedi A, Alizadeh M, Pourian H-R (2018) Sub-lethal concentrations of the entomopathogenic fungus, *Beauveria bassiana* increase fitness costs of *Helicoverpa armigera* (Lepidoptera: Noctuidae) offspring. J Invertebr Pathol 158:32–42

Karami L, Amir-Maafi M, Shahrokhi S, Imani S, Shojai M (2016) Demography of the bird cherry-oat aphid, (*Rhopalosiphum padi* L.) (Hemiptera: Aphididae) on different barley varieties. J Agric Sci Technol 18(5):157–1267

Karimi J, Kharazi-Pakdel A, Hasani-kakhki M (2010) The entomopathogens, *Heterorhabditis bacteriophora* and *Metarhizium anisopliae* (Metch.) Sorokin (Deuteromycota: Hyphomycetes) work additive in controlling white grub, *Polyphylla adspersa* (Col.: Scarabaeidae). Paper presented at the 19th Iranian Plant Protection Congress, Iranian Institute of Plant Protection Research, Tehran, 31 July–3 August 2010

Karimi J, Dara SK, Arthurs S (2018) Microbial insecticides in Iran: history, current status, challenges and perspective. J Invertebr Pathol 165:67–73

Kazemi Yazdi F, Eilenberg J, Mohammadipour A (2011) Biological characterization of *Beauveria bassiana* (Clavicipitaceae: Hypocreales) from overwintering sites of Sunn Pest, *Eurygaster integriceps* (Scutelleridae: Heteroptera) in Iran. Int J Agric Sci Res 2(2):7–16

Khashaveh A, Sakenin Chelav H (2013) Laboratory bioassay of Iranian isolates of Entomopathogenic fungus *Metarhizium anisopliae* (Metsch.) Sorokin (Ascomycota: Hypocreales) for control of two species of storage pest. Agric Conspec Sci 78(1):35–40

Khashaveh A, Safaralizadeh MH, Ghosta Y (2011) Pathogenicity of Iranian isolates of *Metarhizium anisopliae* (Metschinkoff) (Ascomycota: Hypocreales) against *Trogoderma granarium* Everts (Coleoptera: Dermestidae). Biharean Biol 5(1):51–55

Kuhestani K, Askary H, Baghdadi A, Zarrabi M (2010) Evaluation of pathogenicity of *Beauveria bassiana* (Balsamo) Vuillemin and *Lecanicillium muscarium* (Zimmerm.) Zare & Gams on *Bemisia tabaci* Gennadius (Hom.: Aleyrodidae). Paper presented at the 19th Iranian Plant Protection Congress, Iranian Institute of Plant Protection Research, Tehran, 31 July–3 August 2010

Lacey LA, Mesquita LM, Mercadier G, Debire R, Kazmer DJ, Leclant F (1997) Acute and sublethal activity of the entomopathogenic fungus *Paecilomyces fumosoroseus* (Deuteromycotina: Hyphomycetes) on adult *Aphelinus asychis* (Hymenoptera: Aphelinidae). Environ Entomol 26:1452–1460

Latifian M, Soleyman-Nejadian E, Ghazavi M, Hayati J, Mosadegh MS, Nikbakht P (2009) Evaluation of three *Beauveria bassiana* isolates on saw-toothed beetle *Oryzaephilus surinamensis* and the effect of different temperature on their germination and mycelium growth. J Entomol Phytopathol 77(1):151–168

Latifian M, Ghazavi M, Soleimannejadian E (2018) The role of temperature on the pathogenicity of *Beauveria bassiana* in populations of saw toothed grain beetle, *Oryzaephilus surinamensis* (Coleoptera: Silvanidae) fed on stored date fruits. J Crop Prot 7(4):395–402

Lotfalizadeh H, Bayegan Z-A, Zargaran M-R (2016) Species diversity of Chalcidoidea (Hymenoptera) in the rice fields of Iran. J Entomol Res Soc 18(1):99–111

Majidi-Shilsar F, Ershad J, Padasht F (2004) Laboratory evaluation on pathogenicity of six isolates of the fungus *Beauveria bassiana* (Bals.) Vuillemin on the striped stem borer (*Chilo ssuppressalis* Walker). Paper presented at the 16th Iranian Congress of Plant Protection, Tabriz University, Tabriz, 27 August–1 September 2004

Majidi-Shilsar F, Padasht F, Alinia F (2008) Effect of fungus *Beauveria bassiana* on striped stem borer, *Chilo suppressalis* walk. under rice filed condition. Paper presented at the 18th Iranian Plant Protection Congress University of Bu Ali Sina, Hamedan, 26–27 August 2008

Malekan N, Hatami B, Ebadi R, Akhavan A, Radjabi R (2015) Evaluation of entomopathogenic fungi *Beauveria bassiana* and *Lecanicillium muscarium* on different nymphal stages of greenhouse whitefly *Trialeurodes vaporariorum* in greenhouse conditions. Biharean Biol 9 (2):108–112

Masoudi A, Farrokhi N, Parsaeiyan M, Derakhshan Shadmehri A, Mamarabadi M (2013) Virulance of *Beauveria* spp. isolates and the role of conidium count on bicontrol efficacy. Paper presented at the conference of Biological Control in Agriculture and Natural Resources, University of Tehran, Karaj, 27–28 August 2013

Memmott J, Martinez ND, Cohen JE (2000) Predators, parasitoids and pathogens: species richness, trophic generality and body sizes in a natural food web. J Anim Ecol 69(1):1–15

Mesquita ALM, Lacey LA (2001) Interactions among the entomopathogenic fungus, *Paecilomyces fumosoroseus* (Deuteromycotina: Hyphomycetes), the parasitoid, *Aphelinus asychis* (Hymenoptera: Aphelinidae), and their aphid host. Biol Control 22:51–59

Mirabzadeh A, Amir-Sadeghi S, Farrokhi S, Emani B (2000a) Determination the LC50 and LT50 of *Vertiillium lecanii* on cabbage aphid, *Brevicoryne brassicae* and its effect on greenhouse whitefly, *Trialeurodes vaporariorum*. Paper presented at the 14th Iranian Plant Protection Congress, Isfahan University of Technology, Isfahan, 5–8 September 2000

Mirabzadeh A, Moazami N, Amirsadeghi S, Jafari ME (2000b) Determination effect of *Verticillium lecanii* on *Pulvinaria aurantii* and isolation the fungus from pest. Paper presented at the 14th Iranian Plant Protection Congress, Isfahan University of Technology, Isfahan, 5–8 September 2000

Mirshekar A, Kharazi Pakdel A, Ghazavi M, Azmayeshfard P (2004) Comparative and laboratory study of pathogenicity of *Beauveria bassiana* and *Metarhizium anisopliae* on *Chrotogonus trachypterus* (Orth.: Pyrogomorphidae). Paper presented at the 16th Iranian Congress of Plant Protection, Tabriz University, Tabriz, 27 August–1 September 2004

Mohammadipour A, Ghazavi M, Baghdadi A, Sheikhi Garjan A (2009) An investigation of the efficacy of two Iranian isolates of *Metarhizium anisopliae* against Russian wheat aphid, *Diuraphis noxia* (Hemiptera: Aphididae) under laboratory conditions. Iran J Plant Prot Sci 41 (2):353–359

Mohammadyani M, Karimi J, Taheri P, Sadeghi H, Zare R (2016) Entomopathogenic fungi as promising biocontrol agents for the rosaceous longhorn beetle, *Osphranteria coerulescens*. BioControl 61(5):579–590

Moniri VR (1998) Comparison of effect of chemical, microbial and hormonal compositions against *Dericorys albidula* in Isfahan province of Iran. Final report of research project, Research Institute of Forest and Rangeland

Mossavi M, Nouri-Ghanbalani G, Rafiee-Dastjerdi H, Zargarzade F, Rezapanah M (2010) Tha controlling effect of *Lecanicillium muscarium* (Petch) Zara & Gams on *Aphis gossypii* (Glover) & *Myzus persicae* (Sulzer) (Hom: Aphididae). Paper presented at the 19th Iranian Plant Protection Congress, Iranian Institute of Plant Protection Research, Tehran, 31 July–3 August 2010

Mossavi M, Mehrkho F, Ghosta U, Akbari S, Heydarzadeh S (2016) Study the lethal and sublethal effects of the different isolates of fungus *Beauveria bassiana* on adult of *Schizaphis graminum* (Rondani) (Hemiptera: Aphididae) in laboratory condition. Paper presented at the 22rd Iranian Plant Protection Congress, Tehran University, Karaj, 27–30 August 2016

Naseri B, Fathipour Y, Moharramipour S, Hosseininaveh V (2011) Comparative reproductive performance of *Helicoverpa armigera* (Hübner) (Lepidoptera: Noctuidae) reared on thirteen soybean varieties. J Agric Sci Technol 13(1):17–26

Nategh-Jashiran N, Abdollahi M, Hosseinvand M (2013) Effect of *Metarhizium anisopliae* DEMI-001 on black been aphid, *Aphis fabae* (Hem.: Aphididae) at different temperatures. Paper presented at the conference of Biological Control in Agriculture and Natural Resources, University of Tehran, Karaj, 27–28 August 2013

Nouri-Aiin M, Askary H, Imani S, Zare R (2014) Isolation and characterization of entomopathogenic fungi from hibernating sites of Sunn Pest (*Eurygaster integriceps*) on Ilam Mountains. Iran Int J Curr Microbiol App Sci 3(12):314–325

Ortiz-Urquiza A, Keyhani NO (2013) Action on the surface: entomopathogenic fungi versus the insect cuticle. Insects 4:357–37410.3390/insects4030357

Parker BL, Skinner M, Brownbridge M, El-Bohssini M (2000) Control of insect pests with entomopathogenic fungi. Arab J Plant Prot 18:133–138

Parker BL, Skinner M, Costa SD, Gouli S, Reid W, El-Bouhssini M (2003) Entomopathogenic fungi of *Eurygaster integriceps* Puton (Hemiptera: Scutelleridae): collection end characterization for development. Biol Control 27:260–272

Parsi F, Mohammadipour A (2017) Effect of elevation on the presence of *Beauveria bassiana* (Bals.) Vuill Entomopathogen fungus of *Eurygaster integryceps* Puton (Hemiptera: Scutelleridae) in overwintering sites in Center of Iran. Paper presented at the 8th national conference on Biological Control in Agriculture and Natural Resources, University of Guilan, Rasht, 1–2 November 2017

Poor Amiri MN, Alinia F, Imani S, Shayanmehr M, Ahadiyat A (2017) Comparative management of *Chilo suppressalis* (Walker) (Lepidoptera: Crambidae) by convenient pesticides and non-chemical practices in a double rice cropping system. Arthropods 6(4):126–136

Pourali Z, Ajam Hassani M (2017) Cellular defense of the potato tuber moth, *Phthorimaea operculella* (Lepidoptera: Gelechiidae) against the pathogenic fungus, *Beauveria bassiana*. Paper presented at the 8th national conference on Biological Control in Agriculture and Natural Resources, University of Guilan, Rasht, 1–2 November 2017

Rahimzadeh A, Farrokhi S, Ghayourfar R, Tirgari S (2006) Survey on the control possibility of *Microcerotermes gabrielis* (Weidner) using four isolates of *Metarhizium anisopliae* (Metsch.) Sorkok. in the laboratory. Paper presented at the 17th Iranian Congress of Plant Protection, Tehran University, Karaj, 2–5 September 2006

Rahimzadeh A, Rashid M, Sheikhi Garjan A, Naseri B (2012) Laboratory evaluation of *Metarhizium anisopliae* (Metschnikoff) for controlling *Amitermes vilis* (Hagen) and *Microcerotermes gabrielis* (Weidner) (Isoptera: Termitidae). J Crop Prot 1(1):27–34

Ramzi S, Zibaee A (2014) Biochemical properties of different entomopathogenic fungi and their virulence against *Chilo suppressalis* (Lepidoptera: Crambidae) larvae. Biocontrol Sci Tech 24 (5):597–610

Rashki M, Kharazi-Pakdel A (2010) Effect of *Beauveri bassiana* (Ascomycota, Hypocreales) sporulating cadavers of green peach aphids on eggplant colonization and fungus transmission. Paper presented at the 19th Iranian Plant Protection Congress, Iranian Institute of Plant Protection Research, Tehran, 31 July–3 August 2010

Rashki M, Kharazi-Pakdel A, Allahyari H, Van Alphen J (2009) Interactions among the entomopathogenic fungus, *Beauveria bassiana* (Ascomycota: Hypocreales), the parasitoid, *Aphidius matricariae* (Hymenoptera: Braconidae), and its host, *Myzus persicae* (Homoptera: Aphididae). Biol Control 50(3):324–328

Rastegar J, Ghazavi M, Kamali K, Ershad J (2006) Impact of seven native and GHA isolates of *Beauveria bassiana* on adult Sunn pest of Iran. Paper presented at the 17th Iranian Congress of Plant Protection, Tehran University, Karaj, 2–5 September 2006

Rezwani A (2010) Aphids (Hemiptera: Aphidoidea) of herbaceous plants in Iran. Publication of Entomological Society of Iran, Tehran, 557 pp (in Persian)

Roshandel S, Talaei-Hassanloui R, Askary H, Allahyari H (2013) Effect of culture substrates on virulence of *Metarhizium anisopliae* conidia and blastospores against sunn pest, *Eurygaster integriceps*. Iran J Plant Prot Sci 44(2):225–234

Safavi A (2012) Attenuation of the entomopathogenic fungus *Beauveria bassiana* following serial *in vitro* transfers. Biologia 67(6):1062–1068

Safavi S, Kharrazi A, Rasoulian GR, Bandani A (2010) Virulence of some isolates of entomopathogenic fungus, *Beauveria bassiana* on *Ostrinia nubilalis* (Lepidoptera: Pyralidae) larvae. J Agric Sci Technol 12(1):13–21

Schuh RT, Slater JA (1995) True bugs of the world (Hemiptera: Heteroptera). Cornell University Press, New York

Sedighi A, Ghazavi M, Haji Allahverdipour H, Ahadiyat A (2012) Study on the effects of some Iranian isolates of the fungus *Beauveria bassiana* (Balsomo) Vuill. (Deuteromycotina: Hyphomycetes) on the bird cherry-oat aphid, *Rhopalosiphum padi* (Linnaeus) (Hemiptera: Aphididae), under laboratory conditions. Zoology 7(1):267–273

Sedighi N, Askary H, Abbasipour H, Sheikhi Gorjan A, Karimi J (2014) Bioassay with two Iranian isolates of *Metarhizium anisopliae* on eggs, 2nd. nymphal instars and adults of the sunn pest, *Eurygaster integriceps* and the effect of EC oil on pathogenicity. Appl Entomol Phytopathol 81 (2):87–96

Shaabani M, Habibpour B, Mossadegh MS (2015) Compatibility of the entomopathogenic fungus *Metarhizium anisopliae* senso lato with imidacloprid for control of *Microcerotermes diversus* Silvestri (Iso.:Termitidae) in laboratory conditions. Plant Pests Res 5(1):27–36

Shafighi Y, Kazemi MH, Ghosta Y, Akbarian J (2012) Insecticidal efficacy of two isolates of *Beauveria bassiana* (Bals.) (Vuill.), on the second larval stage of *Leptinotarsa decemlineata* (Say) (Col.: Chrysomelidae). Arch Phytopathol Plant Prot 45(15):1852–1860

Shams G, Safaralizadeh MH, Imani S, Shojai M, Aramideh S (2011) A laboratory assessment of the potential of the entomopathogenic fungi *Beauveria bassiana* (Beauvarin) to control *Callosobruchus maculatus* (F.) (Coleoptera: Bruchidae) and *Sitophilus granarius* (L.) (Coleoptera: Curculionidae). Afr J Microbiol Res 5(10):1192–1196

Sharififard M, Mossadegh M, Vazirianzadeh B, Zarei-Mahmoudabadi A (2011) Interactions between Entomopathogenic fungus, *Metarhizium anisopliae* and sublethal doses of spinosad for control of house fly, *Musca domestica*. Iran J Arthropod Borne Dis 5(1):28

Sharififard M, Mossadegh M, Vazirianzadeh B (2012) Effects of temperature and humidity on the pathogenicity of the entomopathogenic fungi in control of the house fly, *Musca domestica* L. (Diptera: Muscidae) under laboratory conditions. J Entomol 9(5):282–288

Sharififard M, Mossadegh MS, Vazirianzadeh B, Latifi SM (2014) Evaluation of conidia-dust formulation of the entomopathogenic fungus, *Metarhizium anisopliae* to biocontrol the brown-banded cockroach, *Supella longipalpa* F. Jundishapur J Microbiol 7(6):6

Sharififard M, Mossadegh MS, Vazirianzadeh B, Latifi SM (2016) Biocontrol of the brown-banded cockroach, *Supella longipalpa* F.(Blattaria: Blattellidae), with entomopathogenic fungus, *Metharhizium anisopliae*. J Arthropod Borne Dis 10(3):335

Shayesteh N, Ranji H, Ziaee M (2015) Abundance and diversity of aphids (Hemiptera: Aphididae) and ladybirds (Coleoptera: Coccinellidae) population in wheat fields of Urmia, Northwestern of Iran. Biharean Biol 9(1):63–65

Skinner M, Parker BL, Gouli S, Reid W, El-Bouhssini M, Amir-Maafi M, Sayyasi Z (2007) Entomopathogenic fungi for Sunn Pest Management: efficacy trials in overwintering sites. In: Parker BL, Skinner M, El-Bouhssini M, Kumari SG (eds) Sunn Pest Management, A decade of progress 1994–2004. Arab Society for Plant Protection, pp 319–328

St. Leger RJ, Screen S (2001) Prospects for strain improvement of fungal pathogens of insects and weeds. In: Butt TM, Jackson C, Morgan N (eds) Fungal biocontrol agents: progress, problems and potential. CAB International, pp 219–238

Tabadkani SM, Mehrasa M, Askary H, Ashouri A (2010) Study on pathogencity effects of the entomophagous fungi *Lecanicillium muscarium* on the greenhouse whitefly *Trialeurodes vaporariorum*. Paper presented at the 19th Iranian Plant Protection Congress, Iranian Institute of Plant Protection Research, Tehran, 31 July–3 August 2010

Talaee L, Talebi A, Fathipour Y, Khajehali J (2017) Performance evaluation of *Spodoptera exigua* (Lepidoptera: Noctuidae) larvae on 10 sugar beet genotypes using nutritional indices. J Agric Sci Technol 19:1103–1112

Talaei-Hassanloui R, Kharazi-Pakdel A, Goettel M, Mozaffari J (2007) Variation in virulence of *Beauveria bassiana* isolates and its relatedness to some morphological characteristics. Biocontrol Sci Tech 16(5):525–534

Talaei-Hassanloui R, Kharazi-Pakdel A, Hedjaroude GA (2009) Transmission possibility of the fungus *Beauveria bassiana* KCF102 by mating behavior between Sunn pest, *Eurygaster integriceps* (Hem.: Scutelleridae) adults. JESI 28(2):1–6

Tork M, Bandani AR, Sayedi M (2008) Pathogenicity and protease production relationship between three isolates of *Metarhizium anisopliae* Metch, on *Eurygaster integriceps* (Hem: Scutelleridae). Paper presented at The International Congress of Entomology, Durban, South Africa, June 2008

Valizadeh B, Jalali Sendi J, Khosravi R, Salehi R (2017) Establishment and characterizations of a new cell line from larval hemocytes of rose sawfly *Arge ochropus* (Hymenoptera; Argidae). JESI 38(2):173–186. 10.22117/jesi.2018.116228.1155

Vega FE, Kaya HK (2012) Insect pathology. Academic Press, London, 508 p

Veys-Behbahani R, Sharififard M, Dinparast-Djadid N, Shamsi J, Fakoorziba MR (2014) Laboratory evolution of the entomopathogenic fungus *Beauveria bassiana* against *Anopheles stephensi* larvae (Diptera: Culicidae). Asian Pac Trop Dis 4(2):S799–S802

Zibaee A, Malagoli D (2014) Immune response of *Chilo suppressalis* Walker (Lepidoptera: Crambidae) larvae to different entomopathogenic fungi. Bull Entomol Res 104(2):155–163

Zibaee I, Bandani AR, Jalali Sendi J (2013) Pathogenicity of *Beauveria bassiana* to fall webworm (*Hyphantria cunea*) (Lepidoptera: Arctiidae) on different host plants. Plant Propt Sci 49 (4):169–176

# Chapter 12
# Entomopathogenic and Insect Parasitic Nematodes

Javad Karimi and Mahnaz Hassani-Kakhki

## 12.1 Introduction

In recent decades, modern agriculture in the way of providing global food demands has challenged by growing problems; these include, but are not limited to, pest damage, especially invasive species, climate change and an increase in the world population. An important challenge ahead of the agricultural sector is annual crop losses due to pests which is expanding worldwide. As a result of the combined effects of globalization, the easier occurrence of pest invasion happens through the exchange of pests/infested plant commodities. Also, climate changes help some insect pest establishes more successfully in the new region. In addition to the effect of climate change on extension of geographical range of pest, it has been demonstrated that this phenomenon contributes to reduction of soil water availability, shifts of thermal and moisture on cropping, soil erosion and desertification, which all means reduction in amount of crop yield (Parry 1992; Kelly and Guo 2007; Kremen et al. 2012; Oliveira et al. 2014; Zabel et al. 2014). These challenges along with an increase in world population growth have resulted in more effort to enhance crop yield per unit of arable area. A common solution is reducing pest losses through using pesticides. Though the consumption of pesticides has offered significant economic benefits, evidence indicates that in the long-term, they not only are producing toxic waste but also induce an increase in soil erosion, development of pesticides resistance, pest resurgence and replacement of a primary pest by a secondary one (Heimlich and Ogg 1982; Sabatier et al. 2014; Yadav et al. 2015).

J. Karimi (✉)
Department of Plant Protection, Ferdowsi University of Mashhad, Mashhad, Iran
e-mail: jkb@um.ac.ir

M. Hassani-Kakhki
Department of Plant Protection, School of Agriculture, Ferdowsi University of Mashhad, Mashhad, Iran

These issues have caused concerns in governments and the general public; as, efforts to combat pests on crops, while maintaining farm profitability and actual crop production levels, is moving toward sustainable agriculture (Kelly and Guo 2007). An eco-friendly strategy in managing pests is Integrated Pest Management (IPM) which provides a healthy offer of agricultural products for consumers. Based on Jacobsen (1997) definition, IPM is a sustainable approach for managing pests by combining biological, cultural, physical, and chemical tools in a way that minimizes economic, health and environmental risks. One important candidate in IPM strategy is control of the pest population using natural enemies, including predators, parasitoids and insect pathogens. Entomopathogens, mostly including viruses, bacteria, fungi, and nematodes, often play an important role in the regulation of insect populations in natural ecosystems.as an alternative method of chemical pesticides, microbial control follows three strategies against a wide variety of insect pests, including classical biological control, augmentation, and conservation.

In Iran, agriculture is pesticide dependent and pests control is often done using chemical products. This huge pesticide use, bring concerns about possible links between the growing levels of cancer incidence and chemical pesticides exposure inside the country, which leads to increased market demand for organic products. on the other hand, sometimes chemical control of certain pests is inefficient. Therefore, there is a tendency within Iranian farmers for using IPM approach by an emphasis on biological control application and insect sex pheromones in recent years. Biological control against agrarian pests in the country is a practice that has initiated in the 1930s (Abivardi 2001), and as a historical point, it has been rise and fall in its own way. Among biocontrol agents, most of the studies have done on predators and parasitoids, and between insect pathogens, *Bacillus thuringiensis* Berliner, 1915 (Bacillaceae) has received more attention than pathogenic fungi and entomopathogenic nematodes, EPNs. While the history of studies on insect pathogenic fungi in the country is much longer, work on EPNs started in the late 1990s, when first attempts were done for isolation and identification of them. Herein, we provide a short overview of IPM programs in Iran and biological control agents with an emphasis on entomopathogenic and entomoparasitic nematodes. We choose selected literature about Iranian entomopathogenic and entomoparasitic nematodes fauna and their possible contribution, challenges, and opportunity, to the biological control of insect pests in IPM programs in the country.

## 12.2 Application of Biological Control Agents in IPM Programs in Iran

From a historical perspective, the attention to biological control in Iran dates back to the Middle Ages, which in literature is mentioned to the benefits of stork and hoopoe in controlling noxious animals and termites. In 1932, however, the first recorded experience in biological control in the country was done through the introduction of *Rodolia cardinalis* Mulsant, 1850 (Coleoptera: Coccinellidae) from France by

Dr. Jalal Afshar for control of *Icerya purchasi* Maskell, 1879 (Hemiptera: Margarodidae). Since then, successful rearing and release of *Trissolcus semistriatus* Nees, 1834 (Hymenoptera: Scelionidae) against Sunn pest, *Eurygaster integriceps* Puton, 1881 (Hemiptera: Scutelleridae) was the most significant efforts for decades (Abiverdi 2001). In current years, Farmers Field School (FFS) approach, which was set up preliminary as a part of the pistachio IPM project in 1999, obviously has been influenced on using various natural enemies on different crops in the country; as studies indicate that farmers who participated in FFS have more tendencies toward biological control and using it than others (Heidari et al. 2011; Moumeni-Helali and Ahmadpour 2013).

According to being new entomopathogenic nematology in Iran, we couldn't find any documentation about the application of EPNs in IPM projects in the fields and most studies were done at the laboratory to investigate the pathogenicity of EPNs on a certain pest. But, there is a report about field application of an entomoparasitic nematode, *Romanomermis culicivorax* (Mermithidae: Nematoda), in controlling of anopheline larvae in southern Iran (Zaim et al. 1998).

## 12.3  An Outline of Studies on Entomopathogenic/ Entomoparasitic Nematodes in Iranian Research Centers

Among entomopathogens, EPNs in the families Steinernematidae and Heterorhabditidae have traits that make them suitable for using against different insect pests in above and below ground as well those with cryptic habitats. During the last three decades, many studies have done for possible control of pest by EPNs, which showed either their successes or failures. These biocontrol agents can potentially be used within an IPM project/sustainable agriculture because of some specific criteria including recycling and persistence of some EPNs species in the environment; possible effects, direct and/or indirect, on plant parasitic nematodes and plant pathogens populations; indirect role in improving soil quality; and compatibility with a wide range of chemical and biological pesticides used in IPM programs (Lacey and Georgis 2012). Currently, more than 90 species of both families are described with several unknown species which among them at least 11 species are commercially formulated and available for large-scale applications. While Steinernematidae and Heterorhabditidae consider historically as EPNs, in recent years, another group of nematodes from the family Rhabditidae including *Oscheius chongmingensis* (Zhang, Liu, Xu, Sun, Yang, An, Gao, Lin, Lai, He, Wu & Zhan) Ye, Torres-Barragan & Cardoza 2010, *Oscheius carolinensis* Ye, Torres–Barragan & Cardoza 2010, *Heterorhabditidoides rugaoensis* Zhang, Liu, Tan, Wang, Qiao, Yedid, Dai, Qiu, Yan, Tan, Su, Lai, & Gao, 2012 and *Caenorhabditis briggsae* Gochnauer & McCoy have reported as entomopathogen (Kaya et al. 2006; Dillman et al. 2012; Lewis and Clarke 2012; Zhang et al. 2012). Another remarkable entomoparasitic nematodes family is Mermithidae which is reported from at least fifteen different insect orders (Kaiser 1991).

The primary research relevant to entomophilic nematodes were done late in the 1980s by Zaim et al. (1988). In the mid-1990s, Rahim Parvizi, one of the pioneer researcher on EPNs, collected the first native isolates of EPNs and assessed their effects on some insect pests. After his works, several studies on EPNs were done by others. Encouraging results have been obtained through these investigations, which we will discuss in two sections: studies on the diversity of EPNs species in the country and their performance on pests. In the last part, we briefly provide information about other groups of entomoparasitic nematodes that were isolated from Iran.

## 12.3.1 The Occurrence of Entomopathogenic Nematodes in Iran

The initial steps for isolation of EPNs in the country goes back to 1990s, when Parvizi and co-workers isolated *Steinernema anomaly*, now *Steinernema arenarium* Wouts, Mráček, Gerdin & Bedding, 1982, and *Heterorhabditis bacteriophora* Poinar, from larvae of *Agrotis ipsilon* Hufnagel, 1766 (Lepidoptera: Noctuidae) and *Agrotis segetum* Denis & Schiffermüller, 1775 (Lepidoptera: Noctuidae), in irrigated cultivations of West Azerbaijan province. These EPNs species were the first EPNs found in Iran. Since then, surveys on the occurrence and distribution of EPNs have been carried out in different regions which resulted in isolation of additional species, including *Steinernema feltiae* (Filipjev) Wouts, Mráček, Gerdin and Bedding, 1982, *Steinernema glaseri* (Steiner) Wouts, Mráček, Gerdin & Bedding, 1982, *Steinernema carpocapsae* (Weiser) Wouts, Mráček, Gerdin & Bedding, 1982, *Steinernema bicornutum* Tallosi, Peters & Ehlers, 1995, *Steinernema kraussei* (Steiner) Travassos, 1927, *O. rugaoensis* as well as unclassified isolates (Darsouei et al. 2014; Karimi and Salari 2015). In 2007, Nikdel et al. (2011a, b) wrote about isolation a new species of steinernematid from Arasbaran forests, near the Kerengan village, East Azarbaijan province. They assigned this species as *Steinernema arasbaranense* and published a work about its description in the Nematologia Mediterranea journal. In studied sites, the most prevalent species is the ubiquitous *S. feltiae*, followed by *H. bacteriophora*. In Table 12.1, we briefly remark the achievements of the Iranian researchers on EPNs isolation and their local distribution.

**Table 12.1** List of isolated species of entomopathogenic and insect parasitic nematodes from sampling sites in Iran

| Nematode species | Isolation source | Locality | References |
|---|---|---|---|
| *Steinernema anomali* and *Heterorhabditis bacteriophora* | *Agrotis ipsilon* *A. segetum* | West Azerbaijan province | Parvizi (1998) |
| *Steinernema* sp. and *H. bacteriophora* | Soil | West Azerbaijan province | Parvizi (2003) |
| *S. feltiae* | Soil | Mazandaran and Tehran provinces | Tanha Ma'afi et al. (2006) |
| *S. carpocapsae* | *P. olivieri* | Tehran province | Karimi (2007) |
| *S. carpocapsae*, *S. glaseri* and *H. bacteriophora* | *Polyphylla olivieri* | Tehran province | Karimi and Kharazi-pakdel (2007) |
| *S. carpocapsae*, *S. feltiae* and *H. bacteriophora* | Soil | Arasbaran forests, northwestern Iran | Nikdel et al. (2008) |
| *S. carpocapsae*, *S. feltiae*, *S. bicornutum* and *H. bacteriophora* | Orchards, alfalfa fields and grasslands | Ardabil, East Azerbaijan and West Azerbaijan provinces | Eivazian Kary et al. (2009) |
| *S. glaseri* | *P. olivieri* | Tehran province | Karimi et al. (2009a) |
| *S. feltiae* | Soil | Tehran province | Karimi et al. (2009b) |
| *S. feltiae* | Soil | Kohgiluyeh and Boyer-Ahmad province | Abdollahi (2010) |
| *H. bacteriophora*, *S. feltiae* and *S. carpocapsae* | Soil of potato fields | East and West Azerbaijan provinces | Agazadeh et al. (2010) |
| *S. bicornutum* | Soil | Marand, East Azerbaijan province | Eivazian Kary et al. (2010a) |
| *H. bacteriophora*, *S. feltiae*, *S. carpocapsae*, *S. bicornutum*, *S. glaseri* and *S. kraussei* | Soil | Arasbaran forests, northwestern Iran | Nikdel et al. (2010a) |
| *H. bacteriophora*, *S. feltiae* and *S. carpocapsae* | Soil of walnut orchards | Arak, Markazi province | Ashtari et al. (2011) |
| *H. bacteriophora* and *S. feltiae* | Soil | Tabriz, East Azerbaijan province | Ebrahimi and Niknam (2011) |
| *S. arasbaranense* | Soil | Kerengan village, East Azerbaijan province | Nikdel et al. (2011a, b) |
| *Hexamermis* cf. *albicans* | *Lymantria dispar Euproctis chrysorrhoea* | Kaleibar Township, East Azerbaijan province | Nikdel et al. (2011a, b) |
| *S. carpocapsae* and *S. feltiae* | Soil | Hamedan province | Saffari et al. (2012) |

(continued)

**Table 12.1** (continued)

| Nematode species | Isolation source | Locality | References |
|---|---|---|---|
| *H. bacteriophora* and *S. feltiae* | Soil | Mashhad, Razavi Khorasan province | Hassani-Kakhki et al. (2013) |
| *H. bacteriophora* and *S. feltiae* | Soil | Bojnourd, North Khorasan province | Kamali et al. (2013) |
| *H. bacteriophora* and *S. feltiae* | Potato fields | Farooj, North Khorasan province | Rahatkhah et al. (2015) |
| *H. bacteriophora* and *S. feltiae* | Soil | Kurdistan province | Abdolmaleki et al. (2016) |
| *H. bacteriophora* | Soil | Kerman province | Seddiqi et al. (2016) |
| *S. feltiae* | Soil | Ardebil | Ebrahimi et al. (2019) |

## 12.3.2 Entomopathogenic Nematodes Experimental Applications in Iran: Case Studies of Laboratory and Greenhouse Experiments in Iran

Despite the knowledge on biological control programs in Iran, the use of EPNs has been restricted to the research level, mainly due to the absence of EPN based products on the Iran market. Only, a limited number of EPN products exist in research centers currently. So, this chapter is a summary of some literature of carried out studies which include laboratory or greenhouse trials about pathogenicity and virulence of native or commercial species/populations of EPNs against certain pests (summarized in Table 12.2).

### 12.3.2.1 The Colorado Potato Beetle, *Leptinotarsa decemlineata* Say, 1824 (Coleoptera: Chrysomelidae)

Among coleopterans, the Colorado potato beetle (CPB), *Leptinotarsa decemlineata* Say, 1824 (Coleoptera: Chrysomelidae) is one of the most economically damaging pests to potato, which developed resistance to the chemical insecticides. To assess the potential of indigenous species of EPN on CPB, Eivazian et al. (2010b) conducted a study using *H. bacteriophora* (four isolates), *S. bicornutum*, *S. carpocapsae* and *S. feltiae* against CPB with three laboratory trials: filter paper (on last larval instar), potato leaf (on last larval instar) and soil (on pre-pupa). In all trials, *H. bacteriophora* caused the highest mortality and S. *bicornutum* had the lowest mortality. In another study, Ebrahimi et al. (2011) assessed the lethal and sublethal effects of two Iranian species of *S. feltiae* and *H. bacteriophora* on the prepupal stage of the CPB, the life stage that EPNs can target in the soil. Results revealed that both species were effective against CPB, but *H. bacteriophora* was

**Table 12.2** Species of target insects and applied entomopathogenic nematode species against them in research and academic centers of Iran (Karimi and Salari 2015)

| Order | Family | Scientific name | Nematode Sp. | References |
|---|---|---|---|---|
| Coleoptera | Curculionidae | *Conorrhynchus brevirostris* | Hb, Ssp. | Parvizi (1998) |
| | | *Curculio glandium* | Hb Sb | Nikdel et al. (2008) |
| | | *Hypera postica* | Sc, Sf | Falahi et al. (2011) |
| | Chrysomelidae | *Leptinotarsa decemlineata* | Hb, Ssp. Sb, Sc, Sf | Parvizi (2000) Eivazian Kary et al. (2010b) Ebrahimi and Niknam (2011) and Ebrahimi et al. (2011) |
| | | *Bruchus lentis* | Sf | Saeidi et al. (2018) |
| | Cerambyciae | *Osphranteria coerulescens* | Hb, Sc | Sharifi et al. (2014) |
| | Melolonthidae | *Polyphylla olivieri* | Hb, Ssp. Sc, Sg | Parvizi (2001) Karimi and Kharazi-pakdel (2007) |
| | | *Polyphylla adspersa* | Hb, Sc, Sf | Karimi (2007) |
| | Scarabaeidae | *Pentadon idiota* | Hb, Sc, Sf | Edraki et al. (2016) |
| Dermaptera | Forficulidae | *Forficula auricularia* | Hb | Kordestani et al. (2013) |
| Diptera | Tephritidae | *Dacus ciliatus* | Hb, Sc | Kamali et al. (2013) |
| | Agromyzidae | *Liriomyza trifolii* *L. sativae* | Sf | Ebrahimi et al. (2012, 2016) |
| Hemiptera | Aleyrodidae | *Trialeurodes vaporariorum* | Hb, Sf | Rezaei et al. (2015) |
| Hymenoptera | Argidae | *Arge ochropus* | Hb, Sc | Sheykhnejad et al. (2014) |
| Lepidoptera | Noctuidae | *Agrotis ipsilon* | Hb | Parvizi (2004) |
| | | *A. segetum* | Sc | Ebrahimi et al. (2019) |
| | | *Helicoverpa armigera* | Hb, Sf Sc | Ebrahimi et al. (2008, 2018) Eivazian Kary et al. (2012) |
| | | *Heliothis viriplaca* | Hb, Ssp. | Parvizi (1998) |
| | | *Spodoptera exigua* | Hb, Ssp. Sc | Parvizi (1998) Aramideh et al. (2005) Karimi et al. (2009b) |
| | Lymantriidae | *Euproctis chrysorrhoea* | Hb, Sc | Nikdel et al. (2010b) |
| | Pieridae | *Pieris rapae* | Hb, Ssp. | Parvizi (1998) |
| | | *P. brassicae* | Hb, Sf | Abdolmaleki et al. (2017) |
| | Gelechiidae | *Phthorimaea operculella* | Sc, Sf, Sg, Hb | Hassani-Kakhki et al. (2013) Eivazian Kary et al. (2018) |
| | | *Tuta absoluta* | Sf Hb, Sc | Abootorabi (2014) Kamali et al. (2017) |
| | | | Sf | Amizadeh et al. (2019) |

(continued)

**Table 12.2** (continued)

| Order | Family | Scientific name | Nematode Sp. | References |
|---|---|---|---|---|
| | Sesiidae | *Synanthedon myopaeformis* | Hb, Ssp. | Parvizi (2002, 2003) |
| | | *Paranthrene diaphana* | Hb, Sc, Sf | Azarnia et al. (2018) |
| | Cossidae | *Zeuzera pyrina* | Hb, Sc | Ashtari et al. (2011) Salari et al. (2014) |
| | Pyralidae | *Ectomyelois ceratoniae* | Hb, Sc, Sf | Memari et al. (2017) |
| | | *Glyphodes pyloalis* | Sc | Mallahi et al. (2019) |
| | Plutellidae | *Plutella xylostella* | Hb, Sc, Sf | Eivazian Kary et al. (2019) |
| | | | Hb, Sc | Zolfagharian et al. (2016) |
| Thysanoptera | Thripidae | *Thrips tabaci* | Hb, Sc, Sf | Kashkouli et al. (2014) Saffari et al. (2013) |

Abbreviations used in Table 2. *Hb H. bacteriophora, Sb S. bicornutum, Sc S. carpocapsae, Sf S. feltiae, Sg S. glaseri, Ssp Steinernema* sp.

more effective at lower concentrations compared to *S. feltiae*. A considerable part of this work was the evaluation of sublethal doses of EPNs on the pest. Some adult insects that survived after exposure to nematodes, 23.5% and 18.2% of insects treated with *H. bacteriophora* and *S. feltiae*, respectively, showed signs like deformed elytra or incomplete formation of wings. Also, in 45.8% of insects treated with *S. feltiae*, delayed eclosion in adult was observed. Sublethal effect of EPN infection on the host would have adverse consequences on the fitness criteria of the target pest, e.g. mating success, fecundity, and host-finding ability. Before this study, the fate of surviving adult insects after infection with EPNs in larval stage have not been reported; these new finding revealed another positive trait of the EPNs ability in controlling insect pest.

### 12.3.2.2    The Potato Tuber Moth, *Phthorimaea operculella* Zeller, 1873 (Lepidoptera: Gelechiidae)

Potato tuber moth (PTM), *Phthorimaea operculella* Zeller, 1873 (Lepidoptera: Gelechiidae) is a primary pest of solanaceous plants which contributes to potato loss in field and storage. For controlling PTM in the field, farmers rely on the repeated application of wide spectrum insecticides. On the other hand, protected tunneling behavior of PTM larvae in foliage and tubers makes it difficult to control, as some studies have found pesticides application can't effectively control PTM when the tubers are infested under the ground. So, the selection of alternative strategies, such as biocontrol agents, for use in PTM control must rely on behavior of its larvae which move in the soil, for pupation or infesting tubers. With regard to the ability of EPNs that can actively search their hosts in a complex environment such as soil, Hassani-Kakhki et al. (2013) assessed the effect of EPNs isolate/species on PTM. The primary objective of their work was to evaluate the susceptibility of different PTM stages, i.e. second (L2) and fourth instar larvae (L4) as well prepupa to four EPN species with different foraging strategies, *S. carpocapsae*, [isolated IJs from commercial product under the trade name Capsanem®], *S. feltiae*, *S. glaseri* and *H. bacteriophora* FUM7 and commercial isolate [under the trade name Larvanem®] in Petri dishes. The initial assessment on filter paper showed that *S. carpocapsae* and *H. bacteriophora* (FUM7 and commercial) species caused the highest mortality on both larval and prepupal stages of PTM. In general, prepupa was the most susceptible stage while pupa had the lowest mortality.

A possible approach for control the destructive stages of PTM that infest tuber or inter to the soil for pupation could be the application of EPNs on the soil surface. Hence, complementary tests in soil columns, by three soil types: loamy, sandy and sandy-loamy, containing infested tuber were established using *S. carpocapsae* and *H. bacteriophora* (FUM7 and commercial) species. Results from this semi-field assay indicated larval mortality induced by *S. carpocapsae* (88%) was higher than those caused either by the commercial population of *H. bacteriophora* (79%) or by *H. bacteriophora* FUM 7 (78%). Also, EPN species had higher performance in lighter soils, which resulted in higher mortality on younger larvae than that in loamy soil (Hassani-Kakhki et al. 2013). The effectiveness of EPN species to infect PTM larvae and prepupa in loamy and sandy-loamy soils are important because potato grows well in these soil textures. On the other hand, during surveys on natural occurrence of EPNs in potato fields of the country, *S. feltiae*, *S. carpocapsae* and *H. bacteriophora* were recovered (Agazadeh et al. 2010; Rahatkhah et al. 2015). In this respect, by providing suitable conditions after EPNs application, IJs can survive, establish and recycle, as one would expect to make a long-term regulation of PTM during the growing season.

### 12.3.2.3   Onion Thrips, *Thrips tabaci* Lindeman, 1889 (Thysanoptera: Thripidae)

Onion thrips, *Thrips tabaci* Lindeman, 1889 (Thysanoptera: Thripidae) is an important pest that has high economic importance. Control of this thrips due to their cryptic behavior and resistance to chemical pesticides has become the major limitation in their management. Thrips pupate in the soil which makes them theoretically reachable targets for EPNs. Hence, Saffari et al. (2013) conducted a study to determine the pathogenicity of a native isolate of *S. feltiae* and two commercial species, *H. bacteriophora* and *S. carpocapsae* against second instar larva, prepupa, and pupa of *T. tabaci* under laboratory conditions. Their results indicated that prepupa was the most vulnerable stage, and the second instar larva showed the least susceptibility to EPNs. *Heterorhabditis bacteriophora* caused the highest mortality in prepupa (54%) while its effect was significantly reduced on second instar larva (25%) at 10,000 IJs/ml. Similarly, *S. carpocapsae* was most effective on prepupal stage (49%) and significantly ineffective against the second instar larva (20%). Also, dissection of cadavers showed that EPNs were unable to reproduce inside immature stages of thrips.

### 12.3.2.4  The Cucurbit Fruit Fly, *Dacus ciliatus* Loew, 1862 (Diptera: Tephritidae)

The cucurbit fruit fly, *Dacus ciliatus* Loew, 1862 (Diptera: Tephritidae) is the most serious pest of cucumber, melon, watermelon and related fruits in Iran (Arghand 1979). This fruit fly is multivoltine without any diapause in the southern part of the country, causing serious damage on fruits from July to September (Arghand 1979; Hancock 1989). Currently, the use of organophosphate insecticides is the main control method for *D. ciliatus*, as authors saw several cases when farmers dip infested fruits into insecticide solutions.

Kamali et al. (2013) examined biocontrol potential of the entomopathogenic nematodes, *H. bacteriophora* and *S. carpocapsae* on third instar larvae, pupae, and adult cucurbit flies under laboratory and greenhouse conditions. In laboratory experiments, filter paper and potting soil, against third instar larvae, *S. carpocapsae* and *H. bacteriophora* exhibited high infectivity and successfully reproduced and emerged from larvae of *D. ciliatus*. Both nematode species induced low mortality on pupae. These EPN species were effective against adult flies, as *S. carpocapsae* caused higher adult mortality in than *H. bacteriophora*. In greenhouse test, they observed that EPNs caused mortality in fly larvae which were located inside the fruit (Fig. 12.1). When EPNs applied directly to the infested cucumber against *D. ciliatus* larvae, the percentage of larval mortality caused with *S. carpocapsae* (27.7 ± 3.1) was significantly greater than that produced with *H. bacteriophora* (12.8 ± 1.8). In

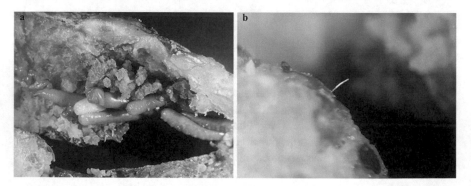

**Fig. 12.1** Application of entomopathogenic nematodes against infested cucumber fruit with cucurbit fruit fly, *Dacus ciliates*. Dead larvae of cucurbit fruit fly infected with *Steinernema carpocapsae* (**a**). Nictation of *S. carpocapsae* larva inside the cucumber fruit (**b**). (Courtesy of Sh. Kamali, Ferdowsi University of Mashhad)

addition, the effect of abiotic factors such as soil texture and the temperature was examined on pathogenicity of these nematodes. The higher concentrations of EPNs induced greater larval mortality of *D. ciliatus* both in sandy loam and sand soil types, whereas all concentrations of EPN species had lower virulence in clay loam. Also, the larval mortality caused by both nematodes at 25 and 30 °C was higher than that at 19 °C. This research highlighted the performance of EPNs on a key pest in Iran and demonstrated the feasibility of using EPNs in *D. ciliatus* IPM programs.

### 12.3.2.5   The Greenhouse Whitefly, *Trialeurodes vaporariorum* Westwood, 1856 (Hemiptera: Aleyrodidae)

The greenhouse whitefly, *Trialeurodes vaporariorum* Westwood, 1856 (Hemiptera: Aleyrodidae) is a cosmopolitan greenhouse pest that poses major problems for farmers. In addition to decrease crop quality through feeding on plant phloem and reduction of photosynthesis due to the development of sooty mold on excreted honeydew of whitefly on leaves, this pest also can transmit plant viruses (Moreau and Isman 2012). Also, there are concerns about the development of resistance to pesticides in *T. vaporariorum*, as prompt researchers to find powerful alternative tactics against this whitefly in IPM program. One scheme for controlling the greenhouse pests, could be the application of EPNs in the greenhouse environment.

Rezaei et al. (2015) evaluated pathogenicity of *S. feltiae* and *H. bacteriophora* on adults and second instar nymphs of *T. vaporariorum* on two different host-plants, sweet pepper, and cucumber, under laboratory and greenhouse conditions. In laboratory tests, both life stages were susceptible to infection by EPNs, but *S. feltiae* had a lower $LC_{50}$ than *H. bacteriophora*. In greenhouse experiments, they observed that *S. feltiae* in lower concentrations than *H. bacteriophora* cause the same mortality in nymphs. Furthermore, application of 250 IJ/cm$^2$ of *S. feltiae* on cucumber resulted in the highest mortality rate (49 ± 1.23%) of the second nymphal instar. In general, in greenhouse trials, the mortality of second nymphal instars was significantly higher on cucumber plant, with hairy leaf, than on pepper plant, with glabrous leaf. Probably, the high density of trichome on the cucumber leaf surface increases humidity, which is critical for enhancing EPN survival. Considering these promising results, probably a foliar application of EPNs with a suitable adjuvant in combination with selective insecticides can cause a profitable control of this greenhouse pest in the short-term.

### 12.3.2.6   The Tomato Leafminer, *Tuta absoluta* Meyrick, 1917 (Lepidoptera: Gelechiidae)

The tomato leaf miner, *Tuta absoluta* Meyrick, 1917 (Lepidoptera: Gelechiidae) is one of the most devastating tomato pests. After its initial tracing in eastern Spain in 2006, it rapidly invaded other European countries and extended throughout the Mediterranean basin. In Iran, *T. absoluta* reported for the first time from Urmia,

West Azerbaijan province (northwestern Iran) in 2010 and distributed in more than 20 various regions before 2012 (Gharekhani and Salek-Ebrahimi 2013). This multivoltine pest mines leaves, fruits, flowers, buds, and stems, which feeding damage causes important yield losses. There are reports regarding insecticide resistance development in populations of *T. absoluta* (Lietti et al. 2005). Hence, using of IPM program with an emphasis on biocontrol agents, like EPNs, against this pest is highlighted, as entomopathogens such as *S. feltiae*, *S. carpocapsae*, and *H. bacteriophora*, have showed efficacy versus *T. absoluta* in laboratory and growth chamber (Batalla-Carrera et al. 2010; Garcia-del-Pino et al. 2013).

Kamali et al. (2017) designed a series of experiments to identify the possible relations among concentration of IJs, soil type, temperature and exposure time, using two species *H. bacteriophora* and *S. carpocapsae* against different life stages of *T. absoluta*. The results showed that the last larval instar of leaf miner was highly susceptible to both *S. carpocapsae* and *H. bacteriophora*. The calculated $LC_{50}$ value after plate assay was 2 IJs/cm$^2$ for *S. carpocapsae* and 1 IJs/cm$^2$ for *H. bacteriophora*. The highest mortality percentage occurred in the coco peat substrate (92.6%) and loamy sand (88.9%), and the lowest level was observed in the sandy loam soil (16.8%). The optimal temperature for the infectivity of *H. bacteriophora* was 25 ± 1 °C, while for *S. carpocapsae* it was 31 ± 1 °C, which caused a maximum mortality rate of 96.0% and 90.6%, respectively. The exposure of last larval instar to *S. carpocapsae* for 240 min was sufficient for an infection of 95.5% of the larvae, whereas 76.9% of the last larval instar infected after 65 min by the same concentration (20 IJs/cm$^2$). In their observations, they found that larvae died inside the galleries after the treatment of tomato plant leaves. Considering the results, they argued that EPNs species such as *S. carpocapsae* need a short time to find and infect the leaf miner larvae within the galleries, as the first 65 min post-application was adequate time for this event. They argued that temperature was not a limiting factor for *S. carpocapsae* and *H. bacteriophora* in their study. Kamali et al. (2017) suggested both species has potential to use in *T. absoluta* management programs.

Based on their experience, it seems these entomopathogens will be effective when pest population is low. In this case, using pheromones for pest monitoring will help to promote the right time for EPNs applications in green house, when pest population is still low.

### 12.3.2.7 The Apple Clearwing Moth, *Synanthedon myopaeformis* Borkhausen, 1789 (Lepidoptera: Sesiidae)

The apple clearwing moth, *Synanthedon myopaeformis* Borkhausen, 1789 (Lepidoptera, Sesiidae) is an economic pest in apple orchards. The larvae exist under tree bark, feed on sap and create galleries between the ross and the cambium layer. The larval feeding behavior leads to a decrease in yield and weakens the host tree, increases susceptibility to bark beetles and fungal diseases. Due to the cryptic habitat of larvae, where they are protected from insecticidal sprays, as well as long oviposition period of female, chemical control is insufficient. So, in an experiment, the efficiency of two native species of *Steinernema* sp. and *H. bacteriophora*, against *S. myopaeformis* was evaluated in a heavily infested apple orchard by Parvizi (2003). A concentration of $10^6$ IJs/300 cm$^3$ of distilled water of *Steinernema* sp. and *H. bacteriophora* were sprayed onto the trunks of infested trees. The outcomes showed that *Steinernema* sp. (up to 60% mortality) is more efficient against *S. myopaeformis* in comparison to *H. bacteriophora* (less than 6% mortality). These observations in combination with additional studies in other countries about efficiency of EPNs on *S. myopaeformis* suggest that EPNs could be a suitable agent in IPM program against this pest.

### 12.3.2.8   The Acorn Weevil, *Curculio glandium* Marsham, 1802 (Coleoptera: Curculionidae)

Oak forests are important ecosystems, which play significant roles in the world. An important pest of the oak forest is the acorn weevil, *Curculio glandium* Marsham, 1802 (Coleoptera: Curculionidae), which its larvae feed on the oak seed. Its damage, for example, in the Hatam-Mashaci forest, located near the city of Mcshkin-Shahr, East Azerbaijan province, had a severe influence on oak regeneration, inducing rotting of almost 70% of oak acorns (Arefipour et al. 2005). The larvae feed and develop within the acorns, and when the ripe acorns fall to the ground, fully developed larvae cease feeding and burrow into the soil to overwinter. Application of chemical pesticide in the soil of oak forests for control this pest has an adverse effect on natural biodiversity and soil biota. Considering the life cycle of pest which live in the soil for overwintering, EPNs can be a potential agent in controlling *C. glandium* in such vulnerable ecosystems.

Nikdel et al. (2008) conducted a study to determine the potential of native species of EPNs, *H. bacteriophora* and *S. bicornutum*, against fifth instar larval of acorn weevil. Bioassays were carried out under laboratory conditions, at two temperature ranges (21–24 °C and 25–28 °C). They observed that the penetration rate of *H. bacteriophora* (1.6%) was higher than *S. bicornutum* (0.55%). Also, the maximum mortality rates caused by *H. bacteriophora* and *S. bicornutum* were 58.3% and 25% (at 21–24 °C) and 63.5% and 30.5% (at 25–28 °C), respectively. Therefore, it seems that the application of EPNs as biocontrol agents against oak pest including *C. glandium* could be an effective recommended strategy to Government Forestry Services.

### 12.3.2.9   The Leopard Moth, *Zeuzera pyrina* L., 1761 (Lepidoptera: Cossidae)

The leopard moth borer, *Zeuzera pyrina* L., 1761 (Lepidoptera: Cossidae) is a polyphagous pest with increasing importance in some countries. The pest attacks more than 150 plant species of up to 20 taxonomic genera, including walnut, apple, pear, peach, cherry and olive trees. Its larvae are wood borers that make deep tunnels, up to 50 cm, in the main branches and trunk of the tree where the main damage occurs at phloem and xylem vessels of the tree. Damage causes weakness in tree and attraction of bark beetles which finally resulted in tree death, especially in the newly established orchards. In recent years, Z. pyrina has become the most damaging pest of walnut trees in Iran, as already led to the uprooting of many walnut groves in some area. The pest has one generation per 2 years, but moths and damages appear every year. On the other hand, the flight period of the moths lasts about 3 months which a female adult can lays about 1000 eggs during this time, after hatching, the larvae bore into twigs, branches, and trunks, so its monitoring and control are extremely difficult (Kutinkova et al. 2006; Gharalari and Kolyai 2014). Among natural enemies, EPNs have adapted for possible use in such hidden habitat.

Ashtari et al. (2011) started investigations toward development of EPNs against the leopard moth borer begin with laboratory bioassays to determine EPNs pathogenicity, followed by comparisons of virulence among different microbial isolates or strains. They used a native isolate of H. bacteriophora and commercial products of S. carpocapsae and H. bacteriophora against the pest larvae in the laboratory bioassays at the rate of 2000 IJs per larva in petri dish. In laboratory tests, S. carpocapsae caused 100% mortality in 2nd, 3rd, and 4th instars larvae at 54, 30 and 36 h after treatment, respectively. Heterorhabditis bacteriophora induced 100% mortality in 2nd, 3rd, and 4th instar larvae at 44, 40 and 52 h after treatment, respectively. For field application, Ashtari et al. (2011) applied same strains/isolate via injection of nematode suspensions, 2000 IJs per active hole, using a 60 cc plastic syringe into the galleries bored in tree stems or branches. Ten days' post application, larvae of Z. pyrina were excerpted from the galleries using a wire, and the number of alive and dead larva was determined. The result showed that both tested nematodes at 2000 IJs/larva proved to be effective on the leopard moth borer larvae (Fig. 12.2).

Subsequently, Salari et al. (2014) evaluated the pathogenicity of S. carpocapsae and H. bacteriophora on Z. pyrina in another study, including the plate and infested branch assay in laboratory condition. The plate assay showed high virulence of both EPNs against the larvae. The calculated values of $LC_{50}$ indicated that S. carpocapsae was comparatively more virulent than H. bacteriophora against larvae. For the branch test, infested branches with active holes were collected from walnut orchards and transferred to the laboratory. Injection of 160 IJs/cm$^2$ of S. carpocapsae and H. bacteriophora to the branch holes induced high mortality (more than 70%) on pest larvae. Also, S. carpocapsae caused higher mortality in the larger larvae than H. bacteriophora.

Currently, pheromone traps and occluding the holes created by the pest, through applying a mixture of glue and fumigating pesticides, are practices that used in IPM programs in walnut orchards. Hidden habitat like galleries inside the trees is considered as a most favorable environment for EPNs activity, though the EPNs delivery method is not practical because it is time consuming and difficult; The female leopard moth lays hundreds eggs which lead to many holes in a tree. On the other hand, usually the height of branches of the walnut trees is a problem for the operator. Maybe using hand-held lances for spraying branches is more practical in

**Fig. 12.2** Field application of entomopathogenic nematodes against the leopard moth borer, *Zeuzera pyrina*. A suspension of 2000 IJs per active hole, using a 60 cc plastic syringe injected into the galleries bored in tree stems or branches (**a**, **b**). Ten days' post application, marked branches were collected and transferred to the laboratory (**c**, **d**). The larvae of *Z. pyrina* were excerpted from the galleries and the number of alive and dead larva was determined. Dead larvae of leopard moth borer infected with *Steinernema carpocapsae* inside the branch (**e**) which dissection of cadaver showed developing nematodes (**f**). (Courtesy of E. Salari, Ferdowsi University of Mashhad)

this case, however this method needs more evaluations to see if nematodes can reach the holes or not before the desiccation. Even if using nematodes can cause 30% mortality in orchards it worth to use, because this pest can destroy a walnut orchard in few years. Using different methods along application of EPNs in IPM programs against this pest is promising.

**12.3.2.10    The Rosaceae Longhorned Beetle,** *Osphranteria coerulescens*
**Redtenbacher, 1850 (Coleoptera: Cerambycidae)**

A serious and economically important pest that causes extensive damage to rosaceous trees in Iran, is the Rosaceae longhorned beetle, *Osphranteria coerulescens*
Redtenbacher, 1850 (Coleoptera: Cerambycidae). Adult beetles lay a single egg into
the upper part of tree twigs. When larva emerges, it bores into the cambium and later
penetrates to hardwood, the center of branches and trunks where feeds for a year.
Due to drought in the country in recent years, some fruit trees, such as almonds, have
been more susceptible to this pest. As similar to *Z. pyrina*, in several area,
*O. coerulescens* damage led to the destruction of many almonds orchards. Efforts
to control this pest are insufficient because of the cryptic behavior of larvae. The
limited movement of *O. coerulescens* larvae, as well as the moist protected environment within their galleries, suggest that EPNs have ability to find and kill larvae
in their habitat. So, Sharifi et al. (2014) evaluated the efficacy of *H. bacteriophora*
and *S. carpocapsae*, against the larvae of this longhorned beetle. The plate assay
showed that the larvae were susceptible to both EPN species but were more
susceptible to *S. carpocapsae* (65–97% mortality) than *H. bacteriophora* (42–88%
mortality). Efficacy of the EPN application against *O. coerulescens* depends, among
other factors, on the ability of the nematodes to find pest in galleries; so these
researchers in the next step, designed laboratory experiments on an infested branch
of apricot, 15–17 cm in length. The tip of the branch, where the larva of
*O. coerulescens* deposits its frass, was placed into a 45 $cm^3$ sponge piece with a
hole inside it. The EPN suspension at a concentration of 10,000 IJs/ml was inoculated into each sponge piece. The result showed that the EPN species located and
killed the larvae within galleries in the branches. These findings highlighted the
ability of EPNs as a potential biocontrol agent of cryptic pests in the IPM program.
    Like the other wood borer pest, the leopard moth borer, the method using
nematodes against *O. coerulescens* is challenging due to complexity of application
method. Although, further field experiments are needed to evaluate their efficacy
under the wide environmental condition.

# 12.4    The Occurrence and Application of Entomoparasitic
##         Nematodes in Iran

The first attempt for the application of entomoparasitic nematodes against an insect
pest in the country was done by Zaim et al. (1988). They investigated the efficacy of
the parasitic nematode *Romanomermis culcivorax* (Nematoda: Mermithidae) in
controlling anopheline larvae, including malaria vectors, in Fars and Baluchistan
provinces. They did not observe any correlation between the level of parasitism and
the density of mosquito larvae present in a site. However, R. culcivorax was

established in the release sites but caused only minor reductions in anopheline larval populations.

Nikdel et al. (2011a, b) reported *Hexamermis* cf. *albicans* (Nematoda: Mermithidae) from infected larvae of two lepidopterans, *Euproctis chrysorrhoea* L. 1785 (Lepidoptera: *Erebidae*) and *Lymantria dispar* L. 1785 (Lepidoptera: *Erebidae*) in the Arasbaran area, northwest of Iran. *Deladenus* species is another group of nematodes which is associated with insect. Three species of this genus including *Deladenus durus* (Jahanshahi Afshar et al. 2014) *D. persicus* sp. n. and *D. apopkaetus* (Miraeiz et al. 2017) were reported from Iran.

Also, the occurrence of some species of the genus *Oscheius* Andrássy, 1976 (Nematoda: Rhabditida) from different regions of Iran were reported. These species include *Oscheius tipulae*, *O. rugaoensis*, *O. necromenus* and *O. onirici* (Hassani-Kakhki et al. 2012; Darsouei et al. 2014; Valizadeh et al. 2017; Karimi et al. 2018).

Salari et al. (2019) surveyed to assess the effect of *Acrobeloides maximus* (Rhabditida: Cephalobidae), collected from Kerman province, against the larvae of the leopard moth. They found that in plate assays, larval mortality was significantly influenced by the concentration of the nematode and after 72 h, the highest mortality of pest larvae was observed at the concentration of 200 IJs per larva.

## 12.5   Conclusion

Based on reports, about 26,000 tons of pesticides including more than 140 types of active ingredients are distributed throughout the country annually. Integrated Pest Management program is a strategy that reduces dependence on chemical control and minimizes the problems associated with pesticides. In recent years, responsible organizations (both governmental and NGOs) in the agriculture sector have contributed to the operation of FAO Regional IPM Project in the country which has led to the acceptance of this procedure among farmers, with an increase in using various natural enemies in different crops. However, there is no doubt that Iranian farmers are at the beginning of the use of biocontrol agents in IPM projects and are still far away from an ideal situation. For example, on account of some issues like import restrictions of entomopathogens based products, like EPNs there is no or little information about their efficiency between growers, which means no demand in the market; hence we could not find any formal record about the application of EPNs in IPM program by Iranian farmers. Nevertheless, there is a developing tendency between scientific workers and we think the prospect of EPNs application in the country is bright.

It seems that for the introduction of these agents for use by farmers, stakeholders, and the agriculture sector, the first step is convincing the government to issue import licenses for EPNs based products. This may be achieved through information obtained from international and national research about the efficacy and safety of EPNs. In the next step, we need to encourage growers to use EPNs. This purpose may be achieved through a network of technical assistance which counsel growers for effective control of pests by EPNs in an IPM program. In this way, growers may even find a more efficient method for EPNs application in their region and share their experience with others via this network; it helps to localization the use of EPNs, considering factors affecting their efficiency in each region. Also, Iranian researchers are interested to collaborate with other researchers in the world and share their experiences to promote the application of EPNs in IPM programs.

Notably, entomopathogenic nematology is a new research field in Iran, as much research input has been directed towards the isolation of indigenous EPN species/ strains and studies about the effect of some of them on insect pests in laboratories condition. Native biocontrol agents are often preferable in biological control programs since they are adapted to local conditions. In this respect, Iran has a diverse climate, from arid or semiarid, to subtropical along the Caspian coast and the northern forests, which provides different vegetation habitats plus a rich fauna of insect species. So, there are opportunities to discover efficacious species/strains, more virulent or tolerant to environmental extreme conditions, and even novel EPN species from the country, especially from natural undisturbed sites.

By considering results obtained from studies, whether in Iran or elsewhere, simultaneously control of several pests that exist at the same within a field/greenhouse, like onion thrips, greenhouse whitefly, tomato leaf miner and even plant parasitic nematodes, by entomopathogenic nematodes is not far-fetched. In this regard, when some EPNs species are applied in soil, IJs have the opportunity to survive, persist and provide long term control of pests during the growing season, or even during several years, which means using less chemical pesticide. It should be kept in mind that selection of an EPN species as biocontrol agents in IPM program must be done based on two criteria: 1- factors related to EPN, including natural enemies and inherent limitations and 2- those related to target pest including host plant, the aimed agroecosystem and also circumstance and facilities available where the program is carried out. Considering obtained data from laboratory/field experiment along with mentioned criteria in above, using EPNs can be an important player in IPM program for controlling of pest.

# References

Abdollahi M (2010) Report of *Steinernema feltiae* in Kohgiluyeh and Boyerahmad province, Iran. In: Manzari SH (ed) Proceedings of the 19th Iranian plant protection congress, Tehran, Aguest 2010. p 589
Abdolmaleki A, Tanha Maafi Z, Rafiee H et al (2016) Isolation and identification of entomopathogenic nematodes and their symbiotic bacteria from Kurdistan province in Iran. J Crop Prot 5(2):259–271
Abdolmaleki A, Maafi Z, Naseri B (2017) Virulence of two entomopathogenic nematodes through their interaction with *Beauveria bassiana* and *Bacillus thuringiensis* against *Pieris brassicae* (Lepidoptera: Pieridae). J Crop Prot 6(2):287–299
Abivardi C (2001) Iranian entomology: an introduction. Springer, Heidelberg
Abootorabi E (2014) Report of native isolate pathogenicity of *Steinernema feltiae* on tomato leafminer, *Tuta absoluta*. Biocontrol Plant Prot 1:107–109

Agazadeh M, Mohammadi D, Eivazian Kary N (2010) Molecular identification of Iranian isolates of the genus *Photorhabdus* and *Xenorhabdus* (Enterobacteriaceae) based on 16S rRNA. Mun Ent Zool 5:772–779

Amizadeh M, Hejazi MJ, Niknam G et al (2019) Interaction between the entomopathogenic nematode, *Steinernema feltiae* and selected chemical insecticides for management of the tomato leafminer. Tuta absoluta BioControl 64:709–721

Aramideh SH, Safarali Zadeh MA, Purmirza AA et al (2005) Study on sensitivity of larvae, pupa and pre–pupa stages of *Spodoptera exigua* H. to *Steinernema carpocapsae* under laboratory condition and on sugar cane plant. J Agric Nat Resour Sci 12:159–166

Arefipour MR, Askary H, Yarmand H et al (2005) Oak forests decline in Iran. In: Villemant C, Lahbib Ben Jamāa M (eds) Integrated protection in oak forests. Proceedings of a meeting at Hammamet, Tunisia, October, 2004. IOBC/wprs Bulletin 28:41–43

Arghand B (1979) Introduction on *Dacus* sp. and preliminary examination in the Hormozgan province. Appl Entomol Phytopathol 51:3–9

Ashtari M, Karimi J, Rezapanah MR et al (2011) Biocontrol of leopard moth, *Zeuzera pyrina* L. (Lep.: Cossidae) using entomopathogenic nematodes in Iran. In: Ehlers RU, Crickmore N, Enkerli N, et al. (eds) Insect pathogens and entomopathogenic nematodes. Preeceedings of the 13th European meeting, Innsbruck, June 2011. IOBC/wprs Bulletin 66:333–335

Azarnia S, Abbasipour H, Saeedizadeh A et al (2018) Laboratory assay of entomopathogenic nematodes against clearwing moth (Lepidoptera: Sesiidae) larvae1. J Entomol Sci 53(1):62 69

Batalla-Carrera L, Morton A, García-del-Pino F (2010) Efficacy of nematodes against the tomato leafminer Tuta absoluta in laboratory and greenhouse conditions. BioControl 55:523–530

Darsouei R, Karimi J, Shokoohi E (2014) *Oschelus rugaoensis* and *Pristionchus maupasi*, two new records of entomophilic nematodes from Iran. Russ J Nematol 22(2):141–155

Dillman AR, Chaston JM, Adams BJ et al (2012) An entomopathogenic nematode by any other name. PLoS Pathog 8:e1002527. https://doi.org/10.1371/journal.ppat.100227

Ebrahimi L, Niknam GH (2011) Detection of thermal preference range of two endemic isolates of entomopathogenic nematodes, *Steinernema feltiae* (Steinernematidae, Tylenchina) and *Heterorhabditis bacteriophora* (Heterorhabditidae, Rhabditina) for application in biological control of insect pests. SAPS 21:77–86

Ebrahimi L, Niknam GR, Nikdel M, Hassanpour M (2008) Study on the response of cotton bollworm, Helicoverpa armigera (Lepidoptera: Noctuidae) to various concentrations of entomopathogenic nematodes, Steinernema feltiae and Heterorhabditis bacteriophora (Rhabditida) under laboratory conditions. In: Proceedings of the 18th Iranian Plant Protection Congress, Hamedan, August 2008, p 7

Ebrahimi L, Niknam GH, Lewis EE (2011) Lethal and sublethal effects of Iranian isolates of *Steinernema feltiae* and *Heterorhabditis bacteriophora* on the Colorado potato beetle, *Leptinotarsa decemlineata*. BioControl 56:781–788

Ebrahimi L, Niknam GH, Askari Saryazdi G (2012) The feasibility of biological control of the vegetable serpentine leafminer, *Liriomyza trifolii* (Diptera: Agromyzidae) by the entomopathogenic nematode, *Steinernema feltiae*. In: Sarafrazi A, Asef M R, Mozhdehi M et al (eds) Proceedings of the 20th Iranian plant protection congress, Shiraz, August 2012, p 64

Ebrahimi L, Shiri MR, Dunphy GB (2016) Efficacy of the entomopathogenic nematode, *Steinernema feltiae* against the vegetable leaf miner, *Liriomyza sativae* Blanchard (Diptera: Agromyzidae). Egypt J Biol Pest Co 26:583–586

Ebrahimi L, Shiri M, Dunphy GB (2018) Effect of entomopathogenic nematode, *Steinernema feltiae*, on survival and plasma phenoloxidase activity of *Helicoverpa armigera* (Hb) (Lepidoptera: Noctuidae) in laboratory conditions. Egypt J Biol Pest Co 28:12

Ebrahimi L, TanhaMaafi Z, Sharifi P (2019) First report of the entomopathogenic nematode, *Steinernema carpocapsae*, from Moghan region of Iran and its efficacy against the turnip moth, *Agrotis segetum* Denis and Schiffermuller (Lepidoptera: Noctuidae), larvae. Egypt J Biol Pest Co 29:66

Edraki V, Fatemi E, Alichi M et al (2016) Biological control of main lawn pests *Pentadon idiota* and Agrotis segetum in shiraz using entomopathogenic nematodes. In: Talaei-Hassanloui R (ed) Proceedings of 22nd Iranian plant protection congress, Karaj, August 2016, pp 27–30

Eivazian Kary N, Niknam GH, Griffin CT (2009) A survey of entomopathogenic nematodes of the families Steinernematidae and Heterorhabditidae (Nematoda: Rhabditida) from north–west of Iran. Nematology 11:107–116

Eivazian Kary N, Niknam GH, Mohammadi SA, Moghaddam M, Nikdel M (2010a) Morphology and molecular study of an entomopathogenic nematode, *Steinernema bicornutum* (Nematoda, Rhabditida, Steinernematidae) from Iran. JESI 29:25–34

Eivazian Kary N, Rafiee HD, Mohammadi D, Afghahi S (2010b) Efficacy of some geographical isolates of entomopathogenic nematodes against *Leptinotarsa decemlineata* (Say) (Col.: Chrysomelidae). Mun ent zool 5:1066–1074

Eivazian Kary N, Golizadeh A, Rafiee HD, Mohammadi D, Afghahi S, Omrani M, Morshedloo MR, Shirzad A (2012) A laboratory study of susceptibility of *Helicoverpa armigera* (Hübner) to three species of entomopathogenic nematodes. Mun ent zool 7:372–379

Eivazian Kary N, Sanatipour Z, Mohammadi D et al (2018) Developmental stage affects the interaction of *Steinernema carpocapsae* and abamectin for the control of *Phthorimaea operculella* (Lepidoptera, Gelechidae). Biol Control 122:18–23

Falahi M, Abdollahi M, Roodaki M, Haghani M (2011) Efficacy of Steinernema carpocapsae for control of the adults of alfalfa weevil, *Hypera postica*. In: Proceedings of National Conference on Modern Agricultural Sciences and Technologies (MAST), Zanjan, September 2011, p 518

Garcia-del-Pino F, Alabern X, Morton A (2013) Efficacy of soil treatments of entomopathogenic nematodes against the larvae, pupae and adults of *Tuta absoluta* and their interaction with the insecticides used against this insect. BioControl 58:723–731

Gharalari AH, Kolyai R (2014) Screening walnut and apple trees against Leopard moth, *Zeuzera pyrina* (Lep.: Cossidae). Appl Entomol Phytopathol 81:11–16

Gharekhani GH, Salek-Ebrahimi H (2013) Evaluating the damage of *Tuta absoluta* (Meyrick) (Lepidoptera: Gelechiidae) on some cultivars of tomato under greenhouse condition. Arch Phytopathol Plant Protect 47:429–436

Hancock DL (1989) Pest status, southern Africa. In: Robinson AS, Hooper G (eds) Fruit flies: their biology, natural enemies and control. Elsevier, Amsterdam, pp 51–58

Hassani-Kakhki M, Karimi J, Hosseini M et al (2012) Efficacy of entomopathogenic nematodes against potato tuber moth, *Phthorimaea operculella* Zeller (Lep.: Gelechiidae). 20th Iranian plant protection congress. Shiraz University, Shiraz, Iran, 25–28 August 2012, 987 pp

Hassani-Kakhki M, Karimi J, Hosseini M (2013) Efficacy of entomopathogenic nematodes against potato tuber moth, *Phthorimaea operculella* (Lepidoptera: Gelechiidae) under laboratory conditions. Biocontrol Sci Tech 23:146–159

Heidari H, Impiglia A, Fathi H, Fredrix M (2011) FAO's FFS approach in Iran: concepts, practical examples and feedbacks. Food and Agriculture Organization, Regional Office in Tehran, Iran

Heimlich RE, Ogg CW (1982) Evaluation of soil-erosion and pesticide-exposure control strategies. J Environ Econ Manag 9:279–288

Hoskins WM, Borden AD, Michelbacher AE (1939) Recommendations for a more discriminating use of insecticides. In: Proceedings of the 6th Pacific science congress. University of California Press, San Francisco, August 1939, p 119

Jacobsen BJ (1997) Role of plant pathology in integrated pest management. Annu Rev Phytopathol 35:373–391

Jahanshahi Afshar F, Pourjam E, Kheiri A (2014) New record of two species belonging to superfamily Sphaerularioidea (Nematoda: Rhabditida) from Iran. J Crop Prot 3(1):59–67

Kaiser H (1991) Insect parasitic nematodes. In: Nickle WR (ed) A manual of agricultural nematology. Marcel Dekker Inc., New York, pp 899–960

Kamali SH, Karimi J, Hosseini M et al (2013) Biocontrol potential of the entomopathogenic nematodes *Heterorhabditis bacteriophora* and *Steinernema carpocapsae* on cucurbit fly, *Dacus ciliatus* (Diptera: Tephritidae). Biocontrol Sci Tech 23:1307–1323

Kamali S, Karimi J, Koppenhöfer AM (2017) New insight into the management of the tomato leaf miner, *Tuta absoluta* (Lepidoptera: Gelechiidae) with entomopathogenic nematodes. J Econ Entomol 111(1):112–119

Karimi J (2007) Taxonomy and pathology of entomopathogenic nematodes on the white grub, *Polyphylla olivieri* (Col., Scarabaeidae) in Tehran province. PhD dissertation, Tehran University, Iran

Karimi J, Kharazi-pakdel A (2007) Incidence of natural infection of the white grub *Polyphylla olivieri* (Coleoptera: Scarabaeidae) with entomopathogenic nematodes in Iran. OILB/srop Bulletin 30:35–39

Karimi J, Salari E (2015) Entomopathogenic nematodes in Iran: research and applied aspects. In: Campos-Herrera R (ed) Nematode pathogenesis of insects and other pests. Springer, Zurich, pp 451–476

Karimi J, Kharazi-pakdel A, Yoshiga T, Koohi-habibi M (2009a) First report of *Steinernema glaseri* (Rhabditida: Steinernematidae) from Iran. Russ J Nematol 17:83–85

Karimi J, Kharazi –p A, Yoshiga T (2009b) Insect pathogenic nematode, *Steinernema feltiae* from Iran. IOBC/wprs Bulletin 45:409–412

Karimi J, Rezaei N, Shokoohi E (2018) Addition of a new insect parasitic nematode, *Oscheius tipulae*, to Iranian fauna. Nematropica 48:45–54

Kary NE, Chahardoli S, Mohammadi D, Dillon AB (2019) Virulence of entomopathogenic nematodes against developmental stages of Plutella xylostella (Lepidoptera: Plutellidae)—effect of exposure time. Nematol 21(3):293–300

Kashkouli M, Khajeali J, Poorjavad N (2014) Effect of entomopathogenic nematodes for controlling the onion thrips, *Thrips tabaci* (Thys.: Thripidae) under semifield condition. In: Proceedings of the 21th Iranian Plant Protection Congress, Urmia, August 2014, p 457

Kaya HK, Aguillera MM, Alumai A et al (2006) Status of entomopathogenic nematodes and their symbiotic bacteria from selected countries or regions of the world. Biol Control 38:134–155

Kelly M, Guo Q (2007) Integrated agricultural pest management through remote sensing and spatial analyses. In: Ciancio A, Mukerji KG (eds) General concepts in integrated pest and disease management. Springer, Dordrecht, pp 191–207

Kordestani M, Karimi J, Modarres Awal M et al (2013) Pathogenecity of entomopathogenic nematode, *Heterorhabditis bacteriophora* Poinar, 1975 on the indigenous population of common earwig, *Forficula auricularia* L. Biol Control Pests Plant Dis 2:35–41

Kremen C, Iles A, Bacon C (2012) Diversified farming systems: an agroecological, systems-based alternative to modern industrial agriculture. Ecol Soc 17(44). https://doi.org/10.5751/ES-05103-170444

Kutinkova H, Andreev R, Arnaoudov V (2006) The leopard moth borer, *Zeuzera pyrina* l. (Lepidoptera: Cossidae) – important pest in Bulgaria. J Plant Prot Res 46:111–115

Lacey LA, Georgis R (2012) Entomopathogenic nematodes for control of insect pests above and below ground with comments on commercial production. J Nematol 44:218–225

Lewis EE, Clarke DJ (2012) Nematode parasites and entomopathogens. In: Vega FE, Kaya HK (eds) Insect pathology, 2nd edn. Elsevier, Amsterdam, pp 395–424

Lietti MMM, Botto E, Alzogaray RA (2005) Insecticide resistance in Argentine populations of *Tuta absoluta* (Meyrick) (Lepidoptera: Gelechiidae). Neotrop Entomol 34:113–119

Mallahi M, Zibaee A, Sendi JJ et al (2019) Effects of *Steinernema carpocapsae* (Weiser) on immunity and antioxidant responses of *Glyphodes pyloalis* Walker. ISJ 16:120–129

Memari Z, Karimi J, Kamali SH et al (2017) Are entomopathogenic nematodes effective biological control agents against the carob moth, *Ectomyelois ceratoniae*? J Nematol 48(4):261–267

Miraeiz E, Heydari R, Golhasan B (2017) A new and a known species of *Deladenus* Thorne, 1941 (Nematoda: Neotylenchidae) from Iran, with an updated species checklist of the genus. Acta Zool Bulgar 69(3):307–316

Moreau TL, Isman MB (2012) Combining reduced-risk products, trap crops and yellow sticky traps for greenhouse whitefly (*Trialeurodes vaporariorum*) management on sweet peppers (Capsicum annum). J Crop Prot 34:42–46

Moumeni-Helali H, Ahmadpour A (2013) Impact of farmers' field school approach on knowledge, attitude and adoption of rice producers toward biological control: the case of Babol township. Iran World Appl Sci J 21:862–868

Nikdel M, Niknam GH, Shojaee M et al (2008) A survey on the response of the last instar larvae of acorn weevil, *Curculio glandium* (Col.: Curculionidae), to entomopathogenic nematodes *Steinernema bicornutum* and *Heterorhabditis bacteriophora* in the laboratory. JESI 28:45–60

Nikdel M, Niknam GH, Griffin CT et al (2010a) Diversity of entomopathogenic nematodes (Nematoda: Steinernematidae, Heterorhabditidae) from Arasbaran forests and rangelands in north–west Iran. Nematology 12:767–773

Nikdel M, Niknam GH, Dordaei AA (2010b) Evaluation of susceptibility of the Brown tail moth, *Euproctis chrysorrhoea* (L.) to entomopathogenic nematode under laboratory condition. Nematol Mediterr 38:3–6

Nikdel M, Kaiser H, Niknam G (2011a) First record of *Hexamermis* cf. *albicans* (Siebold, 1848) (Nematoda: Mermithidae) infecting Lepidopteran larvae from Iran. Nematol Mediterr 39:81–83

Nikdel M, Niknam GH, Ye W (2011b) *Steinernema arasbaranense* sp. n. (Nematoda: Steinernematidae), a new entomopathogenic nematode from Arasbaran forests, Iran. Nematol Mediterr 39:17–28

Oliveira CM, Auad AM, Mendes SM, Frizzas MR (2014) Crop losses and the economic impact of insect pests on Brazilian agriculture. Crop Prot 56:50–54

Parry M (1992) The potential effect of climate changes on agriculture and land use. Adv Ecol Res 22:63–91

Parvizi R (1998) Efficacy of *Steinernema* sp. and *Heterorhabditis bacteriophora* on common pest in West Azerbaijan province. The 13th Iranian plant protection congress, 23–27 Aug., Karaj Junior College of Agriculture, Karaj, Iran, p 206

Parvizi R (2000) Possibility of biological control of Colorado potato beetle, *Leptinotarsa decemlineata*, with entomopathogenic nematodes. In: Proceedings of the 14th Iranian plant protection congress, Esfahan, September 2000, p 66

Parvizi R (2001) Survey on pathogenicity of entomopathogenic nematodes, Steinernema sp. and *Heterorhabditis bacteriophora* infesting larval *Polyphylla olivieri*. JESI 21:63–72

Parvizi R (2003) Efficacy of insect pathogenic nematodes *Steinernema* sp on control of trunk borer butterfly larvae in apple trees *Synanthedon myopaeformis*. Iran Agric Res 34:303–311

Parvizi R (2004) Investigation on efficacy of entomopathogenic nematodes, *Heterorhabditis bacteriophora* against black cutworm *Agrotis ipsilon*. J Agric Sci 14:35–42

Rahatkhah Z, Karimi J, Ghadamyari M (2015) Immune defenses of *Agriotes lineatus* larvae against entomopathogenic nematodes. BioControl 60:641–653. https://doi.org/10.1007/s10526-015-9678-z

Rezaei N, Karimi J, Hosseini M, Goldani M, Campos-Herrera R (2015) Pathogenicity of Two Species of Entomopathogenic Nematodes Against the Greenhouse Whitefly, *Trialeurodes vaporariorum* (Hemiptera: Aleyrodidae), in Laboratory and Greenhouse Experiments. J Nematol 47(1):60–66

Sabatier P, Poulenard J, Fanget B et al (2014) Long-term relationships among pesticide applications, mobility, and soil erosion in a vineyard watershed. PNAS 111:15647–15652

Saeidi K, Adam NA, Omar D et al (2011) Study of some biological aspects and development of integrated pest management program for the safflower fly, *Acanthiophilus helianthi* Rossi (Diptera: Tephritidae) in Iran. Res J Agric Sci 7:1–16

Saeidi K, Pezhman H, Karimipour-Fard H (2018) Efficacy of entomopathogenic nematode *Steinernema feltiae* (Filipjev) as a biological control agent of lentil weevil, *Bruchus lentis*, under laboratory conditions. Not Sci Biol 10(4):503

Saffari T, Karimi J, Madadi H (2012) Characterization some entomopathogenic nematodes from Hamedan using ITS and 28S gene sequences. In: Proceedings of First Conference on DNA Barcoding in the Taxonomy, Mashhad, February 2012, p 199

Saffari T, Madadi H, Karimi J (2013) Pathogenicity of three entomopathogenic nematodes against the onion thrips, *Thrips tabaci* Lind. (Thys.; Thripidae). Arch Phytopathol Plant Protect 46:2459–2468

Salari E, Karimi J, Sadeghi-Nameghi H et al (2014) Efficacy of two entomopathogenic nematodes *Heterorhabditis bacteriophora* and *Steinernema carpocapsae* for control of the leopard moth borer *Zeuzera pyrina* (Lepidoptera: Cossidae) larvae under laboratory conditions. Biocontrol Sci Tech 25:260–275

Salari E, Karimi J, Sadeghi Namaghi H et al (2019) Characterization and evaluation of the pathogenic potential of a native isolate of the insect associated nematode *Acrobeloides maximus* (Rhabditida: Cephalobidae) from Kerman provinces, Iran. The 3rd IICE, Iranian international congress of entomology, 2019-08-31

Seddiqi E, Shokoohi E, Karimi J (2016) New data on *Heterorhabditis bacteriophora* Poinar, 1976 from south eastern Iran. IJAB 12(2):181–190

Sharifi SH, Karimi J, Hosseini M (2014) Efficacy of two entomopathogenic nematode species as potential biocontrol agents against the rosaceae longhorned beetle, *Osphranteria coerulescens,* under laboratory conditions. Nematology 16:729–737

Sheykhnejad H, Ghadamyari M, Koppenhöfer AM et al (2014) Interactions between entomopathogenic nematodes and imidacloprid for rose sawfly control. Biocontrol Sci Tech 24:1481–1486

Tanha Ma'afi Z, Ebrahimi N, Abootorabi E, Spiridonov SE (2006) Record of two Steinernematid species from Iran. In: Proceedings of the 17 th Iranian Plant Protection Congress. University of Tehran, Karaj, Iran, 2-5 Sptember, 482 p

Valizadeh A, Goldasatch SH, Rafiei-Karahroodi Z et al (2017) The occurrence of three species of the genus *Oscheius* Andrássy, 1976 (Nematoda: Rhabditida) in Iran. J Plant Prot Res 57 (3):248–255

Yadav IC, Devi NL, Syed JH et al (2015) Current status of persistent organic pesticides residues in air, water and soil, and their possible effect on neighboring countries: a comprehensive review of India. Sci Total Environ 511:123–137

Zabel F, Putzenlechner B, Mauser W (2014) Global agricultural land resources – a high resolution suitability evaluation and its perspectives until 2100 under climate change conditions. PLoS One 9:e107522. https://doi.org/10.1371/journal.pone.0107522

Zaim M, Ladonni H, Ershadi M (1988) Field application of *Romanomermis culicivorax* (Mermithidae: Nematoda) to control anopheline larvae in southern Iran. J Am Mosq Control Assoc 4(3):351–355

Zhang KY, Liu XH, Tan J et al (2012) *Heterorhabditidoides rugaoensis* n. sp. (Rhabditida: Rhabditidae), a novel highly pathogenic entomopathogenic nematode member of Rhabditidae. J Nematol 44:348–360

Zolfagharian M, Saeedizadeh A, Abbasipour H (2016) Efficacy of two entomopathogenic nematode species as potential biocontrol agents against the diamondback moth, *Plutella xylostella* (L.). JBC 30(2):78–83

# Chapter 13
# Insect Pathogenic Viruses, Microsporidians and Endosymbionts

Mohammad Mehrabadi, Reyhaneh Darsouei, and Javad Karimi

## 13.1 Entomopathogenic Virus

### 13.1.1 Introduction

Annually, herbivorous insect pests are responsible for destroying one-fifth of the total crop production in the world. Therefore, insect pests are a major challenge for food security and provision of food, particularly in developing countries with a rapid annual increase in the human population (2.5–3.0%) compared with a lower rate of increase (1.0%) in food production. Chemical insecticides are often used to control insect pests. However, due to potential risks for human health and environment, adverse effects on non-target organisms, and a significant risk to biodiversity worldwide, their use is gradually becoming more and more restricted (Geiger et al. 2010; Beketov et al. 2013). Therefore, it is necessary to seek safer alternatives to pesticides. To minimize the adverse effects of chemical pesticides, several approaches have been employed including microbial control (Lacey et al. 2001; 2015). In contrast to chemical pesticides that have shown decline on the global market, microbial pesticides have shown long term growth over the past decade; however much of the global market is still dominated by chemical pesticides (Grant et al. 2010).

M. Mehrabadi (✉)
Department of Entomology, Faculty of Agriculture, Tarbiat Modares University, Tehran, Iran
e-mail: m.mehrabadi@modares.ac.ir

R. Darsouei
Department of Plant Protection, Faculty of Agriculture, Ferdowsi University of Mashhad, Mashhad, Iran

J. Karimi
Department of Plant Protection, Ferdowsi University of Mashhad, Mashhad, Iran
e-mail: jkb@um.ac.ir

© The Author(s), under exclusive license to Springer Nature Switzerland AG 2021    505
J. Karimi, H. Madadi (eds.), *Biological Control of Insect and Mite Pests in Iran*,
Progress in Biological Control 18, https://doi.org/10.1007/978-3-030-63990-7_13

Entomopathogens – including insect pathogenic viruses – can be utilized in all three categories of biological control including classical, conservational and augmentative. Moreover, they can be easily used as an ideal tool in sustainable agriculture and integrated pest management (IPM) strategies for the following reasons: (i) they are environmentally safe; (ii) many of them have a narrow host range and are virulent against their insect hosts; (iii) based on their specificity, there is very little effect on non-target organisms; and (iv) they promote biodiversity. Therefore, microbial pesticides are attractive alternatives to chemical pesticides for application in plant protection and pest control.

Insect viruses may contain double or single-stranded DNA (dsDNA and ssDNA, respectively) or double or single-stranded RNA (dsRNA and ssRNA, respectively). Some viruses are contained in a protein matrix called an occlusion body (OB). Baculoviruses are the most important insect viruses due to their applications in biological control of insect pests. Baculoviruses are a family of large rod-shaped viruses that mainly infect insects and have dsDNA circular genomes ranging from 80 to 180 kb in size (Ayres et al. 1994; Grzywacz 2017). The Baculoviridae comprise four genera including *Alphabaculovirus* (the lepidopteran polyhedroviruses), *Betabaculoviruses* (the granoloviruses), *Gammabaculovirus* (the hymenopteran polyhedroviruses) and *Deltabaculovirus* (the dipteran polyhedrovirus) (Jehle et al. 2006).

Baculoviruses have attractive characteristics making them good candidates for use in pest control, particularly in the natural environment (Inceoglu et al. 2006). The most important feature of these viruses is their narrow host range; there are species that infect only one insect host or a few species. This narrow host range makes them suitable for insect pest control without side effects on other non-target insects such as beneficial groups. Baculoviruses are highly pathogenic against their insect hosts; they can kill the insect host within a few days by propagation within the host's body but not through production of toxins.

The study of entomopathogenic viruses in Iran began in 1998 in which case studies on the effect of Nuclear Polyhedrosis Virus (NPVs) on some pests such as *Heliothis armigera* (Hübner 1808), *Spodoptera exigua* (Hübner 1808), *Plutella xylostella* (Linnaeus 1758), and *Cydia pomonella* (Linnaeus 1758) were performed. Issues regarding histopathology and mass rearing were also surveyed during this time.

Although considerable effort has been made in characterization of microbial diversity across the country, there has been little progress in the area of insect pathogenic microorganisms particularly entomopathogenic viruses. Considering the importance of insect pathogens in microbial control programs and their diversity in Iran, more studies need to be done on isolation and characterization of native insect pathogens.

## 13.1.2   Entomopathogenic Viruses in Biocontrol

### 13.1.2.1   *Spodoptera exigua*

The sugar beet armyworm, *Spodoptera exigua* (Lep.: Noctuidae), is a major pest of many crops (Dai et al. 2000). This pest is native to Asia but has been recorded in different regions throughout the world, such as Australia, Europe, and North America (Pathak and Khan 1975). Its larvae feed on the foliage and the upper portion of beet roots and can completely defoliate small plants, frequently causing economic damage. Due to the problem of the pest resistance to insecticides, the adverse effects of insecticides on the environment (Asi et al. 2013), negative effects of insecticides on wildlife and natural enemies (Ordóñez-García et al. 2015), biological control as a pest management strategy is desirable.

Various populations of *S. exigua* are infected by different species/strains of baculoviruses (Virto et al. 2014). Among baculoviruses, *S. exigua multiple nucleopolyhedrovirus* (SeMNPV) was isolated from *S. exigua* larvae (Jakubowsaka et al. 2005; Virto et al. 2013). The SeMNPV is a specific pathogen of *S. exigua* and has the potential to be used as a biocontrol agent as it produces epizooty in larval populations (Virto et al. 2014).

Manzari et al. (1998), surveyed pathogenicity of *Mamestra brassicae* nuclear polyhedrosis virus on *S. exigua*. This virus had a high pathogenicity for *Spodoptera exigua* larvae. In laboratory conditions, a concentration of 12.58 Occlusion bodies (OBs) per $mm^2$ of artificial food led to 50% mortality of 2nd instar larval of *S. exigua* after 6 days. The mortality percentage for 1st, 3rd and 5th larval instars were 100, 69 $\pm$ 6, and 11 $\pm$ 5%, respectively. In surface contamination of eggs with $1 \times 10^5$ OB per $cm^3$ concentration, all larvae died 3 days after hatching (Manzari et al. 1998). Also, Kamali and Pourmirza (2000) surveyed pathogenicity of a NPV strain on 1–5 instars larval of *S. exigua* in laboratory condition. The median lethal does ($LD_{50}$) in 1–5 instars larvae were 14.46, 154.55, 1576.14, 2917.39, and 5902.88 OB per mm, and the mortality value was 93.3, 93.3, 80, 92.9, and 73.3%, respectively. These results indicated that there was a correlation between larval weight and $LD_{50}$. Thus, for virus spraying, the larval weight must be considered. Rabie et al. (2008) performed an experiment about efficacy of MbNPV against different larval stages of *S. exigua* and found that older larvae were more susceptible than younger larvae.

During 2014–2015, a survey was done to identify the natural enemies of sugar beet armyworm, *S. exigua* larva. Different larval stages were collected from beet fields in Mashhad and Chenaran (Northeast of Iran) and kept in laboratory conditions. Finally, several species of parasitoids and pathogens including virus, nematode and fungus were identified. For virus identification, two genes, *lef8* and *Polyhedrin* were amplified, sequenced and analyzed. The reconstructed phylogenetic trees inferred from two loci confirmed that two isolates of SeMNPV were natural enemies of *S. exiguae*. The identified entomopathogenic virus isolates were very important because they had high incidence and virulence (Darsouei and Karimi

508	M. Mehrabadi et al.

**Fig. 13.1** Comparison of healthy (upper) and MNPV-infected larvae (bottom) of *Spodoptera exigua*. Healthy larvae had a greenish color and were able to chew holes in sugar beet leaf. The infected larva had a yellowish color and were only able to scrape on surface of leaf (shown with arrows). (Courtesy of Darsouei R, Ferdowsi University of Mashhad)

2015). The virus-infected larvae were yellow and swollen with negative geotropism (Fig. 13.1). Dissection of the infected larvae showed that their body tissues were liquefied with a thin cuticle layer. Due to cuticle rupture, the larval hemolymph spread on the leaf and the occlusion bodies were released in the emerged liquid, which accelerated horizontal transfer of the pathogen. This isolate (SeMNPV YV) with a 100% vertical transmission had high virulence. Occlusion bodies were polyhedral and irregular in shape (Fig. 13.2). The OB cross-sections from the virus-infected larvae of *S. exigua* indicated the virion contained multiple nucleocapsids arranged randomly within the occlusion matrix (Fig. 13.3) (Darsouei et al. 2017).

### 13.1.2.2 *Heliothis armigera*

The cotton bollworm, *Heliothis armigera* is an important pest in cotton and one of the most polyphagous and cosmopolitan pest specie. Management of *Heliothis* spp. in the past has relied heavily on the use of insecticides, which led to resistance problems in cotton (Fitt 1994). Resistance to pyrethroids amongst *H. armigera* is a serious problem (Trowell et al. 1993). Thus, application of biocontrol agents such as entomopathogenic virus could be useful. The *H. armigera* single nucleopolyhedrovirus (HearNPV) shows high potential for use in a management program of the cotton bollworm. HearNPV has been developed into commercial

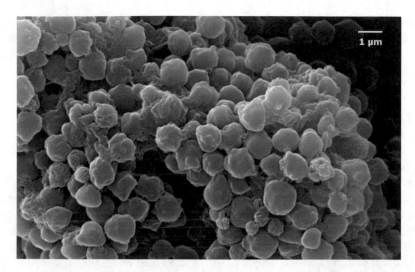

**Fig. 13.2** Scanning Electron Micrographs of polyhedral OBs extracted from infected larvae of *Spodoptera exigua* with a baculovirus (Magnification = 20,000). (Courtesy of Darsouei R, Ferdowsi University of Mashhad)

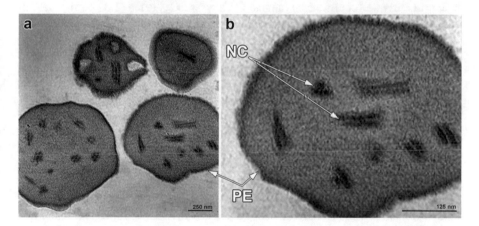

**Fig. 13.3** Transmission electron micrographs from cross section of OBs (**a, b**) obtained from the infected larvae of *Spodoptera exigua* by a baculovirus (NC, nucleocapsids; PE, polyhedral envelope). (Courtesy of Darsouei R, Ferdowsi University of Mashhad)

products for control of *H. armigera* in Europe, Australia and Asia (Moore et al. 2004; Buerger et al. 2007; Sun and Peng 2007).

Despite extensive research on this virus, little attention has been given to HearNPV in Iran. The response of *H. armigera* larvae to nuclear polyhedrosis virus (NPV) was surveyed by Pourmirza (2000). This study estimated that $LD_{50}$ values for the first, second, third as well as early and late fourth larval instars were 5, 141, 1226, 5168, and 24,553 polydera per instar, respectively. The fifth instar larvae were resistant to the infection and there was an inverse relationship between mortality and larval weight. In another study, the pathogenicity of other viral isolates was surveyed on the second instar larvae of *H. armigera*. Then, $LC_{50}$ and $LD_{50}$ values, and yield parameters (yield/larva and productivity ratio) of seven geographic isolates of *H. armigera* NPV were evaluated under laboratory conditions (Mehrvar et al. 2008a). The $LC_{50}$ of seven isolates of NPV including NGM, OTY, PRB, CBE, RHI, MUM and HYD against second instar larvae of *H. armigera* was determined. The highest pathogenicity was observed in the NGM isolate with $0.028$ OB/mm$^2$ which was more virulent than HYD by a factor of 2.75. Also, the $LT_{50}$ of NGM was the lowest among the isolates by 97.8 h.

A few isolates of HearNPV have been collected from tomato fields of the East Azerbaijan province (Mehrvar 2013). Subsequently, a study was designed to evaluate the age-dependent resistance of *H. armigera* larvae against two native virus isolates (EAZ-I and EAZ-II) and a non-native isolate (STD) (Mehrvar 2015a). Results showed that larval resistance to virus isolates increases with larval age. It was shown that the age and weight of the larva as well as ambient temperature had meaningful effects on the larval mortality and the virus replication inside the larva. The optimum larval stage and ambient temperature for HearNPV in vivo production were the early fifth instar larvae and 25 °C, respectively (Mehrvar 2015b). The STD isolate in early second instar larvae had the lowest $LC_{50}$ value ($1.82 \times 10^2$ OB/ml) and the EAZ-II isolate in fifth instar larvae had the highest $LC_{50}$ value ($1.43 \times 10^6$ OB/ml) (Mehrvar 2015b).

Among the entomopathogens, HearNPV is highly virulent in its host. The main limiting factor for application is the required time to kill the insect, which takes at least 3–5 days. Mehrvar (2015a), collected $LC_{50}$ data for a non-native isolate (STD) of HearNPV and two native isolates (EAZ-I and EAZ-II) from the East Azarbaijan province in 2010. All isolates showed a high rate of virulence against the second instar larvae of *H. armigera*. The STD isolate had the lowest $LC_{50}$ value ($7.27 \times 10^2$ OB/ml) and the shortest $LT_{50}$ (97.8 h). EAZ-I and EAZ-II isolates had lower virulence than STD. For fifth instar larvae, the results were similar. When EAZ-II was combined with flufenoxuron, a chitin synthesis inhibitor, the lowest $LC_{50}$ value calculated was $1.83 \times 10^4$ OB/ml. The $LT_{50}$ data showed that flufenoxuron combined with native virus isolates of EAZ-I and EAZ-II, increased speed mortality rate by 19.88 and 19 percent, respectively. Flufenoxuron could effectively enhance virulence by affecting the chitin contents of peritrophic membrane (Mehrvar 2015a).

According to Pour Mirza (1998), efficiency of a viral insecticide is related to various factors such as commune period and larval weight. The necessary time for larval mortality is related to the time required for the virus to proliferate until it reaches a deadly level (Pour Mirza 1998). He bioassayed an isolate of NPV on *Heliothis viriplaca* (Hufnagel 1766) (Lep.: Noctuidae). The results demonstrated that the older larvae have a higher $LT_{50}$. Therefore, one disadvantage of viral insecticides is the long interval between the initial infection and the host's death.

These studies, although limited, show that there are isolates of HearNPV with potential for application against *H. armigera* in Iran which require additional characterization. Therefore, further investigation in other regions of the country is recommended to isolate and characterize native HearNPV isolates.

### 13.1.2.3   *Cydia pomonella*

The codling moth, *Cydia pomonella* (Lep.: Tortricidae) is a well-known agricultural pest, worldwide. In Iran, it has become the primary pest of apples and is usually controlled using chemical pesticides. Frequent spraying increases the risk of resistance developing in the pest population (Pluciennik 2013). Among natural enemies of the pest, a *Betabaculovirus* has high potential for application as a biocontrol agent of codling moth (Miletic et al. 2011).

   The *Cydia pomonella* granulovirus (CpGV) is an effective biological control agent against codling moth, especially for organic apple production. The isolation of CpGV is difficult because infected larvae rupture and are not visible on the trees. In 2002, 11 isolates of CpGV (designated I1, I7, I8, I15, I22, I28, I30, I66, I67, I68, and I70) were collected from northwest and northeast Iran (Rezapanah et al. 1998). Light and electron microscopy studies confirmed the identity of granulovirus in the codling moth populations. Then, these isolates were characterized using bioassays and endonuclease profile analysis (RFLP) (Rezapanah et al. 2008). The isolated CpGVs showed different biological and molecular characteristics which indicated there were various genotypes of this virus in the Iranian strains (Eberle et al. 2009; Rezapanah et al. 2008). Most interestingly, it was shown that one of the isolated CpGVs, known as CpGV-I12, was able to overcome the CpGV resistance issue which occurred in Germany and France (Berling et al. 2009a, b; Eberle et al. 2008). The CpGV-I12 showed great efficacy in laboratory bioassays against all larval instars of a resistant codling moth strain CpR, which has a 100-fold reduced susceptibility to the commercially used isolate of CpGV-M (Berling et al. 2009a, b). Therefore, CpGV-I12 has been suggested as a promising alternative microbial control agent of this insect pest. Further studies should be undertaken to isolate and characterize more isolates of CpGV from the country. The CpGV isolates found in northwest Iran make an important contribution to the known diversity of CpGV strains.

### 13.1.2.4   *Plutella xylostella*

The diamondback moth (DBM), *Plutella xylostella* (Lep.: Plutellidae) is one of the most important pests of crucifers in the world. Fahimi et al. (2008a) surveyed the sensitivity of 2nd, 3rd, and 4th instar larvae, as well as eggs of *P. xylostella* to *Memestra brassicae nucleopolyhedrosis virus* (MbNPV) in laboratory conditions. The leaf surface was infected with viral suspensions containing 47.86 OB/mm$^2$. There was no significant difference in mortality of DBM eggs, or 1st and 2nd instar larvae. However, there was a significant difference in 3rd instar larvae mortality. The mean mortality of infected 2nd, 3rd, and 4th instar larvae were 74.4, 42.77, and 11.11%, respectively. Thus, older larvae have lower sensitivity to MbNPV infection.

In another study, the effect of Taiwanese strain of *Plutella xylostella granulovirus* (PlxyGV) (Baculoviridae), as a natural enemy on *P. xylostella* was surveyed. The experiment was carried out with seven concentrations. The $LC_{50}$ and $LC_{90}$ values of Taiwanese strain of PlxyGV for the second instar larvae of the cabbage moth were calculated as 448.58 and 715.77 granules/mm$^2$, respectively. The $LT_{50}$ values for this PlxyGV strain when applied at concentrations of 749.89 and 1883.65 granule/mm$^2$ were 6.04 and 6.68 days, respectively (Fahimi et al. 2008b).

Magholi et al. (2014), surveyed the insecticidal activity of a baculovirus (HaNPV) against 2nd instar larvae of *P. xylostella*. $LC_{25}$, $LC_{50}$, and $LC_{75}$ were $2.2 \times 10^3$, $3.8 \times 10^4$ and $6.6 \times 10^5$ OB/ml, respectively. Median lethal time ($LT_{50}$) decreased from 114.24 to 106.05 h with decreasing larval age. Larval development time and pupal weight were not affected by the different concentration of HaNPV. There were, however, significant differences between the pupal rate and adult emergence in treated larvae with different concentration.

Dezianian et al. (2015b) evaluated the pathogenicity of *Plutella xylostella Granulovirus* (PlxyGV) on the *P. xylostella* larvae. Bioassay results showed that larval mortality was significantly influenced by larval instar and viral concentrations. The larval mortality increased with increasing viral concentrations and decreased for older instars. The $LC_{50}$ of the virus on the second instar larvae was estimated as $1.39 \times 10^6$ granules/ml. The measured $LT_{50}$ values were between 3.81–6.95, 4.97–9.74 and 5.15–9.41 days for first, second and the third instar larva, respectively. The high specificity and virulence of *PlxyGV* supports its suitability as an appropriate candidate for use within an Integrated Pest Management (IPM) program targeting *P. xylostella*.

### 13.1.2.5 *Bombyx mori*

Grasseri is one of the most important diseases of silkworm, *Bombyx mori* (Linnaeus 1758) (Lep.: Bombycidae) with a significant yield loss caused by nuclear polyhedrosis viruses (NPV) (Etebari et al. 2007). This virus decreases protein content in the infected larvae. Thus, identification of resistant populations of *B. mori* is an important step in lowering the disease damage. Seidavi et al. (2004), performed an experiment to compare 21 new lines of silkworm populations including 11 Japanese and 10 Chinese lines in natural conditions and infected environments against nuclear polyhedrosis virus. The number of surviving larvae, surviving pupae, pupation rate, cocoon weight, shell weight and shell ratio was measured. The results showed that Chinese lines were more susceptible to NPV infection in infected conditions than Japanese lines. Also, the effect of viral infection on silk production potential of individuals was not significant in most lines. Larval resistance to virus could be transmitted to sensitive lines and this characteristic could be applied for improving high productivity lines (Seidavi et al. 2004).

Biabani et al. (2005), attempted to measure the resistance level of 20 new Iranian silkworm hybrids to nuclear polyhedrosis virus. In this study, 3rd instar larvae of *B. mori* were orally inoculated with *B. mori* NPV (550 ppm) and reared up to the spinning stage in summer. The results indicated that lines of 107 K × 124 K, 107 K × 108 K, and 101,433 × 114 had the highest resistance to this virus. Some hybrid had the highest resistance to this virus.

### 13.1.2.6   *Apis mellifera*

At least 18 viruses have been reported in Iranian honeybee colonies, *Apis mellifera* (Linnaeus 1758) (Hym.: Apidae). Of those, six viruses cause severe diseases. These viruses include deformed wing virus (DWV), acute bee paralysis virus (ABPV), chronic bee paralysis virus (CBPV), sacbrood virus (SBV), Kashmir bee virus (KBV), and black queen cell virus (BQCV) (Mosadegh 1990; Ghorani et al. 2017). In addition, CBPV was characterized from different regions including Mazandaran, Razavi Khorasan, Hormozgan, and Kurdistan (Ghorani et al. 2017).

In another study, Sabahi et al. (2018), surveyed 12 apiaries located in the Alborz, Ardabil, and Guilan province and samples of varroa mite, *Varroa destructor* (Anderson and Trueman 2000) (Acari: Varroidae) were taken. Five viruses including DWV, BQCV, ABPV, Israeli acute paralysis virus (IAPV), and KBV were characterized in *V. destructor* using RT-PCR. The results indicated that several mite samples were positive for DWV and BQCV, but all were negative for ABPV, IAPV, and KBV.

## 13.1.3   *Virus and Natural Condition*

Baculoviruses are sensitive to irradiation. Thus, surveying biological activity of NPVs before and after exposure to sunlight is critically important. In a study by Mehrvar et al. (2008b), the effects of different radiation dosage (175, 500, 600, 650, 700, and 750 W/m$^2$) and time exposed to simulated sunlight (30, 60, 90, 120, 180, and 240 minutes) were evaluated under laboratory conditions to monitor NPV virulence in *H. armigera*. Virulence decreased by a factor of 2.63 against second instar larvae of *H. armigera* when exposed to 750 W/m$^2$ dose for 90 min (Mehrvar et al. 2008b).

Moreover, application of baculoviruses in biocontrol programs has three major limitations: they have a high LT$_{50}$, are inactivated by ultraviolet (UV) radiation, and are difficult to mass produce. To address this, Dezianian et al. (2012), surveyed additives (including Tinopal, lignin, molasses and skim milk powder) to a suspension of PlxyGV after UV-B irradiation. The material used as the protectant, and the duration of exposure time (30 or 60 min) significantly affected the larval mortality. Some combinations, including PlxyGV and Tinopal, PlxyGV and molasses, PlxyGV and lignin, and PlxyGV and skim milk, increased the viral suspension's residual activity after exposure to UV radiation by 67.78%, 65.31%, 59.55%, and 31.35%, respectively. These findings suggest that combinations of natural adjuvants including molasses and lignin are as effective as chemical adjuvants such as Tinopal and could improve PlxyGV viral activity (Dezianian et al. 2012).

Microclimate thermal conditions can also affect virulence and viral kill speed. Moshtaghi Maleki et al. (2013), evaluated the effect of temperature (20, 26 and 30 °C) on the $LT_{50}$ and $LD_{50}$ of *Helicoverpa armigera* MNPV in third instar larvae of *H. armigera* under laboratory conditions. The $LD_{50}$ values for three temperatures – 20, 26 and 30 °C – were 3580.83, 2614.82 and 1751.62 OB/larva, respectively. Mortality caused by viral infection increases significantly with temperature, and is especially significant in low doses. The $LT_{50}$ values for $5 \times 10^5$ OB/larva viral concentration in temperatures 20, 26 and 30 °C were 7.15, 6.73 and 6.58 days, respectively. Thus, temperatures in the range of 26–30 °C are suitable conditions for high mortality rate and speed.

Another limitation of viral biopesticides is their persistence in the environment. Considering the issue of resistant formulations, Dezianian et al. (2015a), surveyed the effect of two formulations of PlxyGV on larval mortality of *P. xylostella* under natural conditions. Spray-dried, PlxyGV+ modified food starch combined with lignin and molasses (CLM formulation) considerably increased residual activity of PlxyGV in the environment. The CLM formulation containing $1 \times 10^{13}$ granule/ha after 0.5, 5, 24, 48 and 72 hours exposure in the field condition increased the residual activity of the viral preparation by 13.7, 30.67, 42.93, 52.53 and 21.76% when compared to unformulated virus, respectively. The CLM formulation was significantly more effective on larval mortality than KLM formulation (PlxyGV + kaolin+ lignin + molasses) after exposure to in the field condition (Dezianian et al. 2015a).

### 13.1.4   The Effect of Entomopathogenic Virus on Parasioids

Rabie et al. (2010), investigated the survival of *Habrobracon hebetor* (Say 1836) (Hym.: Braconidae) on nucleopolyhedrovirous (NPV)-infected *S. exigua* larvae. In this study, the second instar larvae of *S. exigua* were exposed to the under-$LD_{50}$, $LD_{50}$ and over-$LD_{50}$ rate of *Mamestra brassicae* NPV (MbMNPV). They were able to be parasitized by *H. hebetor* after 24, 48, and 72 h post-treatment. MbNPV infection of larvae was deleterious to the survival and parasitism of *H. hebetor*. The survival of *H. hebetor* in MbMNPV-infected *S. exigua* was dependent on the interval between viral infection and parasitism. In treated larvae with a 180 PIB/mm$^2$ concentration of MbNPV at 72-h interval treatment, fewer adult parasitoids emerged from infected hosts compared to 24 and 48 h post treatment. Thus, inoculation dosage of MbNPV and timing of parasitoid release has a significant effect on the development of *H. hebetor* within virus-infected hosts. Furthermore, viral biocontrol of *S. exigua* may lead to substantial mortality of immature parasitoids (Rabie et al. 2010).

## 13.1.5   Combination Virus with Other Agents

Marzban et al. (2010), tested *Bacillus thuringiensis* (Bt) toxins combined with cytoplasmic polyhedrosis virus on mortality rate of *H. armigera*. In this study, the effects of interactions between Cry1Ac toxin of *B. thuringiensis* and *Helicoverpa armigera Cytoplasmic polyhedrosis virus* (HaCPV) on the first and third instar larvae of *H. armigera* were evaluated in the laboratory. When the first instar larvae was exposed to a combination of Bt and HaCPV ($6 \times 10^6$, $1 \times 10^7$, and $3 \times 10^7$ OB/ml) on cotton leaf discs, the effect on mortality was additive. When the larvae was exposed to a combination of Cry1Ac (0.3, 0.9, 2.7, or 8.1 µg/g) and the same concentrations of HaCPV, the effect on mortality was additive except for the combination of 0.3 µg/g of Cry1Ac and HaCPV which had a synergistic effect. The third instar larvae had similar results with the first instar larvae. Overal, delayed larval growth and development as well as pupation and pupal weight decreased when larvae fed on an artificial diet containing Cry1Ac and HaCPV or transgenic Bt cotton leaf discs especially for first instar larvae.

In another study, the interaction of *Bacillus thuringiensis* subsp. *kurstaki* and *Helicoverpa armigera* single nucleopolyhedrovirus (HaSNPV) on the survival of second instar *P. xylostella* larvae was evaluated under laboratory conditions. The exposure of second instar larvae to a combination of HaSNPV ($2.3 \times 10^3$, $3.8 \times 10^4$, and $6.6 \times 10^5$ OB/ml) and *B. thuringiensis* subsp. *kurstaki* ($2.4 \times 10^1$, $3.4 \times 10^3$, and $4.7 \times 10^5$ spore/ml), produced an additive effect on the efficiency of most combinations. A synergistic effect was observed when $2.4 \times 10^1$ Bt was combined with $2.3 \times 10^3$ OB/ml HaSNPV. Also, population rate, pupal weight, and emergence rate of adults decreased. The $LT_{50}$ values negatively correlated with *B. thuringiensis* and HaSNPV concentrations (Magholli et al. 2013).

In a similar study, Kalantari et al. (2013) surveyed the effects of *Bacillus thuringiensis* isolates and HaSNPV in combination and individually on *H. armigera* larvae. Combined use of Bt ($3 \times 10^6$ spore/ml) and HaSNPV ($9.2 \times 10^3$ OB/ml) had a synergistic effect on larval mortality. The obtained results indicated a combination of the lowest Bt concentration and the highest HaSNPV concentration had synergistic effect on *H. armigera* second instar larvae.

Assemi et al. (2013a), surveyed the efficacy of different geographical isolates of HaNPV on biological control of *H. armigera*. In this research, two local strains (A, B) of *H. armigera nucleopolyhedrovirus* (HaNPV) were collected from infected *H. armigera* larvae in tobacco fields of north Iran. To maximize the persistence of HaNPV, various adjuvants were evaluated such as neem, henna powder, soybean flour, maize flour, boric acid, and diflubenzuron. Here, the viral concentration was $1 \times 10^5$ OB/ml. The results indicated that a combination of diflubenzuron + virus B and diflubenzuron + virus A caused 85.41 and 82.4% mortality, respectively.

## 13.1.6   Histopathology

Marzban et al. (2013), surveyed midgut histopathology of the cotton bollworm infected with *H. armigera* cytoplasmic polyhedrosis virus. The symptoms of healthy and infected larvae were compared. The bodies of infected larvae were swollen and small milky-white polyhedral inclusion bodies were observed in the columnar cells after 1 or 2 days. Microvilli of infected columnar cells were affected and degenerated immediately prior to rupture of the cells. Some infected columnar cells ruptured to release polyhedral inclusion bodies (PIB) into the gut lumen 3 days after infection. Also, PIB were found in goblet cells, 5 or 6 days after infection. The results showed that cytopathic effects caused in the cotton bollworm larvae midgut by HaCPV were similar to those reported for CPV lepidopteran and dipteran infection.

## 13.1.7   Virus-Insect Interactions

Among the Baculoviruses, the most studied species is *Autographa californica* multiple nucleopolyhedrovirus (AcMNPV). Research on baculovirus-insect interactions has recently begun in Iran. While attempting to find differentially expressed genes of *Spodoptera frugiperda* (Sf9) cells in response to AcMNPV, Mehrabadi et al. (2015) discovered a Dim1-like gene that was upregulated from 4 h post-infection (hpi) up until 24 hpi. The expression level diminished dramatically at 36 hpi up to 120 hpi with no expression detected after 144 hpi. RNAi of SfDim1 resulted in decreased viral DNA levels in comparison to mock-infected cells, which supports the idea that this gene might be involved in virus replication and infection (Mehrabadi et al. 2013a). RNA interference (RNAi) is a highly conserved sequence-specific gene silencing mechanism in diverse eukaryotes, involved in the regulation of many biological processes, including virus-host interactions and antiviral immunity in insects (Mehrabadi et al. 2015). The RNAi-based antiviral response can be mounted through noncoding small RNAs including virus-derived short interfering RNAs (vsiRNAs) and microRNA (miRNA). Using deep sequencing of small RNAs, it was shown that AcMNPV infection induces an RNAi response in Sf9 cells and produces a huge number of vsiRNAs. The vsiRNAs were unevenly mapped onto the AcMNPV genome showing that some regions of AcMNPV genome are targeted more than others (*i.e* hot spots), while there were some regions in the genome with low or no vsiRNA mapping (*i.e* cold spots). Despite the presence of this potent antiviral machinery in Sf9 cells, AcMNPV can infect these cells and replicate within them (Mehrabadi et al. 2015). Therefore, there must be a counter defense strategy in AcMNPV to suppress or avoid this potent cellular antiviral immunity. Furthermore, it was found that AcMNPV encodes potent viral suppressors of RNAi (VSRs) to counter cellular antiviral RNAi. The viral p35 gene, which is well known as an inhibitor of apoptosis, was found to be responsible for suppression of RNAi response in Sf9 cells using a p35-null AcMNPV and ectopic expression of p35. It was also shown that VSR activity of p35 is not linked to the antiapoptotic activity of this gene, as p35 mutant gene with no antiapoptotic activity showed VSR function in Sf9 cells (Mehrabadi et al. 2015). Following this study, it was shown that the transcript levels of core siRNA pathway in Sf9 cells including Dicer-2 (Dcr1), Argonaute-2 (Ago2) increased in response to AcMNPV infection. Silencing of these genes increased viral genomic DNA replication which indicated their role in the host antiviral defense. Moreover, silencing of Dcr2 in Sf9 cells resulted in enhanced expression levels of the selected virus hotspot genes, while the transcript levels of virus cold spots remained unchanged. Replication of p35-null AcMNPV was significantly increased in Sf9 cells with reduced transcript levels of Dcr2 and Ago2. Also, overexpression of p35 as a suppressor of RNAi and anti-apoptosis gene in Sf9 cells increased virus replication (Karamipour et al. 2018). These results indicate the antiviral role of the siRNA pathway in Sf9 cells and show that the AcMNPV p35 suppresses the siRNA pathway.

MicroRNAs (miRNAs) as small non-coding RNAs play important roles in various biological processes including host-virus interactions by regulating host/virus gene expression. Mehrabadi et al. (2013b) identified about 90 conserved and new miRNAs from Sf9 cells using deep sequencing. The majority of the identified miRNAs were differentially expressed in response to AcMNPV infection. Many of the miRNAs were downregulated; however, some of them were upregulated in response to infection (e.g. miR-184, miR-998 and miR-10), highlighting their potential roles in host–virus interaction. Target analysis of miR-184 – one of the most expressed miRNAs that is differentially expressed following AcMNPV infections – showed potential targets on the coding regions of some of host genes. Using miR-184 agonist and antagonist, it was shown that this miRNA targets dimethyl adenosine transferase, glycosyl hydrolase, transmembrane protein 14C and hydroxysteroid dehydrogenase of SF9 cells (Mehrabadi et al. 2013b). This study revealed the potential roles of miRNAs in host–virus interaction. In a follow up study, the role of core components of the miRNA pathway (including Dicer-1 (Dcr1), Argonaute-1 (Ago1), Exportin-5 (Exp5), and Ran) in Sf9-AcMNPV interaction was investigated. Gene expression analyses showed that the expression levels of Dcr1, Ago1, and Exp5 increased following AcMNPV infection; however, Ran expression declined in response to virus infection. The expression levels of cellular miRNAs (i.e. miR-184 and let-7), were also diminished after AcMNPV infection revealed the differential expression of miRNAs. To investigate the role of core miRNA genes in Sf9-AcMNPV interactions, gene silencing was used and the results showed that RNAi of Dcr1, Ago1 and Ran enhanced viral DNA replication and reduced the abundance of miR-184 and let-7. These results provide evidence that the miRNA pathway is involved in the antiviral immunity of Sf9 cells (Karamipour et al. 2019a). It was shown that Toll and Imd pathways played a role in Sf9 cells after AcMNPV infection (Karamipour et al. 2019b).

## 13.1.8  Mass Production of Entomopathogenic Viruses

For the first time, Izadyar et al. (1998) used a Moldavian NPV isolate in laboratory conditions on third instar *H. armigera* larvae. Two concentrations of virus ($3 \times 10^9$ and $5 \times 10^9$ OB/Kg) were added to artificial food. After 10 days, mortality was observed in the fifth instar larvae. Symptoms of the disease were blackening and swelling of the larvae. Dead larvae were collected, frozen and purified for assay the viral virulence on second instar larvae of *H. armigera*.

Rezapanah and Chitic (1998) evaluated mass rearing potential of two baculoviruse isolates of *C. pomonemlla* (CPGV), and *Agrothis segetum* (AgseNPV). For this purpose, *C. pomonella* and *Agrotis segetum* (Denis and Schiffermüller 1775) (Lep.: Noctuidae) larvae were reared. Then, 2nd and 3rd instar larvae were fed artificial diets containing viral suspension. Dead larvae were collected, frozen, homogenized, filtered, and centrifuged, and pellets were formulated with lactose, glycerin, UV protectant, and conservants in liquid form.

In another study, three main factors affecting virus propagation including larval stage, inoculation dose, and incubation temperature were surveyed by Mehrvar (2013). Here, three isolates of *Helicoverpa armigera* nucleopolyhedrovirus – Maragheh (MRG), Nebrin (NBN), and Marand (MRD) – collected from tomato fields in the East Azarbaijan province were used. To select the best larval stage for virus progration, various *H. armigera* larval instars were selected and 10 μl of virus suspension ($5 \times 10^5$ OB/larva) was added to the diet surface. The early 5th instar larvae provided the highest yield of the virus inoculum. The optimal inoculation dose and incubation temperature were 1965.87 OB/mm$^2$ and 25 °C, respectively.

Assemi et al. (2013b) optimized the cost- effective production method of native isolates of *Helicoverpa armigera* nucleopolyhedrovirus. Then, Assemi et al. (2013b), surveyed the effects of seven factors including larval instar, virus isolate, harvesting time, various artificial diets, different generations, and duration of storage on some yield characteristics of *H. armigera* nucleopolyhedrovirus. The maximum viral yield of $9.499 \times 10^9$ OB/larva was obtained when late 5th instar larvae were individually fed a dose of $5 \times 10^4$ OB/ml. Also, a diet without formaldehyde and ascorbate had the greatest impact on *H. armigera* performance and was better for virus production. When the *H. armigera* larvae were fed a diet containing formaldehyde and ascorbate, controlled with 62.38%, but when formaldehyde and ascorbate was removed from the diet, this percentage was 92.02% (overall 32.21% increase).

## 13.2   Insect Endosymbionts

### *13.2.1*   Wolbachia

#### 13.2.1.1   Introduction

In the last decade, insect microbiome research in Iran has begun. Most studies have been related to *Wolbachia* in parasitoids and culicids. The Trichogrammatidae family are egg parasitoids and are widely used as biocontrol agents against the lepidopteran insects. These species have two reproductive systems, including arrhenotoky and thelytokous. Sometimes, thelytoky associated with *Wolbachia* act as master manipulators of arthropods. The application of thelytokous populations increases rate of parasitism in biocontrol programs.

*Wolbachia pipientis* is an intracellular symbiont bacterium belongs to Anaplasmatacae (Class Alphaproteobacteria, Order Rickettsiales) that first described by Hertig (1936). This diverse endosymbiont has subdivided into 13 supergroups named A to F and H to L on the basis of molecular phylogenetic analysis (Casiraghi et al. 2001). *Wolbachia* symbiont of most insects is from supergroups A and B supergroups (Vandekerckhove et al. 1999).

#### 13.2.1.2   *Wolbachia* in *Trichogramma* Wasp

*Wolbachia* induces reproductive manipulations in its host including parthenogenesis, feminization, male killing, and most commonly, cytoplasmic incompatibility (CI). Farrokhi et al. (2008), surveyed thelytoky and its relation to *Wolbachia* in Iranian populations of *Trichogramma* spp. In this work, 22 populations from different provinces of the country were studied. The *Trichogramma* strains were collected from Mazandaran (10 populations), Razavi Khorasan, Ardabil (2 populations), Hormozgan (2 populations), Sistan and Balouchestan, Khuzestan, Golestan (4 populations), and Guilan provinces. All populations were identified using morphological characteristics and DNA sequencing of the ITS gene. In three populations, Baboulsar (Mazandaran), Nava (Mazandaran), and Mashhad (Razavi Khorasan), thelytoky was observed. The obtained results indicated that the *Trichogramma brassicae* (Bezdenko 1968) (Hym.: Trichogrammatidae) from Baboulsar was infected with *Wolbachia*, while thelytoky of Mashhad and Nava populations had a genetic origin. Following these observations, Farrokhi et al. (2010), surveyed the potential of using a thelytokous *Wolbachia*-infected population and an arrhenotokous uninfected population of *T. brassicae* to control the Angoumois grain moth, *Sitotroga cerealella* (Olivier 1789) (Lep.: Gelechidae). The results indicated that there was no significant difference in parasitism rate between the two strains, while handling time was greater in the infected populations than the uninfected population.

*Wolbachia* detection was also performed on some arthropods and nematodes in the Khuzestan province (Pourali et al. 2009). Out of a total of 770 arthropods (of 22 genera) and 41 nematodes (of six genera), 167 samples of arthropods and one nematode sample were positive. All samples positive tested for *Wolbachia* by 16S gene sequencing were placed in the A supergroup.

Poorjavad et al. (2012), identified seven populations of *Trichogramma* species based on PCR-RFLP of ITS2 gene. Those species included *T. brassicae* Bezdenko, *T. cacoeciae* Marchal, *T. embryophagum* Hartig, *T. evanescens* Westwood, *T. euroctidis* Giraults, *T. pintori* Voegele, and *T. tshumakovae* Sorokina. After identification, infection status of wasps with *Wolbachia* was surveyed. Among positive population of *Trichogramma*, a single strain of *Wolbachia* was detected by using MLST in two populations of *T. brassicae*. This strain was classified into the B supergroup. Following this study, a field survey of native *Trichogramma* species was carried out in six provinces including Razavi Khorasan, Tehran, Mazandaran, Guilan, Golestan, and Qom. Based on ITS2 sequence, 14 populations were identified as the species *T. embryophagum*, *T. evanescens*, and *T. brassicae*. *Wolbachia* infection in these Trichogrammatids was detected using *wsp* gene sequence analysis. The highest infection rate of *Wolbachia* was found in *T. evanescens* from Mazandaran and Golestan provinces. The *Wolbachia* strain observed in those species were from the A and B supergroups. In *T. evanescens*, two populations of parasitic wasp were infected with a single strain from the *sib* subgroup and one population had a superinfection (Karimi et al. 2012). In another study, Darsouei et al. (2013) investigated the recombination rate in 19 populations of infected *Trichogramma*. Among them, eight strains of *Wolbachia* were characterized; within these six strains belonged to subgroup *Kue* from A supergroup. The two remaining strains were related to the B supergroup and *Sib* subgroup. Double infection and superinfection were observed in *Trichogramma* populations. Among three species, *T. embryophagum*, *T. brassicae* and *T. evanescens*, the highest prevalence of *Wolbachia* infection was observed in *T. brassicae* with 57% of sampled populations infected.

Previously, Farrokhi et al. (2008) confirmed that a population of *T. brassicae* form Baboulsar is infected with *Wolbachia*. Following this result, Farrokhi et al. (2013) evaluated the effect of *Wolbachia* on olfactory responses and parasitism rate of *T. brassicae* in laboratory conditions. In this study, a series of experiments were conducted to compare the behavioral aspects, dispersal potential, and parasitism of thelytokous (BW+) and bisexual (B) *T. brassicae* strains (Baboulsar ecotype). There was no significant behavioral difference between these two lines. The number of egg parasitized by BW+ and B was 6.01 and 2.88, respectively. Thus, infected *Wolbachia* lines of Baboulsar ecotype had a higher potential for parasitism than non-infected *Wolbachia* lines.

*Wolbachia* can affect the mtDNA diversity of its host. Darsouei and Karimi (2013), surveyed the effect of *Wolbachia* on the genetic diversity of its host. Samples were taken from 37 populations of Trichogrammatid wasps in different areas of Iran, characterized and checked for prevalence of *Wolbachia* infection. Subsequently, the effects of *Wolbachia* infection on mtDNA variation, number of haplotypes, substitution rate, and nucleotide distances in infected and uninfected populations of *T. brassicae* and *T. evanescens* were evaluated by analyzing partial sequences of the COI gene (mtDNA). Seventeen strains of *Wolbachia* were detected among the *Trichogramma* populations. Of these, 94% were identified as members of the A supergroup. Also, a new *Wolbachia* subgroup named *Tbr* belonging to the A supergroup was discovered. The effects of *Wolbachia* infection on mtDNA variation indicated that haplotype frequencies were higher in infected populations than uninfected populations. Nucleotide diversity in infected populations was lower than uninfected populations of *T. brassicae* and *T. evanescens*. Cophylogenetic relationships revealed no congruence between *Wolbachia* strains and mtDNA of their host. Overall, the results showed that *Wolbachia* affect several genomic criteria of *Trichogramma* and therefore impact mtDNA diversity.

Following these studies, the arising question was about relation between status of *Wolbachia* infection and qualitative characteristics of *T. brassicae* reared on cold stored eggs of the host. An increase in host storage time negatively affects some qualitative characteristics of *T. brassicae* such as fecundity, longevity, wing deformity, and sex ratio (Nazeri et al. 2015). However, there was no significant difference between the parasitism rates of thelytokous and arrhenotokous wasps (Nazeri et al. 2015).

*Wolbachia* infected *Trichogrmma* can produce female offspring via gamete diploidization without mating. Nazeri et al. (2015) indicated *Wolbachia* infection in *T. brassicae* resulted in prolonged developmental time, a higher rate of wing deformity, and lower fecundity and emergence rate. Thus, before mass production of the wasp begins, potential negative impacts of *Wolbachia* infection should be evaluated.

Another addressed issue was possible influence of *Wolbachia* on overwintering population of *Trichogramma*. Rahimi et al. (2017) measured the overwintering ability of infected and uninfected populations of *T. brassicae*. The results indicated both strains were able to overwinter but the *Wolbachia* infected wasps had reduced ability. The presence of *Wolbachia* in *T. brassicae* also affects diapause. *Wolbachia* infection makes disturbance on the clock gene expression which consequently reduces the percentage of diapause (Rahimi et al. 2017).

### 13.2.1.3   *Wolbachia* in *Habrobracon hebetor*

*Habrobracon hebetor* (Hym.: Braconidae) is an important parasitoid wasp of some lepidopteran larvae that is commercially produced and used in biological control programs in Iran. Considering the importance of sex ratio in parasitoid wasps and the reproductive manipulations induced by *Wolbachia* that may affect this sex ratio, Bagheri et al. (2019a) screened different populations of *H. hebetor* to see if they were infected with *Wolbachia*. They found a high prevalence of infection in all populations. *Wolbachia* persisted during ontogeny of *H. hebetor* throughout their life cycle. Their initial observations showed that some crosses in *H. hebetor* resulted in high male progeny while others produce more female progeny which supports the

*Wolbachia* reproduction manipulation hypothesis. To test if *Wolbachia* involved in sex ratio modulation in this insect, isolated lines of *Wolbachia*-infected (W+) wasps and uninfected or cured (W−) wasps were generated using tetracycline treatment. Using these lines, different crosses including infected female-infected male, infected female-uninfected male, uninfected female-infected male, and uninfected female-uninfected male, were created and different wasp reproduction parameters were analyzed. The results showed that, with the exception of the cross between infected males and uninfected females – which produced only male progeny, in the other three crosses both male and female progeny, were produced. Considering that sex determination in hymenopterans is based on haplodiploidy system in which males are haploid and females are diploid, the results indicated that *Wolbachia* induced CI in the *H. hebetor*. It was also shown that the presence of *Wolbachia* in both male and female increase the fitness of *H. hebetor*. Moreover, the results provide evidences that *Wolbachia* may modulate sexual behavior of the wasp as the infected females mated with infected males at a higher rate compared to uninfected males (Bagheri et al. 2019a). This study confirmed that *Wolbachia* induces CI and can modulate behavior of *H. hebetor*. Taking the initial evidence of behavioral modulation induced by *Wolbachia*, in a follow up study, Bagheri et al. (2019b) tested whether this behavioral modification persists in a situation like inbreeding that may result in progeny with low fitness. CI strains of *Wolbachia* cause the diploid eggs obtained from infected males and uninfected females not to hatch. Inbreeding in *H. hebetor* leads to the production of homozygous individuals (diploid males). Diploid males have a lower survival rate and do not produce viable offspring. Therefore, female wasps avoid mating with the same brood as themselves. This behavior reduces mating between siblings and, as a result, arrhenotoky increases. It was shown that *sib* subgroup mating and female progeny were higher in *Wolbachia*-infected lines than the uninfected *sib* mates. These results indicate that *Wolbachia* promoted fertile *sib* mating in *H. hebetor*. It was also shown that *Wolbachia* reduces inbreeding depression effects (Bagheri et al. 2019b). By promoting successful sex with siblings and increasing the probability of female progeny, *Wolbachia* enhance their transmission to the next generation and also mitigate inbreeding depression. This is an undescribed effect of *Wolbachia* (symbiont) on the host reproduction.

The presence of prophage WO in the *Wolbachia*-infected *H. hebetor* was also reported (Nasehi et al. 2019). It was shown that prophage WO is present in both males and females from different populations of the wasp. Neither WO nor cif genes were detected in the tetracycline-treated wasps (Nasehi et al. 2019).

### 13.2.1.4  *Wolbachia* in Fruit Flies

The *Wolbachia* has found to be involved in some biological interactions including CI in some species of fruit flies (Zabalou et al. 2004). Karimi and Darsouei (2014), collected the specimens of various populations from several fruit flies species including *Dacus ciliatus* (cucurbit fly), *Rhagoletis cerasi* (cherry fruit fly), *Ceratitis capitata* (Mediterranean fruit fly), *Myiopardalis pardalina* (melon fly) and

*Carypomya vesuviana* (jujube fly) from different geographic regions of the countrytrop. Two species, *R. cerasi* and *C. vesuviana*, showed infection with separate *Wolbachia* strains, namely *wCer6* and *wVes1*, respectively. Here, *C. vesuviana* was introduced as a novel host for *Wolbachia*. Also, using *wsp* sequencing and MLST, the *Wolbachia* strain in *C. vesuviana* was established as a new strain type (ST 277) (Karimi and Darsouei 2014).

### 13.2.1.5   *Wolbachia* in Culicidae

Population differentiation and *Wolbachia* phylogeny in *Aedes scutellaris* (Dip.: Culicidae) was surveyed by Behbahani et al. (2005). Also, *Wolbachia* infection was surveyed among *Culex* mosquitoes in southwest Iran (Behbahani 2012). All samples of *Culex quinquefasciatus* (Dip: Culicidae) had *Wolbachia* infection, while *Culex tritaeniorhynchus* (Dip: Culicidae) and *Culex theileri* (Dip: Culicidae) indicated no infection with *Wolbachia*. Karamin et al. (2016) detected *Wolbachia* in *Culex pipiens* in the northern, central, and southern parts of Iran.

### 13.2.1.6   *Wolbachia* in Other Hosts

A new strain of *Wolbachia* was detected in sand fly, *Paraphlebotomus* sp. (Diptera: Psychodidae) (Parvizi et al. 2013). Later, three new strains of *Wolbachia pipientis* were identified in sandflies in Iran (Bordbar et al. 2014).

Hemmati et al. (2017), identified endosymbionts of *Hishimonus phycitis* (Hem.: Cicadellidae). In this study, *Candidatus* Sulcia meelleri and *Candidatus* Nasuia detocephalinicola were identified as primary endosymbionts. *Wolbachia*, *Arsenophonus* sp., *Diplorichettsia* sp., *Spiroplasma citri*, and *Pantoea agglomerans* were introduced as secondary endosymbionts.

## *13.2.2   Other Endosymbionts*

Mehrabadi et al. (2012) showed that the posterior region of *Eurygaster integriceps* (Hem.: Scutelleridae) has a vacuolated epithelium with many crypts harboring symbiotic bacteria. A follow up study showed that removing these symbiotic bacteria negatively affects fitness of *E. integriceps* (Kafil et al. 2013). Chavsin et al. (2012), surveyed the diversity of bacterial microflora in the midgut of the larvae and adult *Anopheles stephensi* (Dip.: Culicidae) and identified 40 species in 12 genera. In the larval midgut, there were seven gram-negative genera (*Myroides*, *Chryseobacterium*, *Aeromonas*, *Pseudomonas*, *Klebsiella*, *Enterobacter* and *Shewanella*) and five gram-positive genera (*Exiguobacterium*, *Enterococcus*, *Kocuria*, *Microbacterium* and *Rhodococcus*). In the adult midgut, there were 25 gram-negative genera including *Pseudomonas*, *Alcaligenes*, *Bordetella*,

*Myroides* and *Aeromonas*. The crypt-dwelling Gammaproteobacteria were identified in *Graphosoma lineatum* (Hem.: Pentatomidae). It was demonstrated that that these symbiotic bacteria were vertically transmitted and positively affected the fitness of the insect as surface sterilization of eggs resulted in decreased fitness of the offspring compared to the control (Karimpour et al. 2016). Zarei et al. (2017) surveyed the diversity of *Spiroplasma citri* in *Circulifer haematoceps* (Hem.: Cicadellidae), a leafhopper of sesame fields in the Fars province of south Iran. There were six groups –the strains of group 1 could be transmitted from sesame-infected plants to citrus trees by *C. haematoceps*, while strains of group 6 could not infect citrus trees. Hemmati et al. (2017) identified the yeast and yeast-like symbionts from the insect vector of lime witches' broom phytoplasma, *Hishimonus phycitis* (Hem.: Cicadellidae). Thirtheen different populations of *H. phycitis* were collected form Hormozgan, Kerman and Sistan-Baluchestan provinces. Results revealed that the vector harboured yeast-like symbionts of *H. phycitis* (Hp-YLS) and *Candida pimensis*, with high similarity to those reported from the other Cicadellids. Fami-Tafreshi et al. (2017) investigated the microbiota of olive psyllids, *Euphyllura straminea* and *Euphyllura pakistanica* (Hem.: Aphalaridae), using a metagenomic approach. Different populations of *E. straminea* were collected from Tarom and Tehran whereas populations of *E. pakistanica* were collected from Shiraz. The results showed that bacterial taxa within the phylum Proteobacteria were dominant in all populations of *E. straminea* and *E. pakistanica*. The γ-Proteobacteria were the most abundant bacteria in the psyllids followed by α-Proteobacteria, β-Proteobacteria, Actinobacteria, Firmicutes and Bacteroidetes, respectively. Also, the results showed that female psyllids harbored greater populations of gut bacteria compared with males. The bacterial community structure was similar in both species. In another work, the effect of *Arsenophonus* elimination on *Ommatissus lybicus* (Hem.: Tropiduchidae) was surveyed (Karimi et al. 2019). Elimination of *Arsenophonus* increased the developmental time of the hopper immature stages and reduced life-history parameters such as nymphal survival rate and adult longevity in the host. In the aphids, bacterial endosymbionts play important roles in host ecological behaviours. Hosseinzadeh et al. (2019) assessed the endosymbionts and their titers in different tissues of male and female Asian citrus psyllid, *Diaphorina citri* Kuwayama). *D. citri* identified as vector of *Candidatus* Liberibacter asiaticus (CLas). They also compared the level of these endosymbionts in CLas infected psyllids and uninfected psyllids. The highest titer of CLas was observed in the gut while *Wolbachia* titer was highest in the Malpighian tubules. *Wolbachia* and CLas were detected in all psyllid tissues. The titer of *Wolbachia* was higher in the heads of CLas-infected males compared to uninfected males. The titer of two bacterial endosymbionts, *Profftella* and *Carsonella* were detected in specific bacteriome organ and the ovaries. Their titers were significantly reduced in the bacteriome of CLas infected males and increased in the ovaries of CLas infected females. This study showed that CLas modulates the symbionts' titers. In a series of studies, the endosymbionts of pistachio stink bugs *Brachynema germari* (Kashkouli et al. 2019a), *Acrosternum heegeri* (Kashkouli et al. 2020), and *Acrosternum arabicum* (Kashkouli et al. 2019b) (Hemiptera: Pentatomidae) were identified. Pantoea-like endosymbionts were identified from these insects and it was shown that symbiotic bacteria play important roles in the biology of the insects as symbiont removal resulted in impaired growth, development and reproduction. It was shown that high temperature and chemical surface sterilization resulted in symbiont suppression and an overall reduction in insect fitness. Therefore, methods targeting endosymbionts in these insects such as increased temperature and chemical surface sterilization could potentially be used to control pistachio stink bugs (Kashkouli et al. 2019c). Genomic studies of the stinkbug *A. arabicum*'s bacterial symbionts resulted in a draft genomeof the symbiont, which is intermediate in size between free-living *Pantoea* and other highly-reduced obligate symbionts of stink bugs. The genomic annotation revealed that the symbiont retains the predicted capability to perform biosynthesis of amino-acids, vitamins and other metabolites. In addition, approximately one third of total ORFs were classified as pseudogenes. On the basis of this study and the geographic origin of the bacteria, (persica = Persia, Iran), we propose the name "*Candidatus* Pantoea persica" for the midgut symbiont associated with the stinkbug *A. arabicum* (Kashkouli et al. 2019b).

Ayoubi et al. (2020), characterized the bacterial endosymbionts of *Aphis gossypii* (Glover 1877) (Hem.: Aphididae) in Karaj populations of the aphid. In this study, *Buchnera aphidicola, Hamiltonella defense* and *Arsenophonus* sp. were detected. Reduction in *Buchnera* caused prolonged development and no progeny production. Furthermore, secondary symbiont reduction led to a reduction of the total life span, intrinsic rate of natural increase, and emergence of abnormal offspring in treated insects.

## 13.3   Microsporidia

Currently, 93 of the 200 described genera of microsporidia have an insect host. The microsporidian infections (microsporidioses) are classified into two groups, chronic and acute. In infection with microsporidia, there are various signs and symptoms such as tissue manifestations, abnormal development, and behavioral changes. The external effects in infected larvae are changes in color, size, form, or behavior when compared to healthy individuals. The fat body and midgut epithelium are two main tissues in which microsporidia affect them usually (Becnel and Andreadis 2014). Most microsporidian insect pathogens induce sublethal effects on the hosts resulting in reduced fertility, shortened longevity, and loss of vigor (Brooks 1988; Solter et al. 2012). Vertical transmission is a major pathway of transmission for many microsporidian insect pathogens. Contamination of the environment with spores is typically achieved when an infected host dies or when spores are released in fecal excrement that leads to horizontal transmission (Becnel and Andreadis 2014).

## 13.3.1   Apis mellifera

Nosemosis is among the common diseases of adult honeybees in Iran. Two species, namely *Nosema apis* and *Nosema ceranae* – which have been shown to infect honeybee midgut epithelial cells – have been reported by apiaries in various provinces. Infection with *Nosema* induces dysentery and a shorted lifespan in honeybees, delayed fertility of queens, and colony size reduction (Razmaraii et al. 2013). The *Nosema* infection leads to economic and biological damage in colonies. According to Lotfi et al. (2009) and Razmaraii and Karimi 2010, nosemosis in Iran is caused by *N. apis.* However, Nabian et al. (2011) introduced *N. ceranae* to the Mazandaran province for the first time. Infection to *N. apis* was surveyed in North Khorasan province by Moshaverinia et al. (2012). Prevalence of nosemosis in East Azerbaijan was surveyed by Razmaraii et al. (2013). Also, the *N. ceranae* was introduced as a causal agent of nosemosis in East Azerbaijan province (Razmaraii et al. 2013).

Modirrousta et al. (2014) surveyed on *Nosema* incidence in honeybees and 41 positive samples of *Nosema* sp. were collected from five provinces from 2004 to 2013. The samples were examined by multiplex PCR technique and characterized for two species, *N. ceranea* and *N. apis.* The obtained results indicated that all of samples were positive for *N. ceranea*, thus confirming that the *N. ceranea* has been spread in Iran (Modirrousta et al. 2014).

In another study, Aroee et al. (2017), identified *Nosema* species in Isfahan, Fars and Chaharmahal and Bakhtiari provinces. Sequencing the 16S rRNA gene of this microsporid indicated that the *Nosema* infection in southwestern Iran is related to *N. ceranae.* In Kurdistan province, *N. ceranae* was determined as a nosemosis causal agent by 16S rRNA gene data (Khezri et al. 2018). Another study was performed by Mohammadian et al. (2018) that surveyed distribution of *Nosema* spp. in climatic regions of the country. In this study, 183 adult bees were selected and surveyed by microscope and PCR techniques. Infection caused by *N. ceranae* was observed in all regions of the study, and the infection rate was 46.40%. However, infection with *N. apis* was not observed. The prevalence of *N. ceranae* in humid, semi-humid, very humid, arid, semi-arid, and Mediterranean climates was 58.8%, 71%, 68.10%, 29.40%, 34.30%, and 24%, respectively. Here there was a significant difference between different climatic regions. Thus climate can affect the spread of *N. ceranae* (Mohammadian et al. 2018).

## 13.3.2   Culicidae

Omrani et al. (2016), identified a microsporidian in *Anopheles superpictus.* The species was from the genus *Parathelohania* and the family Ambliosporidae. This microsporidian introduced as *Parathelohania iranica* (Omrani et al. 2017).

## 13.3.3   Other Species

The study of microsporidia has a narrow list in Iran. Habibi and Kharazi Pakdel (1989) surveyed horizontal transmission of *Vavaria portugal* (Microsporidia) in *Lymantria dispar* (Linnaeus 1758) (Lep.: Lymantriidae). The larvae of *L. dispar* were infected with $1 \times 10^4$ and $1 \times 10^5$ spore/mL solutions. Then, 3 day-post infection, uninfected larvae were added to the container of infected larvae. Seven days post-infection, spores were produced and horizontal transmission was observed. Ghazavi and Kharazi Pakdel (1995) surveyed histopathology and biology of *Nosema locustae* in desert locust, *Schistocerca gregaria* Forskal. In this research, 3rd, 4th, and 5th nymph stages were given food containing $5.48 \times 10^4$ spores of *N. locustae*. After 10 days, the nymph's tissues were examined. In light infection, fat bodies were the visible infected tissues, but in more developed infections, pericardial cells, ovaries, testes, midgut, malpighian tubes and nervous system were infected.

**Acknowledgment**   We appreciate from Todd Kabaluk and Kyle Mcpherson for reviewing the last version of manuscript.

# References

Anderson DL, Trueman JWH (2000) *Varroa jacobsoni* (Acari: Varroidae) is more than one species. Exp Appl Acarol 24(3):165–189

Aroee F, Azizi H, Shiran B, Pirali Kheirabadi K (2017) Molecular identification of *Nosema* species in provinces of Fars, Chaharmahal and Bakhtiari and Isfahan (Southwestern Iran). Asian Pac J Trop Biomed 7(1):10–13. https://doi.org/10.1016/j.apjtb.2016.11.004

Asi MR, Bashir MH, Afzal M, Zia K, Akram M (2013) Potential of entomopathogenic fungi for biocontrol of *Spodoptera litura* Fabricius (Lepidoptera: Noctuidae). J Anima Plant Sci 3 (3):913–918

Assemi H, Rezapanah MR, Sajjadi A (2013a) Biological control of *Helicoverpa armigera* (Lep.: Noctuidae) with the efficacy of different geographical isolates of HaNPVon tobacco fields. Paper presented at the Conference of Biological Control in Agriculture and Natural Resources, University of Tehran, Tehran, 27–28 August 2013, page 79

Assemi H, Sajjadi A, Rezapanah MR (2013b) Optimization of production of *Helicoverpa armigera nucleopolyhedrovirus* via the Taguchi method and Derringer's desirability function. Paper presented at the Conference of Biological Control in Agriculture and Natural Resources, University of Tehran, Tehran, 27–28 August 2013, page 96

Ayoubi A, Talebi AA, Fathipour Y, Mehrabadi M (2020) Coinfection of the secondary symbionts, *Hamiltonella defensa* and *Arsenophonus* sp. contribute to the performance of the major aphid pest, *Aphis gossypii* (Hemiptera: Aphididae). Insect Sci 27(1):86–98. https://doi.org/10.1111/1744-7917.12603

Ayres MD, Howard SC, Kuzio J, Lopez-Ferber M, Possee RD (1994) The complete DNA sequence of *Autographa californica* nuclear polyhedrosis virus. Virology 202(2):586–605

Bagheri Z, Talebi AA, Asgari S, Mehrabadi M (2019a) *Wolbachia* induce cytoplasmic incompatibility and affect mate preference in *Habrobracon hebetor* to increase the chance of its transmission to the next generation. J Inverteb Pathol 163:1–7

Bagheri Z, Talebi AA, Asgari S, Mehrabadi M (2019b) *Wolbachia* promote successful sex with siblings. bioRxiv 855635

Becnel J, Andreadis T (2014) Chapter 21: Microsporidia in insects. In: Weiss LM, Becnel JJ (eds) Pathogens of opportunity, 1st edn. Wiley, Chichester, pp 521–570

Behbahani A (2012) *Wolbachia* infection and mitochondrial DNA comparisons among *Culex mosquitoes* in South West Iran. Pak J Biol Sci 15(1):54–57

Behbahani A, Dutton T, Davies N, Townson H, Sinkins S (2005) Population differentiation and *Wolbachia* phylogeny in mosquitoes of the *Aedes scutellaris* group. Med Vet Entomol 19 (1):66–71

Beketov MA, Kefford BJ, Schäfer RB, Liess M (2013) Pesticides reduce regional biodiversity of stream invertebrates. PNAS 110:11039–11043

Berling M, Blachere-Lopez C, Soubabere O, Lery X, Bonhomme A, Sauphanor B, Lopez-Ferber M (2009a) *Cydia pomonella* granulovirus genotypes overcome virus resistance in the codling moth and improve virus efficiency by selection against resistant hosts. Appl Environ Microbiol 75:925–930

Berling M, Rey JB, Ondet SJ, Tallot Y, Soubabère O, Bonhomme A, Sauphanor B, Lopez-Ferber M (2009b) Field trials of CpGV virus isolates overcoming resistance to CpGV-M. Virol Sin 24:470

Biabani MR, Seidavi AR, Gholami MR, Etebari K, Matindoost L (2005) Evaluation of reststance to nuclear polyhedrosis virus in 20 commercial hybrids of silkworm (*Bombyx mori*). Formosan Entomol 25:103–112

Bordbar A, Soleimani S, Fardid F, Zolfaghari MR, Parvizi P (2014) Three strains of *Wolbachia pipientis* and high rates of infection in Iranian sandfly species. Bul Entomol Res 104:195–202

Brooks WM (1988) Entomogenous protozoa. In: Ignoffo CM (ed) Handbook of Natural Pesticides, Vol. V, Microbial Insecticides, Part A, Entomogenous Protozoa and Fungi. CRC Press, Inc, Boca Raton, pp 1–149

Buerger P, Hauxwell C, Murray D (2007) Nucleopolyhedrovirus introduction in Australia. Virol Sin 22:173–179

Casiraghi M, Anderson TJC, Bandi C, Bazzocchi C, Genchi C (2001) A phylogenetic analysis of filarial nematodes: comparison with the phylogeny of *Wolbachia* endosymbionts. Parasitology 122:93–103

Chavsin AR, Oshaghi MA, Vatandoost H, Pourmand MR, Raeisi A, Enayati AA, Mardani N, Ghorchian S (2012) Identification of bacterial microflora in the midgut of the larvae and adult of wild caught *Anopheles stephensi*: a step toward finding suitable paratransgenesis candidates. Acta Trop 121:129–134

Dai X, Hajos JP, Hoosyen NN, Oers MM, Ijekel WFJ, Zuidema D, Pang Y, Vlak JM (2000) Isolation of a *Spodoptera exigua* baculovirus recombinant with a 10±6 kbp genome deletion that retains biological activity. J Gen Virol 81:2545–2554

Darsouei R, Karimi J (2013) Effects of *Wolbachia* (endisymbiont of trichogramatid wasps) on genetic diversity of host wasp. Paper presented at the Conference of biological control in agriculture and natural resources, University of Tehran, Karaj, 26–28 August 2013, page 12

Darsouei R, Karimi J (2015). Natural enemies of sugar beet armyworm, *Spodoptera exigua* (Lep.: Noctuidae). Paper presented at 1st Iranian International Congress of Entomology. Tehran University, Karaj, 29–31 August 2015, page 33

Darsouei R, Karimi J, Jahanbakhsh V (2013) *Wolbachia* grouping study, in major *Trichogramma* wasps in Iran. J Entomol Res 5(3):219–236. In Persian

Darsouei R, Karimi J, Ghadamyari M, Hosseini M (2017) Natural enemies of the sugar beet army worm, *Spodoptera exigua* (Lepidoptera: Noctuidae) in north-east Iran. Entomol News 127 (5):446–464. https://doi.org/10.3157/021.127.0508

Dezianian A, Sajap AS, Hong LW, Omar D, Kadir HA, Mohamed R (2012) Evaluation of additives on *Plutella xylostella* granulovirus efficacy after ultraviolet radiation in laboratory conditions. Paper presented at 20th Iranian Plant Protection Congress, Shiraz University, Shiraz, 25–27 August 2012, page 34

Dezianian A, Saeed Sajap A, Lau WH, Dezianian S (2015a) Semi-field evaluation of *Plutella xylostella* Granulovirus microencapsulation formulation on Diamond back Moth (*Plutella xylostella*). Paper presented at Third National Meeting on Biocontrol in Agriculture and Natural Resources of Iran, Ferdowsi University of Mashhad, Mashhad, 2–3 February 2015, page 15

Dezianian A, Said Sajap A, Lau WH, Hussan AK (2015b) Molecular and morphological study and biological activity of *Plutella xylostella* Granulovirus on *Plutella xylostella* (Lepidoptera, Plutellidae) larvae. Paper presented at Third National Meeting on Biocontrol in Agriculture and Natural Resources of Iran, Mashhad. Ferdowsi University of Mashhad, 2–3 February 2015, page 4

Eberle K, Asser-Kaiser S, Sayed S, Nguyen H, Jehle J (2008) Overcoming the resistance of codling moth against conventional *Cydia pomonella* granulovirus (CpGV-M) by a new isolate CpGV-I12. J Inver Pathol 98:293–298

Eberle KE, Sayed S, Rezapanah M, Shojai-Estabragh S, Jehle JA (2009) Diversity and evolution of the *Cydia pomonella* granulovirus. J Gen Virol 90:662–671

Etebari K, Matindoost L, Mirhosseini SZ, Turnbull MW (2007) The effect of BmNPV infection on protein metabolism in silkworm (*Bombyx mori*) larva. ISJ 4:13–17

Fahimi A, Kharazi Pakdel A, Talaei-Hassanloui R, Rezapanah MR, Maleki F (2008a) Sensitivity of different growth stages of cabbage moth, (*Plutella xylostella*: Plutellidae) to MbNPV in laboratory conditions. Paper presented at 18th Iranian Plant Protection Congress, Bu-Ali Sina University, Hemedan, 24–26 August 2008, page 44

Fahimi A, Kharazi Pakdel A, Talaei-Hassanloui R, Rezapanah MR, Maleki F (2008b) Bioassay of PxGV-Taiwanii on cabbage moth *Plutella xylostella* (Lep.: Plutellidae) in laboratory conditions. Paper presented at 18th Iranian Plant Protection Congress, Bu-Ali Sina University, Hemedan, 24–26 August 2008, page 203

Fami-Tafreshi T, Mehrabai M, Askarianzadeh A, Saeedzadeh AA (2017) The gut microbiota of the olive psyllids, *Euphyllura straminea* and *Euphyllura pakistanica* (Hemiptera: Aphalaridae). Paper presented at 2nd Iranian International Congress of Entomology. University of Tehran, Karaj, 2–4 September 2019, page 86

Farrokhi S, Ashouri A, Shirazi J, Huigens M.E, Ebrahimi E, Attaran MR (2008) Thelytoky and its relation to *Wolbachia* in Iranian populations of *Trichogramma* spp. Paper presented at 18th Iranian Plant Protection Congress, Bu-Ali Sina University, Hemedan, 24–26 August 2008, page 50

Farrokhi S, Ashouri A, Shirazi J, Allahyari H, Huigens, M.E (2010) A comparative study on the functional response of *Wolbachia*-infected and uninfected forms of the parasitoid wasp *Trichogramma brassicae*. J Insect Sci 10 (167) DOI: https://doi.org/10.1673/031.010.14127

Farrokhi S, Shirazi J, Attaran MR (2013) *Wolbachia* effect on olfactory responses and parasitism rate of *Trichogramma brassicae* in laboratory conditions. BioControl Plant Protect 1(1):65–79. In Persian

Fitt GP (1994) Cotton pest management: part 3. An Australian perspective. Annu Rev Entomol 39:543–562

Geiger F, Bengtsson J, Berendse F, Weisser WW, Emmerson M, Morales MB, Ceryngier P, Liira J, Tscharntke T, Winqvist C (2010) Persistent negative effects of pesticides on biodiversity and biological control potential on European farmland. Basic Appl Ecol 11:97–105

Ghazavi M, Kharazi-Pakdel A (1995) The Histopathology and biology of *Nosema locustae* canning in desert locust, *Schistocerca gregaria* Forsk. Paper presented at 12th Plant Protection Congress, Tehran University, Karaj, page 286

Ghorani M, Langeroudi A, Madadgar O, Rezapanah MR, Nabian S, Khaltabadi Farahani R, Maghsoudloo H, Forsi M, Abdollahi H, Akbareinn H (2017) Molecular identification and phylogenetic analysis of chronic bee paralysis virus in Iran. Vet Res Forum 8(4):287–292

Ghorbani M, Madadgar O, Langeroudi AG, Rezapanah M, Nabian S, Akbarein H, Farahani RK, Maghsoudloo H, Abdollahi H, Forsi M (2017) The first comprehensive molecular detection of six honey bee viruses in Iran in 2015-2016. Arch Virol 162(8):2287–2291. https://doi.org/10.1007/s00705-017-3370-9

Grant WP, Chandler D, Bailey A, Greaves J, Tatchel M, Prince G (2010) Biopesticides: pest management and regulation. CABI, Wallingford

Grzywacz G (2017) Basic and applied research: Baculovirus. In: Lacey L (ed) Microbial control of insect and mite pests, from theory to practice. Elsevier, Amsterdam, pp 27–46

Habibi J, Kharazi Pakdel A (1989) Horizontal transition *Vavaria portugal* (Microsporidia, Protozoa) in *Lymantria dispar* L. (Lepidoptera: Lymantriidae). Paper presented at 9th Plant Protection Congress. Fersowsi University of Mashhad, Mashhad, p 8

Hemmati C, Moharramipour S, Seyahooei A, Bagheri A, Mehrabadi M (2017) Identification of yeast and yeast-like symbionts associated with *Hishimonus phycitis* (Hemiptera: Cicadellidae), the insect vector of lime witches' broom phytoplasma. J Crop Prot 6(4):439–446

Hertig M (1936) The rickettsia, *Wolbachia pipientis* (gen. et sp. n.) and associated inclusions of the mosquito, *Culex pipiens*. Parasitology 28(4):453–486

Hosseinzadeh S, Shams-Bakhsh M, Mann M, Fattah-Hosseini S, Bagheri A, Mehrabadi M, Heck M (2019) Distribution and variation of bacterial endosymbiont and "*Candidatus* Liberibacter asiaticus" titer in the Huanglongbing insect vector, *Diaphorina citri* Kuwayama. Micro Ecol 78(1):206–222

Inceoglu AB, Kamita SG, Hammock BD (2006) Genetically modified baculoviruses: a historical overview and future outlook. Adv Vir Res. Academic Press, pp 323-360

Izadyar S, Rezapanh MR, Assady HB, Daniali M, Chitic V, Voloshchuk L. (1998) Laboratory production of NPV virus (Helicoviridae-Liquid) and the study of its pathogenicity on the cotton boll worm *in vitro*. Paper presented at 13th Iranian Plant Protection Congress, Tehran University, Karaj, page 61

Jakubowsaka AJ, Vlak J, Ziemnicka J (2005) Characterization of a nucleopolyhedrovirus isolated from the laboratory rearing of the beet armyworm *Spodoptera exigua* (HBN). J Plant Prot Res 45(4):279–286

Jehle JA, Blissard G, Bonning B, Cory J, Herniou E, Rohrmann G, Theilmann D, Thiem S, Vlak J (2006) On the classification and nomenclature of baculoviruses: a proposal for revision. Arch Virol 151:1257–1266

Kafil M, Bandani AR, Kaltenpoth M, Goldansaz SH, Alavi SM, Miller T (2013) Role of symbiotic bacteria in the growth and development of the Sunn pest, *Eurygaster integriceps*. J Insect Sci 13 (99):1–12

Kalantari M, Marzban R, Imani S, Askari H (2013) Effects of *Bacillus thuringiensis* isolates and single nuclear polyhedrosis virus in combination and alone on *Helicoverpa armigera*. Achi Phyto Plant Protec. https://doi.org/10.1080/03235408.2013.802408

Kamali E, Pourmirza AA (2000) The survey of pathogenicity of nucleopolyhedrosis virus against 1–5 instar larvae of *Spodoptera exigua* on laboratory condition. Paper presented at 14th Iranian Plant Protection Congress, Isfahan University, Esfahan, September 2000, page 50

Karamin M, Moosa-Kazemi SH, Oshaghi MA, Vatandoost H, Sedaghat MM, Rajabnia R, Hosseini M, Maleki-Ravasan N, Yahyapour Y, Ferdosi-Shahandishi E (2016) *Wolbachia* Endobacteria in Natural Populations of *Culex pipiens* of Iran and its Phylogenetic Congruence. J Arthropod-Borne Dis 10(3):347–363

Karamipour N, Mehrabadi M, Fathipour Y (2016) Gammaproteobacteria as essential primary symbionts in the striped shield bug, *Graphosoma lineatum* (Hemiptera: Pentatomidae). Sci Rep. https://doi.org/10.1038/srep33168

Karamipour N, Fathipour Y, Talebi AA, Asgari S, Mehrabadi M (2018) Small interfering RNA pathway contributes to antiviral immunity in *Spodoptera frugiperda* (Sf9) cells following *Autographa californica* multiple nucleopolyhedrovirus infection. Insect Biochem Mol Biol 101:24–31

Karamipour N, Fathipour Y, Talebi AA, Asgari S, Mehrabadi M (2019a) The microRNA pathway is involved in *Spodoptera frugiperda* (Sf9) cells antiviral immune defense against *Autographa californica* multiple nucleopolyhedrovirus infection. Insect Biochem Mol Biol 112:103202

Karamipour N, Fathipour Y, Talebi AA, Asgari S, Mehrabadi M (2019b) Toll and Imd pathways in *Spodoptera frugiperda* (Lep.: Noctuidae) cells are induced following *Autographa californica* (Lep.: Noctuidae) multiple nucleopolyhedrovirus infection. Paper presented at 3rd Iranian

International Congress of Entomology. Tabriz University, Tabriz, 17–19 August 2019, page 196

Karimi J, Darsouei R (2014) Presence of the endosymbiont *Wolbachia* among some fruit flies (Diptera: Tephritidae) from Iran: a multilocus sequence typing approach. J Asia Pac Entomol 17:105–112. https://doi.org/10.1016/j.aspen.2013.11.002

Karimi J, Darsouei R, Hosseini M, Stouthamer R (2012) Molecular characterization of Iranian Trichogrammatids (Hymenoptera: Trichogrammatidae) and their *Wolbachia* endosymbiont. J Asia Pac Entomol 15:73–77. https://doi.org/10.1016/j.aspen.2011.08.004

Karimi S, Asgari Seyahooei M, Izadi H, Bagheri A, Khodaygan P (2019) Insect-symbiont interactions effect of *Arsenophonus* endosymbiont elimination on fitness of the date palm hopper, *Ommatissus lybicus* (Hemiptera: Tropiduchidae). Environ Entomol 48(3):614–622. https://doi.org/10.1093/ee/nvz047

Kashkouli M, Fathipour Y, Mehrabadi M (2019a) Heritable gammaproteobacterial symbiont improves the fitness of *Brachynema germari* Kolenati (Hemiptera: Pentatomidae). Environ Entomol 48(5):1079–1087

Kashkouli M, Fathipour Y, Mehrabadi M (2019b) Genome characteristics of the bacterial symbiont of the stinkbug *Acrosternum arabicum* (Hem.: Pentatomidae). Paper presented at 3rd Iranian International Congress of Entomology. Tabriz University, Tabriz, 17–19 August 2019, page 198

Kashkouli M, Fathipour Y, Mehrabadi M (2019c) Potential management tactics for pistachio stink bugs, *Brachynema germari*, *Acrosternum heegeri* and *Acrosternum arabicum* (Hemiptera: Pentatomidae): high temperature and chemical surface sterilants leading to symbiont suppression. J Econ Entomol 112(1):244–254

Kashkouli M, Fathipour Y, Mehrabadi M (2020) Habitat visualization, acquisition features and necessity of the gammaproteobacterial symbiont of pistachio stink bug, *Acrosternum heegeri* (Hem.: Pentatomidae). Bull Entomol Res 110(1):22–33

Khezri M, Moharrami M, Modirrousta H, Torkaman M, Salehi S, Rokhzad B, Khanbabai H (2018) Molecular detection of *Nosema ceranae* in the apiaries of Kurdistan province, Iran. VRF, 9 (3):273–278. https://doi.org/10.30466/vrf.2018.32086

Lacey LA, Frutos R, Kaya H, Vail P (2001) Insect pathogens as biological control agents: do they have a future? Biol Control 21:230–248

Lacey L, Grzywacz D, Shapiro-Ilan D, Frutos R, Brownbridge M, Goettel M (2015) Insect pathogens as biological control agents: back to the future. J Inver Patho 132:1–41

Lotfi A, Jamshidi R, Aghdam Shahryar H, Yousefkhani M (2009) The prevalence of nosemosis in honey bee colonies in Arasbaran region (Northwestern Iran). American-Eurasian. J Agric Environ Sci 5:255–257

Magholi Z, Abbasipour H, Marzban R (2014) Effect of *Helicoverpa armigera Nucleopolyhedrosis* virus (HaNPV) on the larvae of the Diamondback Moth, *Plutella xylostella* (L.) (Lepidoptera: Plutellidae). Plant Prot Sci 50(4):184–189

Magholli Z, Marzban R, Abbasipour H, Shikhi A, Karimi J (2013) Interaction effects of *Bacillus thuringiensis* subsp. *kurstaki* and single nuclear polyhedrosis virus on *Plutella xylostella*. JDPP 120(4):173–178

Manzari S, Safar Alizadeh MH, Kharazi Pakdel A, Pourmirza AA (1998) Effect of MbNPV on different larval instars of *Spodoptera exigua* in laboratory conditions. Paper presented at 13th Iranian Plant Protection Congress. Tehran University, Karaj, August 1998, page 246

Marzban R, He Q, Liu XX, Zhang QW (2010) Interaction of *Bacillus thuringiensis* toxin Cry1Ac and *Helicoverpa armigera* cytoplasmic polyhedrosis virus, against the cotton bollworm larvae, *H. armigera* (Hubner) (Lep.: Noctuidae). Paper presented at 19th Iranian Plant Protection Congress, Institute of Plant Pests and Diseases, Tehran, 31 July-3 August, 2010, page 50

Marzban R, He Q, Zhang Q, Liu XX (2013) Histopathology of cotton bollworm midgut infected with *Helicoverpa armigera* cytoplasmic polyhedrosis virus. Braz J Microbiol 44(4):1231–1236

Mehrabadi M, Bandani AR, Allahyari M, Serrão JE (2012) The Sunn pest, *Eurygaster integriceps* Puton (Hemiptera: Scutelleridae) digestive tract: histology, ultrastructure and its physiological significance. Micron 43(5):631–637

Mehrabadi M, Hussain M, Asgari S (2013a) Cloning and characterization of a Dim1-like mitosis gene of *Spodoptera frugiperda* cells (Sf9) induced by *Autographa californica* multiple nucleopolyhedrovirus. J Inverteb Pathol 113(2):152–159

Mehrabadi M, Hussain M, Asgari S (2013b) MicroRNAome of *Spodoptera frugiperda* cells (Sf9) and its alteration following baculovirus infection. JGMV 94(6):1385–1397

Mehrabadi M, Hussain M, Matindoost L, Asgari S (2015) The baculovirus antiapoptotic p35 protein functions as an inhibitor of the host RNA interference antiviral response. J Virol 89 (16):8182–8192

Mehrvar A (2013) Virus yield parameters in mass production of three Iranina geographic isolates of *Helicoverpa armigera nucleopolyhedrovirus*. Acta Entomol Sin 56(10):1229–1234

Mehrvar A (2015a) Impacts of flufenoxuron on insecticidal activity of *Helicoverpa armigera* nucleopolyhedroviruses. Paper Presented at 1st Iranian International Congress of Entomology, Tehran University, Karaj, page 168

Mehrvar A (2015b) Impact of age-dependently resistance of *Helicoverpa armigera* (H bner) larvae to different geographic isolates of nucleopolyhedroviruses (HearNPV). Paper presented at 1st Iranian International Congress of Entomology. Tehran University, Karaj, p 63

Mehrvar A, Rabindra RJ, Veenakumari K, Narabenchi GB (2008a) Comparative evaluation of seven geographic ioslates of nucleopolyhedrovirus on *Helicoverpa armigera* (Hunber) (Lepidoptera: Noctuidae). Paper presented at 18th Iranian Plant Protection Congress. Bu-Ali Sina University, Hamedan, 24–26 August 2008, page 32

Mehrvar A, Rabindra RJ, Veenakumari K, Narabenchi GB (2008b) Effect of different exposure doses and times of simulated on the sunlight on the virulence of *nucleopolyhedrovirus* on *Helicoverpa armigera* (Hunber) (Lepidoptera: Noctuidae). Paper presented at 18th Iranian Plant Protection Congress, Bu-Ali Sina University, Hamedan, 24–26 August 2008, page 33

Miletic N, Tamas N, Graora D (2011) The control of colding moth (*Cydia pomonella* L.) in apple trees. Agriculture 98(2):213–218

Modirrousta H, Moharrami M, Mamsouri MA (2014) Retrospective study of the *Nosema ceranae* infection of honey bee colonies in Iran (2004–2013). ARI 69(2):197–200

Mohammadian B, Bokaei S, Moharrami M, Nabian S, Forsi M (2018) Distribution of *Nosema* Spp. in climatic regions of Iran. VRF 9(3):259–263. https://doi.org/10.30466/vrf.2018.32082

Moore SD, Pittaway T, Bouwer G, Fourie JG (2004) Evaluation of *Helicoverpa armigera* Nucleopolyhedrovirus (HearNPV) for Control of *Helicoverpa armigera* (Lepidoptera: Noctuidae) on Citrus in South Africa. Biocontrol Sci Tech 14:239–250

Moshaverinia A, Abedi V, Safaei H (2012) A survey of *Nosema apis* infection in apiaries of North Khorasan province, Iran. IJVST 4(2):25–30

Moshtaghi Maleki F, Jalali Sendi J, Rezapanah MR (2013) A laboratory investigation on the effect of environmental temperature on virulence of *Helicoverpa armigera* multiple Nucleopolyhedrovirus. J Ani Res 26(4):327–383. In Persian

Mossadegh MS (1990) Some honey bee virus diseases in Iran. Sci J Agric 13:64–72

Nabian S, Ahmadi K, Nazem Shirazi M, Gerami Sadeghian A (2011) First detection of *Nosema ceranae*, a microsporidian protozoa of European honeybees (*Apis mellifera*) in Iran. Iran J Parasitol 6(3):89–95

Nasehi SF, Fathipour Y, Mehrabadi M (2019) Detection of prophage WO in the *Wolbachia*-infected *Habrobracon hebetor* and its effect on the induction of cytoplasmic incompatibility. Paper presented at 3rd Iranian International Congress of Entomology. Tabriz University, Tabriz, 17–19 August 2019, page 198

Nazeri M, Ashouri A, Hosseini M (2015) Can *Wolbachia* infection improve qualitative characteristics of *Trichogramma brassicae* reared on cold stored eggs of the host? Int J Pest Manage 61 (3):243–249. https://doi.org/10.1080/09670874.2015.1042943

Omrani SM, Moosavi SF, Manouchehri K (2016) Microsporidium Infecting *Anopheles supepictus* (Diptera: Culicidae) Larvae. J Arthropod-Borne Dis 10(3):413–420

Omrani SM, Moosavi SF, Farrokhi E (2017) *Parathelohania iranica* sp. Nov (Microsporidia: Amblyosporidae) in infectiong malaria mosquito *Anopleles superpictus* (Diptera: Culicidae): ultrastructure and molecular characterization. J Invert Path 146:1–6. https://doi.org/10.1016/j.jip.2017.03.011

Ordóñez-Garcia M, Rios-Velasco MC, Berlanga-Reyes D, Acostamuniz CH, Angel Salas-Marina-M, Cambero-Compos OJ (2015) Occurrence of natural enemies of *Spodoptera frugiperda* (Lepidoptera: Noctuidae) in Chihuahua, Mexico. Fla Entomol 98(3):843–847

Parvizi P, Bordbar A, Najafzadeh N (2013) Detection of *Wolbachia pipientis*, including a new strain containing the *wsp* gene, in two sister species of *Paraphlebotomus* sandflies, potential vectors of zoonotic cutaneous leishmaniasis. Mem Inst Oswaldo Cruz, Rio de Janeiro 108(4):414–420

Pathak MD, Khan ZR (1975) Insect pests of rice. International Rice Research Institute, Manilla, 68 pages

Pluciennik Z (2013) The control of codling moth (*Cydia pomonella* L.) population using mating disruption method. J Horticu Res 21(1):65–70

Poorjavad N, Goldansaz SH, Machtelinchx T, Tirry L, Stouthamer R, Leeuwen T (2012) Iranian *Trichogramma*: ITS2 DNA characterization and natural *Wolbachia* infection. BioControl 57:361–374. https://doi.org/10.1007/s10526-011-9397-z

Pourali P, Ardakani R, Jolodar A, Razi Jalali MH (2009) PCR screening of the *Wolbachia* in some arthropods and nematodes in Khuzestan province. IJVR 10(3):216–222. In Persian

Pourmirza AA (1998) Studies on the susceptibility of *Heliothis viriplaca* larvae on Nuclear Polyhedrosis virus. Paper presented at 13th Iranian Plant Protection Congress. Tehran University, Karaj, p 71

Pourmirza AA (2000) Relationship between nuclear polyhedrosis virus susceptibility and larval weight in *Heliothis armigera*. J Agric Sci Tech 2:291–298

Rabie M, Seraj AA, Talaei-Hassanloui R, Rahimi H (2008) A laboratory investigation to effect of MbNPV and indoxacarb against the beet armyworm, *Spodoptera exigua* (Lep.: Noctuidae) larvae. Paper presented at 18th Iranian Plant Protection Congress, Bu-Ali Sina University, Hamedan, 24 26 August 2008, page 2

Rabie MM, Seraj AA, Talaei-Hassanloui R (2010) Interaction between MbMNPV and the braconid parasitoid *Habrobracon hebetor* (Hym., Braconidae) on larvae of beet armyworm, *Spodoptera exigua* (Lep. Noctuidae). Biocontrol Sci Tech 20(10):1075–1078

Rahimi-Kaldeh S, Ashouri A, Bandani A, Tomioka K (2017) The effect of *Wolbachia* on diapause, fecundity, and clock gene expression in *Trichogramma brassicae* (Hymenoptera: Trichogrammatidae). Dev Genes Evol 227(6):401–410. https://doi.org/10.1007/s00427-017-0597-0. Epub 2017 Nov 29

Razmaraii N, Karimi H (2010) A survey of *Nosema* of honeybees (*Apis mellifera*) in East Azarbaijan Province of Iran. J Anim Vet Adv 9(5):879–882

Razmaraii N, Sadegh-Eteghad S, Babaei H, Paykari H, Esmaeilnia K, Froghy L (2013) Molecular identification of *Nosema* species in East-Azerbaijan Province, Iran. Arch Razi Inst 68(1):23–27

Rezapanah MR, Chitic V (1998) Production of two Baculoviruses: *Cydia pomonella granulosis virus* (Cpgv) and *Agrothis segetum baculoviruses*. Paper presented at 13th Iranian Plant Protection Congress. Tehran University, Karaj, p 218

Rezapanah MR, Shojai-Estabragh S, Huber J, Jehle JA (2008) Molecular and biological characterization of new isolates of *Cydia pomonella* granulovirus from Iran. J Pest Sci 81(4):187

Sabahi Q, Morfin N, Nehzati-Pagheleh G, Guzman-Novoa E (2018) Detection and replication of deformed wing virus and black queen cell virus in parasitic mites, *Varroa destructor*, from Iranian honey bee (*Apis mellifera*) colonies. J Apic Res. https://doi.org/10.1080/00218839.2019.1686576

Seidavi A, Mirhosseini SZ, Ghanipoor M (2004) Resistance evaluation of some new lines of silkworm in natural conditions and polluted conditions to Nuclear Polyhedrosis Virus. 16th Iranian Plant Protection Congress, Tabriz University, Tabriz, August 2004, page 62

Solter LE, Becnel JJ, Oi DI (2012) Microsporidian entomopathogens. In: Vega FE, Kaya HK (eds) Insect pathology, 2nd edn. Academic Press, San Diego, pp 221–263

Sun XL, Peng HY (2007) Recent advances in biological control of pest insects by using viruses in China. Virol Sin 22:158–162

Trowell SC, Lang GA, Garsia KA (1993) A *Heliothis* identification kit. In: Corey SA, Dall DJ, Milne WM (eds) Pest control and sustainable agriculture. CSIRO Publishing, Collingwood, pp 176–179

Vandekerckhove TTM, Watteyne S, Willems A, Swing JG, Mertens J, Gillis M (1999) Phylogenetic analysis of the 16S rDNA of the cytoplasmic bacterium *Wolbachia* from the novel host *Folsomia candida* (Hexapoda, Collembola) and its implications for *Wolbachial* taxonomy. FEMS Microbiol Lett 180:279–286

Virto C, Zarate CA, Lopez-Ferber M, Murillo R, Caballero P, Williams T (2013) Gender-mediated differences in vertical transmission of *Spodoptera exigua multiple nucleopolyhedrovirus* (SeMNPV). PLoS One 8:e70932

Virto CD, Navarro D, Tellez M, Herrero S, Williams T, Murillo R, Caballero P (2014) Natural populations of *Spodoptera exigua* are infected by multiple viruses that are transmitted to their offspring. J Invertebr Pathol 122:22–27

Zabalou S, Riegler M, Theodorakopoulou M, Stauffer C, Savakis C, Bourtzis K (2004) *Wolbachia*-induced cytoplasmic incompatibility as a means for insect pest population control. PNAS 101 (42):15042–15045

Zarei Z, Salehi M, Azami Z, Salari Z, Beven L (2017) Stubborn disease in Iran: diversity of *Spiroplasma citri* strains in *Circulifer haematoceps* leafhoppers collected in sesame fields in Fars province. Curr Microbiol 74(2):239–246. https://doi.org/10.1007/s00284-016-1180-z

# Chapter 14
# Microbial Biopecticides: Opportunities and Challenges

Mohammad Reza Moosavi and Vahe Minassian

## 14.1 Introduction

Total losses to agricultural products of Iran caused by pests are estimated to be more than 40% (Ebadzadeh et al. 2016). These pests are traditionally controlled by chemical pesticides and according to latest information released by FAO, 4420 tons of chemicals (1756.1, 1564.2 and 1099.9 tons of insecticides; herbicides; and fungicides plus bactericides, respectively) were used to protect agricultural products of Iran just in the year 2016 (FAO Stat 2019). The mean annual amount of used insecticides; herbicides; and fungicides plus bactericides during 1990 to 2016 was 1844, 2103 and 1687 tons respectively (FAO Stat 2019), however, the application rate fluctuated considerably over the last three decades (Fig. 14.1).

Intensive usage of pesticides causes accumulation of these harmful chemicals in the ecosystem and raises many concerns about the public health and environment safety. Incidence of many chronic diseases like cancers, birth defects, neurodegenerative disorders (such as parkinson, alzheimer, and amyotrophic lateral sclerosis), reproductive disorders, and diabetes has increased unusually in recent years and a lot of evidence has linked the high frequency of these diseases with exposure to pesticides (Mousavi et al. 2009; Mostafalou and Abdollahi 2013). If we cannot find safe alternative pest management methods, the use of pesticides will be increased due to the demand for more food by the increasing population of mankind.

Public awareness about the destructive effects of the chemical and toxic pesticides on environment and human health has generated a high pressure on the

M. R. Moosavi (✉)
Plant Pathology Department, Faculty of Agriculture, Marvdasht Branch, Islamic Azad University, Marvdasht, Iran
e-mail: rmmoosavi@miau.ac.ir

V. Minassian
Department of Plant Protection, Shahid Chamran University, Ahvaz, Iran

© The Author(s), under exclusive license to Springer Nature Switzerland AG 2021
J. Karimi, H. Madadi (eds.), *Biological Control of Insect and Mite Pests in Iran*,
Progress in Biological Control 18, https://doi.org/10.1007/978-3-030-63990-7_14

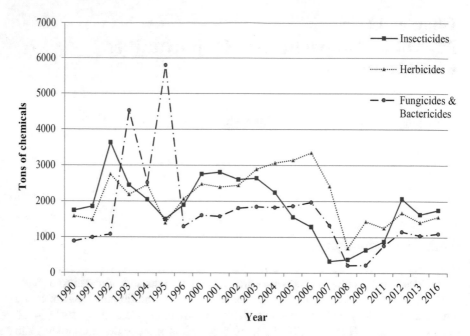

**Fig. 14.1** Application of chemicals in Iran during 1990 to 2016 in terms of tons of chemicals. Data for the years 1997–1999, 2010, 2014 and 2015 were not available (information retrieved from FAO Stat 2019)

authorities to diminish the application of synthetic compounds. The outcome is revealed as some modifications in regulations (Moosavi and Zare 2016). On the other hand, a social rebuttal of chemicals creates an increasing demand for organic- or healthy-products which provide a new market in Iran. The prospect of this market is very bright, so the companies are gradually convinced to invest in new safer pesticides. Microbial pesticides are a proper choice that can suppress the pests in a sustainable manner (Moosavi and Zare 2016). Though the legal enforcement and society views are considered as strong promotional factors, the companies' directions are generally designed according to market-driven demands (Ravensberg 2011d). When there is not enough consumer demand in a reasonable time period, the investment in research and development of a microbial pesticide will be uneconomical (Moosavi and Zare 2015).

Now there is a hope that new safer pesticides can be substituted with chemical ones. Here the history and current status of microbial pesticides are discussed. As well, their success in being launched in the pesticides market of Iran is also reviewed. Then the major challenges in producing and adapting of microbial pesticides by producers and growers are presented. Finally, the prospects of investigation, development and application of these products will be discussed.

## 14.2  History of Biological Control in Iran

Protecting crops had relied on traditional practices for many centuries. It seems that biological control has had a long history in Iran and Iranian natives knew about biocontrol since long ago. The first written document is a book (Zakhira-e-Khârazmi) composed by al-Jurjâni in which is referred to the ability of stork and hoopoe to control harmful animals and termites (Abivardi 2001). Though no document is available to show how farmers protected their crops from pests, agricultural products had been successfully produced for many years. It was probable that growers stimulated naturally occurring biocontrol agents by cultivation practices. These practices included preparing suitable conditions for the proliferation of microbial antagonistic agents; attracting the predators or facilitating their access to damaging pests (Amiri Ardakani and Emadi 2008).

The first attempt to biologically control a pest dates back to 1931 when imported Vedalia beetle (*Rodolia cardinalis*) from France was used to manage cottony cushion scale (*Icerya purchasi*). The second successful example of biocontrol in Iran was the application of *Trissolcus semistriatus* against the Sunn pest (*Eurygaster integriceps*) (Abivardi 2001). Other biocontrol attempts included using of the mealybug destroyer (*Cryptolaemus montrouzieri*) and an Aphelinid (*Prospaltella berlesei*) against citrus mealybug (*Planococcus citri*) and white peach scale (*Pseudaulacaspis pentagona*), respectively. With the establishment of biological control department at the "Iranian Research Institute of Plant Protection" in 1984 and its expansion during 1984–1995, the importance of biocontrol became more obvious. Mass rearing of *Trichogramma* wasps against different pests was one of the most outstanding instances of successful biological control during those years (Moosavi and Zare 2016).

The first national plan to optimize the usage of fertilizers and pesticides was a 10-year program which began in 1995 entitled OUFP (Optimal Utilization of Fertilizers and Pesticides). One of the most important goals of this plan was a reduction of chemical pesticides (about 35% within 5 years) with an outlook for biocontrol; nevertheless, microbial pesticides were not considered a serious candidate in those years. At the beginning of the project, the biocontrol agents were propagated and released in the fields at no cost with the hope that growers would participate and contribute to the expenses in the succeeding years. The program was not welcomed by the farmers and their low participation impeded the plan from reaching its full goals (Sharifi-Moghaddam 2006). However, support by national and international organizations of the program supplied the required foundation for establishing the knowledge, strategy and manpower for the upgrading of IPM/FFS (Integrated Pest Management/Farmers Field School) in Iran (Fathi et al. 2011). The support of FAO Regional IPM Program and GEF/SGP (Global Environment Facility/Small Grants Program) was vital to introducing IPM/FFS to growers. The 252 pilot farms of 37 different crops were selected in 29 provinces to draw the growers' attention to organic agriculture and effectiveness of biopesticides (Razzaghi-borkhani et al. 2010; Etehadi et al. 2011). In spite of these efforts, the organic agriculture needs much more attention to be fully established (Soltani et al. 2014). The area under organic cultivation in Iran was 0.02% of the total cultivable land in 2014 (Willer and Lernoud 2016) which was not a satisfactory share.

As mentioned above, while most of the applied biological agents of the country were predators or parasitoids, a little share was also allocated to microbial agents. However, the market share of microbial biopesticides shows a more or less steady increase during recent years especially for the soil dwelling and greenhouse pests.

## 14.3 Microbial Biopesticides Opportunities in Iran

Biological control of pests, especially those done with microbial agents, has not achieved its proper role in IPM programs in Iran. Biocontrol has now been used sporadically throughout the country but it is not considered as a constant or important part of IPM. However, the increasing pressure of society has enforced a change in the viewpoint of policy makers about the procedure of IPM. For example, an office named "Environment and Food Safety Office" has been established in Ministry of Agriculture-Jahad in 2002. The office is responsible for making policies, planning and supervision toward sustainable agriculture and preserving the environment. Several workgroups (such as Development of Organic Farming, Biosafety, Control and Certification of Agricultural Products, and Green Management Program) have been formed to achieve those goals. The activity of the office has increased several folds in the recent years in response to society's demand for safer foodstuff.

One of the good measures taken by the Environment and Food Safety Office, Ministry of Agriculture-Jahad has been to provide a guideline named "Practical Instruction for Producing Certified and Standard Products". It emphasizes on the procedures of producing agricultural products with minimal hazardous chemical residues. The task of supervision on production process is delegated to Plant Protection Clinics. The required samples will be taken from the products by trustful inspection companies and will be examined by reference laboratories. The product will be certified only if the residues of 105 chemicals (fertilizers and pesticides), nitrate and heavy metals would be below the allowed limit, known as "maximum residue limits (MRLs)". Plant Protection Clinics have the authority to allocate a 16-digit detectable barcode to each product that could pass the process of certification. The first 8 digits represent the province, town, district and village where the farm is located. The next two digits belong to plant protection clinic that supervises the farm. The digits 11 to 13 are allocated to the crop and digits 14 to 16 designate the particulars of the grower that owns the product (the last three digits of grower national code or three last digits of exploitation license of the production unit). Therefore, it is possible to detect the origin of each product by the 16-digit code. A high council consisting of provincial representatives of Ministry of Agriculture-Jahad, Ministry of Health and Medical Education and Iran National Standards

Organization supervises all processes. More recently, volunteer farmers will receive a three-year persuasive standard if the MRLs of their products are below defined limits (Environment and Food Safety Office 2019).

The responsibility of determining the health of products has been delegated to regional health departments according to article 28 of the Iran Plant Protection Implementing Regulations. This article states "regional health departments shall be required to examine fruits, cucurbits, vegetables and other products before they are marketed and if pesticide residues are found, to inform the city hall or other authorities concerned with the matter and prohibit their sale". By law, Ministry of Agriculture-Jahad is responsible for supervising production of safe products and Iran National Standards Organization has a contribution in granting the persuasive standard certification. Obtaining persuasive standard is now voluntary and there is no legal obligation for the growers to follow the instructions for producing safe products. However, it seems that the growers have to adopt this program because of changes in the society trends towards using safer food. This means that the demand for biopesticides including microbial ones is growing and the market size will continuously expand. The old chemical pesticides producers along with many new companies perceive this fact and increase their investment on biopesticides.

In spite of numerous worldwide researches on microbial pesticides, only a small fraction of them pass the commercialization process successfully. These few pesticides have mainly been introduced to the USA, Latin and South America, and Europe (Thakore 2006; Glare et al. 2012; Dutta 2015; Arthurs and Dara 2019) while the share of Iran is very scant (Moosavi and Zare 2016; Karimi et al. 2019). In recent years, country-wide research on microbial biopesticides has increased rapidly but till now only 20 products (occasionally with the same microorganism ingredient) have been produced by Iranian companies which some of them have not completed their registration process yet (Table 14.1). However, there is a hope that the number could increase in the near future.

The leading companies producing microbial pesticides are as follows: Biorun, Greenlife Biotech Co., Green Biotech Co., MehrAsia Biotechnology Company, and Sadra Biotech Co. The active ingredients of the available microbial pesticides (registered or in the final process of registration) in Iran consist mainly of bacteria and fungi.

### 14.3.1  Bacterial Biopesticides

Four groups of bacteria can typically be utilized as biopesticides. These four types consist of I) bacteria that produce crystalliferous spore (like *Bacillus thuringiensis*), II) obligatory parasitic bacteria (like *Bacillus popilliae*), III) facultative parasitic bacteria (like *Pseudomonas aeruginosa*), and IV) opportunistic parasitic bacteria (like *Serratia marcesens*) (Koul 2011).

The most popular bacterium used in Iranian microbial biopesticides is *Bacillus thuringiensis* (Moosavi and Zare 2016). Many subspecies and varieties of *B. thuringiensis* have been identified globally (Osman et al. 2015). This spore

**Table 14.1** Information on the microbial biopesticides in the pesticide market of Iran

| Microorganism | Trade name | Producer/importer | Formulation[a] | Target pest | Made in |
|---|---|---|---|---|---|
| **Bacteria** | | | | | |
| *Bacillus thuringiensis* subsp. *tenebrionis* | Biobeet | Biorun | SC | Coleoptera larvae | Iran |
| *B. thuringiensis* subsp. *kurstaki* | Biolep | Biorun | SC | Lepidaptera larvae | Iran |
| *B. thuringiensis* subsp. *kurstaki* | Biolep | Biorun | WP | Lepidaptera larvae | Iran |
| *B. thuringiensis* subsp. *israliensis* strain MH14 | Bioflash | Biorun | GR/WP | Fly larvae | Iran |
| *B. thuringiensis* subsp. *kurstaki* | Bithurin | MehrAsia Biotechnology Company | SL | Lepidaptera larvae | Iran |
| *B. thuringiensis* subsp. *morrisoni* | Bithiran | MehrAsia Biotechnology Company | SL | Coleoptera larvae | Iran |
| *Bacillus subtilis* | Biosubtil | MehrAsia Biotechnology Company | SL | Plant pathogenic fungi | Iran |
| *Bacillus subtilis* | Subtilin | Talfigh Daneh Agriculture Co. | WP | Plant pathogenic fungi | Iran |
| *Bacillus thuringiensis* | Bt.H | Talfigh Daneh Agriculture Co. | WP | Lepidaptera larvae | Iran |
| *Bacillus* sp. | Nemachit | Sadra Biotech Co. | SL | Plant parasitic nematodes | Iran |
| *Pseudomonas fluorescens* | Pomeg | Sadra Biotech Co. | SL | Plant parasitic nematodes | Iran |
| *Bacillus subtilis* strain BS106 & BS24 | Rouien-1 | Green Biotech Co. | WP | Soil-borne fungi | Iran |
| *Bacillus thuringiensis* subsp. *kurstaki* strain AzLP | Rouien-2 | Green Biotech Co. | WP | Lepidaptera larvae | Iran |
| *B. thuringiensis* subsp. *aizawai* | XenTari & FlorBac | Afrasam Co. | WP | Armyworms and Diamondback Moth larvae | Spain |
| *B. thuringiensis* subsp. *kurstaki* | BioBit & DiPel | Afrasam Co. | WP | Lepidaptera larvae | Spain |

**Fungi**

| | | | | | |
|---|---|---|---|---|---|
| *Trichoderma harzianum* | Trichofarm P | Biorun | Soil-borne fungi | WP | Iran |
| *Trichoderma harzianum* | Trichofarm G | Biorun | Soil-borne fungi | GR | Iran |
| *Trichoderma harzianum* | TricoMix HV | Greenlife Biotech Co. | Plant pathogenic fungi | WP | Iran |
| *Trichoderma harzianum* | Tricodermin B | Talfigh Daneh Agriculture Co. | Plant pathogenic fungi | WP | Iran |
| *Fusarium oxysporum* | Orocide | Greenlife Biotech Co. | Broomrape | L | Iran |
| *Purpureocillium lilacinum* | RootGard | Foroughe Dasht Co. | Plant parasitic nematodes | WP | Spain |
| *Beauveria bassiana* | BootaniGard | Foroughe Dasht Co. | Wide range of insect pest | SC | Spain |
| *Lecanicillium lecanii* | Mycotal | Gyah coroporation | White fly & thrips larvae | WP | Netherland |
| *Trichoderma harzianum* strain T22 | Trianum-P | Gyah corcporation | Soilborn fungi | WP | Netherland |
| *T. harzianum* strain T22 | Trianum-G | Gyah corcporation | Soilborn fungi | GR | Netherland |
| *Beauveria bassiana* | Naturalis | Afrasam Co. | Sucking pests | SC | India |
| *Beauveria bassiana* | Naturalis | Aryan Teb Parto | Sucking pests | SC | Italy |
| **Not stated** | | | | | |
| A group of microorganisms | Probiotect | Biorun | Soil-borne fungi | SC | Iran |
| A group of microorganisms | Pistagaurd | Biorun | *Agonoscena pistaciae* *Cacopsylla pyricola* *Trialeurodes vaporariorum* | SC | Iran |

[a]*GR* granule, *L* liquid, *SL* water soluble liquid, *SC* suspension concentrate, *WP* wettable powder

producing bacterium is an eco-friendly microorganism that can manage the target pest efficiently (Roh et al. 2007; Jouzani et al. 2017). Several microbial pesticides which are currently used in Iran are made from different subspecies or varieties of *B. thuringiensis*. Other bacteria that have been exploited as pesticides are *Bacillus subtilis* and *Pseudomonas fluorecens* (Table 14.1).

## 14.3.2 Fungal Biopesticides

The second most important group of microbial pesticides in Iran consists of fungi. However, the significance of fungi for controlling insect pests, plant-parasitic fungi (Juty and Singh 2016) and phytonematodes (Cumagun and Moosavi 2015) has been rapidly increasing in the recent years. Some fungal characteristics which make them an attractive substrate for research are: host specificity, cosmopolitan dispersal, persistency, occasional production of stress resistant resting spores, the relative simplicity of mass culture and ease of application in the field (Juty and Singh 2016). *Trichoderma*–based products are the most numerous microbial products in Iran that are used against soil borne pathogenic fungi. Other fungal-based products that have been used as microbial biocontrol agents are *Beauveria bassiana*, *Metarhizium anisopliae*, *Purpureocillium lilacinum*, *Lecanicillium lecanii* and *Fusarium oxysporum* (Table 14.1).

## 14.4 Challenges for Growers

Notwithstanding the efforts of the government during successive OUFP national programs, the reliance of growers upon chemical inputs has not decreased significantly. The number of pilot farms in IPM/FFS programs throughout the country was not adequate enough to successfully launch the "biological control" in the farmers' mind as an important part of IPM. Many independent studies have been tried to investigate the barriers to adoption of organic agriculture among Iranian farmers (Veisi et al. 2010; Rezvanfar et al. 2011; Karimi 2011; Soltani et al. 2014; Veisi et al. 2017; Forouzani et al. 2018). Several factors have been involved in hindering the implementation of microbial biopesticides by Iranian growers. The main reasons for this reluctance are almost similar to those in developing countries.

Farmers' mistrust of microbial biopesticides (Mishra et al. 2015), fear of reduced yield (Moosavi and Zare 2016) and weak support of policy network (Kumar and Singh 2015) are among the main reasons. As most Iranian growers are smallholders, they only adopt new controlling methods if they are being completely convinced those methods are risk-free (Rezaei-Moghaddam and Samiee 2019).

In contrast with chemical pesticides, microbial ones lack consistency in their effectiveness. Lack of persistency usually occurs because of the inability to get established in or on the target site due to low competitive capability against the inhabitant microbiota; being affected by the environmental condition; and inferior quality (Moosavi and Zare 2015). Incompatibility of biopesticides application methods with routine delivery systems is another obstacle to farmers' acceptance (Moosavi 2020). Since biopesticides are relatively new for Iranian farmers, there is an essential need to improve their knowledge by experienced trainers. Updating and compatibility of farmers' knowledge and technology is a prerequisite to biopesticides adoption by farmers (Rezaei-Moghaddam and Samiee 2019).

Biological control measures especially those dealing with microbial biocontrol agents cannot be considered as a simple interaction between pests and their antagonists. It is important to contemplate the biological control as a phenomenon which happens in a complicated, incongruous and dynamic environment. Both biotic and abiotic parameters are important effecters in determining the efficacy level of a microbial biocontrol product. That is the key reason why a product shows different efficiency in different times or locations. Deeper studies about the impact of environment and ecological condition on the biocontrol agent-pest-host plant interactions are required to increase the knowledge of how chemical, physical and biological factors affect the relationship between biocontrol agents, their host pest and their host plant (Stirling 2014). This information is necessary for establishing and maintaining of successful and persistent biocontrol systems. Such improvements could help farmers to overcome their uncertainty about biopesticides implementation.

Well-formulated, competitive and high-quality commercial biocontrol agents against plant diseases, insect pests and weeds may reconstruct the trust of growers. However, the farmers should be trained in the processes that microbial pesticides act. According to their mental background, the growers expect that the microbial pesticides act as quickly as chemical ones do. They must be trained during IPM/FFS programs for the principles of biocontrol as well as being informed of adverse effects of chemicals on human life and environment. They should understand that biocontrol agents operate more slowly but with greater longevity. Tolerance of microbial biocontrol agents to chemicals is a desired trait which allows their mixed application. Chemicals provide rapid protection while biocontrol agents supply long-term protection (Desai et al. 2002).

Mixing ability of microbial products with customary agricultural inputs is usually limited which can be considered as a restriction factor for adoption of microbial biocontrol agents. Application of microbial biopesticides sometimes needs special equipment different from essential and current farm machinery (Khan and Anwer 2011). Difficulties and expense of preparing these facilities are deterrent to growers (Bailey et al. 2010).

## 14.5   Challenges for Producers

In spite of many types of research done in Iranian universities or research institutes, the number of commercialized products is not satisfactory. The major problem is a lack of affiliation between researchers and mass producers of biopesticides. The producers rely on their own research and do not use the outcome of academic studies. On the other hand, the academic scientists don't commercialize their isolated robust biocontrol agents due to their unwillingness to act as entrepreneurs or due to inability to afford the expenses of producing processes. Lack of market knowledge is another obstacle for entering the bazaar. A tighter relationship between academic researchers and producer sections is essential for faster developments in the selection, scale-up production, formulation and commercialization of microbial biocontrol agents.

When a competent biocontrol agent is selected, it should be mass produced on an industrial scale. A number of microbial biocontrol agents can be produced in a simple manner while the remaining require special knowledge and technologies (Tormala 1995). Failure in the scale-up production of numerous products causes their rejection (Stewart et al. 2010). In addition to difficulties in mass production, the producers have to be concerned with contamination in their production units and attenuation of pesticidal activity. The microbial BCAs are usually propagated on commercially unselective media and contamination can easily occur if strong sanitation measures are not employed (Desai et al. 2002). Also, the producers should continuously test the number of efficient propagules in the formulation to ensure that the product keeps its specifications up to the end of asserted shelf life (Ravensberg 2011b).

To overcome these barriers, small entrepreneurs require facilities, material, proficiencies and capital that usually cannot be afforded in the developing countries. However, the same situation exists for small businesses in developed countries (Gelernter 2007; Koul 2011; Moosavi and Zare 2016). The required capital for safety tests and registration processes of a biological-based product is appraised from the US \$0.5 to three million that should be invested during a 4–8 years period (Evans 2004; Hokkanen 2007; Marrone 2009; Ehlers 2010; Ravensberg 2011c). This sum of money cannot be easily supplied by small or medium businesses in Iran, while the investment is not completely safe either. The active ingredient of the product is a microorganism which can be reisolated. Therefore business enterprises should protect their product by intellectual property rights which cost them another \$US 50,000–100,000 (Tormala 1995). Though patenting is supportive, its power is debatable and can be challenged with genetic engineering (Ravensberg 2011c).

Commercialization of a product, in addition to biological control knowledge, needs the familiarity with several disciplines like finance, trade, manufacture and personnel management. In other words, a product could be successfully introduced to the market only if a capable team works together. The team members should be experts in research, manufacturing, registration, management and marketing (Ravensberg 2011a).

The process of developing a new microbial pesticide needs many millions of dollars during a 4–8 year period. It is estimated that another 5–8 years is needed till the company reaches to a break-even point (CPL 2006). So, the venture is not a quick profit making one which is deterrent to most investors. Foreign investment or joint venture may be of help in providing the required investment (Karimi et al. 2019).

Another problem with microbial products is their short shelf life. The mean shelf life time for these products is about 1 year. The products that are not made of resting resistance propagules usually need special storage conditions and if not kept in recommended conditions, their efficiency is reduced over time (Yang and del Rio 2002). As the acquisition of these particular storage proficiency or devices is difficult for most wholesalers or retailers, they prefer not to enter the business.

Naturally, a resistance exists against the acceptance of a new product in the market. Therefore, the producers and their distributors' networks need an extensive endeavor even after a product, especially a microbial pesticide, is registered. Considering these matters, few companies decide to enter such business. However widespread IPM/FFS programs could be of help in convincing the growers to integrate robust microbial biopesticides in their IPM measures.

Market instability is another obstacle facing the Iranian producers. The import/export laws and regulations change frequently and the value of the national currency fluctuates occasionally. The interest rates which Iranian banks assign to deposited money is too high which makes the investment on production irrational according to market risks. These factors make the investors (local or foreign) hesitant to venture upon investing as a producer in Iran. The more focused investigation is required to determine the factors slowing down registration, commercialization and adoption of microbial pesticides in Iran.

## 14.6   Future Prospects

Pest management measures need to be revised rapidly because of changing society attitude towards human health and the environment. The demands for various safe foodstuffs have been increasing continuously due to public awareness about the deleterious effects of synthetic chemical pesticides on the health of humans and their surroundings. The microbial biopesticides are a promising alternative which can be used in sustainable farming. Numerous studies in this regard have been accomplished in the recent years (Moosavi et al. 2010, 2011, 2015; Kalantari et al. 2013; Naraghi et al. 2014; Khorramvatan et al. 2014; Damani Zamani et al. 2015; Goudarzi et al. 2015; Salehi Jouzani et al. 2015; Amir-Ahmadi et al. 2017) which increase the hope of introducing new products in the Iran market within a short time. However, it is obvious that the microbial biopesticides need government support to be successfully launched into the market. The government support can be by enforcement of regulations such as forbidding the production or trading of unsafe chemical pesticides; allocating financial assistance or price support systems; and easing the export laws for organic and safe products. These policies can encourage farmers to use biopesticides, enlarge the market size, and direct the capital flow towards investing on microbial biopesticides. Many corporations are still hesitant to enter the business due to the niche market and pricey registration procedures. Government assurance and assistance will certainly persuade them to defeat their uncertainty.

Grower education is an influential and important undertaking which deserves more consideration. Executing the IPM/FFS program in more extensive areas with more experimental farms can be of help. Simultaneous education of consumers about microbial biopesticides is equally important. Both growers and society should be instructed about the key role that microbial pesticides can play in sustainable agriculture.

The active companies in this venture especially those newly engaged should also be instructed about the procedures that a product must pass to be commercialized. Science and technology parks can satisfactorily accomplish this task. Science and technology parks have been now established in most provinces of Iran and are ready to provide required services to applicant knowledge-base companies. These services include technical, accounting, market and legal consultation. Installing efficient networks to obtain assistance from international agricultural institutes is another possible service rendered by the science and technology parks which improve the prospects for continued acceptance of microbial pesticides.

The process that needs the most help is formulation. Many efficient microorganisms for controlling pests have been identified in Iran that can be commercialized if they are suitably and stably formulated. In other words, we need to adopt the knowledge which helps us to commercialize the isolated indigenous biocontrol agents. Genetic engineering is another strategy that can be used to enhance virulence and efficiency of microorganisms or to tailor their host range as desired (Narayanan 2001; Kumari et al. 2014; Moshi and Matoju 2017).

However microbial pesticides will be confronted with the last obstacle, the grower acceptance, when finally marketed. Therefore the focus and emphasis should be on the extensive pilot testing to represent the potential of these kinds of pesticides. The microbial pesticide producer should manufacture a well-formulated product that can be easily used with conventional farm machinery, and government authorities enact policies which facilitate their use.

The introduction of relatively harmless and efficient chemical pesticides will be a serious challenge for microbial pesticides in the future. Therefore the microbial products can successfully be accepted by the growers only if they can emulate the agrochemicals. The long term prospects for microbial pesticides seem very brilliant despite slow adoption by growers. The pressure of NGOs as well as public awareness will surely direct the path toward the use of safer products.

**Acknowledgement** The authors thank "Knowledge-base Plant Protection Clinic" (Shiraz-Iran) for its help in collecting information on microbial biopesticides available in the market of Iran. We also thank Dr. Anahita Yazdanpak from "Fars Environment and Food Safety Office" for sharing her information about "Practical Instruction for Producing Certified and Standard Products".

# References

Abivardi C (2001) Iranian entomology–an introduction. Springer, New York

Amir-Ahmadi N, Moosavi MR, Moaf-Poorian GR (2017) Investigating the effect of soil texture and its organic content on the efficacy of *Trichoderma harzianum* in controlling *Meloidogyne javanica* and stimulating the growth of kidney bean. Biocontrol Sci Tech 27(1):115–127

Amiri Ardakani M, Emadi MH (2008) Traditional knowledge of Iranian farmers on biological pest management. Indian J Tradit Know 7(4):676–678

Arthurs S, Dara SK (2019) Microbial biopesticides for invertebrate pests and their markets in the United States. J Invertebr Pathol 165:13–21

Bailey A, Chandler D, Grant WP et al (2010) The economics of making the switch in technologies. In: Bailey A, Chandler D, Grant WP, Creaves J, Prince G, Tatchell M (eds) Biopesticides, pest management and regulation. CABI Publishing, Wallingford, pp 131–147

CPL (2006) A how to do it guide to biopesticides. Biopesticides 2007, vol 5. CPL Business Consultants, Wallingford

Cumagun CJR, Moosavi MR (2015) Significance of biocontrol agents of phytonematodes. In: Askary TH, Martinelli PRP (eds) Biocontrol agents of phytonematodes. CABI Publishing, Wallingford, pp 50–78

Damani Zamani F, Moosavi MR, Asadi R (2015) Efficacy of entomopathogenic nematode in biological control of tomato leafminer, *Tuta absoluta* in laboratory and greenhouse conditions. Biocontrol Plant Prot 3:19–32

Desai S, Reddy MS, Kloepper JW (2002) Comprehensive testing of biocontrol agents. In: Gnanamanickam SS (ed) Biological control of crop diseases. Marcel Dekker, New York, pp 391–424

Dutta S (2015) Biopesticides: an ecofriendly approach for pest control. World J Pharm Pharm Sci 4 (6):250–265

Ebadzadeh HR, Ahmadi K, Mohammadnia Afroozi S et al (eds) (2016) Agricultural statistics, vol 2. Ministry of Agriculture-Jahad Publishing, Tehran

Ehlers R-U (2010) REBECA – EU-policy support action to review regulation of biological control agents. In: Gisi U, Chet I, Gullino ML (eds) Recent developments in management of plant diseases, plant pathology in the 21st century. Springer Science + Business Media, Dordrecht, pp 147–161

Environment and Food Safety Office (2019) Practical instruction for producing certified and standard products, 2nd revision. Agricultural organization of Fars protocol, Shiraz

Etehadi M, Rusta K, Gholi-Nia MJ (2011) Investigating the effectiveness of FFS approach in disseminating IPM practices from farmers' overview, case study Sistan and Baluchestan Province. Iran Agr Exten Edu J 7:41–52

Evans J (2004) Shifting perceptions on biopesticides. Agrow No 455, September 3rd, 2004: 18–21

FAO Stat (2019) Food and agriculture organization of the United Nations, statistics division. http://www.fao.org/faostat/en/#compare. Accessed 26 Dec 2019

Fathi H, Heidari H, Impiglia A et al (2011) History of IPM/FFS in Iran. GTFS/REM/070/ITA regional integrated pest management (IPM) programme in the near east. FAO Publishing, Rome

Forouzani M, Merdasi G, Nargesi Z et al (2018) Factors affecting attitude of farmers toward organic farming in Khuzestan, Iran. Agric For 64(1):89–96. https://doi.org/10.17707/AgricultForest.64.1.11

Gelernter WD (2007) Microbial control in Asia: a bellwether for the future? J Invertebr Pathol 95:161–167

Glare T, Caradus J, Gelernter W et al (2012) Have biopesticides come of age? Trends Biotechnol 30:250–258

Goudarzi M, Moosavi MR, Asadi R (2015) Effects of enthomopathogenic nematodes, *Heterorhabditis bacteriophora* and *Steinernema carpocapsae*, in biological control of *Agrotis segetum*. Turk Entomol Derg-TU 39:239–250

Hokkanen H (2007) Survey among industry concerning biological plant protection products. In Hokkanen H, Menzler-Hokkanen I (eds) Proceedings of the REBECA (Regulation of Biological Control Agents) workshop on "balancing the benefits and costs of regulating biological plant protection products", Porvoo, 2007

Jouzani GS, Valijanian E, Sharafi R (2017) *Bacillus thuringiensis*: a successful insecticide with new environmental features and tidings. Appl Microbiol Biotechnol 101:2691–2711

Juty S, Singh DP (2016) Fungi as biocontrol agents in sustainable agriculture. In: Singh JS, Singh DP (eds) Microbes and environmental management. Studium Press, New Delhi, pp 172–194

Kalantari M, Marzban R, Imani S et al (2013) Effects of *Bacillus thuringiensis* isolates and single nuclear polyhedrosis virus in combination and alone on *Helicoverpa armigera*. Arch Phytopathol Plant Prot 47:1–9

Karimi E (2011) Investigating the barriers of organic agriculture development. J Iran Agric Econ Dev 42:231–242

Karimi J, Dara SK, Arthurs S (2019) Microbial insecticides in Iran: history, current status, challenges and perspective. J Invertebr Pathol 165:67–73

Khan MR, Anwer MA (2011) Fungal bioinoculants for plant disease management. In: Ahmad I, Ahmad F, Pichtel J (eds) Microbes and microbial technology: agricultural and environmental applications. Springer Science + Business Media, Dordrecht, pp 447–488

Khorramvatan S, Marzban R, Ardjmand M et al (2014) The effect of polymers on the stability of microencapsulated formulations of *Bacillus thuringiensis* subsp. *kurstaki* (Bt-KD2) after exposure to ultra violet radiation. Biocontrol Sci Tech 24:462–472

Koul O (2011) Microbial biopesticides: opportunities and challenges. CAB Rev Prospect Agric Vet Sci Nutr Resour 6:056. https://doi.org/10.1079/PAVSNNR20116056

Kumar S, Singh A (2015) Biopesticides: present status and the future prospects. J Fertil Pestic 6:2. https://doi.org/10.4172/jbfbp.1000e129

Kumari BR, Vijayabharathi R, Srinivas V et al (2014) Microbes as interesting source of novel insecticides: a review. Afr J Biotechnol 13:2582–2592

Marrone PG (2009) Barriers to adoption of biological control agents and biological pesticides. In: Radcliffe EB, Hutchison WD, Cancelado RE (eds) Integrated pest management. Cambridge University Press, Cambridge, pp 163–178

Mishra J, Tewari S, Singh S, Arora NK (2015) Biopesticides: where we stand? In: Arora NK (ed) Plant microbes symbiosis: applied facets. Springer India, New Delhi, pp 37–75

Moosavi MR (2020) Efficacy of microbial biocontrol agents in integration with other managing methods against phytonematodes. In: Ansari RA, Rizvi R, Mahmood I (eds) Management of phytonematodes – recent advances and future challenges. Springer, Cham, pp 333–384

Moosavi MR, Shakeri S, Mohammadi S (2015) The ability of separate and combined application of five nematopathogenic fungi against *Meloidogyne javanica*. Iran J Plant Prot Sci 46:179–190

Moosavi MR, Zare R (2015) Factors affecting commercial success of biocontrol agents of phytonematodes. In: Askary TH, Martinelli PRP (eds) Biocontrol agents of phytonematodes. CABI Publishing, Wallingford, pp 423–445

Moosavi MR, Zare R (2016) Present status and the future prospects of microbial biopesticides in Iran. In: Singh HB, Sarma BK, Keswani C (eds) Agriculturally important microorganisms: commercialization and regulatory requirements in Asia. Springer, Singapore, pp 293–305

Moosavi MR, Zare R, Zamanizadeh HR et al (2010) Pathogenicity of *Pochonia* species on eggs of *Meloidogyne javanica*. J Invertebr Pathol 104:125–133

Moosavi MR, Zare R, Zamanizadeh HR et al (2011) Pathogenicity of *Verticillium epiphytum* isolates against *Meloidogyne javanica*. Int J Pest Manage 57:291–297

Moshi AP, Matoju I (2017) The status of research on and application of biopesticides in Tanzania. Review. Crop Prot 92:16–28

Mostafalou S, Abdollahi M (2013) Pesticides and human chronic diseases: evidences, mechanisms, and perspectives. Toxicol Appl Pharmacol 268:157–177

Mousavi SM, Gouya MM, Ramazani R et al (2009) Cancer incidence and mortality in Iran. Ann Oncol 20:556–563

Naraghi L, Heydari A, Askari H et al (2014) Biological control of *Polymyxa betae*, fungal vector of rhizomania disease of sugar beets in greenhouse conditions. J Plant Prot Res 54:109–114

Narayanan K (2001) Microbial control of insect pests: role of genetic engineering and tissue culture. In: Koul O, Dhaliwal GS (eds) Microbial biopesticides. Taylor & Francis, London, pp 117–180

Osman GEH, Already R, Assaeedi ASA et al (2015) Bioinsecticide *Bacillus thuringiensis* a comprehensive review. Egypt J Biol Pest Control 25:271–288

Ravensberg WJ (2011a) Critical factors in the successful commercialization of microbial pest control products. in: Ravensberg WJ (ed) A roadmap to the successful development and commercialization of microbial pest control products for control of arthropods. Progress in biological control, vol 10. Springer, Dordrecht, pp 295–356

Ravensberg WJ (2011b) Quality control. in: Ravensberg WJ (ed) A roadmap to the successful development and commercialization of microbial pest control products for control of arthropods. Progress in biological control, vol 10. Springer, Dordrecht, pp 129–170

Ravensberg WJ (2011c) Registration of microbial pest control agents and products and other related regulations. in: Ravensberg WJ (ed) A roadmap to the successful development and commercialization of microbial pest control products for control of arthropods. Progress in biological control, vol 10. Springer, Dordrecht, pp 171–233

Ravensberg WJ (2011d) Roadmap to success and future perspective. In: Ravensberg WJ (ed) A roadmap to the successful development and commercialization of microbial pest control products for control of arthropods. Progress in biological control, vol 10. Springer Science + Business Media, Dordrecht, pp 357–376

Razzaghi-borkhani F, Rezvanfar A, Shabanali Fami H (2010) The role of educational and communicational factors on the knowledge of Integrated Pest Management (IPM) among Paddy farmers in Sari county. J Agric Edu Manag Res 13:2–17

Rezaei-Moghaddam K, Samiee S (2019) Adoption of integrated pest management (IPM): the case of Iranian farmers. Eur Online J Nat Soc Sci 8(2):269–284

Rezvanfar A, Eraktan G, Olhan E (2011) Determine of factors associated with the adoption of organic agriculture among small farmers in Iran. Afr J Agric Res 6(13):2950–2956

Roh JY, Choi JY, Li MS et al (2007) *Bacillus thuringiensis* as a specific, safe, and effective tool for insect pest control. J Microbiol Biotechnol 17:547–549

Salehi Jouzani G, Abbasalizadeh S, Moradali MF et al (2015) Development of a cost effective bioprocess for production of an Iranian anti-coleoptera *Bacillus thuringiensis* strain. J Agric Sci Tech 17:1183–1196

Sharifi-Moghaddam M (2006) Report of the activities in IPM/FFS sites in Iran. Ministry of Agriculture, Agricultural Extension and Farming System Department, Tehran

Soltani S, Azadi H, Mahmoudi H et al (2014) Organic agriculture in Iran: farmers' barriers to and factors influencing adoption. Renew Agric Food Syst 29:126–134

Stewart A, Brownbridge M, Hill RA et al (2010) Utilizing soil microbes for biocontrol. in: Dixon GR, Tilston EL (eds) Soil microbiology and sustainable crop production. Springer, Dordrecht, pp 315–371

Stirling GR (2014) The soil environment and the soil–root interface. In: Stirling GR (ed) Biological control of plant-parasitic nematodes, soil ecosystem management in sustainable agriculture, 2nd edn. CABI Publishing, Wallingford, pp 15–47

Thakore Y (2006) The biopesticide market for global agricultural use. Indust Biotech 2:194–208

Tormala T (1995) Economics of biocontrol agents: an industrial view. In: Hokkanen HMT, Lynch JM (eds) Biological control: benefits and risks. Plant and microbial biotechnology research series 4. Cambridge University Press, Cambridge, pp 277–282

Veisi H, Carolan MS, Alipour A (2017) Exploring the motivations and problems of farmers for conversion to organic farming in Iran. Int J Agric Sustain 15:303–320

Veisi H, Mahmoudi H, Sharifi-Moghaddam M (2010) Identifying farmers' adoption of integrated pest management technologies. J Iran Agric Econ Dev 41:481–490

Willer H, Lernoud J (eds) (2016) The world of organic agriculture – statistics and emerging trends 2016. Research Institute of Organic Agriculture (FiBL) and IFOAM – Organics International.

Available via https://shop.fibl.org/fileadmin/documents/shop/1698-organic-world-2016.pdf. Accessed 24 Jan 2017

Yang XB, del Rio L (2002) Implementation of biological control of plant diseases in integrated pest-management systems. In: Gnanamanickam SS (ed) Biological control of crop diseases. Marcel Dekker, New York, pp 343–358

# Part IV
# Other Approaches and Analyses of Current States of Biological Control in Iran

# Chapter 15
# Biological Control of Greenhouse Pests in Iran

Zahra Tazerouni and Ali Asghar Talebi

## 15.1   Introduction

The changes in the weather condition and amount of rainfall affect the types of crops grown in different parts of the world. Climate change and decreasing of precipitation increased the greenhouse cultivation in Iran in order to use the available resources. The greenhouse productions are important because besides supplying the local markets, they are greatly valued for their export potential and play an important role in the foreign trade balance of several national economies. However, the intensification of greenhouse crop production has created favorable conditions for many devastating pests and diseases. In greenhouse conditions, pests and diseases develop rapidly. Therefore, early detection and diagnosis of pest insects are necessary to make control decisions before the problem gets out of hand and suffer economic loss (FAO 2013).

Biological control as a main part of integrated pest management (IPM) is one of the best options for controlling pests in greenhouses. Natural enemies (parasitoids, predators or pathogens) are best used early in the crop growth cycle when plants are small, pest numbers are low and damage is not yet observed. Also, a detailed action plan needed to ensure success (Perdikis et al. 2008).

In this chapter, main pests on important greenhouse production and their effective biological control agents in Iran is discussed.

Z. Tazerouni · A. A. Talebi (✉)
Department of Entomology, Faculty of Agriculture, Tarbiat Modares University, Tehran, Iran
e-mail: talebia@modares.ac.ir

## 15.2    The Main Greenhouse Crops

Greenhouse crop production is now a growing reality throughout Iran with an estimated about 9000 ha of greenhouses. The vegetable crops (approximately 4000 ha) are the main greenhouse crops in Iran. Cucumber, tomato, pepper, cherry tomato, beans, broad beans and eggplant are the most commonly grown in greenhouses. Strawberry is one of the most important fresh fruit that grows in greenhouses. Also, many flowers (mainly rose, chrysanthemum, carnation, gladiolus, tuberose, alstromeria, saintpaulias, gerbera, orchid, anthurium and lilium) and ornamental plants are cultivated in greenhouses in Iran (Pahlevan et al. 2012; FAO 2013).

## 15.3    The Main Pests of Greenhouse Crops

Four groups of insects (include aphids, whiteflies, thrips and leaf miners) and plant mites are common and important pests on greenhouse crops in Iran. Several species are most serious pests of the greenhouse-grown crops that economic importance for greenhouse, are discussed here.

### 15.3.1    Aphids

Aphids are one of the most serious pests of vegetable and ornamental crops in commercial greenhouses. Almost all species of plants grown in greenhouses are susceptible to some aphid species (van Emden and Harrington 2007). Aphids can damage crops in different ways: (1) direct damage by sucking the sap of plants causing stunted and distorted plant tissue, new foliage and flowers are often malformed and (2) indirect damage by excreting honeydew and by transmission of pathogens. The cucumber mosaic virus (CMV), potato virus Y (PVY) and bean yellow mosaic virus (BYMV) are the most common and economically important viruses of greenhouse crops (van Emden and Harrington 2007; Pinto et al. 2008) that transmit by aphids. These damages result in significant yield reduction and economic loss (van Emden and Harrington 2007; Karami et al. 2018a). The most important aphid pests on greenhouse crops in Iran are given in Table 15.1.

*Aphis gossypii* is a polyphagous aphid, attacks the wide range of greenhouse crops (Table 15.1). The estimated intrinsic rate of increase of this aphid ranged from 0.173 day$^{-1}$ (on chrysanthemum) (Rostami et al. 2012) to 0.493 day$^{-1}$ (on cucumber, Super Sultan variety) (Tazerouni et al. 2016a). Pepper, eggplant, chrysanthemum and cucumber are most important host plants for *M. persicae* in greenhouses in Iran. The maximum intrinsic rate of increase of *M. persicae* was determined on pepper (0.389 day$^{-1}$) (Ghorbanian et al. 2019).

**Table 15.1** The most important aphid pests (Hemiptera: Aphididae) on greenhouse crops in Iran

| Aphid species | Host plants | References |
|---|---|---|
| *Aphis gossypii* Glover (cotton or melon aphid) | cucumber, pepper, broad bean, chrysanthemum | Zamani et al. (2006), Rostami et al. (2012), Valizadeh et al. (2013), Alizadeh et al. (2014), Hajiramezani-Chaleshtori et al. (2014), Makareminia et al. (2014), Keshavarz et al. (2015), Seyedebrahimi et al. (2015), Rahsepar et al. (2016), Tazerouni et al. (2016a) |
| *Myzus persicae* (Sulzer) (green peach aphid) | pepper, eggplant, chrysanthemum, cucumber | Zamani et al. (2007), Ghadamyari et al. (2008), Rashki et al. (2009), Valizadeh et al. (2013), Mardani-Talaee et al. (2015), Tazerouni et al. (2016c), Ghorbanian et al. (2019) |
| *Aphis fabae* Scopoli (black bean aphid) | broad been | Sabahi et al. (2011), Keshavarz et al. (2015) |
| *Macrosiphum rosae* (L.) (rose aphid) | Rose | Mehrparvar and Hatami (2007), Modarres Najafabadi (2014), Jafarinasab et al. (2015), Jalilian (2015) |
| *Macrosiphoniella sanborni* (Gillette) (chrysanthemum aphid) | chrysanthemum | Madjdzadeh and Mehrparvar (2009), Valizadeh et al. (2013) |

*Aphis fabae* is reported as one of the major pests of broad been (Sabahi et al. 2011; Keshavarz et al. 2015) in Iran greenhouses.

The rose aphid, *M. rosae* is one of the most important pests of rose which reducing the quality and quantity of products by attacking and hurting the leaves and twigs. The fecundity of females (progeny/female) increased with increasing temperature from 15 °C to 22 °C. The highest fecundity of *M. rosae* was reported 38.88 (progeny/female) at 22 °C on rose (Mehrparvar and Hatami 2007). *Macrosiphoniella sanborni* is a serious pest of chrysanthemum in Iran, high population densities of this aphid result in significant economic damage to chrysanthemum by decreasing its beauty and the value of cut flowers (Madjdzadeh and Mehrparvar 2009; Valizadeh et al. 2013).

## 15.3.2  Whiteflies

Whiteflies are one of the economic pests of a wide range of greenhouse crops in the world and Iran. They may cause economically important damages on host plants. Direct damage occurs by feeding plant phloem (Malumphy 2003) and indirect damage occurs by producing high amounts of honeydew (Perumal and Marimuthu 2009) which serves as a growing medium for sooty mold fungus. They are also vectors of some of the plant viruses such as tomato yellow leaf curl virus (TYLCV) and cucumber vein yellowing virus (CVYV) (Makkouk 1978; Nuez et al. 1999).

Z. Tazerouni and A. A. Talebi

The greenhouse whitefly, *Trialeurodes vaporariorum* (Westwood) and sweet potato whitefly, *Bemisia tabaci* (Gennadius) (Hemiptera: Aleyrodidae) are the most common whitefly species found in greenhouses in Iran. These pests can cause a high rate of economic losses on host plants.

*Trialeurodes vaporariorum* is the most harmful pest of a wide range of greenhouse crops specifically cucumber (Dehghani and Ahmadi 2013; Pirmoradi Amozegarfard et al. 2013; Amiri et al. 2014; Mirzamohammadzadeh et al. 2014), tomato (Ebrahimifar et al. 2015) and gerbera (Seddigh and Kiani 2012). Cucumber is the main host plants for *T. vaporariorum* and the highest and lowest total fecundity of this pest was 203.82 and 132.26 eggs on cucumber cultivars Royal Sluis and Soltan, respectively. Also, the mortality rate of this pest in pre-immaginal period was reported low (maximum 7%) (Mirzamohammadzadeh et al. 2014). Cucumber (Baniameri and Sheikhi 2006; Yarahmadi et al. 2013) and tomato (Samareh Fekri et al. 2013; Jafarbeigi et al. 2014) are also reported as the most important host plants for *B. tabaci* in greenhouses in Iran.

Some species of thrips are serious pests on greenhouse crops in the world and Iran. They can directly damage to its host plant through the feeding on cell sap and indirectly through the transmission of harmful plant viruses such as tomato spotted wilt virus (TSWV), iris yellow spot virus (IYSV) and impatiens necrotic spot virus (INSV) (Kirk 2001; Doi et al. 2003; Ghotbi and Baniameri 2006; Thungrabeab et al. 2006). Damages by thrips on host plant decrease production marketable and also they can result in significant yield losses (Mahr et al. 2001). Control of thrips is difficult because they have a small size and commonly hide in flowers, buds and leaf axils.

Two species of thrips include western flower thrips, *Frankliniella occidentalis* (Pergande) and onion thrips, *Thrips tabaci* Lindeman, (Thysanoptera: Thripidae) are the most dangerous species in greenhouses in Iran.

*Frankliniella occidentalis* was first detected in Iran in 2004 (Jalili Moghadam and Azmayeshfard 2004). It is a major pest of cucumber (Eshrati et al. 2014; Gholami et al. 2015), strawberry (Kiani et al. 2012; Golmohammadi and Mohammadipour 2015), green bean (Koupi et al. 2016), saintpaulias, gerbera and anthurium (Jalili Moghadam and Azmayeshfard 2004) in Iran. Intrinsic rate of increase ($r_m$), gross reproductive rate and net reproductive rate of *F. occidentalis* on cucumber, Viola variety were calculated $0.1902 \pm 0.008$ day$^{-1}$, 170.02 and 48.96 females/female/ generation, respectively (Eshrati et al. 2014). Damages by *T. tabaci* on greenhouse cucumber (Modarres Awal 2001; Pourian et al. 2009; Rajabpour et al. 2011; Jafari et al. 2013) are important economically. Also, this pest causes a serious damage on cut flower products especially carnation, gladiolus, chrysanthemum and tuberose in the greenhouse in Iran (Hosseininia and Malkeshi 2009).

## 15.3.4  Leaf Miners

Several species of leafminers include vegetable leaf miner, *Liriomyza sativae* Blanchard and serpentine leaf miner, *Liriomyza trifolii* Burgess (Diptera: Agromyzidae) attack greenhouse crops. Both larvae and adults damage the host plants. The larvae feed on leaf mesophyl and reduce chlorophyll content, which it result to decrease photosynthesis (Parrella 1983). The adults also damage the leaves by puncturing them for feeding and oviposition (Parrella and Keil 1985). Infested leaves are favorable habitats for invading bacterial and fungal plant pathogens (Capinera 2001). Yield losses, in general, can be considerable (Waterhouse and Norris 1987).

*Liriomyza sativae* has seriously damages on tomato (Baniameri et al. 2012) and cucumber (Asadi et al. 2006; Fathipour et al. 2006; Haghani et al. 2006; Asghari-Tabari et al. 2009; Haghani et al. 2007; Tavanapour et al. 2009; Basij et al. 2011; Namvar et al. 2011; Dashtbani et al. 2013). Effect of temperature (10, 15, 20, 25, 30, 35 and 40 °C) on life table parameters of *L. sativae* on cucumber, Negin variety revealed that the intrinsic rate of natural increase ($r_m$) and net reproductive rate ($R_0$) were significantly higher at 25 °C (0.196 day$^{-1}$ and 52.452 females/female/generation time, respectively) (Haghani et al. 2006). Among greenhouse crops, tomato (Baniameri and Cheraghian 2012) and bean (Askari Saryazdi et al. 2012) are reported as the most important host plants for *L. trifolii*.

The tomato leaf miner, *Tuta absoluta* (Meyrick) (Lepidoptera: Gelechiidae) has been introduced into Iran in 2010 (Baniameri and Cheraghian 2012). Larvae of *T. absoluta* mine the leaves, flowers, shoots, and fruit of tomato (Pastrana 2004). After hatching, larvae penetrate apical buds, flowers, new fruit, leaves or stems. Conspicuous irregular mines and galleries, as well as dark feces of larvae, make infestations relatively easy to spot. Fruits can be attacked soon after they have formed, and the galleries made by the larvae can be colonized by pathogens that cause fruit rot. The damage caused by this pest is severe, especially in young plants (Vargas 1970). *Tuta absoluta* is one of the economic pests of tomato (Hashemi Tasuji et al. 2014; Javadi Khederi et al. 2014; Sohrabi et al. 2014; Ahmadipour et al. 2015; Dehghani et al. 2015; Nemati et al. 2015; Rostami et al. 2015; Sharifian et al. 2015; Tamoli Torfi et al. 2016), but also attacks eggplant (Hashemi Tasuji et al. 2014; Tamoli Torfi et al. 2016) and pepper (Hashemi Tasuji et al. 2014) in greenhouses in Iran. High temperature (above 25 °C) has a negative effect on the biological parameters of tomato leaf miner moths (Gharekhani et al. 2014). The mean number of larvae varied significantly on different host plants. The highest number of larvae was determined on tomato (26.33 ± 4.48) and the lowest on the pepper plant (1.33 ± 0.88) (Hashemi Tasuji et al. 2014).

## 15.3.5 Mites

Two spotted spider mite, *Tetranychus urticae* Koch (Acari: Tetranychidae) is a very destructive pest of many crops in greenhouses around the world. This mite attacks almost all greenhouse crops and cause damage by sucking plant juices. Spider mites remove chlorophyll from plant cells and reduce photosynthesis. This damage produces the characteristic stippling or mottling of foliage and sometimes causes leaf drop (Abdel-Wali 2013). When populations of this pest are low, the mites prefer to attack the underside of mature leaves, but they may move upward of leaves even on fruits when populations increase. In severe infestations, the plants may be covered with webbing, which is why they are referred to as spider mites. Feeding and silk production of this mite affects the quality and quantity of crop yield (Zhang 2003). *Tetranychus urticae* has been considered as a major pest on cucumber (Kheradpir et al. 2006, 2009; Jafari et al. 2012; Shoorooei et al. 2012; Motahari et al. 2014; Maleknia et al. 2016a), bean (Hamedi et al. 2010; Ganjisaffar et al. 2011; Kouhjani Gorji et al. 2012; Modarres Najafabadi 2012; Motazedian et al. 2012), tomato (Kheradpir et al. 2009; Saeidi et al. 2012), eggplant (Khanjani 2005; Kheradpir et al. 2006, 2009; Khanamani et al. 2013), strawberry (Rezaei et al. 2013; Mortazavi et al. 2015), pepper (Kheradpir et al. 2009) and rose (Bidarnamani et al. 2015). The estimated intrinsic rate of increase of this mite on 12 greenhouse cucumber cultivars (Royal, Amitral, Ariya, Sultan, Vida, Bahman, Storm, Nasim, Negin, PS-29, Tornado and Caspian) ranged from 0.249 day$^{-1}$ (on Caspian) to 0.331 day$^{-1}$ (on Vida) (Maleknia et al. 2016a). The total fecundity of *T. urticae* varied from 82.45 to 142.05 eggs on five bean genotype (Modarres Najafabadi 2012). The positive relation exists between the density of glandular trichomes on different cultivars of tomato and their resistance to the two spotted spider mite (Saeidi et al. 2012). The intrinsic rate of increase of *T. urticae* on different cultivars of eggplant (Isfahan, Dezful, Shend-Abad, Neishabour, Bandar-Abbas, Jahrom and Borazjan) varied from 0.022 day$^{-1}$ on Neishabour to 0.157 day$^{-1}$ on Isfahan (Khanamani et al. 2013). A comparative study on the life table parameters of *T. urticae* on seven strawberry cultivars (Marak, Yalova, Aliso, Gaviota, Sequoia, Camarosa and Chandler) revealed that the higher intrinsic rate of increase found on Chandler (0.28 day$^{-1}$), that may be due to the higher nutritional quality and low density of trichomes (Rezaei et al. 2013).

The strawberry spider mite, *Tetranychus turkestani* (Ugarov and Nikolskii) (Acari: Tetranichidae) are reported as an important pest of greenhouse cucumber (Mohammadi et al. 2015) and eggplant in Iran (Soleimannejadian et al. 2006). The life table parameters of this mite were studied on six greenhouse cucumber cultivars (Puia, Hedieh, Milad Jadid, Milad Ghadim, Khasib and Negin). The result showed that the highest total fecundity was observed on the Hedieh cultivar (29.784 eggs/female) and the lowest on the Puia cultivar (15.773 egg/ female). Moreover, *T. turkestani* gross hatching rate on the six cucumber cultivars differed significantly, varying from 0.18 on Negin to 0.48 on Hedieh (Mohammadi et al. 2015).

The broad mite, *Polyphagotarsonemus latus* (Banks) (Acari: Tarsonemidae) has also been reported on tomatoes, eggplants, cucumbers and strawberries, in Iran, although the economic damage caused has been brought under control (Baker and Arbabi 2015). Its attack is confined mostly to new growths resulting in curling of leaf margins, firmness of infested leaves, necrosis of growing points, aborted buds, malformed fruits and growth inhibition (Grinberg et al. 2005).

## 15.4   The Main Biological Control Agents of Greenhouse Crop Pests

Biological control is an important component of integrated pest management (IPM) in greenhouses. Augmentaion biological control involves the inoculative and inundative releases of natural enemies (predators, parasitoids and pathogens) for the control of insect and mite pests (Koul and Dhaliwal 2003).

Greenhouses are isolated units therefore they are suitable for the use of biological control agents, because: (1) its closed system prevent to dispersal of natural enemies, (2) the greenhouse conditions may be more favorable for the natural enemies than pests, (3) the protected environment of greenhouses also make possible application of biological control agents in low pest population However, biological control can be also effective under high pest population, but in these cases, the costs increase significantly (Perdikis et al. 2008).

In this part, effective biological control agents against pests of greenhouse crops in Iran are discussed.

### 15.4.1   Biological Control of Aphids

Natural enemies evaluated against aphid pests in greenhouse crops in Iran include predators, parasitoids and pathogens. The most important species are listed in Table 15.2.

#### 15.4.1.1   Predators

The efficiency of several species of predators as potential candidates for suppression of greenhouse aphids populations are evaluated in Iran. Most of them belong to the families Coccinellidae (Coleoptera), Syrphidae, Cecidomyiidae (Diptera), Chrysopidae (Neuroptera), Miridae and Anthocoridae (Hemiptera) (Table 15.2).

The population of *C. septempunctata* on *A. gossypii* was able to multiply 3.45 times per week ($r_w$) (Mollashahi et al. 2009). The production indices, 41.3 and 43.12 were calculated for *H. variegata* and *C. septempunctata* fed on *A. gossypii* at 30 °C, respectively (Mollashahi et al. 2012). Each individual ladybeetle (*H. variegata*) during their life, the average of 432 green peach aphid is consumption (Hassankhani

**Table 15.2** The main natural enemies evaluated against aphids in greenhouse crops in Iran

| Natural enemies | Host aphids | Crops | References |
|---|---|---|---|
| **Predators** | | | |
| Coleoptera: Coccinellidae | | | |
| *Coccinella septempunctata* Linnaeus | *A. gossypii* | cucumber | Mollashahi et al. (2009) |
| | *M. rosae* | Rose | Amin Afshar et al. (2014) |
| *Hippodamia variegata* (Goeze) | *A. gossypii* | Cucumber | Mojib Hagh Ghadam and Yousefpour (2012), Mollashahi et al. (2012) |
| | *M. persicae* | | |
| | *M. rosae* | | |
| *Adalia bipunctata* (Linnaeus) | *M. persicae* | | Jalali et al. (2010) |
| *Propylea quatuordecimpunctata* (Linnaeus) | *A. gossypii* | broad bean | Keshavarz et al. (2015) |
| | *A. fabae* | | |
| Diptera: Syrphidae | | | |
| *Scaeva albomaculata* (Macquart) | *M. persicae* | different greenhouse crops | Fathipour et al. (2005) |
| | *M. rosae* | Rose | Jalilian (2015) |
| *Eupeodes corollae* (Fabricius) | *M. persicae* | different greenhouse crops | Fathipour et al. (2005) |
| *Episyrphus balteatus* (De Geer) | *M. persicae* | different greenhouse crops | Fathipour et al. (2005) |
| Diptera: Cecidomyiidae | | | |
| *Aphidoletes aphidimyza* (Rondani) | *A. gossypii* | Cucumber | Hosseini et al. (2010), Madahi et al. (2013), Moayeri et al. (2013), Malkeshi et al. (2015) |
| Neuroptera: Chrysopidae | | | |
| *Chrysoperla carnea* (Stephens) | *A. gossypii* | | Takalloozadeh (2015) |
| | *M. persicae* | | |
| Hemiptera: Miridae | | | |
| *Deraeocoris lutescens* (Schilling) | *M. persicae* | sweet pepper | Azimizadeh et al. (2012) |
| Hemiptera: Anthocoridae | *A. gossypii* | Cucumber | Hosseini et al. (2010), Moayeri et al. (2013) |
| *Orius laevigatus* (Fieber) | | | |
| **Parasitoids** | | | |
| Hymenoptera: Braconidae | | | |
| *Aphidius matricariae* Haliday | *M. persicae* | cucumber and pepper | Zamani et al. (2007), Rashki et al. (2009), Norouzi (2011), Tazerouni et al. (2016c) |
| | *A. gossypii* | cucumber and pepper | Talebi et al. (2006), Zamani et al. (2012), Tazerouni et al. (2017) |

(continued)

**Table 15.2** (continued)

| Natural enemies | Host aphids | Crops | References |
|---|---|---|---|
| | *A. fabae* | | Tahriri Adabi et al. (2010), Pourtaghi et al. (2014) |
| *Aphidius colemani* Viereck | *A. gossypii* | Cucumber | Talebi et al. (2006), Zamani et al. (2007, 2012), Malkeshi et al. (2015) |
| *Lysiphlebus fabarum* (Marshall) | *A. fabae* | | Mohammadi et al. (2014), Mohseni et al. (2014) |
| *Praon volucre* (Haliday) | *M. persicae* | Sweet pepper | Tazerouni et al. (2016c) |
| **Pathogens** | | | |
| *Beauveria bassiana* (Balsamo-Crivelli) | *A. gossypii* | cucumber, eggplant | Rashki et al. (2009), Rashki and Shirvani (2013) |
| | *M. persicae* | | |

et al. 2014). In the three hoverfly species, about 70% consumptions occurred during the third larval instars (Fathipour et al. 2005). Each larva of *S. albomaculata* ate 516.167 ± 73.17 aphids (*M. rosae*) during larval period. There were significant differences (P < 0.01) between daily feeding rates of larval instars. The third instar larvae had an important role in feeding rate and 78.32% of total larval feeding was due to this instar (Jalilian 2015).

These predators are suggested to control mentioned aphid pests in greenhouses in Iran. Intrinsic rate of increase ($r_m$) of *A. aphidimyza* on *A. gossypii* ranged from $0.110 ± 0.016$ to $0.166 ± 0.014$ day$^{-1}$ with increasing prey density (5, 10, 20, 40, 60, 80) (Madahi et al. 2013).

The maximum mean fecundity per female of *C. carnea* was 478.50 ± 8.38 eggs when larvae fed on *M. persicae* followed by 409.33 ± 8.16 eggs on *A. gossypii* (Takalloozadeh 2015).

Theoretical maximum attack rates ($T/T_h$) of *Orious laevigatus* on *A. gossypii* on cucumber were estimated to be 55.58, 54.74, 61.98 and 64.64 aphids/day in 1st, 2nd, 3rd and 4th days of the adult lifetime, respectively (Rostamian et al. 2014).

### 15.4.1.2  Parasitoids

All members of the subfamily Aphidiinae (Hymenoptera: Braconidae) are solitary endoparasitoids of aphids (Starý 1970). They are among the most important natural enemies of aphids, which can effectively regulate the aphid populations and prevent serious outbreaks (Hagvar and Hofsvang 1991; Karami et al. 2018b). Several species have shown potential to control aphids in greenhouses in Iran (Table 15.2).

The highest value of the mean fertile eggs per day for *A. matricariae* and *A. colemani* on *A. gossypii* on greenhouse cucumber were recorded at 25 °C and 30 °C, respectively (Zamani et al. 2012). Talebi et al. (2006) reported that the body size of the progeny of *A. colemani* and *A. matricariae* at emergence increased with host (*A. gossypii*) stage at the time of parasitization. The greenhouse release of parasitoids would be best timed to coincide with the period when third and fourth nymphal instars of cotton aphid are most abundant.Tazerouni et al. (2016b) were expressed that both *A. matricariae* and *P. volucre* are potential biological control agents to decreasing population of *A. gossypii* and *M. persicae*. The maximum parasitism rate of *A. matricariae* and *P. volucre* on *A. gossypii* was reported 34.28 and 24.74 nymphs, respectively (Tazerouni et al. 2017). The percentage of parasitism of *A. gossypii* and *M. persicae* by *A. matricariae* was 41% and 47.6%, respectively and by *A. colemani* was obtained 56.4% and 50.6%, respectively (Zamani et al. 2007). The $r_m$ value for *A. matricariae* on *M. persicae* was reported 0.3 day$^{-1}$ (Rashki et al. 2009). *Aphidius matricariae* caused reasonable mortality of the *M. persicae* by parasitism of 52.17 host aphids, in 24 h (Tazerouni et al. 2016c). Searching efficiency, handling time and maximum attack rate of *P. volucre* on *M. persicae* were $0.020 \pm 0.003$ h$^{-1}$, $0.507 \pm 0.160$ h and 47.34 nymphs per 24 h, respectively (Tazerouni et al. 2016c). Other braconid wasp, *Aphidius rosae* Haliday is beneficial parasitoid wasp against *A. rosae* (Mehrparvar et al. 2005).

*Aphidius matricariae* and *A. colemani* produce commercially and they use as successful biological control agent against *M. persicae* and *A. gossypii*, respectively in greenhouses.

*Beauveria bassiana* strain DEBI008 can reduce the melon aphid fitness or repel the aphid (Rashki and Shirvani 2013). After 48 hours, to use concentration of $10^6$ of *Metarhizium anisopliae* (Metschnikoff) (DEMI-001 isolate), mortality of *M. rosae* was 52.1% and 74.2% at 10 and 20 °C, respectively (Jafarinasab 2015).

## 15.4.2 Biological Control of Whiteflies

There are a few effective parasitoid wasps against whiteflies in Iran. Also, there are some predators and pathogens that they are used single or in combination with parasitoids to improve control of whiteflies.

### 15.4.2.1 Predators

*Amblyseius swirskii* Athias-Henriot (Acari: Phythoseiidae) as a general predator has the effective influence to decreasing *T. vaporariorum* populations, particularly pupal stage of this pest on greenhouse cucumber (Ardeh et al. 2014). Also, this predator is useful for control of *B. tabaci* (especially on eggs and first nymphal instars) on bean (Soleymani et al. 2015a) and cucumber (Soleymani et al. 2015b) in greenhouses.

### 15.4.2.2 Parasitoids

Several parasitoid wasps attack whiteflies, which have been used successfully for biological control of these pests in greenhouses. The parasitic wasps of the family

Aphelinidae as effective natural enemies of whiteflies are used under greenhouses condition. *Encarsia inaron* (Walker) (Hymenoptera: Aphelinidae) significantly increased the mortality rate of immature stages of *T. vaporariorum* (Hosseini and Pourmirza 2010). In mentioned above research, pre-immaginal mortality in the presence of the parasitoid was 93.9%. The release of *Encarsia formosa* Gahan (Hymenoptera: Aphelinidae) against greenhouse whitefly (*T. vaporariorum*) on white and pink gerbera can decrease the population of this pest beneath the economic damage threshold (Seddigh and Kiani 2012). *Encarisa formosa* and *Eretmocerus eremicus* Rose and Zolnerowich are efficient parasitoids to control of *T. vaporariorum* and *B. tabaci* on greenhouse cucumber in Iran (Ardeh et al. 2014). Ebrahimifar et al. (2015) reported *Eretmocerus delhiensis* Mani as major parasitoid on greenhouse whitefly, *T. vaporariorum* on tomato in greenhouses in Iran. The number of eggs deposited by a single parasitoid (*Encarsia acaudaleyrodis* Hayat), on *B. tabaci* on cucumber depended on the number of host individuals, the maximum being 44.4 when the number of nymphs was 100 (Shishehbor and Zandi-Sohani 2011).

### 15.4.2.3  Pathogens

The application of two species of entomopathogenic nematodes includes *Steinernema feltiae* (Filipjev) (Nematoda: Steinernematidae) and *Heterorhabditis bacteriophora* Poinar (Nematoda: Heterorhabditidae) as a biological control agents on second nymphal instars of *T. vaporariorum* on greenhouse cucumber result significant mortality of this pest. Therefore, use both species of nematodes, *S. feltiae* and *H. bacteriophora* was suggested against greenhouse whitefly on cucumber in greenhouses in Iran (Rezaei et al. 2015).

## 15.4.3  Biological Control of Thrips

The efficiency of several important natural enemies has been evaluated to suppress the populations of thrips on greenhouse crops in Iran. Also, there are a few parasitoids and pathogens attack thrips in greenhouses. Some pathogens have also potential to reduce thrips population.

### 15.4.3.1  Predators

*Orius niger* (Wolff) (Hemiptera: Anthocoridae) is an ideal predator to use as a biological agent against *T. tabaci* in greenhouse crops which have enough pollen, or when they are mixed with banker plants (Baniameri et al. 2006). *Neoseiulus barkeri* Hughes and *Neoseiulus cucumeris* (Oudemans) (Acari: Phytoseiidae) are predators of first instars larvae of thrips (*F. occidentalis* and *T. tabaci*) and to a smaller extent, second instar larvae on ornamental crops in greenhouses (Malkeshi and Hosyni 2014). *Neoseiulus barkeri* is an indigenous and potential biological control agent against *T. tabaci* on greenhouse cucumber in Iran. It successfully developed on 1st instars larvae of *T. tabaci*. The intrinsic rate of increase ($r_m$) and the net reproductive rate ($R_0$) of *N. barkeri* on *T. tabaci* were 0.252 day$^{-1}$ and 18.70 female offspring, respectively. This predator can prevent the outbreak of *T. tabaci* (Jafari et al. 2013).

#### 15.4.3.2 Pathogens

Entomopathogenic fungi such as *Metarhizium anisopliae* is an important biocontrol agent against thrips in greenhouses. All mobile stages of thrips are susceptible to this fungus. The results of research by Koupi et al. (2016) showed that *M. anisopliae* isolate 'DEMI002' can be considered as a promising tool in biological control programs of the *F. occidentalis* on bean in greenhouses in Iran. Biocontrol potential of entomopathogenic nematodes includes *Steinernema carpocapsae* (Weiser) and *H. bacteriophora* were investigated on *F. occidentalis* on the bean. The results showed that the highest thrips mortality was obtained by using these biological control agents at $2 \times 10^4$ IJs ml$^{-1}$ with 65 and 61% thrips mortality, respectively (Kashkouli et al. 2014).

### 15.4.4 Biological Control of Leaf Miners

Biological control of leaf miners is possible on many greenhouse crops. There are few predators of leaf miners in the world and Iran, however numerous parasitoid wasps are effective against leaf miners that they commonly found in greenhouses; their potential for successful biological control is high on vegetable crops.

#### 15.4.4.1 Predators

*Nesidiocoris tenuis* (Reuter) and *Macrolophus pygmaeus* (Rambur) (Hemiptera: Miridae) and *Nabis pseudoferus* Remane (Hemiptera: Nabidae) are reported as predators of *T. absoluta* on tomato (Mahdavi and Madadi 2015; Sohrabi and Hosseini 2015; Sharifian et al. 2015). They have high predatory capacity on this pest on tomato. The attack rate and handling time of *M. pygmaeus* on *T. absoluta* eggs were 0.2251 h$^{-1}$ and 2.7415 h, respectively (Sharifian et al. 2015). The mean daily feeding rate of 3rd, 4th and 5th instar nymphs, adult males and females of *N. pseudoferus* on *T. absoluta* egg was 35.6, 51.6, 61.7, 70.2 and 63.15 and total predation rate of the same stages were 117.95, 221.6, 381.3, 983.2 and 2510.45, respectively. The mean daily feeding rate of the same stages by feeding on the 1st instar larvae of *T. absoluta* were 23.7, 27.8, 59.6, 37.7 and 62.7. The total predation rate was 76.85, 116.05, 304.85, 749.05 and 2217.85, respectively. It has been suggested that due to the larger size and greater energy requirements to produce eggs, predator females have the highest feeding rate. Oviposition rate of *N. pseudoferus* by feeding on eggs and 1st instar larvae of tomato leaf miner was 54.4 and 127.85 eggs per female which were differed significantly (Mahdavi and Madadi 2015).

### 15.4.4.2   Parasitoids

Three eulophid parasitoid species including *Diglyphus isaea* (Walker), *Hemiptarsenus zilahisebessi* Erdos and *Neochrysocharis formosus* (Westwood) (Hymenoptera: Eulophidae) are common parasitoid of *L. sativae* larvae on cucumber greenhouse. *Diglyphus isaea* was reported as most important and common parasitoid of *L. sativae* on cucumber in greenhouses in Iran (Asadi et al. 2006; Fathipour et al. 2006; Haghani et al. 2007; Dashtbani et al. 2013). The activity of this parasitoid was increased following the increase of ambient temperature and host population (Dashtbani et al. 2013). Parasitism of *L. sativae* on cucumber was found to reach 39% with *D. isaea* as the most common species in Iran in unsprayed greenhouses (Fathipour et al. 2006).

In addition to above mentioned three parasitoid species, *Cirrospilus vittatus* Walker and *Diglyphus crassinervis* Erdos (Hymenoptera: Eulophidae) are effective parasitoid wasps on *L. satiave* and *L. trifolii* on different host plants, in particular on cucumber (Asadi et al. 2006). Some species of *Trichogramma* (Hymenoptera: Trichogrammatidae) as the egg parasitoids, are potential candidates for biologocal control of *T. absoluta*. The efficiency of some *Trichogramma* species includes *T. brassicae* Bezdenko (Baboulsar strain), *T. evanescens* Westwood, *T. principium* Sugonjaev and Sorokina and *T. pintoi* Voegele are investigated against *T. absoluta* on tomato, the highest parasitism rate achieved as 47.5 ± 0.85% for *T. brassicae* (Ahmadipour et al. 2015).

## 15.4.5   Biological Control of Mites

Predatory insects and mites are the only effective natural enemies of spider mites (Tetranychidae). There are many species of predatory mites that used successfuly to control of spider mites. Several species of the family Phytoseiidae are the most efficient predators of pest mites and commonly used to control of spider mites.

### 15.4.5.1   Predators

*Phytoseiulus persimilis* Athias-Henriot is known as a very efficient biological control agent of several species of spider mites (Helle and Sabelis 1985; Maleknia et al. 2016b). Fathipour et al. (2018) revealed that *P. persimilis* can show both type II and type III functional responses depending on the age of the predator. Maximum prey consumption at the highest prey density (128 eggs) was 39.3, 41.7, 39.3 and 38.1 eggs/day on 15, 20, 25 and 30 days of ptredatory mite age, respectively (Fathipour et al. 2018).

The predatory mite, *Neoseiulus barkeri* Hughes has been reported from many parts of the world and Iran. *Neoseiulus barkeri* has an inherent potential for the control of two-spotted spider mite on cucumber at higher temperatures especially at temperatures between 30 and 35 °C (Jafari et al. 2010). This predator has been used in the augmentative biological control of *T. urticae*. The highest intrinsic rate of increase ($r_m$) of *N. barkeri* on *T. urticae* was 0.256 day$^{-1}$ at 30 °C. During the oviposition period, the total prey consumption by *N. barkeri* increased with increasing temperature from 15 (160.43 preys) to 30 °C (286.71 preys) and then declined and reached to 191.57 preys at 37 °C (Jafari et al. 2011). *Neoseiulus californicus* (McGregor) as an introduced species in Iran have the high predation potential particularly on eggs of *T. urticae*. It has been suggested to control of *T. urticae* on cucumber (Farazmand et al. 2012) and on strawberry (Rezaei 2015) in greenhouses. *Neoseiulus californicus* had the highest predation potential particularly on eggs (14.59 per day) of *T. urticae* (Farazmand et al. 2012). In a recent study, effect of spirotetramat on the life table parameters of *Neoseiulus californicus* was studies under laboratory conditions. The result showed that total fecundity of *N. californicus* on *T. urticae* was between 29.45 and 39.46 (offspring/individual) in different treatments (Sangak Sani Bozhgani et al. 2018). They suggested that spirotetramat can be introduced as a compatible pesticide alongside the predatory mite, *N. californicus*.

*Amblyseius swirskii* Athias-Henriot against *T. urticae* (red form) on bean induce the egg retention phenomena in prey in the presence of a predator (Askarieh Yazdi et al. 2015). *Phytoseius plumifer* (Canestrini and Fanzago) has been reported as effective biological control agent of *T. urticae*. The total female prey consumption was reported 426.98 preys at different life stages of *P. plumifer* at 25 °C (Kouhjani Gorji et al. 2009).

*Typhlodromus bagdasarjani* is indigenous and widespread species in Iran. It is well adapted to high temperatures, therefore they can use against two-spotted spider mite in greenhouses (Ganjisaffar et al. 2011). The maximum attack rate of *T. bagdasarjani* on two spotted spider mite on strawberry was estimated to be 12.56 eggs per 24 h. This predator may be used as an efficient predator for population management of *T. urticae* on strawberry in greenhouses (Mortazavi et al. 2015).

Some of mite predators such as *T. bagdasarjani* and *A. swirskii* have been commercially used for biological control of *T. urticae*. There are several studies about their mass rearing on artificial diets. The results of studies showed that *Ephestia kuehniella* eggs, cysts of *Artemia franciscana* and *Tyrophagus putrescentiae* is recommendable for mass rearing of *T. bagdasarjani* and utilizing on crop as alternative food (Riahi et al. 2018). Also almond pollen can be used to amplify the experimental and commercial mass rearing programs of this predator and maize plant can be recommended as banker plants in greenhouses (Riahi et al. 2016).

Accoring to the results, *A. swirskii* required an average of 17.69, 32.33, 27.87, and 18.04 prey to produce a single predator egg on date+*T. urticae*, bee pollen +*T. urticae*, *T. urticae*, and almond+*T. urticae*, respectively. Overall, *A. swirskii* had greater control potential of *T. urticae* in the presence of almond pollen than other diets (Riahi et al. 2017). The intrinsic rate of increase ($r_m$) of minute pirate bug, *O. niger* a predator of *T. urticae* was reported 0.1184 day$^{-1}$ (Nojumian et al. 2015). *Scolothrips longicornis* Priesner (Thysanoptera: Thripidae) is a native beneficial thrips in Iran. It is considered to be an important predator of numerous spider mite species such as *T. urticae* (Pakyari et al. 2008). This predator achieved higher predation (16.1 preys/day) at a higher temperature (35 °C). Therefore it may be more effective for biological control of two spotted spider mite on bean in warmer conditions. The efficiency of this predator was evaluated against *T. urticae* on four host plants (cucumber, tomato, sweet pepper and eggplant) (Kheradpir et al. 2009). Predacious thrips preferred *T. urticae* on host plants (such as eggplant) with rational dense trichomes, which are arranged in clusters and leave some space for the predator to walk, search prey and also oviposition.

The ladybird beetles of the genus *Stethorus* (Coleoptera: Coccinellidae) are one the best candidates for controlling *T. urticae* in the world and Iran. In Iran, *Stethorus*

**Table 15.3** The main natural enemies evaluated against mites in greenhouse crops in Iran

| Natural enemies | Host mite | Crops | References |
|---|---|---|---|
| **Predators** | | | |
| Acari: Phytoseiidae | | | |
| *Phytoseiulus persimilis* Athias-Henriot | *T. urticae* | cucumber | Fathipour et al. (2018) |
| *Neoseiulus barkeri* Hughes | *T. urticae* | cucumber | Jafari et al. (2010) |
| | | strawberry | Rezaei and Askari (2015) |
| *Neoseiulus californicus* (McGregor) | *T. urticae* | cucumber | Farazmand et al. (2012) |
| | | | Khanamani et al. (2017) |
| | | strawberry | Rezaei et al. (2015) |
| | | Bean | Sangak Sani Bozhgani et al. (2018) |
| *Amblyseius swirskii* Athias-Henriot | *T. urticae* (red form) | Bean | Askarieh Yazdi et al. (2015) |
| *Phytoseius plumifer* (Canestrini and Fanzago) | *T. urticae* | | Kouhjani Gorji et al. (2009) |
| *Typhlodromus bagdasarjani* | *T. urticae* | | Ganjisaffar et al. (2011) |
| Wainstein and Arutunjan | | strawberry | Mortazavi et al. (2015) |
| Hemiptera: Anthocoridae | | | |
| *Orius niger* (Wolff) | *T. urticae* | | Nojumian et al. (2015) |
| Thysanoptera: Thripidae | | | |
| *Scolothrips longicornis* Priesner | *T. urticae* | | Pakyari et al. (2008) |
| Coleoptera: Coccinellidae | | | |
| *Stethorus gilvifrons* (Mulsant) | *T. urticae* | | Kheradpir et al. (2006) |
| | | bean | Taghizadeh et al. (2008) |

*gilvifrons* (Mulsant) is the most common species, especially around Tehran. Female predators have the highest feeding rate on adult mites and larvae had the lowest feeding rate, especially on mite eggs (Kheradpir et al. 2006). The highest values of $r_m$ (0.240 day$^{-1}$) and net reproductive rate (59.27 females/female/generation time) of *S. gilvifrons* on *T. urticae* on bean were reported at temperature of 35 °C. The temperature greatly affected fecundity, survivorship and life table parameters of *S. gilvifrons*, and that 35 °C is a suitable temperature for population growth of this predator (Taghizadeh et al. 2008) (Table 15.3).

# References

Abdel-Wali M (2013) Integrated pest management and plant hygiene under protected cultivation. In: FAO (ed) Good agricultural practices for greenhouse vegetable crops: principles for mediterranean climate areas. Food and Agriculture Organization of the United Nations, Rome, pp 399–426

Ahmadipour R, Farrokhi S, Shakarami J et al (2015) Selection of *Trichogramma* native strain for biological control of tomato leafminer, *Tuta absoluta* (Meyrick). In: Abstract presented at the 1st Iranian International Congress of Entomology, Iranian Research Institute of Plant Protection, Tehran, 29–31 August 2015, p 362

Alizadeh Z, Haghani M, Sedaratian A (2014) Reproductive parameters of two populations of *Aphis gossypii* Glover (Hem.: Aphididae) on different sweet pepper *Capsicum annuum* L. cultivars. In: Abstract presented at the 21st Plant Protection Congress, University of Urmia, Urmia, 23–26 August 2014, p 501

Amin Afshar E, Khanjani M, Zahiri B (2014) Evaluation of life table parameters of *Coccinella septempunctata* (L.) fed on *Macrosiphum rosae* (L.). In: Abstract presented at the 21st Plant Protection Congress, University of Urmia, Urmia, 23–26 August 2014, vol I. Pests, p 536

Amiri F, Hassan Poor A, Shirzadi MH (2014) The effect of vermicompost extract on the white fly population on green house cucumber (*Cucumis sativus* L.) in 2 varieties. J Nov Appl Sci 3 (8):883–885

Ardeh MJ, Farrokhi S, Valiollah B (2014) Efficiency of two biocontrol agents to control cucumber whiteflies under greenhouses. In: Abstract presented at the 21st Plant Protection Congress, University of Urmia, Urmia, 23–26 August 2014, vol I. Pests, p 471

Asadi R, Talebi AA, Fathipour Y et al (2006) Identification of parasitoids and seasonal parasitism of the Agromyzid Leafminers genus *Liriomyza* (Dip.: Agromyzidae) in Varamin. J Agr Sci Tech 8:293–303

Asghari-Tabari B, Sheikhi Gorjan A, Shojaei M et al (2009) Susceptibility of three developmental stages of *Liriomyza sativae* Blanchard (Dip.: Agromyzidae) to biorational insecticides in vitro conditions. J Entomol Res 1(1):23–34

Askari Saryazdi G, Hejazi MJ, Saber M (2012) Residual Toxicity of Abamectin, Chlorpyrifos, Cyromazine, Indoxacarb and Spinosad on *Liriomyza trifolii* (Burgess) (Diptera: Agromyzidae) in Greenhouse Conditions. Pestic Phytomed 27(2):107–116

Askarieh Yazdi S, Zahedi Golpayegani A, Saboori A et al (2015) Different forms of *Tetranychus urticae* Koch (Acari: Tetranychidae) and their plasticity in retaining eggs in the presence of predatory mites *Amblyseius swirskii* (Acari: Phytoseiidae). In: Abstract presented at the 1st Iranian International Congress of Entomology, Iranian Research Institute of Plant Protection, Tehran, 29–31 August 2015, p 312

Azimizadeh N, Ahmadi K, Imani S et al (2012) Evalution of oviposition-site preference behavior in predatory bug *Deraeocoris lutescens* Schilling (Hemiptera: Miridae). Munis Entomol Zool 7 (1):506–515

Baker RA, Arbabi M (2015) The broad mite, *Polyphagotarsonemus latus* (Banks), a résumé of a widely distributed invasive species and plants pest. J Entomol Res 7(2):23–29

Baniameri V, Cheraghian A (2012) The first report and control strategies of *Tuta absoluta* in Iran. EPPO Bull 42:322–324

Baniameri V, Sheikhi A (2006) Imidoclopride as soil application against whitefly *bemisai tabaci* in greenhouse cucumber. IOBC Bull 29:101–102

Baniameri V, Soleyman-nejadian E, Mohaghegh J (2006) The predatory bug *Orius niger*: its biology and potential for controlling *Thrips tabaci* in Iran. IOBC/wprs Bull 29:207–209

Baniameri V, Bagheri MR, Farrokhi Sh (2012) Evaluation of biological control agents to control of leafminers in greenhouse tomato. Retreived from: http://agris.fao.org/agris-search/search.do? recordID=IR2015000474

Basij M, Askarianzaeh A, Asgari S et al (2011) Evalution of resistance of cucumber cultivars to the vegetable leafminer (*Liriomyza sativae* Blanchard) (Diptera: Agromyzidae) in greenhouse. Chil J Agr Res 71(3):395–400

Bidarnamani F, Sanatgar E, Shabanipoor M (2015) Spatial distribution pattern of *Tetranychus urticae* Koch (Acari: Tetranychidae) on different rosa cultivars in greenhouse Tehran. J Ornam Plants 5(3):175–182

Capinera JL (2001) Handbook of vegetable pests. Academic Press, New York

Dashtbani KH, Baniameri V, Shojaei M et al (2013) Evaluation of *Diglyphus isaea* (Hymenoptera: Eulophidae) (Miglyphus®) for biological control of *Liriomyza sativae* (Dip.: Agromyzidae) on Greenhouse cucumber. Appl Entomol Phytopathol 81(1):31–41

Dehghani M, Ahmadi K (2013) Anti-oviposition and repellence activities of essential oils and aqueous extracts from five aromatic plants against greenhouse whitefly *Trialeurodes vaporariorum* Westwood (Homoptera: Aleyrodidae). Bulg J Agric Sci 19(4):691–696

Dehghani S, Talebi AA, Hajiqanbar HR (2015) Comparison of life table parameters of tomato leafminer, *Tuta absoluta* (Lepidoptera: Gelechiidae), on three tomato cultivar in laboratory conditions. In: Abstract presented at the 1st Iranian International Congress of Entomology, Iranian Research Institute of Plant Protection, Tehran, 29–31 August 2015, p 328

Doi M, Zen S, Okuda M et al (2003) Leaf necrosis disease of lisianthus (*Eustoma grundiflorum*) caused by Iris yellow spot virus. Japan J Phytopathol 69:181–188

Ebrahimifar J, Jamshidnia A, Allahyari H (2015) Observations on host-feeding and fecundity of *Eretmocerus delhiensis* Mani on greenhouse whitefly, *Trialeurodes vaporariorum* Westwood. In: Abstract presented at the 1st Iranian International Congress of Entomology, Iranian Research Institute of Plant Protection, Tehran, 29–31 August 2015, p 357

Eshrati M, Asgari S, Moeeni Naghade N et al (2014) Demography of western flower thrips, *Frankliniella occidentalis* Pergande (Thysanoptera, Thripidae), on greenhouse cucumber, Viola variety. In: Abstract presented at the 21st Plant Protection Congress, University of Urmia, Urmia, 23–26 August 2014, vol. I. Pests, p 675

FAO (2013) Good Agricultural Practices for Greenhouse Vegetable Crops: Principles for Mediterranean Climate Areas, 217th edn. Food and Agriculture Organization of United Nations, Rome

Farazmand A, Fathipour Y, Kamali K (2012) Functional response and mutual interference of *Neoseiulus californicus* and *Typhlodromus bagdasarjani* (Acari: Phytoseiidae) on *Tetranychus urticae* (Acari: Tetranychidae). Int J Acarol 38(5):369–376

Fathipour Y, Jalilian F, Talebi AA et al (2005) Voracity of larvae of three hoverfly species (Dip.: Syrphidae) as potential biological control agents of *Myzus persicae* (Hom.: Aphididae) on greenhouse crops. Integrated Control in Protected Crops, Temprate Climate. IOBC/wprs Bull 28(1):91–94

Fathipour Y, Haghani M, Talebi AA et al (2006) Natural parasitism of *Liriomyza sativae* (Diptera: Agromyzidae) on cucumber under field and greenhouse conditions. IOBC/WPRS Bull 29:155–160

Fathipour Y, Karimi M, Farazmand A et al (2018) Age-specific functional response and predation capacity of *Phytoseiulus persimilis* (Phytoseiidae) on the two-spotted spider mite. Acarologia 58 (1):31–40

Ganjisaffar F, Fathipour Y, Kamali K (2011) Temperature-dependent development and life table parameters of *Typhlodromus bagdasarjani* (Phytoseiidae) fed on two-spotted spider mite. Exp Appl Acar 55:259–272

Ghadamyari M, Hiroshi M, Suenghyup OH et al (2008) Studies on pirimicarb resistance mechanisms in Iranian populations of the peach-potato aphid, *Myzus persicae*. Appl Entomol Zool 43 (1):149–157

Gharekhani Gh, Hakimi H, Eshan N (2014) Evaluation of the stage mortality, survival rate and life expectancy of tomato leaf miner *Tuta absoluta* (Meyric) (Lepidoptera: Gelechiidae) under different temperatures. In: Abstract presented at the 21st Plant Protection Congress, University of Urmia, Urmia, 23–26 August 2014, vol. I. Pests, p 476

Gholami Z, Sadeghi A, Sheikhi Garjan A et al (2015) Susceptibility of western flower thrips *Frankliniella occidentalis* (Thysanoptera: Thripidae) to some synthetic and botanical insecticides under laboratory conditions. J Crop Prot 4:627–632

Ghorbanian M, Fathipour Y, Talebi AA et al (2019) Different pepper cultivars affect performance of second (*Myzus persicae*) and third (*Diaeretiella rapae*) trophic levels. J Asia-Pac Entomol 22:194–202

Ghotbi T, Baniameri V (2006) Identification and determination of transmission ability of thrips species as vectors of two tospovirus, tomato spotted wilt virus (TSWV) andimpatiens necrotic spot virus (INSV) on ornamental plants in Iran. Integrated Control in Protected Crops, Mediterranean Climate. IOBC/wprs Bull 29(4):297

Golmohammadi G, Mohammadipour A (2015) Efficacy of herbal extracts and synthetic compounds against strawberry thrips, *Frankliniella occidentalis* (Pergande) under greenhouse conditions. J Entomol Zool Stud 3(4):42–44

Grinberg M, Perl-Treves R, Palevsky E et al (2005) Interaction between cucumber plants and the broad mite, *Polyphagotarsonemus latus* from damage to defense gene expression. Entomol Exp Appl 115(1):135–144

Haghani M, Fathipour Y, Talebi AA et al (2006) Comparative demography of *Liriomyza sativae* Blanchard (Diptera: Agromyzidae) on cucumber at seven constant temperatures. Insect Sci 13:477–483

Haghani M, Fathipour Y, Talebi AA et al (2007) Thermal requirement and development of *Liriomyza sativae* (Diptera: Agromyzidae) on cucumber. J Econ Entomol 100(2):350–356

Hagvar EB, Hofsvang T (1991) Aphid parasitoids (Hymenoptera: Aphidiidae): biology, host selection, and use in biological control. Biocontrol News Inform 12:13–41

Hajiramezani-Chaleshtori MR, Nouri-Ganbalani G, Razmjoue J et al (2014) Comparison of the intrinsic rate of increase and mean relative growth rate of cotton aphid, *Aphis gossypii* Glover on two cultivars of cucumber under the laboratory conditions. In: Abstract presented at the 21st Plant Protection Congress, University of Urmia, Urmia, 23–26 August 2014, vol. I. Pests, p 780

Hamedi N, Fathipour Y, Saber M (2010) Sublethal effects of fenpyroximate on life table parameters of the predatory mite *Phytoseius plumifer*. BioControl 55:271–278

Hashemi Tasuji Z, Safaralizadeh MH, Aramidae S et al (2014) Host preference of tomato leafminer moth *Tuta absoluta* (Meyrick) (Lepidoptera: Gelechiidae) on the foure plant of Solanacea family, in the greenhouse conditions. In: Abstract presented at the 21st Plant Protection Congress, University of Urmia, Urmia, 23–26 August 2014, vol. I. Pests, p 667

Hassankhani Kh, Allahyari H, Saei Dehghan M (2014) Predation rate of *Hippodamia variegata* (Col.: Coccinellidae) preying on densities of *Myzus persicae* (Hem.: Aphididae) under laboratory conditions. In: Abstract presented in the 3rd Integrated Pest Management Conference (IPMC), University of Kerman, Kerman, 21–22 January 2014, p 441

Helle W, Sabelis MW (1985) Spider mites, their biology, natural enemies and control, vol 1. Elsevier, Amsterdam, p 458

Hosseini SA, Pourmirza AA (2010) Impacts of pyriproxyfen on the efficacy of *Encarsia inaron* Walker (Hym: Aphelinidae) on control of *Trialeurodes vaporariorum* Westwood (Hom: Aleyrodidae). Munis Entomol Zool 5:1119–1124

Hosseini M, Ashouri A, Enkegaard A et al (2010) Plant quality effects on intraguild predation between *Orius laevigatus* and *Aphidoletes aphidimyza*. Entomol Exp Appl 135:208–216

Hosseininia A, Malkeshi SH (2009) Comparison of some control methods of *Thrips tabaci* Lindeman on carnation under greenhouse condition. Res Reconst Agron Hort 81:140–146

Jafarbeigi F, Samih MA, Zarabi M et al (2014) Sublethal effects of some synthetic and botanical insecticides on Bemisia tabaci (Hemiptera: Aleyrodidae). Anthropods 3(3):127–137

Jafari S, Fathipour Y, Faraji F et al (2010) Demographic response to constant temperatures in *Neoseiulus barkeri* (Phytoseiidae) fed on *Tetranychus urticae* (Tetranychidae). Syst Appl Acarol 15:83–99

Jafari S, Fathipour Y, Faraji F (2011) Re-description of *Amblyseius Meghriensis* arutunjan and *Typhlodromus haiastanius* (Arutunjan) With discussion on using preanal pores as a character in the subgenus *Anthoseius* (Mesostigmata: Phytoseiidae). Int J Acarol 37(3):244–254

Jafari S, Fathipour Y, Faraji F (2012) Temperature-dependent development of *Neoseiulus barkeri* (Acari: Phytoseiidae) on *Tetranychus urticae* (Acari: Tetranychidae) at seven constant temperatures. Insect Sci 19:220–228

Jafari S, Abassi N, Bahirae F (2013) Demographic parameters of *Neoseiulus barkeri* (Acari: Phytoseiidae) fed on *Thrips tabaci* (Thysanoptera: Thripidae). Pers J Acarol 2(2):287–296

Jafarinasab B (2015) The effect of DEMI-001 isolate of *Metarhizium anisopliae* on Aphis Rose (Hemiptera: Aphididae) at different temperatures. Int J Farm Allied Sci 4(6):496–498

Jafarinasab B, Rajabi R, Gholamian E (2015) Comparison of neonicotinoid with commonly insecticides effect on population mortality of Rose Aphid (*Macrosiphum rosae* L.). Adv Environ Biol 9(3):874–879

Jalali MA, Tirry L, Arbab A et al (2010) Temperature-dependent development of the two-spotted ladybeetle, *Adalia bipunctata,* on the green peach aphid, *Myzus persicae,* and a factitious food under constant temperatures. J Insect Sci 10(124):1–14

Jalili Moghadam M, Azmayeshfard P (2004) Thrips of ornamental plants in Tehran and Mahallat. In: Abstract presented in the 16th Iranian plant protection congress, University of Tabriz, Tabriz, 28-1 August-September 2004, vol. I. Pests, p 16

Jalilian F (2015) Development and feeding capacity of *Scaeva albomaculata* (Macqaurt) (Diptera: Syrphidae) fed with rose aphid, *Macrosiphum rosae* (Homoptera: Aphididae) development and feeding capacity of *Scaeva albomaculata* (Macqaurt) (Diptera: Syrphidae) fed with rose aphid, *Macrosiphum rosae* (Homoptera: Aphididae). Biolo Forum – Int J 7(1):1377–1381

Javadi Khederi S, Hosseini MA, Khanjani M et al (2014) Role of different trichome style in the resistance of various tomato genotypes to tomato leaf miner *Tuta absoluta* (Meyrick) (Lepidoptera: Gelechiidae) in greenhouse condition. In: Abstract presented at the 21st Plant Protection Congress, University of Urmia, Urmia, 23–26 August 2014, vol. I. Pests, p 432

Karami A, Fathipour Y, Talebi AA et al (2018a) Canola quality affects second (*Brevicoryne brassicae*) and third (*Diaeretiella rapae*) trophic levels. Arthropod-Plant Inte 12:291–301

Karami A, Fathipour Y, Talebi AA et al (2018b) Parasitism capacity and searching efficiency of *Diaeretiella rapae* parasitizing *Brevicoryne brassicae* on susceptible and resistant canola cultivars. J Asia-Pac Entomol 21(4):1095–1101

Kashkouli M, Khajehali J, Poorjavad N (2014) Effect of entomopathogenic nematodes for controlling the onion thrips, *Thrips tabaci* (Thysanoptera: Thripidae) under semi-field condition. In: Abstract presented at the 21st Plant Protection Congress, University of Urmia, Urmia, 23–26 August 2014, vol. I. Pests, p 457

Keshavarz M, Seiedy N, Allahyari H (2015) Preference of two population of *Popylea quatuordecimounctata* (Coleoptera: Coccinellidae) for *Aphis fabae* and *Aphis gossypii* (Homoptera: Aphididae). Eur J Entomol 112(3):560–563

Khanamani M, Fathipour Y, Hajiqanbar HR (2013) Population growth response of *Tetranychus urticae* to eggplant quality: application of female age-specific and age-stage, two-sex life tables. Int J Acarol 39(8):638–648

Khanamani M, Fathipour Y, Talebi AA et al (2017) Quantitative analysis of long-term mass rearing of *Neoseiulus californicus* (Acari: Phytoseiidae) on Almond Pollen. J Econ Entomol 110 (4):1442–1450

Khanjani M (2005) Field crop pests (insects and mites) in Iran. Abu-Ali Sina University Press, Hamedan

Kheradpir N, Khalghani J, Ostovan H et al (2006) Feeding rate of *Stethorus gilvifrons* on *Tetranychus urticae* in three greenhouse cucumber cultivars with different resistance levels. Integrated Control in Protected Crops, Mediterranean Climate. IOBC/wprs Bull 29(4):139–143

Kheradpir N, Rezapanah MR, Kamali K et al (2009) Physical structures of host plants affect preference behaviour of predaceous thrips, *Scolothrips longicornis* (Thysanoptera: Thripidae). J Biol Control 23(2):131–135

Kiani L, Yazdanian M, Tafaghodinia B et al (2012) Control of western flower thrips, *Frankliniella occidentalis* (Pergande) (Thysanoptera: Thripidae), by plant extracts on strawberry in greenhouse conditions. Munis Entomol Zool 7(2):857–866

Kirk WDJ (2001) The pest and vector from the west: *Frankliniella occidentalis*. In: Paper presented in the the the 7Th International Symposium on Thysanoptera, University of Reggio Calabria, Reggio Calabria, 2–7 July 2001, p 33–42

Kouhjani Gorji M, Fathipour Y, Kamali K (2009) The effect of temperature on the functional response and prey consumption of *Phytoseius plumifer* (Acari: Phyto seiidae) on the two-spotted spider mite. Acarina 17(2):231–237

Kouhjani Gorji M, Fathipour Y, Kamali K (2012) Life table parameters of *Phytoseius plumifer* (Phytoseiidae) fed on two-spotted spider mite at different constant temperatures. Int J Acarol 38 (5):377–385

Koul O, Dhaliwal GS (2003) Predators and parasitoids. Taylor & Francis, London

Koupi N, Ghazavi M, Kamali K et al (2016) Virulence of Iranian isolates of *Metarhizium anisopliae* on western flower thrips, *Frankliniella occidentalis* (Pergande) (Thysanoptera: Thripidae). J Crop Prot 5(1):131–138

Madahi K, Sahragard A, Hosseini R (2013) Influence of *Aphis gossypii* Glover (Hemiptera: Aphididae) density on life table parameters of *Aphidoletes aphidimyza* Rondani (Diptera: Cecidomyiidae) under laboratory conditions. J Crop Prot 2(3):355–368

Madjdzadeh SM, Mehrparvar M (2009) Morphological discrimination of geographical populations of *Macrosiphoniella sanborni* (Gillette, 1908) (Hem.: Aphididae) in Iran. North-West J Zool 5 (2):338–348

Mahdavi T, Madadi H (2015) Determination predation and oviposition rate of *Nabis pseudoferus* Remanefed on egg and 1st instar larvae of *Tuta absoluta* Meyrick. In: Abstract presented at the 1st Iranian International Congress of Entomology, Iranian Research Institute of Plant Protection, Tehran, 29–31 August 2015, p 287

Mahr SER, Cloyd RA, Mahr DL et al (2001) Biological control of insects and other pests of greenhouse crops. North Central Regional Publication 581, University of Wisconsin, Madison

Makareminia G, Jalalizand A, Khajehali J et al (2014) The effect of some biorational insecticides on cotton aphid (*Aphis gossypii*) in laboratory and greenhouse conditions. Int J Agric Innov Res 2 (5):2319–1473

Makkouk KM (1978) A study on tomato viruses in the Jordan Valley with special emphasis on tomato yellow leaf curl. Plant Dis Rep 62:259–262

Maleknia B, Fathipour Y, Soufbaf M (2016a) How greenhouse cucumber cultivars affect population growth and two-sex life table parameters of *Tetranychus urticae* (Acari: Tetranychidae). Int J Acarol 42(2):70–78

Maleknia B, Fathipour Y, Soufbaf M (2016b) Intraguild predation among three phytoseiid species, *Neoseiulus barkeri*, *Phytoseiulus persimilis* and *Amblyseius swirskii*. Sys Appl Acarol 21:417–426

Malkeshi SH, Hosyni A (2014) Biological control of thrips on ornamental crops. In: Paper presented in the 1st National Ornamental Plants Congress, Seed and Plant Improvement Institute, Karaj, 21–22 October 2014, p 1–8

Malkeshi SH, Shahrokhi S, Nori H (2015) A study on the efficiency of *Aphidoletes aphidimyza* and *Aphidius colemani* for controlling *Aphis gossypii* on commercial greenhouse cucumber. In: Abstract presented at the 1st Iranian International Congress of Entomology, Iranian Research Institute of Plant Protection, Tehran, 29–31 August 2015, p 296

Malumphy CP (2003) The status of *Bemisia afer* (Priesner and Hosny) in Britain (Homoptera: Aleyrodidae). Entomol Gaz 54:191–196

Mardani-Talaee M, Nouri-Ganblani G, Razmjou J et al (2015) Effects of chemical, organic and biofertilizers on life table parameters of the green peach aphid, *Myzus persicae* (Sulzer) (Hem.: Aphididae), on bell pepper, *Capsicum annuum* L. In: Abstract presented at the 1st Iranian International Congress of Entomology, Iranian Research Institute of Plant Protection, Tehran, 29–31 August 2015, p 363

Mehrparvar M, Hatami B (2007) Effect of temperature on some biological parameters of an Iranian population of the rose aphid, *Macrosiphum rosae* (Hemiptera: Aphididae). Eur J Entomol 104:631–634

Mehrparvar M, Hatami B, Stary P (2005) Report of *Aphidius rosae* (Hym.: Braconidae), a parasitoid of rose aphid, *Macrosiphum rosae* (Hom.: Aphididae) from Iran. J Entomol Soc Iran 25(1):63–64

Mirzamohammadzadeh S, Iranipour S, Lotfalizadeh H et al (2014) Biological parameters of *Trialeurodes vaporariorum* (Hem.: Aleyrodidae) in four greenhouse cucumber cultivars. J Entomol Soc Iran 34(4):53–67

Moayeri HRS, Mohandesi AR, Ashouri A (2013) The effect of interaguild predation on avoidance behavior of the aphidophagous midge, *Aphidoletes aphidimyza* (Dip.: Cecidomyiidae) on its encounter with the predatory bug *Orius laevigatus* (Het.: Anthocoridae). J Entomol Soc Iran 32 (2):35–48

Modarres Awal M (2001) List of agricultural pests and their natural enemies in Iran. Ferdowsi University Press, Mashhad

Modarres Najafabadi SS (2012) Resistance to *Tetranychus urticae* Koch (Acari: Tetranychidae) in *Phaseolus vulgaris* L. Middle-East J Sci Res 11(6):690–701

Modarres Najafabadi S (2014) Effect of various vermicompost-tea concentrations on life table parameters of *Macrosiphum rosae* L. (Hemiptera: Aphididae) on rose (Rosa hybrida L.) flower. J Ornam Plants 4(2):81–92

Mohammadi Z, Rasekht A, Kocheilil F et al (2014) Evaluation of biological characteristics of sexeual population of *Lysiphlebus fabarum* (Marshall), reared in different instar hosts (the black bean aphid, *Aphis fabae* Scopoli). In: Abstract presented at the 21st Plant Protection Congress, University of Urmia, Urmia, 23–26 August 2014, vol. I. Pests, p 480

Mohammadi S, Seraj AA, Rajabpour A (2015) Effects of six greenhouse cucumber cultivars on reproductive performance and life expectancy of *Tetranychus turkestani* (Acari: Tetranychidae). Acarologia 55(2):231–242

Mohseni L, Rasekh A, Kocheili F, et al (2014) The effect of host density on superparasitism and larval competition in a thelytokous population of *Lysiphlebus fabarum* Marshall (Hym., Braconidae). In: Abstract presented at the 21st Plant Protection Congress, University of Urmia, Urmia, 23–26 August 2014, vol. I. Pests, p 626

Mojib Hagh Ghadam Z, Yousefpour M (2012) Effects of feeding from different hosts on biological parameters of the lady beetle *Hippodamia variegata* (Goeze) in the laboratory conditions. Int J Agric Crop Sci 4(12):755–759

Mollashahi M, Sahragard A, Hosseini R (2009) A comparative study on the population growth parameters of *Coccinella septempunctata* (Col.: Coccinellidae) and melon aphid, *Aphis gossypii* (Hem.: Aphididae) under laboratory conditions. J Entomol Soc Iran 29(1):1–12

Mollashahi M, Sabouri H, Sedghi M (2012) A comparative study on product index (PI) of lady beetles, *Hippodamia variegata* and *Coccinella septempunctata* feeding on wheat green aphid, cabbage aphid and melon aphid under laboratory conditions. J Iran Plant Pests Res 1(1):1–16

Mortazavi N, Fathipour Y, Talebi AA (2015) Functional response of *Typhlodromus bagdasarjani* (Acari Phytoseiidae) on eggs of the twospotted spider mite strawberry. In: Abstract presented at the 1st Iranian International Congress of Entomology, Iranian Research Institute of Plant Protection, Tehran, 29–31 August 2015, p 319

Motahari M, Kheradmand K, Roustaee AM et al (2014) The impact of cucumber nitrogen nutrition on life history traits of Tetranychus urticae (Koch) (Acari: Tetranychidae). Acarol 54 (4):443–452

Motazedian N, Ravan S, Bandani AR (2012) Toxicity and repellency effects of three essential oils against Tetranychus urticae Koch (Acari: Tetranychidae). J Agr Sci Tech 14:275–284

Namvar P, Safaralizadeh MH, Baniameri V et al (2011) Spatial distribution and fixed-precision sequential sampling of Liriomyza sativae Blanchard (Diptera: Agromyzidae) on cucumber greenhouse. Middle-East J Sci Res 10(2):157–163

Nemati A, Zahiri B, Khanjani M (2015) The effect of Jasmonic acid and Salicylic acid on induced resistance of three different tomato cultivars to Tuta absoluta. In: Abstract presented at the 1st Iranian International Congress of Entomology, Iranian Research Institute of Plant Protection, Tehran, 29–31 August 2015, p 373

Nojumian F, Sabahi Q, Talaei-Hassanloui R, Darvishzadeh A (2015) Sublethal effects of spirodiclofen on life table parameters of minute pirate bug Orius niger Wolff (Hemiptera: Anthocoridae). J Entomol Zool Stud 3(1):227–232

Norouzi Y (2011) Compatibility of the parasitoid Aphidius matricariae with BotaniGard for the control of greenhouse aphids. M. Sc.Dissertation, University of Simon Fraser, Canada, p 158

Nuez F, Pico B, Iglesias A et al (1999) Genetics of melon yellows virus resistance derived from Cucumis melo ssp. agrestis. Eur J Plant Pathol 105(5):453–464

Pahlevan R, Omid M, Akram A (2012) The relationship between energy inputs and crop yield in greenhouse basil production. J Agr Sci Tech 14:1243–1253

Pakyari H, Fathipour Y, Rezapanah KM et al (2008) Prey-stage preference in Scolothrips longicornis Priesner (Thysanoptera: Thripidae) on Tetranychus urticae Koch (Acari: Tetranychidae). Integrated Control in Protected Crops. Temperature Climate IOBCΙ\vprs Bull 32:167–169

Parrella MP (1983) Intraspecific competition among larvae of Liriomyza trifolii (Diptera., Agromyzidae): effects on colony production. Environ Entomol 12:1412–1414

Parrella MP, Keil CB (1985) Toxicity of methamidophos to four species of Agromyzidae. J Agric Entomol 2:234–237

Pastrana JA (2004) Los Lepidopteros Argentinos – sus plantas hospederas yotros sustratos alimenticios. South American Biological Control Laboratory USDA-ARS and Sociedad Entomológica Argentina, Buenos Aires

Perdikis D, Kapaxidi E, Papadoulis G (2008) Biological control of insect and mite pests in greenhouse solanaceous crops. Eur J Plant Sci Biotech 2(1):125–144

Perumal Y, Marimuthu M (2009) Host plant mediated population variations of cotton whitefly Bemisia tabaci Gennadius (Aleyrodidae: Homoptera) characterized with random DNA markers. Am J Biochem Biotechnol 5:40–46

Pinto ZV, Rezende JAM, Yuki VA et al (2008) Ability of Aphis gossypii and Myzus persicae to transmit cucumber mosaic virus in single and mixed infection with two potyviruses to zucchini squash. Summa Phytopathol 34(2):183–185

Pirmoradi Amozegarfard N, Sheikhigarjan A, Baniameri V et al (2013) Evaluation of susceptibility of the first instar nymphs and adults of Trialeurodes vaporariorum (Hemiptera: Aleyrodidae) to neonicotinoid insecticides under laboratory conditions. J Entomol Scoc Iran 31(1):13–24

Pourian HR, Mirab-balou M, Alizadeh M et al (2009) Study on biology of onion the, Thrips tabaci Lindeman (Thysanoptera: Thripidae) on cucumber (Var. Sultan) in laboratory conditions. J Plant Prot Res 49(4):390–394

Pourtaghi E, Rashki M, Shirvani A (2014) Study of life table parameters Aphidius matricariae (Hym.: Aphidiidae) on Aphis fabae Scopoli (Hem.:Aphididae). In: Abstract presented at the 21st Plant Protection Congress, University of Urmia, Urmia, 23–26 August 2014, vol. I. Pests, p 527

Rahsepar A, Haghani M, Sedaratian-Jahromi A et al (2016) Different cucumber (Cucumis sativus) varieties could affects biological performance of cotton aphid, Aphis gossypii Glover (Hemiptera: Aphididae), a case study at laboratory condition. Entomofauna 37(21):353–364

Rajabpour A, Seraj AA, Allahyari H et al (2011) Evaluation of *Orius laevigatus* Fiber (Heteroptera: Anthocoridae) for biological control of *Thrips tabaci* Lindeman (Thysanoptera: Thripidae) on greenhouse cucumber in south of Iran. Asian J Biol Sci 4:457–467

Rashki M, Shirvani A (2013) The effect of entomopathogenic fungus, *Beauveria bassiana* on life table parameters and behavioural response of *Aphis gossypii*. Bull Insecology 66(1):85–91

Rashki M, Kharazmi A, Pakdel A et al (2009) Interactions among the entomopathogenic fungus, *Beauveria bassiana* (Ascomycota: Hypocreales), the parasitoid, *Aphidius matricariae* (Hymenoptera: Braconidae), and its host, *Myzus persicae* (Homoptera: Aphididae). Biol Control 50:324–328

Rezaei M (2015) Comparison of life table parameters of the predatory mite, *Neoseiulus californicus* (McGregor) fed on two-spotted spider mite and western flower thrips. In: Abstract presented at the 1st Iranian International Congress of Entomology, Iranian Research Institute of Plant Protection, Tehran, 29–31 August 2015, p 349

Rezaei M, Askari S (2015) Effect of feeding with different plant pollens on biological parameters of *Neoseiulus barkeri* (Hughes) (Acari: Phytoseiidae). In: Abstract presented at the 1st Iranian International Congress of Entomology, Iranian Research Institute of Plant Protection, Tehran, 29–31 August 2015, p 350

Rezaei M, Saboori A, Baniameric V et al (2013) Susceptibility of *Tetranychus uticae* Koch (Acari: Tetranychidae) on seven strawberry cultivars. Int Res J Appl Basic Sci 4(9):2455–2463

Rezaei N, Karimi J, Hosseini M et al (2015) Pathogenicity of two species of entomopathogenic nematodes against the greenhouse whitefly, *Trialeurodes vaporariorum* (Hemiptera: Aleyrodidae), in laboratory and greenhouse experiments. J Nematol 47(1):60–66

Riahi E, Fathipour Y, Talebi AA et al (2016) Pollen quality and predator viability: life table of *Typhlodromus bagdasarjani* on seven different plant pollens and two-spotted spider mite. Syst Appl Acarol 21(10):1399–1412

Riahi E, Fathipour Y, Talebi AA et al (2017) Linking life table and consumption rate of *Amblyseius swirskii* (Acari: Phytoseiidae) in presence and absence of different pollens. Ann Entomol Soc Am 110(2):244–253

Riahi E, Fathipour Y, Talebi AA et al (2018) Factitious prey and artificial diets: do they all have the potential to facilitate rearing of *Typhlodromus bagdasarjani* (Acari: Phytoseiidae)? Int J Acarol 44(2–3):121–128

Rostami M, Zamani AA, Goldasteh S et al (2012) Influence of nitrogen fertilization on biology of *Aphis gossypii* (Hemiptera: Aphididae) reared on *Chrysanthemum indicum* (Asteraceae). J Plant Prot Res 52(1):118–121

Rostami E, Abbasipour H, Allahyari H (2015) Life table parameter of tomato leaf miner, *Tuta absoluta* (Meyrick) (Lepidoptera: Gelechiidae) on the whole plant under laboratory conditions. In: Abstract presented at the 1st Iranian International Congress of Entomology, Iranian Research Institute of Plant Protection, Tehran, 29–31 August 2015, p 311

Rostamian P, Hassanpour M, Rafiee-Dastjerdi H et al (2014) Age-dependent functional response of the predatory bug, *Orius laevigatus* (Fieber) to the melon aphid, *Aphis gossypii* Glover. In: Abstract presented at the 21st Plant Protection Congress, University of Urmia, Urmia, 23–26 August 2014, vol. I. Pests, p 799

Sabahi Q, Rasekh A, Michaud JP (2011) Toxicity of three insecticides to *Lysiphlebus fabarum*, a parasitoid of the black bean aphid, *Aphis fabae*. J Insect Sci 11(104):1–8

Saeidi Z, Mallik B, Nemati A et al (2012) Resistance of 14 accessions/cultivars of *Lycopersicon* spp. to two-spotted spider mite, *Tetranychus urticae* (Acari: Tetranychidae), in laboratory and greenhouse. J Entomol Soci Iran 32(1):93–108

Samareh Fekri M, Samih MA, Imani S et al (2013) Study of host preference and the comparison of some biological characteristcs of *Bemisia tabasi* (Genn.) on tomato varieties. J Plant Prot Res 53 (2):137–142

Sangak Sani Bozhgani N, Kheradmand K, Talebi AA (2018) The effects of spirotetramat on the demographic parameters of *Neoseiulus californicus* (Phytoseiidae). Syst Appl Acarol 23 (10):1952–1964

Seddigh S, Kiani L (2012) Greenhouse whitefly (*Trialeurodes vaporariorum* Westwood) control by *Encarsia formosa* Gahan and its color preference in commercial gerbera greenhouses in Iran. Ann Biol Res 3(5):2414–2418

Seyedebrahimi SS, Talebi Jahromi K, Imani S et al (2015) Characterization of imidacloprid resistance in *Aphis gossypii* (Glover) (Hemiptera: Aphididae) in Southern Iran. Turk J Entomol 39(4):413–423

Sharifian I, Sabahi Q, Khoshabi J (2015) Investigation on functional response type and predatory indices of *Macrolophus pygmaeus* (Rambur) feeding on *Ephestia kuehniella* (Zeller) and *Tuta absoluta* (Meyrick). In: Abstract presented at the 1st Iranian International Congress of Entomology, Iranian Research Institute of Plant Protection, Tehran, 29–31 August 2015, p 279

Shishehbor P, Zandi-Sohani N (2011) Investigation on functional and numerical responses of *Encarsia acaudaleyrodis* parasitizing *Bemisia tabaci* on cucumber. Biocontrol Sci Technol 21 (3):271–280

Shoorooei M, Nasertorabi M, Soleimani A et al (2012) Screening of some cucumber accessions to two-spotted spider mite (*Tetranychus urticae*). Int Res J Appl Basic Sci 3(8):1580–1584

Sohrabi F, Hosseini R (2015) *Nesidiocoris tenuis* (Reuter) (Heteroptera: Miridae), a predatory species of the tomato leafminer, *Tuta absoluta* (Meyrick) (Lepidoptera: Gelechiidae) in Iran. J P Prot Res 55(3):322–323

Sohrabi F, Lotfalizadeh H, Salehipour H (2014) Report of two larval parasitoids of *Tuta absoluta* (Meyrick) (Lepidoptera: Gelechiidae) from Iran. In: Abstract presented at the 21st Plant Protection Congress, University of Urmia, Urmia, 23–26 August 2014, vol. I. Pests, p 726

Soleimannejadian E, Nemati A, Shishehbor P et al (2006) Biology of the two spotted spider mite, *Tetranychus turkestani* (Acari: Tetranychidae) on four common varieties of eggplant in Iran. Bulletin OILB/SROP 29(4):115–119

Soleymani S, Hakimitabar M, Seiedy M (2015a) The predation rate of *Amblyseius swirskii* Athias-Henriot (Acari: Phytoseiidae) on *Tetranychus urticae* Koch (Acari: Tetranychidae) and *Bemisia tabaci* Gennadius (Hem: Aleyrodidae). In: Abstract presented at the 1st Iranian International Congress of Entomology, Iranian Research Institute of Plant Protection, Tehran, 29–31 August 2015, p 329

Soleymani S, Seiedy M, Hakimitabar M (2015b) Evaluation of life table parameters of *Amblyseius swirskii* Athias-Henriot (Acari: Phytoseiidae) fed on *Tetranychus urticae* Koch (Acari: Tetranychidae) and *Bemisia tabaci* Gennadius (Hem: Aleyrodidae). In: Abstract presented at the 1st Iranian International Congress of Entomology, Iranian Research Institute of Plant Protection, Tehran, 29–31 August 2015, p 330

Starý P (1970) Biology of aphid parasites (Hymenoptera: Aphidiidae) with respect to integrated control. Series entomologica, Hague

Taghizadeh R, Fathipour Y, Kamali K (2008) Influence of temperature on life-table parameters of *Stethorus gilvifrons* (Mulsant) (Coleoptera: Coccinellidae) fed on *Tetranychus urticae* Koch. J Appl Entomol 132:638–645

Tahriri Adabi S, Talebi AA, Fathipour Y et al (2010) Life history demographic parameters of *Aphis fabae* (Hemiptera: Aphididae) and its parasitoid, *Aphidius matricaria* (Hymenoptera: Aphidiidae) on four sugar beet cultivars. Acta Entomol Serbica 15(1):61–73

Takalloozadeh HM (2015) Effect of different prey species on the biological parameters of *Chrysoperla carnea* (Neuroptera: Chrysopidae) in laboratory conditions. J Crop Prot 4 (1):11–18

Talebi AA, Zamani AA, Fathipour Y et al (2006) Host stage preference by *Aphidius colemani* and *Aphidius matricariae* (Hymenoptera: Aphidiidae) as parasitoids of *Aphis gossypii* (Hemiptera: Aphididae) on greenhouse cucumber. Integrated Control in Protected Crops, Mediterranean Climate. IOBC/Wprs Bull 29:173–177

Tamoli Torfi E, Seraj AA, Rajabpour A (2016) Biological characteristics and population parameters of tomato leaf miner, *Tuta absoluta* (Lep.: Gelechiidae) on potato and tobacco plants under laboratory conditions. Plant Prot 38(4):79–88

Tavanapour SH, Emami MS, Behdad E (2009) Study on feeding and oviposition preference of *Liriomyza sativae* Blanchard (Dip.: Agromyzidae) on greenhouse cucumber cultivars. J Entomol Res 1(1):57–66

Tazerouni Z, Talebi AA, Fathipour Y et al (2016a) Bottom-up effect of two host plants on life table parameters of *Aphis gossypii* (Hemiptera: Aphididae). J Agr Sci Tech 18:179–190

Tazerouni Z, Talebi AA, Fathipour Y et al (2016b) Interference competition between *Aphidius matricariae* and *Praon volucre* (Hymenoptera: Braconidae) attacking two common aphid species. Biocontrol Sci Technol 26(11):1552–1564

Tazerouni Z, Talebi AA, Fathipour Y et al (2016c) Age-specific functional response of *Aphidius matricariae* and *Praon volucre* (Hymenoptera: Braconidae) on *Myzus persicae* (Hemiptera: Aphididae). Neotrop Entomol 45:642–651

Tazerouni Z, Talebi AA, Fathipour Y et al (2017) Age-specific functional response of *Aphidius matricariae* and *Praon volucre* (Hym.: Braconidae) on *Aphis gossypii* (Hem.: Aphididae). J Entomol Soc Iran 36(4):239–248

Thungrabeab M, Blaeser P, Sengonca C (2006) Possibilities for biocontrol of the onion thrips *Thrips tabaci* Lindeman (Thysan., Thripidae) using different entomopathogenic fungi from Thailand. Mitt Dtsch Ges Allg Angew Entomol 15:299–304

Valizadeh M, Deraison C, Kazemitabar SK et al (2013) Aphid resistance in florist's chrysanthemum (*Chrysanthemum morifolium* Ramat.) induced by sea anemone equistatin overexpression. Afr J Biotechnol 12(50):6922–6930

van Emden HF, Harrington R (eds) (2007) Aphids as crop pests. CABI, Wallingford

Vargas H (1970) Observaciones sobre la biologia enemigos naturales de las polilla del tomate, *Gnorimoschema absoluta* (Meyrick). Idesia 1.75–110

Waterhouse DF, Norris KR (1987) Biological control: Pacific prospects. Inkata Press, Melbourne

Yarahmadi F, Rajabpour A, Zandi Sohani N et al (2013) Investigating contact toxicity of *Geranium* and *Artemisia* essential oils on *Bemisia tabaci* Gen. Avicenna J Phytomed 3(2):106–111

Zamani AA, Talebi AA, Fathipour Y et al (2006) Effect of temperature on biology and population growth parameters of *Aphis gossypii* Glover (Hom. Aphididae) on greenhouse cucumber. J Appl Entomol 130(8):453–460

Zamani AA, Talebi AA, Fathipour Y et al (2007) Effect of temperature on life history of *Aphidius colemani* and *Aphidius matricariae* (Hymenoptera: Braconidae), two parasitoids of *Aphis gossypii* and *Myzus persicae* (Homoptera: Aphididae). Environ Entomol 36(2):263–271

Zamani AA, Haghani M, Kheradmand K (2012) Effect of temperature on reproductive parameters of *Aphidius colemani* and *Aphidius matricariae* (Hymenoptera: Braconidae) on *Aphis gossypii* (Hemiptera: Aphididae) in laboratory conditions. J Crop Prot 1(1):35–40

Zhang ZQ (2003) Mites of greenhouses: identification, biology and control. CABI Publishing, Oxon

# Chapter 16
# Biocontrol for Arthropods of Medical and Veterinary Importance in Iran

Hana Haji Allahverdipour and Kamran Akbarzadeh

## 16.1 Introduction

Having been situated at Palaearctic region and under influence of Oriental and Afrotropical zoogeographical regions, Iran has diverse fauna of medically important arthropods and consequently, different kinds of arthropod-borne diseases.

Diseases of medical and veterinary importance in the country include malaria, leishmaniasis, myiasis, West Nile virus, Crimean-Congo hemorrhagic fever, pappataci fever, Q fever, scabies, some diseases with silent foci such as tularemia and recently reported diseases such as dengue fever (Askarian et al. 2012; Alizadeh et al. 2014; Farhadpour et al. 2016; Seyedi Arani et al. 2016; Soleimani-Ahmadi et al. 2017; Dehghani et al. 2018; Nejati et al. 2018; Esmaeili et al. 2019).

Historic milestones of malaria vector biocontrol include first introduction of exotic larvivorous fish in the North (1920), establishment and effectiveness of mosquito fish in the South (1966) and successful field trials of Bti (1986) in the country.

The development of insecticide resistance in pest and vector population, the hazards on wildlife and non-target organisms and the realization of other environmental concerns of chemical insecticides as well as stability of vector-borne disease have led to an increasing interest in biological control within integrated pest management (IPM) programs (Moazami 2009).

H. H. Allahverdipour
Biological Control Department, Iranian Research Institute of Plant Protection, Agricultural Research, Education and Extension Organization, Tehran, Iran
e-mail: h.hajiallahverdi@areeo.ac.ir

K. Akbarzadeh (✉)
Department of Medical Entomology and Vector Control, School of Public Health, Tehran University of Medical Sciences, Tehran, Iran
e-mail: kakbarzadeh@tums.ac.ir

© The Author(s), under exclusive license to Springer Nature Switzerland AG 2021
J. Karimi, H. Madadi (eds.), *Biological Control of Insect and Mite Pests in Iran*,
Progress in Biological Control 18, https://doi.org/10.1007/978-3-030-63990-7_16

## 16.2 Arthropod-Borne Diseases

### 16.2.1 Malaria and West Nile Virus

Malaria is one of the oldest parasitic diseases of the country where it is recorded as endemic (Vatandoost et al. 2019). The disease was pandemic more than half a century ago but now is common in some parts in South and Southeast of the country. Its control measures was launched since 1949 and a malaria eradication campaign was initiated in 1951 (Berger 2019). Shortly, malaria was eradicated from majority parts of the land. However, Southeastern Iran is still characterized by "refractory malaria" affected by socioeconomic, geographical, climatic factors and vector diversity (Vatandoost et al. 2019; Hanafi-Bojd et al. 2011).

There are 31 anopheline species in Iran while 8 of these species are proved to be malaria vectors (Azari-Hamidian 2007; Hanafi-Bojd et al. 2011). Herein, the malaria vectors are composed of *Anopheles stephensi, An. culicifacies, An. fluviatilis, An. superpictus, An. d'thali, An. sacharovi, An. maculipennis* and *An. pulcherrimus*. All these species show endo/exophilic and endo/exophagic behaviour (Mehravaran et al. 2012; Schapira et al. 2018).

Vector control is the key strategy to reduce malaria transmission which is primarily based on chemical control. Malaria vectors control in the country is focused on Indoor residual spraying (IRS) using Lambda-cyhalothrin and Deltamethrin, targeting larval breeding sites using Chlorpyrifos-methyl, application of *Bacillus thuringiensis israelensis*, employment of larvivorous fish including *Aphanius dispar* (Cyprinodontidae) and *Gambusia affinis* (Poeciliidae) and distributing long-lasting insecticidal nets (LLINs) (Hanafi-Bojd et al. 2010). IRS increases the direct contact of humans and animals with insecticides (Gyalpo et al. 2012). Among many disadvantages of chemical control is insecticide resistance in vectors of malaria (Abbasi et al. 2019). Main malaria vectors of the country are resistant to some of the pyrethroids (Vatandoost and Hanafi-Bojd 2012). Reviews on insecticide resistance of *Anopheles* mosquitoes indicate *An. stephensi* is resistant to Dieldrin and Malathion and tolerant to Deltamethrin, Bendiocarb and Fenthion (Mogaddam et al. 2016; Hanafi-Bojd et al. 2012). *An. culicifacies* act as the main vector in the areas which are under pressure of IRS while the main vector of malaria in the country is *An. stephensi* (Pakdad et al. 2017).

However, ineffective chemical control in addition to environmental and health hazards of this strategy inspired scientists to find some applicable biological control agents. More than two hundred pathogens and parasites and more than five hundred predators have been listed as natural enemies of mosquitoe larvae and adults (Bay et al. 1976). Biological control of malaria vectors continues to be a component of Integrated Vector Management (IVM).

### 16.2.1.1   Predators of Malaria Vectors

The main invertebrate predators of mosquito larvae are *Toxorhynchites* (Diptera: Culicidae), Dytiscidae (Coleoptera), Belastomatidae (Hemiptera), Libellulidae and Coenagrionidae (Odonata). Potential of the predators for reducing the mosquito populations as natural regulators is highly varied with some ineffective predator species. While their establishment in mosquito habitats (Service 1981) and successful mass rearing are quite challenging, their conservation is of main priority.

#### 16.2.1.1.1   Larvivorous Fishes

Larvivorous fishes are the most famous and applicable predators of mosquito larvae. Fish families recommended for biological control of mosquitoes are Cyprinidac, Cyprinodontidae and Cichlidae (WHO 2013a). These predators have got more attention among natural enemies of mosquito larvae to use in biological control programs in Iran. Release of larvivorous fish, *G. affinis*, *Carassius auratus* (Cyprinidae) in larval breeding sites integrated with other control tactics in Kermanshah, Fars and Bushehr provinces, decreased the populations of *Anopheles* larvae and malaria transmission dramatically in 1969 (Tabibzadeh et al. 1970). *Gambusia* sp., *A. dispar*, *C. auratus*, *Colisa lalia* (Osphronemidae), *Danio rerio* (Cyprinidae), *Poecilia reticulata* (Poeciliidae) and *Oreochromis mossambica* (Cichlidae) are present in different regions of the land (Jafari et al. 2019). There were three species of *A. dispar, Aphanius* sp. and *Gambusia holbrooki* mosquito fish in southern Iran which seem to have important role in malaria vector control. Malaria control in Iran by biological methods, included the highly successful method of transporting fish in wooden barrels by trucks during the night. *Gambusia* sp. has been used for mosquito control the last 40 years in Sistan and Baluchistan province and *A. dispar* is active in Hormozgan and Kerman (Kahnouj) provinces (WHO 2003).

Tooth-carp fish, *Aphanius* spp. are among native larvivorous fish with ten known species inside the genus in Iran (Shahi et al. 2015) which can live under diverse ecological conditions from freshwater to marine ecosystems (Esmaeili and Shiva 2006). Conservation of *A. dispar* is more sensible strategy than introduction of *G. holbrooki* as an alien species in mosquito breeding sites (Shahi et al. 2015).

*G. affinis*, native to Southeastern United States, is the best known larvivorous fish (Bay et al. 1976). This fish is viviparous and does not possess any food or economic value. It has been transplanted progressively to various countries. *G. affinis was* introduced from Italy into the Ghazian marshes, Caspian littoral in Iran during 1922–1930. After solving initial technical problems, this fish was successfully used in malaria eradication. After several years, this species was established through the country (Jouladeh Roudbar et al. 2015).

*Gambusia* have been adapted to various climates and ecological conditions of the south and the southeastern parts of Iran and act as a natural predator on mosquito larvae without mass rearing and release (Tabibzadeh et al. 1970). However, investigation into native larvivorous fish in southern Iran showed that the presence of *G. holbrooki* resulted in reduced biodiversity in breeding sites. The use of *Gambusia* fish in malaria control programs is a serious threat for native species like *Aphanius* in its distribution area and employment of native mosquito fish is much more recommended than non-native species in malaria control (Shahi et al. 2015). Ecology and behavior of larvivorous fish and mosquito larvae should be considered in every release strategy (Arthington and Marshall 1999).

*Gambusia* species are usually not very effective in reducing malaria transmission, although they are more likely to reduce malaria in arid areas in countries such as Iran, Afghanistan, Somalia and Ethiopia where larval habitats are discrete and permanent (Service and Service 2012).

Four species of carnivorous plant, *Utricularia* (Lentibulariaceae) including *U. ochroleuca*, *U. minor*, *U. vulgaris* and *U. australis*, collected from wetlands and lakes situated in various climates of Iran, have been reported (Dinarvand 2012). This genus can trap and ingest mosquito larvae.

### 16.2.1.2 Parasites and Pathogens of Malaria Vectors

#### 16.2.1.2.1 Nematodes

Mermithids have received attention as biological alternative to chemical insecticides due to lethality to their hosts and potential for mass rearing (Alavo et al. 2015). Because of a host range that includes mosquitoes of public health importance, *Romanomermis iyengari*, *Strelkovimermis spiculatus* and *Romanomermis culicivorax* (Nematode: Mermithidae) have been massively released across the world to control vector mosquitoes (Alavo et al. 2015; Achinelly and Micieli 2009; Platzer 2007).

The only release of mermithids in Iran dates back to inundative release of an imported *R. culicivorax* in anopheline and culicine habitats in Fars and Sistan and Baluchistan provinces during 1984–1985. Inconsistent level of parasitism in different sites hindered the development of this biocontrol agent for control of mosquito vectors, although the average parasitism ranged between 56% and 69% obtained 24 h post-inoculation. This mermithid established and was recovered from 2 release sites after 2 years (Zaim et al. 1988).

In a recent study, release of artificially-parasitized *Culex pipiens* (Diptera: Culicidae) adults by *Strelkovimermis spiculatus* (Fig. 16.1) was suggested to control the vector through autodissemination strategy (Haji Allahverdipour et al. 2017). It is assumed that these parasitized adult mosquitoes prefer water rather than blood-feeding and offer a means for mermithids to colonise new host larval habitats (Haji Allahverdipour et al. 2019).

**Fig. 16.1** *Culex* mosquito parasitized by mermithid *Strelkovimermis spiculatus*. (Permission from Elsevier, license number 4797780366164)

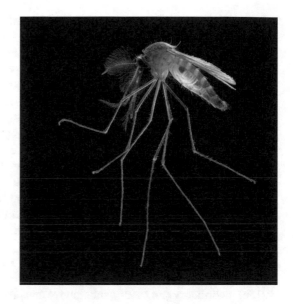

### 16.2.1.2.2   Insect Pathogens

Immature stages of Anopheline mosquitoes suffer from some microbial pathogens such as fungi, viruses and some protozoans (Walker and Lynch 2007).

Vast array of insect pathogens have been collected from naturally infected mosquitoes but some of them aren't amenable in applied biological control. The low rate of infection, difficult mass rearing and application are main reasons for few number of mosquitoes' microbial control agents. Some of the pathogens can kill mosquitoes in laboratory but they are not effective in wild. Factors which limit the effectiveness of these microbial agents under natural conditions include organic contamination of larval breeding sites, water hardness, water pH, rate of dissolved oxygen, light intensity, temperature and draught (Basseri 1988). Interaction between environmental conditions and mosquito pathogens remains largely unknown.

### 16.2.1.2.3   Entomopathogenic Bacteria

Ishiwata (1902) isolated an air-borne *bacillus* from diseased silkworm larvae and Iwabuch called the bacterium *Bacillus sotto* Ishiwata (Tanada and Kaya 1993) but the name was later ruled invalid. Berliner (1915) found similar *Bacillus* in Thuringia region of Germany from larvae of *Ephestia kuehniella* (Lepidoptera: Pyralidae) which he named *Bacillus thuringiensis* (Bt) (Milner 1994).

*B. thuringiensis* subsp. *israelensis* (Bti) (Bacillaceae) a soil bacterium has been found in Neger desert of Israel in 1976 and has been tested on the various genus of Culicine (Goldbery et al. 1977). This strain act as a natural pathogen of immature stages of mosquitoes (Walker and Lynch 2007). Spores of Bti produce crystal toxins which can damage digestive tract of the mosquito larvae (Sanahuja et al. 2011). Research on Bt in the context of mosquito vector control started in 1986 with isolation of *B. thuringiensis* strain from dead larvae of *An. stephensi* in Lorestan Province, Iran (Gezelbash et al. 2014). Following up, field assay of Teknar[®], Bt-H14 based product, was carried out on culicine and anopheline larvae in rice field of Kazeroon, Fars. The Teknar[®] was more effective on culicine than anopheline larvae (Motabar et al. 1987). Anophelines usually rest on the water surface while the aforementioned Bt formulation settled to the bottom of breeding sites. Fayaz and Moazami (1987) and Fayaz et al. (1988) could develop a media and formulation of Bt-H14 based on molasses and cornsteep liquor which proved very effective against *C. pipiens* and *An. stephensi*. Despite of high efficacy of locally produced Bt, e.g. Teknar[®], Bactimos[®] in field (Motabar et al. 1987; Zaim et al. 1992; Kasiri and Zaim 1997; Mousakazemi et al. 2000), Bioflash[®] proved low to moderate effectiveness (Gezelbash et al. 2014). Resistance of mosquito vectors to Bt has not been reported (Shahi et al. 2013).

Bioflash[®] (10% a.i.) slow release corn grit formulation remains effective for 3 weeks in various breeding habitats at concentration of 1800 ITU/ml (WHO 2013b). Laboratory and field evaluations of this product obtained inconsistent results concerning biological control of anopheline vectors of malaria (Shahi et al. 2013; Gezelbash et al. 2014). Nevertheless, biological control using Bt continues to be integrated with chemical control and personal protection measures against malaria vectors in Iran (Hanafi-Bojd et al. 2010, Fekri et al. 2013). Around $14 \times 10^6 \, m^2$ of mosquito larval habitats have been treated by Bioflash[®] in the past two years in the country (personal communication).

Although hosts of *Lysinibacillus sphaericus* (Bacillaceae) are not as diverse as Bt strains (Brown et al. 2004), high persistence of this aerobic bacterium in wild due to recycling from dead mosquito larvae makes it more effective for combating the Culicid larvae (Ganushkina et al. 2000).

There are few studies regarding field efficacy of *L. sphaericus* against mosquito vectors in Iran (Hayat-gheib 1988) and knowledge gaps still remain in successful application of *L. sphaericus* against culicidae larvae.

## 16.2.1.2.4  Entomopathogenic Fungi

The first record of entomopathogenic fungus, *Coelomomyces irani* (Coelomomycetaceae) (Fig. 16.2) was reported from *An. maculipennis*, the main malaria vector in Europe and Mediterranean countries, from Gilan province in 1988 (Weiser et al. 1991). *Coelomomyces indicus* and *Coelomomyces* sp. have been isolated from *An. culicifacies* and *An. superpictus* in Southeastern and West Central Iran, respectively (Azari-Hamidian and Abai 2019).

Iranian isolates of *Beauveria bassiana* (Cordycipitaceae) and *Metarhizium anisopliae* (Clavicipitaceae) were evaluated against *An. stephensi* Chabahr strain. Both *B. bassiana* and *M. anisopliae* isolates had high virulence on 1st and 2nd instar larvae and caused 97–100% mortality at 7 days post-inoculation with $LT_{50}$ range of

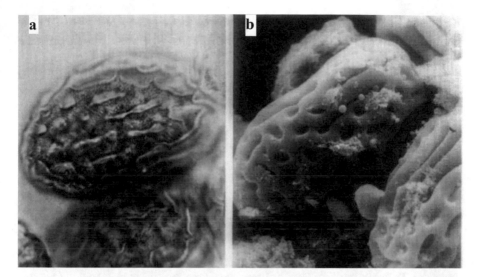

**Fig. 16.2** *Coelomomyces irani* resting sporangium (**a**) sporangium (**b**). (Permission from Elsevier, License number 4797780124271)

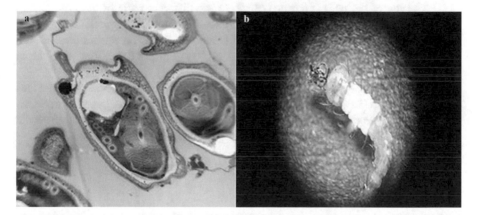

**Fig. 16.3** Parathelohania iranica spores infecting Anopheles superpictus (Left) Heavy infection of three abdominal segments (Right). (Permission from Elsevier, Licence number 4797780898620 and Courtesy from publisher of Journal of Arthropod-Borne Diseases)

0.63–2.53 days. Among these fungal isolates, *M. anisopliae* Iran 1018C isolate, possessing some favourable characteristics in microbial control of mosquitoes, was selected for further investigation (Fakoorziba et al. 2014). Among the aforementioned isolates, there were several isolates which also caused 100% mortality in larvae and adults of house fly, *Musca domestica* (Muscidae) (Sharififard et al. 2011).

Natural infection of *An. superpictus*, malaria vector in Middle East, by microsporidium, *Parathelohania iranica* (Amblyosporidae) (Fig. 16.3) was reported from West Central Iran (Omrani et al. 2017). Role of this microsporidium as natural population regulator and its potential in biocontrol of Anophelines are still obscure.

### 16.2.1.2.5   Entomopathogenic Oomycetes

*Lagenidium giganteum* (Lagenidiaceae) was tested against *An. stephensi* and *C. pipiens*. $LC_{50}$ of $1.57 \times 10^5$ and $1.78 \times 10^5$ spores/ml were obtained for *An. stephensi* and *C. pipiens*, respectively (Vatandoost et al. 1995). But, commercial strains of *L. giganteum* are currently unregistered and no longer available due to public health concern (Vilela et al. 2015).

### 16.2.1.2.6   Parasitic Mites

Parasitic mites of mosquitoes also known as water mites (Hydrachnidia) belong to the order Trombidiformes. They follow parasitengona life cycle. Their larvae are parasites while their nymph and adults are predators of insect eggs and larvae (Proctor et al. 2015).

The larvae of the parasitic mites attach to the different parts of the host mostly pre-abdominal regions (Kirkhoff et al. 2013). Parasitic mites can affect longevity and fertility of mosquitoes but not the rate of fecundity. Thus, parasitic mites are not suggested candidate for biological control of mosquitoes (Rajendran and Prasad 1992).

Water mites of the genus *Arrenurus* (Arrenuridae) have been reported as parasites of three species of anopheline, *An. maculipennis*, *An. pseudopictus* and *An. hyrcanus*, from North of Iran (Karami et al. 2019). However, recommending water mites in biological control of mosquitoes is still in doubt.

### 16.2.1.3   Symbiotic Bacteria

Characterization of *Wolbachia* (Anaplasmataceae) as master manipulator in mosquitoes of Northwest, Southwest, Northern, Central, and Southern Iran is limited to a couple of studies (Bozorg-Omid et al. 2019; Behbahani 2012; Karami et al. 2016). In determination of *Wolbachia* infection in probable arbovirus vectors in northwest Iran, results of semi-nested PCR using *Wolbachia surface protein gene*) showed that *Wolbachia* infection was present in *Aedes caspius* (Diptera: Culicidae), *Cx. pipiens*, *Culex theileri* and *Culiseta longiareolata* (Diptera: Culicidae), where the highest infection rate was observed in *Cx. pipiens*. The infection rates of mosquitoes with *Wolbachia* in *Cx. pipiens*, *Cs. longiareolata*, *Cx. theileri*, and *Ae. caspius* were 96.9%, 11.5%, 5.2% and 0%, respectively. The *Wolbachia* strain found in this study belonged to supergroup B. Among the 17 supergroups of *Wolbachia*, supergroups A and B have been described to frequently cause changes in the reproductive system of the host and are ubiquitous bacteria occurring in many arthropods (Bozorg-Omid et al. 2019). Prevalence of infection in different populations of *Cx. pipiens* in the North, Center, and South of Iran showed that all populations of the mentioned species were infected with *Wolbachia* (Karami et al. 2016; Bozorg-Omid et al. 2019).

Screening and identification of midgut bacterial microbiota from wild-caught *An. stephensi* larvae and adults in southern Iran for generation of paratransgenic mosquitoes refractory to transmission of malaria resulted in isolation of *Pseudomonas* (Pseudomonadaceae), *Aeromonas* (Aeromonadaceae) and *Myroides* (Flavobacteriaceae) genera from both larvae and adult stages indicating possible trans-stadial transmission (Chavshin et al. 2012). Among the isolated genera, *Pseudomonas* possessing characteristics like successful trans-stadial transmission, survival after several molting or ecdysis events during larval stages as well as colonisation of malpighian tubules and midgut (Fig. 16.4), proliferation following blood meals and persistence in the lumen of adult mosquitoes more than 21 days after eclosion have been nominated as suitable candidate for Incompatible Insect Technique. Further study is needed to engineer the *Pseudomonas* isolate with some effector molecules and to test the ability of the recombinant bacterium to block parasite transmission (Chavshin et al. 2015). Majority of midgut microbiota of *An. maculipennis* and *An. stephensi* from Southwest and North of Iran belonged to genera of *Psuedomonas* sp. and *Aeromonas* sp.. Additionally, *Pantoea* (Erwiniaceae), *Acinetobacter* (Moraxellaceae), *Brevundimonas* (Caulobacteraceae), *Bacillus*, *Sphingomonas* (Sphingomonadaceae) and *Lysinibacillus* were isolated as well (Dinparast Djadid et al. 2011).

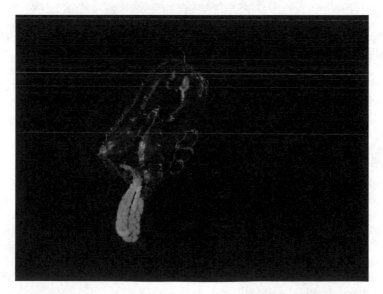

**Fig. 16.4** Colonisation of *Pseudomonas*-GFP in malpighian tubules. (Courtesy of Alireza Chavshin)

Rami et al. (2018) also introduce native *Asaia* sp. (Acetobacteraceae), isolated from five *Anopheles* species for malaria vector control via paratransgenesis technology. They conclude the fact that the vector populations were collected from different zoogeographical zones in the South, East, and North of Iran suggest that this candidate can tolerate the complicated environmental conditions of the vector-borne diseases endemic regions.

On the other hand, Soltani et al. (2017) indicate correlation between the presence of symbiotic bacteria in *An. stephensi* and resistance to temephos. Laboratory removal of bacteria from symbiotic organs could result in manipulation of resistant population to susceptible.

## 16.2.2 Leishmaniasis

There are two forms of leishmaniasis in Iran, cutaneous and visceral. From the transmission point of view, cutaneous leishmaniasis has two subgroups: Zoonotic Cutaneous Leishmaniasis (ZCL) and Anthroponotic Cutaneous Leishmaniasis (ACL) (Yaghoobi-Ershadi 2012). Herein, around 80% of reported leishmaniasis cases were recorded as ZCL (Akhavan et al. 2007).

Nevertheless, a number of studies proved the efficacy of entomopathogenic fungi like *B. bassiana* and *M. anisopliae* on sand flies (Reithinger et al. 1997; Amora et al. 2009; Ngumbi et al. 2011), these agents have not been employed in control of sand flies in the country.

Sant'Anna et al. (2014) discovered colonisation of sand fly gut with yeast and bacteria like *Asaia* reduces the potential for *Leishmania* (Trypanosomatidae) to establish within the sand fly vector. On the other hand, six fungal genera including *Penicillium, Aspergillus, Acremonium, Fusarium, Geotrichum* and *Candida* have been found in wild-caught phlebotomine sand flies in northwest Iran. Presence of some bacteria such as *Asaia* sp. was recognized as well (Akhoundi et al. 2012). This possibility that microbial composition of female sand fly may influence *Leishmania* transmission necessitates further studies on the role of this microbiome in regulating sandfly competence.

Parvizi et al. (2013a) detected two strains of *Wolbachia pipientis* (Turk54 and Turk07) belonging to group A, in natural populations of *Phlebotomus mongolensis* (Diptera: Psychodidae) and *Ph. caucasicus*, potential vectors of ZLC, from Turkmen Sahra, Northern Iran. Strain Turk54 had previously been isolated from *Phlebotomus papatasi* but strain Turk07 was a new record for these vector species. The opportunity of using these endosymbionts to drive transgenes through wild sand flies and reduce their natural population is promising. Moreover, *Phlebotomus perfiliewi transcaucasicus*, potential vector of visceral leishmaniasis, infected with *W. pipientis* haplotypes pave the way for release of genetically-modified *Wolbachia* within the vector populations (Parvizi et al. 2013b).

Various groups of rodents are main natural reservoirs for leishmaniasis especially for ZCL in Iran (Yaghoobi-Ershadi 2012). It has been assumed that control of rodents can reduce the incidence of ZCL. However, biological control of rodents for leishmaniasis prevention has both its pros and cons. One of the disadvantages is that some predators such as foxes are reservoirs of visceral leishmaniasis parasite in Iran (Nilforoushzadeh et al. 2014). Additionally, some predators of rodents work effectively under certain conditions (Wodzicki 1973). There are some evidence on decrease in populations of *Rattus* (Muridae) in South part of the country, Abu-Musa Island by mongoose (Herpestidae, Carnivora) as predator. However, its sole impact has not been confirmed yet (Khoobdel et al. 2016). Further research on the potential and threats of Phlebotomine natural enemies is needed.

### 16.2.3   Tick-Borne Diseases

Crimean-Congo hemorrhagic fever (CCHF) has been documented as the most frequent tick-borne viral infection in the country with more than 50 cases annually (Khakifirouz et al. 2018). The first incidence of CCHF in Iran (Khorasan Province) dates back in 1978 and 871 confirmed cases of CCHF were reported during 1999–2012 (Berger 2019). The main mode of transmission for CCHF virus is tick bites but contact to infected blood and livestock meat and nosocomial infection can also cause CCHF in human (Farhadpour et al. 2016). CCHF has been documented in almost all provinces of the country, with highly endemic areas including Sistan and Baluchistan, Razavi Khorasan, Fars, Esfahan, and Kerman (Khakifirouz et al. 2018). Main local vectors of this disease belong to *Hyalomma* and *Rhipicephalus* genera (Ixodidae) (Farhadpour et al. 2016; Champour et al. 2016). The unique feeding behavior to find their host in each developmental stage may result in more difficulties for their control. Indiscriminate use of acaricides is not sensible due to their environmental hazards, milk contamination and resistance development in target species (Onofre et al. 2001).

According to Kalsbeek et al. (1995) entomopathogenic fungi play an important role in regulating the population of *Ixodes ricinus* (Ixodidae) and females are the most frequently infected stage by *B. bassiana*, *Beauveria brongniartii*, *Paecilomyces fumosoroseus* (Cordycipitaceae) and *Verticillium aranearum* (Plectosphaerellaceae). Additionally, application of entomopathogenic fungi for tick burden on cattle in South America has obtained promising results (Kaya and Hassan 2000). Gindin et al. (2002) emphasize that biological control of the tick vectors with entomopathogenic fungi is feasible as there is a high chance of finding a fungal strain pathogenic for all tick stages of several tick species. To enhance the efficacy of entomopathogenic fungi in field, selection of highly lethal strain, host-targeted application, dose optimization, delivery medium and timing have been suggested (Ostfeld et al. 2006).

Effective concentration of *B. bassiana* to prevent eclosion of *Rhipicephalus microplus* Canestrini, 1887 (Acari: Ixodidae) has to be $10^7$–$10^8$ conidia/ml for field application (Bittencourt et al. 1997). In a similar study, minimum concentration of $1.5 \times 10^6$ spores/ml of *B. bassiana* was estimated to induce infection and can be

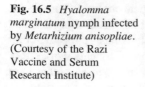

**Fig. 16.5** *Hyalomma marginatum* nymph infected by *Metarhizium anisopliae*. (Courtesy of the Razi Vaccine and Serum Research Institute)

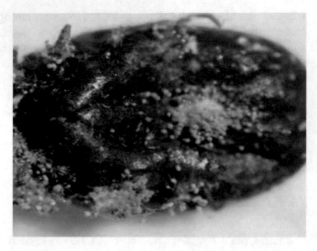

applied in biological control of *Hyalomma anatolicum anatolicum* and *H. marginatum* adult stages (Abdigoudarzi et al. 2009). Several native isolates of *M. anisopliae* show great potential in control of *H. anatolicum anatolicum*, *H. marginatum* (Fig. 16.5) and *Haemaphysalis punctata*, nymphs (Tavassoli et al. 2009; Rivaz et al. 2016).

Exposure of *Rhipicephalus annulatus* to Iranian isolates of *M. anisopliae*, DEMI 001 and IRAN 437C showed the isolates potential in biological control of engorged females. Mortality rate of 90–100% was recorded 6–11 days post treatment. *Lecanicillium psalliotae* (Cordycipitaceae) was also isolated from *R. annulatus* in this study (Pirali-Kheirabadi et al. 2007). Also, another work indicated up to 50% reduction in egg clutches and decrease in body mass of *Ixodes scapularis* engorging females exposed to *M. anisopliae* were observed (Hornbostel et al. 2004).

Soft ticks species studied for their susceptibility to *M. anisopliae* native isolates include *Dermanyssus gallinae* (Acari: Gamasidae) and *Argas persicus* (Acari: Argasidae). The fungal isolated showed promising results especially in case of *A. persicus* (Tavassoli et al. 2008, 2011; Pourseyed et al. 2010). However, The use of parasitic fungi as a way to control soft ticks infestation could generate downstream environmental disequilibrium, since entomopathogenic fungi are generally not specific for soft ticks and may affect other naturally existing insect populations (Pritchard et al. 2015).

Concerning endosymbionts of CCHF vectors, *Wolbachia* group A was found in *Rhipicephalus sanguineus* and *R. annulatus* in Iran (Pourali et al. 2009).

### 16.2.4 Myiasis

Despite significant and useful role in forensic entomology and maggot therapy, Muscomorpha are well known for their potential in mechanical transmission of their larvae to human and animals called myiasis (Kassiri et al. 2012). Myiasis is sometimes regarded as an occupational disease because of its prevalence in feedlots and traditional ranches (Akbarzadeh et al. 2012).

Different forms of human and animal myiasis are common in Iran. *Wohlfahrtia magnifica* (Diptera: Sarcophagidae) is the main agent of wound myiasis in animal. Oral myiasis as the dominant type of human myiasis is caused by *Oestrus ovis* (Diptera: Oestridae) (Akbarzadeh et al. 2012; Alizadeh et al. 2014). Other fly families are also involved in transmission of myiasis in the country (Alizadeh et al. 2014).

Parasitic wasps in the families Pteromalidae and Chalcididae parasitize Muscomorpha in various regions of the country. The dominant species of Pteromalidae collected from naturally parasitized housefly pupae was *Nasonia vitripennis* Walker, 1836 (Hymenoptera: Pteromalidae) with parasitic rate of 26.7% (Akbarzadeh et al. 2017). This species was reported in 1991 as a suitable biological control agent in reduction of *Musca domestica* (Diptera: Muscidae) population in Iran (Iranpour et al. 1991). Presence of *N. vitripennis* was reconfirmed in the checklist by Lotfalizadeh and Gharali (2008). *N. vitripennis* is a cosmopolitan parasitie of Calliphoridae and Sarcophagidae pupae (Van den Assem and Werren 1994).

One of the widely distributed genera of Chalcididae is *Brachymeria*. Members of this genus are parasitoids of muscoid flies (Marchiori 2001). Species of *Brachymeria* have been reported in Iran (Lotfalizadeh et al. 2012).

*N. vitripennis* and *Brachymeria podagrica* have been collected from medically important flies in Brazil (Marchiori, 2004) and Argentina (Horenstein and Salvo 2012). Natural host preference of three species of parasitic wasps including *N. vitripennis*, *Pachycrepoide vindemmiae* (Hymenoptera: Pteromalidae) and *Spalangia nigroaenea* (Hymenoptera: Pteromalidae) has been observed on three medical important fly species including *M. domestica*, *Lucilia sericata* (Diptera: Calliphoridae) and *Sarcophaga haemorrhoidalis* (Diptera: Sarcophagidae) showing high preference of *S. nigroaenea* for *M. domestica* and potential of *N. vitripennis* for control of synanthropic flies (Khoobdel et al. 2019). There is also compatibility between *N. vitripennis* and cyromazine as an Insect Growth Regulator (IGR). Deployment of 1.1% or 0.9% Cyromazine treated cloths in addition to release of *N. vitripennis* is suggested for control of housefly in poultry coops and livestock farms (Vazirianzadeh et al. 2008).

## 16.2.5 Diseases Transmitted by Cockroaches

Cockroaches can transmit parasites, bacteria, fungi and viruses of public health importance. The prevalent species in Iran including *Periplaneta americana* (Blattodea: Blattidae), *Blattella germanica* (Blattodea: Ectobiidae), *B. orientalis* and *Supella longipalpa* (Blattodea: Ectobiidae) (Arfa Rahimian et al. 2019) are mostly suppressed by chemical insecticides mainly applied in sewage systems.

However, several attempts have been made to deploy entomopathogenic fungi for cockroaches' control. Entomopathogenic fungi of *B. bassiana*, *Lecanicillium muscarium* and *M. anisopliae* proved lethal to different stages of *B. germanica*; albeit, their virulence was highly dependant on the methods of application and formulations (Davari et al. 2015; Abedi and Dayer 2006).

## 16.3   Open End Story

Limited use of biocontrol agents for insects of medical importance in Iran could be generally explained by usually temporary habitats of pests, complexity in setting their tolerance level like other countries but it has some other domestic restrications as well. Despite extensive effort in collection and identification of indigenous natural enemies of medically important vectors so far, progress in the development of biological control agents for vectors has been hampered due to socioeconomic factors and authority's insufficient support. Hard and fast rules for product registration/regulation, low profit margin and turnover discourage stakeholders to invest in biological control market.

Several biological measures against vector mosquitoes have been exploited in the country with few being taken for large scale mosquito control. There were a few parasites and pathogens of mosquito vectors e.g. *Romanomermis culicivorax*, *L. sphaericus* which their efficiency in mosquito control in the past have been neglected. Further investigation on these agents are of priority. Techniques like Paratransgenesis and Incompatible Insect Technique have not been implemented for vector control yet. However, due to the special attention of governmentsectors and decision makers to environment pollution with chemical insecticides and threat of extinction of endangered species, increase in biocontrol of pests of medical and veterinary importance are expected in near future. Various climates and rich biodiversity confers Iran diverse fauna of macrobials and microbials to be used in vector control. Use of biological control agents against anopheline vectors as a component of integrated vector management is a promising example to be adapted in control of other life-threatening vector-borne diseases in the country. All in all, each vector control program including biological control as integral part has to be done at regional level; Collective efforts of Iran and bordering countries to prevent outbreaks is necessary.

**Acknowledgement**   The authors Randy Gaugler for his comments and Javad Karimi for review and edit the draft. The authors appreciate Seyed-Mohammad Omrani and Ali Reza Chavshin for sharing their original images and Shokoofeh Kamali for images processing.

# References

Abbasi M, Hanaf-Bojd AA, Yaghoobi-Ershadi MR, Vatandoost H, Oshaghi MA, Hazratian T, Hazratian T, Sedaghat MM, Fekri S, Mojahedi AR, Salari Y (2019) Resistance status of main malaria vector, *Anopheles stephensi* Liston (Diptera: Culicidae) to insecticides in a malaria Endemic Area, Southern Iran. Asian Pac J Trop Med 12:43–48

Abdigoudarzi M, Esmaeilnia K, Shariat N (2009) Laboratory study on biological control of ticks (Acari: Ixodidae) by entomopathogenic indigenous fungi (*Beauveria bassiana*). Iran J Arthropod Borne Dis 3:36–43

Abedi A, Dayer MS (2006) Evaluation of the effect of the fungus *Metarhizium anisopliae*, as a biological control agent, on german cockroaches *Blattella germanica*. Modares J Med Sci 8 (1):31–36

Achinelly MF, Micieli MV (2009) Experimental releases of *Strelkovimermis spiculatus* (Nematoda: Mermithidae) against three mosquito species in Argentina. Nematol 11:151–154

Akbarzadeh K, Rafinejad J, Alipour H, Biglarian A (2012) Human myiasis in Fars Province, Iran. Southeast Asian J Trop Med Public Health 43:1205–1211

Akbarzadeh K, Mirzakhanlou AA, Lotfalizadeh H, Malekian A, Hazratian T, Rezaei Talarposhti K, Babapour Darzi R, Radi E, Aghaei Afshar A (2017) Natural parasitism associated with species of Sarcophagidae family of Diptera in Iran. Ann Trop Med Public Health 10(1):134–137

Akhavan AA, Yaghoobi-Ershadi MR, Hasibi F, Jafari R, Abdoli H, Arandian MH, Soleimani H, Zahraei-Ramazani AR, Mohebali M, Hajjaran H (2007) Emergence of cutaneous leishmaniasis due to *Leishmania major* in a new focus of southern Iran. Iran J Arthropod Borne Dis 1.1–8

Akhoundi M, Bakhtiari R, Guillard T, Baghaei A, Tolouei R, Sereno D, Toubas D, Depaquit J, Razzaghi Abyaneh M (2012) Diversity of the bacterial and fungal microflora from the midgut and cuticle of phlebotomine sand flies collected in north-western Iran. PLoS One 7(11):1–10

Alavo TB, Abagli AZ, Perez-Pacheco R, Platzer EG (2015) Large scale production of the malaria vector biocontrol agent *Romanomermis iyengari* (Nematoda: Mermithidae) in Benin, West Africa. Malar J 6:1–5

Alizadeh M, Mowlavi G, Kargar F, Nateghpour M, Akbarzadeh K, Hajenorouzali-Tehrani M (2014) A review of myiasis in Iran and a new nosocomial case from Tehran. Iran J Arthropod Borne Dis 8:124–131

Amora SSA, Bevilaqua CML, Feijo FMC, Silva MA, Pereira RHMA, Silva SD, Alves ND, Freire FAM, Oliveira DM (2009) Evaluation of the fungus *Beauveria bassiana* (Deuteromycotina: Hyphomycetes), a potential biological control agent of *Lutzomyia longipalpis* (Diptera, Psychodidae). Biol Control 50:329–335

Arfa Rahimian A, Hanafi-Bojd AA, Vatandoost H, Zaim M (2019) A review on the insecticide resistance of three species of cockroaches (Blattodea: Blattidae) in Iran. J Econ 112(1):1–10

Arthington AH, Marshall CJ (1999) Diet of the exotic mosquitofish, *Gambusia holbrooki*, in an Australian lake and potential for competition with indigenous fish species. Asian Fish Sci 12 (1):1–8

Askarian M, Mansour Ghanaie R, Karimi A, Habibzadeh F (2012) Infectious diseases in Iran: a bird's eye view. Clin Microbiol Infect 18:1081–1088

Azari-Hamidian S (2007) Checklist of Iranian mosquitoes (Diptera: Culicidae). J Vector Ecol 32:235–242

Azari-Hamidian S, Abai MR (2019) The first record of the fungal pathogen *Coelomomyces* (Blastocladiales: Coelomomycetaceae) in the Malaria Vector *Anopheles culicifacies* s.l. (Diptera: Culicidae) in Iran. In: Abstracts of the 11th National and 4th International Congress of Parasitology and Parasitic Diseases, Urmia, Iran, 9–11 Oct 2019

Basseri HR (1988) Study on distribution of *Aphanius dispar* in Baluchestan region and study on its application in malaria control programs. A thesis for Master Science in Public Health, School of Public Health, Tehran University of Medical Sciences. No 30732

Bay EC, Berg CO, Chapman HC, Legner EF (1976) Biological control of medical and veterinary pest. In: Huffaker CB, Messenger PS (eds) Theory and practice of biological control. Academic Press, New York, pp 457–479

Behbahani A (2012) *Wolbachia* infection and mitochondrial DNA comparisons among *Culex mosquitoes* in South West Iran. Pak J Biol Sci 15:54–57

Berger S (2019) Infectious diseases of Iran. GIDEON Informatics, Inc, California

Berliner E (1915) Über die Schlaffsucht der Mehlmottenraupe (*Ephestia kühniella* Zell.) und ihren Erreger *Bacillus thuringiensis* n. sp. Z. Angew. Entomol. 2: 29–56

Bittencourt V, Souza EJ, Peralva S, Mascarenhas A (1997) Evaluation of the in-vitro efficacy of two isolates of the entomopathogenic fungus *Beauveria bassiana* (Bals.) vuill. In engorged females of *Boophilus microplus* (Canestrini, 1887) (Acari: Ixodidae). Rev Bras Parasitol Vet 6(1):49–52

Bozorg-Omid F, Oshaghi MA, Vahedi M, Karimian F, Seyyed-Zadeh SJ, Chavshin AR (2019) *Wolbachia* infection in West Nile virus vectors of Northwest Iran. Appl Entomol Zool 55:105–113

Brown MD, Watson TM, Carter J, Purdie DM, Kay BH (2004) Toxicity of VectoLex (*Bacillus sphaericus*) products to selected Australian mosquito and nontarget species. J Econ Entomol 97 (1):51–58

Champour M, Chinikar S, Mohammadi G, Razmi G, ShahHosseini N, Khakifirouz S, Mostafavi E, Jalali T (2016) Molecular epidemiology of Crimean–Congo Hemorrhagic Fever virus detected from ticks of one humped camels (*Camelus dromedarius*) population in Northeastern Iran. J Parasit Dis 40(1):110–115

Chavshin AR, Oshaghi MA, Vatandoost H, Pourmand MR, Raeisi A, Enayati AA, Mardani N, Ghoorchian S (2012) Identification of bacterial microflora in the midgut of the larvae and adult of wild caught *Anopheles stephensi*: a step toward finding suitable paratransgenesis candidates. Acta Trop 121:129–134

Chavshin AR, Oshaghi MA, Vatandoost H, Yakhchali B, Zarenejad F, Terenius O (2015) Malpighian tubules are important determinants of *Pseudomonas* transstadial transmission and long time persistence in *Anopheles stephensi*. Parasit Vectors 8:36–43

Davari B, Limoee M, Khodavaisy S, Zamini G, Izadi S (2015) Toxicity of entomopathogenic fungi, *Beauveria bassiana* and *Lecanicillium muscarium* against a field-collected strain of the German cockroach *Blattella germanica* (L.) (Dictyoptera: Blattellidae). Trop Biomed 32(3):463–470

Dehghani R, Charkhloo E, Seyyedi-Bidgoli N, Chimehi E, Ghavami-Ghameshlo M (2018) A review on scorpionism in Iran. Iran J Arthropod Borne Dis 12(4):325–333

Dinarvand M (2012) A taxonomic revision of *Utricularia* (Lentibulariaceae) for aqua flora of Iran. Iran J Bot 18(2):191–195

Dinparast Djadid N, Jazayeri H, Raz A, Favia G, Ricci I, Zakeri S (2011) Identification of the midgut microbiota of *An. stephensi* and *An. maculipennis* for their application as a paratransgenic tool against malaria. PLoS One 7(1):10.1371

Esmaeili HR, Shiva AH (2006) Reproductive biology of the Persian tooth-carp, *Aphanius persicus* (Jenkins, 1910) (Cyprinodontidae), in Southern Iran. Zool Middle East 37:39–46

Esmaeili S, Ghasemi A, Naserifar R, Jalilian A, Molaeipoor L, Maurin M, Mostafav IE (2019) Epidemiological survey of Tularemia in Ilam Province, west of Iran. BMC Infect Dis 19:502

Fakoorziba MR, Veys-Behbahani R, Dinparast Djadid N, Azizi K, Sharififard M (2014) Screening of the entomopathogenic fungi, *Metarhizium anisopliae* and *Beauveria bassiana* against early larval instars of *Anopheles stephensi* (Diptera: Culicidae). J Entomol 11:87–94

Farhadpour F, Telmadarraiy Z, Chinikar S, Akbarzadeh K, Moemenbellah-Fard MD, Faghihi F, Fakoorziba MR, Jalali T, Mostafavi E, Shahhosseini N, Mohammadian M (2016) Molecular detection of Crimean–Congo haemorrhagic fever virus in ticks collected from infested livestock populations in a new endemic area, South of Iran. Tropical Med Int Health 21(3):340–347

Fayaz F, Moazami N (1987) Three different media formulated from molassess and cornsteep liquor for the production of *Bacillus thuringiensis* serotype H-14. J Pasteur Inst Iran 3:54–60

Fayaz F, Moazami N, Zaiem M, Motabar M, Holakouei NK (1988) Local production of primary powder of *Bacillus thuringiensis* serotype H-14 in Iran and determination of its insecticidal properties against *Culex pipiens* and *Anopheles stephensi*. Med J Iran 2(3):229–236

Fekri S, Vatandoost H, Daryanavard A, Shahi M, Safari R, Raeisi A, Sheikh Omar A, Sharif M, Azizi A, Ahmad Ali A, Nasser A, Hasaballah I, Hanafi–Bojd AA (2013) Malaria situation in an endemic area, Southeastern Iran. Iran J Arthropod Borne Dis 8(1):82–90

Ganushkina LA, Lebedeva NN, Azizbekian RR, Sergiev VP (2000) The duration of the larvicidal effects of spore crystalline mass of the bacteria *Bacillus thuringiensis* spp. *israelensis* and *Bacillus sphaericus* in the laboratory setting. Med Parazitol (Mosk) 4:25–29

Gezelbash Z, Vatandoost H, Abai MR, Raeisi A, Rassi Y, Hanafi-Bojd AA, Jabbari H, Nikpoor F (2014) Laboratory and field evaluation of two formulations of *Bacillus thuringiensis* M-H-14 against mosquito larvae in the Islamic Republic of Iran, 2012. East Mediterr Health J 20 (4):229–235

Gindin G, Samish M, Zangi G, Mishoutchenko A, Glazer I (2002) The susceptibility of different species and stages of tick to entomopathogenic fungi. Exp Appl Acarol 28:283–288

Goldbery LJ, Goldbery EM, Margalit J (1977) Potential application of a bacterial spore, CNR 60A to mosquito larval control demonstrated rapid larvicidal activity against *Anopheles sergenti*, *Uranotaenia unguicolata*, *Culex univitatus*, *Aedes aegypti* and *Culex pipiens* (Complex). Nimeographed document. WHO/VBC/77, 662

Gyalpo T, Fritsche L, Bouwman H, Bornman R, Scheringer M, Hungerbühler K (2012) Estimation of human body concentrations of DDT from indoor residual spraying for malaria control. Environ Pollut 169:235–241

Haji Allahverdipour H, Talaei-Hassanloui R, Karimi J, Wang Y, Gaugler R (2017) Production of *Culex pipiens* (Dip.: Culicidae) adults infected by *Strelkovimermis spiculatus* (Nematoda: Mermithidae) in autodissemination control strategy. J E S I 37(1):125–134. [In Persian]

Haji Allahverdipour H, Talaei-Hassanloui R, Karimi J, Wang Y, Rochlin I, Gaugler R (2019) Behavior manipulation of mosquitoes by a mermithid nematode. J Invertebr Pathol 168:107273

Hanafi-Bojd AA, Vatandoost H, Philip E, Stepanova E, Abdi AI, Safari R, Mohseni GH, Bruhi MI, Peter A, Abdulrazag SH, Mangal G (2010) Malaria situation analysis and stratification in Bandar Abbas county, Southern Iran, 2004–2008. Iran J Arthropod Borne Dis 4(1):31–41

Hanafi-Bojd AA, Azari-Hamidian S, Vatandoost H, Charrahy Z (2011) Spatio-temporal distribution of malaria vectors (Diptera: Culicidae) across different climatic zones of Iran. Asian Pac J Trop Med 4(6):498–504

Hanafi-Bojd AA, Vatandoost H, Oshaghi MA, Haghdoost AA, Shahi M, Sedaghat MM (2012) Entomological and epidemiological attributes for malaria transmission and implementation of vector control in southern Iran. Acta Trop 121:85–92

Hayat-gheib D (1988) Efficacy and preliminary field evaluation of *Bacillus sphaericus* 1593 against anopheline larvae in southern Iran, Kazeroon, Fars province. MSc Thesis. Tehran University of Medical Science [In Persian]

Horenstein MB, Salvo A (2012) Community dynamics of carrion flies and their parasitoids in experimental carcasses in central Argentina. J Insect Sci 12(8):1–10

Hornbostel V, Ostfeld LR, Zhioua SE, Benjamin A (2004) Sublethal effects of *Metarhizium anisopliae* (Deuteromycetes) on engorged larval, nymphal, and adult *Ixodes scapularis* (Acari: Ixodidae). J Med Entomol 41:922–929

Iranpour M, Tirgari S, Shayeghi M (1991) First attempt on the study of the biology and mass rearing of two Iranian parasitoids of house fly pupae. In: Abstracts of the 10th Iranian Plant Protection Congress, Kerman, Iran, 1–5 Sep 1991

Ishiwata S (1902) Sule bacilli appele, Sitto. Japan Bull Asso Seri 114:1–5

Jafari A, Enayati A, Jafari F, Motevalli Haghi F, Hosseini-Vasoukolaei N, Sadeghnezhad R, Azarnoosh M, Fazeli-Dinan M (2019) A narrative review of the control of mosquitoes by larvivorous fish in Iran and the World. Iran J Health Sci 7(2):49–60

Jouladeh Roudbar A, Eagderi S, Esmaeili HR (2015) Fishes of the Dasht-e Kavir basin of Iran: an updated checklist. Int J Aquat Biol 3(4):263–273

Kalsbeek V, Frandsen F, Steenberg T (1995) Entomopathogenic fungi associated with *Ixodes ricinus* ticks. Exp Appl Acarol 19:45–51

Karami M, Moosa-Kazemi SH, Oshaghi MA, Vatandoost H, Sedaghat MM, Rajabnia R, Hosseini M, Maleki-Ravasan N, Yahyapour Y, Ferdosi-Shahandashti E (2016) *Wolbachia* endobacteria in natural populations of *Culex pipiens* of Iran and its phylogenetic congruence. J Arthropod Borne Dis 10(3):347–363

Karami M, Saboori A, Asadi M, Moosa-Kazemi SH, Gorouhi MA, Motevali Haghi F, Maleki-Ravasan N, Zahedi Golpayegani A (2019) Parasitism of mosquitoes (Diptera: Culicidae) by water mite larvae (Acari: Hydrachnidia) in Amol, Mazandaran Province, northern Iran. Syst Appl Acarol 24:423–434

Kasiri H, Zaim M (1997) Field assessment of *Bacillus thuringiensis* serotype H-14 (Bactimos WP), Abate and oil against *Anopheles* and *Culex* in south of Iran. Iran J Public Health 26(3–4):69–76. [In Persian]

Kassiri H, Akbarzadeh K, Ghaderi A (2012) Isolation of pathogenic bacteria on the house fly, *Musca domestica* L. (Diptera: Muscidae), body surface in Ahwaz hospitals, outhwestern Iran. Asian Pac J Trop Biomed 2:1116–1119

Kaya GP, Hassan S (2000) Entomogenous fungi as promising biopesticides for tick control. Exp Appl Acarol 24:913–926

Khakifirouz S, Mowla SJ, Baniasadi V, Fazlalipour M, Jalali T, Mirghiasi SM, Salehi-Vaziri M (2018) No detection of Crimean Congo Hemorrhagic Fever (CCHF) virus in ticks from Kerman province of Iran. J Med Microbiol Infect Dis 6(4):108–111

Khoobdel M, Jafari H, Firouzi F (2016) Evaluation of biological control of *Rattus* population by mongoose (Herpestidae, Carnivora) in Abu Musa Island, Iran. Asian Pac J Trop Dis 6 (10):802–806

Khoobdel M, Sobati H, Dehghan O, Akbarzadeh K, Radi E (2019) Natural host preferences of parasitoid wasps (Hymenoptera: Pteromalidae) on synanthropic flies. Eur J Transl Myol 29:118–123

Kirkhoff CJ, Simmons TW, Hutchinson M (2013) Adult mosquitoes parasitized by larval water mites in Pennsylvania. J Parasitol 99(1):31–39

Lotfalizadeh HA, Gharali B (2008) Pteromalidae (Hymenoptera: Chalcidoidea) of Iran: new records and a preliminary checklist. Entomofauna 29(6):93–120

Lotfalizadeh H, Ebrahimi E, Delvare G (2012) A contribution to the knowledge of the family Chalcididae (Hym.: Chalcidoidea) in Iran. J E S I 31(2):67–100

Marchiori CH (2001) Ocorrência de *Brachymeria podagrica* (Fabricius) (Hymenoptera: Chalcididae) como parasitóide de *Peckia chrysostoma* (Wiedemann) (Diptera: Sarcophagidae) no Brasil. Entomol vectores 8:513–517

Marchiori CH (2004) Parasitoids of *Chrysomya megacephala* (Fabricius) collected in Itumbiara, Goias, Brazil. Rev Saude Publica 38(2):1–2

Mehravaran A, Vatandoost H, Oshaghi M, Abai M, Edalat H, Javadian E, Mashayekhi M, Piazak N, Hanafi-Bojd A (2012) Ecology of *Anopheles stephensi* in a malarious area, Southeast of Iran. Acta Med Iran 50(1):61–65

Milner RJ (1994) History of *Bacillus thuringiensis*. Agric Ecosyst Environ 49(1):9–13

Moazami N (2009) Biopesticide production. In: Doelle HW, Rockem S, Berovie M (eds) Biotechnology: special processes for products, fuel and energy, vol VI. Encyclopedia of life support systems (EOLSS), Oxford, pp 1–52

Mogaddam MY, Motevalli Haghi F, Fazeli-Dinan M, Hosseini-Vasoukolaei N, Enayati AA (2016) A review of insecticide resistance in malaria vectors of Iran. J Mazandaran Univ Med Sci 26 (134):394–411. [In Persian]

Motabar M, Ladoni H, Zaim M (1987) Study the larvicidal effect of *Bacillus thuringiensis* (Teknar®) on mosquito larvae in rice field of Kazeroon, Fars in 1984. Iran J Public Health 15:21–29. [In Persian]

Mousakazemi SH, Motabar M, Moazami N, Kamali F (2000) Field evaluation of Bactimos wettable powder and corn cob granule of *Bacillus thuringiensis* var H-14 formulations for the control of Anopheline larvae in Bandar Abbas and Kazerun, the south of Iran. Iran South Med J 3 (1):16–19. [In Persian]

Nejati J, Keyhani A, Tavakoli Kareshk A, Mahmoudvand H, Saghafipour A, Khoraminasab M, Tavakoli Oliaee R, Mousavi SM (2018) Prevalence and risk factors of pediculosis in primary school children in south west of Iran. Iran J Public Health 47(12):1923–1929

Ngumbi PM, Irungu LW, Ndegwa PN, Maniania NK (2011) Pathogenicity of *Metarhizium anisopliae* (Metch) Sorok and *Beauveria bassiana* (Bals) Vuill to adult *Phlebotomus duboscqi* (Neveu-Lemaire) in the laboratory. J Vector Borne Dis 48:37–40

Nilforoushzadeh MA, Shirani-Bidabadi L, Saberi S, Hosseini SM (2014) Effect of integrated pest management on controlling zoonotic cutaneous leishmaniasis in Emamzadeh Agha Ali Abbas (AS) District, Isfahan province, 2006–2009. Adv Biomed Res 3:104–110

Omrani SM, Moosavi SF, Farrokhi E (2017) *Parathelohania iranica* sp. nov. (Microsporidia: Amblyosporidae) infecting malaria mosquito *Anopheles superpictus* (Diptera: Culicidae): ultrastructure and molecular characterization. J Invertebr Pathol 146:1–6

Onofre SB, Miniuk CM, de Barros NM, Azevedo JL (2001) Pathogenicity of four strains of entomopathogenic fungi against the bovine tick *Boophilus microplus*. Am J Vet Res 62 (9):1478–1480

Ostfeld RS, Price A, Hornbostel RL, Benjamin MA, Keesing M (2006) Controlling ticks and tick-borne zoonoses with biological and chemical agents. J Biosci 56(5):384–394

Pakdad K, Hanafi-Bojd AA, Vatandoost H, Sedaghat MM, Raeisi A, Moghaddam AS, Foroushani AR (2017) Predicting the potential distribution of main malaria vectors *Anopheles stephensi, An. culicifacies s.l.* and *An. fluviatilis s.l.* in Iran based on maximum entropy model. Acta Trop 169:93–99

Parvizi P, Bordbar A, Najafzadeh N (2013a) Detection of *Wolbachia pipientis* including a new strain containing the wsp gene in two sister species of *Paraphlebotomus* sandflies potential vectors of zoonotic cutaneous leishmaniasis. Mem Inst Oswaldo Cruz 108:414–420

Parvizi P, Fardid F, Soleimani S (2013b) Detection of a new strain of *Wolbachia pipientis* in *Phlebotomus perfiliewi* transcaucasicus a potential vector of visceral leishmaniasis in North West of Iran by targeting the major surface protein gene. Iran J Arthropod Borne Dis 7(1):46–55

Pirali-Kheirabadi K, Haddadzadeh H, Razzaghi-Abyaneh M, Bokaie S, Zare R, Ghazavi M, Shams-Ghahfarokhi M (2007) Biological control of *Rhipicephalus* (*Boophilus*) *annulatus* by different strains of *Metarhizium anisopliae, Beauveria bassiana* and *Lecanicillium psalliotae* fungi. Parasitol Res 100:1297–1302

Platzer EG (2007) Mermithid Nematodes. In T. G. Floore, ed. Biorational control of mosquitoes. J Am Mosquito Contr 7(Suppl. 2):58–64

Pourali P, Roayaei Ardakani M, Jolodar A, Razi Jalali M (2009) PCR screening of the *Wolbachia* in some arthropods and nematodes in Khuzestan province. Iran J Vet Res 10:216–222

Pourseyed SH, Tavassoli M, Bernousi I, Mardani K (2010) *Metarhizium anisopliae* (Ascomycota: Hypocreales): An effective alternative to chemical acaricides against different developmental stages of fowl tick *Argas persicus* (Acari: Argasidae). Vet Parasitol 172:305–310

Pritchard J, Kuster T, Sparagano O, Tomley F (2015) Understanding the biology and control of the poultry red mite *Dermanyssus gallinae*: a review. Avian Pathol 44(3):143–153

Proctor HC, Smith IM, Cook DR, Smith BP (2015) Subphylum Chelicerata, class Arachnida. In: Thorp J, Rogers DC (eds) Ecology and general biology: Thorp and Covich's freshwater invertebrates. Academic, pp 599–660

Rajendran R, Prasad RS (1992) Influence of mite infestation on the longevity and fecundity of the mosquito *Mansonia uniformis* (Diptera: Insecta) under laboratory conditions. J Biosci 17 (1):35–40

Rami A, Raz A, Zakeri S, Dinparast Djadid N (2018) Isolation and identification of *Asaia* sp. in *Anopheles* spp. mosquitoes collected from Iranian malaria settings: steps toward applying paratransgenic tools against malaria. Parasit Vectors 11(1):367–374

Reithinger R, Davies CR, Cadena H, Alexander B (1997) Evaluation of the fungus *Beauveria bassiana* as a potential biological control agent against Phlebotomine sand flies in Colombian coffee plantations. J Invertebr Pathol 70(2):131–135

Rivaz SH, Pirali Kheirabadi KH, Abdigoudarzi M, Karimi GH (2016) A survey on virulence of different isolates of *Metarhizium anisopliae* for biological control of *Hyalomma marginatum* nymphal stage. Vet Res 29(2):45–55. [In Persian]

Sanahuja G, Banakar R, Twyman RM, Capell T, Christou P (2011) *Bacillus thuringiensis*: a century of research, development and commercial applications. Plant Biotech J 9(3):283–300

Sant'Anna MRV, Hector D-A, Aguiar-Martins K, Al Salem WS, Cavalcante RR, Dillon VM, Bates PA, Genta FA, Dillon RJ (2014) Colonisation resistance in the sand fly gut: *Leishmania* protects *Lutzomyia longipalpis* from bacterial infection. Parasit Vectors 7:329–338

Schapira AM, Zaim M, Raeisi A, Ranjbar M, Kolifarhood G, Nikpour F, Amlashi M, Faraji L (2018) History of the successful struggle against malaria in the Islamic Republic of Iran. Ministry of Health and Medical Education-Center for Disease Control, 1st edn. Shayan gostar, Tehran

Service MW (1981) Ecological consideration in biocontrol strategies against mosquitoes. In: Laird K (ed) Biocontrol of medical and veterinary pests. Praeger Publication, New York, pp 173–195

Service MW, Service M (2012) Medical entomology for students. Cambridge University Press, Cambridge

Seyedi Arani HR, Dehghani R, Ghannaee Arani M, Zarghi I (2016) Scabies contamination status in Iran: a review. Int J Epidemiol Res 3(1):86–94

Shahi M, Hanafi-Bojd AA, Vatandoost H, Soleimani Ahmadi M (2013) Susceptibility status of *Anopheles stephensi* Liston the main malaria vector, to Deltamethrin and *Bacillus thuringiensis* in the endemic malarious area of Hormozgan province, Southern Iran. J Kerman Univ Med Sci 20(1):87–95. [In Persian]

Shahi M, Kamrani E, Salehi M, Habibi R, Hanafi Bojd AA (2015) Native larvivorous fish in an endemic malarious area of southern Iran, a biological alternative factor for chemical larvicides in malaria control program. Iran J Public Health 44(11):1544–1549

Sharififard M, Mossadegh MS, Vazirianzadeh B, Zarei Mahmoudabadi A (2011) Laboratory pathogenicity of entomopathogenic fungi, *Beauveria bassiana* (Bals.) Vuill. And *Metarhizium anisoplae* (Metch.) Sorok. To larvae and adult of house fly, *Musca domestica* L. (Diptera: Muscidae). Asian J Biol Sci 4:128–137

Soleimani-Ahmadi M, Zare M, Abtahi SM, Khazeni A (2017) Species identification and prevalence of house dust mites as respiratory allergen in kindergartens of the Bandar Abbas city. Iran J Allergy Asthma Immunol 16(2):133–139

Soltani A, Vatandoost H, Oshaghi MA, Enayati AA, Chavshin AR (2017) The role of midgut symbiotic bacteria in resistance of *Anopheles stephensi* (Diptera: Culicidae) to organophosphate insecticides. Pathog Glob Health 111(6):289–296

Tabibzadeh I, Behbehani G, Nakhai R (1970) Use of *Gambusia* fish in the malaria eradication programme of Iran. Bull World Health Org 43(4):623–626

Tanada Y, Kaya HK (1993) Insect pathology. Academic Press, San Diego

Tavassoli M, Ownag A, Pourseyed SH, Mardani K (2008) Laboratory evaluation of three strains of the entomopathogenic fungus *Metarhizium anisopliae* for controlling *Dermanyssus gallinae*. Avian Pathol 37(3):259–263

Tavassoli M, Ownag A, Meamari R, Rahmani S, Mardani K, Butt T (2009) Laboratory evaluation of three strains of the entomopathogenic fungus *Metarhizium anisopliae* for controlling *Hyalomma anatolicum anatolicum* and *Haemaphysalis punctata*. Int J Vet Res 3(1):11–15

Tavassoli M, Allymehr M, Pourseyeda SH, Ownag A, Bernousi I, Mardani K, Ghorbanzadegana M, Shokrpoor S (2011) Field bioassay of *Metarhizium anisopliae* strains to control the poultry red mite *Dermanyssus gallinae*. Vet Parasitol 178(3–4):374–378

Van den Assem J, Werren JH (1994) A comparisons of the courtship and mating behaviors of three species of *Nasonia* (Hymenoptera: Pteromalidae). J Insect Behavior 7(1):53–66

Vatandoost H, Hanafi-Bojd AA (2012) Indication of pyrethroid resistance in the main malaria vector, *Anopheles stephensi* from Iran. Asian Pac J Trop Med 5(9):722–726

Vatandoost H, Zaim M, Moazami N, Asmar M, Djavadian E (1995) Use of *Lagenedium giganteum* for biological control of mosquito larvae in Iran. Ann Trop Med Parasitol 89(2):213. [Abstract]

Vatandoost H, Raeisi A, Saghafipour A, Nikpour F, Nejati J (2019) Malaria situation in Iran: 2002–2017. Malar J 18:200–207

Vazirianzadeh B, Kidd NAC, Moravvej SA (2008) Side effects of IGR Cyromazine on *Nasonia vitripennis* (Hymenoptera: Pteromalidae), a parasitic wasp of house fly pupae. Iran J Arthropod Borne Dis 2(2):1–6

Vilela R, Taylor JW, Walker ED, Mendoza L (2015) *Lagenidium giganteum* pathogenicity in mammals. Emerg Infect Dis 21(2):290–297

Walker K, Lynch M (2007) Contributions of Anopheles larval control to malaria suppression in tropical Africa: review of achievements and potential. Med Vet Entomol 21(1):2–21

Weiser J, Zaim M, Saebi E (1991) *Coelomomyces irani* sp. n. Infecting *Anopheles maculipennis* in Iran. J Invertebr Pathol 57(2):290–291

Wodzicki K (1973) Prospects for biological control of rodent populations. Bull World Health Org 48:461–467

World Health Organization (WHO) (2003) Use of fish for mosquito control. Regional Office for Eastern Medditeranean, Cairo

World Health Organization (WHO) (2013a) Larval source management: a supplementary measure for malaria vector control: an operational manual. WHO, Geneva

World Health Organization (WHO) (2013b) Malaria entomology and vector control. Participant's guide. WHO, Geneva

Yaghoobi-Ershadi MR (2012) Phlebotomine sand flies (Diptera: Psychodidae) in Iran and their role on *Leishmania* transmission. Iran J Arthropod Borne Dis 6(1):1–17

Zaim M, Ladonni H, Ershadi MR, Manouchehri AV, Sahabi Z, Nazari M, Shahmohammadi H (1988) Field application of *Romanomermis culicivorax* (Mermithidae: Nematoda) to control anopheline larvae in southern Iran. J Am Mosq Control Assoc 4(3):351–355

Zaim M, Kasiri H, Motabar M (1992) Efficacy of a flowable concentrate formulation of *Bacillus thuringiensis* (H-14) against larval mosquitoes in Southern Iran. J Am Mosq Control Assoc 8 (2):156–158

# Chapter 17
# Analytical Approach to Opportunities and Obstacles of Iranian Biological Pest Control

**Javad Karimi, Shokoofeh Kamali, Hossein Madadi,
Seyed Hossein Goldansaz, Soleiman Ghasemi, Heshmatollah Saadi,
and Mohammadreza Attaran**

## 17.1 Introduction

The history of biological control (BC) in Iran goes back more than 85 years. This history began with the importation of three biological control agents: biological control agents (BCA) for three imported scales, (Hem., Coccoidea) in classical BC approach. The BC of the Sunn pest, *Eurygaster integriceps* using parasitic wasp, *Trissolcus grandis* which continued for several years, was stopped around 1951 (Farahbakhsh 1961; Zomorodi 2003). The reason behind this was likely the introduction of chemical pesticides, especially DDT, into the country. Although the devastating effects of chemical pesticides on wildlife and environment appeared in

J. Karimi (✉)
Department of Plant Protection, Ferdowsi University of Mashhad, Mashhad, Iran
e-mail: jkb@um.ac.ir

S. Kamali
Department of Plant Protection, Faculty of Agriculture, Ferdowsi University of Mashhad, Mashhad, Iran

H. Madadi
Department of Plant Protection, Faculty of Agriculture, Bu-Ali Sina University, Hamedan, Iran

S. H. Goldansaz
Department of Plant Protection, College of Agriculture and Natural Resources, University of Tehran, Karaj, Iran

S. Ghasemi
Research and Development Unit, Nature Biotechnology Company (Biorun), Karaj, Iran

H. Saadi
Department of Agricultural Extension Education, Bu- Ali Sina University, Hamedan, Iran

M. Attaran
Biological Control Research Station, Iranian Research Institute of Plant Protection, Agricultural Research, Education and Extension Organization (AREEO), Amol, Iran

the 1940s, this issue suppressed to run of a new and highly profitable industry for the production and sale of chemical pesticides (Behdad 2002).

The BC in the country has remained on the margin until the 1960s for various reasons. Following implementing plans for reducing the use of pesticides in Europe, scenarios for chemical pesticides reduction initiated in Iran using appropriate techniques including BC and other non-chemical approaches over two decades. This plan aimed at mass rearing and augmentative release of BCAs. However, field studies proved that improper planning and poor management of the crops are the main causes of the failure of the plan. During the implementation of the plan, there was an issue with some farmers not participating, but this was never addressed. Farmers were expected to pay some of the BC costs. This didn't happen when it was decided that some costs of producing or introducing BCAs should be paid by them. Literature shows that the use and release of certain agents, mainly *Trichogramma* have been very similar to chemical pesticides' patterns regarding decision levels including the Economic Injury Level (EIL) without consistency with the ecological nature of BC (Heidari et al. 2011). While reports indicate a reduction in pesticide use in the country, a realistic view implies that this reduction in pesticide use was not due to simplistic BC practices. After three decades of using BCAs, there is still no reliable level of croplands that have reached ecological balance and sustainability. Furthermore, the continuous use of chemical pesticides in crops, especially in the Caspian region, had the significant pesticides reduction plan that was hindered and resulted in the emergence of fake chemical pesticides and trafficking of various un-authorized pesticides with improper application techniques. Despite four decades of efforts to use BC in the country, there are still many problems faced with planning. History shows that many socioeconomic issues have prevented the technology from being implemented. The remaining risks of pesticides at high levels in a variety of crops still threaten consumers' health. This requires cooperation and planning to change the current situation.

In the following part, we will analyze the current state of production and use of macro BCA as well as microbial control agent (MCA). Then we will discuss government support, mass rearing of BCA for greenhouses and open fields, the role of academic elites on the advancement of BC plans, and the ability of farmers for BC. In the end, some strategies are mentioned to improve the process of using BCA in the country.

## 17.2   The Current Situation of BC in the Country

Both natural and applied BC strategies including introduction, augmentation and conservation have been involved within pest management plans in the country. Natural BC has a significant role in regulating arthropods density in various regions of the country. There are reports about the infection of overwintering populations of the Sunn pest with hypocrealean fungi during winter in various provinces through the years. Another case is the natural regulation of the gypsy moth (*Lymantria dispar*) population in north forests by viral and nosemose infection. The information about the vital role of conservation BC has neglected mostly.

From classic biocontrol, while the initial of biocontrol was started with the introduction of Novius beetle, followed by mealybug destructor and *Prospatella* wasp, but the last importation was those of Phytosiid mite, *Phytoseiulus persimlis* in 1988 and then, there is no more new introduce (Table 17.1) (Shirazi et al. 2011). The current situation of using BCAs in the country including mostly augmentation programs of eight BCAs used within 119,000 ha. Those crops consist of field crops (97,000 ha), fruit orchards (2000 ha) and greenhouses (21,000 ha) (Fig. 17.1). Eighty-three insectaries are active for the mass rearing of macro BCAs in 24 provinces. The available agents from predators are *Amblyseius swirskii, Cryptolaemus montrouzieri, Macrolophus pygmaeus, Neoseiulus californicus, Orius laevigatus* and *Phytoseiulus persimilis*. The prevalent parasitoids are *Trichogramma* spp. (mostly *Trichogramma brassicae*) and *Bracon hebetor* (Fig. 17.2). The total crop area which used microbial pesticides was 11,597 ha (6030 kg of products based on *Bacillus thuringiensis*). Also during this time, 2380 kg of mosquitocidal products based on *B. thuringiensis is*raliensis* was distributed in the country (PPO 2020).

The total area of croplands that use BCA has fluctuations through years but near the total area with BC in 10 years ago (105,000 ha in 2009–2010 which included 0.68% of agricultural land area) (Askary and Alinia 2011).

Here, we will explore the main groups of BCAs in three sections. First, *Trichogramma* as the key parasitoid will be review in terms of historical progress and obstacles for its improvement. Then, there is a glance over usable predators as

**Table 17.1** List of classic biocontrol projects in Iran (Shirazi et al. 2011)

| Agent | Year | Target species | Target common name | Performed by | Result | Current status |
|---|---|---|---|---|---|---|
| *Novius cardinalis* | 1934 | *Icerya purchasi* | Cottony cushion scale | Jalal Afshar and Mohammad Kaussari | Successful | Active |
| *Crptolaemus montrouzieri* | 1966 | Pseudococcidae | Mealybugs | Mohammad Safavi | Successful | Active |
| *Prospaltella perniciosi* | 1971 | *Quadraspidiotus perniciosus* | San Jose scale | IRIPP | NA | NA |
| *Trichogramma* spp. | 1974 | *Chilo suppressalis* | Striped rice stem borer | Firouz Nikkho | No | Absent |
| *Prospaltella berlesi* | 1977 | *Pseudaulacaspis pentagona* | White peach scale | Mohammad Safavi | Successful | NA |
| *Phytoseiulus persimilis* | 1988 | *Tetranychus urticae* | Two-spotted spider mites | Hooshang Daneshvar | Successful | Active |

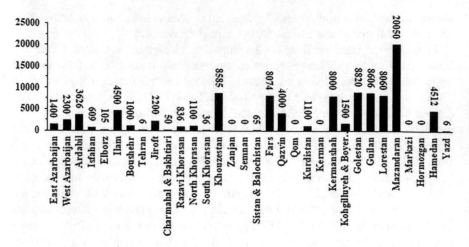

**Fig. 17.1** Implementation level (ha) of biological control in different provinces of Iran during 2018–19. (Data from PPO 2020)

BCAs in the biocontrol plans of the country. The last part is the general view on the progress of MCA.

### 17.2.1    *Parasitoids:* **Trichogramma** *as the Master Agent*

In the 1950s, with the invasion of the rice stem borer, *Chilo suppressalis* into the Northern part of the country, two species of *Trichogramma* from two origins, former the Soviet Union and Germany were introduced by Firouz Nikkho of the Iranian Research Institute of Plant Protection (IRIPP). In 1974, without considering the rich fauna of Trichogrammatidae of the wider Iranian plateau, they were released in the Northern part of the country after preliminary rearing during 1975–1977. These activities were developing year by year since 1981 with the identification of important Trichogrammatid species in the central plateau and Northern areas of the country by researchers of the IRIPP and the Iranian Research Organization for Science and Technology (IROST) (Shojai et al. 1990; Kharazi-Pakdel et al. 1993; Shojai et al. 1998). In 1984, by a collaboration of relevant research and executive organizations, and conducting research projects to study the efficiency, mass production and use of *Trichogramma*, these agents have been used to control insect pests, including *Chilo suppressalis, Ectomyelois ceratoniae, Ostrinia nubilalis, Helicoverpa armigera* and *Cydia pomonella.*

Mass production technology of *Trichogramma*, developed in the public sector has been transferred to the private sector and more than 40 insectariums have been active throughout the country. In general, by 2010, the release level of *Trichogramma* on seven different crops was occurred on 187,896 ha (Plant

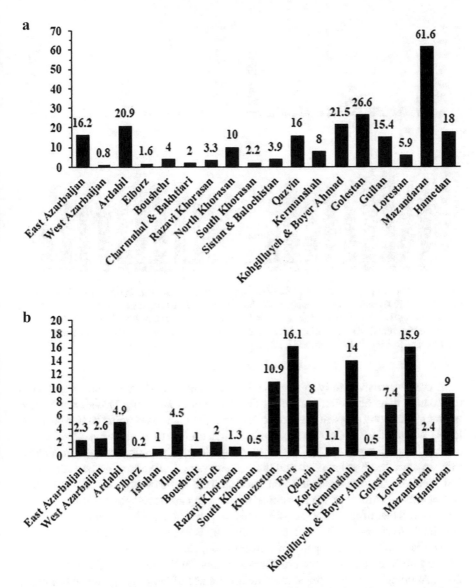

**Fig. 17.2**  Annual amount of released number (million individuals) of (**a**) *Trichogramma* species in 18 provinces (kg) and, (**b**) *Harbobracon hebetor* in 20 provinces through 2018–2019. (Data from PPO 2020)

Protection Organization, PPO 2011). The speed of *Trichogramma* development in crops such as tomato and maize was very high, while after initial decline on other crops, it reached a steady level (such as rice) and in other crops was negative (such as apple) (Fig. 17.3). Nevertheless, the last data showed a decline in the application area for the *Trichogramma*.

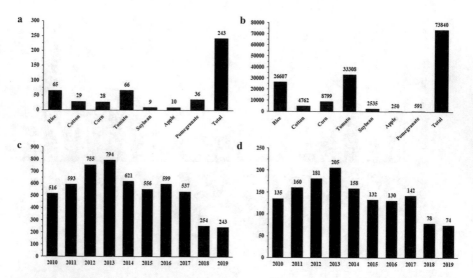

**Fig. 17.3** The crop area under inundative release of *Trichogramma*. (**a**) Augmentative release amount (kg) of *Trichogramma* in main crops during 2018–19, (**b**) augmentative release area (ha) of *Trichogramma* in various main crops through 2018–19, (**c**) application level (kg) of *Trichogramma* during recent years, (**d**) fluctuations level of crops area (ha) under use of *Trichogramma* during recent year. (Retrieved data from PPO, 2020)

A comprehensive study of changes in mass rearing and augmentation tactics in the country can help better realize the development of BC using *Trichogramma* and overcome the obstacles and enhance its strengths. The key point shown at these levels is that the application rates are set according to a plan and whether the anticipated infrastructure has been met, the levels have increased, but finally, infrastructure has had its effect and the annual usage of the agent has been changed. This kind of planning has caused the method itself to be questioned. The voluntary acceptance of farmers at the beginning of the plan caused a rapid development, but due to the effect of many factors, it has led to a lack of acceptance by many farmers and finally to some who believe that this method is ineffective. For example, experts in the agro-industrial company of Dasht-e-Naz, Sari, Mazandaran province, concluded failure of the plan for use *Trichogramma* in the farms of this company in 2010. Another example is the immediate acceptance of the *Trichogramma* use in the pomegranate orchards of the Fars province over the past few years, followed by incomplete implementation and finally, distrust farmers imply biocontrol development occurred without having the infrastructure and the required criteria for mass rearing and release program. In a review by Attaran and Dadpour Moghanloo (2011), they had an analysis on the use of *Trichogramma* as BCA of *Chilo suppressalis*, and varied evolutionary processes were reviewed.

About forty years ago, biocontrol of the rice stem borer was initiated. The *Trichogramma* wasps were first introduced and released at a low level. The basic tasks of collecting, identifying, releasing and evaluating the process of

*Trichogramma* use have been developed during 1980–1990. During 1991–1994, the *Trichogramma* was considered as an effective BCA in pest control. The results of BC using *Trichogramma* were successful or at least promising (Ebrahimi et al. 1998). In 1995–2001, the development of BC in rice fields and subsequent crops began. Numerous insectaries were developed, the pesticide subsidies were allocated for the development of the BC plan, mass production is practically refined experienced, but on the other hand, the release methods were not optimized and the storage method was the same as before, except for some details. The issue of quality control (QC) began to emerge and the private sector gradually started collaboration during which problems appeared. During this time, efforts were made to maintain and enhance the BC in the country. The plan to reduce pesticide use was turned into a program for the Optimal Utilization of Fertilizers and Pesticides (OUFP), and this change in attitude, as well as decrease of subsidies to pesticides, were combined with the use of BC. However, there were problems with financial allocation. It was determined that a percentage of mass rearing and application costs of the agents would come from farmers with an annual increase until it covers the costs, and the producer be in direct contact with the farmer. Despite these problems, development continued. While the formal appearance of the BC program seemed good more serious problems arose and the process was further refined

Since 2001, the BC development in the rice fields has stopped; the two Ministries of Agriculture and Jihad have been merged, a High Council for Pesticides and Chemical Fertilizers use Reduction has been abolished and so the coordination center was dismantled. Thus the oversight, planning, and procurement of BC were lost; credits shifted and allocation changed; research and implementation were separated with BC measures continued only with the efforts of a few true proponents of its development. Some rice farmers have taken a negative insight into BC approach as research budgets related to methods of developing and removing the bottlenecks were insignificant and in fact, credits did not move to the relevant research plans.

The problems with the use of *Trichogramma* are not limited to mass production, and there are various other issues, including planning and monitoring of essential barriers. We don't have *Trichogramma* post-release data in quantitative form, and we don't know where we succeed or failed or by how much (Attaran and Dadpour Moghanloo 2011).

Organizing a single and specific planner and supervisor, pushing subsidies to the production of healthy crop rather than producer, removing government purchases of BCAs, maximizing producer responsibilities (*i.e.* mass production, release, and packaging), and the necessity of forming major producers and/or aggregating small producers into a large subset are among the most important issues which need thoroughly examining in the *Trichogramma* application.

### 17.2.1.1 More Technical Issues About *Trichogramma*

Regarding the necessity of mass rearing of *Trichogramma* and its hosts mentioned above, it is necessary to address issues like mass rearing of the alternative host, Trichocard type (cardboard sheet to the size of $10 \times 10$ cm containing parasitized egg of alternative host which adult *Trichogramma* emerge 8–10 days from the date of parasitization), distribution, evaluation, storage, and QC.

**Fig. 17.4** Mass rearing of *Trichogramma* and its host. (Courtesy of PPO)

### 17.2.1.1.1   Mass Production of Alternative Hosts

The major factitious host used in the mass rearing of *Trichogramma* is the Angou-mois grain moth, *Sitotroga cereallela* (Fig. 17.4) The quality of barley can affect the quality of the moth as a host, but this issue is not much considered in mass rearing because it is not easy to find the suitable cultivar of barley and producers have to use existing ones that are sometimes inappropriate. The average production efficiency of *S. cereallela* in the country is about 1: 4–5, which means that 1 g of *S. cereallela* eggs introduced into the rearing system will lead finally to 4 to 5 g of *S. cereallela* eggs, each moth laying about 100–150 eggs. Some insectaries also report production efficiency up to 1: 12; if the producers are able to increase the efficiency rate from 1: 5 to about 1: 6, then producers who use one ton of barley grain for the infestation in a rearing cycle will produce about 1 kg more eggs, which means cheaper agent and more rearing of *Trichogramma* in a given period. This issue is important when the price of BCA is a limiting factor in their use. Apart from the cultivar and size of barley grains and problems related to the degree of infestation of barley grains with *S. cereallela* larvae, the process of host moth egg production also faces concerns. Storage conditions of the collected eggs and the storage time of the eggs which use for initiation of rearing are the factors affecting production efficiency. The health issue of insectaries, in terms of mites and the employees health impact on the efficiency of production of *S. cereallela* eggs and finally the *Trichogramma* wasps. It seems that many production units in the country lack the necessary standards in this regard. Having a well-developed insectarium can solve many of these problems. On the other hand, research in this field is mostly unheard of, and governor budgets have almost completely shifted to plan implementation, while the research section has neglected some main questions about mass rearing (Attaran and Dadpour Moghanloo 2011).

### 17.2.1.1.2   Mass Production of *Trichogramma*

During 2018–19, more than 240 kg of the host eggs containing *Trichogramma* pupae have been purchased in the country under contracts between the PPO and the producers. Perhaps the most important issue in cost-effective production of *Trichogramma* is production efficiency, which should not diminish the importance of *Trichogramma* strain selection and other qualitative issues. In the current mass-rearing system in Iran, the efficiency of production is about 1 to 3 because a single female *Trichogramma* in the mass rearing only provides six parasitized eggs. Comparing this to the wasp fertility of about 100 eggs from each female under normal conditions means wasting of some production inputs by taking into account the efficiency of production of *S. cereallela* eggs, from each kilogram of barley; thus about 1 g of the *Trichogramma* is released. To increase the production efficiency of *Trichogramma*, a high-quality host should be produced, and by selecting the appro-priate strain of the wasp, the rearing process has to be optimized. Proper planning for mass production management is another factor that plays an important role in production efficiency. Unfortunately, some economic problems, limited government contracts and the lack of a specific long-term plan for some producers have created conditions where the producer uses the previous system of mass rearing and looks for short-term economic benefits. Establishing a large production unit for host egg production is one of the solutions to overcome these problems (Dadpour Moghanlou 2002; Attaran and Dadpour Moghanloo 2011).

### 17.2.1.1.3 Trichocards, Storage, and QC

The current Trichocards, with minor changes, are the same as those introduced at the beginning of the development of mass rearing and releases plan of the *Trichogramma*. Recently sporadic efforts has been made to change this method. This present method affords to advantages ease of release, wasps protection from adverse conditions and natural enemies at a low cost. The predation of ants on the wasp within the cards is a challenge that needs to be considered and resolved (Attaran and Dadpour Moghanloo 2011).

An important issue for mass rearing and use of the *Trichogramma* which usually is underestimated, despite its importance, is storage. The storage method available in production units that directly transfer the *Trichogramma* to the refrigerator on days five and six post- parasitism has many disadvantages and imposes high loss and high-quality reduction of the *Trichogramma* efficacy. Recently, there is an increasing trend for using the cold storage of pupae in the mass rearing of the wasp in insectaries.

The QC considered the major concerns facing BC in the country. The QC is a process that runs from the beginning to produce the final product and ends with the evaluation of the product's efficiency. Monitoring this process requires technical or necessary information and a comprehensive system. The system has not yet been launched and seems to be the main part of BC implementation in Iran. The quality of the product is of great importance for its acceptance and use by end users. Among the factors predicted in the QC guidelines during the last two decades, the target pest egg acceptance and the life span of the *Trichogramma* have not been evaluated, and the selection of criteria due to the lack of evaluation of these factors may be indicative of low quality of Trichogrammatid wasps. The postproduction guidelines approved by the International Organization for Biological Control (IOBC) contain main criteria for measuring the wasp traits (van Lenteren et al. 2018).

## 17.2.2   Predators

Long time, predators have been considered as the most prominent biocontrol agents. 45 years after the emblematic case of classic BC by *Rodolia cardinalis,* this predator imported to Iran and substantially reduced the cottony cushion scale population in Northern provinces. After this great achievement, a promising prospect seemed for

**Fig. 17.5** (**a**, **c** and **d**) Inundative use and, (**b**) mass rearing of predators in norther provinces. (Courtesy of PPO)

use of predators but it seems that success was forgotten and the use of predators abandoned at the expense of using parasitoids. The resumption of predator application begins with releasing mealybug destroyer, *Cryptolaemus montrouzieri* in Northern provinces of the country against mealybugs and other scales on citrus trees and tea more than 50 years ago (Fig. 17.5). This lady beetle imported from Spain in 1966 and reared in Tonekabon insectarium, Mazandaran Province for the first time. The most recent statistics showed 800,000 individuals of this predator has been produced by private insectarium and released during 2018–19 (PPO 2019). However, it is likely that this species not being supported by the government anymore.

Predatory bugs are amongst the other common macro BCAs. However, their role in BC has not been well appreciated by many growers. Different native predatory bugs are naturally dispersed in fields and greenhouses of the country but first of all, the minute pirate bugs, Anthocoridae, have been considered and planned to be reared commercially in 20 years ago. The commercial rearing of *Orius* bugs against thrips and spider mites was one of the main subtitles of the OUFP plan. Accordingly, 50,000 *Orius* species (mainly *O. albidipennis* as native species of Iran) have been produced commercially through 2018–19. However, this amount of production seems too low to be efficient and the optimization of rearing, storage and releasing procedure required. The other predatory bug is *Macrolophus pygmaeus* which

---

105,000 individuals have been produced mainly against South American tomato leafminer and whiteflies at the same time. This predatory species proved to be effective to suppress *Tuta absoluta* population within the greenhouse solitary or in combination with *Trichogramma* wasps and needs to be considered more.

Predatory mites have been considered as one of the most prominent predatory groups which have high potential to reduce the pest population in the greenhouse mainly. The most important families used practically are Phytoseiidae and Laelapidae. Different phytoseiids used as BCAs worldwide but in Iran, only three species, *Amblyseius swirskii*, *Neoseiulus californicus* and *Phytoseiulus persimilis* produced commercially which among those species, the *A. swirskii* has the highest production (ca. 21 million individuals) and two others produced as 100,000 and 200,000 individuals per year, respectively. These species largely used in augmentative release within greenhouses and against greenhouse whiteflies and two-spotted spider mite. Regarding their simple mass rearing procedure and high rate of producing cost to benefit makes this group of the predator as ideal candidates for rearing and releasing at least in greenhouses.

The result of the above discussion about commercial production of predator clear that the suppressive potential of this group has been ignored largely by many growers and farmers who at least in some cases aware of parasitoids and MCA potential. Compare to parasitoids, few insectaries produced predators and their production rate is not comparable at all. This problem somewhat arouses from polyphagous habits of most predators. Thereupon, they do not respond to pest population outbreaks quickly and in some cases, they seem inefficient. Moreover, the mass rearing of many predator species needs sophisticated methods and requirements which could not affordable by private insectaries and the end product would be expensive. In addition to the above deficiencies, in recent years, the governmental subsidies contribution decreased from 100% to 55% which has not paid timely (The third Author, Personal communications with private producers). This, in turn, might be causing that some minor private producers discarded their predator production. Fortunately, direct contract between predators producer and farmers have recorded recently which promise the BC plan using these agents in the future.

Besides the above species, it sounds that green lacewings have a high potential to be produced commercially. Still, research on optimization of predator mass rearing, cold storage, packaging and releasing should be a high priority of research institutes and Universities especially regarding effective native predators which are acclimatized with the weather of Iran. Defective knowledge in each of these steps prevents the widespread use of predators as BCAs.

## 17.2.3 Microbial Control Agents (MCA)

Currently, the diversity of available MCA in the country is restricted and hence has reduced the range of farmers' choice. The reasons for this should be sought in the processes and rules for the registration and regulation of MCAs. There may be a variety of issues noted about the obstacles of registration, but there seems to be a point at the top of these problems that if the issue is resolved, the flow of production and introduction of new products will change, and that is changing the attitudes of regulatory authorities such as the PPO in the field of BC. This administrative, executive, and legislative structure of the country in the field of pest control, stemming from the belief in the chemical pesticides is the key control method and even the only way to manage the pests that has led to an attitude of biological inputs in harmony and as the same as chemical inputs. Currently, the most important and widely used MCA is *Bacillus thuringiensis* (Bt). This bacterium is currently produced in high quality in bulk in the country, is competitively priced with its chemical counterparts, and most importantly, kills the target pest rapidly, unlike other types of BCAs. However, there is no widespread use of Bt products in the country.

Another reason for the low number and types of microbial pesticides comes from the fact that large and reputable producers in the country, with dozen of chemical pesticides, lack the motivation to follow the research and development of new products. One example of this is the production of MCAs with high-inflation and free foreign currency prices in the country, while the government has subsidized currency for synthetic chemicals and hence imports are high. This makes no sense except to demand non-production domestically. Therefore, it can be said that the number, variety, quality and quantity of microbial pesticides in the country are

**Fig. 17.6** Microbial control facilities in Karaj. (**a** and **d**) main building, (**b**) fermentation unit, (**c**) control system room. (Courtesy of Biorun)

affected by the question of the status of successful existing products that have all characteristics that are acceptable. The obvious answer is that if the national and friendly goals and motives of some of these producers were not valid, companies producing MCAs would have been shut down soon after their establishment.

Currently, unlike parasitoids and predators, the country's ability to produce MCAs is well known (Fig. 17.6) and can be used to reduce dependence on chemical pesticide importation. For example, about 30,000 tonnes of chemical pesticides are currently imported into the country from official sources; about 20,000 tonnes of these are insecticides. Of this amount, a significant percentage is related to the pesticides used against lepidopterans pest that can be replaced or combined with or alternated in field and greenhouse applications with a Bt-based microbial insecticide. This requires a firm and decisive decision by the relevant authorities in the Ministry of Agriculture and the Ministry of Health and Medical Education. It seems that only by activating this potential, the percentage of chemical pesticide imports could fall below the current level.

## 17.3   Government Support of BC

Currently, agricultural pesticides are inexpensive, but their final cost is several folds more than the BC cost when taking into account the comprehensive and environmental costs. Accordingly, developed countries usually apply a combined plan with a combination of BCAs and other tactics to lead the community to higher use of BC via accompanying additional costs to the market price of chemical pesticides through taxation and the creation of real prices, and subsidizing BC plans. For support payments, it should cover the crop, not alone the agent.

Since a restricted number of BCAs registered for users recorded in previous years, various products are not available for recommendation to users. Of course, with the establishment of the Expert Committee for Supervision and Registration of Biological Control and Review of Registration Requests, more BCAs are in the queue of being registered, however, there are still problems. Regarding the registration of BCAs according to Soroush et al. (2011), it is necessary to carefully study the technical information of the products and perform QC analyzes in well-equipped laboratories. For example, each isolate of a microbial agent has a completely separate and unique characteristic and full characterization of each microbial strain required a wide range of data which can only be obtained in a well equipped laboratory and at relatively high costs. Regarding the approved criteria as standard levels asked from registrants, it is difficult for companies, especially domestic producers, to provide technical characterization and QC analysis. Most companies that produce BCAs and even import their products are suffering from insufficient financial resources to afford the high costs of performing QC tests and studying the efficiency regarding the existing tariffs that have increased recently as well. Therefore, a solution should be thought to facilitate the registration of MCAs, due to their significant potential role. Considering the high cost of studying the efficiency of each MCA for each target/product which is borne by the emerging private sector, it is suggested to assist the producers financially or accelerate the registration procedures. To support the mass production, it is also necessary to provide credits for the establishment and equipment of the R &D units, which supports all stages relating to production, package, storage, transport and field persistence and efficacy. One recommends the implementation a coherent plan through the involvement of non-governmental organizations (NGOs) to enhance the technical and technological knowledge of the private sector.

Today, microbial pesticides play an important role in pest management as well organic farming and because of their application manner, fewer limitations with their similarity to conventional pesticides application, they are readily accepted by farmers and can play a significant role in the development of BC in the country. The Expert Committee for Supervision and Registration of Biological Pest Control Products in the PPO could provide more options for increasing the use of microbial pesticides, especially under greenhouse conditions as well as open fields.

The need to interact more fully with the scientific community of developed and developing countries through the bilateral cooperation agreements to use the achievements and experiences of other countries will make a significant contribution to accelerating the development of these agents. Participation of experts, researchers, and managers of the private sector in relevant internationally accredited fields will be effective in transferring these experiences to the country. We hope that in the future, we will be able to establish regional reference laboratories to facilitate and accelerate the QC evaluation of the microbial pesticides.

## 17.4  Target Crops of BC: Greenhouses and Outdoor Area

Applying biocontrol in greenhouses is much easier and feasible than other ecosystems. The closed environment of the greenhouses provides a chance for the farmers to use various agents for each pest type that allows the greenhouse manager to take action to control the pest population of crops according to conditions/products. Hence, more than ten agents have been registered for use in these agrosystems in Iran so far (PPO 2019). In this regard, due to water deficiency as a great threat and the necessity for increasing greenhouse and closed cultivation, the mass rearing of agents for these systems has some challenges.

In the current state, mass rearing of all the required and approved agents for greenhouses in the country is neither possible nor rational, though it may be objectionable at first glance. In recent years, these agents were from the imported source. This issue has some concerns as the introduction of the exotic populations of an agent, its possible establishment and release can disrupt the native fauna of the ecosystem. To overcome this challenge, it is recommended to introduce available and native species first and foremost, and also precision risk analysis procedures should be carried out before importations (Ashouri 2011).

There are two strategies herein considered. The first one supports mass rearing and the use of *Trichogramma* as an inundative agent on a large scale, but the second one is the mass production of native agents like predatory mites and lacewings. Therefore, the capacity of mass production, which lies in the private sector, should be improved to modify the methods of mass rearing/production and the use of BCAs in open fields, such as with *Trichogramma*. The country's greenhouse area is about 15,000 hectares, while open field crops cover more than 14 million hectares (Ministry of Agriculture Statistics, 2018–2019). Of this amount, about 1,200,000 hectares are allocated to seven main crops on which currently biocontrol plans are implemented using *Trichogramma*. It is possible to cover the whole of this area only by mass rearing of qualified *Trichogramma* species to the amount of more than 21 tons of *Trichogramma*. Therefore, if this agent is required in the country, it is necessary to start large scale mass production of the wasp for wide range use. *Trichogramma* have been mass-reared in the country, millions of areas are covered by this wasp. Technical knowledge for mass production of single agent will always be easier than producing several agents. The private sector should be encouraged to invest in establishing modern insectaries, not the small and local ones, which come from changing the old houses. Skilled insectaries should be equipped with R & D unit. Despite initial costs incurred in setting up this type of insectary, it is not only the product itself that will produce the quality but also the cost of the final product will be reduced. With this policy, the BCAs are produced according to the desired standards and the BC will be cost-effective.

## 17.5    Role of Academia on the Development of BC

Irrespective of the greenhouse owners, currently the Iranian growers have restricted management tactics. This reflects the scientific, administrative and economic development of the country being focused predominantly on chemical control. This should be brought to the attention of government officials by biocontrol experts. In this regard, Ashouri (2011) acknowledged that elites should make greater efforts to convince decision-making centers to change their attitudes and minds. In many countries, this has been well done. To achieve the goal of implementing BC plans, there is a need for two- way interaction. Moreover, the elites are trying to convince the statesmen, and on the other hand, the statesmen have put the both methods of supporting research and the elites in the field of BC in their plans so that they can reciprocally fulfill their mission. Soroush et al. (2011), also emphasized the role of the academic elite in the development of BC plans. Currently, domestic mass production companies are low-tech and require efficient and specialized production technology and manpower. In general, the lack of sufficient technical knowledge in the country, mass production of various BCAs and numerous producers of the agents are among the most important parameters which required to use the scientific potential of elites to improve mass-production and commercialization of BCAs. Assigning and providing the necessary credits for research, creating a responsible and independent structure for research (grant body) and development of BC as a bridge for transfer of technical knowledge to the production sector will make a significant difference in this regard. The development of technology for microbial agents requires training and education of the public and private sectors.

## 17.6  Farmers Issues and Their Ability for BC

The lack of awareness about natural enemies and their potential in regulating herbivore populations is important for using BC. The farmers need skills to make their own farm decisions. In this regard, Heidari et al. (2011) acknowledged that there is no other way than to rely on empowering farmers to solve the problem of protection for agricultural products and the reduction in chemical pollutants that endanger human health. Any decision on production is made only by farmers and plans should be tailored to the socio-economic and cultural characteristics of different communities in ways that help farmers make better decisions. The articles related to agricultural sector plans for using IPM and BC will be realized if the decision-makers in crop management are farmers. For such decision-making, they need the training enhancing their skills and make efforts as researchers rather than just message recipients. Case studies in many countries, including Iran, which will be referred to below, have shown that farmers are highly motivated to engage in participatory research and acquire the skills to manage crop production and protection. However, poor management of different sectors has prevented farmers from providing the necessary input for their farms to participate.

For example, rice farmers in Fereidoonkar, Mazandaran Province, North part of Iran with the combination of rice-duck cultivation, in addition to removing other weeds, herbicides and other pesticides improved the diversity of useful species in the field and a better understanding of farmers contributed to the conservation of ecosystems (Fig. 17.7) (Heidari et al. 2011; Osku et al. 2012). In many developing countries, farmer field school (FFS) has been used for many years as the main method to promote IPM. But without a full understanding of the effects of its deployment at the national level, it has hardly been expanded from the pilot stage to the broader levels. Studying the effects of FFS on its graduated farmers and comparing the results with other farmers in the same villages and farmers in other villages who were not aware of FFS at all showed that FFS participating farmers were less likely to use chemical pesticides and had a practical knowledge of natural enemies and the importance of not rushing into the spraying, but not transferring this knowledge to other farmers as they should and perhaps. Silk producers in the North of the country are also partially involved in the BC plan of the mulberry scale, *Pseudaulacaspis pentagona*, with much of mulberry scale inhibitory activity captured by parasitoid wasp of *Prospaltela berlesi* from the insectarium producing this species at Iran Silk Company (Fathi et al. 2011; Heidari et al. 2011).

**Fig. 17.7** Biocontrol in rice field using *Trichogramma*. (**a**) The inundative release of *Trichogramma brassicae*, (**b**, **c** and **d**) implementation of trichocards by a farmer in Mazanadarn, Northern region of Iran. (Courtesy of PPO)

## 17.7    Final Analysis and Solutions

Supplying healthy food and protecting the environment perhaps is the main goals of today's human. To achieve this, the effectiveness of IPM aimed to reduce the use of chemical pesticides is inevitable, which is possible with the focus on eco-friend manners including BC. The desired performance of IPM, relying on BC, is the result of the disproportionately complex interaction of various sociopolitical and even cultural factors that can only be developed through a specific set of factors. The development of BC, not only as part of sustainable agricultural development, but also as an important stage of the development through changing the attitude of officials and experts in addition to helping to preserve the environment contributes to the achievement of advanced communication technologies, relating to the large markets and the modern organization of the private sector, and finally improves the health of the household food and the welfare of the community. The first question is, how to develop production, market and use of BCAs and/or, in short, how to improve BC plans in the face of the existing problems? Is the answer might be to take advantage of other countries' experiences and combine their achievements with our knowledge. Among these experiences could be the EU, which adoption of IPM and considering BC as the cornerstone, interests for BCA application increased and a large number of chemical pesticides discontinued. Another model is the policy strategy of China which launched a plan for the reduction of chemicals and development of non-chemical manner including BC with 340 million US$. The main recommendations for improving the BC implementation suggest here.

1. Considering IOBC recommendations for dealing with important pests in the country as well as compatibility and capabilities of natural enemies to Iranian conditions, and planning and applications of experiences in other countries.
2. Define the updated standards required and recognized for BC to apply only those high-quality BCAs into the market.
3. Revise the guidelines on the production and application of chemical pesticides, strict application of pesticides and more attention to behavior-based tactics including pheromones.
4. For accelerating BC plans, especially open space plans, there should be a maternal product line (*Ephestia kuehniella* egg production line) for adequate, timely, and high-quality supply in the country. For this purpose, it is better to acquire the technical knowledge required by the private sector and transfer it to the country.
5. Specializing production units (insectaries) to focus on qualified agent production so that by enhancing their technical knowledge, they can produce and store throughout the year so, the need for instantaneous production capacities is met.
6. Monitoring the final result of BC plans. This monitoring should focus more on performance and result rather than on inputs. Here, it is imperative to comply with the residual limit on chemicals required to encourage farmers to use BC as much as possible.
7. Encouraging private sector investment and removing obstacles are among the other requirements for the development of production, business and use of BC in the country. An important factor in developing the agricultural section and therefore, BC is the active presence of the private sector while high economic risk, lack of rapid return of capital, administrative bureaucracy and the legal gap in the field of license and permission prevent this active presence.
8. Encourage the involving NGOs with BC to work on side effects and residue levels of pesticides and their illegal application to increase public knowledge, restricted use of chemicals and promote the BC activities (van Lenteren et al. 2018).
9. Use of startups capability for interdisciplinary and multidisciplinary programs, suitable for engaging a new generation of graduates.

**Acknowledgment** Finalizing this chapter about biocontrol challenges and status was so harder than we thought it. Many people helped us during preperation this job. The authors appreciate providing data and information about the biocontrol plan by "*Plant Protection Organization*" of Iran. We thank Gary B Dunphy for his comments and edit the final draft. Ahmad Hemet Abadi helped the first author with his experience about the biocontrol producers and had some suggests for the work which acknowledges him. Also Hossein Ranjbar Aghdam and Ahmad Ashouri provided some basic data for the first author who acknowledges him. Ali Rezaei helped us by sharing his information and images of parasitoids and predators and Marzieyh Aminzadeh by edit some images which appreciate them.

# References

Ashouri A (2011) Production and commercial use of biological control agents in worldwide and Iran. Paper presented at the 1st Biological Control Development Congress in Iran, Iranian Research Institute of Plant Protection, Tehran, 27–28 July 2011

Askary H, Alinia F (2011) The necessity to developed biological control in forest and rangelands protection. Paper presented at the 1st Biological Control Development Congress in Iran, Iranian Research Institute of Plant Protection, Tehran, 27–28 July 2011

Attaran MR, Dadpour Moghanloo H (2011) An analytical review of present status and future prospective in utilization of *Trichogramma* wasps for biological control of agricultural pests in Iran. Paper presented at the 1st Biological Control Development Congress in Iran, Iranian Research Institute of Plant Protection, Tehran, 27–28 July 2011

Behdad E (2002) Introductory entomology and important plant pests in Iran. Nashre Yadbood Publication, Isfahan, p 824. (in Persian)

Dadpour Moghanlou H (2002) An investigation on the host-parasitoid system between *Trichogramma pintoi* (Voegele) and the Mediterranean flour and Angoumois grain moth, in laboratory conditions. MSc thesis, Tarbiat Modares University, Tehran, Iran

Ebrahimi E, Pintureau B, Shojai M (1998) Morphological and enzymatic study of the genus *Trichogramma* in Iran (Hym. Trichogrammatidae). Appl Entomol Phytopathol 66:122–141

Farahbakhsh G (1961) Family Pentatomidae (Heteroptera). In: Farahbakhsh G (ed) A checklist of economically important insects and other enemies of plants and agricultural products in Iran, vol 1. Department of Plant Protection, Ministry of Agriculture, Tehran, pp 25–28

Fathi H, Heidari H, Impiglia A, Fredrix M (2011) History of IPM/FFS in Iran. Food and Agriculture Organization, Regional Office in Tehran, Iran, p 24

Heidari H, Impiglia A, Mirzaee F (2011) Farmers' empowerment; a key to implement biological control. Paper presented at the 1st Biological Control Development Congress in Iran, Iranian Research Institute of Plant Protection, Tehran, 27–28 July 2011

Kharazi-Pakdel A, Heydari M, Vakili A, Arab M, Alavipour MS, Valipour M (1993) Non-chemical methods for control of rice stem borer, *Chilo suppressalis* walk (Lep. Pyralidae) in Mazandaran province. In: Abstracts of the 11th Iranian Plant Protection Congress, Guilan University, Rasht, 28 August–2 September 1993

Osku T, Omrani M, Zare L (2012) The expansion of the management approach of *Chilo suppressalis* (Walker) in Mazandaran agricultural fields. J Plant Prod 35:61–68. (in Persian with English summary)

Plant Protection Organization (2011) Annual Report. Available online at https://ppo.ir. Accessed 10 May 2018

Plant Protection Organization (2019) Annual Report. Available online at: https://b2n.ir/g23976. Accessed 10 June 2019

Plant Protection Organization (2020) Announcement of biological control plan in crop, orchard and greenhouse. Available online at: https://ppo.ir/_douranportal/documents/eblaghiyeh1399-bio.pdf. Accessed 15 March 2019

Shirazi J, Attaran M, Farrokhi Sh, Dadpour H, Nouri H (2011) An analytical review on the classical biological control of pests in Iran and the world. Paper presented at the 1st Biological Control Development Congress in Iran, Iranian Research Institute of Plant Protection, Tehran, 27–28 July 2011

Shojai M, Tirgari S, Azma M, Nasrollahi AA (1990) Faunistic study of beneficial parasitoid wasps *Trichogramma* and prospect for their application in agricultural field in Iran. J Iran Res Org Sci Technol 9:33–47. (in Persian with English summary)

Shojai M, Ostovan H, Khodaman AR, Hosseini M, Daniali M, Seddighfar M, Nasrollahi AA, Labbafi Y, Ghavam F, Honarbakhsh S (1998) An investigation on beneficial species of *Trichogramma* spp. (Hym., Trichogrammatidae), active in apple orchards, and providing optimum conditions for mass production in laboratory cultures. J Agric Sci 16:5–39. (in Persian with English summary)

Soroush MJ, Heydari Sh, Jalali Nia M, Marzban R (2011) The trend for registration of biological control agents in Iran. Paper presented at the 1st Biological Control Development Congress in Iran, Iranian Research Institute of Plant Protection, Tehran, 27–28 July 2011

van Lenteren JC, Bolckmans K, Köhl J, Ravensberg WJ, Urbaneja A (2018) Biological control using invertebrates and microorganisms: plenty of new opportunities. BioControl 63:39–59

Zomorodi A (2003) History of Iranian plant protection. Agriculture Education Press, p 698

Printed in the United States
by Baker & Taylor Publisher Services